Cyst Nematodes

Cyst Nematodes

Edited by

Roland N. Perry

University of Hertfordshire, Hatfield, Hertfordshire, UK and Ghent University, Ghent, Belgium

Maurice Moens

Flanders Research Institute for Agriculture, Fisheries and Food, Merelbeke, Belgium and Ghent University, Ghent, Belgium

and

John T. Jones

The James Hutton Institute, Invergowrie, Dundee, UK, University of St Andrews, St Andrews, UK and Ghent University, Ghent, Belgium

CABI is a trading name of CAB International

CABI
Nosworthy Way
Wallingford
Oxfordshire OX10 8DE
UK

CABI
745 Atlantic Avenue
8th Floor
Boston, MA 02111
USA

Tel: +44 (0)1491 832111
Fax: +44 (0)1491 833508
E-mail: info@cabi.org
Website: www.cabi.org

Tel: +1 (617)682-9015
E-mail: cabi-nao@cabi.org

© CAB International 2018. All rights reserved. No part of this publication may be reproduced in any form or by any means, electronically, mechanically, by photocopying, recording or otherwise, without the prior permission of the copyright owners.

A catalogue record for this book is available from the British Library, London, UK.

Library of Congress Cataloging-in-Publication Data

Names: Perry, Roland N., editor. | Moens, Maurice, editor. | Jones, John T., Professor, editor. | C.A.B. International, issuing body.
Title: Cyst nematodes / edited by Roland N. Perry, Maurice Moens, and John T. Jones.
Description: Oxfordshire, UK ; Boston, MA : CAB International, [2018] | Includes bibliographical references and index.
Identifiers: LCCN 2017050571 (print) | LCCN 2017051611 (ebook) | ISBN 9781786390844 (ePDF) | ISBN 9781786390851 (ePub) | ISBN 9781786390837 (hbk : alk. paper)
Subjects: | MESH: Nematoda--microbiology | Plants--microbiology | Disease Resistance | Host-Pathogen Interactions
Classification: LCC QL391.N4 (ebook) | LCC QL391.N4 (print) | NLM QX 203 | DDC 592/.57--dc23
LC record available at https://lccn.loc.gov/2017050571

ISBN-13: 978 1 78639 083 7

Commissioning editor: Rachael Russell
Editorial assistant: Emma McCann
Production editor: Tim Kapp

Typeset by SPi, Pondicherry, India
Printed and bound in the UK by CPI Group (UK) Ltd, Croydon, CR0 4YY

To Clare, Monique and Bridget, with grateful thanks for their patience and support during the preparation of this book; also to colleagues who have continually stimulated our interest in the fascinating biology of nematodes.

Contents

About the Editors	xvii
Contributors	xxi
Preface	xxiii

1 Cyst Nematodes – Life Cycle and Economic Importance — 1
Maurice Moens, Roland N. Perry and John T. Jones
- 1.1 Introduction — 1
- 1.2 Impact — 2
- 1.3 History of the Genus — 2
- 1.4 Distribution — 3
- 1.5 Identification — 3
  - 1.5.1 Traditional identification — 3
    - 1.5.1.1 Cysts — 3
    - 1.5.1.2 Second-stage juveniles — 4
  - 1.5.2 Molecular identification — 4
- 1.6 Life Cycle — 4
- 1.7 Syncytium — 7
- 1.8 Effect of Abiotic Factors — 8
- 1.9 Important Species — 8
  - 1.9.1 Sugar beet cyst nematode *Heterodera schachtii* — 8
  - 1.9.2 Cereal cyst nematodes *Heterodera avenae, H. latipons* and *H. filipjevi* — 8
  - 1.9.3 Soybean cyst nematode *Heterodera glycines* — 10
  - 1.9.4 Pigeon pea cyst nematode *Heterodera cajani* — 10
  - 1.9.5 Sugarcane cyst nematode *Heterodera sacchari* — 11
  - 1.9.6 Rice cyst nematodes *Heterodera oryzae* and *H. oryzicola* — 11
  - 1.9.7 Potato cyst nematodes *Globodera rostochiensis, G. pallida* and *G. ellingtonae* — 12
    - 1.9.7.1 *Globodera rostochiensis*, the golden or yellow potato cyst nematode — 12
    - 1.9.7.2 *Globodera pallida*, the pale potato cyst nematode — 13
    - 1.9.7.3 *Globodera ellingtonae*, the Ellington potato cyst nematode — 13
- 1.10 Pathotypes and Races — 13
- 1.11 Symptoms — 14

	1.12 Management	17
	1.13 References	18

2 Genomics and Transcriptomics – a Revolution in the Study of Cyst Nematode Biology   27
Sebastian Eves-van den Akker and John T. Jones

- 2.1 Introduction   27
- 2.2 A Note of Caution   29
- 2.3 Current Status of Genome and Transcriptome Projects for Cyst Nematodes and other Plant-parasitic Nematodes   29
- 2.4 Key Findings from Genome/Transcriptome Projects   30
  - 2.4.1 Identifying genes of interest from genome/transcriptome resources   30
  - 2.4.2 Sequence similarity searches: gene birth and death   30
  - 2.4.3 Temporal transcriptional profiling   32
  - 2.4.4 Spatial transcriptomic profiling   33
  - 2.4.5 Promoter motifs and their associated predictive power   34
- 2.5 Population Genetics and Metagenetics   34
- 2.6 Identification of Key Biochemical Pathways and Targets for Control   35
- 2.7 Mitochondrial Genomes   36
- 2.8 Horizontal Gene Transfer   36
- 2.9 Accessibility   38
- 2.10 Conclusions and Future Prospects   38
- 2.11 Acknowledgements   39
- 2.12 References   39

3 Hatch, Survival and Sensory Perception   44
Edward P. Masler and Roland N. Perry

- 3.1 Introduction   44
- 3.2 Biology of Hatching   44
  - 3.2.1 Abiotic influences on hatching   45
  - 3.2.2 Root diffusate activity   45
  - 3.2.3 Chemistry of root diffusates   46
  - 3.2.4 The effect of microorganisms on hatching   47
  - 3.2.5 Hatching mechanism   47
  - 3.2.6 Hatching of species with multiple generations during a season   49
  - 3.2.7 Durability of hatching and developmental arrest   50
- 3.3 Survival   53
  - 3.3.1 Diapause   54
  - 3.3.2 Osmotic regulation   56
  - 3.3.3 Cold tolerance   56
- 3.4 Sensory Perception   57
  - 3.4.1 Sensilla   57
  - 3.4.2 Plant signals   58
  - 3.4.3 Sex pheromones   60
  - 3.4.4 Ascaroside and peptide signals   61
  - 3.4.5 Blocking sensory perception   62
- 3.5 Conclusions and Future Prospects   63
- 3.6 References   63

4 Biology of Effectors   74
John T. Jones and Melissa G. Mitchum

- 4.1 Introduction   74
  - 4.1.1 What are effectors?   74

		4.1.2	Sources of effectors	75
		4.1.3	Where do effectors go in plants?	75
	4.2	Degradation of the Plant Cell Wall		77
		4.2.1	Cellulases	77
		4.2.2	Degradation of pectin	78
		4.2.3	Non-enzymatic modifications of the plant cell wall	78
	4.3	Suppression of Host Defences		79
	4.4	Manipulation of Host Biochemistry for Feeding Site Induction		82
	4.5	Conclusions and Future Prospects		83
	4.6	Acknowledgements		83
	4.7	References		83

5 Biochemistry — 89
David J. Chitwood and Edward P. Masler

	5.1	Introduction		89
	5.2	Lipids		90
		5.2.1	Content and utilization	90
		5.2.2	Lipid class composition and fatty acids	91
		5.2.3	Phospholipids	91
		5.2.4	Hydrocarbons	92
		5.2.5	Sterols	92
		5.2.6	Ascarosides	92
	5.3	Carbohydrates		93
	5.4	Proteins		93
		5.4.1	Lipid-binding proteins	94
		5.4.2	Feeding, hatching and parasitism	94
		5.4.3	Cell integrity	95
		5.4.4	Neuropeptides	95
	5.5	Conclusions and Future Prospects		96
	5.6	References		96

6 Role of Population Dynamics and Damage Thresholds in Cyst Nematode Management — 101
George W. Bird, Inga A. Zasada and Gregory L. Tylka

	6.1	Introduction		101
	6.2	*Heterodera schachtii* (Sugar Beet Cyst Nematode)		103
		6.2.1	Population dynamics	104
		6.2.2	Crop damage thresholds	105
		6.2.3	Information-based sugar beet cyst management	106
	6.3	*Globodera* spp. (Potato Cyst Nematodes)		107
		6.3.1	Population dynamics	108
		6.3.2	Crop damage thresholds	110
		6.3.3	Application of PCN population dynamics and damage models	110
	6.4	*Heterodera glycines* (Soybean Cyst Nematode)		112
		6.4.1	Population dynamics	112
		6.4.2	Damage thresholds	116
	6.5	The Rest of the Heteroderinae		117
	6.6	Conclusions and Future Prospects		118
	6.7	References		119

7 Quarantine, Distribution Patterns and Sampling — 128
Jon Pickup, Adrian M.I. Roberts and Loes J.M.F. den Nijs

| | 7.1 | Introduction | | 128 |

	7.2	Quarantine	129
		7.2.1 Soybean cyst nematode *Heterodera glycines*	131
		7.2.2 Potato cyst nematodes *Globodera pallida* and *G. rostochiensis*	131
	7.3	Phytosanitary Status	132
		7.3.1 Pest free areas	132
		7.3.2 PCN eradication and containment	132
		7.3.3 PCN management	135
	7.4	Soil Sampling	135
		7.4.1 Detection of cysts/juveniles	136
		7.4.2 Distributions of PCN within fields	137
		7.4.3 Sampling for detection	139
		7.4.4 Sampling for population density estimation	140
		7.4.5 Sampling for quarantine	144
	7.5	Laboratory Diagnosis	144
		7.5.1 Extraction of nematodes	144
		7.5.1.1 Free-living stages from soil	145
		7.5.1.2 Endoparasitic and sedentary stages from roots	145
		7.5.1.3 Cysts from soil	146
		7.5.2 Identification and viability assessment	146
		7.5.3 Optimization for statutory or commercial purposes	147
		7.5.4 Quantitative assessments	148
		7.5.5 Laboratory diagnostics	150
	7.6	Conclusions and Future Prospects	150
	7.7	References	151
8	**Mechanisms of Resistance to Cyst Nematodes**		**154**
	Aska Goverse and Geert Smant		
	8.1	Introduction	154
	8.2	Basal Immunity to Cyst Nematodes	155
		8.2.1 Recognition of cyst nematodes during host invasion	155
		8.2.2 Activation of basal immune responses upon root invasion	155
		8.2.3 Modulation of basal immune responses	156
	8.3	Host-specific Immunity to Cyst Nematodes	156
		8.3.1 *R* gene-mediated cyst nematode resistance	156
		8.3.2 Extra- and intracellular immune receptors encoded by *R* genes	157
		8.3.3 Effector recognition by extra- and intracellular R proteins	158
		8.3.4 Activation of effector-triggered immune responses	159
		8.3.5 Modulation of effector-triggered immune responses	160
	8.4	Quantitative Aspects of Host-specific Immunity	161
		8.4.1 NB-LRR gene clusters linked to quantitative cyst nematode resistance	161
		8.4.2 Resistance based on copy number variation and natural mutations	162
	8.5	Transcriptional Reprogramming and Defence Gene Expression	162
		8.5.1 Local induction of defence gene expression in susceptible host plants	162
		8.5.2 Local induction of defence gene expression in resistant host plants	163
	8.6	Hormone-mediated Defence Responses by Cyst Nematodes	164
		8.6.1 Local induced hormone-mediated defence responses	164
		8.6.2 Systemic induced hormone-regulated defence responses	165
		8.6.3 Modulation of local and systemic hormone-mediated defence responses	166
	8.7	Conclusions and Future Prospects	166
	8.8	References	168

9	**Resistance Breeding**	**174**
	Vivian C. Blok, Gregory L. Tylka, Richard W. Smiley, Walter S. de Jong and Matthias Daub	
9.1	Introduction	174
9.2	Resistance and Tolerance	176
9.3	Sources of Resistance and Genetics	178
9.4	Virulence, Pathotypes and Races	178
9.5	Screening (Phenotyping)	179
	9.5.1 Inoculum	180
	9.5.2 Marker assisted selection	180
9.6	Cereal Cyst Nematodes	181
	9.6.1 Scale of the problem	181
	9.6.2 Sources of resistance	181
	9.6.3 Selection of nematode populations for inoculum	182
	9.6.3.1 Naturally infested soil	182
	9.6.3.2 Cysts as inoculum	182
	9.6.3.3 Juveniles as inoculum	182
	9.6.3.4 Preparation of a new inoculum	182
	9.6.4 Screening assays	183
	9.6.4.1 Field trials	183
	9.6.4.2 Pot trials	183
	9.6.4.3 Miniaturized trials	184
	9.6.5 Interpretation of 'resistance'	185
	9.6.6 Development of resistant cultivars	185
	9.6.7 Future challenges	186
9.7	Potato Cyst Nematodes	187
	9.7.1 Scale of the problem	187
	9.7.2 Sources of resistance	188
	9.7.3 Selection of nematode populations	188
	9.7.3.1 Inoculum	189
	9.7.3.2 Screening protocols	189
	9.7.4 Development of resistant cultivars	190
	9.7.5 Future challenges	191
9.8	Soybean Cyst Nematodes	191
	9.8.1 Scale of the problem	191
	9.8.2 Sources of resistance	192
	9.8.3 Selection of nematode populations	194
	9.8.4 Screening protocols	194
	9.8.5 Developing resistant cultivars	196
	9.8.6 Future challenges	196
9.9	Sugar Beet Cyst Nematodes	197
	9.9.1 Scale of the problem	197
	9.9.2 Sources of resistance	198
	9.9.3 Selection of nematode populations	199
	9.9.4 Screening protocols	199
	9.9.5 Development of resistant cultivars	199
	9.9.6 Future challenges	200
9.10	Other Cyst Nematodes	201
	9.10.1 Rice cyst nematodes	201
	9.10.2 Pea cyst nematode	202
	9.10.3 Pigeon pea cyst nematode	202
	9.10.4 Clover cyst nematode	202
	9.10.5 Carrot cyst nematode and corn cyst nematodes	202
9.11	Conclusions and Future Prospects	202
9.12	References	203

10	**Plant Biotechnology Approaches: From Breeding to Genome Editing**	**215**
	Godelieve Gheysen and Catherine J. Lilley	
	10.1 Introduction	215
	10.2 PCR and Next Generation Sequencing	216
	10.3 Molecular Research into Plant–Nematode Interactions: Using Molecular Biology to Understand the Parasite	217
	10.4 Genetic Engineering: The Basic Principles	217
	10.5 Breeding or Genetic Engineering with Natural Resistance Genes	218
	10.6 Genetic Engineering Using Genes that Interfere with Nematode Feeding and Digestion	219
	10.7 Secretion of Peptide Repellents from Root Tips Inhibits Nematode Invasion	220
	10.8 RNAi Acting across Kingdom Borders	221
	10.9 Susceptibility Genes as an Alternative Target for Nematode Resistance	222
	10.10 Genome Editing Can Adapt Susceptibility Genes with Surgical Precision	224
	10.11 Strategic Development of Resistance Technology	225
	10.11.1 Field trials of GM nematode-resistant crops	225
	10.11.2 Biosafety considerations	227
	10.11.3 Promoter choice can target transgene expression for enhanced biosafety	228
	10.11.4 The route to uptake of future GM crops for nematode control	229
	10.12 Conclusions and Future Prospects	229
	10.13 References	230
11	**Biological Control of Cyst Nematodes through Microbial Pathogens, Endophytes and Antagonists**	**237**
	Keith G. Davies, Sharad Mohan and Johannes Hallmann	
	11.1 Introduction	237
	11.2 Nematophagous and Antagonistic Fungi	238
	11.2.1 Nematophagous fungi	238
	11.2.1.1 Nematode–trapping fungi	238
	11.2.1.2 Endoparasitic fungi	240
	11.2.1.3 Fungi parasitizing eggs and females	241
	11.2.1.4 Toxin-producing fungi	242
	11.2.2 Antagonistic fungi	242
	11.2.2.1 Endophytic fungi	242
	11.2.2.2 Arbuscular mycorrhiza fungi	243
	11.3 Rhizobacteria and Root Endophytic Bacteria	245
	11.3.1 *Pseudomonas* species	245
	11.3.2 *Bacillus* species	246
	11.3.3 *Rhizobium etli*	247
	11.3.4 Chitinase-producing bacteria	247
	11.3.5 *Pasteuria* species	248
	11.3.6 *Brevibacillus laterosporus*	248
	11.3.7 Others	248
	11.4 Microbial–Nematode Molecular Interactions	249
	11.4.1 Insights from *Caenorhabditis elegans*	249
	11.4.2 Microbial diversity and molecular interactions	250
	11.4.2.1 Microbial inter- and intra-specific extracellular enzyme diversity	251
	11.4.2.2 Chitinases	251
	11.4.2.3 Proteinases	251
	11.4.3 Attachment and surface recognition	252

		11.5	Biological Control Products	253

		11.5	Biological Control Products	253
		11.6	Managing Biological Control	254
			11.6.1 Effect of tillage	254
			11.6.2 Organic matter content	256
			11.6.3 C:N ratio	256
		11.7	Conclusions and Future Prospects	257
		11.8	References	258

12 Interactions with Other Pathogens — 271
Horacio D. Lopez-Nicora and Terry L. Niblack

- 12.1 Introduction — 271
- 12.2 Defining Interactions — 272
- 12.3 Methodologies for Investigation of Interactions between Nematodes and Other Organisms — 272
- 12.4 Biological and Statistical Evidence of Interactions — 274
- 12.5 Mechanisms of Interactions — 276
  - 12.5.1 Mechanisms of synergistic interactions — 276
    - 12.5.1.1 The 'wounding agent' effect — 277
    - 12.5.1.2 The 'physiological changes' effect — 278
    - 12.5.1.3 The 'modified rhizosphere' effect — 279
    - 12.5.1.4 The 'resistance breaker' effect — 279
  - 12.5.2 Mechanisms of antagonistic interactions — 280
- 12.6 Interactions between Cyst Nematodes and Other Organisms — 281
  - 12.6.1 Cyst nematodes – oomycete and fungal pathogen interactions — 281
    - 12.6.1.1 Interactions between *Heterodera glycines* and fungal pathogens — 281
    - 12.6.1.2 Interactions between *Heterodera schachtii* and fungal pathogens — 284
    - 12.6.1.3 Interactions between *Globodera* spp. and fungal pathogens — 288
    - 12.6.1.4 Interactions between cereal cyst nematodes and fungal pathogens — 290
    - 12.6.1.5 Interactions between other cyst nematodes and fungal pathogens — 291
  - 12.6.2 Interactions between cyst nematodes and other nematodes — 291
  - 12.6.3 Interactions between cyst nematodes and symbionts — 292
    - 12.6.3.1 Interactions between cyst nematodes and rhizobia — 292
    - 12.6.3.2 Interactions between cyst nematodes and mycorrhizae — 293
- 12.7 Conclusions and Future Prospects — 294
- 12.8 References — 295

13 Field Management and Control Strategies — 305
Matthew A. Back, Laura Cortada, Ivan G. Grove and Victoria Taylor

- 13.1 Introduction — 305
- 13.2 Cultural Control Methods for Cyst Nematodes — 306
  - 13.2.1 Crop rotation — 306
  - 13.2.2 Soil cultivations (tillage) — 308
  - 13.2.3 Anaerobic soil disinfection, soil amendments and inundation — 309
  - 13.2.4 Solarization — 309
  - 13.2.5 Weed management — 310
- 13.3 Biofumigation — 311
- 13.4 Trap Cropping and Natural Resistance — 313
- 13.5 Use of Plant Biomass, Oils and Extracts — 314

	13.6	Application of Biological Control Agents	315
		13.6.1 Direct antagonism through the application of biological control agents	316
		13.6.2 Antagonists and endophytes: Suppressing cyst nematodes through the enhancement of the soil food web	319
	13.7	Agrochemical Control of Cyst Nematodes	320
		13.7.1 Active substances currently in use	321
		13.7.2 Nematicide persistence	325
		13.7.3 Nematicide registration and stewardship	326
		13.7.4 Precision aspects of nematode management	326
	13.8	Conclusions and Future Prospects	327
	13.9	References	328

14 General Morphology of Cyst Nematodes 337
James G. Baldwin and Zafar A. Handoo

14.1	Introduction	337
14.2	Egg and Embryo	338
14.3	Second- to Fourth-stage Juveniles	340
14.4	Males	346
14.5	Females	347
14.6	Cysts	350
14.7	Techniques	352
	14.7.1 Light microscopy (LM)	352
	14.7.2 Scanning electron microscopy (SEM)	353
	14.7.3 Transmission electron microscopy (TEM)	355
14.8	Minimal Standards for Species Descriptions	357
14.9	Conclusions and Future Prospects	359
14.10	References	360

15 Taxonomy, Identification and Principal Species 365
Zafar A. Handoo and Sergei A. Subbotin

15.1	Introduction	365
	15.1.1 History	365
	15.1.2 Major reference sources	366
15.2	Identification	366
	15.2.1 General techniques	367
	15.2.2 Cone mounts	368
	15.2.3 Root staining	368
	15.2.4 Scanning electron microscopy	368
	15.2.5 Diagnostic characters	368
15.3	Systematic Position	368
15.4	Subfamily Heteroderinae Diagnosis	369
15.5	Genus *Heterodera* Schmidt, 1871	369
	15.5.1 List of species and synonyms	369
	15.5.2 Principal species	373
	15.5.2.1 European cereal cyst nematode, *H. avenae*	373
	15.5.2.2 Yellow beet cyst nematode, *H. betae*	374
	15.5.2.3 Pigeon pea cyst nematode, *H. cajani*	376
	15.5.2.4 Carrot cyst nematode, *H. carotae*	377
	15.5.2.5 Cabbage cyst nematode, *H. cruciferae*	379
	15.5.2.6 Japanese cyst nematode, *H. elachista*	379
	15.5.2.7 Fig cyst nematode, *H. fici*	380

		15.5.2.8	Filipjev cereal cyst nematode, *H. filipjevi*	380
		15.5.2.9	Soybean cyst nematode, *H. glycines*	380
		15.5.2.10	Pea cyst nematode, *H. goettingiana*	381
		15.5.2.11	Barley cyst nematode, *H. hordecalis*	382
		15.5.2.12	Hop cyst nematode, *H. humuli*	382
		15.5.2.13	Mediterranean cereal cyst nematode, *H. latipons*	383
		15.5.2.14	Sugar beet cyst nematode, *H. schachtii*	383
		15.5.2.15	Clover cyst nematode, *H. trifolii*	384
		15.5.2.16	Maize cyst nematode, *H. zeae*	384

15.6 Subfamily Punctoderinae Diagnosis — 384
15.7 Genus *Globodera* Skarbilovich, 1959 — 385
    15.7.1 List of species and synonyms — 385
    15.7.2 Principal species — 386
        15.7.2.1 Golden potato cyst nematode, *G. rostochiensis* — 386
        15.7.2.2 Ellington potato cyst nematode, *G. ellingtonae* — 387
        15.7.2.3 Pale potato cyst nematode, *G. pallida* — 388
        15.7.2.4 Tobacco cyst nematode, *G. tabacum* — 389
15.8 Genus *Punctodera* Mulvey & Stone, 1976 — 390
    15.8.1 List of species and synonyms — 390
    15.8.2 Principal species — 390
        15.8.2.1 Grass cyst nematode, *P. punctata* — 390
15.9 Genus *Cactodera* Krall & Krall, 1978 — 390
    15.9.1 List of species and synonyms — 391
    15.9.2 Principal species — 391
        15.9.2.1 Cactus cyst nematode, *Cactodera cacti* — 391
15.10 Genus *Dolichodera* Mulvey & Ebsary, 1980 — 391
15.11 Genus *Betulodera* Sturhan, 2002 — 392
15.12 Genus *Paradolichodera* Sturhan, Wouts & Subbotin, 2007 — 393
15.13 Genus *Vittatidera* Bernard, Handoo, Powers, Donald & Heinz, 2010 — 393
15.14 Conclusions and Future Prospects — 393
15.15 Acknowledgements — 393
15.16 References — 394

16 Molecular Taxonomy and Phylogeny — 399
Sergei A. Subbotin and Andrea M. Skantar

16.1 Introduction — 399
16.2 Nuclear Ribosomal RNA Genes — 400
16.3 Nuclear Protein-coding Genes — 400
    16.3.1 Heat shock protein 90 — 400
    16.3.2 Actin — 401
    16.3.3 Fructose-bisphosphate aldolase — 402
    16.3.4 β-tubulin gene — 403
16.4 Mitochondrial DNA Genome Organization — 403
    16.4.1 Mitochondrial DNA genes — 406
16.5 Origin and Phylogeny of Heteroderidae — 407
    16.5.1 Phylogeny of Heteroderinae — 407
16.6 Phylogeny and Phylogeography of Punctoderinae — 409
16.7 Phylogeny and Phylogeography of *Globodera* — 410
16.8 Co-evolution of Cyst Nematodes with their Host Plants — 412
16.9 Conclusions and Future Prospects — 413
16.10 References — 414

17	**Biochemical and Molecular Identification**			**419**
	Lieven Waeyenberge			
	17.1	Statutory and Non-Statutory Issues		419
	17.2	Biochemical and Molecular Identification		421
		17.2.1	Introduction	421
		17.2.2	Protein-based identification	422
			17.2.2.1 Isoelectric focusing	422
			17.2.2.2 Antibody mediated detection and identification	424
		17.2.3	DNA-based identification and detection	425
			17.2.3.1 rDNA markers	425
			17.2.3.2 mtDNA markers	427
			17.2.3.3 sDNA markers	430
			17.2.3.4 Other DNA markers	432
			17.2.3.5 Detection in soil	433
			17.2.3.6 Barcoding	433
	17.3	Conclusions and Future Prospects		434
	17.4	References		436

Genes Index	**443**
Nematodes Index	**445**
General Index	**451**

About the Editors

Roland N. Perry Professor Roland Perry is based at the University of Hertfordshire, UK. He graduated with a BSc (Hons) in zoology from Newcastle University, UK, where he also obtained a PhD in zoology on physiological aspects of desiccation survival of *Ditylenchus* spp. After a year's postdoctoral research at Newcastle, he moved to Keele University, UK, where he taught parasitology; after 3 years at Keele, he was appointed to Rothamsted Experimental Station (now Rothamsted Research). His research interests centred primarily on plant-parasitic nematodes, especially focusing on nematode hatching, sensory perception, behaviour and survival physiology, and several of his past PhD and postdoctoral students are currently involved in nematology research. He remained at Rothamsted until 2014, when he moved to the Department of Biological and Environmental Sciences, University of Hertfordshire.

He co-edited *The Physiology and Biochemistry of Free-living and Plant-parasitic Nematodes* (1998), *Root-knot Nematodes* (2009), *Molecular and Physiological Basis of Nematode Survival* (2011) and the first (2006) and second (2013) editions of the text book, *Plant Nematology*. He is author or co-author of over 40 book chapters and refereed reviews and over 100 refereed research papers. He is Editor-in-Chief of *Nematology* and Chief Editor of the *Russian Journal of Nematology*. He co-edits the book series Nematology Monographs and Perspectives. In 2010, he was elected Fellow of the Society of Nematologists (USA) in recognition of his research achievements; in 2008 he was elected Fellow of the European Society of Nematologists for outstanding contribution to the science of nematology; and in 2011 he was elected Honorary Member of the Russian Society of Nematologists. He is a visiting professor at Ghent University, Belgium, where he lectures on nematode biology, focusing on physiology and behaviour.

Maurice Moens Professor Maurice Moens is Honorary Director of Research at the Flanders Research Institute for Agriculture, Fisheries and Food (ILVO) at Merelbeke, Belgium, and an honorary professor at Ghent University, Belgium, where he gave a lecture course on agro-nematology at the Faculty of Bioscience Engineering. He is a past director of the postgraduate international nematology course (MSc nematology) and coordinator of the Erasmus Mundus – European Master of Science in nematology, where he gave five lecture courses on plant nematology. The MSc course is organised in the Faculty of Sciences of Ghent University.

He graduated as an agricultural engineer from Ghent University and obtained a PhD at the same University on the spread of plant-parasitic nematodes and their management in hydroponic cropping systems. Within the framework of the Belgian Cooperation, he worked from 1972 to 1985 as a researcher in crop protection, including nematology, at two research stations in Tunisia. Upon his return to Belgium, he was appointed as senior nematologist at the Agricultural Research Centre (now ILVO). There, he expanded the research in plant nematology over various areas covering molecular characterization, biology of host–parasite relationships, biological control, resistance and other forms of non-chemical control. He was appointed Head of the Crop Protection Department in 2000 and became Director of Research in 2006. He retired from both ILVO and Ghent University in 2012 but continued to supervise PhD students until 2017. In 2001, he was elected Fellow of the Society of Nematologists (USA) for outstanding contributions to nematology; in the same year he was elected Fellow of the European Society of Nematologists for outstanding contribution to the science of nematology. In 2012 he was elected Honorary Fellow of the Chinese Society for Plant Nematology, and in 2013 he became Honorary Member of the Russian Society of Nematologists. He supervised 27 PhD students, who are active in nematology all over the world. He is past president of the European Society of Nematologists (2010–2014). He co-edited *Root-knot Nematodes* (2009) and the first (2006) and second (2013) editions of the text book, *Plant Nematology*. He is author or co-author of ten book chapters and refereed reviews and over 150 refereed research papers. He is a member of the editorial board of the *Russian Journal of Nematology*.

John Jones Professor John Jones is Head of the Cell and Molecular Sciences Department at the James Hutton Institute and holds a joint appointment as Professor of Biology at The University of St Andrews. He is also a guest professor at Ghent University, Belgium. John graduated with a BSc (Hons) in Zoology from Newcastle University and obtained a PhD from University College of Wales, Aberystwyth, on the structure and function of nematode sense organs. This project was undertaken jointly with Rothamsted Research. Following a 2-year postdoctoral position at the University of South Carolina, John moved to the Scottish Crop Research Institute in 1993 and has remained there since then. The Scottish Crop Research Institute merged with the Macauley Land Use Research Institute in 2011 to form the James Hutton Institute. John's research is focused on understanding the molecular basis of the interactions between plant-parasitic nematodes and their hosts, and has included extensive analysis of nematode genomes and transcriptomes. John is author or co-author of over 85 refereed research papers and has co-edited two books. John is a member of the editorial board of *Nematology* and a senior editor for *Molecular Plant Pathology*.

Contributors

Matthew A. Back Centre for Integrated Pest Management (CIPM), Harper Adams University, Newport, Shropshire TF10 8NB, UK (mback@harper-adams.ac.uk)

James G. Baldwin Department of Nematology, University of California, 900 University Avenue, Riverside, CA 92521, USA (james.baldwin@ucr.edu)

George W. Bird Department of Entomology, Michigan State University, 288 Farm Lane, East Lansing, MI 48824, USA (birdg@msu.edu)

Vivian C. Blok Cell and Molecular Sciences Group, The James Hutton Institute, Dundee, DD2 5DA, UK (vivian.blok@hutton.ac.uk)

David J. Chitwood, Mycology and Nematology Genetic Diversity and Biology Laboratory, USDA, ARS, Building 010A, BARC-West, 10300 Baltimore Avenue, Beltsville, MD 20705, USA (david.chitwood@ars.usda.gov)

Laura Cortada Soil Health (Nematology), International Institute of Tropical Agriculture, icipe campus, Kasarani, PO Box 30772-00100, Nairobi, Kenya (l.cortada-gonzalez@cgiar.org)

Matthias Daub Julius Kühn-Institut, Federal Research Centre for Cultivated Plants, Institute for Plant Protection in Field Crops and Grassland – Field station Elsdorf-Dürener Str. 71, D-50189, Elsdorf, Germany (matthias.daub@julius-kuehn.de)

Keith G. Davies Department of Biological and Environmental Sciences, School of Life and Medical Sciences, University of Hertfordshire, Herts AL10 9AB, UK (k.davies@herts.ac.uk)

Walter S. de Jong School of Integrative Plant Science, College of Agriculture and Life Sciences, Section of Plant Breeding and Genetics, Cornell University, 240 Emerson Hall, Ithaca, NY 14853, USA (walter.dejong@cornell.edu)

Loes J.M.F. den Nijs National Reference Centre, Netherland Food and Consumer Product Safety Authority, Geertjesweg 15, 6706 EA Wageningen, The Netherlands (l.j.m.f.dennijs@nvwa.nl)

Sebastian Eves-van den Akker Division of Plant Sciences, College of Life Sciences, University of Dundee, Dundee, DD1 5EH, UK (s.evesvandenakker@dundee.ac.uk) and Department of Biological Chemistry, The John Innes Centre, Norwich, Norfolk NR4 7UH, UK (sebastian.eves-vandenakker@jic.ac.uk)

Godelieve Gheysen Department of Biotechnology, Ghent University, Coupure Links 653, B-9000 Ghent, Belgium (godelieve.gheysen@ugent.be)

Aska Goverse Laboratory of Nematology, Wageningen University, 6708PD Wageningen, The Netherlands (aska.goverse@wur.nl)

Ivan G. Grove Centre for Integrated Pest Management (CIPM), Harper Adams University, Newport, Shropshire TF10 8NB, UK (igrove@harper-adams.ac.uk)

Johannes Hallmann Julius Kühn-Institut, Federal Research Centre for Cultivated Plants, Institute for Epidemiology and Pathogen Diagnostics, Toppheideweg 88, 48161 Münster, Germany (johannes.hallmann@julius-kuehn.de)

Zafar A. Handoo Mycology and Nematology Genetic Diversity and Biology Laboratory, USDA, ARS, Northeast Area, Bldg. 010A, BARC-West, 10300 Baltimore Avenue, Beltsville, MD 20705, USA (zafar.handoo@ars.usda.gov)

John T. Jones Cell and Molecular Sciences Group, Dundee Effector Consortium, The James Hutton Institute, Dundee, DD2 5DA, UK (john.jones@hutton.ac.uk)

Catherine J. Lilley Centre for Plant Sciences, School of Biology, University of Leeds, Leeds, LS2 9JT, UK (c.j.lilley@leeds.ac.uk)

Horacio D. Lopez-Nicora Department of Plant Pathology, 248 Kottman Hall, 2021 Coffey Road, The Ohio State University, Columbus, OH 43210, USA (lopez-nicora.1@osu.edu)

Edward P. Masler Mycology and Nematology Genetic Diversity and Biology Laboratory, USDA, ARS, Building 010A, BARC-West, 10300 Baltimore Avenue, Beltsville, MD 20705, USA (edward.masler@ars.usda.gov)

Melissa G. Mitchum Division of Plant Sciences and Bond Life Sciences Center, University of Missouri, Columbia, Missouri, MO 65211, USA (goellnerm@missouri.edu)

Maurice Moens ILVO, Flanders Research Institute for Agriculture, Fisheries and Food, Burg. Van Gansberghelaan 92, B-9820 Merelbeke, Belgium (maurice.moens@ilvo.vlaanderen.be)

Sharad Mohan Division of Nematology, Indian Agricultural Research Institute, New Delhi 110012, India (sharad@iari.res.in)

Terry L. Niblack College of Food, Agricultural, and Environmental Sciences, 140 Agricultural Administration Building, 2120 Fyffe Road, The Ohio State University, Columbus, OH 43210, USA (niblack.2@osu.edu)

Roland N. Perry Department of Biological and Environmental Sciences, School of Life and Medical Sciences, University of Hertfordshire, Herts AL10 9AB, UK (r.perry2@herts.ac.uk)

Jon Pickup Virology and Zoology, Science and Advice for Scottish Agriculture, Edinburgh, EH12 9FJ, UK (jon.pickup@sasa.gsi.gov.uk)

Adrian M.I. Roberts Biomathematics & Statistics Scotland, JCMB, King's Buildings, Peter Guthrie Tait Road, Edinburgh EH9 3FD, UK (adrian.roberts@bioss.ac.uk)

Andrea M. Skantar Mycology and Nematology Genetic Diversity and Biology Laboratory, USDA, ARS, Building 010A, BARC-West, 10300 Baltimore Avenue, Beltsville, MD 20705, USA (andrea.skantar@ars.usda.gov)

Geert Smant Laboratory of Nematology, Wageningen University, 6708PD Wageningen, The Netherlands (geert.smant@wur.nl)

Richard W. Smiley Columbia Basin Agricultural Research Center, Oregon State University, Pendleton, OR 97801, USA (richard.smiley@oregonstate.edu)

Sergei A. Subbotin Plant Pest Diagnostic Center, California Department of Food and Agriculture, 3294 Meadowview Road, Sacramento, CA 95832, USA (sergei.a.subbotin@gmail.com)

Victoria Taylor Arcis Biotechnology, Daresbury Innovation Centre, Sci-Tech Daresbury, Cheshire WA4 4FS, UK (victoria.taylor@arcisbio.com)

Gregory L. Tylka Department of Plant Pathology and Microbiology, Iowa State University, 321 Bessey Hall, Ames, IA 50011-1020, USA (gltylka@iastate.edu)

Lieven Waeyenberge ILVO, Flanders Research Institute for Agriculture, Fisheries and Food, Burg. Van Gansberghelaan 96, B-9820, Merelbeke, Belgium (lieven.waeyenberge@ilvo.vlaanderen.be)

Inga A. Zasada Horticultural Crops Research Laboratory, USDA, ARS, 3420 NW Orchard Ave., Corvallis, OR 97330, USA (inga.zasada@ars.usda.gov)

Preface

Cyst nematodes are an amazing example of a successful parasite group and the damage they cause to crops has considerable detrimental economic and social impacts world-wide, and several species are listed as quarantine organisms. Although cyst nematodes were originally considered to be largely a pest of temperate regions, many species are now known to be present in tropical and sub-tropical regions, making them globally important agricultural pests. Several aspects of their biology, from their ability to induce complex feeding sites in host plants, through the development of the cyst as a protective feature, to the sophisticated hatching and survival physiology of the infective juveniles set them apart from other parasites. The combination of the economic impact and their unique biology makes them a fascinating and challenging subject for research.

Advances in our understanding of the molecular biology of cyst nematodes have contributed to a deeper knowledge of host–parasite interactions. This has been driven by the availability of genome and transcriptome sequences for some of the most economically important species. As well as contributing to the possibility of novel control options, the information is interesting from a purely scientific aspect. However, the driving impetus has more recently been engendered by the need to develop environmentally acceptable ways to manage and control these pests. An enormous volume of literature has accumulated since the first description of a cyst nematode and it is important to distil and integrate this information. To ensure that older and directly relevant work is not ignored, in this volume we aimed to include some of the earlier research and link it to the more recent advances facilitated by molecular investigations.

We are grateful to the chapter authors for their time and dedication in contributing to this volume. Their expertise is essential in ensuring as complete an overview as possible of cyst nematodes.

Roland N. Perry
Maurice Moens
John T. Jones
May 2017

1 Cyst Nematodes – Life Cycle and Economic Importance

Maurice Moens[1,2], Roland N. Perry[2,3] and John T. Jones[2,4,5]

[1]*Flanders Research Institute for Agriculture, Fisheries and Food, Merelbeke, Belgium;* [2]*Ghent University, Ghent, Belgium;* [3]*University of Hertfordshire, Hatfield, Hertfordshire, UK;* [4]*The James Hutton Institute, Invergowrie, Dundee, UK;* [5]*University of St Andrews, St Andrews, UK*

1.1	Introduction	1
1.2	Impact	2
1.3	History of the Genus	2
1.4	Distribution	3
1.5	Identification	3
1.6	Life Cycle	4
1.7	Syncytium	7
1.8	Effect of Abiotic Factors	8
1.9	Important Species	8
1.10	Pathotypes and Races	13
1.11	Symptoms	14
1.12	Management	17
1.13	References	18

1.1 Introduction

Cyst nematodes are remarkable parasites. They have highly specialized interactions with plants and induce the formation of a unique feeding structure, the syncytium, within the roots of their hosts. Cyst nematodes are of enormous economic importance throughout the world and the various species infect all of the world's most important crops (Jones *et al.*, 2013); they also share a unique feature: the ability of the female to turn her cuticle into a durable, protective capsule for her eggs. Cyst nematodes are classified within eight genera of the subfamily Heteroderinae: *Heterodera*, *Globodera*, *Cactodera*, *Dolichodera*, *Paradolichodera*, *Betulodera*, *Punctodera* and *Vittatidera* (Subbotin *et al.*, 2010a, b). However, not all genera of the Heteroderinae are cyst forming; for example, *Atalodera*, *Bellodera*, *Meloidodera* and *Verutus* (Evans and Rowe, 1998). The most economically important species occur within the genera *Heterodera* and *Globodera*.

All cyst nematodes are obligatory endoparasites feeding within the roots of their hosts. Following fertilization and production of eggs, the body wall of the female tans and desiccates. This generates a long-lasting cyst that holds a large number of embryonated eggs, which can survive for long periods until a suitable host is available. This persistence is one of the characteristics that explain the economic importance of this group. At low nematode densities, above-ground symptoms of cyst nematode damage may be minimal. When nematode populations increase, host plants may become stunted and wilt. These symptoms are often misattributed to other abiotic factors, such as soil characteristics, mineral nutrition and water availability, or diseases.

1.2 Impact

The most damaging species include soybean cyst nematodes (SCN) (*Heterodera glycines*), potato cyst nematodes (PCN) (including *Globodera pallida* and *G. rostochiensis*) and cereal cyst nematodes (CCN) (including *Heterodera avenae* and *H. filipjevi*) (Jones *et al.*, 2013). Losses caused by these and other cyst nematodes are difficult to define. PCN have been estimated to cause losses of 9% of total potato production worldwide (Turner and Subbotin, 2013). SCN are thought to be responsible for losses in excess of US$1.5 billion each year in the USA alone (Chen *et al.*, 2001). Losses caused by CCN largely depend on environmental conditions but may sometimes exceed 90% (Nicol *et al.*, 2011).

In addition to this direct impact on crop yield, cyst nematodes also have an indirect effect due to increased costs incurred by growers for their control and due to a reduction of product quality. Control of cyst nematodes uses a number of strategies including host resistance, prolonged rotation, chemicals, the action of biological antagonists (which may be stimulated through solarization) or trap crops (see Chapters 9 and 13, this volume). To reduce potential costs for control, quarantine status is assigned to some species. For example, PCN, a group of global importance, are native to South America (Plantard *et al.*, 2008; Grenier, 2010) from where they were introduced in Europe along with potatoes (Turner and Evans, 1998). Because of their substantial impact on potato production and their ease of dissemination, PCN are listed as quarantine nematodes in many countries around the world (CABI, 2016a; Chapter 7, this volume). The rapid expansion in the area of soybean production in the USA that started in the mid-1900s served to establish a huge reservoir of hosts that permitted a similarly large increase in SCN (Davis and Tylka, 2000). A federal quarantine for SCN was established in 1957; however, the quarantine was lifted in 1972 because it was ineffective. It has been suggested that the movement of SCN-infested soil and plant material into new soybean production areas had already occurred before the quarantine could be established.

1.3 History of the Genus

Wouts and Baldwin (1998) wrote an excellent review of the taxonomy and identification of the cyst nematodes, including historical aspects. The following is partly inspired by this review. The first cyst-forming nematode was reported by Schacht (1859) from sugar beets that showed poor development. The nematode was later described by Schmidt (1871) as *Heterodera schachtii* in honour of its discoverer. During the years that followed, further cyst nematodes were reported to cause similar effects on other crops; it was believed that *H. schachtii* was the causal agent in all of these cases. However, when cyst nematodes collected from peas did not infect oats, a well-known host of *H. schachtii*, it became clear that other cyst-forming species were involved. Eventually, Liebscher (1892) described the species as *H. goettingiana*. Based on the host specificity of cyst nematodes, Wollenweber described the potato cyst nematode (*G. rostochiensis*) and oat cyst nematode (*H. avenae*) in 1923 and 1924, respectively (Wollenweber, 1923, 1924). All of these species were detected in Europe; *H. punctata* was the first cyst nematode species to be described from North America (Thorne, 1928). Being unaware of the description of *H. rostochiensis*, a species with a spherical cyst, the author considered the spherical shape of the cyst the basis of *H. punctata*. In his review of the genus *Heterodera*, Filipjev (1934) recognized seven cyst-forming species. Before restoration of the earlier generic name *Meloidogyne* by Chitwood (1949), the genus *Heterodera* contained both cyst-forming species of Heteroderidae and various root-knot species under the name *H. marioni* (Luc, 1986; Moens *et al.*, 2009). With her monograph on the genus *Heterodera*, Franklin (1951) enabled the identification of ten species. However, the host range remained for several years the standard for the identification of *Heterodera* species. In 1959, Skarbilovich erected the subgenus *Globodera* to group the round cyst nematode species, including the PCN. Later, Behrens (1975) raised *Globodera* to generic level.

On the basis of solid morphological characters the former genus *Heterodera* was eventually split into the genera *Heterodera*, *Globodera*, *Punctodera* and *Cactodera* (Wouts and Baldwin, 1998). The genera *Dolichodera*, *Betulodera*, *Paradolichodera* and *Vittatidera* (each represented by one species) were added later (Turner and Subbotin, 2013). In 2017, eight genera and a total of 121 species are recognized within the cyst nematodes (see Chapter 15, this volume).

1.4 Distribution

For a long time, cyst nematodes were thought to occur only in temperate areas (Luc, 1986) or to be largely a pest of temperate regions. The first report of a cyst nematode in a tropical crop was on sugarcane in Hawaii (Muir and Henderson, 1926). When Luc (1961) reported two *Heterodera* species, one from swamp rice in the Ivory Coast and another from sugarcane in the Congo, it became clear that cyst nematodes could occur in the tropics. The species were described as *H. oryzae* and *H. sacchari*, respectively. Later, several other species were described from tropical crops and weeds (Fourie *et al.*, 2017; Chapter 15, this volume). Their host range may be large (e.g. *H. cajani*) or restricted to a small number of plants (e.g. *H. sacchari*).

Within the genus *Globodera*, three species are of major importance: *G. pallida*, *G. rostochiensis* and *G. tabacum*. They are found worldwide in temperate regions or in temperate areas of tropical regions where their hosts are grown. The range of their economically important hosts is restricted to Solanaceae. The genus *Heterodera* contains many more species of economic importance. Species important in temperate regions are (major hosts): *H. avenae* (cereals), *H. filipjevi* (cereals), *H. latipons* (cereals), *H. cruciferae* (Cruciferae), *H. glycines* (legumes), *H. schachtii* (various) and *H. trifolii* (various). The following species are important in tropical regions: *H. cajani* (legumes), *H. oryzicola* (rice, banana), *H. sacchari* (rice, sugarcane), *H. sorghi* (cereals) and *H. zeae* (cereals).

The most important CCN, that is, *H. avenae*, *H. filipjevi* and *H. latipons*, differ in their distribution (reviewed by Toumi *et al.*, 2017). *Heterodera avenae* was first reported in Germany and subsequently detected in most European countries. It has also been detected in Asia and South Africa, New Zealand, Peru, Canada, the Middle East, North Africa and the USA. *Heterodera latipons* is distributed in the Mediterranean region, but has also been detected in the former Soviet Union, Iran, Japan and Canada. *Heterodera filipjevi* was first reported in the Sverdlovsk region (Russia). Later, it was found in the former Soviet Union, Europe (Norway, Germany, Poland, Spain and Sweden), Turkey, Iran, Syria, India, Tajikistan and the USA. *Heterodera glycines* is a major pest of soybean in regions of the USA, particularly semi-arid areas. The nematode has also been found as a pest of soybean outside the USA (e.g. Argentina, Brazil, Colombia, China, Egypt, Indonesia, Iran, Italy, Japan, Korea, Paraguay and the former Soviet Union) (CABI, 2016b). Both *H. schachtii* and *H. trifolii* have a cosmopolitan distribution. Other species have a rather limited distribution: *H. cajani* is restricted to India and neighbouring countries (Pakistan and Myanmar; CABI, 2016c); *H. oryzicola* has only been identified in India; *H. sacchari* in several West African countries, India and Pakistan; *H. sorghi* in India and Pakistan; whilst *H. zeae* has been identified in India, Pakistan, Egypt and the USA. The patchy distributions of some species may reflect a restricted presence in a relatively small geographical area but may also reflect redistribution due to human activities as well as the lack of thorough sampling in some areas.

1.5 Identification

1.5.1 Traditional identification

Species of the Heteroderinae share similar morphology and are often distinguished from each other only by small details (Turner and Subbotin, 2013). Traditional identification is mainly based on the morphology and morphometrics of both females (cysts) and second-stage juveniles (J2) (Chapters 14–15, this volume). These stages are found in soil extracts and possess useful species-discriminating features in both morphology and morphometrics.

1.5.1.1 Cysts

The presence or absence of a vulval cone is an important character separating genera (*Heterodera* is the only genus with a prominent vulval cone; the cone is reduced in size in *Cactodera*; no vulval cone is present in *Globodera*, *Dolichodera* and *Punctodera*, hence the round-shaped cyst). A thin-walled area surrounds the vulva. The thin wall can be lost; the opening that is formed is called the fenestra. The degree of fenestration (presence or absence; shape) is used in the diagnosis of genera and species. Measurements of the fenestral area (fenestral length and width, vulval bridge width and length, length of underbridge, length of vulval slit) as well as vulval features (e.g. the presence of bullae) are very useful for species and genus diagnostics.

1.5.1.2 Second-stage juveniles

Both morphometrics (body length and width, hyaline tail length, true tail length, stylet length, stylet knob width) and morphology (stylet knobs shape, number of lines in lateral field, number head annules) of J2 are used to separate genera and species.

1.5.2 Molecular identification

Because identification of cyst nematodes is time consuming and difficult, especially when the sample contains more than one species, biochemical techniques allowing a clear-cut identification have been developed. It was shown that isoelectric focusing (IEF) could be used to identify *G. rostochiensis* and *G. pallida* (Fleming and Marks, 1982) and to separate *Heterodera* species of the *Avenae* group (Subbotin et al., 1996) on the basis of different protein profiles. IEF is used as a routine diagnostic technique for *G. rostochiensis* and *G. pallida* (Karssen et al., 1995). Esterase isoenzyme patterns of the white females have been used reliably to identify *H. cajani*, *H. graminis*, *H. sorghi* and *H. zeae* (Meher et al., 1998).

Compared with IEF, DNA-based identification techniques have several advantages. DNA profiles can be obtained rapidly from a few or even single nematodes and the clarity of the results enables species to be identified very easily. In addition, no problems are encountered with the effects of environmental and developmental variation (Subbotin et al., 2013). Polymerase chain reaction-restriction fragment length polymorphism (PCR-RFLP) and PCR with species-specific primer(s) presently are used for diagnostics of many cyst nematode species (Chapter 17, this volume). PCR-RFLP of the internal transcribed spacer (ITS) region of the ribosomal RNA (rRNA) gene has been used for identifying species from the genus *Heterodera* (Subbotin et al., 1999, 2000; Zheng et al., 2000; Madani et al., 2004). Species-specific primers have been developed for major species, such as *G. rostochiensis* and *G. pallida* (Mulholland et al., 1996; Bulman and Marshall, 1997), *H. schachtii* (Amiri et al., 2001, 2002), *H. glycines* (Subbotin et al., 2001), *H. latipons* (Toumi et al., 2013a), *H. avenae* and *H. filipjevi* (Toumi et al., 2013b). Real-time PCR was used for the detection of single J2 of *G. pallida* and *H. schachtii* (Madani et al., 2005) or *H. glycines* (Ye, 2012); a qPCR protocol was developed for *H. avenae* and *H. latipons* (Toumi et al., 2015). Van den Elsen et al. (2012) developed a qualitative viability test based on the detection of trehalose (a disaccharide sugar present in the perivitelline fluid between the unhatched J2 and the eggshell; see Chapter 3, this volume) in viable eggs of potato cyst nematodes. In accordance with these findings, Ebrahimi et al. (2015) found a relationship between egg viability and the trehalose content, and developed a trehalose-based method to quantify viable eggs of *G. rostochiensis* and *G. pallida*.

The phylogeny of *Heterodera* spp. can be inferred from sequences of ITS ribosomal DNA (rDNA) (Tanha Maafi et al., 2003; Chapter 16, this volume). The combination of ITS with morphological data enabled several groups within *Heterodera* to be recognized: *Afenestra*, *Avenae*, *Bifenestra*, *Cardiolata*, *Cyperi*, *Goettingiana*, *Humuli*, *Sacchari* and *Schachtii* groups (Subbotin et al., 2010b). Close relationships were revealed between the *Avenae* and *Sacchari* groups and between the *Humuli* group and the species *H. turcomanica* and *H. salixophila* (Subbotin et al., 2013).

1.6 Life Cycle

The life cycle of cyst nematodes is summarized in Figure 1.1. In cyst nematodes the eggs are retained in a cyst, which is the dead body wall of the female. Within these eggs, the embryo develops into the first-stage juvenile, which moults to the J2. This stage is equipped with a stylet. The stylet is used to cut a slit in the eggshell allowing

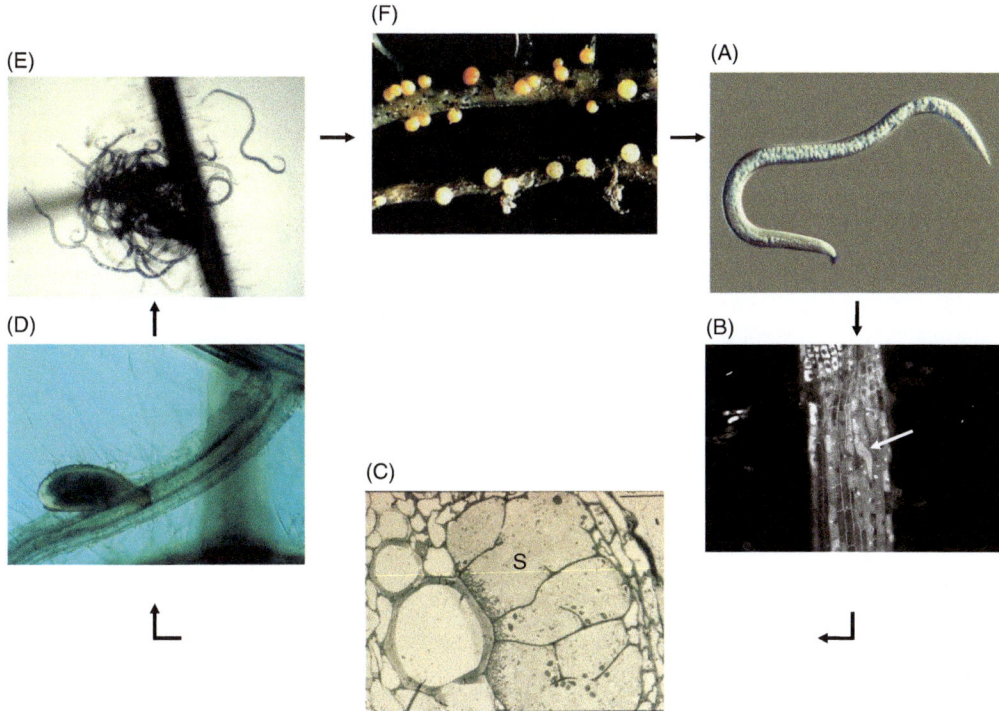

Fig. 1.1. Life cycle of a cyst nematode. Cysts contain up to approximately 400 eggs, each one containing a second-stage juvenile (J2). After hatch (A), the J2 moves through the soil, invades a host root (B; arrowed) and moves through the root to establish a feeding site (C; syncytium; S) on which it feeds and develops. Juveniles develop either into females, which become saccate and rupture the root (D), or into vermiform males, which leave the root, locate the female and mate (E). The female then dies to form the cyst (F). (From Turner and Subbotin, 2013.)

the J2 to hatch from the egg. The unhatched J2 may be able to survive within the egg for a long time, depending on the species and environmental conditions. This stage is the dormant stage of the life cycle, a stage of arrested development. Cyst nematodes have two types of dormancy: diapause and quiescence (Chapter 3, this volume). Diapause has a time component that enables the J2 to overcome seasonal environmental conditions that are unfavourable for hatch, such as extreme temperatures or drought. *Globodera rostochiensis*, *G. pallida* and *H. avenae* usually have an obligate diapause during their first season of development. In *G. rostochiensis* and *G. pallida*, diapause is terminated in late spring, when the combination of rising soil temperature and adequate soil moisture is favourable for infection of the new potato crop. Facultative diapause is initiated by external factors from the second season onwards. Once diapause is completed, the J2 may enter into quiescence, a spontaneous reversible response to unpredictable unfavourable environmental conditions (Perry et al., 2013). This requires various environmental cues to effect further development. In temperate regions this usually occurs with an increase in soil temperature together with specific hatching stimuli produced by the host root system, termed root diffusate or root exudate (Perry, 2002). Cyst nematodes can be classified into four categories based on their hatching responses to water and root diffusate of the host: (i) high J2 root diffusate hatch, low J2 water hatch (*G. rostochiensis*, *G. pallida*, *H. cruciferae*, *H. carotae*, *H. goettingiana*, *H. humuli*); (ii) high J2 root diffusate hatch, moderate J2 water hatch (*H. trifolii*, *H. galeopsidis*, *H. glycines*); (iii) high J2 root diffusate hatch, high J2 water hatch (*H. schachtii*, *H. avenae*); and (iv) high J2 root diffusate hatch of later generations, large J2 hatch in all generations (*H. cajani*, *H. sorghi*). The degree to which cyst nematodes are dependent on root diffusates to initiate hatch is related to host range: species that have a narrow host range tend to have strong dependency on the presence of diffusates from host species in order to hatch, whereas those with broader host range hatch more readily in the absence of such cues. Not all juveniles hatch at the same time – a proportion of J2 are retained either within the cyst body or in external egg masses. *Globodera* species do not produce eggsacs. In some *Heterodera* species (*H. avenae*) all eggs are retained in the cyst, whilst in others (e.g. *H. cajani*, *H. glycines*, *H. cruciferae*) eggs are also deposited in an eggsac, which is attached to the cyst. Within a species, eggsac production can vary according to environmental conditions; for example, *H. glycines* produces more eggs in eggsacs under favourable conditions (Ishibashi *et al.*, 1973). Upon hatch, J2 leave the cyst either via the fenestrae or the neck. Once in the soil, they search for a suitable host. The J2 locate the host by exploiting gradients of chemicals released by the root system of the host (Chapter 3, this volume).

The J2, the infective stage of cyst nematodes, penetrates the root system of its host near the root tip. Inside the root, the nematode migrates intracellularly, using its stylet to cut through cell walls, to the pericycle, where it selects a suitable cell used to form a feeding site, termed a syncytium (Chapter 4, this volume). The hollow stylet pierces the wall of this cell and saliva is injected from the pharyngeal glands. If the protoplast collapses or if the stylet becomes covered with a layer of callose-like material, the stylet is retracted. This behaviour is repeated until a suitable cell that does not respond adversely to J2 probing is found. This cell becomes the initial syncytial cell (Golinowski *et al.*, 1997). Cell wall openings are formed between the initial syncytial cell and its neighbours by widening of plasmodesmata followed by controlled breakdown of the plant cell wall in these regions. The cytoplasm of the initial syncytial cell proliferates, the central vacuole breaks down and the nucleus becomes enlarged. Similar changes are observed in the cells surrounding the initial syncytial cell as they become connected to the initial syncytial cell. Eventually the protoplasts of the initial syncytial cell and its neighbours fuse at the cell wall openings and this process is repeated with further layers of cells until 200–300 cells are incorporated into the syncytium. Cell wall ingrowths form where the syncytium is adjacent to the vascular tissues that increase the internal surface area and facilitate the passage of nutrients into the syncytium. If the J2 is able to induce and maintain syncytia (compatible relationship), the juvenile will develop into a male or female adult (Chapter 8, this volume).

About 7 days after penetration, the J2 moults to the third-stage juvenile (J3). The J3 has a well-developed genital primordium and rectum; the male has a single testis and the female has paired ovaries. At this stage female nematodes become saccate. The J3 moults to the fourth-stage juvenile; eventually, the nematode will moult a final time to the adult stage. At this (fourth) moult, the females are saccate and their posterior ends protrude through the root cortex, ready for mating. Males develop in the same root as the females. When they emerge at the fourth moult they are still enclosed in the J3 cuticle. Males revert to the vermiform body shape, are non-feeding and leave the roots; they live for about 10 days after first leaving the root (Evans, 1970). They are attracted to females, which exude sex pheromones (Chapter 3, this volume). Females may mate with different males. When a population is exposed to poor environmental conditions (e.g. competition for feeding sites due to overcrowding or a resistant response; see section 1.7) more males are present (Trudgill, 1967).

After mating, the embryo develops within the egg, which is still within the female's body. Eventually, the female dies and her cuticle undergoes polyphenol oxidase tanning to form a tough protective cyst containing several hundred embryonated eggs. The cysts become detached from the roots as the plant dies and remain dormant in the soil until the next suitable host grows in the vicinity (Turner and Evans, 1998).

The number of generations per year varies between cyst nematode species. Under field conditions most temperate species of cyst nematodes will complete one or two generations, corresponding to the natural life cycle of its host combined with the length of the optimal temperature range. For example, temperature was crucial in determining the number of generations *H. schachtii* completed on oil seed rape during the growing season (Kakaire *et al.*, 2015). However, in tropical regions where favourable environmental conditions are more constant throughout the year, multiple generations are present, with up to 11 generations being reported for *H. oryzicola* (Jayaprakash and Rao, 1983). The time needed to complete the life cycle of a cyst nematode depends upon the co-evolution of the species with its host range and the environmental conditions. In temperate regions, life cycles are completed in about 30 days, but this may be reduced in warmer climates.

1.7 Syncytium

One of the most remarkable adaptations of cyst nematodes for parasitism is the ability to induce the formation of a syncytium in the roots of their host (Fig. 1.2). Cyst nematodes depend on the syncytium for all nutrients that are required for development to the adult stage. In addition, each nematode can only induce a single syncytium, meaning that this structure must be kept alive for a period of several weeks throughout the duration of feeding. The syncytium is a large, multinucleate and metabolically active structure. The central vacuole present in other root cells is absent or greatly reduced and the cytoplasm is enriched in rough endoplasmic reticulum and other subcellular organelles (Sobczak and Golinowski, 2009, 2011). These changes in plant cell structure are underpinned by profound changes in host gene expression in the syncytium. Microarrays have been used to probe changes in gene expression that occur in the syncytia induced by *H. schachtii* in *Arabidopsis* (Puthoff *et al.*, 2003; Szakasits *et al.*, 2009) and *H. glycines* in *Glycine max* (e.g. Ithal *et al.*, 2007; Klink *et al.*, 2007). These and similar studies have shown that expression profiles of thousands of plant genes are changed in the syncytia. Changes in nuclear structure within the syncytium are likely to be due to induction of the endocycle (De Almeida Engler *et al.*, 2011). Although the precise mechanisms that underlie syncytium induction remain unknown, effectors have been identified from cyst nematodes that may interact with auxin transport proteins or that are similar to peptide ligands that control host developmental processes (see Chapter 4, this volume).

Resistance against biotrophic pathogens is associated with a hypersensitive reaction. A small number of resistance responses against cyst nematodes are characterized by this type of strong, early response targeted against the syncytium. For example, the response of *Sinapsis alba* 'Maxi' to *H. schachtii* features a strong necrotic response surrounding the J2, which is unable to induce a syncytium (Soliman *et al.*, 2005). However, most resistance against cyst nematodes operates by restricting the development of the syncytium, preventing it from growing to a stage where it meets the vascular tissues. In these cases invasion, migration and the early stages of syncytium induction proceed as seen in the susceptible host. However, after this initial developmental stage the syncytium itself, or more often the cells surrounding the syncytium, collapse. This degeneration of the surrounding tissues restricts the development of the syncytium, leading either to the death of the nematode or to production of a much larger proportion of males, which tend to develop from smaller syncytia (Sobczak and Golinowski, 2011).

There are some superficial similarities between syncytia induced by cyst nematodes and the giant cells induced by root-knot nematodes (Moens *et al.*, 2009). Both are large multinucleate and highly metabolically active structures that show enriched cytoplasm compared to the

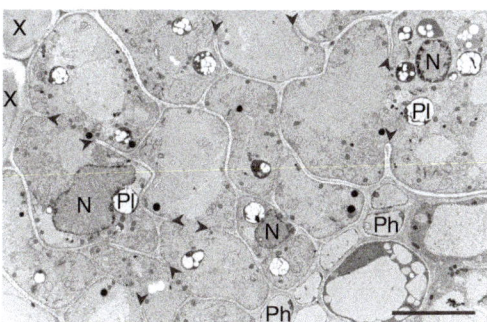

Fig. 1.2. Transmission electron micrograph of a syncytium induced by *Globodera rostochiensis* in root of tomato 'Moneymaker' (7 days post infection), showing several broken cell walls (arrowheads) and enlarged syncytial elements. Abbreviations: N: nucleus; Ph: phloem; Pl: plastid with starch grains; X: xylem. Scale bar = 10 µm. (Courtesy of Miroslaw Sobczak, Dept Botany, Warsaw University of Life Sciences, Poland.)

surrounding tissues. However, phylogenetic analysis shows that the ability to induce feeding structures has evolved independently in cyst and root-knot nematodes (Baldwin *et al.*, 2004). Giant cells also have a completely different ontogeny to syncytia and are formed as a result of repeated rounds of nuclear division in the absence of cytokinesis.

1.8 Effect of Abiotic Factors

Cyst nematodes exhibit considerable variation in optimum temperature for hatching *in vitro*; for example, *G. pallida* is adapted to lower temperatures than *G. rostochiensis* (16°C and 20°C, respectively) (Turner and Subbotin, 2013). Kaczmarek *et al.* (2014) observed the greatest cumulative percentage hatch of J2 occurring between 15 and 27°C for *G. rostochiensis* and 13 and 15°C for *G. pallida*. In addition to hatching, soil temperature also influences mobility, infectivity and lipid utilization of J2 of PCN (Robinson *et al.*, 1987; Ebrahimi *et al.*, 2014). Low optimum temperatures for hatching are characteristic of cyst nematodes that can invade during winter or early spring, such as *H. cruciferae*. As expected, nematodes adapted to warmer climates exhibit higher temperature optima, for example, 30°C for *H. zeae*.

Soil type can also affect rates of hatch. In general, coarse-textured soils favour hatching and subsequent invasion of root systems, providing suitable conditions for aeration and nematode migration. Maximum hatch usually occurs in soil at field capacity, whilst drought and waterlogging inhibit hatch (Turner and Subbotin, 2013).

1.9 Important Species

Out of the eight genera of cyst nematodes, only two, *Heterodera* and *Globodera*, contain economically important species. The species discussed in this section are distributed in temperate regions, in temperate areas of tropical regions, sub-tropical regions and tropical areas and, because of their importance, have been studied intensively. It is clear that there is substantial overlap between these climatic groupings.

1.9.1 Sugar beet cyst nematode *Heterodera schachtii*

The sugar beet cyst nematode was the first cyst-forming nematode detected when it was found associated with stunted and declining sugar beet in Germany (see section 1.3). *Heterodera schachtii* is now found in all major sugar beet production areas of the world; it is mainly present in temperate regions but is occasionally established in hot climates. The species is widespread in most European countries, the USA, Canada, the Middle East, Africa, Australia and South America (Baldwin and Mundo-Ocampo, 1991; Evans and Rowe, 1998; CABI, 2016d). The nematode causes serious yield reductions and decreases sugar content of sugar beet wherever the crop is grown. In European countries the annual yield losses were estimated at *ca* €90 million (Müller, 1999). The optimum temperature for development is around 25°C. In some climates, three to five generations may complete development on sugar beet in one season (Franklin, 1972). *Heterodera schachtii* belongs to the *Schachtii* species group. Host plants of *H. schachtii* are numerous and belong mainly to the families Amaranthaceae and Brassicaceae, but also to Polygonaceae, Scrophulariaceae, Caryophillaceae and Solanaceae (Turner and Subbotin, 2013).

A second cyst nematode parasitizes sugar beet, *H. betae*, described by Wouts *et al.* (2001). Before its description it used to be called the *forma specialis beta* or race of *H. trifolii*. *Heterodera betae* is not easily distinguished morphometrically from *H. schachtii*. However, unlike cysts of *H. schachtii*, cysts of *H. betae* mature through a yellow stage. However, it is readily distinguished by RFLP analysis of the ITS regions of rDNA (Amiri *et al.*, 2002). It has been reported from The Netherlands, France, Switzerland, Italy, Germany, Sweden and Morocco. Both *H. schachtii* and *H. betae* belong to the *Schachtii* group.

1.9.2 Cereal cyst nematodes *Heterodera avenae*, *H. latipons* and *H. filipjevi*

The CCN form a complex of several closely related species, which are distributed worldwide, mainly on plants of the Poaceae (Rivoal and Cook, 1993; Nicol and Rivoal, 2007). Smiley *et al.* (2017) published an excellent, comprehensive review of CCN. Among CCN, *H. avenae* was the first species to be described (Wollenweber, 1924), followed by the Mediterranean *H. latipons* (Franklin, 1969), the north European *H. hordecalis* (Andersson, 1974), the east European *H. filipjevi* (Madzhidov, 1981) and several other species (Wouts *et al.*, 1995). So far, 12 species of the CCN group have been described. However, *H. avenae*, *H. latipons* and *H. filipjevi* are considered the most economically important species in cereals worldwide (Rivoal and Cook, 1993; Nicol and Rivoal, 2007).

Heterodera avenae (common name: cereal or oat cyst nematode) is the main nematode species on cereals in temperate regions. It has been reported in most European countries and also in North Africa, South Africa, Asia (China, India, Iran, Japan, Pakistan, Syria, Saudi Arabia and Turkey), New Zealand, Peru, Canada and the USA. In Europe, more than 50% of the fields in major cereal-growing areas were found to be infested by *H. avenae* (Rivoal and Cook, 1993); annual yield losses are estimated at £3 million (Nicol and Rivoal, 2008). In the USA, the annual loss is estimated at US$3.4 million in wheat production in the states of Idaho, Oregon and Washington. *Heterodera avenae* belongs to the *Avenae* group.

In the 1960s, another cyst nematode was detected in the Mediterranean region (Israel and Libya) on the roots of stunted wheat plants. It was described as *H. latipons* (common name: Mediterranean cereal cyst nematode) based on the morphological characteristics of the Israel population (Franklin, 1969). *Heterodera latipons* has a wide distribution and is essentially distributed in the Mediterranean region and the Middle East (Toumi *et al.*, 2017), but was also detected in more or less temperate continental climates of the former USSR (Mulvey and Golden, 1983; Subbotin *et al.*, 1996), Japan (Momota, 1979) and Canada (Sewell, 1973). *Heterodera latipons* belongs to the *Avenae* group; it often occurs in mixed populations with *H. avenae* in cereal cropping systems. *Heterodera latipons* is believed to cause less damage to cereals compared with *H. avenae* (Mor *et al.*, 1992, 2008). However, in Cyprus, *H. latipons* was reported to decrease barley yield by 50%. The loss was greatest under severe drought conditions and monoculture systems (Philis, 1988, 1997). In Syria, the nematode causes average yield losses of 20 and 30% in barley and durum wheat, respectively, and the nematode was more damaging under water stress conditions (Schölz, 2001).

The third species of the CCN complex is *H. filipjevi* (common name: Filipjev cyst nematode; previously called Gotland strain of *H. avenae*, pathotype 3 of *H. avenae* or race 3 of *H. avenae*). Currently, its distribution seems to be restricted to countries in Asia (China, India, Iran, Syria, Tajikistan and Turkey), Europe (Germany, Norway, Poland, Spain, Sweden and the former USSR) and the USA. On winter wheat under rain-fed conditions in Turkey, the average yield loss caused by *H. filipjevi* was between 42 and 50% (Nicol *et al.*, 2006). In Iran, the yield loss due to *H. filipjevi* on winter wheat in monoculture was estimated at 48% (Hajihasani *et al.*, 2010). In Norway, the occurrence of *H. filipjevi* caused damage to *Secale cereale* (winter rye) (Holgado *et al.*, 2005).

There is considerable information on hatching and dormancy of the three major species of CCN. Environmental stresses on the female of *H. avenae* initiate a facultative diapause (Wright and Perry, 2006; see Chapter 3, this volume). In West Australia, *H. avenae* hatches optimally at a temperature between 10 and 15°C, with the hatching peak of 80% under the field conditions in late May (Banyer and Fisher, 1971; Meagher, 1977; Stanton and Eyres, 1994). Similar results were reported from France for *H. avenae*, although differences were observed between southern and northern ecotypes (Rivoal, 1986). In the south of France, an obligatory diapause acts during the summer and autumn and is disrupted by low temperatures, which explains the winter hatching of this ecotype. By contrast, for the northern ecotype, a facultative diapause acts during winter and is broken by an increase of temperature, which leads to hatching in spring (Rivoal, 1983). *Heterodera latipons* hatched well at 10°C in Syria (Schölz and Sikora, 2004) and Jordan (Al Abed *et al.*, 2009) with the maximum hatching not more than 33% of viable cyst contents, and with one hatching peak at the end of January–early February. Hatching of *H. filipjevi* in root diffusate of susceptible barley was similar to that in water and, overall, hatch of *H. latipons* was similar to that of the southern (Mediterranean) ecotype of *H. avenae* (Chapter 3, this volume). The optimal hatch for Turkish populations of *H. filipjevi* was between 10 and 15°C; in *in vitro* and field conditions 94% of the J2 hatched, with two peaks recorded at early October and between the end of January to early March (Sahin *et al.*, 2010).

The infection process of CCN differs between species (Mor *et al.*, 1992). J2 of *H. avenae* attacked the root tip region inducing typical branching and swelling of roots with ensuing adherence of soil particles; J2 of *H. latipons*, however, penetrated at sites along roots more distant from the root tip. Hence, *H. latipons* did not produce clearly visible root symptoms in the early infection period or the seedling stage. Mor *et al.* (1992, 2008) further observed differences between *H. latipons* and *H. avenae* in the infection process and in the feeding cell structures in cereals. The growth inhibition caused by *H. avenae* was more severe compared with *H. latipons* (Mor *et al.*, 1992).

1.9.3 Soybean cyst nematode *Heterodera glycines*

A cyst nematode parasitizing soybean plants, *Glycine max* was found in Japan in 1915. This was initially known as *H. schachtii*. Eventually, Ichinohe (1952) named the species *H. glycines*, and gave a brief description of this nematode.

The species occurs in most countries of the world where soybean is produced, that is, the USA, South America (Argentina, Brazil, Colombia and Paraguay), Asia (China, Indonesia, Iran, Japan and Korea), Egypt, Italy and the former Soviet Union. It is a pest in temperate areas and does not develop below 15°C or above 33°C. *Heterodera glycines* probably evolved either in China or Japan, from where it has been spread to the New World (see Chapter 9, this volume). In a study of losses in ten soybean-producing countries together accounting for 97% of the world crop, *H. glycines* appeared to be the most important constraint on yield and caused damage estimated at US$1960 million (Wrather *et al.*, 2001). Wrather *et al.* (2003) found that in the period 1999–2002, the highest yield losses on soybean were caused by *H. glycines* in both the USA and Canada, the reduction in yield in the USA in 2002 amounting to US$784 million. In Japan, yield losses have been estimated at 10–70% (Ichinohe, 1988). *Heterodera glycines* has a broad host range, especially within the Fabaceae (Turner and Subbotin, 2013). Riggs (1992) compiled the most comprehensive host list for SCN: it included 22 plant families and 286 species. Evaluating weed species from the Northern Great Plains as hosts of soybean cyst nematode, Poromarto *et al.* (2015) identified 26 weed species from 11 plant families as new hosts of SCN.

Heterodera glycines interferes with nodulation and causes early yellowing of soybean plants. The above-ground symptoms of damage are not specific. *Heterodera glycines* has been classified into a large number of races, which are distinguished using differential hosts (see Chapter 9, this volume). *Heterodera glycines* belongs to the *Schachtii* group.

1.9.4 Pigeon pea cyst nematode *Heterodera cajani*

In 1964, Swarup *et al.* reported a cyst nematode species of the *Schachtii* group under the name of *H. trifolii* from a pigeon pea field at IARI, New Delhi, India. The species was later briefly described from roots of pigeon pea, *Cajanus cajan*, by Koshy (1967) and named *H. cajani*. Later, a more detailed description was provided by Koshy *et al.* (1971) and Koshy and Swarup (1971).

Heterodera cajani is widely spread in India (CABI, 2016c); it has also been reported in Egypt (Aboul-Eid and Ghorab, 1974). Primary hosts include: *C. cajan*, *Vigna unguiculata*, *V. mungo*, *V. radiata*, *V. aconitifolia*, *Phaseolus* species, *Pisum sativum* and *Phyllanthus maderaspatensis* (Evans and Rowe, 1998) and *S. indicum* and *Cyamopsis tetragonolobus* from Haryana (India) (Bhatti and Gupta, 1973). The most important hosts are species of Fabaceae and Pedaliaceae.

The nematode completes its life cycle in 16 days at 29°C but during cooler conditions (10–25°C), the life cycle takes 45–80 days to complete (Koshy and Swarup, 1971). On one crop of pigeon pea, eight or nine generations were recorded. Although amphimixis is the rule, females sometimes reproduce without males. Eggs may be retained inside the female body but many are laid in a gelatinous matrix forming eggsacs. Emergence from eggsacs is higher and more rapid than from white (young) or brown (mature) cysts (Sharma and Swarup, 1984; Sharma and Sharma, 1998). Juvenile emergence is greater from cysts produced on 30-day-old pigeon pea plants than from cysts produced on older plants. The pattern of J2 emergence is complex and temperature is a major, but not the only, important factor. Some of the encysted juvenile population undergoes diapause (Singh and Sharma, 1996). Root leachates of host plants stimulate J2 hatch of *H. cajani*. Leachates collected from 2-, 3- or 4-week-old plants or soil are more stimulatory than those from 1-week-old plants (Yadav and Walia, 1988).

Survival is greater at 20 and 25°C than at 15 and 30°C. Eggs within cysts are able to withstand extremes of desiccation. Exposing nine different hosts to 14 populations of *H. cajani* from different origin revealed the presence of three races (Siddiqui and Mahmood, 1993).

1.9.5 Sugarcane cyst nematode *Heterodera sacchari*

Heterodera sacchari was originally described on sugarcane in the Niari Valley, Congo (Luc and Merny, 1963). Later, it was reported in the Ivory Coast, where it was detected in flooded rice fields (Merny, 1970), in Nigeria on sugarcane (Jerath, 1968) and wild grasses (Odihirin, 1975), on flooded rice fields of Casamance Province, Senegal and in Gambia (Fortuner and Merny, 1973) and on sugarcane in Burkina Faso (Cadet and Merny, 1978). The species is also reported on *Saccharum spontaneum* in India (Swarup *et al.*, 1964). Next to sugar cane and rice, various wild Graminaceae are considered as hosts (Odihirin, 1975). The species has also been reported in Trinidad. However, the veracity of *H. sacchari* records from Asian countries, as well as from Trinidad, may need confirmation (Tanha Maafi *et al.*, 2007).

The morphology of *H. sacchari* has been described by Luc and Merny (1963) and the original description was supplemented by Luc (1974) and several others (e.g. Vovlas *et al.*, 1986; Nobbs *et al.*, 1992; Tanha Maafi *et al.*, 2007). *Heterodera sacchari* is a triploid parthenogenetic species (Netscher, 1969). The nonfunctional male is rare and has been described by Netscher *et al.* (1969). The hatching of juveniles from cysts is similar in sugarcane root diffusate and tap water (Garabedian and Hague, 1984).

In western areas of Africa, *H. sacchari* may constitute a danger for sugarcane, and both upland and swamp rice. Jerath (1968) observed that infested sugarcane plants are stunted and thin, and secondary roots are less abundant than healthy plants. In sugarcane plantations, cysts of *H. sacchari* may be transported by the water in irrigation canals for at least 5 to 8 km (Odihirin, 1977). On rice, experiments show (Babatola, 1983a) that infested plants are chlorotic and their growth retarded, roots are necrotic and blackened, tiller numbers are reduced and the grain yield is lower. *Heterodera sacchari* attacks both swamp and upland rice but the latter appears more susceptible to damage (Babatola, 1983a). Rice cultivars react very differently to infestation by the nematode (Babatola, 1983b).

Heterodera sacchari belongs to the *Sacchari* group and to the *H. sacchari* species complex, together with *H. leuceilyma* and *H. goldeni* (Tanha Maafi *et al.*, 2007). The three species are not easily separated. Several restriction enzymes distinguish *H. sacchari* from *H. goldeni* (Subbotin *et al.*, 2010b).

1.9.6 Rice cyst nematodes *Heterodera oryzae* and *H. oryzicola*

Heterodera oryzae and *H. oryzicola*, both called rice cyst nematodes, are parasites of rice and plantains. Luc and Berdon Brizuela (1961) described *H. oryzae* from swamp rice fields in the central part of Côte d'Ivoire. The species was also reported from Asian countries. However, Luc (1986) and Nobbs *et al.* (1992) suggest these findings need to be confirmed at species level because of possible confusion with *H. elachista* and *H. oryzicola*. All of these species belong to the *Cyperi* species group and are morphologically very similar and difficult to separate on basis of morphology. However, non-specific esterase banding patterns allowed separation of all four species (Nobbs *et al.*, 1992). Rao and Jayaprakash (1978) described *H. oryzicola* on roots of upland rice in the state of Kerala, India. Later it was detected on rice and banana in several states in India (Kaushal *et al.*, 2007).

Females of both species retain eggs and produce large egg masses. *Heterodera oryzicola* is dependent on root diffusates to induce substantial hatch (Ibrahim *et al.*, 1993). The dependence of *H. sacchari* on diffusates is less easily defined; in the study by Ibrahim *et al.* (1993) it was only with cysts from the last two polyphenol extractions that a small proportion of eggs were dependent on root diffusates for hatch and the total percentage hatch from these cysts was considerably less than from cysts collected from younger plants.

Hosts for *H. oryzae* are rice, plantain (*Musa paradisiaca*) and maize. Rice is the primary host for *H. oryzicola*. Other hosts include plantain and a few grasses. The damage is not quantified for both species. *Heterodera oryzae* is less aggressive than *H. sacchari* on rice (Luc, 1986).

1.9.7 Potato cyst nematodes *Globodera rostochiensis*, *G. pallida* and *G. ellingtonae*

Globodera rostochiensis and *G. pallida* are cosmopolitan pests in both temperate countries and temperate regions of tropical countries. They have been reported on all continents where potatoes are grown (OEPP/EPPO, 2014). They have been detected in 71 (*G. rostochiensis*) and 55 (*G. pallida*) countries (CABI, 2016a). They are native to South America (Grenier *et al.*, 2010), where they are the principal pest of Andean potato crops. In this region, the species are mainly found between 2000 and 4000 m.a.s.l., with the heaviest infestations between 2900 and 3800 m.a.s.l. *Globodera rostochiensis* and *G. pallida* are differently distributed in the Andes. The demarcation line between the two species is near 15.6°S. With few exceptions, populations north of this line are mainly *G. pallida*. Those from areas around Lake Titicaca and further south are predominantly *G. rostochiensis* with few *G. pallida* or mixtures of both species. The most southerly populations, from the east side of the Andes in Bolivia, are mixtures of *G. rostochiensis* and *G. pallida*. In 2008 cysts were isolated from a field in Oregon and in two fields in Idaho, all three having a history of growing potatoes; the cysts belonged to a new species, which was named and described in 2012 as *G. ellingtonae* (Handoo *et al.*, 2012).

PCN do not cause specific symptoms of infection. At low nematode densities crops display patches of poor growth and affected plants may show chlorosis and wilting; tuber sizes are reduced, whereas at higher densities both number and size of tubers can be reduced. *Globodera rostochiensis* and *G. pallida* are responsible for annual potato tuber losses of up to 9% in Europe. By contrast, in Oregon, *G. ellingtonae* caused minimal damage to potato.

1.9.7.1 Globodera rostochiensis, the golden or yellow potato cyst nematode

The golden cyst nematode was first reported in 1881 when it was found associated with potato plants, *Solanum tuberosum*, in Rostock, Germany. For a long time, the nematode was referred to as *H. schachtii*, because this was the only known species of cyst nematode at that time. Following its first detection, *G. rostochiensis* became more widely known throughout Europe and, eventually, was described in 1923. Its common name derives from the fact that as the female dies, her body wall (cuticle) undergoes polyphenol oxidase tanning from white through golden yellow to brown. The dead female cyst contains the eggs, often more than 300, and is effective in protecting the unhatched J2 from environmental extremes (see Chapter 3, this volume) and predation by mites etc.

The development of one generation requires 6–10 weeks. The J2 can enter diapause and remain viable for 20 or more years (Perry *et al.*, 2013). Every year a small proportion of J2 hatch spontaneously in the absence of a host and die of starvation. This natural decline has been estimated to be 20–40% per year for *G. rostochiensis* and 10–30% per year for *G. pallida* (Whitehead, 1995). On average, *G. rostochiensis* has a 40% greater spontaneous hatch than *G. pallida* (Evans and Haydock, 2000). Substantial hatch depends on stimulation from hatching factors, found in host root diffusates (see Chapter 3, this volume). The optimum temperature for the hatch of *G. rostochiensis* is about 15°C, with the largest proportion of adults in a population at 650–830 day degrees over a basal temperature of 4.4°C (Evans, 1968). Crops that are attacked by *G. rostochiensis* are tomato, eggplant and potato (Subbotin *et al.*, 2010a); the number of weed hosts is limited, although numerous wild *Solanum* species from South America have been described as hosts.

1.9.7.2 Globodera pallida, the pale potato cyst nematode

Globodera pallida was originally considered to be a pathotype of *H. rostochiensis* (= *G. rostochiensis*); during tanning, the female cuticle does not pass through a yellow stage, hence the common name. The species was described from two localities: Epworth in Lincolnshire, England, representing the Pa3 pathotype, and Duddingston, Scotland, representing the Pa1 pathotype (see Chapter 9, this volume).

Globodera pallida is a major pest of potato crops in cool temperate climates. Using the mtDNA gene, cytochrome b (cytb) sequences and microsatellite loci, Plantard *et al.* (2008) showed that the majority of *G. pallida* presently distributed in Europe derived from a single restricted area in the extreme south of Peru, located between the north shore of Lake Titicaca and Cusco.

Globodera pallida usually develops one generation per vegetation season. It is adapted to cool temperatures and is able to hatch earlier in the year and develop at temperatures 2°C cooler than *G. rostochiensis* (Langeslag *et al.*, 1982). *Globodera pallida* hatches at around 10°C or less and is adapted to develop at cool temperatures between 10 and 18°C, whereas *G. rostochiensis* seems to be adapted to a temperature range of 15 to 25°C (Franco, 1979). Day length also influences hatch, which is faster where the host has continuous light rather than prolonged hours of darkness (Hominick, 1986; see Chapter 3, this volume).

1.9.7.3 Globodera ellingtonae, the Ellington potato cyst nematode

The third species of PCN, *G. ellingtonae*, is currently the subject of detailed investigation to determine its biology and, especially, its pathogenicity. Although not strictly a 'principal species' worldwide, it is of considerable interest. Populations were isolated from soil collected from a research farm near Powell Butte, Oregon, USA, and from two farmers' fields in Idaho, USA in 2008. The isolates were characterized both morphologically, using cysts and hatched J2, and molecularly, and it was described as a new species by Handoo *et al.* (2012). The USA isolates of *G. ellingtonae* fell into the same group as isolates from Antofagasta, Chile, but morphological measurements of the isolates from South America are not available. The Attacama desert region of Chile, which spans Bolivia, Chile and Argentina, is likely to be where the nematode originates (Inga Zasada, Oregon, 2017, personal communication).

Globodera ellingtonae is a restricted pathogen in the USA. The Animal and Plant Health Inspection Service, USA, sampled 300,000 additional fields in Idaho and 100,000 fields in other states, and no additional *G. ellingtonae* cysts were found (Zasada *et al.*, 2015). Potato varieties resistant to *G. rostochiensis* pathotype Ro1 are resistant to *G. ellingtonae* (Zasada *et al.*, 2015). Development of *G. ellingtonae* is similar to that reported for *G. rostochiensis* and *G. pallida*, and in bare soil the maximum reduction of eggs per cyst was 55 to 73%, which is similar to that for *G. pallida* and *G. rostochiensis* (Phillips *et al.*, 2017). As with the other species of PCN, hatch of *G. ellingtonae* depends on stimulation by host root diffusates (Zasada *et al.*, 2015).

1.10 Pathotypes and Races

Several species of cyst nematodes can be controlled by the use of resistant cultivars (see Chapter 9, this volume). Within this context, the term 'resistance' refers to the genetics of traits in host plants that interact with nematode (a)virulence genes. Resistant cultivars inhibit reproduction of a nematode population. When such a population is able to reproduce and increase in number on a resistant cultivar, the population is said to be virulent. The use of resistant varieties has demonstrated the genetic variation between virulent populations (Cook and Rivoal, 1998). Interactions between the (plants and nematode) genetic systems are the basis for the identification of pathotypes. These are usually recognized by the effect of the virulence phenotype in experiments with host plant differentials. Pathotypes are regarded as a group of individual nematodes with common gene(s) for (a)virulence and differing from gene or gene combinations found in other groups. Pathotype schemes were proposed for the major cyst nematodes, PCN (*G. rostochiensis* and *G. pallida*), CCN (*H. avenae*, *H. filipjevi* and *H. australis*) and SCN (*H. glycines*). They are all based on the ability (or inability) of populations within each species to reproduce on a range of 'differential' host plants.

Pathotype schemes for *G. rostochiensis* and *G. pallida* were proposed by Kort *et al.* (1977) and Canto-Saenz and de Scurrah (1977); the schemes described the virulence of populations from Europe and South America, respectively (Table 1.1). In the pathotype/differential clone interactions, susceptible (+) indicates a multiplication rate $(P_f/P_i) > 1.0$, and resistant (−) indicates a $P_f/P_i < 1.0$, where P_i and P_f are the initial and final population sizes, respectively. The schemes standardized contrasting national schemes, especially those used within European countries, but it soon became clear that environmental influences and the extensive heterogeneity of some populations, especially those of *G. pallida*, caused problems (Turner and Subbotin, 2013). Populations in the centres of origin of the two species in South America are more heterogeneous in virulence characteristics than those introduced and dispersed in the rest of the world. Some populations are relatively homozygous for virulence, for example, Ro1 (R1A) and Pa1 (P1A). Others, including most other *G. pallida* populations, are heterogeneous and give varying results; thus, these populations cannot reliably be described as pathotypes and are increasingly referred to as virulence groupings (Trudgill, 1985). Potato clones with the gene *H1* are resistant to *G. rostochiensis* pathotypes Ro1 and Ro4, which have been combined as virulence group Ro1.

The pathotype scheme for CCN developed by Anderson and Anderson (1982) describes the pathotypes of species of CCN (*H. avenae*, *H. filipjevi* and *H. australis*) based on their multiplication on host differentials of barley, oats and wheat. The separation into three pathotype groups is based on reactions of the barley cultivars with the known resistance genes *Rha1* ('Ortolan'), *Rha2* ('Siri' and 'KVL191') and *Rha3* ('Morocco'). Each pathotype group is further subdivided according to their reactions on other differentials (Table 1.2). Resistance is defined as fewer than 5% new females compared with numbers on susceptible control. As with the PCN pathotype schemes, because the genetics of field populations are largely unknown and variability exists within them, the term 'virulence phenotype' has been proposed (Turner and Subbotin, 2013). Evidence suggests that the aforementioned species of CCN have populations with different virulence phenotypes. There is limited evidence for loss of effectiveness of resistance genes used in widely grown cultivars.

Differences in virulence between populations of *H. glycines* are termed races rather than pathotypes. Such differences were first observed during breeding programmes in the USA for resistant soybean varieties (Golden *et al.*, 1970). The term 'races' was used to define field populations with different abilities to reproduce on plant lines carrying various sources of resistance and on resistant cultivars. Four soybean differentials ('Pickett', 'Peking', 'PI 88788' and 'PI 90763') were used in this race test, with 'Lee' as the susceptible standard. A resistant response (avirulence) was defined as a Female Index of <10% of that obtained on 'Lee'. However, it was soon found that the scheme inadequately described the genetic diversity of SCN (Epps and Duclos, 1970). When Riggs *et al.* (1981) increased the number of differentials to 12 resistant soybean lines, they identified 25 different 'races' (Table 1.3). Because *H. glycines* populations vary in genetic diversity, and this variation has implications for management strategies, a mechanism was needed for documenting and discussing population differences. Niblack *et al.* (2002) proposed an HG Type Test to describe population variation better and to expand the flexibility of the race classification system. The HG Type system uses three of the four resistant soybean genotypes used as indicator hosts ('Peking' (= 'PI 548402'), 'PI 88788' and 'PI 90763'). Reproduction of 10% or more on a resistant cultivar, when compared with the susceptible 'Lee 74', results in a designation of compatibility. Virulence is measured as a Female Index using numbers of females on both 'Lee 74' and the test cultivar (Niblack *et al.*, 2009).

More information is provided in the excellent review by Cook and Rivoal (1998). Turner and Subbotin (2013) conclude: 'Despite the limitations of the various pathotype/race schemes for cyst nematodes, providing their limitations are recognized, they continue to give a useful indication of the virulence characteristics of particular nematode gene pools. As such, they can provide critical information necessary for effective management and the emergence of new virulent strains.'

1.11 Symptoms

The above-ground symptoms of cyst nematodes are not specific. The only unique symptom of

Table 1.1. Pathotype groups of potato cyst nematodes, *Globodera rostochiensis* and *G. pallida*. (Adapted from Cook and Noel, 2002.)

Species		*G. rostochiensis*					*G. pallida*						
Globodera species virulence group[a]		Ro1		Ro2	Ro3	Ro5	Pa1		–	–	Pa2	Pa2/3	Pa3
European pathotypes[b]		Ro1	Ro4	Ro2	Ro3	Ro5	Pa1				Pa2		Pa3
South American pathotypes[c]		R1A	R1B	R2A	R3A	–	P1A	P1B	P2A	P3A	P4A	P5A	P6A
Species and accession	Ploidy, resistance gene												
Solanum tuberosum ssp. *tuberosum*	4x, (minor)	+/−	+	+	+	–	+	+	–	–	–	–	–
S. tuberosum spp. *andigena* CPC 1673	4x, *H1* on chromosome 5	–	–	+	+	+	+	"	+	+	+	+	"
S. kurtzianum KTT 60.21.19	2x, K1 K2 A & B	–	(+)	–	(+)	(+)	+	"	+	+	+	+	"
S. vernei GLKS 58.1642.4	2x, Quantitative	–	+	–	–	+	+	"	+	–	–	+	+
S. vernei Vt 62.33.3	2x, Quantitative	+	–	+	+	+	+	+	–	–	+	+	+
ex. *S. multidissectum* hyprid P55/7	2x, 1 + polygenes *H2*	+	+	+	+	+	–	−/+	+	+	+	+	+
S. t. ssp. *andigena*	H3 + polygenes	"	"	"	"	"	(−)	"	"	"	"	(−)	(−)	"
CIP 280090.10	Quantitative	–	–	–	–	–	–	"	+	–	–	–	–	+
S. vernei hybrid 69.1377/94	2x, Polygenes	–	–	+	–	–	+	"	+	+	+	+	"
S. vernei hybrid 63.346/19	2x, Polygenes	–	–	+	+	+	–	"	–	–	+	+	"
S. spegazzinii	2x, Fa = H1	–	–	–	–	–	–	"	–	–	–	–	"
S. spegazzinii	2x, Fb+2 minor	+	–	+	–	+	+	"	"	"	"	"	"
	Glo1 on chromosome 7	(−)	+	+	+	+	+	"	"	"	"	"	"

[a]Trudgill (1985); [b]Kort *et al.* (1977); [c]Canto-Saenz and de Scurrah (1977).

Note: + = compatible interaction: nematode multiplication, potato susceptible; − − = incompatible interaction: nematode no multiplication, potato resistant ; ( ) = partial or uncertain interaction ; " = no information.

Table 1.2. Pathotype groups of cereal cyst nematodes, *Heterodera avenae*, *H. filipjevi* and *H. australis*. (Adapted from Cook and Rivoal, 1998.)

Species				*H. avenae*						*H. australis*		*H. filipjevi*	
Pathotype group	Ha1 group						Ha2 group		Ha3 group				
Pathotype	Ha11	Ha21	Ha31	Ha41	Ha51	Ha61	Ha71	Ha12	Ha13	Ha23	Ha33			
Different species and cultivar														
Barley														
Varde	+	"	"	+	"	+	+	+	+	+	+			
Emir	+	+	"	+	"	–	+	+	+	+	+			
Ortolan/Drost	–	–	–	–	–	–	–	+	–	–	–			
Morocco	–	–	–	–	–	–	–	–	–	–	–			
Siri	–	–	–	+	+	–	–	–	+ "	+ "	+ "			
KVL 191	–	–	"	"	+ "	+	+	–	+	+	+			
Bajo Aragon	–	"	–	–	–	"	–	–	+	+ "	+ "			
Herta	+	+ "	–	–	–	–	–	+	+	+	+			
Martin 403	–	"	"	–	"	–	–	–	–	+	–			
Dalmatische	(–)	"	–	+ "	"	–	(+)	+ "	+ "	(–)	+ "			
La Estanzuela	"	"	–	"	"	"	+	"	"	(–)	+ "			
Hartian 43	–	"	"	–	"	"	–	–	–	–	+			
Oat														
Nidar	+	"	"	(+)	"	+	–	+	+	+	+			
Sol II	+	–	–	–	–	+	–	+	+	–	+			
Pura Hybrid BS1	–	–	"	–	–	–	–	–	+	–	–			
Avena sterilis 1376	–	"	"	–	"	"	–	–	–	–	+			
Silva	(–)	"	"	–	"	(–)	–	(–)	(–)	(–)	+			
IGV.H. 72-646	–	"	"	–	"	–	–	–	+	+	+			
Wheat														
Capa	+	+ "	"	+	"	+	+	+	+	+	+			
AUS10894	–	"	"	–	"	–	+	–	(–)	+	+			
Loros	–	–	"	–	"	(–) "	"	–	(–)	+	–			
Psathias	"	"	"	+	"	"	"	+	+	+	+			
Iskamish K-2-light	+	"	"	–	"	(–)	"	+	+	+	+			

Note: + = susceptible; – = resistant (< 5% new females compared to numbers on susceptible control); () = intermediate; " = no observation.

Table 1.3. Races of soybean cyst nematode *Heterodera glycines*. (Adapted from Cook and Rivoal, 1998.)

Differential cultivar	Race Virulence phenotype	3 0	6 1	13 2	9 3	1 4	5 5	11 6	2 7	8 8	10 9	12 10	14 11	7 12	15 13	1 14	4 15
'Pickett'		+	−*	+	−	+	−	+	−	+	−	+	−	+	−	+	−
'Peking'		+	−	−	+	+	−	−	+	+	−	−	+	+	−	−	+
'PI 88788'		+	−	−	−	−	+	+	+	+	−	−	−	−	+	+	+
'PI 90763'		+	−	−	−	−	−	−	−	−	+	+	+	+	+	+	+

* − resistant (female index <10% 'Lee'); + susceptible (female index >10% that of susceptible control 'Lee').

infection by cyst nematodes is the presence of adult female nematodes and cysts on the host roots. Symptoms of cyst nematode damage to host crops may include the appearance in the field of circular- or oval-shaped areas of stunted, yellowed and less vigorous plants. Infested patches vary in size and often show a clear-cut border between stunted and apparently healthy plants. PCN causes growth retardation and, at very high population densities, damage to the roots. Potato plants may show chlorosis and wilting, resulting in early senescence of plants. Crops infected with PCN show patches of poor growth. PCN cause yield loss and tubers will be smaller. Like PCN, symptoms caused by SCN infection are not unique. They are easily confused with nutrient deficiency, particularly iron deficiency, stress from drought or herbicide injury. Roots infected with SCN are dwarfed, yellowing and stunted. *Heterodera glycines* interferes with nodulation and may decrease the number of nitrogen-fixing nodules on the roots. CCN cause stunting and chlorosis on wheat and barley. Symptoms in the foliage are generally assumed to be associated with irregularities in soil depth, soil texture, soil pH, mineral nutrition, water availability or diseases. The roots branch excessively at sites where *H. avenae* females have established a syncytium, resulting in a knotted appearance on the root. On oats, *H. avenae* does not cause this knotted symptom. *Heterodera schachtii* can parasitize roots of plants of all ages. Seedlings may be severely injured or killed. When infected with beet cyst nematodes, outer leaves of plants usually wilt during the hot period of the day or when soil moisture becomes limited. Leaves of parasitized plants also may have pronounced yellowing. Infected plants have small storage roots that are severely branched with excess fibrous roots and are often referred to as bearded.

1.12 Management

Management of nematodes involves the reduction of nematode densities to non-damaging threshold levels using several measures in relation to the whole production system; control implies the use of a single measure to reduce or eliminate nematode numbers (Brown and Kerry, 1987). Several characteristics of cyst nematodes are of essential importance in management or control (Riggs and Schuster, 1998): (i) J2 are protected within eggs, which in turn are protected in a cyst with a hardened protective wall; (ii) unhatched J2 may remain dormant for many years; (iii) in many cases (e.g. *Globodera* spp.) substantial hatch will only occur in the presence of hatching factors produced by a potential host; and (iv) some cyst nematodes have a relatively narrow host range.

Fundamental to the prevention of cyst nematodes spreading into non-infested regions is the use of certified planting material, and strict legislation for those commodities being traded both internationally and locally. This policy has been the pillar for controlling several major cyst nematodes such as *G. rostochiensis*, *G. pallida* and *H. schachtii* (Hockland *et al.*, 2013). Cysts are easily spread on farm machinery and may be present in soil adhering to planting material or in water run-off. Cleaning machinery before and after use is recommended as are restricting movement of soil outside the field boundary and construction of natural windbreaks.

Restriction of the risk of movement of important pests is regulated via quarantine measures. In 2011, four species of cyst nematodes were on the quarantine list of different countries: *G. rostochiensis* (119), *G. pallida* (80), *H. glycines* (55) and *H. schachtii* (14) (Hockland *et al.*,

2013). Crop rotation can be an important component in managing cyst nematode densities when their host range is narrow. Cyst nematodes that have a small number of cultivated host plants include G. rostochiensis, G. pallida, H. avenae, H. zeae and H. carotae. Non-host crops can safely be cultivated, during which time a combination of spontaneous hatch and natural mortality will reduce the field population to below threshold levels. However, at high infestation levels the time between susceptible crops required to reduce populations to non-damaging levels may render control by rotation uneconomic.

Resistance of major crop hosts to G. rostochiensis, G. pallida, G. tabacum tabacum, G. tabacum solanacearum, H. avenae, H. glycines, H. schachtii and H. cajani has been found and attempts made to incorporate it into commercial cultivars. Only low level, or no, resistance is known in the major crop hosts of H. cruciferae, H. oryzae, H. sacchari and H. oryzicola (Riggs and Schuster, 1998). In many cases resistance is found only in wild species, with the accompanying inherent difficulties of transferring into commercial cultivars (see Chapters 8 and 9, this volume). The inappropriate continuous planting of resistant cultivars increases the selection pressure for virulent populations (e.g. potatoes and G. rostochiensis and G. pallida), limiting the durability of resistance. This is demonstrated by the repeated use of cultivars containing the *H1* gene against G. rostochiensis leading to selection for G. pallida, which has now become the principal species infecting potato in the UK.

Eggs of cyst nematodes are grouped, either inside the female's body/cyst or in a gelatinous sac. In these conditions they represent an easy target for parasitism by fungi or bacteria living in the rhizosphere. Numerous studies using nematophagous fungi and bacteria against economically important cyst nematodes have been undertaken with varying degrees of success and failure. The best example of suppressiveness of soils towards cyst nematodes is in the control of H. avenae by the fungi *Nematophthora gynophila* and *Pochonia chlamydosporia* (see Chapter 11, this volume).

Nematicides are very effective in controlling cyst nematodes (Whitehead and Turner, 1998; see Chapter 13, this volume) but several of the most effective have now been withdrawn because of health, safety and environmental concerns. The effectiveness of nematicides is reduced by their biological degradation by soil organisms, a process that may be increased by repeated use (Karpouzas and Giannakou, 2002). Nematicides have been extensively used as a control strategy for G. rostochiensis and G. pallida on potatoes, H. goettingiana on peas and H. avenae on cereals (all one or two generations a year). Species that produce several generations in a year (e.g. H. glycines) appear to be more difficult to control. Inadequate control of G. pallida increase on susceptible potatoes by an oximecarbamate nematicide of short persistence, such as oxamyl, is primarily due to the slow rate of J2 emergence in most populations of G. pallida, with a second generation and the upward migration of J2 from deeper untreated soil later in the growing season as potential contributory factors (Whitehead, 1992).

The repeated use of a single control measure is likely to fail, sooner or later, because of the selection of individuals unaffected by any control measure that may be applied. The potential for managing cyst nematodes by combining two or more control strategies in an integrated programme has been widely demonstrated (Roberts, 1993). Usually some level of crop rotation is practised, alongside additional measures. The advantages of this approach include the use of partially effective strategies and protection of highly effective ones that are vulnerable to nematode adaptation or environmental risk; examples include integrated control of G. rostochiensis and G. pallida in Europe, and H. glycines in the USA.

1.13 References

Aboul-Eid, H.Z. and Ghorab, A.I (1974) Pathological effects of *Heterodera cajani* on cowpea. *Plant Disease Reporter* 58, 1130–1133.

Al Abed, A., Al-Momany, A. and Al Banna, L. (2009) *Heterodera latipons* on barley in Jordan. *Phytopathologia Mediterranea* 43, 311–317. DOI: 10.14601/Phytopathol_Mediterr-1767

Amiri, S., Subbotin, S.A. and Moens, M. (2001) An efficient method for identification of the *Heterodera schachtii sensu stricto* group using PCR with specific primers. *Nematologia Mediterranea* 29, 241–246.

Amiri, S., Subbotin, S.A. and Moens, M. (2002) Identification of the beet cyst nematode *Heterodera schachtii* by PCR. *European Journal of Plant Pathology* 108, 497–506. DOI: 10.1023/A:1019974101225

Andersen, S. and Andersen, K. (1982) Suggestions for determination and terminology of pathotypes and genes for resistance in cyst forming nematodes, especially *Heterodera avenae*. *Bulletin OEPP* 12(4), 379–386.

Andersson, S. (1974) *Heterodera hordecalis* n.sp. (Nematoda: Heteroderidae) a cyst nematode of cereals and grasses in southern Sweden. *Nematologica* 20, 445–454. DOI: 10.1163/187529274X00078

Babatola, J.O. (1983a) Pathogenicity of *Heterodera sacchari* on rice. *Nematologia Mediterranea* 11, 21–25.

Babatola, J.O. (1983b) Rice cultivars and *Heterodera sacchari*. *Nematologia Mediterranea* 11, 103–105.

Baldwin, J.G. and Mundo-Ocampo, M. (1991) Heteroderinae, cyst- and non cyst-forming nematodes. In: Nickle, W.R. (ed.) *Manual of Agricultural Nematology*. Marcel Dekker, New York, pp. 275–362.

Baldwin, J.G., Nadler, S.A. and Adams, B.J. (2004) Evolution of plant parasitism among nematodes. *Annual Review of Phytopathology* 42, 83–105. DOI: 10.1146/annurev.phyto.42.012204.130804

Banyer, R.J. and Fisher, J.M. (1971) Seasonal variation in hatching of eggs of *Heterodera avenae*. *Nematologica* 17, 225–236. DOI: 10.1163/187529271X00071

Behrens, E. (1975) Taxonomically useful characters for the differentiation of *Heterodera* species. *Probleme der Phytonematologie. Vortrage anlässlich der 10 Tagung über Probleme der Phytonematologie im Institut für Pflanzenzuchtung Gross–Lusewitz der Deutschen Akademie der Landwirtschaftswissenschaften zu Berlin am 11 Juni 1971*, Berlin, Germany, pp. 122–142.

Bhatti, D.S. and Gupta, D.C. (1973) Guar an additional host of *Heterodera cajani*. *Indian Journal of Nematology* 3, 160.

Brown, R.H. and Kerry, B.R. (1987) *Principles and Practice of Nematode Control in Crops*. Academic Press, Melbourne, Australia.

Bulman, S.R. and Marshall, J.W. (1997) Differentiation of Australasian potato cyst nematode (PCN) populations using the polymerase chain reaction (PCR). *New Zealand Journal of Crop and Horticultural Science* 25, 123–129. DOI: 10.1080/01140671.1997.9513998

CABI (2016a) Invasive species compendium; *Globodera rostochiensis*. Available at: www.cabi.org/isc/datasheet/27034 (accessed 1 July 2016).

CABI (2016b) Invasive species compendium; *Heterodera glycines*. Available at: www.cabi.org/isc/datasheet/27027 (accessed 1 July 2016).

CABI (2016c) Invasive species compendium. *Heterodera cajani*. Available at: www.cabi.org/isc/datasheet/27023 (accessed 1 July 2016).

CABI (2016d) Invasive species compendium. *Heterodera schachtii*. Available at: www.cabi.org/isc/datasheet/27036 (accessed 1 July 2016).

Cadet, P. and Merny, G. (1978) Premiers essais de traitements chimiques contre les nématodes parasites de la canne à sucre en Haute-Volta. *Revue de Nématologie* 1, 53–62.

Canto-Saenz, M. and de Scurrah, M.M. (1977) Races of potato cyst nematode in the Andean region and a new system of classification. *Nematologica* 23, 340–349. DOI: 10.1163/187529277X00066

Chen, S.Y., Porter, P.M., Orf, J.H., Reese, C.D., Stienstra, W.C., Young, N.D., Walgenbach, D.D., Schaus, P.J., Arlt, T.J. and Breitenbach, F.R. (2001) Soybean cyst nematode population development and associated soybean yields of resistant and susceptible cultivars in Minnesota. *Plant Disease* 85, 760–766. DOI: 10.1094/PDIS.2001.85.7.760

Chitwood, B.J. (1949) Root-knot nematodes – Part I. A revision of the genus *Meloidogyne* Goeldi, 1887. *Proceedings of the Helminthological Society of Washington* 16, 90–104.

Cook, R. and Noel, G.R. (2002) Cyst nematodes: *Globodera* and *Heterodera* species. In Starr, J.L., Cook, R. and Bridge, J. (eds) *Plant Resistance to Parasitic Nematodes*. CAB International, Wallingford, UK, pp. 71–105.

Cook, R. and Rivoal, R. (1998) Genetics of resistance and parasitism. In: Sharma, S.B. (ed.) *The Cyst Nematodes*. Kluwer Academic Publishers, Dordrecht, The Netherlands, pp. 322–352.

Davis, E.L. and Tylka, G.L. (2000) Soybean cyst nematode disease. *The Plant Health Instructor*. DOI: 10.1094/PHI-I-2000-0725-01

De Almeida Engler, J., Engler, G. and Gheysen, G. (2011) Unravelling the plant cell cycle in nematode induced feeding sites. In: Jones, J., Gheysen, G. and Fenoll, C. (eds) *Genomics and Molecular Genetics of Plant–Nematode Interactions*. Springer, Dordrecht, The Netherlands, pp. 349–368. DOI 10.1007/978-94-007-0434-3_17

Ebrahimi, N., Viaene, N., Demeulemeester, K. and Moens, M. (2014) Observations on the life cycle of potato cyst nematodes *Globodera rostochiensis* and *G. pallida,* on early potato cultivars. *Nematology* 16, 937–952. DOI: 10.1163/15685411-00002821

Ebrahimi, N., Viaene, N. and Moens, M. (2015) Optimising trehalose-based quantification of live eggs in potato cyst nematodes (*Globodera rostochiensis* and *G. pallida*). *Plant Disease* 99, 947–953. DOI: 10.1094/PDIS-09-14-0940-RE

Epps, J.M. and Duclos, L.A. (1970) Races of soybean cyst nematode in Missouri and Tennessee. *Plant Disease Reporter* 54, 319–320.

Evans, K. (1968) The influence of some factors on the reproduction of *Heterodera rostochiensis*. PhD Thesis. London University, London.

Evans, K. (1970) Longevity of males and fertilisation of females of *Heterodera rostochiensis*. *Nematologica* 16, 369–320. DOI: 10.1163/187529270X00054

Evans, K. and Haydock, P.P.J. (2000) Potato cyst nematode management – present and future. *Aspects of Applied Biology* 59, 91–98.

Evans, K. and Rowe, J.A. (1998) Distribution and economic importance. In: Sharma, S.B. (ed.) *The Cyst Nematodes*. Kluwer Academic Publishers, Dordrecht, The Netherlands, pp. 1–30.

Filipjev, I.N. (1934) *Harmful and Useful Nematodes in Rural Economy*. Moscow and Leningrad, Russia (in Russian).

Fleming, C.C. and Marks, R.J. (1982) A method for the quantitative estimation of *Globodera rostochiensis* and *Globodera pallida* in mixed-species samples. *Record of Agricultural Research* 30, 67–70.

Fortuner, R. and Merny, G. (1973) Les nématodes parasites des racines associés au riz en Basse-Casamance (Sénégal) et en Gambie. *Cahiers ORSTOM Série Biologie* 21, 4–43.

Fourie, H., Spaull, V.W., Jones, R.K., Daneel, M.S. and De Waele, D. (2017) *Nematology in South Africa: A View from the 21st Century*. Springer International Publishing, Basel, Switzerland.

Franco, J. (1979) Effect of temperature on hatching and multiplication of potato cyst nematodes. *Nematologica* 25, 237–244. DOI: 10.1163/187529279X00253

Franklin, M.T. (1951) *The Cyst-forming Species of* Heterodera. *Commonwealth Agricultural Bureau Technical Communication* 29. Commonwealth Agricultural Bureau, Farnham Royal, UK.

Franklin, M.T. (1969) *Heterodera latipons* n. sp., a cereal cyst nematode from the Mediterranean region. *Nematologica* 15, 535–542. DOI: 10.1163/187529269X00867

Franklin, M.T. (1972) *Heterodera schachtii*. In: *C.I.H. Descriptions of Plant-Parasitic Nematodes*. Set 1, No. 1. CAB International, Wallingford, UK.

Garabedian, S. and Hague, N.G. (1984) The effect of weekly exposures to non-volatile nematicides and sugarcane root diffusate on the hatching of *Heterodera sacchari*. *Revue de Nématologie* 7, 95–96.

Golden, A.M., Epps, J.M., Riggs, R.D., Duclos, L.A., Fox, J.A. and Bernard, R.L. (1970) Terminology and identity of infraspecific forms of the soybean cyst nematode (*Heterodera glycines*). *Plant Disease Reporter* 54, 544–546.

Golinowski, W., Sobczak, M., Kurek, W. and Grymaszewska, G. (1997) The structure of syncytia. In: Fenoll, C., Grundler, F.M.W. and Ohl, S. (eds) *Cellular and Molecular Aspects of Plant–Nematode Interactions*. Kluwer Academic Publishers, Dordrecht, The Netherlands, pp. 80–97.

Grenier, E., Fournet, S., Petit, E. and Anthoine, G. (2010) A cyst nematode 'species factory' called the Andes. *Nematology* 12, 163–169. DOI: 10.1163/138855409X12573393054942

Hajihasani, A., Tanha, M.Z., Nicol, J.M. and Rezaee, S. (2010) Effect of the cereal cyst nematode, *Heterodera filipjevi*, on wheat in microplot trials. *Nematology* 12, 357–363. DOI: 10.1163/138855409X12548945788321

Handoo, Z.A., Carta, L.K., Skantar, A.M. and Chitwood, D.J. (2012) Description of *Globodera ellingtonae* n. sp. (Nematodea: Heteroderidae) from Oregon. *Journal of Nematology* 44, 40–57.

Hockland, S., Inserra, R. and Kohl, L.M. (2013) International Plant health – putting legislation into practice. In: Perry, R.N. and Moens, M. (eds) *Plant Nematology*, 2nd edn. CAB International, Wallingford, UK, pp. 359–382.

Holgado, R., Andersson, S., Rowe, J. and Magnusson, C. (2005) Management of rye cyst nematode *Heterodera filipjevi* (Madzhidov, 1981) Stelter 1984 in Norway. *Proceedings of the Confererence on Advances in Nematology*, London.

Hominick, W.M. (1986) Photoperiod and diapause in the potato cyst nematode, *Globodera rostochiensis*. *Nematologica* 32, 408–418. DOI: 10.1163/187529286X00291

Ibrahim, S.K., Perry, R.N., Plowright, R.A. and Rowe, J. (1993) Hatching behaviour of the rice cyst nematodes *Heterodera sacchari* and *H. oryzicola* in relation to age of host plant. *Fundamental and Applied Nematology* 16, 23–29.

Ichinohe, M. (1952) On the soybean nematode, *Heterodera glycines* n. sp., from Japan (trans.). *Oyo-Dobutsugaku-Zasshi* 17, 1–4.

Ichinohe, M. (1988) Current research on the major nematode problems in Japan. *Journal of Nematology* 20, 184–190.

Ishibashi, N., Kondo, E., Muraoka, M. and Yokoo, T. (1973) Ecological significance of dormancy in plant parasitic nematodes. I. Ecological difference between eggs in gelatinous matrix and cysts of *Heterodera glycines* Ichinohe (Tylenchida: Heteroderidae). *Applied Entomology and Zoology* 8, 53–63.

Ithal, N., Recknor, J., Nettleton, D., Maier, T., Baum, T.J. and Mitchum, M.G. (2007) Developmental transcript profiling of cyst nematode feeding cells in soybean roots. *Molecular Plant-Microbe Interactions* 20, 510–525. DOI: doi:10.1094/MPMI-20-5-0510

Jayaprakash, A. and Rao, Y.S. (1983) Hatching behaviour of the cyst-nematode *Heterodera oryzicola*. *Indian Journal of Nematology* 13, 117–118.

Jerath, M.L. (1968) *Heterodera sacchari*, a cyst nematode pest of sugar cane new to Nigeria. *Plant Disease Reporter* 52, 237–239.

Jones, J.T., Haegeman, A., Danchin, E.G.J. et al. (2013) Top 10 plant-parasitic nematodes in molecular plant pathology. *Molecular Plant Pathology* 14, 946–961. DOI: 10.1111/mpp.12057

Kaczmarek, A., Mackenzie, K., Kettle, H. and Blok, V.C. (2014) Influence of soil temperature on *Globodera rostochiensis* and *Globodera pallida*. *Phytopathologia Mediterranea* 53, 396–405. DOI: 10.14601/Phytopathol_Mediterr-13512

Kakaire, S., Grove, I.G. and Haydock, P.J. (2015) The number of generations of *Heterodera schachtii* completed on oilseed rape (*Brassica napus* L.) during the UK growing season. *Nematology* 17, 557–565. DOI: 10.1163/15685411-00002889

Karpouzas, D.G. and Giannakou, I.O. (2002) Biodegradation and enhanced biodegradation: a reason for reduced biological efficacy of nematicides in soil. *Russian Journal of Nematology* 10, 59–78.

Karssen, G., Van Hoenselaar, T., Verkerk-Bakker, B. and Janssen, R. (1995) Species identification of cyst and root-knot nematodes from potato by electrophoresis of individual females. *Electrophoresis* 16, 105–109. DOI: 10.1002/elps.1150160119

Kaushal, K.K., Srivastava, A.N., Pankaj, Chawla, G. and Singh, K. (2007) Cyst forming nematodes in India – a review. *Indian Journal of Nematology* 37, 1–7.

Klink, V., Overall, C., Alkharouf, N., MacDonald, M. and Matthews, B. (2007) A time-course comparative microarray analysis of an incompatible and compatible response by *Glycine max* (soybean) to *Heterodera glycines* (soybean cyst nematode) infection. *Planta* 226, 1423–1447. DOI: 10.1007/s00425-007-0581-4

Kort, J., Ross, H., Rumpenhorst, H.J. and Stone, A.R. (1977) An international scheme for the identification of pathotypes of potato cyst nematodes *Globodera rostochiensis* and *G. pallida*. *Nematologica* 23, 333–339. DOI: 10.1163/187529277X00057

Koshy, P.K. (1967) A new species of *Heterodera* from India. *Indian Phytopathology* 20, 272–274.

Koshy, P.K. and Swarup, G. (1971) On the number of generations of *Heterodera cajani*, the pigeon-pea cyst nematode in a year. *Indian Journal of Nematology* 1, 88–90.

Koshy, P.K., Swarup, G. and Sethi, C.L. (1971) Further notes on the pigeon-pea cyst nematode, *Heterodera cajani*. *Nematologica* 16, 477–482. DOI: 10.1163/187529270X00658

Langeslag, M., Mugniery, D. and Fayet, G. (1982) Développement embryonnaire de *Globodera rostochiensis* et *G. pallida* en fonction de la température, en conditions contrôlées et naturelles. *Revue de Nématologie* 5, 103–109.

Liebscher, G. (1892) Beobachtungen über das Auftreten eines Nematoden an Erbsen. *Journal für Landwirtschaft* 40, 357–368.

Luc, M. (1961) Nématodes du genre *Heterodera* parasites de cultures tropicales en Afrique. *Comptes rendus des Séances de l'Académie d'Agriculture de France* 47, 940.

Luc, M. (1974) *Heterodera sacchari*. CIH Descriptions of Plant-Parasitic Nematodes, Set 4, No. 48. Commonwealth Institute of Helminthology, St Albans, UK.

Luc, M. (1986) Cyst nematodes in equatorial and hot tropical regions. In: Lamberti, F. and Taylor, C.E. (eds) *Cyst Nematodes*. Plenum Press, New York, pp. 355–372.

Luc, M. and Berdon Brizuela, R. (1961) *Heterodera oryzae* n. sp. (Nematoda: Tylenchoidea) parasite du riz en Côte d'Ivoire. *Nematologica* 6, 272–279. DOI: 10.1163/187529261X00135

Luc, M. and Merny, G. (1963) *Heterodera sacchari* n. sp. (Nematoda: Tylenchoidea) parasite de la canne à sucre au Congo-Brazzaville. *Nematologica* 9, 31–37. DOI: 10.1163/187529263X00089

Madani, M., Vovlas, N., Castillo, P., Subbotin, S.A. and Moens, M. (2004) Molecular characterization of cyst nematode species (*Heterodera* spp.) from the Mediterranean basin using RFLPs and sequences of ITS-rDNA. *Journal of Phytopathology* 152, 229– 234. DOI: 10.1111/j.1439-0434.2004.00835.x

Madani, M., Subbotin, S.A. and Moens, M. (2005) Quantitative detection of the potato cyst nematode, *Globodera pallida*, and the beet cyst nematode, *Heterodera schachtii*, using real-time PCR with SYBR green I dye. *Molecular and Cellular Probes* 29, 81–86. DOI: 10.1016/j.mcp.2004.09.006

Madzhidov, A.R. (1981) [Bidera filipjevi n. sp. (Heteroderina: Tylenchida) in Tadzhikistan.] *Izvestiya Akademii Nauk Tadzhiksköï SSR. Biologicheskie Nauki* 2, 40–44.

Meagher, J.W. (1977) World dissemination of the cereal cyst nematode (*Heterodera avenae*) and its potential as a pathogen of wheat. *Journal of Nematology* 9, 9–15.

Meher, H.C., Kaushal, K.K., Khan, E. and Naved, S.H. (1998) Use of esterase phenotypes of females for precise diagnosis of four *Heterodera* species. *Indian Journal of Nematology* 28, 81–84.

Merny, G. (1970) Les nématodes phytoparasites des rizières inondées de Côte d'Ivoire. 1. Les espèces observées. *Cahiers ORSTOM Série Biologie* 11, 3–43.

Moens, M., Perry, R.N. and Starr, J.L. (2009) *Meloidogyne* species – a diverse group of novel and important plant parasites. In: Perry, R.N., Moens, M. and Starr, J.L. (eds) *Root-knot Nematodes*. CAB International, Wallingford, UK, pp. 1–17.

Momota, Y. (1979) The first report of *Heterodera latipons* Franklin, 1969 in Japan. *Japanese Journal of Nematology* 9, 73–74.

Mor, M., Cohn, E. and Spiegel, Y. (1992) Phenology, pathogenicity and pathotypes of cereal cyst nematodes, *Heterodera avenae* and *H. latipons* (Nematoda: Heteroderidae) in Israel. *Nematologica* 38, 494–501. DOI: 10.1163/187529292X00469

Mor, M., Spiegel, Y. and Oka, Y. (2008) Histological study of syncytia induced in cereals by the Mediterranean cereal cyst nematode, *Heterodera latipons*. *Nematology* 10, 279–287. DOI: 10.1163/156854108783476340

Muir, F. and Henderson, G. (1926) Nematodes in connection with sugar cane root rot in the Hawaiian islands. *Hawaii Plant Records* 30, 233.

Mulholland, V., Carde, L., O'Donnell, K.J., Fleming, C.C. and Powers, T.O. (1996) Use of the polymerase chain reaction to discriminate potato cyst nematode at the species level. In: Marshall, G. (ed.) *BCPC Proceedings No. 65, Diagnostics in Crop Production*. The British Crop Protection Council, Farnham, UK, pp. 247–252.

Müller, J. (1999) The economic importance of *Heterodera schachtii* in Europe. *Helminthologia* 36, 205–213.

Mulvey, R.H. and Golden, A.M. (1983) An illustrated key to the cyst-forming genera and species of Heteroderidae in the western hemisphere with species morphometrics and distribution. *Journal of Nematology* 15, 1–59.

Netscher, C. (1969) L'ovogénèse et la reproduction chez *Heterodera oryzae* et *H. sacchari* (Nematoda: Heteroderidae). *Nematologica* 15, 10–14. DOI: 10.1163/187529269X00038

Netscher, C., Luc, L. and Merny, G. (1969) Description du mâle d'*Heterodera sacchari* Luc & Merny, 1963. *Nematologica* 15, 156–157. DOI: 10.1163/187529269X00218

Niblack, T.L., Arelli, P.R., Noel, G.R., Opperman, C.H., Orf, J.H., Schmitt, D.P., Shannon, J.G. and Tylka, G.L. (2002) A revised classification scheme for genetically diverse populations of *Heterodera glycines*. *Journal of Nematology* 34, 279–288.

Niblack, T., Tylka, G.L., Arelli, P. *et al.* (2009) A standard greenhouse method for assessing soybean cyst nematode resistance in soybean: SCE08 (standardized cyst evaluation 2008). *Plant Health Progress* (available online). DOI: 10.1094/PHP-2009-0513-01-RV.

Nicol, J. and Rivoal, R. (2008) Global knowledge and its application for the integrated control and management of nematodes on wheat. In: Ciancio, A. and Mukerji, K.G. (eds) *Integrated Management and Biocontrol of Vegetable and Grain Crops Nematodes*. Springer, The Netherlands, pp. 251–294.

Nicol, J.M., Bolat, N., Sahin, E., Tülek, A., Yıldırım, A.F., Yorgancılar, A., Kaplan, A. and Braun, H.J. (2006) The cereal cyst nematode is causing economic damage on rain-fed wheat production systems of Turkey. *Phytopathology* 96, S169. [abstract].

Nicol, J.M., Turner, S.J., Coyne, D.L., den Nijs, L., Hockland, S. and Maafi, Z.T. (2011) Current nematode threats to world agriculture. In: Jones, J., Gheysen, G. and Fenoll, C. (eds) *Genomics and Molecular Genetics of Plant–Nematode Interactions*. Springer, Dordrecht, The Netherlands, pp. 21–43.

Nobbs, J.M., Ibrahim, S.K. and Rowe, J. (1992) A morphological and biochemical comparison of the four cyst nematode species, *Heterodera elachista*, *H. oryzicola*, *H. oryzae* and *H. sacchari* (Nematoda: Heteroderidae) known to attack rice (*Oryza sativa*). *Fundamental and Applied Nematology* 15, 551–562.

Odihirin, R.A. (1975) Occurrence of *Heterodera* cysts nematode (Nematoda: Heteroderidae) on wild grasses in southern Nigeria. *Occasional Publications of Nigerian Society of Plant Protection* 1, 24.

Odihirin, R.A. (1977) Irrigation water as a means for the dissemination of the sugar cane cyst nematode *Heterodera sacchari* at Bacita Sugar Estate. *Nigerian Society for Plant Protection* 2, 58.
OEPP/EPPO (2014) PQR database. Paris, France: European and Mediterranean Plant Protection Organization. Available at: www.eppo.int/DATABASES/pqr/pqr.htm (accessed 5 October 2017).
Perry, R.N. (2002) Hatching. In: Lee, D.L. (ed.) *The Biology of Nematodes*. Taylor and Francis, London, pp. 147–169.
Perry, R.N., Wright, D. and Chitwood, D. (2013) Reproduction, physiology and biochemistry. In: Perry, R.N. and Moens, M. (eds) *Plant Nematology*, 2nd edn. CAB International, Wallingford, UK, pp. 219–245.
Philis, I. (1988) Occurrence of *Heterodera latipons* on barley in Cyprus. *Nematologia Mediterranea*, 16, 223.
Philis, I. (1997) *Heterodera latipons* and *Pratylenchus thornei* attacking barley in Cyprus. *Nematologia Mediterranea* 25, 305–309.
Phillips, W.S., Kitner, M. and Zasada, I.A. (2017) Developmental dynamics of *Globodera ellingtonae* in field-grown potato. *Plant Disease* DOI: org/10.1094/PDIS-10-16-1439-RE
Plantard, O., Picard, D., Valette, S., Scurrah, M., Grenier, E. and Mugniéry, D. (2008) Origin and genetic diversity of Western European populations of the potato cyst nematode (*Globodera pallida*) inferred from mitochondrial sequences and microsatellite loci. *Molecular Ecology* 17, 2208–2218. DOI: 10.1111/j.1365-294X.2008.03718.x
Poromarto, S.H., Gramig, G.G., Nelson, B.D., Jr and Jain, S. (2015) Evaluation of weed species from the Northern Great Plains as hosts of soybean cyst nematode. *Plant Health Progress* DOI: 10.1094/PHP-RS-14-0024
Puthoff, D.P., Nettleton, D., Rodermel, S.R. and Baum T.J. (2003) *Arabidopsis* gene expression changes during cyst nematode parasitism revealed by statistical analyses of microarray expression profiles. *Plant Journal* 33, 911–921.
Rao, Y.S. and Jayaprakash, A. (1978) *Heterodera oryzicola* n. sp. (Nematoda: Heteroderidae) a cyst nematode on rice (*Oryza sativa* L.) from Kerala State, India. *Nematologica* 24, 341–346. DOI: 10.1163/187529278X00461
Riggs, R.D. (1992) Host range. In: Riggs, R.D. and Wrather, J.A. (eds) *Biology and Management of the Soybean Cyst Nematode*. St. Paul, Minnesota, American Phytopathological Society, pp. 107–114.
Riggs, R.D. and Schuster, R.P. (1998) Management. In: Sharma, S.B. (ed.) *The Cyst Nematodes*. Kluwer Academic Publishers, Dordrecht, The Netherlands, pp. 388–416.
Riggs, R.D., Hamblen, M.L. and Rakes, L. (1981) Infra-species variation in reactions to hosts in populations. *Journal of Nematology* 13, 171–179.
Rivoal, R. (1983) Biologie d'Heterodera avenae Wollenweber en France. III. Evolution des diapauses des races Fr1 et Fr4 au cours de plusieurs années consecutives: influence de la température. *Revue de Nématologie* 6, 157–164.
Rivoal, R. (1986) Biology of *Heterodera avenae* Wollenweber in France. IV. Comparative study of the hatching cycles of two ecotypes after their transfer to different climatic conditions. *Revue de Nématologie* 9, 405–410.
Rivoal, R. and Cook, R. (1993) Nematode pests of cereals. In: Evans, K., Trudgill, D.L. and Webster, J.M. (eds) *Plant-parasitic Nematodes in Temperate Agriculture*. CAB International, Wallingford, UK, pp. 259–303.
Roberts, P.A. (1993) The future of nematology: integration of new and improved management strategies. *Journal of Nematology* 25, 383–394.
Robinson, M.P., Atkinson, H.J. and Perry, R.N. (1987) The influence of temperature on the hatching, activity and lipid utilisation of second stage juveniles of the potato cyst nematodes *Globodera rostochiensis* and *G. pallida*. *Revue de Nématologie* 10, 349–354.
Sahin, E., Nicol, J.M., Elekcioglu, I.H. and Rivoal, R. (2010) Hatching of *Heterodera filipjevi* in controlled and natural temperature conditions in Turkey. *Nematology* 12, 193–200. DOI: 10.1163/156854109X463738
Schacht, H. (1859) Über den Rübennematoden. *Zeitschrift für die Rübenzuckerindustrie im Zollverein* 21, 1–19.
Schmidt, A. (1871) Über einige Feinde der Rübenfelder. *Zeitschrift für die Rübenzuckerindustrie im Zollverein* 9, 175–179.
Schölz, U. (2001) Biology, pathogenicity and control of the cereal cyst nematode *Heterodera latipons* Franklin on wheat and barley under semiarid conditions, and interactions with common root rot *Bipolaris sorokiniana* (Sacc.) Shoemaker [Teleomorph: *Cochliobolus sativus* (Ito et Kurib.) Drechs. ex Dastur.]. PhD thesis. Bonn University, Bonn, Germany.

Schölz, U. and Sikora, R.A. (2004) Hatching behaviour and life cycle of *Heterodera latipons* Franklin under Syrian agro-ecological conditions. *Nematology* 6, 245–256. DOI: 10.1163/1568541041217924

Sewell, R. (1973) Plant-parasitic nematodes from Canada and abroad. *Canadian Plant Disease Survey* 53, 34–35.

Sharma, S.B. and Sharma, R. (1998) Hatch and emergence. In: Sharma S.B. (ed.) *The Cyst Nematodes*. Kluwer Academic Publishers, Dordrecht, The Netherlands, pp. 191–216.

Sharma, S.B. and Swarup, G. (1984) *Cyst Forming Nematodes of India*. Cosmo Publications, New Delhi, India.

Siddiqui, Z.A. and Mahmood, I. (1993) Occurrence of races of *Heterodera cajani* in Uttar Pradesh, India. *Nematologia Mediterranea* 21, 185–186.

Singh, M. and Sharma, S.B. (1996) Emergence of *Heterodera cajani* juveniles from cysts and egg sacs. *Indian Journal of Plant Protection* 24, 90–97.

Skarbilovich, T.S. (1959) On the structure of the nematodes of the order Tylenchida Thorne, 1949. *Acta Parasitologica Polonica* 15, 117–132.

Smiley, R.W., Dababat, A.A., Iqbal, S., Jones, M.G.K., Tanha Maafi, Z., Peng, D., Subbotin, S.A. and Waeyenberge, L. (2017) Cereal cyst nematodes: a complex and destructive group of Heterodera species. *Plant Disease* 101, 1692–1720. DOI: 10.1094/PDIS-03-17-0355-FE

Sobczak M. and Golinowski W. (2009) Structure of cyst nematode feeding sites. In: Berg R.H. and Taylor C.G. (eds) *Cell Biology of Plant Nematode Parasitism*. Plant Cell Monographs, vol. 15. Springer, Dordrecht, The Netherlands, pp. 153–187.

Sobczak, M. and Golinowski, W. (2011) Cyst nematodes and syncytia. I In: Jones, J., Gheysen, G. and Fenoll, C. (eds) *Genomics and Molecular Genetics of Plant–Nematode Interactions*. Springer, Dordrecht, The Netherlands, pp. 61–82. DOI 10.1007/978-94-007-0434-3_4

Soliman, A.H., Sobczak, M. and Golinowski, W. (2005) Defence responses of white mustard, *Sinapis alba*, to infection with the cyst nematode *Heterodera schachtii*. *Nematology* 7, 881–889. DOI: 10.1163/156854105776186389

Stanton, J.M. and Eyres, M. (1994) Hatching of Western Australian populations of cereal cyst nematode, *Heterodera avenae*, and effects of sowing time and method of sowing on yield of wheat. *Australasian Plant Pathology* 23, 1–7.

Subbotin, S.A., Rumpenhorst, H.J. and Sturhan, D. (1996) Morphological and electrophoretic studies on populations of *Heterodera avenae* complex from the former USSR. *Russian Journal of Nematology* 4, 29–38.

Subbotin, S.A., Waeyenberge, L., Molokanova, I.A. and Moens, M. (1999) Identification of species from the *Heterodera avenae* group by morphometrics and ribosomal DNA RFLPs. *Nematology* 1, 195–207. DOI: 10.1163/156854199508018

Subbotin, S.A., Waeyenberge, L. and Moens, M. (2000) Identification of cyst forming nematodes of the genus *Heterodera* (Nematoda: Heteroderidae) based on the ribosomal DNA-RFLPs. *Nematology* 2, 153–164. DOI: 10.1163/156854100509042

Subbotin, S.A., Peng, D. and Moens, M. (2001) A rapid method for the identification of the soybean cyst nematode *Heterodera glycines* using duplex PCR. *Nematology* 3, 365–370. DOI: 10.1163/156854101317020286

Subbotin, S.A., Mundo-Ocampo, M. and Baldwin, J.G. (2010a) Systematics of cyst nematodes (Nematoda: Heteroderinae). *Nematology Monographs and Perspectives 8A* (series editors: Hunt, D.J. and Perry, R.N.). Brill, Leiden, The Netherlands.

Subbotin, S.A., Mundo-Ocampo, M. and Baldwin, J.G. (2010b) Systematics of cyst nematodes (Nematoda: Heteroderinae). *Nematology Monographs and Perspectives 8B* (series editors: Hunt, D.J. and Perry, R.N.). Brill, Leiden, The Netherlands.

Subbotin, S.A., Waeyenberge, L. and Moens, M. (2013) Molecular systematics. In: Perry, R.N. and Moens, M. (eds) *Plant Nematology*, 2nd edn. CAB International, Wallingford, UK, pp. 40–72.

Swarup, G., Prasad, S.K. and Raski, D.J. (1964) Some *Heterodera* species from India. *Plant Disease Reporter* 48, 235.

Szakasits, D., Heinen, P., Wieczorek, K., Hofmann, J., Wagner, F., Kreil, D.P., Sykacek, P., Grundler, F.M.W. and Bohlmann, H. (2009) The transcriptome of syncytia induced by the cyst nematode *Heterodera schachtii* in *Arabidopsis* roots. *Plant Journal* 57, 771–784. DOI: 10.1111/j.1365-313X.2008.03727.x.

Tanha Maafi, Z., Subbotin, S.A. and Moens, M. (2003) Molecular identification of cyst-forming nematodes (Heteroderidae) from Iran and a phylogeny based on the ITS sequences of rDNA. *Nematology* 5, 99–111. DOI: 10.1163/156854102765216731

Tanha Maafi, Z., Sturham, D., Handoo, Z., Mor, M., Moens, M. and Subbotin, S.A. (2007) Morphological and molecular studies on *Heterodera sacchari*, *H. goldeni*, and *H. leuceilyma* (Nematoda: Heteroderidae). *Nematology* 9, 483–497. DOI: 10.1163/156854107781487242

Thorne, G. (1928) *Heterodera punctata* n. sp. a nematode parasite on wheat roots from Saskatchewan. *Scientific Agriculture* 8, 707–711.

Toumi, F., Waeyenberge, L., Viaene, N., Dababat, A., Nicol, J.M., Ogbonnaya, F. and Moens, M. (2013a) Development of a species-specific PCR to detect the cereal cyst nematode, *Heterodera latipons*. *Nematology* 15, 709–717. DOI: 10.1163/15685411-00002713

Toumi, F., Waeyenberge, L., Viaene, N., Dababat, A. Nicol, J.M., Ogbonnaya, F. and Moens, M. (2013b) Development of two species-specific primer sets to detect the cereal cyst nematodes *Heterodera avenae* and *Heterodera filipjevi*. *European Journal of Plant Pathology* 136, 613–624. DOI: 10.1007/s10658-013-0192-9

Toumi, F., Waeyenberge, L., Viaene, V., Dababat, A.A., Nicol, J.M., Ogbonnaya, F.C. and Moens, M. (2015) Development of qPCR assays for quantitative detection of *Heterodera avenae* and *H. latipons*. *European Journal of Plant Pathology*, 143, 305–316. DOI: 10.1007/s10658-015-0681-0

Toumi, F., Waeyenberge, L., Viaene, N., Dababat, A.A., Nicol, J.M., Ogbonnaya, F. and Moens, M. (2017) Cereal cyst nematodes: importance, distribution, identification, quantification, and control. *European Journal of Plant Pathology* 1–20. DOI: 10.1007/s10658-017-1263-0

Trudgill, D.L. (1967) The effect of environment on sex determination in *Heterodera rostochiensis*. *Nematologica* 13, 263–272. DOI: 10.1163/187529267X00120

Trudgill, D.L. (1985) Potato cyst nematodes: a critical review of the current pathotyping scheme. *EPPO Bulletin* 15, 273–279. DOI: 10.1111/j.1365-2338.1985.tb00228.x

Turner, S.J. and Evans, K. (1998) The origins, global distribution and biology of potato cyst nematodes (*Globodera rostochiensis* (Woll.) and *Globodera pallida* (Stone)). In: Marks, R. and Brodie, B. (eds) *Potato Cyst Nematodes. Biology, Distribution and Control*. CAB International, Wallingford, UK, pp. 7–26.

Turner, S.J. and Subbotin, S.A. (2013) Cyst nematodes. In: Perry, R.N. and Moens, M. (eds) *Plant Nematology*, 2nd edn. CAB International, Wallingford, UK, pp. 109–143.

van den Elsen, S., Ave, M., Schoenmakers, N., Landeweert, R., Bakker, J. and Helder, H. (2012) A rapid, sensitive and cost-efficient assay to estimate viability of potato cyst nematodes. *Phytopathology* 102, 140–146. DOI: 10.1094 / PHYTO-02-11-0051

Vovlas, N., Lamberti, F. and Tuopay, D.K. (1986) Observations on the morphology and histopathology of *Heterodera sacchari* attacking rice in Liberia. *FAO Plant Protection Bulletin* 34, 153–156.

Whitehead, A.G. (1992) Emergence of juvenile potato cyst-nematodes *Globodera rostochiensis* and *G. pallida* and the control of *G. pallida*. *Annals of Applied Biology* 120, 471–486. DOI: 10.1111/j.1744-7348.1992.tb04907.x

Whitehead, A.G. (1995) Decline of potato cyst nematodes, *Globodera rostochiensis* and *G. pallida*, in spring barley microplots. *Plant Pathology* 44, 191–195. DOI: 10.1111/j.1365-3059.1995.tb02728.x

Whitehead, A.G. and Turner, S.J. (1998) Management and regulatory control strategies for potato cyst nematodes (*Globodera rostochiensis* and *G. pallida*). In: Marks, R.J. and Brodie, B.B. (eds) *Potato Cyst Nematodes: Biology, Distribution and Control*. CAB International, Wallingford, UK, pp. 135–152.

Wollenweber, H.W. (1923) Krankheiten und Beschädigungen der Kartoffel. *Arbeiten des Forschungsinstitutes für Kartoffelbau Berlin* 7, 1–56.

Wollenweber, H.W. (1924) Zur Kenntnis der Kartoffel-Heteroderen. *Illustrierte landwirtschaftliche Zeitung* 44, 100–101.

Wouts, W.M. and Baldwin, J.G. (1998) Taxonomy and identifcation. In: Sharma, S.B. (ed.) *The Cyst Nematodes*. Kluwer Academic Publishers, Dordrecht, The Netherlands, pp. 83–122.

Wouts, W.W., Schoemaker, A., Sturhan, D. and Burrows, P.R. (1995) *Heterodera spinicauda* sp. n. (Nematoda: Heteroderidae) from mud flats in the Netherlands, with a key to the species of the *H. avenae* group. *Nematologica*, 41, 575–583. DOI: 10.1163/003925995X00512

Wouts, W.M., Rumpenhorst, H.J. and Sturhan, D. (2001) *Heterodera betae* sp. n., the yellow beet cyst nematode (Nematoda: Heteroderidae). *Russian Journal of Nematology* 9, 33–42.

Wrather, J.A., Anderson, T.R., Arsyad, D.M., Tan, Y., Ploper, L.D., Porta-Puglia, A., Ram, H.H. and Yorinori, J.T. (2001) Soybean disease loss estimates for the top ten soybean-producing countries in 1998. *Canadian Journal of Plant Pathology* 23, 115–121. DOI: 10.1080/07060660109506918

Wrather, J.A., Koenning, S.R. and Anderson, T.R. (2003) Effect of diseases on soybean yields in the United States and Ontario (1999 to 2002). *Plant Health Progress*. DOI: 10.1094/PHP-2003-0325-01-RV

Wright, D.J. and Perry, R.N. (2006) Reproduction, physiology and biochemistry. In: Perry, R.N. and Moens, M. (eds) *Plant Nematology*. CAB International, Wallingford, UK, pp. 187–209.

Yadav, U.S. and Walia, R.K. (1988) On the biology of pigeon pea cyst nematode, *Heterodera cajani* Koshy, 1967. *Indian Journal of Nematology* 18, 35–39.

Ye, W. (2012) Development of PrimeTime-Real-Time PCR for species identification of soybean cyst nematode (*Heterodera glycines* Ichinohe, 1952) in North Carolina. *Journal of Nematology* 44, 284–290.

Zasada, I.A., Phillips, W.S. and Ingham, R.E. (2015) Biological insights into *Globodera ellingtonae*. In: *4th Symposium of Potato Cyst Nematode Management (including other nematode parasites of potatoes), Aspects of Applied Biology 130*. Harper Adams University, Newport, Shropshire, UK.

Zheng, J., Subbotin, S.A., Waeyenberge, L. and Moens, M. (2000) Molecular characterisation of Chinese *Heterodera glycines* and *H. avenae* populations based on RFLPs and sequences of rDNA-ITS regions. *Russian Journal of Nematology* 8, 109–113.

2 Genomics and Transcriptomics – a Revolution in the Study of Cyst Nematode Biology

Sebastian Eves-van den Akker[1,2] and John T. Jones[3,4,5]

[1]*Division of Plant Sciences, College of Life Sciences, University of Dundee, Dundee, UK;* [2]*Department of Biological Chemistry, John Innes Centre, Norwich Research Park, Norwich, UK;* [3]*The James Hutton Institute, Invergowrie, Dundee, UK;* [4]*The University of St Andrews, North Haugh, St Andrews, UK;* [5]*Ghent University, Ghent, Belgium*

2.1	Introduction	27
2.2	A Note of Caution	29
2.3	Current Status of Genome and Transcriptome Projects for Cyst Nematodes and Other Plant-parasitic Nematodes	29
2.4	Key Findings from Genome/Transcriptome Projects	30
2.5	Population Genetics and Metagenetics	34
2.6	Identification of Key Biochemical Pathways and Targets for Control	35
2.7	Mitochondrial Genomes	36
2.8	Horizontal Gene Transfer	36
2.9	Accessibility	38
2.10	Conclusions and Future Prospects	38
2.11	Acknowledgements	39
2.12	References	39

2.1 Introduction

This is a fascinating time to be a biologist. Our field is in the middle of a revolutionary change brought about by the availability of massive high throughput and inexpensive sequencing capacity. The ability to generate genome or transcriptome sequence from any organism has facilitated a step change in the manner in which we approach biology. In the past the first, and often the most time consuming, part of a molecular biology study was cloning a gene that could subsequently be used for functional studies. This is exemplified by the process that led to the cloning of the first cyst nematode effector – a β1,4 endoglucanase from *Globodera rostochiensis* (Smant *et al.*, 1998). Cloning this gene took many years and included purification of nematode proteins, raising of antibodies against these proteins, identification of the antibodies that recognized gland cells followed by immunopurification of the antigenic protein, N-terminal sequencing of purified proteins and subsequent cloning of the full length cDNA by rapid amplification of cDNA ends polymerase chain reaction (RACE

PCR). Only after this process was complete was it possible to start the key studies that allowed the role of this gene in the biology of *G. rostochiensis* to be determined. One of the huge advantages of access to genomic resources is the ability to start work with a comprehensive list of genes present in your organism of choice and to move directly to functional studies of genes deemed to be of interest. A genome or transcriptome therefore offers a blueprint for functional studies.

The recent explosion in genome data has been underpinned by new developments in sequencing technology. Figure 2.1 shows how the costs of generating raw DNA sequence (at a large genome sequencing centre) have changed since 2001. Between 2001 and 2006 costs dropped in line with what might be expected in terms of Moores law – a derivation of the observation that computing power could be expected to double every 2 years to an expectation that for a highly successful field the costs of a key operation would be anticipated to halve every 2 years. From 2006 onwards costs dropped dramatically as new sequencing technologies, primarily Illumina, became widely used. It is likely that future developments in the field will lead to further improvements, particularly in terms of the length of individual sequencing reads that can be obtained.

The familiarity of many researchers working on plant-parasitic nematodes with the *Caenorhabditis elegans* project meant that the plant-parasitic nematode community was keenly aware of the potential benefits of genomic scale approaches to biology. Consequently, our discipline was among the first to apply these tools outside the well-characterized model organisms. A comprehensive review of the huge amount of information generated from cyst nematode genome and transcriptome projects is beyond the scope of this chapter;

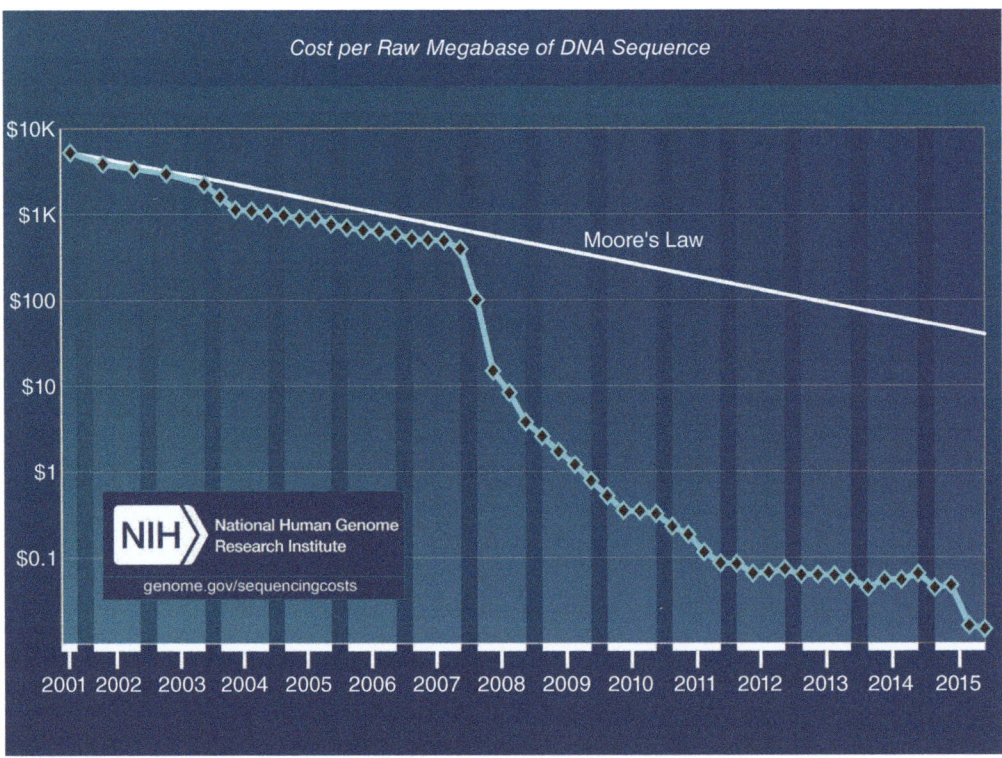

Fig. 2.1. Costs of raw DNA sequencing between 2001 and 2015, with anticipated costs in keeping with Moore's law plotted for comparison. Note the huge drops in costs after 2006, co-incident with the development of 454 and Illumina sequencing. (Taken from https://www.genome.gov/sequencingcostsdata/)

selected illustrative examples of the techniques and associated discoveries are provided instead.

2.2 A Note of Caution

> Any genome or transcriptome assembly is a hypothesis.
> Leighton Prichard, The James Hutton Institute

Genomics has had, and will continue to have, a profound impact on all areas of biology. Although it is tempting to view any published genomic resource as the final end product of a complex and lengthy process, it is important to recognize that there may be several limitations to the genome or transcriptome that is produced. It is more difficult to generate a meaningful assembly for an organism with a large, repeat-rich genome than from a less complex sample. Similarly, a sequence generated from one individual, or from a clonal line of an asexually reproducing organism, is likely to be easier to assemble than a sequence generated from large numbers of sexually reproducing organisms that have not been through a genetic bottleneck. A practical example of this difference from within cyst nematodes is given in section 2.4.2. For transcriptome projects, where the focus is often on identifying differentially expressed transcripts, sequencing a heterogeneous population is less problematic. Here, replication is critical, as discussed in detail in section 2.4.3. For both genome and transcriptome assemblies, there are common criteria that ought to be met, but are frequently overlooked. Most important is the removal of potential contaminant reads (Koutsovoulos *et al.*, 2016), often of fungal, bacterial or host origin. With sufficient coverage, every next generation sequencing project is essentially a metagenome. Even if the original sample was prepared in sterile culture, biofilms introduced during the library preparation stage guarantee the sequencing of non-target organisms at low coverage. Removal of contamination is particularly important for faithful reporting of horizontal gene transfer in plant-parasitic nematodes, and almost invariably improves assembly of the genome of the target organism (Kumar *et al.*, 2013).

2.3 Current Status of Genome and Transcriptome Projects for Cyst Nematodes and Other Plant-parasitic Nematodes

In 2014, the draft genome of *G. pallida* was completed and with it the first cyst nematode genome was made available to the scientific community (Cotton *et al.*, 2014). This paved the way for several advances and set the trend for the future of cyst nematode genomics. In 2016, the second cyst nematode genome project, for *G. rostochiensis*, was completed (Eves-van den Akker *et al.*, 2016a). The low genetic variation present in the UK *G. rostochiensis* used for sequencing (Eves-van den Akker *et al.*, 2014a) is reflected in the minimal pseudo-duplication of conserved eukaryote genes (Parra *et al.*, 2007), and indeed of all genes, suggesting that the *G. rostochiensis* genome assembly is a more accurate representation of a *Globodera* genome. The differences between these two genomes emphasize the need to minimize heterogeneity in starting material as far as possible.

Within the cyst nematodes, coverage of *Globodera* is set to increase in the near future with the addition of the *G. ellingtonae* genome and transcriptome (D. Denver and I. Zasada, 2017, pers. comm.). However, we currently lack data for several key species. These include the most economically important cyst nematode, *Heterodera glycines* (although genome and transcriptome projects are actively being pursued by the community (T. Baum, M. Goellner Mitchum, pers. comm.)) and *H. schachtii*, a species that is widely used as a model for cyst nematode biology due to its ability to infect *Arabidopsis thaliana*.

Outside the cyst nematodes, draft genomes are available for the root-knot nematodes *Meloidogyne incognita* (Abad *et al.*, 2008), *M. hapla* (Opperman *et al.*, 2008) and *M. floridensis* (Lunt *et al.*, 2014), with *M. arenaria* and *M. javanica* due to follow in the near future (E. Danchin, pers. comm.). In terms of other plant-parasitic nematodes, the small size and difficulty of culturing *Trichodorus* and *Paratrichodorus* mean that no genome or transcriptome data of any description are available for a Clade 1 plant-parasitic nematode. Transcriptome projects using next generation sequencing are in progress for both *Xiphinema index* and *Longidorus elongatus*

(E. Danchin, J. Jones and S. Eves-van den Akker, pers. comm.) but at present the only genomic resource for any Clade 2 plant-parasitic nematode is a small-scale expressed sequence tag (EST) project on *X. index* (Furlanetto *et al.*, 2005), and raw reads of low coverage *X. americanum* genome skimming (Denver *et al.*, 2016). Table 2.1 summarizes the genomic and transcriptomic resources available for cyst nematodes, and other plant-parasitic nematodes, at the time of writing. This table does not include small-scale EST projects but focuses on genome and comprehensive transcriptome projects. It should also be noted that one implication of a steady reduction in the cost of sequencing is that any such list is likely to become rapidly outdated. As costs continue to decline, and techniques become refined to allow their use with smaller samples, it is likely that genomic resources will become available for all economically important nematodes as well as for phylogenetically informative species that would not previously have been considered.

2.4 Key Findings from Genome/Transcriptome Projects

Once a genome sequence is deemed sufficiently well assembled for analysis, and the likely gene complement present has been identified, researchers are able to interrogate this information to address the areas of biology of greatest interest to them. This may include specific biochemical pathways or genes involved in the parasitic process.

2.4.1 Identifying genes of interest from genome/transcriptome resources

Often the 'problem' associated with a genome or transcriptome study is the volume of data produced: there are too many genes to work on. Homing in on particular classes of genes of interest is therefore critical. Several different approaches have been employed for plant-parasitic nematodes in general and cyst nematodes in particular, each of which has produced novel insights into nematode biology. Many of these approaches are broadly applicable to all areas of interest, but have primarily been applied to

identification of effectors (see Chapter 4, this volume) due to their role in dictating the outcome of the cyst nematode–plant interaction and their scope for informing nematode management (see section 2.5).

2.4.2 Sequence similarity searches: gene birth and death

Simple sequence similarity searches are a rapid method to identify and superficially classify genes of interest. However, this approach is biased in that it identifies candidate genes on the basis of similarity to genes that are deemed to be of interest in another system. Analysis of the SPRY domain-containing gene family by similarity searches raises questions about gene birth and gene death in cyst nematodes, and simultaneously serves to highlight why inferring function based on similarity can be misleading, as well as demonstrating why less heterogeneous starting material is paramount for a genome assembly.

A family of effectors (SPRYSECs) was identified from cyst nematodes and are characterized by the predicted presence of a SPRY domain and a signal peptide for secretion. These SPRYSECs were abundant in EST datasets (Jones *et al.*, 2009) and one was shown to be a determinant of avirulence in cyst nematodes (RBP1) (Sacco *et al.*, 2009). The first large-scale analysis of SPRY domain-containing proteins in a cyst nematode was carried out as part of the *G. pallida* genome project. This analysis revealed that 299 SPRY domain-containing proteins were predicted. However, it seems likely that the pseudo-duplication of genes in the *G. pallida* assembly led to an over-estimation of the number of SPRY domain-containing proteins. The more accurate *G. rostochiensis* assembly estimates just 66 SPRY domain-containing proteins, highlighting the value of comparative genomics and of using genetically homogeneous starting material. Regardless of the total number present, the SPRY domain-containing family in cyst nematodes is massively expanded compared to that in other organisms. Comparative work by Thorpe *et al.* (2014) demonstrates that although a small number of SPRY domain-containing proteins are present in most animals (including humans), a significant lineage-specific expansion exists in

Table 2.1. Current state of genome and transcriptome projects for cyst nematodes and other plant-parasitic nematodes.

Species	Genome data			Transcriptome data		
	Resources available	Reference	Technology	Life stages sampled	Reference
Globodera rostochiensis	Full download of assemblies, gene calls, cDNA and protein sequences. BLAST server	(Eves-van den Akker *et al.*, 2016a)	Illumina	Egg, J2, 14 dpi parasitic	(Eves-van den Akker *et al.*, 2016a)
G. pallida	Full download of assemblies, gene calls, cDNA and protein sequences. BLAST server	(Cotton *et al.*, 2014)	Illumina	Egg, J2, 7 dpi, 14 dpi, 21 dpi, 28 dpi, 35 dpi parasitic, male	(Cotton *et al.*, 2014)
G. ellingtonae		(Denver *et al.*, 2016; Phillips *et al.*, 2017b)			(Phillips *et al.*, 2017a)
Heterodera schachtii			FLX	J2	
H. avenae			Illumina	J2, females	
Rotylenchulus reniformis		(Showmaker *et al.*, in progress)	Illumina	J2, sedentary females	(Eves-van den Akker *et al.*, 2016b)
Nacobbus aberrans			Illumina	J2, migratory parasitic, sedentary parasitic	(Eves-van den Akker *et al.*, 2014b)
Meloidogyne incognita	BLAST server, individual proteins	(Abad *et al.*, 2008)			
M. hapla	Full download of assembly, gene calls, proteins, BLAST server (through Wormbase ParaSite)	(Opperman *et al.*, 2008)			
M. floridensis	Assembly, gene calls, BLAST server (through nematode.net)	(Lunt *et al.*, 2014)			
M. arenaria		(Blanc-Mathieu *et al.*, 2017)			
M. javanica		(Blanc-Mathieu *et al.*, 2017)			
M. graminicola			454 FLX; Illumina	J2, J2, parasitic J2, J3–J4, female (on susceptible and resistant plants)	(Haegeman *et al.*, 2013, Petitot *et al*, 2016)

Continued

Table 2.1. Continued.

Species	Genome data		Transcriptome data		
	Resources available	Reference	Technology	Life stages sampled	Reference
Pratylenchus coffeae	Raw reads	(Burke et al., 2015)	454 FLX	Mixed	(Haegeman et al., 2011a)
P. penetrans	Genome skim raw reads	(Denver et al., 2016)	Illumina	Mixed (including parasitic stages)	(Vieira et al., 2015)
P. zeae			454 FLX	Mixed	(Fosu-Nyarko et al., 2015)
P. thornei	Genome skim raw reads	(Denver et al., 2016)	454 FLX	Mixed	(Nicol et al., 2012)
P. neglectus	Genome skim raw reads	(Denver et al., 2016)			
Ditylenchus destructor	Not clear	(Zheng et al., 2016)			
Hirschmanniella oryzae			454FLX	Mixed	(Bauters et al., 2014)
Bursaphelenchus xylophilus	Full download of assembly, gene calls, proteins, BLAST server (through GeneDB)	(Kikuchi et al., 2011)	Illumina	Fungal feeding and parasitic	(Espada et al., 2016)
Aphelenchoides besseyi			454 FLX; Illumina	Mixed	(Kikuchi et al., 2014; Wang et al., 2014)
Xiphinema index			Illumina	Stress + standard conditions	(Eves-van den Akker et al., 2016b)
Xiphinema americanum	Genome skim raw reads	(Denver et al., 2016)			
Longidorus elongatus			Illumina	Mixed	(Eves-van den Akker et al., 2016b)

J2, J3, J4 = second-, third- and fourth-stage juveniles, respectively; dpi = days post-inoculation.

the cyst nematodes. The SPRYSECs form part of this lineage-specific expansion, yet they are the exception not the rule. In *G. pallida* and *G. rostochiensis*, 10% and 25% of SPRY domain-containing proteins, respectively, are predicted to be SPRYSECs. Given that SPRY domain-containing proteins are present in most animals, and that SPRYSECs are in the minority, caution is therefore required in interpreting the output of similarity searches alone when identifying effectors or assigning function more generally. The SPRY family also raises an interesting question: if ~25% of all SPRYs in cyst nematodes are SPRYSECs, what is or was the function of the remaining 75% of the sequences? Interestingly,

the presence of a signal peptide is an almost perfect predictor of gene expression specifically at the early stages of parasitism (Mei et al., 2015), whereas SPRY domain-containing proteins lacking a signal peptide tend not to be expressed at any of the life stages tested (Cotton et al., 2014). It is therefore unclear if this apparent functional dichotomy represents a reservoir of diversity or an effector graveyard.

2.4.3 Temporal transcriptional profiling

Sequence-unrelated genes can be subdivided into categories on the basis of their temporal expression patterns. This approach is based on the premise that genes expressed at a particular stage are required for process/es that occur in that stage. When compared to similarity searches, the lack of bias in temporal transcriptional profiling provides an advantage for discovery of novel genes.

By far the most widely adopted method for transcriptional profiling is Illumina RNA sequencing (RNAseq). In an ideal situation, RNAseq data are mapped to a pre-existing genome assembly in order to mitigate some of the redundancy issues that can be encountered during transcriptome assembly (Eves-van den Akker et al., 2014b). Nevertheless, with sufficient coverage, the RNAseq reads themselves can be assembled into a *de novo* reference transcriptome, and the differential expression analysis can then be performed by mapping reads from individual life stages to the reference transcriptome. This approach is an efficient and inexpensive method to analyse non-model organisms, and has been effectively deployed for the study of the false root-knot nematode *Nacobbus aberrans* (Eves-van den Akker et al., 2014b). Biological replication is central to the success of this approach. As the costs of Illumina sequencing have decreased, the expectations of the number of replications that are required have steadily increased. Recently, a comprehensive 48-replicate experiment sought to provide recommendations on the number of replications required. This analysis showed that six replicates is the minimum required for all experiments, with 12 biological replicates required reliably to identify most differentially expressed genes (Schurch et al., 2016).

Nevertheless, prior to the publication of these recommendations, experiments using fewer replicates have successfully identified important differentially expressed genes. In 2014, the duplicated life stage-specific transcriptome of *G. pallida* was used to identify a gene family of unprecedented variability that is involved in sustained biotrophy. Based on the naïve hypothesis that the most highly expressed, and most highly differentially expressed genes specific to the sedentary parasitic stages would contain genes required for parasitism, a pipeline was developed (Eves-van den Akker et al., 2014a). This pipeline identified a novel group of hyper-variable extracellular effectors, termed HYPs. Genetically, HYP effectors are a large and complex gene family, with a modular structure similar to the oomycete RXLR and CRN effectors and a tandem repeat structure with variable di-residues similar to the bacterial transcription activator-like (TAL) effectors. Most notably, HYP effectors show unparalleled genetic diversity between individuals of the same population: no two nematodes tested had the same genetic complement of HYP effectors. Individuals vary in the number, size and type of effector subfamilies. To date, there is still no known genetic mechanism that could account for such drastic genetic variation. HYP effectors highlight well how much there is still to learn in cyst nematode genomics, and also illustrate the scope for interesting discoveries based on relatively simple temporal transcriptomic profiling.

2.4.4 Spatial transcriptomic profiling

Genes that are important in some biological processes can be identified on the basis of the tissues in which they are expressed. Such genes lend themselves well to spatial transcriptomic analysis. For example, effectors from cyst and other plant-parasitic nematodes can be identified on the basis of their expression in, and secretion from, the pharyngeal gland cells. In the first relatively large-scale spatial transcriptomic profiling experiment, the contents of pharyngeal gland cells of *H. glycines* were aspirated, RNA extracted and cDNA libraries were generated (Gao et al., 2001). ESTs generated from these libraries were screened for gland cell expression by *in situ* hybridization resulting in a list of 53 novel effectors (Wang et al., 2001; Gao et al., 2003). A similar approach was used by Noon et al. (2015, 2016) with the addition of 18 new gland cell genes. Depending on the filtering criteria used, this approach is also unbiased. The main distinction between temporal and spatial transcriptomic profiling is that the former will identify genes of all sorts that are involved in the parasitism process, while the latter is specific for effector identification. More recently, Maier et al. have developed a method for the purification of whole gland cells from nematodes, and the subsequent extraction of RNA (Maier et al., 2013). These samples have been subjected to Illumina RNAseq and *de novo* assembly. This approach shows great promise in effector identification, in particular for species with accompanying genomes or transcriptomes.

Spatial and temporal transcriptome studies have been carried out in a range of plant-parasitic nematodes. However, there is still no individual species for which both approaches have been used. This combination of approaches would provide a powerful resource to subdivide effectors into functional groups.

2.4.5 Promoter motifs and their associated predictive power

One of the benefits of a genome sequence over a transcriptome is the ability to explore the mechanisms underlying biological traits. Given the highly tissue-specific expression pattern of effectors, a recent study aimed to identify promoter elements that were associated with genes expressed in the gland cells. By employing a differential motif discovery algorithm, Eves-van den Akker et al. (2016a) identified a six-base pair DOrsal Gland Box (termed DOG box) that is highly enriched in the promoter region of representatives from 26 of the 28 experimentally validated dorsal gland cell effector families. Importantly, genes with multiple iterations of the DOG box in their promoter regions were more likely to encode proteins with a signal peptide for secretion: a required feature of an effector. This DOG box was used to predict a superset of putative effectors and experimentally validated gland cell expression of two novel genes by in situ hybridization. The DOG box was subsequently used to catalogue DOG effectors from both cyst nematode genomes. This finding represents a major turning point: for the first time, we are able to predict cyst nematode effectors in silico.

The ability for rapid identification of effectors in other pathosystems (e.g. the RXLR peptide motif of Phytophthora effectors (Whisson et al., 2007) or the bacterial type secretion systems) was a prerequisite for major advances in understanding. The trend is set to exploit the ever increasing genomic information to identify vastly more effectors than previously known by using promoter motifs descriptive of other glands (subventral, amphids, etc.) or from other nematodes.

Given the almost complete lack of overlap in effector repertoires between nematodes with independent evolutionary origins of sedentary endo-parasitism (Eves-van den Akker et al., 2016a), expanding these in silico analyses will likely reveal many novel effectors and effector functions, ultimately giving us a better understanding of how these nematodes modify plant development and immunity. The extensive gland cell transcriptome libraries for H. glycines and M. incognita could serve as excellent training sets for this approach in these species.

2.5 Population Genetics and Metagenetics

Several cyst nematode species have been characterized to pathotype based on various sources of host resistance (Kort et al., 1977; see Chapters 1 and 9, this volume). Genomics allows us to understand the genetic basis of these differences, with profound implications for the management of cyst nematodes, particularly in terms of deployment of available resistance sources.

Towards this aim, 23 populations of G. rostochiensis from nine countries covering five G. rostochiensis pathotypes (Ro1–5) were genotyped by sequencing (GBS) in an effort to study the relationships between populations of different origins (Mimee et al., 2015). This Pool-Seq approach provided good evidence that the pathotype system is not consistent with unique genotypes, but instead points towards multiple independent origins of the same pathotypes in different genotypes. As expected, no individual polymorphisms covering the five pathotypes was descriptive of virulence on H1 resistance (Eves-van den Akker et al., 2016a). Convergent evolution of the same phenotype by independent mutations may be explained by identifying genes that contain at least one, but not necessarily the same, predicted loss or change of function variant in populations virulent on H1, and always absent for any predicted loss or change of function variants in populations avirulent on H1. When allowing for some heterozygosity in populations virulent on H1, 190 such genes were identified, two of which correspond to known effectors (Eves-van den Akker et al., 2016a).

Although these predictions remain untested, there is promise that identifying such a marker descriptive at the level of pathotype could be rapidly deployed to combat cyst nematodes through pathogen-informed cultivar choice in the field. In 2015, a proof of concept study identified a region of mitochondrial DNA descriptive of three main groups of *G. pallida* present in the UK and adopted a metagenetic approach to the sequencing and analysis of nearly 1000 field samples simultaneously (Eves-van den Akker *et al.*, 2015). This study was able to map relative abundance of each type at the level of the country, the field and the individual nematode, providing three different landscapes of diversity. Properly exploiting three aspects of the genomics revolution (effector identification, comparative genomics and population genetics) to identify a marker descriptive of pathotypes would have direct and immediate implications for agriculture.

2.6 Identification of Key Biochemical Pathways and Targets for Control

Knowledge of the full gene complement of a pathogen allows the details of its biochemistry to be inferred and potential vulnerabilities that can be exploited in terms of new control strategies to be identified. Such analyses will frequently use the pathways that have been functionally validated in *C. elegans* as a starting point to determine whether orthologues are present in the species being analysed. For example, genes involved in the RNAi pathway were examined in both *G. pallida* and *M. incognita*. Although putative orthologues of many functionally characterized *C. elegans* sequences were identified, several genes involved in spread of double stranded RNA between cells that are important for systemic RNAi were absent in both species (Abad *et al.*, 2008; Cotton *et al.*, 2014). The ability to generate RNAi by feeding dsRNA to both species (Urwin *et al.*, 2002; Rosso *et al.*, 2005) implies the existence of an alternative system for spread of the RNAi signal in these nematodes. Caution needs to be applied when interpreting the outputs of these analyses. Too often no attempts are made to seek true orthologues and the only analysis that is done is to examine whether similar genes are present. Functional validation is required to allow biological inferences to be made from a genome project; identification of putative orthologues is a starting point rather than the end product.

The ability to explore the complete gene complement of an organism allows potentially important genes and gene families to be identified in a way that would not be possible without such information. For example, as part of the *G. pallida* genome analysis, the presence of a hugely expanded family of glutathione synthetases (GS) was noted (Cotton *et al.*, 2014). All nematodes contain one GS gene, plant parasites often contain between one and four, but 52 are present in the *G. pallida* genome. More surprisingly, a significant number of these have a signal peptide for secretion, something that is not observed in any other species. The function of the *G. pallida* GS gene family remains uncertain. However, this is clearly an important gene family in terms of *G. pallida* biology and without the availability of the genome sequence this would never have come to light.

The availability of genomes from several plant-parasitic nematodes allows inferences about biology to be tested across species and provides some level of reassurance that hypotheses that are being made reflect the biology of the species under investigation. For example, the *M. incognita* genome was noted to contain reduced numbers of genes associated with immune responses to pathogens; it was suggested that the endoparasitic lifestyle of this nematode allowed it a degree of protection from biotic stresses (Abad *et al.*, 2008). Reassuringly, this pattern of reduction in immune system was also observed in the unrelated sedentary endoparasite *G. pallida* (Cotton *et al.*, 2014). The ability of plant-parasitic nematode genomes to respond (in evolutionary terms) to environmental conditions encountered during their life cycles is further illustrated by *Bursaphelenchus xylophilus*. This nematode spends much of its life cycle living within pine trees and its genome contains larger numbers of genes involved in various detoxification processes compared to other free-living and plant-parasitic nematodes (Kikuchi *et al.*, 2011).

One potentially productive area from genomics projects that has yet to be fully exploited in studies on plant-parasitic nematodes is pathway mapping. Pathway mapping allows interrogation of genome or transcriptome information to identify key pathways that may underlie biological traits. For example, tools such as MapMan (Thimm et al., 2004) can be used in conjunction with transcriptomic data to examine global changes in expression profiles of genes associated with various metabolic pathways in response to biotic or abiotic stresses (Urbanczyk-Wochniak et al., 2006; Rotter et al., 2007). Pathway mapping has also been used to compare the genomes of 25 *Dickeya* species and strains (https://dx.doi.org/10.6084/m9.figshare.767275.v2). This analysis revealed a correlation between the metabolic capacity of these bacterial species and their host range. The predicted metabolic capabilities were also predictive of the ability of the strains to grow on specific single carbon sources. In the future, pathway mapping offers the prospect of identifying vulnerabilities in an organism that can subsequently be used as targets for control.

2.7 Mitochondrial Genomes

Mitochondria are present in almost every observed eukaryote (Karnkowska et al., 2016), exceptions being only where they are secondarily lost. The highly conserved canonical mitochondrial genome is typically represented as a single circular molecule (Sloan, 2013), and contains a specific set of genes, the order of which is often largely conserved across whole phyla (Mindell et al., 1998).

A striking exception to this rule is multipartism – the subdivision of the canonical mitochondrial gene content across multiple circles. There are relatively few examples of multipartite mitochondrial genomes across the tree of life, yet examples exist for representatives of most major groups: protists (Vlcek et al., 2010), rotifers (Suga et al., 2008), plants (Palmer and Shields, 1984), lice (Wei et al., 2012; Chen et al., 2014) and nematodes (Hunt et al., 2016; Phillips et al., 2016). In the case of the latter, two distantly related genera of nematodes have been discovered with multipartite mitochondrial genomes, the free-living *Rhabditophanes* and the plant-parasitic *Globodera*.

In 2000, the multipartite genome of *G. pallida* was reported to contain at least six circles and unclosed fragments (Armstrong et al., 2000).

In 2007, the multipartite genome of *G. rostochiensis* was also reported to contain six sub-genomes (Gibson et al., 2007). It was noted that the arrangement is remarkably similar between these two *Globodera* species and indicated that this may be a common feature of the *Globodera*. In contrast to this assertion, Phillips et al. (2016) described the mitochondrial genome of *G. ellingtonae* to contain a bi-partite arrangement, providing the first indication that the mitochondrial genomes of *Globodera* are not as similar as previously thought. The sequencing of a nuclear genome to hundreds of fold coverage invariably sequences the mitochondrial genome to thousands of fold coverage. The genomics revolution has thus further strengthened this case, by uncovering and correcting historical misannotations of *G. rostochiensis* circles. By exploiting the whole genome sequence data for *G. rostochiensis* (Eves-van den Akker et al., 2016a) and *G. pallida* (Cotton et al., 2014), it was shown that five of the six mitochondrial chromosomes thought to originate from *G. rostochiensis* actually originate from *G. pallida* (S. Eves-van den Akker, 2017, pers. comm.). Furthermore, four new species-unique mitochondrial chromosomes were identified for *G. rostochiensis* and, together with the remaining *G. rostochiensis* sequence, these five chromosomes collectively contain all genes usual to nematode mitochondrial genomes. Thus, there are three species of cyst nematodes within the same genus each with a different number of mitochondrial chromosomes. Exploring the unusual evolutionary pressures that gave rise to, and result from, such organizations will be fascinating.

2.8 Horizontal Gene Transfer

One of the most remarkable findings from cyst nematode genome and transcriptome projects has been the discovery of large numbers of genes that have been acquired by a process of horizontal gene transfer (HGT). Horizontal gene transfer seems to have occurred on several different occasions and from several different sources. The best characterized examples of horizontally acquired genes are the plant cell wall modifying proteins. Many different classes of these proteins have been identified in Clade 12 plant-parasitic nematodes, including cellulases, pectate lyases, xylanases, polygalacturonases, arabinogalactan galactosidases, arabinanases and expansin-like proteins. The phylogeny of many of these sequences has been examined in detail by Danchin et al. (2010) and a bacterial origin has been hypothesized. In the genome of *G. pallida*, 40 genes from six different protein families involved in plant cell wall degradation are present, with a similar range present in *M. incognita* (Abad et al., 2008) and *M. hapla* (Opperman et al., 2008). Intriguingly, the Clade 10 pinewood nematode, *B. xylophilus*, which is unrelated to the Clade 12 nematodes, also contains cell wall modifying proteins. While the pectate lyase present in this nematode is from the same enzyme class as that in Clade 12 plant-parasitic nematodes (Kikuchi et al., 2006), the cellulases present in *B. xylophilus* are entirely unrelated to those in the Clade 12 nematodes and appear to have been acquired from fungi (Kikuchi et al., 2004). Several reviews of the evolution and function of the cell wall modifying proteins present in plant-parasitic nematodes have been published (e.g. Haegeman et al., 2011b) and this topic is not covered further here.

Other genes have also been acquired by HGT that play different roles in the biology of cyst nematodes. Chorismate mutase is present in Clade 12 plant-parasitic nematodes including cyst and root-knot nematodes as well as several migratory endoparasitic species (Lambert et al., 1999; Jones et al., 2003; Bauters et al., 2014). While a role in induction of the giant cells induced by root-knot nematodes was originally proposed for chorismate mutase (Doyle and Lambert, 2003), the presence of this gene in nematodes with such diverse feeding strategies argues against this role. Chorismate can be converted to the defence signalling molecule salicylic acid via an isochorismate intermediate (Wildermuth et al., 2001) and therefore it is possible that chorismate mutase has a role in suppressing host defences. More recently, two studies have identified genes that encode invertases in cyst nematodes (Danchin et al., 2016; Noon and Baum, 2016). The *G. pallida* invertases are expressed in the digestive system and are biochemically active in metabolizing sucrose, suggesting that HGT has enhanced the ability of the nematode to exploit the major translocation carbohydrate of their host plants (Danchin et al., 2016).

Surprisingly, genes acquired by HGT in nematodes appear to have all the features associated with 'regular' nematode genes. Almost all horizontally acquired genes contain multiple spliceosomal introns. Given the bacterial origin of most of these sequences, and the fact that bacterial genes lack introns, these were most likely gained after the HGT event. An analysis of six different families of plant cell wall-degrading enzymes has shown that all have at least one intron position conserved between different lineages of plant-parasitic nematodes (Danchin et al., 2010), supporting the idea that the introns were gained before the separation of the different lineages and therefore soon after their transfer. However, it should be noted that this observation can also be explained by the existence of 'hot spots' that are predisposed for intron insertion (Danchin et al., 2010).

The presence of introns suggests that bacterial genes gained eukaryotic features as part of an adaptation process to their host genome after their acquisition by HGT. Analysis of guanine-cytosine (GC) content and codon usage of horizontally acquired genes further supports this idea. The GC content of the *NodL* gene from *M. incognita* is more similar to that of other *Meloidogyne* genes than to the GC content of genes of rhizobial bacterial species (Scholl et al., 2003). A more comprehensive analysis of the GC contents of three genes present in a range of plant-parasitic nematode species showed that the majority of these genes had a GC content similar to other sequences in the recipient genome and significantly different to that of the potential donor bacteria (Noon and Baum, 2016). A similar pattern is seen for codon usage in these two studies, with codon usage patterns much more similar to those of the nematode genomes than of the donor bacteria. Similar analysis on a wider range of horizontally acquired genes showed that it was not possible to discriminate genes acquired by HGT from other nematode sequences on the basis of GC content or codon usage patterns (Danchin et al., 2010). Genes acquired by HGT have therefore undergone a process of homogenization within their recipient genomes and are now indistinguishable from other nematode sequences.

One question that genome sequencing has yet to address is the mechanism by which these genes were acquired by plant-parasitic nematodes.

Various phylogenetic studies have been undertaken that have attempted to identify the likely origins of the horizontally acquired genes (e.g. Danchin et al., 2010; Palomares-Rius et al., 2014). Unfortunately, sequence motifs that might provide clues as to how the sequences were transferred into nematodes are not readily identified. However, analysis of the *G. rostochiensis* genome showed that genes acquired by HGT are significantly closer to transposable elements when compared to other *G. rostochiensis* genes (Eves-van den Akker et al., 2016a). Given the length of time since many of the HGT events have most likely occurred (in many cases before the split of the major Clade 12 plant-parasitic nematodes) and the homogenization that has occurred within the nematode genomes, this may be as close as it is possible to get to an explanation as to the mechanism underlying HGT. Regardless, it appears that HGT has played a major role in the adaptation to parasitism in plant-parasitic nematodes. In the *G. rostochiensis* genome project, an attempt was made to identify all genes likely to have been acquired via HGT. Support for an HGT origin was found for 519 genes (3.5%), with strong support for a subset of 91 (Eves-van den Akker et al., 2016a). This suggests that we have not yet uncovered the full repertoire of biological functions that cyst nematodes have acquired via HGT.

2.9 Accessibility

For the scientific community to make full use of genomics and transcriptomics, and to facilitate advances towards successful management and control of cyst nematodes, accessibility of the data is critical. All reputable scientific journals require that researchers make their raw data available to the community before publication is permitted. However, if all that is made available to the community from a sequencing project is the raw sequence reads, anyone wishing to interrogate the genome will need to undertake a significant amount of work before being able to undertake any biological investigations. The two cyst nematode genomes sequenced to date are examples of good practice in this area. The full *G. pallida* genome assembly and gene calls can be interrogated or downloaded through the Sanger centre website (www.sanger.ac.uk/resources/downloads/helminths/globodera-pallida.html; Cotton et al., 2014). Both *G. rostochiensis* and *G. pallida* data are similarly accessible at the WormBase ParaSite (http://parasite.wormbase.org/), and will soon be available through the permanent National Center for Biotechnology Information (NCBI) repository – submission to this database can be a lengthy process.

Whilst genome assemblies and gene calls are sometimes, but not often, deposited in permanent databases, this is very rarely the case for other useful synthesized data (e.g. normalized expression values, genome wide single nucleotide polymorphisms (SNPs)). Several of these data have no natural home in standard journal publication format, yet they contain some of the most valuable information for the community. Dryad, a general purpose repository that allows data behind scientific publications to be accessed by the wider community, was created to address this disparity (http://datadryad.org). Importantly, each Dryad submission is curated, linked to a publication and receives its own unique DOI. Much of the data underlying the *G. rostochiensis* genome project, including normalized expression data, BLAST2GO annotation, clustering of expression data, interproscan information and information on SNPs are available, accessible and free to download (http://dx.doi.org/10.5061/dryad.4s5r6). To realize the full potential of genomics and transcriptomics, the sharing of data is necessary.

2.10 Conclusions and Future Prospects

New sequencing technologies deliver the prospect of improved, and indeed perfect, genome assemblies. While Illumina sequencing was largely responsible for the genomics revolution in cyst nematodes, the short read lengths and non-random errors that are introduced during sequencing mean that it cannot be used to generate a perfect genome assembly. Perfect assemblies are well within the reach of new long-read technologies (Oxford Nanopore, but in particular PacBio). PacBio sequencers produce long reads with truly random errors: regardless of the error rates, perfect genomes are therefore simply a matter of coverage.

The accessibility of high throughput sequencing tools is likely to increase further in coming years. It is also likely that as sample preparation techniques are refined, the required quantity of starting material for sequencing (currently around 1 μg for most sequencing centres) will decline. In addition, methods for amplification of nucleic acids have been developed that may allow sequencing of nucleic acids derived from individual nematodes. These factors are likely to mean that an even broader range of cyst nematodes and other plant-parasitic nematodes will have their genomes or transcriptomes explored. Economically important species will clearly remain a focus for these studies but it is also likely that a more eclectic selection of species will be examined. Taxonomically important species that may be of little economic importance and that may be extremely difficult to collect but that reside at the basal positions in important plant-parasitic nematode clades will be able to be analysed. Such studies may reveal a great deal about the evolution of plant parasitism.

2.11 Acknowledgements

The James Hutton Institute receives funding from the Rural and Environmental Science and Analytical Services division of the Scottish Government. This work benefited from interactions funded through COST Action FA 1208. SE-vdA is supported by the Biotechnology and Biological Sciences Research Council grant BB/M014207/1.

2.12 References

Abad, P., Gouzy, J., Aury, J.-M. *et al*. (2008) Genome sequence of the metazoan plant-parasitic nematode *Meloidogyne incognita*. *Nature Biotechnology* 26, 909–915. DOI: 10.1038/nbt.1482

Armstrong, M.R., Blok, V.C. and Phillips, M.S. (2000) A multipartite mitochondrial genome in the potato cyst nematode *Globodera pallida*. *Genetics* 154, 181–192.

Bauters, L., Haegeman, A., Kyndt, T. and Gheysen, G. (2014) Analysis of the transcriptome of *Hirschmanniella oryzae* to explore potential survival strategies and host–nematode interactions. *Molecular Plant Pathology* 15, 352–363. DOI: 10.1111/mpp.12098

Blanc-Mathieu, R., Perfus-Barbeoch, L., Aury, J.M. *et al*. (2017) Hybridization and polyploidy enable genomic plasticity without sex in the most devastating plant-parasitic nematodes. *PLoS Genetics* 13(6), e1006777.

Burke, M., Scholl, E.H., Bird, D.M., Schaff, J.E., Colman, S.D., Crowell, R., Diener, S., Gordon, O., Graham, S. and Wang, X. (2015) The plant parasite *Pratylenchus coffeae* carries a minimal nematode genome. *Nematology* 17, 621–637. DOI: 10.1163/15685411-00002901

Chen, S.-C., Wei, D.-D., Shao, R., Shi, J.-X., Dou, W. and Wang, J.-J. (2014) Evolution of multipartite mitochondrial genomes in the booklice of the genus *Liposcelis* (Psocoptera). *BMC Genomics* 15, 861. DOI: 10.1186/1471-2164-15-861

Cotton, J.A., Lilley, C.J., Jones, L.M. *et al*. (2014) The genome and life-stage specific transcriptomes of *Globodera pallida* elucidate key aspects of plant parasitism by a cyst nematode. *Genome Biology* 15, R43. DOI: 10.1186/gb-2014-15-3-r43

Danchin, E.G.J., Rosso, M.-N., Vieira, P., de Almeida-Engler, J., Coutinho, P.M., Henrissat, B. and Abad, P. (2010) Multiple lateral gene transfers and duplications have promoted plant parasitism ability in nematodes. *Proceedings of the National Academy of Sciences of the United States of America* 107, 17651–17656. DOI: 10.1073/pnas.1008486107

Danchin, E.G., Guzeeva, E.A., Mantelin, S., Berepiki, A. and Jones, J.T. (2016) Horizontal gene transfer from bacteria has enabled the plant-parasitic nematode *Globodera pallida* to feed on host-derived sucrose. *Molecular Biology and Evolution* 33, 1571–1579. DOI: 10.1093/molbev/msw041

Denver, D.R., Brown, A.M., Howe, D.K., Peetz, A.B. and Zasada, I.A. (2016) Genome skimming: a rapid approach to gaining diverse biological insights into multicellular pathogens. *PLoS Pathogens* 12, e1005713.

Doyle, E.A. and Lambert, K.N. (2003) *Meloidogyne javanica* chorismate mutase 1 alters plant cell development. *Molecular Plant-Microbe Interactions* 16, 123–131. DOI: 10.1094/MPMI.2003.16.2.123

Espada, M., Silva, A.C., Eves van den Akker, S., Cock, P.J., Mota, M. and Jones, J.T. (2016) Identification and characterization of parasitism genes from the pinewood nematode *Bursaphelenchus xylophilus* reveals a multilayered detoxification strategy. *Molecular Plant Pathology* 17(2), 286–295.

Eves-van den Akker, S., Lilley, C.J., Jones, J.T. and Urwin, P.E. (2014a) Identification and characterisation of a hyper-variable apoplastic effector gene family of the potato cyst nematodes, *PLoS Pathogens* 10, e1004391.

Eves-van den Akker, S., Lilley, C., Danchin, E., Rancurel, C., Cock, P., Urwin, P. and Jones, J. (2014b) The transcriptome of *Nacobbus aberrans* reveals insights into the evolution of sedentary endoparasitism in plant-parasitic nematodes. *Genome Biology and Evolution* evu171.

Eves-van den Akker, S., Lilley, C.J., Reid, A., Pickup, J., Anderson, E., Cock, P.J., Blaxter, M., Urwin, P.E., Jones, J.T. and Blok, V.C. (2015) A metagenetic approach to determine the diversity and distribution of cyst nematodes at the level of the country, the field and the individual. *Molecular Ecology* 24, 5842–5851. doi: 10.1111/mec.13434

Eves-van den Akker, S., Laetsch, D.R., Thorpe, P. et al. (2016a) The genome of the yellow potato cyst nematode, *Globodera rostochiensis*, reveals insights into the basis of parasitism and virulence. *Genome Biology* 17, 1–23. DOI: 10.1186/s13059-016-0985-1

Eves-Van Den Akker, S., Lilley, C.J., Yusup, H.B., Jones, J.T. and Urwin, P.E. (2016b) Functional C-TERMINALLY ENCODED PEPTIDE (CEP) plant hormone domains evolved *de novo* in the plant parasite *Rotylenchulus reniformis*. *Molecular Plant Pathology* 17, 1265–1275. DOI: 10.1111/mpp.12402

Fosu-Nyarko, J., Tan, J.A.C., Gill, R., Agrez, V.G., Rao, U. and Jones, M.G. (2015) *De novo* analysis of the transcriptome of *Pratylenchus zeae* to identify transcripts for proteins required for structural integrity, sensation, locomotion and parasitism. *Molecular Plant Pathology* 17, 532–552. DOI: 10.1111/mpp.12301

Furlanetto, C., Cardle, L., Brown, D.J. and Jones, J.T. (2005) Analysis of expressed sequence tags from the ectoparasitic nematode *Xiphinema index*. *Nematology* 7, 95–104. DOI: 10.1163/1568541054192180

Gao, B., Allen, R., Maier, T., Davis, E.L., Baum, T.J. and Hussey, R.S. (2001) Identification of putative parasitism genes expressed in the esophageal gland cells of the soybean cyst nematode *Heterodera glycines*. *Molecular Plant-Microbe Interactions* 14, 1247–1254. DOI: 10.1094/MPMI.2001.14.10.1247

Gao, B.L., Allen, R., Maier, T., Davis, E.L., Baum, T.J. and Hussey, R.S. (2003) The parasitome of the phytonematode *Heterodera glycines*. *Molecular Plant-Microbe Interactions* 16, 720–726. DOI: 10.1094/MPMI.2003.16.8.720

Gibson, T., Blok, V.C. and Dowton, M. (2007) Sequence and characterization of six mitochondrial subgenomes from *Globodera rostochiensis*: multipartite structure is conserved among close nematode relatives. *Journal of Molecular Evolution* 65, 308–315. DOI: 10.1007/s00239-007-9007-y

Haegeman, A., Joseph, S. and Gheysen, G. (2011a) Analysis of the transcriptome of the root lesion nematode *Pratylenchus coffeae* generated by 454 sequencing technology. *Molecular and Biochemical Parasitology* 178, 7–14. DOI: 10.1016/j.molbiopara.2011.04.001

Haegeman, A., Jones, J.T. and Danchin, E.G.J. (2011b) Horizontal gene transfer in nematodes: a catalyst for plant parasitism? *Molecular Plant-Microbe Interactions* 24, 879–887. DOI: 10.1094/MPMI-03-11-0055

Haegeman, A., Bauters, L., Kyndt, T., Rahman, M.M. and Gheysen, G. (2013) Identification of candidate effector genes in the transcriptome of the rice root knot nematode *Meloidogyne graminicola*. *Molecular Plant Pathology* 14, 379–90. DOI: 10.1111/mpp.12014

Hunt, V.L., Tsai, I.J., Coghlan, A. et al. (2016) The genomic basis of parasitism in the *Strongyloides* clade of nematodes. *Nature Genetics* 48, 299–307. DOI: 10.1038/ng.3495

Jones, J.T., Furlanetto, C., Bakker, E., Banks, B., Blok, V., Chen, Q., Phillips, M. and Prior, A. (2003) Characterization of a chorismate mutase from the potato cyst nematode *Globodera pallida*. *Molecular Plant Pathology* 4, 43–50. DOI: 10.1046/j.1364-3703.2003.00140.x

Jones, J.T., Kumar, A., Pylypenko, L.A., Thirugnanasambandam, A., Castelli, L., Chapman, S., Cock, P.J., Grenier, E., Lilley, C.J., Phillips, M.S. and Blok, V.C. (2009) Identification and functional characterization of effectors in expressed sequence tags from various life cycle stages of the potato cyst nematode *Globodera pallida*. *Molecular Plant Pathology* 10, 815–828. DOI: 10.1111/j.1364-3703.2009.00585.x

Karnkowska, A., Vacek, V., Zubáčová, Z. et al. (2016) A eukaryote without a mitochondrial organelle. *Current Biology* 26, 1274–1284. DOI: http://dx.doi.org/10.1016/j.cub.2016.03.053

Kikuchi, T., Jones, J.T., Aikawa, T., Kosaka, H. and Ogura, N. (2004) A family of glycosyl hydrolase family 45 cellulases from the pine wood nematode *Bursaphelenchus xylophilus*. *FEBS Letters* 572, 201–205. DOI.org/10.1016/j.febslet.2004.07.039

Kikuchi, T., Shibuya, H., Aikawa, T. and Jones, J.T. (2006) Cloning and characterization of pectate lyases expressed in the esophageal gland of the pine wood nematode *Bursaphelenchus xylophilus*. *Molecular Plant-Microbe Interactions* 19, 280–287. DOI: 10.1094/MPMI-19-0280

Kikuchi, T., Cotton, J.A., Dalzell, J.J., Hasegawa, K., Kanzaki, N., McVeigh, P., Takanashi, T., Tsai, I.J., Assefa, S.A. and Cock, P.J. (2011) Genomic insights into the origin of parasitism in the emerging plant pathogen *Bursaphelenchus xylophilus*. *PLoS Pathogens* 7, e1002219.

Kikuchi, T., Cock, P.J., Helder, J. and Jones, J.T. (2014) Characterisation of the transcriptome of *Aphelenchoides besseyi* and identification of a GHF 45 cellulase. *Nematology* 16, 99–107. DOI: 10.1163/15685411-00002748

Kort, J., Ross, H., Rumpenhorst, H. and Stone, A. (1977) An international scheme for identifying and classifying pathotypes of potato cyst-nematodes *Globodera rostochiensis* and *G. pallida*. *Nematologica* 23, 333–339. DOI: 10.1163/187529277X00057

Koutsovoulos, G., Kumar, S., Laetsch, D.R., Stevens, L., Daub, J., Conlon, C., Maroon, H., Thomas, F., Aboobaker, A.A. and Blaxter, M. (2016) No evidence for extensive horizontal gene transfer in the genome of the tardigrade *Hypsibius dujardini*. *Proceedings of the National Academy of Sciences of the United States of America* 113, 5053–5058. DOI: 10.1073/pnas.1600338113

Kumar, S., Jones, M., Koutsovoulos, G., Clarke, M. and Blaxter, M. (2013) Blobology: exploring raw genome data for contaminants, symbionts and parasites using taxon-annotated GC-coverage plots. *Frontiers in Genetics* 4, 273. DOI: 10.3389/fgene.2013.00237

Lambert, K.N., Allen, K.D. and Sussex, I.M. (1999) Cloning and characterization of an esophageal-gland-specific chorismate mutase from the phytoparasitic nematode *Meloidogyne javanica*. *Molecular Plant-Microbe Interactions* 12, 328–336. DOI: 10.1094/MPMI.1999.12.4.328

Lunt, D.H., Kumar, S., Koutsovoulos, G. and Blaxter, M.L. (2014) The complex hybrid origins of the root knot nematodes revealed through comparative genomics, *PeerJ* 2, e356.

Maier, T.R., Hewezi, T., Peng, J. and Baum, T.J. (2013) Isolation of whole esophageal gland cells from plant-parasitic nematodes for transcriptome analyses and effector identification. *Molecular Plant-Microbe Interactions* 26, 31–35. DOI: 10.1094/MPMI-05-12-0121-FI

Mei, Y., Thorpe, P., Guzha, A., Haegeman, A., Blok, V.C., MacKenzie, K., Gheysen, G., Jones, J.T. and Mantelin, S. (2015) Only a small subset of the SPRY domain gene family in *Globodera pallida* is likely to encode effectors, two of which suppress host defences induced by the potato resistance gene *Gpa2*. *Nematology* 17, 409–424. DOI: 10.1163/15685411-00002875

Mimee, B., Duceppe, M.O., Véronneau, P.Y., Lafond-Lapalme, J., Jean, M., Belzile, F. and Bélair, G. (2015) A new method for studying population genetics of cyst nematodes based on Pool-Seq and genome wide allele frequency analysis. *Molecular Ecology Resources* 15, 1356–1365. DOI: 10.1111/1755-0998.12412

Mindell, D.P., Sorenson, M.D. and Dimcheff, D.E. (1998) Multiple independent origins of mitochondrial gene order in birds. *Proceedings of the National Academy of Sciences of the United States of America* 95, 10693–10697.

Nicol, P., Gill, R., Fosu-Nyarko, J. and Jones, M.G. (2012) de novo analysis and functional classification of the transcriptome of the root lesion nematode, *Pratylenchus thornei*, after 454 GS FLX sequencing. *International Journal for Parasitology* 42, 225–237. DOI: 10.1016/j.ijpara.2011.11.010

Noon, J.B. and Baum, T.J. (2016) Horizontal gene transfer of acetyltransferases, invertases and chorismate mutases from different bacteria to diverse recipients. *BMC Evolutionary Biology* 16, 74. DOI: 10.1186/s12862-016-0651-y

Noon, J.B., Hewezi, T.A.F., Maier, T.R., Simmons, C., Wei, J.-Z., Wu, G., Llaca, V., Deschamps, S., Davis, E. and Mitchum, M. (2015) Eighteen new candidate effectors of the phytonematode *Heterodera glycines* produced specifically in the secretory esophageal gland cells during parasitism. *Phytopathology* 17, 832–844. DOI: 10.1111/mpp.12330.

Noon, J.B., Qi, M., Sill, D.N., Muppirala, U., Eves-van den Akker, S., Maier, T.R., Dobbs, D., Mitchum, M.G., Hewezi, T. and Baum, T.J. (2016) A plasmodium-like virulence effector of the soybean cyst nematode suppresses plant innate immunity. *New Phytologist*, 212, 444–460. DOI: 10.1111/nph.14047

Opperman, C.H., Bird, D.M., Williamson, V.M., Rokhsar, D.S., Burke, M., Cohn, J., Cromer, J., Diener, S., Gajan, J. and Graham, S. (2008) Sequence and genetic map of *Meloidogyne hapla*: a compact nematode genome for plant parasitism. *Proceedings of the National Academy of Sciences of the United States of America* 105, 14802–14807. DOI: 10.1073/pnas.0805946105

Palmer, J.D. and Shields, C.R. (1984) Tripartite structure of the *Brassica campestris* mitochondrial genome. *Nature* 307, 437–440. DOI: 10.1038/307437a0

Palomares-Rius, J.E., Hirooka, Y., Tsai, I.J., Masuya, H., Hino, A., Kanzaki, N., Jones, J.T. and Kikuchi, T. (2014) Distribution and evolution of glycoside hydrolase family 45 cellulases in nematodes and fungi. *BMC Evolutionary Biology* 14, 69. DOI: 10.1186/1471-2148-14-69

Parra, G., Bradnam, K. and Korf, I. (2007) CEGMA: a pipeline to accurately annotate core genes in eukaryotic genomes. *Bioinformatics* 23, 1061–1067. DOI: 10.1093/bioinformatics/btm071

Petitot, A.S., Dereeper, A., Agbessi, M., Da Silva, C., Guy, J., Ardisson, M. and Fernandez, D. (2016) Dual RNA-seq reveals *Meloidogyne graminicola* transcriptome and candidate effectors during the interaction with rice plants. *Molecular Plant Pathology* 17, 860–874. DOI: 10.1111/mpp.12334

Phillips, W.S., Brown, A.M.V., Howe, D.K., Peetz, A.B., Blok, V.C., Denver, D.R. and. Zasada, I.A (2016) The mitochondrial genome of *Globodera ellingtonae* is composed of two circles with segregated gene content and differential copy numbers. *BMC Genomics* 17, 1–14. DOI: 10.1186/s12864-016-3047-x

Phillips, W.S., Howe, D.K., Brown, A.M. et al. (2017a) The draft genome of *Globodera ellingtonae*. *Journal of Nematology* 49(2), 127–128.

Phillips, W.S., Eves-Van Den Akker, S. and Zasada, I.A. (2017b) Draft transcriptome of *Globodera ellingtonae*. *Journal of Nematology* 49(2), 129.

Rosso, M.-N., Dubrana, M., Cimbolini, N., Jaubert, S. and Abad, P. (2005) Application of RNA interference to root-knot nematode genes encoding esophageal gland proteins. *Molecular Plant-Microbe Interactions* 18, 615–620. DOI: 10.1094/MPMI-18-0615

Rotter, A., Usadel, B., Baebler, Š., Stitt, M. and Gruden, K. (2007) Adaptation of the MapMan ontology to biotic stress responses: application in solanaceous species, *Plant Methods* 3, 1. DOI: 10.1186/1746-4811-3-10.

Sacco, M.A., Koropacka, K., Grenier, E., Jaubert, M.J., Blanchard, A., Goverse, A., Smant, G. and Moffett, P. (2009) The cyst nematode SPRYSEC protein RBP-1 elicits Gpa2-and RanGAP2-dependent plant cell death. *PLoS Pathogens* 5, e1000564.

Scholl, E.H., Thorne, J.L., McCarter, J.P. and Bird, D.M. (2003) Horizontally transferred genes in plant-parasitic nematodes: a high-throughput genomic approach. *Genome Biology* 4, R39. DOI: 10.1186/gb-2003-4-6-r39

Schurch, N.J., Schofield, P., Gierliński, M., Cole, C., Sherstnev, A., Singh, V., Wrobel, N., Gharbi, K., Simpson, G.G. and Owen-Hughes, T. (2016) How many biological replicates are needed in an RNA-seq experiment and which differential expression tool should you use? *RNA* 22, 839–851. DOI: 10.1261/rna.053959.115

Sloan, D.B. (2013) One ring to rule them all? Genome sequencing provides new insights into the 'master circle' model of plant mitochondrial DNA structure. *New Phytologist* 200, 978–985. DOI: 10.1111/nph.12395

Smant, G., Stokkermans, J., Yan, Y.T. et al. (1998) Endogenous cellulases in animals: Isolation of beta-1, 4-endoglucanase genes from two species of plant-parasitic cyst nematodes. *Proceedings of the National Academy of Sciences of the United States of America* 95, 4906–4911.

Suga, K., Welch, D.B.M., Tanaka, Y., Sakakura, Y. and Hagiwara, A. (2008) Two circular chromosomes of unequal copy number make up the mitochondrial genome of the rotifer *Brachionus plicatilis*. *Molecular Biology and Evolution* 25, 1129–1137. DOI: 10.1093/molbev/msn058

Thimm, O., Bläsing, O., Gibon, Y., Nagel, A., Meyer, S., Krüger, P., Selbig, J., Müller, L.A., Rhee, S.Y. and Stitt, M. (2004) Mapman: a user-driven tool to display genomics data sets onto diagrams of metabolic pathways and other biological processes. *The Plant Journal* 37, 914–939.

Thorpe, P., Mantelin, S., Cock, P.J. et al. (2014) Genomic characterisation of the effector complement of the potato cyst nematode *Globodera pallida*. *BMC Genomics* 15, 923. DOI: 10.1186/1471-2164-15-923

Urbanczyk-Wochniak, E., Usadel, B., Thimm, O. et al. (2006) Conversion of MapMan to allow the analysis of transcript data from Solanaceous species: effects of genetic and environmental alterations in energy metabolism in the leaf. *Plant Molecular Biology* 60, 773–792. DOI: 10.1007/s11103-005-5772-4

Urwin, P.E., Lilley, C.J. and Atkinson, H.J. (2002) Ingestion of double-stranded RNA by preparasitic juvenile cyst nematodes leads to RNA interference. *Molecular Plant-Microbe Interactions* 15, 747–752. DOI: 10.1094/MPMI.2002.15.8.747

Vieira, P., Eves-van den Akker, S., Verma, R., Wantoch, S., Eisenback, J.D. and Kamo, K. (2015) The *Pratylenchus penetrans* transcriptome as a source for the development of alternative control strategies: mining for putative genes involved in parasitism and evaluation of in planta RNAi. *PLoS ONE* 10, e0144674.

Vlcek, C., Marande W., Teijeiro S., Lukeš J. and Burger G. (2010) Systematically fragmented genes in a multipartite mitochondrial genome. *Nucleic Acids Research* 39, 979–988. DOI: 10.1093/nar/gkq883

Wang, F., Li, D., Wang, Z., Dong, A., Liu, L., Wang, B., Chen, Q. and Liu, X. (2014) Transcriptomic analysis of the rice white tip nematode, *Aphelenchoides besseyi* (Nematoda: Aphelenchoididae). *PLoS ONE* 9, e91591.

Wang, X., Allen, R., Ding, X., Goellner, M., Maier, T., de Boer, J.M., Baum, T.J., Hussey, R.S. and Davis, E.L. (2001) Signal peptide-selection of cDNA cloned directly from the esophageal gland cells of the soybean cyst nematode *Heterodera glycines*. *Molecular Plant-Microbe Interactions* 14, 536–544. DOI: 10.1094/MPMI.2001.14.4.536

Wei, D.-D., Shao, R., Yuan, M.-L., Dou, W., Barker, S.C. and Wang, J.-J. (2012) The multipartite mitochondrial genome of *Liposcelis bostrychophila*: insights into the evolution of mitochondrial genomes in bilateral animals. *PloS ONE* 7, e33973.

Whisson, S.C., Boevink, P.C., Moleleki, L. *et al.* (2007) A translocation signal for delivery of oomycete effector proteins into host plant cells. *Nature* 450, 115–118. DOI:10.1038/nature06203

Wildermuth, M.C., Dewdney, J., Wu, G. and Ausubel, F.M. (2001) Isochorismate synthase is required to synthesize salicylic acid for plant defence. *Nature* 414, 562–565. DOI: 10.1038/35107108

Zheng, J., Peng, D., Chen, L., Liu, H., Chen, F., Xu, M., Ju, S., Ruan, L. and Sun, M. (2016) The *Ditylenchus destructor* genome provides new insights into the evolution of plant parasitic nematodes. *Proceedings of the Royal Society B*, 283. DOI: 10.1098/rspb.2016.0942

3 Hatch, Survival and Sensory Perception

Edward P. Masler[1] and Roland N. Perry[2,3]
[1]USDA-ARS, Beltsville, Maryland, USA; [2]University of Hertfordshire, Hatfield, UK; [3]Ghent University, Ghent, Belgium

3.1 Introduction 44
3.2 Biology of Hatching 44
3.3 Survival 53
3.4 Sensory Perception 57
3.5 Conclusions and Future Prospects 63
3.6 References 63

3.1 Introduction

The need to understand the functional biology of nematodes is central both to fundamental science and to practical aspects. Research on the biology of economically important cyst nematodes frequently has the declared aim of identifying novel control targets based on disruption of the nematode life cycle. The knowledge from such research also provides a fascinating exposition of the complex aspects of host–parasite interactions and the sophisticated mechanisms involved. In this chapter we focus on three aspects, hatching, survival and sensory perception, and explore the links among them and the associated interfaces between the cyst nematodes and their hosts.

3.2 Biology of Hatching

The eggshell and cyst protect the unhatched second-stage juveniles (J2) from predation and environmental extremes but, once hatched, the now vulnerable J2 need to find a host to commence feeding. In some species, particularly *Globodera* spp., this vulnerability is offset by a sophisticated host–parasite interaction whereby the J2 does not hatch unless stimulated by chemicals emanating from host roots. These root emanations have variously been termed exudates, diffusates or leachates; for consistency throughout this chapter the term diffusates will be used. The host root diffusate effectively synchronizes hatch with the availability of nearby host plants, and is an important component of the survival attributes of some species of cyst nematodes.

The reliance on host root diffusates to cause hatch varies considerably between species of cyst nematode (Table 3.1). Species such as *G. rostochiensis* and *G. pallida* that have a host range restricted to members of the Solanaceae family are almost completely dependent on host diffusates for hatch. By contrast, *Heterodera schachtii*, for example, hatches well in water and its survival is correlated with a very wide host range (over 200 plant species, including many

Table 3.1. Grouping of some species of cyst nematodes into four broad categories, based on their hatching response to host root diffusates. (From Perry, 2002.)

Group 1	Very large numbers of juveniles hatching in response to host root diffusates; few hatching in water	e.g. *Globodera rostochiensis*, *G. pallida*, *Heterodera cruciferae*, *H. carotae*, *H. goettingiana*, *H. humuli*
Group 2	Very large numbers of juveniles hatching in response to host root diffusates; moderate hatch in water	e.g. *H. trifolii*, *H. galeopsidis*, *H. glycines*
Group 3	Very large numbers of juveniles hatching in response to host root diffusates; large hatch in water	e.g. *H. schachtii*, *H. avenae*
Group 4	Hatching of juveniles induced by diffusates only in later generations produced during the host growing season; very large hatch in water for all generations	e.g. *H. cajani*, *H. sorghi*

weeds). *Heterodera avenae* also has a large hatch in water but a relatively narrow host range; however, the hosts (mainly grasses and cereal crops, including barley, oats, wheat and rye) are very common. Some species of cyst nematodes, such as *H. schachtii*, that hatch freely in water exhibit an enhanced rate of hatching when exposed to host root diffusates.

3.2.1 Abiotic influences on hatching

Under suitable environmental conditions, including appropriate temperature, oxygen availability and soil water levels, and an absence of physiological barriers, such as diapause (see section 3.3.1), hatch of most species of cyst nematodes proceeds without requiring specific host cues. Maximum hatch usually occurs when soil water content is at field capacity, although this can vary with different soil types. Extremes, such as drought and water-logging, hinder hatching. Soil type is an important abiotic factor; in general, coarse-textured soils favour hatching and subsequent invasion of roots, providing suitable conditions for aeration and nematode migration.

Nematodes are poikilotherms and so temperature is an important influence in modifying their hatching activities. Perry (2002) pointed out that the majority of research on the effects of temperature on hatch examined the *in vitro* hatch during exposure to constant temperatures, which do not relate to conditions experienced by the nematode *in vivo*. The optimum temperature for *in vitro* hatch may differ markedly from that likely to be experienced *in vivo*. For example, the optimum temperature for *in vitro* hatch of *G. rostochiensis* is 20°C (Robinson *et al.*, 1987a), a temperature that cysts at root depth in the soil are unlikely to experience. In general, low optimum temperatures for hatching are characteristic of those species that invade during winter or early spring, such as *H. cruciferae*, whereas nematodes that are adapted to warmer climates, such as *H. zeae*, exhibit higher temperature optima. Variation in the *range* of temperatures at which hatching can occur is an important adaptive characteristic. For example, the hatching of *G. pallida* is adapted to lower temperatures than that of *G. rostochiensis* (Robinson *et al.*, 1987a).

Temperature extremes are also relevant to hatching biology. High temperatures can suppress hatch or are lethal to nematodes. For example, Stoyanov and Trifonova (1995) reported 30°C as the upper limit for hatch of *G. rostochiensis*, whilst no hatch was observed from *H. zeae* at 40°C (Hashmi and Krusberg, 1995). Temperature may be involved in the induction of diapause and the influence of diapause on hatching has been discussed by Evans and Perry (1976) and Jones *et al.* (1998), and is briefly presented in section 3.3.1.

3.2.2 Root diffusate activity

Diffusate is produced along the entire root length, but it has been shown by Rawsthorne and Brodie (1986) that cells near the root tip of potato plants produced a more active diffusate than cells located elsewhere. The root tip is metabolically very active, being associated with elongation and differentiation. It is the preferred invasion site for infective J2 of cyst nematodes and the more active diffusate may act as a possible local attractant for J2 (see section 3.4).

Shepherd and Clarke (1971) and Clarke and Perry (1977) reviewed early work on the effects of diffusates on hatching of cyst nematodes. The efficacy of root diffusates is usually assessed in laboratory-based hatching bioassays by determining the percentage hatch from cysts over a given period of time, with diffusate being replaced with fresh solution each week. Over a period of 3–5 weeks, depending on species, hatch of 60–90% of J2 from cultured populations (usually single generation cysts) can be expected; the percentage will vary and may be substantially less with field populations containing cysts of different ages. However, in several plant species the production of active chemicals in root diffusate, the hatching factors (HF), is confined to a short period of plant growth. *Heterodera goettingiana* hatched only in diffusates from 4- and 6-week-old pea plants (Perry *et al.*, 1980a) and from 6- and 8-week-old bean plants (Beane and Perry, 1983), whilst hatch of *H. carotae* was largely restricted to diffusates from 5- and 7-week-old carrots (Greco and Brandonisio, 1986). Root diffusate activity in relation to plant age also influenced hatching of *H. sacchari* and *H. oryzicola* (Ibrahim *et al.*, 1993). Maximum hatching activity of potato root diffusate (PRD) was reached 2 weeks after planting potatoes and was maintained for a further 2 weeks (Twomey, 1995). Hatching of *Globodera* spp. in potato fields occurs over an 8-week period (Trudgill *et al.*, 1996), although LaMondia and Brodie (1986) reported that most J2 that successfully invaded the host had hatched within the first 3 weeks; the utilization of energy reserves by late-hatching J2 (Robinson *et al.*, 1985) may adversely affect invasion.

Diffusion of root diffusate in the soil, and thus the zone of root diffusate influence, is influenced by soil moisture, temperature and microbial activity, among other factors. Diffusion and hatching activity are reduced in soils with high organic content and HF are rapidly inactivated at pH > 8. Hatching activity towards *G. rostochiensis* and *G. pallida* has been detected up to 80 cm from potato roots (Rawsthorne and Brodie, 1987; Malinowska, 1996). Although Fenwick (1956) considered that PRD was highly labile in the soil, this has since been refuted for *G. rostochiensis* (Tsutsumi, 1976; Perry *et al.*, 1981) and *H. goettingiana* (Perry *et al.*, 1980a); hatching activity towards *G. rostochiensis*, for example, was detected in soil for up to 100 days after removal of the plants (Tsutsumi, 1976).

3.2.3 Chemistry of root diffusates

The first HF isolated was the pentanor-triterpene, glycinoeclepin A, which induced hatch of *H. glycines* (Masamune *et al.*, 1982) and the complex structure was subsequently elucidated (Fukuzawa *et al.*, 1985a; Corey and Houpis, 1990; Watanabe and Mori, 1991). Later, two more triterpene HF, glycinoeclepins B and C, were purified (Fukuzawa *et al.*, 1985b). Masamune *et al.* (1987a) proposed that HF are derivatives of an essential plant biosynthetic pathway from a cycloartane such as acerinol or cimigenol. These glycinoeclepins were obtained from macerated roots rather than root diffusates. Although early work synthesized artificial compounds, with simpler partial structures of glycinoeclepin A, that had hatching activity towards *H. glycines* (Kraus *et al.*, 1994, 1996), it seems that, in general, natural and artificial HF would be prohibitively expensive for commercial use. The fractionation of PRD using mass spectrometry has revealed the presence of more than ten HF of two classes of terpenoid molecules with similar chemical profiles and a molecular weight of 530.5 Da (Devine and Jones, 2000a). HF are present only in trace amounts; for example, 1058 kg dry weight of bean roots yielded 1.25 mg of glycinoeclepin A as its bisphenacyl ester (Masamune *et al.*, 1982), and a hatching factor isolated by Devine and Jones (2000b) represented less than $2.9 \times 10^{-5}\%$ of the organic material recovered from PRD. However, HF have very high specific activities, stimulating hatch of *H. glycines* and *G. rostochiensis* at concentrations as low as 10^{-10} g ml^{-1} (Devine and Jones, 2000b). The maximum hatching activity of the HF isolated by Mulder *et al.* (1992) and Devine and Jones (2000b) occurred at approximately 10^{-8} M. Several empirical formulae have been given for *Globodera* HF, including $C_{18}H_{24}O_8$ (Marrian *et al.*, 1949), $C_{19}H_{28}O_8$ (Johnson, 1952), $C_{13}H_{12}O_3$ (Hartwell *et al.*,

1959), $C_{11}H_{16}O_4$ (Clarke, 1970) and $C_{27}H_{30}O_9$ (Mulder et al., 1992).

HF from potato roots include the potato steroid glycoalkaloids, α-solanine and α-chaconine (Byrne et al., 1998; Būda and Čėsulytė-Rakauskienė, 2015). *Globodera rostochiensis* hatches faster in response to these two glycoalkaloids and this may relate to the adaptation of *G. rostochiensis* to the bitter (i.e. high glycoalkaloid) potatoes of the Andes. Solanoeclepin A is a hatching stimulant for *G. rostochiensis* and *G. pallida* and its *de novo* chemical synthesis has been reported (Tanino et al., 2011). *De novo* synthesis of glycinoeclepin A has also been achieved (Masamune et al., 1987a, b; Miwa et al., 1987; Murai et al., 1988; Mori and Watanabe, 1989; Corey and Houpis, 1990; Watanabe and Mori, 1991) and the structure resembles that of gibberellins.

Two other classes of hatching chemicals have been identified in PRD: hatch inhibitors (HI) and hatching factor stimulants (HS). HI are produced earlier than the HF, resulting in an initial net inhibition of hatch while the roots of the new plants become established. Subsequently, the J2 of *G. rostochiensis* hatched in response to a rise in the HF:HI ratio (Byrne et al., 1998). HS synergize the effect of HF to increase the number and/or rate of hatching but are inactive by themselves. HS are more plentiful in diffusate produced later in the host growth (Byrne et al., 2001). Hearne (2016 and Hearne et al., 2017) studied the population dynamics of *G. pallida* and *G. rostochiensis* in various mixed- and single-species competition assays and found that *G. pallida* showed greater multiplication in mixed- than single-species populations, with *G. rostochiensis* showing the opposite. Hearne (2016) attributed the greater *G. pallida* competitiveness to its later hatch – a later peak in hatching activity and more prolonged hatch – relative to *G. rostochiensis*. Interestingly, hatch of *G. pallida* was greater when it was induced with PRD containing *G. rostochiensis*-specific compounds, indicating that *G. pallida* hatch is stimulated upon perception of *G. rostochiensis*-derived compounds.

3.2.4 The effect of microorganisms on hatching

As well as plant-derived HF, there is increasing evidence for a role for microbial HF in cyst nematode hatching in soil, suggesting a tritrophic (host–microbe–cyst nematode) interaction (Ryan and Jones, 2003; Ryan et al., 2003). Tsutsumi (1976) showed that soil microorganisms stimulate hatch of potato cyst nematodes, and Ryan and Jones (2003) demonstrated that spontaneous hatch in the absence of diffusate or host plants was significantly greater in non-sterile sand compared to that *in vitro*, confirming a role for microorganisms. Root diffusate from conventionally grown potato plants contains more HF activity and induces greater hatch compared to diffusate from aseptically grown plants, and Ryan and Jones (2004a,b) concluded that HF production is partially mediated by root-associated microorganisms, particularly arbuscular mycorrhizal fungi (AMF) and plant growth promoting rhizobacteria (PGPR), such as *Bacillus* spp. Lettice and Jones (2015) demonstrated that isolated rhizobacteria were capable of inducing hatch of *G. pallida* in the absence of the host plant, and Lettice and Jones (2016) compared the hatching activity of soil from the ridge of potato fields with bulk soil from the same field. Soil samples and soil washes were incubated with PRD produced under sterile conditions in order to measure the effect that soil microorganisms had on hatching activity. As in other studies, the conclusion was that soil microorganisms played a role in hatching and may be responsible, at least in part, for 'spontaneous' hatch, that is, the hatch that occurs without stimulation by host root diffusates.

3.2.5 Hatching mechanism

Most information on hatching mechanisms derives from research on potato–*Globodera* spp. interactions and detailed information, especially of the early work, is presented by Jones et al. (1998) and Perry (2002), and provides considerable information about the direct and indirect effects on the eggshell and the unhatched J2. The overt response to hatch stimulation is eclosion, or hatch from the egg, but detailed analysis of the hatching response has demonstrated that a sequence of inter-related events occurs between stimulation and hatch.

Perry (2002) divided the hatching process of cyst nematodes into three phases: changes in the eggshell, activation of the J2 and eclosion.

In contrast to species of *Meloidogyne*, where activation of the J2 appears to precede or even cause changes in eggshell structure, the alteration of eggshell permeability characteristics in *G. rostochiensis* is a necessary prerequisite for metabolic changes in the J2. The following details of the cascade of events leading to eclosion (Fig. 3.1) are based primarily from research on *G. rostochiensis*.

Inside the egg, the J2 is surrounded by perivitelline fluid, which contains trehalose at a concentration of 0.34 M in *G. rostochiensis* (Clarke *et al.*, 1978) and 0.5 M in *H. goettingiana* (Perry *et al.*, 1980a). Osmotic pressure generated by trehalose reduces water content of the unhatched J2 from the normal fully hydrated state of 72% (Ellenby and Perry, 1976) to 67.3% in *G. rostochiensis* and 66.5% in *H. goettingiana*.

This partial dehydration inhibits J2 movement because the turgor pressure is insufficient to antagonize the longitudinal musculature. In *H. schachtii*, the osmotic pressure of the perivitelline fluid is lower and tolerance of osmotic stress by J2 is greater (Perry *et al.*, 1980b), perhaps explaining why this species hatches well in water without the need for diffusate stimulation. To 'activate' the unhatched J2 of *G. rostochiensis*, the osmotic pressure needs to be removed. This is achieved by a change in permeability of the inner lipid layer of the eggshell, a tetralaminate layer of lipoprotein membranes. This lipid layer is the primary permeability barrier to liquids, although allowing passage of oxygen. The permeability change is caused by HF binding or displacing internal Ca^{2+}. Three classes of Ca^{2+}-binding sites have been distinguished in the

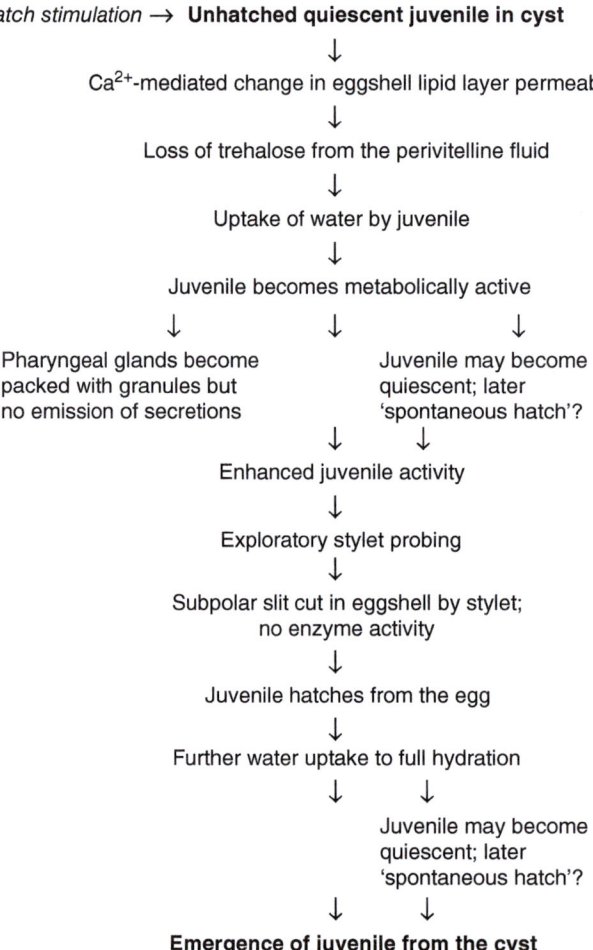

Fig. 3.1. Flow diagram showing events in the hatching process of second-stage juveniles of *Globodera rostochiensis* after stimulation with potato root diffusate. The main components are in chronological order but there will be overlap in the sequence of some events. Secretions in the pharyngeal glands do not appear to be involved in hatching but their accumulation is a preparation for the subsequent invasion and feeding phases of the life cycle. (Modified from Perry, 2002.)

G. rostochiensis eggshell (Clarke and Perry, 1985): (i) sites in the lipoprotein layer, from which Ca^{2+} can be removed by HF (these are proposed to be the sites associated with HF-stimulated hatch); (ii) sites on the lipoprotein layer that bind additional Ca^{2+} ions in the presence of HF; and (iii) sites in the outer layers that bind Ca^{2+} tightly but are not involved in the hatching process. Atkinson and Taylor (1983) reported a sialoglycoprotein with high Ca^{2+} affinity that was involved in the hatching process, which may correspond to Ca^{2+}-binding site class i.

The change in eggshell permeability is a necessary precursor to hatch of both G. rostochiensis and G. pallida, and in both species a 5-min exposure to PRD is sufficient to stimulate hatch (Perry and Beane, 1982). This suggests the involvement of a receptor–ligand interaction between the HF and the eggshell lipoprotein membrane. A change in chemistry of the lipoprotein membranes induced by HF is also possible. For example, potato steroidal glycoalkaloids, α-solanine and α-chaconine, induce hatch of G. rostochiensis (Devine et al., 1996); glycoalkaloids are known to destabilize lipid membranes during which leakage of trehalose is possible. The change in eggshell permeability results in trehalose leaving the egg, releasing the osmotic stress on the J2 and permitting an influx of water and subsequent rehydration of the J2 to a water content commensurate with restored metabolic activity and movement. Martin (1994) detected trehalose efflux from G. rostochiensis eggs within 8 h of exposure to PRD. The presence of trehalose in eggs containing viable J2 has been used as the basis for a viability assay for G. rostochiensis and G. pallida (van den Elsen et al., 2012) and further evaluation by Ebrahimi et al. (2015) has shown this viability assessment to be a robust, rapid technique capable of a minimum detection level of five viable eggs.

A Zn^{2+}-dependent enzyme mediates hatching of H. glycines and leucine aminopeptidase was found in the egg supernatant (Tefft and Bone, 1985a), although root diffusate does not increase its activity (Tefft and Bone, 1985b). However, the eggshell of G. rostochiensis remains rigid during the hatching process and there is no evidence of enzyme involvement (Perry et al., 1992).

Activation by diffusate and rehydration induce a series of physiological and behavioural changes in the unhatched J2, including changes in the cuticle and ultrastructure of the amphids (Jones et al., 1994). Within 24 h of exposure of unhatched J2 of G. rostochiensis to root diffusate, the adenylate energy charge falls (Atkinson and Ballantyne, 1977a), oxygen consumption commences (Atkinson and Ballantyne, 1977b) and the content of cyclic adenosine monophosphate (cAMP) (a possible secondary messenger in receptor–ligand interactions) rises (Atkinson et al., 1987). Changes in gene expression appear to occur during or immediately after the hatching process (Blair et al., 1999).

Vigorous movement of the J2 does not begin until at least 3 days after initial exposure to root diffusate, when the J2 starts local exploration of the inner surface of the egg, using its lips and stylet, and then begins thrusting movements with the stylet. This causes a regular pattern of perforations in the subpolar region of the rigid eggshell, which the J2 extends to a slit through which it hatches (Doncaster and Seymour, 1973). The rigidity of the eggshell imposes a limit on the water uptake and associated expansion of the J2, which only achieves full hydration after hatching (Ellenby, 1974).

These effects of host root diffusate on the rehydrated J2 are due in part to removal of osmotic pressure and hydration and in part to direct stimulation of the J2 by root diffusate; thus, diffusate has a bimodal action affecting eggshell structure as well as directly activating the J2.

3.2.6 Hatching of species with multiple generations during a season

Species such as G. rostochiensis and G. pallida in temperate regions may have only a single generation during the host growing season. However, other species of cyst nematodes, especially those in the tropics, are polycyclic, completing several generations during the host growing season. Species such as H. schachtii may have only one or two generations in cooler temperate regions but in warmer areas several annual generations can occur. Zheng and Ferris (1991) found that there were four types of unhatched J2 of H. schachtii: (i) J2 that hatched rapidly in water; (ii) J2 that hatched rapidly when stimulated by host root diffusates; (iii) J2 that hatched over a long period in water; and (iv) J2 that hatched over a long period in diffusates. By exposing cysts to various temperature regimes, Zheng and Ferris (1991) confirmed a seasonal diapause (see section 3.3) in *H. schachtii*. Later studies on the durability of hatching types have focused on three phases linked to development (see section 3.2.7).

Hatching patterns of polycyclic cyst nematodes are part of the panoply of cyst nematode survival adaptations. As well as retaining eggs inside the cyst, some species of *Heterodera* extrude a proportion of their eggs into an eggsac, which remains attached to the cyst. Working with *H. glycines*, Ishibashi *et al.* (1973) were the first to note that under favourable conditions, *H. glycines* produced most eggs in the eggsacs and J2 from these eggs hatched in water, providing a secondary inoculum for rapid re-infestation of the host plant, thus allowing a rapid population increase. However, under less favourable conditions, more eggs were retained within the cysts and a large proportion of these encysted eggs required host root diffusates (Ishibashi *et al.*, 1973) or artificial HF (Thompson and Tylka, 1997) to stimulate hatch of J2. Similar phenomena have been reported for *H. carotae* (Greco, 1981; Aubert, 1986) and *H. goettingiana* (Greco *et al.*, 1986). Later generations of some species of polycyclic cyst nematodes show an increased dependence on root diffusates for hatch, reflecting a change of priority during the host plant growing season from rapid re-infection and population increase to survival after host senescence (Perry and Gaur, 1996). Hatch from cysts of *H. sacchari* extracted from young plants was not dependent on root diffusate but a percentage of J2 in cysts from older plants required root diffusate for hatch (Ibrahim *et al.*, 1993). *Heterodera sorghi* produced three types of eggs: (i) eggs from which J2 hatched freely in soil leachate; (ii) that required root diffusate to stimulate hatch; and (iii) a large percentage of eggs from which J2 did not hatch immediately. The proportions of these three types of eggs changed during the host growing season, with a trend towards increased persistence in the later generations (Gaur *et al.*, 1995). For the first four generations of *H. cajani*, produced in glasshouse pot cultures, J2 in cysts hatched well in water with no enhancement of hatch by root diffusates, but in the fifth and sixth generations on senescing plants, 18–22% of the eggs required root diffusate to stimulate hatch (Gaur *et al.*, 1992). Also, in the final generation, the encysted J2 contained more lipid (energy) reserves than those in eggsacs. Thus, a large proportion of J2 from the later generations remained protected by the egg and cyst during the period between crops and had enhanced energy reserves. In this research, cysts were exposed to diffusate with maximum hatching activity; however, the decline in activity of diffusate as plants age is an additional factor ensuring that J2 do not hatch when host plants are not available. The factors causing the change in hatching response are unknown. Perry and Gaur (1996) suggested that changes in the feeding site with age and the onset of senescence of the host plant may affect the feeding female and trigger biochemical changes in the J2 to enhance survival; a comparison of feeding site chemistry at different stages of plant growth would be useful.

3.2.7 Durability of hatching and developmental arrest

Most of the experiments detailed above utilized cysts from populations grown in culture, usually on plants in a glasshouse. It is important to determine how consistent the hatch is from such cultures maintained over many years, and how this relates to field populations. Experiments with *H. glycines* examined these aspects. Hatching of *H. glycines* obtained from cultures maintained on an optimized *Glycine max* rearing system (Sardanelli and Kenworthy, 1997) can be profuse in water, consistently ≥60%, and was characterized by a three-phase rate profile. Eggs removed from the cyst, rinsed with water and incubated in bulk on Baermann funnels at 27°C always yielded some J2 within 24 h (phase 1). Phase 2, from 1 to ~12 days was characterized by a linear rate of 5% hatch day^{-1}, which dropped sharply and ended after 12 days (phase 3). Phase 2 accounted for >80% of total hatch (Masler *et al.*, 2008a, b). Importantly, even under these favourable conditions, significant numbers of eggs (30–40%) yielded no hatched J2 and exhibited some form of dormancy (Masler *et al.*, 2008a, b; section 3.3). Hatch dynamics were rigorously maintained, and cultures continuously produced these patterns through many (>100) generations. Thus, even in culture conditions that provide constant access to host and minimized environmental variation, *H. glycines* maintained heterogeneous populations of eggs (embryos).

In contrast to eggs from culture, hatching from *H. glycines* eggs collected from the field was very poor in water (<1%; Masler *et al.*, 2008a). However, when such eggs were used to inoculate *G. max* in the Sardanelli and Kenworthy (1997) rearing system, the level of infection was greater than would have been possible from a ~1% hatch rate. Thus, it was probable that hatch from these field eggs was stimulated by exposure to *G. max*. Hatch rates and infectivity levels of the field samples increased rapidly in *G. max* culture. Within nine generations, rates, hatching profiles, maximum percentage hatch and infectivity levels were identical to those of the long-term laboratory culture (Masler *et al.*, 2008a). Laboratory rearing conditions on *G. max* evidently selected for populations of *H. glycines* that hatch readily in water, but retained a significant fraction of the population that remained refractory to hatch in water alone. Thus, selection merely changed the proportions of *H. glycines* with different hatch characteristics, while some form of dormancy (Zheng and Ferris, 1991; Yen *et al.*, 1995; Jones *et al.*, 1998) persisted. By contrast, attempts to culture *G. ellingtonae* in the glasshouse were constrained by low hatch rates relative to the field, and various manipulations of temperature and moisture increased hatch only modestly and remained dependent upon PRD (Ingham *et al.*, 2015). Curiously, no differences in reproduction on field microplot potato plants exposed to glasshouse or field sourced eggs were observed. Each population apparently responded to field conditions similarly. Despite the differences in hatch requirements and plasticity of *H. glycines* and *G. ellingtonae*, their behaviours illustrate the durability of hatch biology, its underlying developmental mechanisms, and the complex interactions among nematode, host plant and environment.

Cyst nematode dormancy is essential to survival of the species (see section 3.3 for definitions of types of dormancy). It can be influenced by temperature, an important non-biotic environmental factor that affects metabolic rate and development (Van Gundy, 1965; Alston and Schmitt, 1988; Tzortzakakis and Trudgill, 2005; Phillips *et al.*, 2015). Incubation of *H. glycines* eggs at 5°C for 1 week *in vitro* caused a 40% decrease in hatch, relative to controls, after eggs were returned to normal rearing temperature (Masler *et al.*, 2008b). Suppression was increased to as much as 70% with extended exposure to 5°C, but was never complete (Masler *et al.*, 2008b). The phase 1 group was unaffected by low temperature, with a persistent small percentage hatch. Such early hatch was also reported in *H. schachtii* (Zheng and Ferris, 1991). Embryos in this first group require no further development to hatch and are fully hatch-competent. The phase 2 group comprises a continuum of developmental stages, each affected by low temperature and reduced metabolism. Although members of this group require further development to become biologically competent for hatching, they must have reached a critical developmental stage before exposure to low temperature, since their return to rearing temperature restored development and resulted in qualitatively normal hatch. The phase 3 group comprises embryos fundamentally affected by low temperature; they had not yet reached a critical developmental stage and were unable to recover from depressed metabolism. Normal development beyond the critical stage requires the action of internal molecular signals that cannot be issued without additional factors, which include those provided by environmental conditions, host plants or time. These dormant embryos account for essentially all of the 'missing' hatch, and are considered to be in diapause (Sommerville and Davey, 2002; Masler and Rogers, 2011; Masler *et al.*, 2013). The influence of dormancy, and the associated physiological states of diapause and quiescence, on survival are examined in detail in section 3.7.

The low temperature-induced dormancy in *H. glycines* is initiated in early embryos, which have not developed to at least the first-stage juvenile (J1) by the time low temperature is applied (Masler and Rogers, 2011). However, the effects of this induction are not apparent until the J1 stage (Fig. 3.2). The control plots (Fig. 3.2A) illustrate the phased hatch curve (J2), the decline in early embryos (EE) and periodicity of the J1 embryos. The plots from treated eggs (Fig. 3.2B) demonstrate decreased hatch with retention of the phased hatch curve and decrease in EE, but a strikingly different J1 profile. The increase in J1 at Day 5 was followed by only a mild decline. Day 0 values for EE, J1 and J2 are the same for control and treated groups, but only the EE values are the same at Day 14. All of

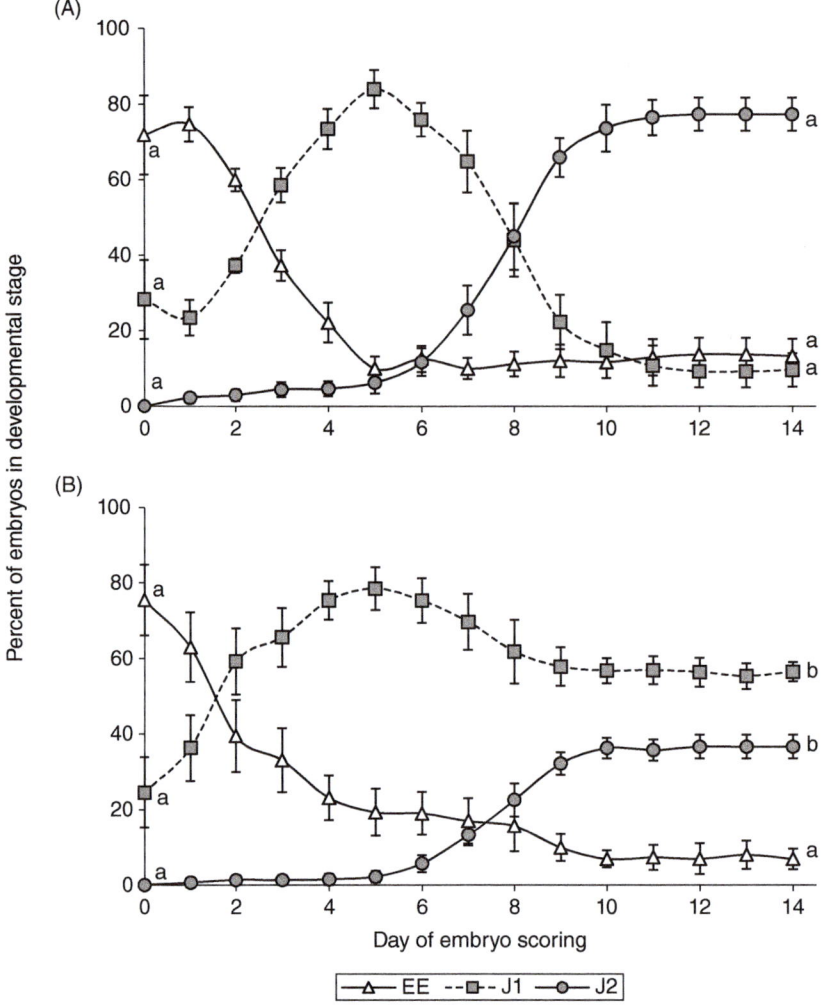

Fig. 3.2. *Heterodera glycines* embryonic development in response to low temperature. Eggs were collected from cysts and washed free of cyst content. Development of embryos in individual eggs was scored daily for 14 days using a simplified four-stage system (details in Masler and Rogers, 2011). Embryos from multicellular to the appearance of some tissue formation were scored as early (EE). Embryos with clearly vermiform structure were scored as first-stage juveniles (J1). Hatched juveniles were scored as second-stage juveniles (J2). The percentage of embryos at each stage ((number of individual embryos in one of three stages/total individuals of all stages) × 100) was determined daily. Control eggs (A) were incubated at 28°C, whereas treated eggs (B) were first incubated at 5°C, then transferred to 28°C. Each data point represents the mean ± SEM of daily percentage values across multiple (five to six) experiments. Means for analogous stages (EE, J1, J2) were compared between control and treated groups at Day 0 and Day 14. Means with different letters are significantly different ($P < 0.01$; Student's *t*-test). (Modified from Masler and Rogers, 2011.)

the reduced J2 hatch can be accounted for by the increase in J1 (Fig. 3.2B). J1 completely recover from low temperature with no effect on hatch (Masler and Rogers, 2011). Thus, the presence of J1 on Day 14 must be due to effects on EE. Since EE stages decline, all must have gone on with development, but all or most ceased development at the J1 stage. Low temperature, then, caused a delayed developmental arrest in *H. glycines* that was initiated early in development but not expressed until J1.

Retention of dormancy after transfer to laboratory conditions, and the induction of developmental arrest in laboratory populations, attest to the fundamental importance of this survival strategy. In addition, the stable and characteristic hatching patterns and the response of embryos to temperature stress indicate that the developmental programming of hatch in *H. glycines* regulates a durable but plastic mechanism.

In addition to the essential function of protection of unhatched J2 against adverse environmental conditions, the cyst contains materials that can influence hatch (Okada, 1972, 1974; Charlson and Tylka, 2003; Pridannikov et al., 2007). Thompson and Tylka (1997) reported that differences in hatching from eggs obtained from *H. glycines* cysts and external eggsacs could not be explained by differences in development, and Masler and Rogers (2011) observed different effects of low temperature exposure on hatch if the exposed eggs were free or encysted. The mechanisms and materials involved with these effects remain unclear, and biochemical exploration of cyst content is essential. In this context, proteases represent one functional molecular class worth attention (Masler and Chitwood, 2016, 2017; Nonaka et al., 2016).

3.3 Survival

It is clear that the hatching responses of cyst nematodes are closely linked not only to the capacity to survive in the absence of a host, but also to survive environmental extremes. Once hatched the J2 have limited survival; for example, J2 of *G. rostochiensis* and *G. pallida* can survive for less than 2 weeks without feeding (Robinson et al., 1987a). Although they are sibling species, *G. rostochiensis* and *G. pallida* show marked physiological differences, especially in rate of development and hatching, which are directly linked to survival. *Globodera pallida* has a faster rate of postembryonic development and eggshell structural development than *G. rostochiensis* (Perry, 2002). The two species have different rates of lipid utilization (Robinson et al., 1987b; see Chapter 5, this volume), with *G. pallida* utilizing its lipid reserves more slowly than *G. rostochiensis*, and hatching more slowly and over a longer period of time than *G. rostochiensis* (Robinson et al., 1985; Den Nijs and Lock, 1992). These physiological differences are central to the superior ability of *G. pallida* to survive and remain infective for periods much longer than *G. rostochiensis*.

Results from studies on competition between *G. rostochiensis* and *G. pallida* differ, perhaps due to different experimental conditions. Marshall (1986, 1989) found *G. rostochiensis* to be the more competitive species in outdoor and glasshouse pot experiments in New Zealand, whereas Den Nijs (1992) in The Netherlands and Ryan et al. (2005), Lettice (2014) and Hearne et al. (2017) in outdoor pot trials in Ireland found *G. pallida* to be the more competitive species.

The cyst and eggshell afford protection to the unhatched J2, which is in a state of osmotically induced dormancy. Dormancy is a suspension of development associated with lowered metabolism and can be conveniently subdivided into quiescence and diapause. The term dauer, which describes an alternative developmental stage for surviving unfavourable conditions (Grant and Viney, 2011), may not be relevant to J2 of cyst nematodes as they do not represent an 'alternative' stage of the life cycle.

Quiescence is a spontaneous reversible response to unpredictable unfavourable environmental conditions and release from quiescence occurs when favourable conditions return. Quiescence can be facultative or obligate. Obligate quiescence occurs when the environmental cue affects a specific receptive stage, such as the J1 or J2. By contrast, facultative quiescence is not stage-specific. Adverse environmental conditions and the types of quiescence they induce include cooling (cryobiosis), high temperatures (thermobiosis), lack of oxygen (anoxybiosis), osmotic stress (osmobiosis) and dehydration, or desiccation (anhydrobiosis). Cryptobiosis is a further term that has been used in connection with quiescence and is defined as a state where no metabolism can be detected. However, it is difficult to separate states on the basis of metabolic activity and the cause of the arrest in development is a more relevant criterion; on this basis cryptobiosis can be viewed as the same kind of phenomenon as quiescence. Evans (1987) further distinguished between dormancy affecting ontogenetic development and that affecting somatic development. In contrast to quiescence, diapause is a state of arrested development whereby development does not continue until specific requirements have been satisfied, even if favourable conditions return. It is either programmed into the life cycle (obligatory diapause) or is triggered by environmental stimuli (facultative diapause), such as day length. Separating different types of quiescence is somewhat artificial as many of the environmental stresses involve removal or immobilization of water. For example, desiccation concentrates body solutes and increases internal osmotic stress, exposure to hyperosmotic conditions causes partial dehydration of a nematode and freezing may involve dehydration through sublimation of water from the solid phase.

The ability to withstand dehydration for periods considerably in excess of the duration of the normal life cycle is a feature associated with a dispersal phase, such as the cysts of *Globodera* and *Heterodera* species. When exposed to desiccation, the permeability characteristics of the surface layers of the cyst wall and of the eggshell of *G. rostochiensis* change; the resulting control of water loss is a major factor in the survival of this species. The ability of cyst nematodes to survive severe desiccation varies considerably between species and long-term anhydrobiosis seems to be associated primarily with those species, such as *G. rostochiensis*, that have a very restricted host range. In addition to its role in desiccation survival of unhatched J2, the eggshell enables J2 of *G. rostochiensis* to supercool in the presence of ice as a freeze avoidance strategy for cryobiotic survival (Wharton *et al.*, 1993; see Chapter 5, this volume).

3.3.1 Diapause

Diapause has been documented for cyst nematodes as a strategy to overcome cyclic long-term conditions, such as seasonal conditions and/or the absence of the host, that are not conducive to hatch and infection. The incidence of diapause varies greatly between species and between populations of the same species.

Obligate diapause is initiated by endogenous factors and can be ended by the J2 receiving exogenous stimuli for a required period of time. Nematodes can undergo obligate diapause only once in their life. Temperature is the most important environmental cue for the termination of obligate diapause, with a fixed period of exposure to low temperatures relieving the arrested development. The Fr 1 ecotype of *H. avenae* has an obligate diapause (Rivoal, 1979). Diapause in *G. rostochiensis* and *G. pallida* can be circumvented by avoiding desiccation of the cysts and in pot cultures three to five generations can be produced (Janssen *et al.*, 1987).

Facultative diapause is initiated by exogenous, rather than endogenous, stimuli and terminated by endogenous factors after a critical period of time. This type of diapause is illustrated by the predictable periods of non-responsiveness to root diffusates of cyst nematode J2. Initiation of facultative diapause in *G. rostochiensis* is affected by both day length and low temperatures. Hominick *et al.* (1985) and Hominick (1986) examined the influence of photoperiod on diapause and hatch of *G. rostochiensis*. Cysts grown in canisters in the dark produced J2 with suppressed hatching and prolonged diapause, whereas a large proportion of J2 in cysts grown on plants exposed to constant light hatched immediately on exposure to diffusate. The conclusion was that the signal initiating diapause was triggered by photoperiod acting on the host potato plant to the female nematode and thence to the developing J2, and that the length of the diapause was negatively correlated with the amount of light to which the host plant was exposed (Hominick, 1986).

In early work with *H. avenae* in Australia, Banyer and Fisher (1971) divided hatching into two phases: the first had an optimum of about 10°C and was associated with juvenile development; the second, of relatively short duration, had an optimum of about 20°C and was associated with eclosion. Both phases occurred over the range 5 to 20°C and rate of hatching at constant temperature was the result of the different times taken to complete each phase. Banyer and Fisher (1971) noted that dormancy was induced by warmth following a period of cold and was modified by time at the low temperature and magnitude of the temperature rise. In China, Jing *et al.* (2014) noted a summer diapause in *H. avenae* (from May to October). *Heterodera avenae* is polymorphous (Cook and Rivoal, 1998) and induction of diapause varies. In Mediterranean climates, the diapause is obligate and durable, occurring when the climate is hot and dry and ending when the soil temperature falls and moisture rises (Rivoal and Cook, 1993). Rivoal (1979) separated *H. avenae* populations in France into two ecotypes, Fr 1 and Fr 4. Fr 1 is termed the Mediterranean ecotype and is found in the south of France and the Australian populations correspond to this ecotype. The Fr 4 ecotype is characteristic of populations from areas with a temperate Oceanic climate. Mokabli *et al.* (2001) found that diapause in two Algerian populations was induced by high temperature treatments (20 and 25°C) and broken subsequently by lower temperatures (3 and 7°C); this is typical of the Mediterranean Fr 1 ecotype. The induction temperature is much higher than those inducing diapause in Oceanic Fr 4 populations of *H. avenae* from France (Rivoal, 1978, 1979, 1982, 1983, 1986). It is clear from these studies that hatching of *H. avenae* and the presence and influence of diapause vary according to regional adaptation and ecotypes.

Information on gene expression that may differentiate diapause from quiescence is limited. Palomares-Rius *et al.* (2016) compared the gene expression changes of anhydrobiotic unhatched J2 of *G. pallida* and J2 in hydrated cysts exposed to tomato root diffusate (a hatching stimulant). Microarray gene expression analysis showed that there were differences between hydrated J2 in quiescence and diapause (Fig. 3.3), indicating difference in adaptation to short- (quiescence) and long- (diapause) term survival. With quiescent J2, most gene expression was associated with ion transport, whilst the J2 in diapause showed several transporters (e.g. amino acid and ion transport).

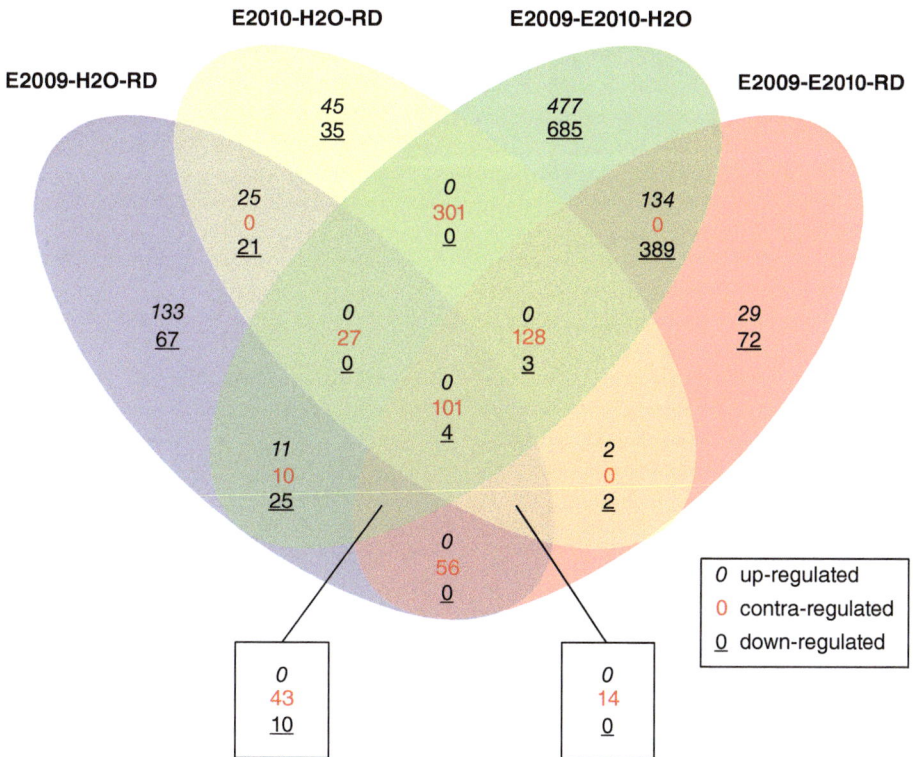

Fig. 3.3. Venn diagram for genes of *Globodera pallida* differentially expressed in comparisons of different treatments. E2009-H2O-RD: comparison between hydrated and hydrated-diffusate soaked quiescent unhatched J2; E2010-H2O-RD: comparison between hydrated and hydrated-diffusate soaked diapaused unhatched J2; E2009-E2010-H2O and E2009-E2010-RD: comparison between hydrated quiescent and diapaused unhatched J2 exposed to RD. RD = root diffusate. (From Palomares-Rius *et al.*, 2016.)

The different types of dormancy provide nematodes with a range of strategies with which to synchronize hatching with unpredictable and seasonal environmental changes. For example, the newly formed unhatched J2 of *G. rostochiensis* immediately enter obligate diapause, which is broken by the chilling stimulus of autumn and winter. In early spring, the unhatched J2 then enter obligate quiescence, which will be terminated by increasing soil temperature and PRD to stimulate hatching of the J2. If no host is present, quiescence continues followed by facultative diapause as the unstimulated eggs enter their second winter. This combination of diapause and quiescence enables *G. rostochiensis* to persist in the soil for more than 20 years in the absence of the host plant.

3.3.2 Osmotic regulation

Osmotic and ionic regulation in nematodes has been reviewed by Wright and Newall (1976, 1980), Wright (1998, 2004), Thompson and Geary (2002) and Wharton and Perry (2011), plus Wormbook and Google searches will provide considerable information on *Caenorhabditis elegans*, including osmotic avoidance assays and information on the genes involved in osmotic stress responses. The small size of plant-parasitic nematodes obviates extracting pseudocoelomic fluid for analysis, so studies on osmoregulation have been based on indirect methods, such as measuring changes in water content and length and/or volume changes in response to osmotic stress. However, many studies of nematode osmoregulation have used non-ionic or single salt solutions as incubation media, so interpretation of osmoregulatory abilities has to be done with caution.

Nematodes rely on a high internal turgor pressure to antagonize the longitudinal musculature for locomotion. If the nematode was hyposmotic to its surroundings water would be lost and the reduced turgor pressure would be incompatible with movement. This is the situation with unhatched J2 of *G. rostochiensis*, for example, where the concentration of trehalose (0.34 M) in the perivitelline fluid causes an osmotic stress and a reduction of water content with concomitant inactivity and dormancy, which is only reversed with the release of osmotic pressure during the hatching process (see section 3.2.5). The hatched J2 of *G. rostochiensis* and *H. schachtii* maintain their body fluids as hyperosmotic in relation to the external medium (Clarke *et al.*, 1978; Perry *et al.*, 1980b).

Once hatched, the J2 experience fluctuating osmotic pressures in the soil. Soil water has salts and other solutes dissolved in it and the J2 will need to regulate its internal ionic and osmotic composition to maintain its internal solutes at different concentrations to those in the surrounding medium. When entering the plant root, migrating through plant tissue and setting up a feeding site the J2 will experience different physiochemical conditions. Newall (quoted in Wright and Newall, 1976) found that J2 of *G. rostochiensis* can regulate their sodium content across a range of NaCl concentrations but they lost internal sodium ions rapidly in solutions containing less than 16 mM NaCl. Newall found that J2 of *G. rostochiensis* had an internal osmotic pressure equivalent to about 0.1 M NaCl. Clarke *et al.* (1978) showed that the water content of J2 decreased in 0.2 M sucrose but was little altered in 0.1 M sucrose; 0.1 and 0.2 M sucrose are osmotically equivalent to 0.05 and 0.1 M NaCl, respectively. Thus, the internal osmotic pressure of *G. rostochiensis* J2 is likely to be equivalent to about 0.05 M NaCl.

3.3.3 Cold tolerance

The two main cold tolerance strategies of organisms are freezing tolerance and freeze avoidance (Wharton, 2011). Freezing-tolerant animals survive the freezing of at least part of their body fluids. Freeze-avoiding animals reduce their risk of freezing by preventing fatal ice nucleation into internal tissues and survive in a supercooled state (where their body fluids are still liquid at temperatures below their melting point) but die if their body fluids freeze. Once hatched, J2 of *G. rostochiensis* cannot survive freezing (Perry and Wharton, 1985) but the unhatched J2 is protected by the eggshell and cyst wall, both of which prevent inoculative freezing and allow the J2 to supercool to temperatures as low as −38°C, before the J2 freezes and dies (Wharton *et al.*, 1993; Wharton and Ramløv, 1995; Devine, 2010). Wharton and Ramløv (1995) and Devine (2010) consider that ice-nucleating agents are present within the cyst walls of both *G. rostochiensis* and *G. pallida*. In addition, the synthesis of unsaturated fatty acids (Gibson *et al.*, 1995) and the high concentration of the antifreeze, trehalose, in the perivitelline fluid may enhance their supercooling capacity (Wharton and Ramløv, 1995).

3.4 Sensory Perception

The cyst nematode life cycle comprises three distinctly different environments: cyst/egg, rhizosphere and host plant. Each presents a different set of physical and chemical challenges to perception and response by the postembryonic nematode. Chemical and tactile cues are the primary signals used by soil-dwelling nematodes for orientation, communication and guidance (Jones, 2002; Rasmann *et al.*, 2012).

The infective J2 depends upon a sophisticated neurosensory system to interpret the complex environments of the rhizosphere and within the plant. Such perception and response are important to all cyst nematode life processes including egg laying, hatching, locomotion, host seeking and location, attraction and mating, toxin avoidance, social behaviour, cuticular changes related to infection, interactions with the host plant, and other behaviours necessary for survival and reproduction (Perry, 1996, 1997; Lilley *et al.*, 2005; Haegeman *et al.*, 2012).

In the context of cyst nematode hatch, locomotion and host location, chemical signals (semiochemicals) comprise the primary stimuli. Chemicals for communication between different organisms (allelochemicals), such as between host plant and hatched juvenile, and between members of the same species (pheromones), form the core of this signal system. Chemical perception is of particular importance to plant-parasitic nematode survival during the high-risk period between hatch and infection. Chemical signals in root diffusates have fundamental effects on the induction of cyst nematode hatch (see section 3.2), and are essential to the attraction of infective J2 to, and entry into, the host. However, the rhizosphere presents an exceedingly complex perception environment. Chemical signals from bacteria, fungi, plants and soil fauna, including other nematodes, and soil chemistry such as pH and CO_2, create this complexity, and cyst nematodes are required to negotiate such noisy signal conditions to survive. They are equipped, as are other nematodes, with a system of cuticular sense organs (sensilla) to deal with this task. Sensilla comprise neuronal and structural components that are responsible for the detection and transmission of sensory information, and this system perceives, filters and integrates cues to evoke proper responses.

3.4.1 Sensilla

Nematode sensilla are located both externally and internally, with external sensilla assigned primarily to chemical and physical (mechanical, temperature) stimuli, whereas internal sensilla are primarily involved with mechanoreception. External sensilla comprise the anterior amphids and posterior phasmids. This head-to-tail placement of sensilla allows the nematode to orient within a chemical field (Hilliard *et al.*, 2002; Curtis, 2008). All cuticular sensilla are constructed with the basic components of sheath cell, socket cell and neuronal dentrites (Jones, 2002; Bargmann, 2006; Perry and Curtis, 2013; Chapter 14, this volume). Dendrites are supported and externally exposed through a canal comprising the sheath and socket cells. The sheath cell forms the initial, more internal, portion of the amphidial canal and joins the socket cell towards the exterior. The dendrites within the amphidial canal are bathed in secretions from these cells, although structural analysis suggests that the socket cell may be the sole contributor. Nevertheless, the secretions bathe and protect the neuronal dendrites and form a protective external cap – the latter may facilitate signal transduction. Secretions vary by species and by developmental stage (Perry and Curtis, 2013). With multiple sensory neurons and variable secretory materials, amphids are sensitive and selective signal transducers (Bargmann, 2006; Perry and Curtis, 2013). In males, phasmids may be involved with mating.

While the basic structural and functional features of amphids are present across nematodes, there are anatomical, biochemical and functional differences between species and between developmental stages within species. Jones *et al.* (1994) observed that amphids present in unhatched J2 of *G. rostochiensis* lack amphidial secretions and are apparently inactive, whereas the amphids of hatched J2 contain obvious secretions and appear functional. This appearance is maintained through the juvenile stage. Eves-van den Akker *et al.* (2014) characterized the expression of genes encoding a hyper-variable effector (HYP) family in *G. pallida*. The effectors are associated with host invasion and infectivity, and are expressed by, and secreted from, the amphids. However, expression is detected only in parasitic stages of *G. pallida* and not in infective stage J2. In addition, the authors demonstrated effector expression variation between *G. pallida* and *G. rostochiensis*. Jones *et al.* (2000) reported the product of a gene in *G. rostochiensis* (*gr-ams-1*) as an exclusively amphidial protein (AMS). Although function was not determined, the authors suggest that, given its exclusive expression and lack of homologues outside the Nematoda, it might be a target worth exploring. In fact, Chen *et al.* (2005) demonstrated that silencing *gr-ams-1* in infective stage J2 almost completely suppressed the ability to locate and invade a host. Curiously, *gr-ams-1* is expressed in both J2 and feeding females (Jones *et al.*, 2000). A homologue of *gr-ams-1* has recently been reported in *H. schachtii*, but no *H. schachtii* transcripts with homology to HYP effectors were found (Fosu-Nyarko *et al.*, 2016).

Winter *et al.* (2002) were able to stain the amphidial neurons of *H. glycines* J2 using fluorescein isothiocyanate, and stain nuclei of the cell bodies with bisbenzimide. By contrast, Han *et al.* (2016) reported that *H. glycines* J2 amphids failed to dye-fill with DiI (1,1′-Dioctadecyl-3,3,3′,3′-tetramethylindocarbocyanine perchlorate). Differences in methodologies, including chemically different dyes, may explain the contradictory results. It is also possible that there were physiological differences between the *H. glycines* J2 used in the two studies. It is worth noting that amphidial neuronal and receptor composition is quite variable in *C. elegans*. Each *C. elegans* amphidial neuron expresses a set of 12 candidate receptor genes and detects a set of attractants, repellents and pheromones (Bargmann, 2006). Collectively, these observations illustrate the potential for extensive variation among what are fundamental and conserved nematode sensory structures.

3.4.2 Plant signals

The infective J2 responds to chemicals released from the root at both long-range and short-range distances (Perry, 2005; Farnier *et al.*, 2012). Around actively growing roots there exist several gradients of volatile and non-volatile compounds, including amino acids, ions, pH, temperature and CO_2. It is evident that nematodes orientate towards the roots using at least some of these gradients (Curtis *et al.*, 2009; Reynolds *et al.*, 2011). Perry (2005) classified attractants into three types: 'long distance attractants' that enable nematodes to move to the root area, 'short distance attractants' that enable the nematode to orientate to individual roots and 'local attractants' that are used by endoparasitic nematodes, such as cyst nematodes, to locate the preferred invasion site behind the root tip. Since cyst nematodes hatch close to their potential hosts, long distance attractants, such as respiratory gases (e.g. O_2, CO_2), may have little specific effect but they are important as non-specific signals to help orient the nematode within the rhizosphere.

Radiant heat from the sun results in temperature gradients within the soil that may aid the nematode in vertical positioning (Perry, 2005) that has biological significance. Robinson (1994) used temperature gradients, created to mimic actual field conditions, in observing nematode response behaviours. Two root plant-parasitic nematodes, *Rotylenchus reniformis* and *Meloidogyne incognita*, consistently responded by either moving towards (*M. incognita*) or away from (*R. reniformis*) the heat source in search of their preferred temperatures. Different behaviours in response to the same stimulus are suggested to be related to survival, such as *M. incognita* dispersal and the relative vertical distributions of the two species in the crop fields that the experiments were designed to mimic (Robinson, 1994; Perry and Curtis, 2013). The extent to which such general signals might affect cyst nematode orientation may vary by species and circumstance, but it is the more specific and targeted chemical signal that profoundly affects reproduction and survival.

The roles of short distance attractants, primarily root chemicals, in nematode orientation, chemotaxis and host location are broadly documented (Rasmann *et al.*, 2012; Gang and

Hallem, 2016), but few specific attractants from plants have been chemically identified (Būda and Čėsulytė-Rakauskienė, 2011, 2015). However, sophisticated *in vitro* bioassays reveal complex and specific behavioural responses to root signals. Grundler *et al.* (1991) demonstrated temporal effects of mustard (*Sinapis alba*) root signals on the locomotion of *H. schachtii*. Root diffusate applied to an agarose disc on an agar plate rapidly (~1 h) caused infective J2 to aggregate at the source disc. As early as ~2 h the aggregation began to decrease and was absent by 11 h. The aggregation appeared to involve the continual arrival and leaving of J2 rather than a subset of J2 remaining at the site. In addition, not all J2 in the assay dish were attracted to the root stimulus. Some could be seen passing through but not stopping at the stimulus disc site. It appeared that J2 were not specifically attracted from a distance to the stimulus site *per se*, but upon encountering the site by chance some were induced to remain (local attraction?). Significantly, others were not attracted. The diffusate also stimulated J2 stylus thrusting similar to that exhibited during root exploration and penetration.

Farnier *et al.* (2012) demonstrated the attraction of *G. pallida* J2 to fresh PRD in a sand-based assay comprising an arena with a narrow channel connecting two rectangular sides. J2 placed in the connection channel could freely migrate to either the water (control) or diffusate sides. After 24 h, ~33% of J2 were in the application channel. The proportion of the remaining J2 was significantly greater in the diffusate side than in the control side. The same fresh diffusate also stimulated J2 hatch. Freeze drying the diffusate removed any attraction to J2, suggesting that volatiles were involved. *Heterodera glycines* J2 were differentially attracted to root extracts from a variety of plant species in agar plate assays (Masler *et al.*, 2017), with evidence for both volatile and non-volatile attractants in *G. max* and *Tagetes patula*. Dalzell *et al.* (2011) demonstrated attraction of *G. pallida* to PRD, but also demonstrated a strong aversion of *G. pallida* to non-host root preparations.

In one of the more complete *in vitro* analyses of cyst nematode and root diffusate interactions, Devine and Jones (2003) demonstrated clear differences between *Globodera* spp. Chromatographic fractions of PRD were tested for effects on hatch and attraction of infective juveniles. Hatching was stimulated in *G. rostochiensis* and *G. pallida* by some, but not all, of the fractions. The array of stimulatory fractions was different for the two species. Both *G. rostochiensis* and *G. pallida* were also attracted to a subset of fractions, and again the fraction subsets were different for each species. Interestingly, the attraction of J2 of either species to their respective PRD fraction sets differed depending upon whether the J2 were hatched in water or in diffusate.

Chemoreception is essential for suitable food selection, and analysis of responses of cyst nematodes to specific plant compounds will provide information about feeding deterrents and stimulants that may relate to plant host suitability (Perry, 1996). Using the electrophysiology technique to analyse the neuronal spike activity, Rolfe *et al.* (2000) examined responses of J2 of *G. rostochiensis* to some of the many known phagostimulatory compounds of insects; for example, in insects, the D-isomers of many amino acids usually elicit a phagostimulatory response, whilst many L-amino acids are feeding deterrents (Mullin *et al.*, 1994). Exposure of J2 of *G. rostochiensis* to D-glutamic acid resulted in a significant increase in spike activity but there was no response on exposure to L-glutamic acid. This is the reverse of the responses of males of *G. rostochiensis* and *G. pallida* to these chemicals (Riga *et al.*, 1997a), which may indicate concentration-specific responses that differ between stages. However, as males are considered to be non-feeding (see Chapter 14, this volume), the results with J2 probably more accurately reflect the responses to phagostimulants. Stimulation of J2 and males of *G. rostochiensis* with 10 mM D-tryptophan resulted in a large increase in spike activity but males of *G. pallida* failed to respond (Riga *et al.*, 1997a; Rolfe *et al.*, 2000). Males of *G. rostochiensis* also responded to 1 mM D-tryptophan (Jones *et al.*, 1991).

Glycine is an alpha amino acid, usually eliciting phagostimulatory responses in insects (Ma, 1972), but the amino acid concentration of plant parts differs. The lack of increase in spike activity of J2 of *G. rostochiensis* on exposure to glycine (Rolfe *et al.*, 2000) may correlate with low levels of glycine in root diffusates or within the host plant. Thus, orientation of hatched J2 may not be mediated by glycine.

Males of *G. rostochiensis* and *G. pallida* also did not respond to 10 mM glycine (Riga *et al.*, 1997a).

There was a significant increase in spike activity of J2 of *G. rostochiensis* when exposed to PRD (Rolfe *et al.*, 2000) and it is likely that this response is associated with the additional role of diffusate in enhancing the rate of movement of infective J2 through soil and attracting them to host roots (Perry, 1997). There was no response from J2 on exposure to root diffusate from sugar beet, which is not a host for *G. rostochiensis*. This indicates that responses to diffusates may be host-specific. The active fractions in PRD responsible for attraction of *G. rostochiensis* and *G. pallida* J2 are different from those known to stimulate hatching (see section 3.2.3). Other chemicals are hatch-stimulatory but are not involved in attraction. For example, Voinilo (1976) considered that citric acid stimulated hatching of J2 of *G. rostochiensis* but there is no evidence that citric acid is an attractant and J2 and males gave no electrophysiological response on exposure to 10 mM citric acid (Riga *et al.*, 1997a; Rolfe *et al.*, 2000). In contrast to J2, males of *G. rostochiensis* and *G. pallida* did not respond to PRD (Riga *et al.*, 1996b). Males emerge from the roots and probably remain close to the root surface, requiring only sex pheromones to locate the females.

3.4.3 Sex pheromones

In contrast to the situation in insects, where a panoply of pheromones have been investigated, in cyst nematodes only sex pheromones (Greet *et al.*, 1968; Greet and Perry, 1992; Perry and Aumann, 1998) have been studied in any detail. Following the first demonstration of pheromone-mediated sex attraction in *Panagrolaimus rigidus* (Greet, 1964), pioneering work by Green and co-workers, summarized in Green (1980), examined sex attraction between some species of cyst nematodes. Green and Plumb (1970) investigated the inter- and intraspecific attractiveness to males of the secretions from females of species of cyst nematodes. They suggested a division into three groups: (i) *H. schachtii, H. glycines* and *H. trifolii*; (ii) *H. carotae, H. cruciferae* and *H. goettingiana*; and (iii) *H. mexicana, H. tabacum* and possibly *H. avenae*. Males of species within these groups are always attracted to females within the same group, thus sharing common sex pheromone components. Green and Plumb (1970) considered that the interactions between groups can be explained only by postulating at least six distinct attractive substances. *Heterodera* (= *Globodera*) *rostochiensis* was included in the third group, but it is not clear whether this was *G. rostochiensis sensu stricto* or *G. pallida*.

Subsequently, electrophysiological recordings were obtained of the intra- and interspecific responses of *G. rostochiensis* and *G. pallida* males to secretory-excretory product (containing sex pheromones) from virgin females (Riga *et al.*, 1996a). The spike frequency produced by *G. rostochiensis* males increased significantly after the application of their homospecific pheromone, but there was no response to sex pheromones from female *G. pallida*. However, there was a significant increase in spike frequency from males of *G. pallida* in response to sex pheromones from females of both *G. pallida* and *G. rostochiensis*. Thus, only *G. rostochiensis* males exhibit specific mate recognition and sex pheromones of *G. rostochiensis* females can elicit both inter- and intraspecific responses. Although *G. pallida* males responded to the sex pheromones from females of both species, the homospecific response was much greater and may be dominant in the soil environment (Riga *et al.*, 1996a). These data, and those of Green and Plumb (1970), showing interspecific attraction are contrary to the strict definition of a pheromone. The results with *G. rostochiensis* and *G. pallida* may reflect their sibling status, and that the evolutionary separation into two 'biological species' is not perfect. It is known that crosses between these two species produce viable F1 offspring and may occur under natural conditions (Thiéry *et al.*, 1996).

J2 of *G. rostochiensis* showed no change in spike activity in response to sex pheromones from virgin females (Rolfe *et al.*, 2000); J2 hatch and need to find host roots, whereas the sole function of adult males is to fertilize the females. The difference in responses of J2 and males of *G. rostochiensis* to certain semiochemicals could be mediated by changes in expression of the corresponding receptors in the amphids.

Using reverse-phase high-performance liquid chromatography (HPLC), the secretory-excretory solution from *G. rostochiensis* females was separated into four fractions, only two of which had sex pheromone activity (Riga *et al.*, 1997b). The sex pheromone of this species is composed of several polar compounds, which are less polar than PRD and are weakly basic (Riga et al., 1997b). One male attractant has been identified for a plant-parasitic nematode: vanillic acid for *H. glycines* (Jaffe et al., 1989). Vanillic acid did not attract *G. rostochiensis* males and concentrations of 10 µmol l^{-1}, 1 µmol l^{-1} and 100 nmol l^{-1} were not attractive for *H. schachtii* males (Aumann and Hasheem, 1993). These data indicate species specificity for vanillic acid.

Green and Greet (1972) considered that the sex pheromones of *G. rostochiensis* and *H. schachtii* were secreted through the cuticle but it seems more likely that they are secreted through the vulva (Aumann and Hasheem, 1993).

3.4.4 Ascaroside and peptide signals

Golden and Riddle (1982, 1985) demonstrated that *C. elegans* dauer formation (developmental arrest) is induced by a nematode-produced pheromone constitutively produced throughout the life cycle. Pheromone production increased in response to environmental stress, leading to entry into the protective dauer state, and the term 'daumone' was used to identify this pheromone. Their conclusions, based upon experiments with crude preparations, were prescient. They reported that that the pheromone is a fatty acid-like compound, that such pheromones are not confined to *C. elegans*, that they are nematode species-specific, and that other 'similar compounds' may be involved in parasitic nematode development.

The structures of active dauer-inducing pheromones were determined (Jeong et al., 2005; Butcher et al., 2007, 2008; Pungaliya et al., 2009) to be fatty acid derivatives of a dideoxy sugar. The dauer-inducing pheromone is in fact a mixture of glycoside derivatives of the dideoxy sugar ascarylose, and the structure of the derivative reported by Jeong et al. (2005) was called daumone-1. An enormously varied population of such molecules has now been identified in nematodes (Ludewig and Schroeder, 2013; Schroeder, 2015; von Reuss and Schroeder, 2015). They are structurally similar to compounds described from the animal parasitic nematode *Ascaris lumbricoides* and termed ascarosides (Jezyk and Fairbairn, 1967). Consequently, ascaroside has been adopted as a naming convention for these nematode signalling molecules. Daumone-1, for example, is also known as ascaroside #1 (ascr#1).

Nematodes produce a suite of ascaroside and ascaroside-related molecules that function as signals in a broad range of developmental and behavioural events (Srinivasan et al., 2008; Braendle, 2012; Kaplan et al., 2012; Ludewig and Schroeder, 2013; Aprison and Ruvinsky, 2015; Schroeder, 2015). Ascaroside biosynthetic pathways provide high levels of chemical flexibility and specificity, producing chemical structural features that may be unique to nematodes (Srinivasan et al., 2012; Artyukhin et al., 2013; Schroeder, 2015). Ascarosides often function as mixtures (e.g. daumones), providing additional specificity (von Reuss and Schroeder, 2015), and comprise an expanding area of interest in nematode biology.

Ascaroside pheromones are widely distributed and conserved within the Phylum Nematoda (Braendle, 2012; Choe et al., 2012; Schroeder, 2015) and show some effects across species (Choe et al., 2012; Hollister et al., 2013; Schroeder, 2015; Srinivasan et al., 2008). They have been identified in the exo-metabolomes of infective juveniles from five plant-parasitic nematodes (Manosalva et al., 2015). *Meloidogyne hapla*, *M. incognita* and *M. javanica* each secreted seven ascarosides (ascr#s 10, 16, 18, 20, 22, 24, 26), with ascr#18 in the most abundance. *Heterodera glycines* and *Pratylenchus brachyurus* each secreted ascr#18, but no other ascaroside was detected. It is notable that none of the ascarosides found in the *C. elegans* daumone mixture (ascr#s 1, 2, 3, 5; Ludewig and Schroeder, 2013) was detected in the plant parasite samples.

Manosalva et al. (2015) tested ascr#18 in plant exposure assays, eliciting immunity to a number of plant pathogens. In addition, ascr#18 treatment resulted in reduced infectivity of *Arabidopsis thaliana* seedlings by *H. schachtii* and *M. incognita*. These data, along with analysis of the signalling pathways induced in plants by exposure to ascr#18, suggest that ascarosides are recognized by plants as a pathogen-associated molecular pattern (PAMP) (Manosalva et al., 2015). Guo et al. (2016) synthesized a number of daumone pheromones and found that hatch of *H. glycines* J2 was stimulated by daumone-3 (ascr#3) but not by daumone-1 (ascr#1). In the same study, ascr#3 was more effective than ascr#1 at inducing *C. elegans* dauer formation, as expected (Butcher *et al.*, 2008).

Two major bioactive peptide families in nematodes are the FMRFamide-like peptides (FLPs; Li, 2005; Li and Kim, 2008, 2014; Peymen *et al.*, 2014) and the insulin-like peptides (ILPs, insulins; Pierce *et al.*, 2001; Li *et al.*, 2003; Ritter *et al.*, 2013). FLPs are abundant and varied regulators of neuromuscular, behavioural and other fundamental processes (Johnston *et al.*, 2010; Holden-Dye and Walker, 2011; Chang *et al.*, 2015; Yu *et al.*, 2016), and there are at least 14 *flp* genes in the soybean cyst nematode *H. glycines* (Li and Kim, 2014; Peymen *et al.*, 2014). Different *flp* genes may encode multiple copies of a single peptide sequence, single copies of multiple sequences or combinations of these patterns (McVeigh *et al.*, 2005; Masler *et al.*, 2013), generating significant biochemical variety.

Some *C. elegans flp* genes are expressed in chemosensory neurons (Li and Kim, 2008; Peymen *et al.*, 2014; Chang *et al.*, 2015) and their products interact with chemoreceptors (Li and Kim, 2014). Atkinson *et al.* (2013) demonstrated that *G. pallida flp-32* is widely expressed in the nervous system, including brain and ventral nerve, and that silencing *G. pallida flp-32* increased infective J2 locomotion. In addition, infectivity on potato was enhanced. The authors speculate that *G. pallida* FLP-32 (AMRNALVRFa) may have a suppressive effect on locomotion, which is lost by silencing. Yang *et al.* (2017) reported enrichment of KEGG pathways associated with environmental perception and enhanced expression of *flp* genes in *H. avenae* infective J2. FLPs clearly have important roles in environmental sensory perception, its integration with the neuromuscular system and the regulation of behaviour.

Insulins figure prominently in studies of the *C. elegans* survival strategy of developmental arrest (dauer state). Ascaroside pheromones are detected by nematode chemoreceptors which are coupled to a complex signalling pathway that includes ILP ligands of DAF-2, a key regulator of dauer formation (Li and Kim, 2008; Crook, 2014). However, DAF-2 is the only *C. elegans* insulin receptor known, while there are ~ 40 predicted *C. elegans* ILP ligands (Li and Kim, 2008; Cornils *et al.*, 2011; Matsunaga *et al.*, 2016). Thus, only a subset of ILPs has been associated with dauer regulation (Li and Kim, 2008; Cornils *et al.*, 2011), although others may prove to be involved (Matsunaga *et al.*, 2016). Nevertheless, ILPs not known to be involved with dauer regulation may function in other pathways, including thermotaxis (Kodama *et al.*, 2006), chemotaxis (Tomioka *et al.*, 2006), olfaction (Chalasani *et al.*, 2010) and mechanosensation (Campbell *et al.*, 2015), and through different receptors. In cyst nematodes, developmental arrest or dormancy can occur in the late embryo stages (Zheng and Ferris, 1991; Niblack *et al.*, 2006; Masler and Rogers, 2011) as a protective response to the environment and as a means of synchronizing hatch with availability of host plants. Although it is tempting to equate the functional similarities of *C. elegans* dauer and cyst nematode dormancy survival strategies, their molecular components are not fully shared (Elling *et al.*, 2007; Palomares-Rius *et al.*, 2013; Cotton *et al.*, 2014). However, *ins* genes have been identified in the *H. avenae* transcriptome (Kumar *et al.*, 2014). It is also relevant to note that the presence of octopamine ascaroside (Artyukhin *et al.*, 2013) suggests additional mechanisms involving neurotransmitters, neuropeptides and ascarosides in nematode chemosensation and behaviour. Thus, both ILPs and FLPs should be explored in cyst nematodes given their potential for transmitting environmental cues to pathways regulating metabolism, development and behaviour.

3.4.5 Blocking sensory perception

Perturbation of chemoreception before the J2 invade the host may be the basis for a novel nematode management strategy. The possibility of blocking sensory perception in order to disrupt host location would result in nematodes moving randomly and using up energy reserves, thus compromising infectivity. Attention has focused on blocking the amphids. Molecules of chemical substances initially come into contact with the amphidial secretions. The secretions may protect the dendritic ending of the nerve cells from desiccation or microbial attack, or they may maintain electrical contact between the tips and the bases of the dendritic processes. It is likely that the proteins in the secretions have an important role in chemoreception, perhaps similar to that of odorant-binding proteins and odorant-degrading enzymes that are present in insect sense organs (Perry and Aumann, 1998; Jones, 2002).

Stewart et al. (1993a, b) showed that a polyclonal antibody specific to the amphids of species of *Meloidogyne* recognized a glycoprotein (GP32) that was involved in chemoreception of the active stages of the life cycle. Binding of the antibody to the amphids of J2 prevented the nematodes from locating host roots. Subsequent studies on cyst nematodes showed the same phenomenon. Experiments with a monoclonal antibody showing immunoreactivity to amphidial secretions of *Globodera* (Fioretti, 2000) demonstrated that the antibody blocked electrophysiological responses of J2 of *G. rostochiensis* to PRD (Rolfe, 1999). Electrophysiological tests with *G. rostochiensis* demonstrated that exposing J2 for 1 h to 1% DiTera® (Valent BioSciences Corp.), a bio-nematicide obtained from fermentation with a hyphomycete fungus, prevented subsequent response to PRD, probably by blocking the amphidial pores or binding to the amphidial secretions (Twomey et al., 2002). Similarly, Fioretti et al. (2002) found that antibodies recognizing the amphids of *G. pallida* adversely affected nematode movement and delayed nematode penetration of roots. However, in all these experiments the nematodes were not permanently desensitized, presumably because turnover of amphidial secretions 'unblocks' the amphids and restores sensory perception.

Rolfe and Perry (2001) demonstrated that electropharyngeogram measurements from the anterior end of live, individual J2 of *G. rostochiensis* were correlated with stylet protractor muscle activity. Electropharyngeogram measurements of stylet activity showed that exposure of J2 of *G. rostochiensis* to 1% DiTera® significantly reduced stylet thrusting when nematodes were subsequently transferred to a stimulus, such as 10 mM serotonin (Twomey et al., 2002). Blocking of the amphids by DiTera® may be the explanation for the decreased stylet activity of *G. rostochiensis*, assuming that stylet thrusting is, in part at least, a response to chemosensory stimuli. Hu et al. (2014) reported the development of a microfluidic chip, which they used for electrophysiological recording of stylet activity of *G. pallida*.

3.5 Conclusions and Future Prospects

This is an exciting period for studies on nematode biology and host–parasite interactions. Examining the linkages among sensory perception, signalling and messenger molecules, metabolic responses and development will have profound effects on our understanding of cyst nematodes. Exploiting the tools of genome sequences, transcriptome sequences and RNAi applications in the context of fundamental biology provides immense potential for functional genomic studies.

3.6 References

Alston, D.G. and Schmitt, D.P. (1988) Development of *Heterodera glycines* life stages as influenced by temperature. *Journal of Nematology* 20, 366–372.

Aprison, E.Z. and Ruvinsky, I. (2015) Sex pheromones of *C. elegans* males prime the female reproductive system and ameliorate the effects of heat stress. *PLoS Genetics* 11, e1005729.

Artyukhin, A.B., Schroeder, F.C. and Avery, L. (2013) Density dependence in *Caenorhabditis* larval starvation. *Scientific Reports* 3, 2777. DOI: 10.1038/srep02777

Atkinson, H.J. and Ballantyne, A.J. (1977a) Changes in the adenine nucleotide content of cysts of *Globodera rostochiensis* associated with the hatching of juveniles. *Annals of Applied Biology* 87, 167–174. DOI: 10.1111/j.1744-7348.1977.tb01872.x

Atkinson, H.J. and Ballantyne, A.J. (1977b) Changes in the oxygen consumption of cysts of *Globodera rostochiensis* associated with the hatching of juveniles. *Annals of Applied Biology* 87, 159–166. DOI: 10.1111/j.1744-7348.1977.tb01871.x

Atkinson, H.J. and Taylor, J.D. (1983) A calcium binding sialoglycoprotein associated with an apparent egg shell membrane of *Globodera rostochiensis*. *Annals of Applied Biology* 102, 345–354. DOI: 10.1111/j.1744-7348.1983.tb02704.x

Atkinson, H.J., Taylor, J.D. and Fowler, M. (1987) Changes in the second stage juveniles of *Globodera rostochiensis* prior to hatching in response to potato root diffusate. *Annals of Applied Biology* 110, 105–114. DOI: 10.1111/j.1744-7348.1987.tb03237.x

Atkinson, L.E., Stevenson, M., McCoy, C.J., Marks, N.J., Fleming, C., Zamanian, M., Day, T.A., Kimber, M.J., Maule, A.G. and Mousley, A. (2013) *flp-32* ligand/receptor silencing phenocopy faster plant pathogenic nematodes. *PLoS Pathogens* 9(2), e1003169. DOI: 10.1371/journal.ppat.1003169

Aubert, V. (1986) Hatching of the carrot cyst nematodes. In: Lamberti F. and Taylor C.E. (eds) *Cyst Nematodes*. Plenum Press, New York, USA, pp. 347–348.

Aumann, J. and Hasheem, M. (1993) Studies on substances with sex pheromone activity produced by *Heterodera schachtii* females, *Revue de Nématologie* 16, 43–46.

Banyer, R.J. and Fisher, J.M. (1971) Effect of temperature on hatching of eggs of *Heterodera avenae*. *Nematologica* 17, 519–534. DOI: 10.1163/187529271X00242

Bargmann, C.I. (2006) Chemosensation in *C. elegans*. *WormBook* (The *C. elegans* Research Community Ed.) DOI/10/1985/wormbook.1.123.1

Beane, J. and Perry, R.N. (1983) Hatching of the cyst nematode *Heterodera goettingiana* in response to root diffusate from bean (*Vicia faba*). *Nematologica* 29, 361–363. DOI: 10.1163/187529283X00113

Blair, L., Perry, R.N., Oparka, K and Jones, J.T. (1999) Activation of transcription during the hatching process of the potato cyst nematode *Globodera rostochiensis*. *Nematology* 1, 103–111. DOI: 10.1163/156854199507910

Braendle, C. (2012) Pheromones: evolving language of chemical communication in nematodes. *Currrent Biology* 22, R294. DOI: 10.1016/j.cub.2012.03.035

Būda, V. and Čeśulytė-Rakauskienė, R. (2011) The effect of linalool on second-stage juveniles of the potato cyst nematodes *Globodera rostochiensis* and *G. pallida*. *Journal of Nematology* 43, 149–151.

Būda, V. and Čeśulytė-Rakauskienė, R. (2015) The effects of of α-solanine and zinc sulphate on the behavior of potato cyst nematodes *Globodera rostochiensis* and *G. pallida*. *Nematology* 17, 1105–1111. DOI: 10.1163/15685411-00002927

Butcher, R.A., Fujita, M., Schroeder, F.C. and Clardy, J. (2007) Small-molecule pheromones that control dauer development in *Caenorhabditis elegans*. *Nature Chemical Biology* 3, 420–422.

Butcher, R.A., Ragains, J.R., Kim, E. and Clardy, J. (2008) A potent dauer pheromone component in *Caenorhabditis elegans* that acts synergistically with other components. *Proceedings of the National Academy of Sciences of the United States of America* 105, 14288–14292. DOI: 10.1073pnas.0806676105

Byrne, J., Twomey, U., Maher, N., Devine, K.J. and Jones, P.W. (1998) Detection of hatching inhibitors and hatching factor stimulants for golden potato cyst nematode, *Globodera rostochiensis*, in potato root leachate. *Annals of Applied Biology* 132, 463–472. DOI: 10.1111/j.1744-7348.1998.tb05222.x

Byrne, J.T., Maher, N.J. and Jones, P.W. (2001) Comparative responses of *Globodera rostochiensis* and *G. pallida* to hatching chemicals. *Journal of Nematology* 33, 195–202.

Campbell, J.C., Chin-Sang, I.D. and Bendena, W.G. (2015) Mechanosensation circuitry in *Caenorhabditis elegans*: a focus on gentle touch. *Peptides* 68, 164–174. DOI: 10.1016/j.peptides.2014.12.004

Chalasani, S.H., Kato, S., Albrecht, D.R. *et al.* (2010) Neuropeptide feedback modifies odor-evoked dynamics in *C. elegans* olfactory neurons. *Nature Neuroscience* 13, 615–621. DOI: 10.1038/nn.2526

Chang, Y.-J., Burton, T., Ha, L., Huang, Z. and Olajubelo, A. (2015) Modulation of locomotion and reproduction by FLP neuropeptides in the nematode *Caenorhabditis elegans*. *PLoS ONE* 10, e0135164. DOI: 10.1371/journal.pone.0135164

Charlson, D.V. and Tylka, G.L. (2003) *Heterodera glycines* cyst components and surface disinfestants affect *H. glycines* hatching. *Journal of Nematology* 35, 458–464.

Chen, Q., Rehman, S., Smant, G. and Jones, J.T. (2005) Functional analysis of pathogenicity proteins of the potato cyst nematode *Globodera rostochiensis* using RNAi. *Molecular Plant Microbe Interaction* 18, 621–625. DOI: 10.1094/MPMI-10-0621

Choe, A., von Reuss, S.H., Kogan, D. *et al.* (2012) Ascaroside signaling is widely conserved among nematodes. *Current Biology* 22, 772–780. DOI 10.1016/j.cub.2012.03.024

Clarke, A.J. (1970) Hatching factors and sex attractants of cyst nematodes. In: *Report of Rothamsted Experimental Station for 1969*. Rothamsted Research Station, Harpenden, UK, p. 179.

Clarke, A.J. and Perry, R.N. (1977) Hatching of cyst-nematodes. *Nematologica* 23, 350–368. DOI: 10.1163/187529277X00075

Clarke, A.J. and Perry, R.N. (1985) Egg-shell calcium and the hatching of *Globodera rostochiensis*. *International Journal of Parasitology* 15, 511–516. DOI: 10.1016/0020-7519(85)90046-3

Clarke, A.J., Perry, R.N. and Hennessy, J. (1978) Osmotic stress and the hatching of *Globodera rostochiensis*. *Nematologica* 24, 384–392. DOI: 10.1163/187529278X00506

Cook, R. and Rivoal, R. (1998) Genetics of resistance and parasitism. In: Sharma, S.B (ed.) *The Cyst Nematodes*. Kluwer Academic Publishers, Dordrecht, The Netherlands, pp. 322–352.

Corey, E.J. and Houpis, I.N. (1990) Total synthesis of glycinoeclepin A. *Journal of the American Chemical Society* 112, 8997–8998.

Cornils, A., Gloeck, M., Chen, Z., Zhang, Y. and Alcedo, J. (2011) Specific insulin-like peptides encode sensory information to regulate distinct developmental processes. *Development* 138, 1183–1193. DOI: 10.1242/dev.060905

Cotton, J.A., Lilley, C.J., Jones, L.M. *et al.* (2014) The genome and life-stage specific transcriptomes of *Globodera pallida* elucidate key aspects of plant parasitism by a cyst nematode. *Genome Biology* 15, R43. Available at: http://genomebiology.com/2014/15/3/R43 (accessed 13 February 2018).

Crook, M. (2014) The dauer hypothesis and the evolution of parasitism: 20 years on and still going strong. *International Journal for Parasitology* 44, 1–8. DOI: 10.1016/j.ijpara.2013.08.004

Curtis, R.H.C. (2008) Plant-nematode interactions: environmental signals detected by the nematode's chemosensory organs control changes in the surface of the cuticle and behaviour. *Parasite* 15, 310–316. DOI: 10.1051/parasite/2008153310

Curtis, R.H.C., Robinson, A.F. and Perry, R.N. (2009) Hatch and host location. In: Perry, R.N., Moens, M. and Starr, J.L. (eds) *Root-knot Nematodes*. CAB International, Wallingford, UK, pp. 139–162.

Dalzell, J.J., Kerr, R., Coebett, M.D., Fleming, C.C. and Maule, A.G. (2011) Novel bioassays to examine the host-finding ability of plant-parasitic nematodes. *Nematology* 13, 211–220. DOI: 10.1163/138855410X516760

Den Nijs, L.J. (1992) Interaction between *Globodera rostochiensis* and *G. pallida* in simultaneous infections on potatoes with different resistance properties. *Fundamental and Applied Nematology* 15, 173–178.

Den Nijs, L.J. and Lock, C. (1992) Differential hatching of the potato cyst nematodes *Globodera rostochiensis* and *G. pallida* in root diffusates and water of differing ionic composition. *European Journal of Plant Pathology* 98, 117–128.

Devine, K.J. (2010) Comparison of the effects of freezing and thawing on the cysts of the two potato cyst nematode species, *Globodera rostochiensis* and *G. pallida* using differential scanning calorimetry. *Nematology* 12, 81–88. DOI: 10.1163/156854109X448357

Devine, K.J. and Jones, P.W. (2000a) Response of *Globodera rostochiensis* to exogenously applied hatching factors in soil. *Annals of Applied Biology* 137, 21–29. DOI: 10.1111/j.1744-7348.2000.tb00053.x

Devine, K.J. and Jones, P.W. (2000b) Purification and partial characterisation of hatching factors for the potato cyst nematode *Globodera rostochiensis* from potato root leachate. *Nematology* 2, 231–236. DOI: 10.1163/156854100508971

Devine, K.J. and Jones, P.W. (2003) Investigations into the chemoattraction of the potato cyst nematodes *Globodera rostochiensis* and *G. pallida* towards fractionated potato root leachate. *Nematology* 5, 65–75. DOI: 10.1163/156854102765216704

Devine, K.J., Byrne, J., Maher, N. and Jones, P.W. (1996) Resolution of natural hatching factors for the golden potato cyst nematode, *Globodera rostochiensis*. *Annals of Applied Biology* 129, 323–334. DOI: 10.1111/j.1744-7348.1996.tb05755.x

Doncaster, C.C. and Seymour, M.K. (1973) Exploration and selection of penetration sites by Tylenchida. *Nematologica* 19, 137–145. DOI: 10.1163/187529273X00277

Ebrahimi, N., Viaene, N. and Moens, M. (2015) Optimising trehalose-based quantification of live eggs in potato cyst nematodes (*Globodera rostochiensis* and *G. pallida*). *Plant Disease* 99, 947–953. DOI: 10.1094/PDIS-09-14-0940-RE

Ellenby, C. (1974) Water uptake and hatching in the potato cyst nematode, *Heterodera rostochiensis* and the root knot nematode, *Meloidogyne incognita*. *Journal of Experimental Biology* 61, 773–779.

Ellenby, C. and Perry, R.N. (1976) The influence of the hatching factor on the water uptake of the second stage larva of the potato cyst nematode, *Heterodera rostochiensis*. *Journal of Experimental Biology* 64, 141–147.

Elling, A.A., Mitreva, M., Recknor, J. *et al.* (2007) Divergent evolution of arrested development in the dauer stage of *Caenorhabditis elegans* and the infective stage of *Heterodera glycines*. *Genome Biology* 8, R211. DOI: 10.1186/gb-2007-8-10-r211

Evans, A.A.F. (1987) Diapause in nematodes as a survival strategy. In: Veech, J.A. and Dickson, D.W. (eds) *Vistas on Nematology*. Society of Nematologists Inc., Hyattsville, Maryland, pp. 180–187.

Evans, A.A.F. and Perry, R.N. (1976) Survival strategies in nematodes. In: Croll, N.A. (ed.) *The Organization of Nematodes*. Academic Press, London, pp. 383–424.

Eves-van den Akker, S., Lilley, C.J., Jones, J.T. and Urwin, P.E. (2014) Identification and characterisation of a hyper-variable apoplastic effector gene family of the potato cyst nematodes. *PLoS Pathogens* 10, e1004391. DOI: 10.1371/journal.ppat.1004391

Farnier, K., Bengtsson, M., Becher, P.G., Witzell, J., Witzgall, P. and Manduric, S. (2012) Novel bioassay demonstrates attraction of the white potato cyst nematode *Globodera pallida* (Stone) to non-volatile and volatile host plant cues. *Journal of Chemical Ecology* 38, 795–801. DOI: 10.1007/s10886-012-0105-y

Fenwick, D.W. (1956) The hatching of cyst-forming nematodes. In: *Report of Rothamsted Experimental Station for 1955*. Rothamsted Experimental Station, Harpenden, UK, pp. 202–209.

Fioretti, L. (2000) Identification and characterisation of potato cyst nematode excretory/secretory products suitable for the plantibody approach and production of functional single chain antibody fragment. PhD thesis. Open University, Milton Keynes, UK.

Fioretti, L., Porter, A., Haydock, P.J. and Curtis, R. (2002) Monoclonal antibodies reactive with secreted-excreted products from the amphids and the cuticle surface of *Globodera pallida* affect nematode movement and delay invasion of potato roots. *International Journal for Parasitology* 32, 1709–1718. DOI: 10.1016/S0020-7519(02)00178-9

Fosu-Nyarko, J., Nicol, P., Naz, F., Gill, R. and Jones, M.G.K. (2016) Analysis of the transcriptome of the infective stage of the beet cyst nematode, *H. schachtii*. *PLoS ONE* 11(1), e0147511. DOI: 10.1371/journal.pone.0147511

Fukuzawa, A., Furusaki, A., Ikura, M. and Masamune, T. (1985a) Glycinoeclepin A, a natural hatching stimulant for the soybean cyst nematode. *Journal of the Chemical Society* 4, 222–224.

Fukuzawa, A., Matsue, H., Ikura, M. and Masamune, R. (1985b) Glycinoeclepins B and C, nortriterpenes related to glycinoeclepin A. *Tetrahedron Letters* 26, 5539–5542.

Gang, S.S. and Hallem, E.A. (2016) Mechanisms of host seeking by parasitic nematodes. *Molecular and Biochemical Parasitology* 208, 23–32. DOI: 10.1016/j.molbiopara.2016.05.007

Gaur, H.S., Perry, R.N. and Beane, J. (1992) Hatching behaviour of six successive generations of the pigeon-pea cyst nematode, *Heterodera cajani*, in relation to growth and senescence of cowpea, *Vigna unguiculata*. *Nematologica* 38, 190–202. DOI: 10.1163/187529292X00162

Gaur, H.S., Beane, J. and Perry, R.N. (1995) Hatching of four successive generations of *Heterodera sorghi* in relation to the age of sorghum, *Sorghum vulgare*. *Fundamental and Applied Nematology* 18, 599–601.

Gibson, D.M., Moreau, R.A., McNeil, G.P. and Brodie, B.B. (1995) Lipid composition of cyst stages of *Globodera rostochiensis*. *Journal of Nematology* 27, 304–311.

Golden, J.W. and Riddle, D.L. (1982) A pheromone influences larval development in the nematode *Caenorhabditis elegans*. *Science* 218, 578–580.

Golden, J.W. and Riddle, D.L. (1985) A gene affecting production of the *Caenorhabditis elegans* dauer-inducing pheromone. *Molecular Genetics and Genomics* 198, 534–536. DOI: 10.1007/BF00332953

Grant, W. and Viney, M. (2011) The dauer phenomenon. In: Perry, R.N. and Wharton, D.A. (eds) *Molecular and Physiological Basis of Nematode Survival*. CAB International, Wallingford, UK, pp. 99–125.

Greco, N. (1981) Hatching of *Heterodera carotae* and *H. avenae*. *Nematologica* 27, 366–371. DOI: 10.1163/187529281X00377

Greco, N. and Brandonisio, A. (1986) The biology of *Heterodera carotae*. *Nematologica* 32, 447–460. DOI: 10.1163/187529286X00327

Greco, N., Vito, M.D. and Lamberti, F. (1986) Studies on the biology of *Heterodera goettingiana* in southern Italy. *Nematologia Mediterranea* 14, 23–29.

Green, C.D. (1980) Nematode sex attractants. *Helminthological Abstracts Series B* 49, 81–93.

Green, C.D. and Greet, D.N. (1972) The location of the secretions that attract male *Heterodera schachtii* and *H. rostochiensis* to their females. *Nematologica* 18, 347–52. DOI: 10.1163/187529272X00610

Green, C.D. and Plumb, S.C. (1970) The interrelationships of some *Heterodera* spp. indicated by the specificity of the male attractants emitted by their females. *Nematologica* 16, 39–46. DOI: 10.1163/187529270X00441

Greet, D.N. (1964) Observations on sexual attraction and copulation in the nematode *Panagrolaimus rigidus* (Schneider). *Nature* 204, 96–97.

Greet, D.N. and Perry, R.N. (1992) Sexual differentiation and behaviour: Nematoda and Nematomorpha. In: Adiyodi, K.G. and Adiyodi, R.G. (eds) *Reproductive Biology of Invertebrates, Vol. 5*. Oxford and IBH, New Delhi, India, pp. 147–173.

Greet, D.N., Green, C.D. and Poulton, M.E. (1968) Extraction, standardization and assessment of the volatility of the sex attractants of *Heterodera rostochiensis* Woll. and *H. schachtii* Schm. *Annals of Applied Biology* 61, 511–519. DOI: 10.1111/j.1744-7348.1968.tb04553.x

Grundler, F., Schnibbe, L. and Wyss, U. (1991) *In vitro* studies on the behavior of second-stage juveniles of *Heterodera schachtii* (Nematoda: Heteroderidae) in response to host plant root diffusates. *Parasitology* 103, 149–155. DOI: https://doi.org/10.1017/S0031182000059394

Guo, H., La Clair, J.J., Masler, E.P., O'Doherty, G.A. and Xing, Y. (2016) De novo asymmetric synthesis and biological analysis of the daumone pheromones in *Caenorhabditis elegans* and in the soybean cyst nematode *Heterodera glycines*. *Tetrahedron* 72, 2280–2286. DOI: 10.1016/j.tet.2016.03.033

Haegeman, A., Mantelin, S., Jones, J.T. and Gheysen, G. (2012) Functional roles of effectors of plant-parasitic nematodes. *Gene* 492, 19–31. DOI: 10.1016/j.gene.2011.10.040

Han, Z., Boas, S. and Schroeder, N.E. (2016) Unexpected variation in neuroanatomy among diverse nematode species. *Frontiers in Neuroanatomy* 9, 162. DOI: 10.3389/fnana.2015.00162

Hartwell, W.V., Dahlstrom, R.V. and Neal, A.L. (1959) Crystallisation of a natural hatching factor for the larvae of the golden nematode. *Phytopathology* 49, 540–541.

Hashmi, S. and Krusberg, L.R. (1995) Factors influencing emergence of juveniles from cysts of *Heterodera zeae*. *Journal of Nematology* 27, 362–369.

Hearne, R. (2016) Interspecific competition between the potato cyst nematode species *Globodera pallida* and *G. rostochiensis*. PhD thesis. University College Cork, Cork, Ireland.

Hearne, R., Lettice, E.P. and Jones, P.W. (2017) Interspecific and intraspecific competition in potato cyst nematodes *Globodera pallida* and *G. rostochiensis*. *Nematology* 19, 463–475. DOI: 10.1163/15685411-00003061

Hilliard, M.A., Bargmann, C.I. and Bazzaicalupo, P. (2002) *C. elegans* responds to chemical repellents by integrating sensory inputs from head and tail. *Current Biology* 12, 730–734.

Holden-Dye, L. and Walker, R.J. (2011) Neurobiology of plant parasitic nematodes. *Invertebrate Neuroscience* 11, 9–19. DOI: 10.1007/s10158-011-0117-2

Hollister, K.A., Conner, E.S., Zhang, X., Spella, M., Bernarda, G.M., Patela, P., de Carvalhob, A.C.G.V., Butcher, R.A. and Ragainsa, J.R. (2013) Ascaroside activity in *Caenorhabditis elegans* is highly dependent on chemical structure. *Bioorganic and Medicinal Chemistry* 21, 5754–5769. DOI: 10.1016/j.bmc.2013.07.018

Hominick, W.M. (1986) Photoperiod and diapause in the potato cyst-nematode, *Globodera rostochiensis*. *Nematologica* 32, 408–418. DOI: 10.1163/187529286X00291

Hominick, W.M., Forrest, J.M.S. and Evans, A.A.F.E. (1985) Diapause in *Globodera rostochiensis* and variability in hatching trials. *Nematologica* 31, 159–170. DOI: 10.1163/187529285X00210

Hu, C., Kearn, J., Urwin, P., Lilley, C., O' Connor, V., Holden-Dye, L. and Morgan, H. (2014) StyletChip: a microfluidic device for recording host invasion behaviour and feeding of plant parasitic nematodes. *Lab Chip* 14, 2447–2455. DOI: 10.1039/c4lc00292j

Ibrahim, S.K., Perry, R.N., Plowright, R.A. and Rowe, J. (1993) Hatching behaviour of the rice cyst nematode *Heterodera sacchari* and *H. oryzicola* in relation to age of host plant. *Fundamental and Applied Nematology* 16, 23–29.

Ingham, R.E., Kroese, D. and Zasada, I.A. (2015) Effect of storage environment on hatching of the cyst nematode *Globodera ellingtonae*. *Journal of Nematology* 47, 45–51.

Ishibashi, N., Kondo, E., Muraoka, M. and Yokoo, T. (1973) Ecological significance of dormancy in plant parasitic nematodes. I. Ecological difference between eggs in gelatinous matrix and cysts of *Heterodera glycines* Ichinohe (Tylenchida: Heteroderidae). *Applied Entomology and Zoology* 8, 53–63. DOI: 10.1303/aez.8.53

Jaffe, H., Huettel, R.N., Demilo, A.B., Hayes, D.K. and Rebois, R.V. (1989) Isolation and identification of a compound from soybean cyst nematode, *Heterodera glycines*, with sex pheromone activity. *Journal of Chemical Ecology* 15, 2031–2043. DOI: 10.1007/BF01207435

Janssen, R., Bakker, J. and Gommers, F.J. (1987) Circumventing the diapauses of potato cyst nematodes. *Netherlands Journal of Plant Pathology* 83, 107–113.

Jeong, P.-Y., Jung, M., Yim, Y.-H., Kim, H., Park, M., Hong, E., Lee, W., Kim, Y.H., Kim, K. and Paik, Y-K. (2005) Chemical structure and biological activity of the *Caenorhabditis elegans* dauer-inducing pheromone. *Nature* 433, 541–545. DOI: 10.1038/nature03201

Jezyk, P.F. and Fairbairn, D. (1967) Ascarosides and ascaroside esters in *Ascaris lumbricoides* (Nematoda). *Comparative Biochemistry and Physiology* 23, 691–705.

Jing, B.X., He, Q., Wu, H.Y. and Peng, D.L. (2014) Seasonal and temperature effects on hatching of *Heterodera avenae* (Shandong population, China). *Nematology* 16, 1209–1217. DOI: 10.1163/15685411-00002847

Johnson, A.W. (1952) The eelworm problem: biological aspects. The potato eelworm hatching factor. *Chemistry and Industry* 40, 998–999.

Johnston, M.J.G., McVeigh, P., McMaster, S., Fleming, C.C. and Maule, A.G. (2010) FMRFamide-like peptides in root knot nematodes and their potential role in nematode physiology. *Journal of Helminthology* 84, 253–265. DOI: 10.1017/S0022149X09990630

Jones, J. (2002) Nematode sense organs. In: Lee, D.L. (ed.) *The Biology of Nematodes*. Taylor & Francis, London, pp. 353–368.

Jones, J.T., Perry, R.N. and Johnston, M.R.L. (1991) Electrophysiological recordings of electrical activity and responses to stimulants from *Globodera rostochiensis* and *Syngamus trachea*. *Revue de Nématologie* 14, 467–473.

Jones, J.T., Perry, R.N. and Johnston, M.R.L. (1994) Changes in the ultrastructure of the amphids of the potato cyst nematode, *Globodera rostochiensis*, during development and infection. *Fundamental and Applied Nematology* 17, 369–382.

Jones, J.T., Smant, G. and Blok, V.C. (2000) SXP/RAL-2 proteins of the potato cyst nematode *Globodera rostochiensis*: secreted proteins of the hypodermis and amphids. *Nematology* 1, 887–893. DOI: 10.1163/156854100750112833

Jones, P.W., Tylka, G.L. and Perry, R.N. (1998) Hatching. In: Perry, R.N. and Wright, D.J. (eds) *The Physiology and Biochemistry of Free-Living and Plant-Parasitic Nematodes*. CAB International, Wallingford, UK, pp. 181–212.

Kaplan, F., Alborn, H.T., von Reuss, S.H. *et al.* (2012) Interspecific nematode signals regulate dispersal behavior. *PLoS ONE* 7(6), e38735. DOI: 10.1371/journal.pone.0038735

Kodama, E., Kuhara, A., Mohri-Shiomi, A., Kimura, K.D., Okumura, M., Tomioka, M., Iino, Y. and Mori, I. (2006) Insulin-like signaling and the neural circuit for integrative behavior in *C. elegans*. *Genes and Development* 20, 2955–2960. DOI: 10.1101/gad.1479906

Kraus, G.A., Johnson, B., Kongsjahju, A. and Tylka, G.L. (1994) Synthesis and evaluation of compounds that affect soybean cyst nematode egg hatch. *Journal of Agricultural and Food Chemistry* 42, 1839–1840. DOI: 10.1021/jf00045a001

Kraus, G.A., Vander Louw, S.J., Tylka, G.L. and Soh, D.H. (1996) The synthesis and testing of compounds that inhibit soybean cyst nematode egg hatch. *Journal of Agricultural and Food Chemistry* 44, 1548–1550. DOI: 10.1021/jf9505382

Kumar, M., Gantasala, N.P., Roychowdhury, T., Thakur, P.K., Banakar, P., Shukla, R.N., Jones, M.G.K. and Rao, U. (2014) De novo transcriptome sequencing and analysis of the cereal cyst nematode, *Heterodera avenae*. *PLoS ONE* 9(5), e96311. DOI: 10.1371/journal.pone.0096311

LaMondia, J.A. and Brodie, B.B. (1986) The effects of potato trap crops and fallow on decline of *Globodera rostochiensis*. *Annals of Applied Biology* 108, 347–352. DOI: 10.1111/j.1744-7348.1986.tb07656.x

Lettice, E. (2014) Interactions between nematodes, potato and soil-borne microorganisms and their effect on the management of potato cyst nematodes. PhD thesis. University College Cork, Cork, Ireland.

Lettice, E.P. and Jones, P.W. (2015) Evaluation of rhizobacterial colonisation and the ability to induce *Globodera pallida* hatch. *Nematology* 17, 203–212. DOI: 10.1163/15685411-00002863

Lettice, E.P. and Jones, P.W. (2016) Effect of soil and soil bacteria on hatching activity towards potato cyst nematodes (*Globodera* spp.). *Nematology* 18, 803–810. DOI: 10.1163/15685411-00002994

Li, C. (2005) The ever-expanding neuropeptide gene families in the nematode *Caenorhabditis elegans*. *Parasitology* 131, S109–127.

Li, C. and Kim, K. (2008) Neuropeptides. In: *WormBook*, ed. The *C. elegans* Research Community, WormBook, doi/10.1895/wormbook.1.142.1. Available at: www.wormbook.org (accessed 23 January 2018).

Li, C. and Kim, K. (2014) Family of FLP peptides in *Caenorhabditis elegans* and related nematodes. *Frontiers in Endocrinology* 5, Art. 150. DOI: 10.3389/fendo.2014.00150

Li, W., Kennedy, S.G. and Ruvkin, G. (2003) *daf-28* encodes a *C. elegans* insulin superfamily member that is regulated by environmental cues and acts in the DAF-2 pathway. *Genes and Development* 17, 844–858. DOI: 10.1101/gad.1066503

Lilley, C.J., Atkinson, H.J. and Urwin, P.E. (2005) Molecular aspects of cyst nematodes. *Molecular Plant Pathology* 6, 577–588. DOI: 10.1111/J.1364 3703.2005.00306.X

Ludewig, A.H. and Schroeder, F.C. (2013) Ascaroside signaling in *C. elegans*. In: *WormBook*, ed. The *C. elegans* Research Community, WormBook, DOI/10.1895/wormbook.1.155.1. Available at: www.wormbook.org (accessed 23 January 2018).

Ma, W.-C. (1972) Dynamics of feeding responses in *Pieris brassicae* Linn as a function of chemosensory input: a behavioural, ultrastructural and electrophysiological study. *Mededelingen Landbouwhogeschool Wageningen* 72, 1–162.

Malinowska, E. (1996) The range of stimulatory action of the nematode resistant cultivar Tarpan on hatching of juveniles from the cysts of golden potato cyst nematode (Ro1 pathotype) under field conditions. *Biuletyn Instytutu Ziemniaka* 46, 104.

Manosalva, P., Manohar, M., von Reuss, S.H. *et al.* (2015) Conserved nematode signaling molecules elicit plant defenses and pathogen resistance. *Nature Communications* 6, 7795. DOI: 10.1038/ncomms8795

Marrian, D.H., Russell, P.B., Todd, A.R. and Waring, R.S. (1949) The potato eelworm hatching factor. 3. Concentration of the factor by chromatography. Observations on the nature of eclepic acid. *Biochemical Journal* 45, 524–528.

Marshall, J.M. (1986) Competition between two sibling species of plant parasitic nematodes (*Globodera rostochiensis* and *G. pallida*) on a susceptible potato plant. *New Zealand Journal of Zoology* 13, 219–223.

Marshall, J.W. (1989) Changes in relative abundance of two potato nematode species *Globodera rostochiensis* and *G. pallida* in one generation. *Annals of Applied Biology* 115, 79–87. DOI: 10.1111/j.1744-7348.1989.tb06814.x

Martin, B. (1994) Development of a reliable assay for potato cyst nematode hatching factors. MSc thesis. University of Aberdeen, Aberdeen, UK.

Masamune, T., Anetai, M., Takasugi, M. and Katsui, N. (1982) Isolation of a natural hatching stimulus, glycinoeclepin A, for the soybean cyst nematode. *Nature* 297, 495–496. DOI: 10.1038/297495a0

Masamune, T., Anetai, M., Fukuzawa, A., Takasugi, M., Matsue, H., Kabayashi, K., Ueno, S. and Katsui, N. (1987a) Glycinoeclepins, natural hatching stimuli for the soybean cyst nematode, *Heterodera glycines*. I. Isolation. *Bulletin of the Chemical Society of Japan* 60, 981–999.

Masamune, T., Fukuzawa, A., Furusaki, A., Ikura, M., Matsue, H., Kaneko, T., Abiko, A., Sakamoto, N., Tanimoto, N. and Murai, A. (1987b) Glycinoeclepins, natural hatching stimuli for the soybean cyst nematode, *Heterodera glycines*. II. Structural elucidation. *Bulletin of the Chemical Society of Japan* 60, 1001–1014.

Masler, E.P. (2013) Free-living nematodes. In: Kastin, A.J. (ed.) *Handbook of Biologically Active Peptides*. Academic Press/Elsevier, San Diego, USA, pp. 247–254.

Masler, E.P. and Chitwood, D.J. (2016) *Heterodera glycines* cysts contain an extensive array of endoproteases as well as inhibitors of proteases in *H. glycines* and *Meloidogyne incognita* infective juveniles. *Nematology* 18, 4894–4899. DOI: 10.1163/156854110-0002972

Masler, E.P. and Chitwood, D.J. (2017) Evaluation of proteases and protease inhibitors in *Heterodera glycines* cysts obtained from laboratory and field populations. *Nematology* 19, 1091–1020. DOI: 10.1163/156854110-0003035

Masler, E.P. and Rogers, S.T. (2011) Effects of cyst components and low temperature exposure of *Heterodera glycines* eggs on juvenile hatching *in vitro*. *Nematology* 13, 837–844. DOI: 10.1163/138855410X552689

Masler, E.P., Donald, P.A. and Sardanelli, S. (2008a) Stability of *Heterodera glycines* (Tylenchida: Heteroderidae) juvenile hatching from eggs obtained from different sources of soybean, *Glycine max*. *Nematology* 10, 271–278. DOI: 10.1163/156854108783476322

Masler, E.P., Zasada, I.A., and Sardanelli, S.S. (2008b) Hatching behavior in *Heterodera glycines* in response to low temperature. *Comparative Parasitology* 75, 76–81. DOI: 10.1654/4292.1

Masler, E.P., Rogers, S.T. and Chitwood, D.J. (2013) Effects of catechins and low temperature on embryonic development and hatching in *Heterodera glycines* and *Meloidogyne incognita*. *Nematology* 15, 653–663. DOI: 10.1163/15685411-00002708

Masler, E.P., Rogers, S.T. and Hooks, C.R.R. (2017) Behavioral differences of *Heterodera glycines* and *Meloidogyne incognita* infective juveniles exposed to root extracts *in vitro*. *Nematology* 19, 175–183. DOI.org/10.1163/15685411-00003038

Matsunaga, Y., Honda, Y., Honda, S., Iwasaki, T., Qadota, H., Benian, G.M. and Kawano, T. (2016) Diapause is associated with a change in the polarity of secretion of insulin-like peptides. *Nature Communications* 7, 10573. DOI: 10.1038/ncomms10573

McVeigh, P., Leech, S., Mair, G., Marks, N., Geary, T. and Maule, A. (2005) Analysis of FMRFamide-like peptide (FLP) diversity in phylum Nematoda. *International Journal for Parasitology* 35, 1043–1060. DOI: 10.1016/j.ijpara.2005.05.010

Miwa, A., Nii, Y., Okawara, H. and Sakakibara, M. (1987) Synthetic study on hatching stimuli for the soybean cyst nematode. *Agricultural and Biological Chemistry* 51, 3459–3461.

Mokabli, A, Valette, S., Gauthier, J.-P. and Rivoal, R. (2001) *Nematology* 3, 171–178. DOI: 10.1163/156854101750236303

Mori, K. and Watanabe, H. (1989) Recent results in the synthesis of semiochemicals: synthesis of glycinoeclepin A. *Pure and Applied Chemistry* 61, 543–546.

Mulder, J.G., Diepenhorst, P., Plieger, P. and Brüggemann-Rotgans, I.E.M. (1992) Hatching agent for the potato cyst nematode. Patent application PCT/NL92/00126.

Mullin, C.A., Chyb, S., Eichenseer, H., Hollister, B. and Frazier, J.L. (1994) Neuroreceptor mechanisms in insect gustation: a pharmacological approach. *Journal of Insect Physiology* 40, 913–931. DOI: 10.1016/0022-1910(94)90130-9

Murai, A., Tanimoto, H., Sakamoto, H. and Masamune, T. (1988) Total synthesis of glycinoeclepin A. *Journal of the American Chemical Society* 110, 1985–1986.

Niblack, T.L., Lambert, K.N. and Tylka, G.L. (2006) A model plant pathogen from the kingdom Animalia: *Heterodera glycines*, the soybean cyst nematode. *Annual Review of Phytopathology* 44, 283–303. DOI: 10.1146/annurev.phyto.43.040204.140218

Nonaka, S., Katsuyama, T., Kondo, T., Sasaki, Y., Asami, T., Yajima, S. and Ito, S. (2016) 1,10-Phenanthroline and its derivatives are novel hatching stimulants for soybean cyst nematodes. *Bioorganic and Medicinal Chemistry Letters* 26, 5240–5243. DOI: 10.1016/j.bmcl.2016.09.052

Okada, T. (1972) Hatch inhibitory factor in the cyst contents of the soybean cyst nematode, *Heterodera glycines* Ichinhe (Tylenchida: Heteroderidae). *Applied Entomology and Zoology*, 99–102.

Okada, T. (1974) Effects of hatching stimulants obtained from the cyst contents of *Heterodera species* (Tylenchida: Heteroderidae) on the hatching of other species. *Applied Entomology and Zoology* 9, 49–51.

Palomares-Rius, J.E., Jones, J.T., Cock, P.J., Castillo, P. and Blok, V.C. (2013) Activation and hatching in diapaused and quiescent *Globodera pallida*. *Parasitology* 140, 445–454. DOI: 10.1017/S0031182012001874

Palomares-Rius, J.E., Hedley, P., Cock, P.J.A., Morris, J.A., Jones, J.T. and Blok, V.C. (2016) Gene expression changes in diapause or quiescent potato cyst nematode, *Globodera pallida*, eggs after hydration or exposure to tomato root diffusate. *PeerJ* DOI: 10.7717/peerj.1654

Perry, R.N. (1989) Root diffusates and hatching factors. *Aspects of Applied Biology* 22, 121–128.

Perry, R.N. (1996) Chemoreception in plant parasitic nematodes. *Annual Review of Phytopathology* 24, 181–199.

Perry, R.N. (1997) Plant signals in nematode hatching and attraction. In: Fenoll, C., Grundler, F.M.W. and Ohl, S.A. (eds) *Cellular and Molecular Aspects of Plant-Nematode Interactions*. Kluwer Academic Publishers, Dordrecht, The Netherlands, pp. 38–50.

Perry, R.N. (2002) Hatching. In: Lee, D.L. (ed.) *The Biology of Nematodes*. Taylor and Francis, London.

Perry, R.N. (2005) An evaluation of types of attractants enabling plant-parasitic nematodes to locate plant roots. *Russian Journal of Nematology* 13, 83–88.

Perry, R.N. and Aumann, J. (1998) Behaviour and sensory responses. In: Perry, R.N. and Wright, D.J. (eds) *The Physiology and Biochemistry of Free-living and Plant-parasitic Nematodes*. CAB International, Wallingford, UK, pp. 75–102.

Perry, R.N. and Beane, J. (1982) The effect of brief exposures to potato root diffusate on the hatching of *Globodera rostochiensis*. *Revue de Nématologie* 5, 221–224.

Perry, R.N. and Curtis, R.H.C. (2013) Behaviour and sensory perception. In: Perry, R.N. and Moens, M. (eds) *Plant Nematology*, 2nd edn. CAB International, Wallingford, UK, pp. 246–273.

Perry, R.N. and Gaur, H.S. (1996) Host plant influences on the hatching of cyst nematodes. *Fundamental and Applied Nematology* 19, 505–510.

Perry, R.N. and Wharton, D.A. (1985) Cold tolerance of hatched and unhatched second stage juveniles of the potato cyst nematode *Globodera rostochiensis*. *International Journal for Parasitology* 15, 441–445. DOI 10.1016/0020-7519(85)90031-1

Perry, R.N., Clarke, A.J. and Beane, J. (1980a) Hatching of *H. goettingiana in vitro*. *Nematologica* 26, 493–495. DOI: 10.1163/187529280X00422

Perry, R.N., Clarke, A.J. and Hennessy, J. (1980b) The influence of osmotic stress on the hatching of *Heterodera schachtii*. *Revue de Nématologie* 3, 3–9.

Perry, R.N., Hodges, J.A. and Beane, J. (1981) Hatching of *Globodera rostochiensis* in response to potato root diffusate persisting in soil. *Nematologica* 27, 349–352. DOI: 10.1163/187529281X00584

Perry, R.N., Knox, D.P. and Beane, J. (1992) Enzymes released during hatching of *Globodera rostochiensis* and *Meloidogyne incognita*. *Fundamental and Applied Nematology* 15, 283–288.

Peymen, K., Watteyne, J., Frooninckx, L., Schoofs, L. and Beets, I. (2014) The FMRFamide-like peptide family in nematodes. *Frontiers in Endocrinology* 5, 90. DOI: 10.3389/fendo.2014.00090

Phillips, W.S., Kieran, S.R. and Zasada, I.A. (2015) The relationship between temperature and development in *Globodera ellingtonae*. *Journal of Nematology* 47, 283–289.

Pierce, S.B., Costa, M., Wisotzkey, R. *et al*. (2001) Regulation of DAF-2 receptor signaling by human insulin and *ins-1*, a member of the unusually large and diverse *C. elegans* insulin gene family. *Genes and Development* 15, 672–686. DOI: 10.1101/gad.867301

Pridannikov, M.V., Petelina, G.G., Palchuk, M.V., Masler, E.P. and Dzhavakhiya, V.G. (2007) Influence of components of *Globodera rostochiensis* cysts on the in vitro hatch of second-stage juveniles. *Nematology* 9, 837–844. DOI: 10.1163/156854107782331126

Pungaliya, C., Srinivasan, J., Fox, B.W., Malik, R.U., Ludewig, A.H., Sternberg, P.W. and Schroeder, F.C. (2009) A shortcut to identifying small molecule signals that regulate behavior and development in

Caenorhabditis elegans. *Proceedings of the National Academy of Sciences of the United States of America* 106, 7708–7713. DOI: 10.1073pnas.0811918106

Rasmann, S., Ali, J.G., Helder, J. and van der Putten, W.H. (2012) Ecology and evolution of soil nematode chemotaxis. *Journal of Chemical Ecology* 38, 615–628. DOI: 10.1007/s10886-012-0118-6

Rawsthorne, D. and Brodie, B.B. (1986) Relationship between root growth of potato, root diffusate production, and hatching of *Globodera rostochiensis*. *Journal of Nematology* 18, 379–384.

Rawsthorne, D. and Brodie, B.B. (1987) Movement of potato root diffusate through the soil. *Journal of Nematology* 19, 119–122.

Reynolds, A.M., Dutta, T.K., Curtis, R.H.C., Powers, S.J., Gaur, H.S. and Kerry, B.R. (2011) Chemotaxis can take plant-parasitic nematodes to the source of chemo-attractant via the shortest possible routes. *Journal of the Royal Society Interface* 8, 569–577. DOI: 10.1098/rsif.2010.0417

Riga, E., Perry, R.N., Barrett, J. and Johnston, M.R.L. (1996a) Electrophysiological responses of males of the potato cyst nematodes, *Globodera rostochiensis* and *G. pallida*, to their sex pheromones. *Parasitology* 112, 239–246. DOI: https://doi.org/10.1017/S0031182000084821

Riga, E., Perry, R.N. and Barrett, J. (1996b) Electrophysiological analysis of the response of males of *Globodera rostochiensis* and *G. pallida* to their female sex pheromones and to potato root diffusate. *Nematologica* 42, 493–498. DOI: 10.1163/004525996X00091

Riga, E., Perry, R.N., Barrett, J. and Johnston, M.R.L. (1997a) Electrophysiological responses of males of the potato cyst nematodes, *Globodera rostochiensis* and *G. pallida*, to some chemicals. *Journal of Chemical Ecology* 23, 417–428. DOI: 10.1023/B:JOEC.0000006368.52520.07

Riga, E., Holdsworth, D.R., Perry, R.N., Barrett, J. and Johnston, M.R.L. (1997b) Electrophysiological analysis of the response of males of the potato cyst nematode, *Globodera rostochiensis*, to fractions of their homospecific sex pheromone. *Parasitology* 115, 311–316.

Ritter, A.D., Shen, Y., Bass, J.F. *et al*. (2013) Complex expression dynamics and robustness in *C. elegans* insulin networks. *Genome Research* 23, 954–965. DOI: 10.1101/gr.150466

Rivoal, R. (1978) Biologie d'*Heterodera avenae* Wollenweber en France I. Différences dans les cycles d'éclosion et de développement des deux races Fr1 et Fr4. *Revue de Nématologie* 1, 171–179.

Rivoal, R. (1979) Biologie d'*Heterodera avenae* Wollenweber en France II. Etude des différences dans les conditions thermiques d'éclosion des races Fr1 et Fr4. *Revue de Nématologie* 2, 233–248.

Rivoal, R. (1982) Caractérisation de deux écotypes d'*Heterodera avenae* en France par leurs cycles et conditions thermiques d'éclosion. *Bulletin OEPP* 12, 353–359.

Rivoal, R. (1983) Biologie d'*Heterodera avenae* Wollenweber en France. III. Evolution des diapauses des races Fr1 et Fr4 au cours de plusieurs années consecutives: influence de la température. *Revue de Nématologie* 6, 157–164.

Rivoal, R. (1986) Biology of *Heterodera avenae* Wollenweber in France. IV. Comparative study of the hatching cycles of two ecotypes after their transfer to different climatic conditions. *Revue de Nématologie* 9, 405–410.

Rivoal, R. and Cook, R. (1993) Nematode pests of cereals. In: Evans, K., Trudgill, D.L. and Webster, J.M. (eds) *Plant Parasitic Nematodes in Temperate Agriculture*. CAB International, Wallingford, UK, pp. 259–303.

Robinson, A.F. (1994) Movement of five nematode species through sand subjected to natural temperature gradient fluctuations. *Journal of Nematology* 26, 46–58.

Robinson, M.P., Atkinson, H.J. and Perry, R.N. (1985) The effect of delayed emergence on the subsequent infectivity of second stage juveniles of the potato cyst nematode *Globodera rostochiensis*. *Nematologica* 31, 171–178. DOI: 10.1163/187529285X00229

Robinson, M.P., Atkinson, H.J. and Perry, R.N. (1987a) The influence of temperature on the hatching, activity and lipid utilization of the potato cyst nematodes *Globodera rostochiensis* and *G. pallida*. *Revue de Nématologie* 10, 349–354.

Robinson, M.P., Atkinson, H.J. and Perry, R.N. (1987b) The influence of soil moisture and storage time on the motility, infectivity and lipid utilization of second stage juveniles of the potato cyst nematodes *Globodera rostochiensis* and *G. pallida*. *Revue de Nématologie* 10, 343–348.

Rolfe, R.N. (1999) Electrophysiological analysis of nematode responses. PhD thesis. University of Wales, Aberystwyth, UK.

Rolfe, R.N. and Perry, R.N. (2001) Electropharyngeograms and stylet activity of second stage juveniles of *Globodera rostochiensis*. *Nematology* 3, 31–34. DOI: 10.1163/156854101300106865

Rolfe, R.N., Barrett, J. and Perry, R.N. (2000) Analysis of chemosensory responses of second stage juveniles of *Globodera rostochiensis* using electrophysiological techniques. *Nematology* 2, 523–533. DOI: 10.1163/156854100509448

Ryan, A.N. and Jones, P.W. (2003) Effect of tuber-borne micro-organisms on hatching activity of potato root leachate towards potato cyst nematodes. *Nematology* 5, 55–63.

Ryan, A.N., Deliopoulos, T., Jones, P. and Haydock, P.P.J. (2003) Effects of a mixedisolate mycorrhizal inoculum on the potato-potato cyst nematode interaction. *Annals of Applied Biology* 143, 111–119.

Ryan, N.A. and Jones, P. (2004a) The effect of mycorrhization of potato roots on the hatching chemicals active towards the potato cyst nematodes, *Globodera pallida* and *G. rostochiensis*. *Nematology* 6, 335–342. DOI: 10.1163/1568541042360456

Ryan, N.A. and Jones, P. (2004b) The ability of rhizosphere bacteria isolated from nematode host and non-host plants to influence the hatch in vitro of the two potato cyst nematode species, *Globodera rostochiensis* and *G. pallida*. *Nematology* 6, 375–387. DOI: 10.1163/1568541042360528

Ryan, A.N., Jones, P.W. and Devine, K. (2005) The effect of competition between *G. rostochiensis* and *G. pallida* on PCN multiplication rates on non-resistance potato cultivars. *Proceedings of Advances in Potato Cyst Nematode Management*. Harper Adams University, Newport, UK, p. 1.

Sardanelli, S. and Kenworthy, W.L. (1997) Soil moisture control and direct seeding for bioassay of *Heterodera glycines* on soybean. *Journal of Nematology* 29, supplement 4, 625–634.

Schroeder, F.C. (2015) Modular assembly of primary metabolic building blocks: a chemical language in *C. elegans*. *Chemical Biology* 22, 7–16. DOI: 10.1016/J.CHEMBIOL.2014.10.012

Shepherd, A.M. and Clarke, A.J. (1971) Molting and hatching stimuli. In: Zuckerman, B.M., Mai, W.F. and Rohde, R.A. (eds) *Plant Parasitic Nematodes, Volume II*. Academic Press, London, pp. 267–287.

Sommerville, R.I. and Davey, K.G. (2002) Diapause in parasitic nematodes: a review. *Canadian Journal of Zoology* 80, 1817–1840. DOI: 10.1139/z02-163

Srinivasan, J., Kaplan, F., Ajredini, R. et al. (2008) A blend of small molecules regulates both mating and development in *Caenorhabditis elegans*. *Nature* 454, 1115–1119. DOI: 10.1038/nature07168

Srinivasan, J., von Reuss, S.H., Bose, N. et al. (2012) A modular library of small molecule signals regulates social behaviors in *Caenorhabditis elegans*. *PLoS Biology* 10, e1001237. DOI: 10.1371/journal.pbio.1001237

Stewart, G.R., Perry, R.N., Alexander, J. and Wright, D.J. (1993a) A glycoprotein specific to the amphids of *Meloidogyne* species. *Parasitology* 106, 405–412. DOI: https://doi.org/10.1017/S0031182000067159

Stewart, G.R., Perry, R.N. and Wright, D.J. (1993b) Studies on the amphid specific protein gp32 in different in different life cycle stages of *Meloidogyne* species. *Parasitology* 107, 573–578. DOI: 10.1017/S0031182000068165

Stoyanov, D. and Trifonova, Z. (1995) Hatching dynamics of golden potato cyst nematode larvae *Globodera rostochiensis* Woll. *Bulgarian Journal of Agricultural Science* 1, 241–246.

Tanino, K., Takahashi, M., Tomata, Y., Tokura, H., Uehara, T., Narabu, T. and Miyashita, M. (2011) Total synthesis of solanoeclepin A. *Nature Chemistry* 3, 484–488.

Tefft, P.M. and Bone, L.W. (1985a) Leucine aminopeptidase in eggs of the soybean cyst nematode *Heterodera glycines*. *Journal of Nematology* 17, 270–274.

Tefft, P.M. and Bone, L.W. (1985b) Plant-induced hatching of the soybean cyst nematode *Heterodera glycines*. *Journal of Nematology* 17, 275–279.

Thiéry, M., Mugniéry, D., Fouville, D. and Schots, A. (1996) Hybridations naturelles entre *Globodera rostochiensis* et *G. pallida*. *Fundamental and Applied Nematology* 19, 437–442.

Thompson, D.P. and Geary, T.G. (2002) Excretion/secretion, ionic and osmotic regulation. In: Lee, D.L. (ed.) *The Biology of Nematodes*. Taylor & Francis, London, pp. 291–320.

Thompson, J.M. and Tylka, G.L. (1997) Differences in hatching of *Heterodera glycines* egg-mass and encysted eggs in vitro. *Journal of Nematology* 29, 315–321.

Tomioka, M., Adachi, T., Suzuki, H., Kunitomo, H., Schafer, W.R. and Iino, Y. (2006) The insulin/PI 3-kinase pathway regulates salt chemotaxis learning in *Caenorhabditis elegans*. *Neuron* 51, 613–625. DOI: 10.1016/j.neuron.2006.07.024

Trudgill, D.L., Phillips, M.S. and Hackett, C.A. (1996) The basis of predictive modelling for estimating yield loss and planning potato cyst nematode management. *Pesticide Science* 47, 89–94. DOI: 10.1002/(SICI)1096-9063(199605)47:1<89::AID-PS389>3.0.CO;2-S

Tsutsumi, M. (1976) Conditions for collecting the potato root diffusate and the influence on the natural hatching of potato cyst nematode. *Japanese Journal of Nematology* 6, 10–13.

Twomey, U. (1995) Hatching chemicals involved in the interaction between potato cyst nematodes, host and non-host plants. PhD thesis. The National University of Ireland, Cork, Ireland.

Twomey, U., Rolfe, R.N., Warrior, P. and Perry, R.N. (2002) Effects of the biological nematicide, DiTera®, on movement and sensory responses of second stage juveniles of *Globodera rostochiensis*, and stylet activity of *G. rostochiensis* and fourth stage juveniles of *Ditylenchus dipsaci*. *Nematology* 4, 909–915. DOI: 10.1163/156854102321122520

Tzortzakakis, E.A. and Trudgill, D.L. (2005) A comparative study of the thermal time requirements for embryogenesis in *Meloidogyne javanica* and *M. incognita*. *Nematology* 7, 313–315. DOI: 10.1163/1568541054879467

van den Elsen, S., Ave, M., Schoenmakers, N., Landeweert, R., Bakker, J. and Helder, H. (2012) A rapid, sensitive and cost-efficient assay to estimate viability of potato cyst nematodes. *Phytopathology* 102, 140–146. DOI: 10.1094 / PHYTO-02-11-0051

van Gundy, S.D. (1965) Factors in survival of nematodes. *Annual Review of Phytopathology* 3, 43–68.

Voinilo, V.A. (1976) The effect of organic acids on the issue from the cyst of potato cyst nematode larva. *Zashchita Rastenii Minsk* 1, 147–150.

von Reuss, S.H. and Schroeder, F.C. (2015) Combinitorial chemistry in nematodes: modular assembly of primary metabolism-derived building blocks. *Natural Products Reports* 32, 994–1006. DOI: 10.1039/c5np00042d

Watanabe, H. and Mori, K. (1991) Triterpenoid total synthesis. Part 2. Synthesis of glycinoeclepin A, a potent hatching stimulus for the soybean cyst nematode. *Journal of the Chemical Society Perkin Transactions* 1, 2919–2934.

Wharton, D.A. (2011) Cold tolerance. In: Perry, R.N. and Wharton, D.A. (eds) *Molecular and Physiological Basis of Nematode Survival*. CAB International, Wallingford, UK, pp. 182–204.

Wharton, D.A. and Perry, R.N. (2011) Osmotic and ionic regulation. In: Perry, R.N. and Wharton, D.A. (eds) *Molecular and Physiological Basis of Nematode Survival*. CAB International, Wallingford, UK, pp. 256–281.

Wharton, D.A. and Ramløv, H. (1995) Differential scanning calorimetry studies on the cysts of the potato cyst nematode *Globodera rostochiensis* during freezing and melting. *Journal of Experimental Biology* 198, 2551–2555.

Wharton, D.A., Perry, R.N. and Beane, J. (1993) The role of the eggshell in the cold tolerance mechanisms of the unhatched juveniles of *Globodera rostochiensis*. *Fundamental and Applied Nematology* 16, 425–431.

Winter, M.D., McPherson, M.J. and Atkinson, H.L. (2002) Neuronal uptake of pesticides disrupts chemosensory cells of nematodes. *Parasitology* 125, 561–565. DOI: 10.1017/S0031182002002482

Wright, D.J. (1998) Respiratory physiology, nitrogen excretion and osmotic and ionic regulation. In: Perry, R.N. and Wright, D.J. (eds) *The Physiology and Biochemistry of Free-living and Plant-parasitic Nematodes*. CAB International, Wallingford, UK, pp. 103–131.

Wright, D.J. (2004) Osmoregulatory and excretory behaviour. In: Gaugler, R. and Bilgrami, A.R. (eds) *Nematode Behaviour*. CAB International, Wallingford, UK, pp. 177–196.

Wright, D.J. and Newall, D.R. (1976) Nitrogen excretion, osmotic and ionic regulation in nematodes. In: Croll, N.A. (ed.) *The Organisation of Nematodes*. Academic Press, London, pp. 163–210.

Wright, D.J. and Newall, D.R. (1980) Osmotic and ionic regulation in nematodes. In: Zuckerma, B.M. (ed.) *Nematodes as Biological Models. Volume 2, Aging and other Model Systems*. Academic Press, London, pp.143–164.

Yang, D., Chen, C., Liu, Q. and Jian, H. (2017) Comparative analysis of pre- and post-parasitic transcriptomes and mining pioneer effectors of *Heterodera avenae*. *Cell and Bioscience* 7, 11. DOI: 10.1186/s13578-017-0138-6

Yen, J.H., Niblack, T.L. and Wiebold, W.J. (1995) Dormancy of *Heterodera glycines* in Missouri. *Journal of Nematology* 27, 153–163.

Yu, Y., Zhi, L., Guan, X., Wang, D. and Wang, D. (2016) FLP-4 neuropeptide and its receptor in a neuronal circuit regulate preference choice through functions of ASH-2 trithorax complex in *Caenorhabditis elegans*. *Scientific Reports* 6, 21485. DOI: 10.1038/srep21485

Zheng, L. and Ferris, H. (1991) Four types of dormancy exhibited by eggs of *Heterodera schachtii*. *Revue de Nématologie* 14, 419–426.

4 Biology of Effectors

John T. Jones[1,2,3] and Melissa G. Mitchum[4]

[1]The James Hutton Institute, Invergowrie, Dundee, UK; [2]The University of St Andrews, North Haugh, St Andrews, UK; [3]Ghent University, Ghent, Belgium; [4]University of Missouri, Columbia, Missouri, USA

4.1 Introduction	74
4.2 Degradation of the Plant Cell Wall	77
4.3 Suppression of Host Defences	79
4.4 Manipulation of Host Biochemistry for Feeding Site Induction	82
4.5 Conclusions and Future Prospects	83
4.6 Acknowledgements	83
4.7 References	83

4.1 Introduction

4.1.1 What are effectors?

A wide range of definitions of effectors has been proposed by various authors, with varying degrees of inclusivity. For the purposes of this article we will use a deliberately broad definition as suggested by Hogenhout *et al.* (2009): 'all pathogen (i.e. nematode) proteins and small molecules that alter host-cell structure and function'. Effectors in cyst nematodes need to fulfil a variety of functional roles, and each of these is considered in detail below. Effectors allow nematodes to overcome the physical barrier presented by the plant cell wall so that they can invade their hosts and migrate to a place suitable for induction of the feeding site. Effectors are required to suppress host defence responses and to protect the nematode from toxic compounds produced as a part of these responses. Many of these functional requirements are shared with effectors from other plant pathogens, including bacteria, fungi and oomycetes. However, effectors from sedentary endoparasitic nematodes (including cyst nematodes) are also required to induce the formation of the feeding structure (syncytium) that will provide the developing nematode with the food it requires to complete its life cycle, a feature unique to these nematodes.

The increasing accessibility of genomics tools and the technical developments that have enabled the deployment of these techniques to plant-parasitic nematodes have allowed large numbers of effectors to be identified from a wide range of species. The approaches that have enabled these developments are described in detail by Eves-van den Akker and Jones (Chapter 2, this volume) but for cyst nematodes they have included analysis of expressed sequence tags (e.g. Popeijus *et al.*, 2000a; Jones *et al.*, 2009), sequencing of gland cell libraries (Gao *et al.*, 2003; Maier *et al.*, 2013) and, more recently, bioinformatic analysis of complete genome sequences of cyst nematodes (Thorpe et al., 2014; Eves-van den Akker et al., 2016). Researchers in this field have therefore been able to move on from work aimed at identifying candidate effectors to studying the function of these proteins. This chapter is focused specifically on effectors that have been characterized from cyst nematodes, although detailed studies have also been performed on effectors from root-knot nematodes (reviewed by Hewezi and Baum, 2013; Mitchum et al., 2013; and Truong et al., 2015). Although some effectors are common to root-knot and cyst nematodes, including the cell wall degrading enzymes (CWDEs) and chorismate mutase, recent genomic analyses of cyst and root-knot nematodes have shown that the vast majority of effectors are specific to each group (Davis et al., 2004; Cotton et al., 2014; Thorpe et al., 2014). This is likely to reflect the independent origin of biotrophic (sedentary) parasitism in cyst and root-knot nematodes (e.g. Baldwin et al., 2004).

4.1.2 Sources of effectors

The main sources of cyst nematode effectors are the dorsal and subventral pharyngeal gland cells. Proteins produced in these structures are secreted through the stylet into the host. The dorsal and subventral pharyngeal gland cells show developmental profiles that suggest distinct roles for the effectors produced in each gland cell. The two subventral gland cells are large and full of secretory granules in the pre-parasitic second-stage juvenile (J2), decrease in size and activity during the sedentary parasitic stages but become active again in adult male nematodes, which leave the root in order to locate females. By contrast, the dorsal gland cell is relatively small in the J2 but increases in size and activity in the sedentary stages (Hussey and Mimms, 1990; Mitchum et al., 2013). These morphological differences are reflected to some extent in temporal expression profiles of effectors identified from *Globodera pallida*. The expression profiles of most of the identified effectors from this species were clustered and formed three distinct groups (Fig. 4.1; Thorpe et al., 2014). Some effectors were expressed solely at the pre-parasitic J2 stage with others showing expression at J2 and adult male. These clusters included sequences such as CWDEs that are important in migration and known to be restricted to the subventral gland cells. A further set of effectors were expressed only at parasitic stages (Fig. 4.1) and included many sequences expressed in the dorsal gland cells. These observations suggest that effectors produced in the subventral gland cells are important for invasion, migration and processes occurring at the early stages of parasitism, whereas those produced in the dorsal gland cell play a role in the later stages of parasitism.

Although the majority of effectors are produced in the pharyngeal gland cells, other nematode tissues produce effectors and these may play important roles in the host–parasite interaction. A variety of antioxidant proteins, including glutathione peroxidase and peroxiredoxin, are produced in the hypodermis and secreted onto the surface of the nematode cuticle (reviewed in Smant and Jones, 2011). These enzymes may protect the nematode from reactive oxygen species produced as part of host defence responses. More recently, a highly variable group of effectors (HYPs) has been identified, the members of which are secreted from the secretory cells surrounding the main anterior sense organs of *G. pallida* (Eves-van den Akker et al., 2014). Although the precise role of these proteins remains to be established, silencing the genes encoding HYP proteins had a profound negative effect on parasitism.

4.1.3 Where do effectors go in plants?

Secretion of effectors to both apoplastic and symplastic compartments of plant cells, as well as the observed differences in subcellular localization, reflects the diverse array of functions being attributed to cyst nematode effectors (described below).

CWDEs were the first effector proteins shown to be secreted into plant tissues during the intracellular migration of cyst nematodes through root tissues (Wang et al., 1999; Goellner et al., 2001). Since then, secretion of cyst nematode effector proteins from sedentary life stages to both the apoplast and cytoplasm of feeding cells has also been observed. Several effectors

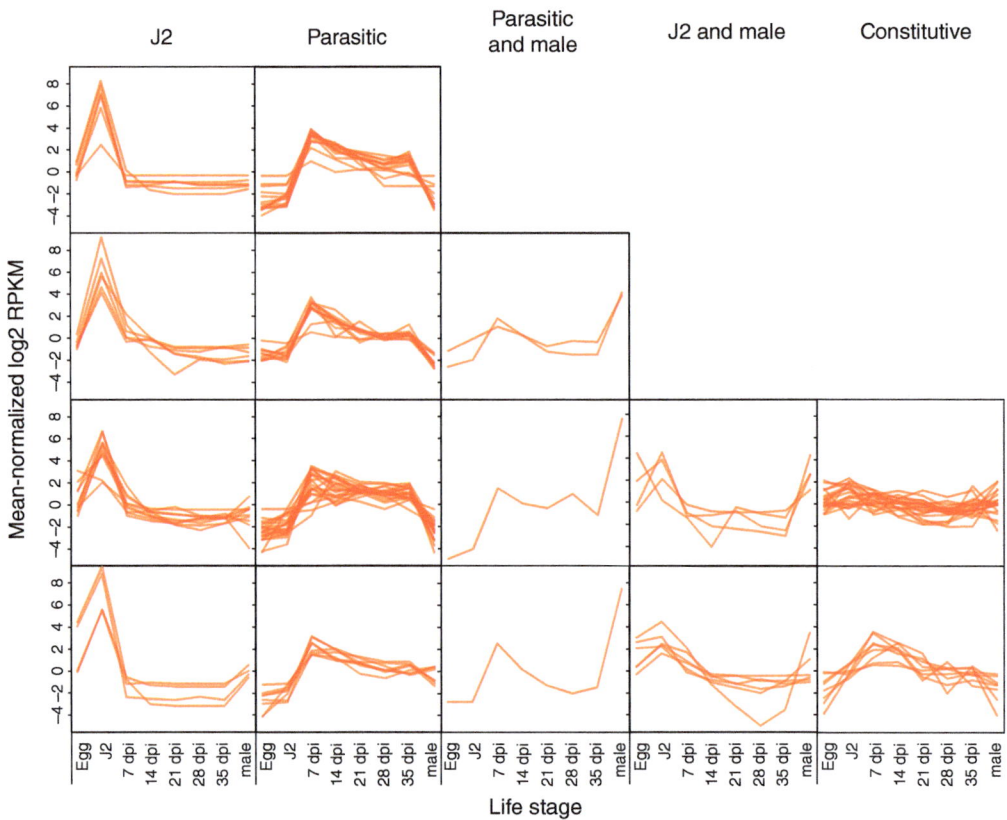

Fig. 4.1. Cluster analysis of co-regulated candidate effectors during *Globodera pallida* life cycle. The *y*-axis represents fold change in expression values, determined by calculating fold changes over mean expression values across all samples from RNAseq data. Clusters of effectors are seen that share temporal expression profiles across the *G. pallida* life cycle. J2 = second-stage juvenile; dpi = days post-inoculation. (From Thorpe *et al.*, 2014.)

function in the apoplast. Venom-allergen proteins (VAPs) are probably secreted to the apoplast during early stages of infection to suppress plant defences (Lozano-Torres *et al.*, 2014), whereas the HYP effectors are secreted directly into the apoplast surrounding the established syncytium (Eves-van den Akker *et al.*, 2014). In a unique case, cyst nematode effectors that mimic plant CLAVATA3/ENDOSPERM SURROUNDING REGION (CLE) peptides were observed within the cytoplasm of the developing syncytium, yet found to function in the apoplast (Wang *et al.*, 2010a). The redirection of cyst nematode CLE peptide effectors from the cytoplasm to the apoplast of plant cells has been attributed to a unique trafficking function of the N-terminal portion of the variable domain (VD) of these proteins that is absent from plant CLEs (Wang *et al.*, 2010b). Although the trafficking mechanism remains to be determined, *G. rostochiensis* CLE1 was shown to be glycosylated when expressed in potato callus cultures (Chen *et al.*, 2015), suggesting that cyst nematode CLE effectors may in fact be targeted to the plant cell secretory pathway. How nematode CLE peptide effectors might enter the plant secretory pathway to acquire these modifications remains a mystery considering that the typical route is via co-translational translocation into the endoplasmic reticulum. However, more recent findings in yeast and mammalian systems indicate that proteins can also enter the secretory pathway after a protein is fully formed: the expected scenario of a nematode secreted CLE protein delivered to a plant cell. While such pathways probably exist in plants, they remain be characterized, underscoring the potential significance of elucidating the VD trafficking function of this class of effectors.

Numerous cyst nematode effectors also function within host cells. Transient expression of these effectors fused to green fluorescent protein (GFP) show a range of subcellular localizations including the cytoplasm, peroxisomes, nucleus and nucleolus of plant cells (Elling et al., 2007; Jones et al., 2009; Chronis et al., 2013; Thorpe et al., 2014), potentially conferring the ability to modulate a wide variety of host cellular processes.

4.2 Degradation of the Plant Cell Wall

The plant cell wall represents the first line of defence that any organism wishing to feed on plants will need to overcome. Cyst nematodes have evolved strategies that allow them to penetrate and migrate through plant tissues, and effectors that metabolize the plant cell wall are a key component of these strategies. The plant cell wall is a complex, rigid structure formed from a network of cross-linked cellulose microfibrils, hemicelluloses, pectins and proteins. Secondary cell walls also contain lignins, in addition to the components listed above, that help provide mechanical stability to cells. Degradation of the plant cell wall therefore requires a mixture of enzymatic and non-enzymatic components that metabolize the various ingredients.

Until the first report of a β-1,4-endoglucanase (cellulase) from cyst nematodes (Smant et al., 1998), it was generally accepted that animals did not produce CWDEs. Animals with a need to degrade the plant cell wall evolved a symbiotic relationship with other organisms, most frequently bacteria, that produced the enzymes required. So, although it had been known for decades that homogenates of plant-parasitic nematodes contained cellulase activity (e.g. Dropkin, 1963), it was assumed that – as seen in other animals – this activity would have a bacterial origin. However, it has become clear that the CWDEs present in plant-parasitic nematodes, including cyst nematodes, are endogenous nematode genes and that these have been acquired by a process of horizontal gene transfer (HGT) (Yan et al., 1998). The phenomenon of HGT in plant-parasitic nematodes is discussed in Chapter 2 of this volume, has been reviewed extensively (Jones et al., 2005; Danchin et al., 2010; Haegeman et al., 2011) and is not considered further here. Following on from the initial discovery of the cyst nematode cellulases (Smant et al., 1998), the completion of the G. pallida and G. rostochiensis genome sequences (Cotton et al., 2014; Eves-van den Akker et al., 2016) has subsequently allowed the full complement of CWDEs present in a cyst nematode to be characterized. These proteins are summarized in Table 4.1.

4.2.1 Cellulases

Cellulase was the first CWDE identified in any plant-parasitic nematode (Smant et al., 1998). All cyst nematode cellulases identified to date are from glycosyl hydrolase family (GHF) 5. These enzymes degrade cellulose through hydrolysis of the 1,4-β linkages within the cellulose chains. GHF5 cellulases are also present in many other Clade 12 plant-parasitic nematodes (van Megen et al., 2009), including root-knot nematodes, as well as many migratory endoparasitic nematodes. This suggests that the HGT event that led to the presence of GHF5 cellulases in plant-parasitic nematodes occurred in a common ancestor of the entire clade (Danchin

Table 4.1. Numbers of cell wall degrading and modifying proteins present in the genomes of *Globodera pallida* and *G. rostochiensis*.

Species	Cellulase (GHF5)	Pectate lyase	Arabinase (GHF43)	Arabinogalactan endo-1,4-beta-galactosidase (GHF53)	Expansin	Carbohydrate binding module (CBM) domain
G. pallida	16	7	1	1	9	6
G. rostochiensis	11	3	0	1	7	5

et al., 2010). Interestingly, cellulases in Clade 10 of plant-parasitic nematodes, such as *Aphelenchoides* and *Bursaphelenchus*, are from GHF45 and are entirely unrelated to the GHF5 cellulases, suggesting independent HGT within the two clades. GHF5 cellulases in cyst nematodes may consist of the catalytic domain alone or may be composed of the catalytic domain fused to a family 2 carbohydrate binding module (CBM), with the CBM found at the C-terminus of the protein. The CBM may aid in the binding of the protein to the substrate (Boraston et al., 2004), and is also found on other cell wall modifying proteins present in cyst nematodes such as expansins (below). The importance of cellulases to the nematode is illustrated in a variety of studies using RNA interference (RNAi); silencing of cellulase expression frequently results in a greatly reduced capacity to infect plants (e.g. Chen et al., 2005).

4.2.2 Degradation of pectin

Pectate lyases (PELs) play a critical role in degradation of pectin within cell walls. These enzymes cleave α-1,4-galacturonan, the major component of pectin backbone, by β-elimination rather than by hydrolysis (reviewed by Davis et al., 2011). All nematode pectate lyases identified to date are from polysaccharide lyase (PL) family 3. Like the cellulases, pectate lyases are widespread within Clade 12 of plant-parasitic nematodes, and have been identified in several species of cyst nematodes (e.g. Popeijus et al., 2000b; de Boer et al., 2002; Vanholme et al., 2007) as well as root-knot nematodes (e.g. Huang et al., 2005) and several migratory nematode species (e.g. Peng et al., 2013). In addition, similar sequences have also been found within the Clade 10 plant-parasitic nematode *Bursaphelenchus xylophilus* (Kikuchi et al., 2006). However, a detailed phylogenetic analysis of nematode pectate lyases suggested that more than one HGT event may have given rise to the sequences present in plant-parasitic nematodes (Danchin et al., 2010). Functional studies of pectate lyases have confirmed that cyst nematode pectate lyase sequences are biochemically active (Kudla et al., 2007) and that RNAi of these pectate lyases results in reduced infection levels (Vanholme et al., 2007).

Two other enzymes that may participate in degradation of pectin have been identified from cyst nematodes (Table 4.1). A GHF53 arabinogalactan endo-1,4-β-galactosidase from *Heterodera schachtii* has been characterized (Vanholme et al., 2009). The gene encoding this protein was expressed in the subventral gland cells, with expression peaking at early parasitic stages. Although the cyst nematode enzymes have not been biochemically characterized, if they have the same mode of action as other proteins of this type they are likely to hydrolyse β-1,4-galactan in branched regions of pectin, making the molecule more accessible to pectate lyases. Similarly, a GHF43 arabinase has been identified in the *G. pallida* genome sequence. Although no characterization of this protein has been carried out, it may hydrolyse α-1,5 linkages in arabinan polysaccharides, which are present as side chains of pectin and thus make the pectin backbone accessible to pectate lyase activity (Davis et al., 2011).

4.2.3 Non-enzymatic modifications of the plant cell wall

In addition to the enzymes described above, cyst nematodes produce effectors that encode proteins that may catalyse other changes to the plant cell wall or that act in a way that allows the CWDEs to operate more efficiently. Expansins are thought to act by disruption of non-covalent bonds between cell wall components, thus allowing better access for CWDEs. The first animal expansin was identified from *G. rostochiensis* (Qin et al., 2004) with similar sequences subsequently identified from other plant-parasitic nematodes. Although the *G. rostochiensis* sequence was shown to be biochemically active in terms of loosening cell walls, a recent study has suggested that expansins targeted to the apoplast are also able to suppress host defence responses induced by an elicitor from *Phytophthora infestans* and by the Bs2 NB-LRR resistance gene (Ali et al., 2015).

Many plant-parasitic nematode CWDEs are multi-domain proteins consisting of a catalytic domain fused to a CBM2 domain. Such domains may be present on cyst nematode cellulases and expansins but are also present as solitary CBM domain proteins not fused to any catalytic domain. Expression patterns of some of these CBM sequences in *G. pallida* are consistent with a role in facilitating activity of CWDEs (Thorpe *et al.*, 2014) but it has been shown that at least one CBM from *H. schachtii* has a different functional role (Hewezi *et al.*, 2008). This CBM interacts with a host pectin methyl esterase (PME) and overexpression of either the CBM or the PME increased infection by the nematode. These data suggest that the PME is a susceptibility factor for cyst nematodes and that the CBM interacts with this protein to promote its activity. This interaction may facilitate targeted modifications of the plant cell wall within the nematode feeding site. Analysis of expression data from *G. pallida* shows that one CBM in this nematode is upregulated at parasitic stages and, therefore, it may have a similar functional role (Thorpe *et al.*, 2014). The data for expansins and CBMs from cyst nematodes demonstrate that proteins involved in plant cell wall modification may become adapted for different functional roles in the context of plant parasitism.

4.3 Suppression of Host Defences

Cyst nematodes are biotrophic pathogens. Each nematode can only induce a single syncytium and, once formed, this structure needs to be kept alive while the nematode feeds, a period that can be as long as 6 weeks for some cyst nematodes. This prolonged biotrophic phase is extremely unusual in a plant pathogen and is comparable to that of the classical biotrophic rusts and powdery mildews.

The function and evolution of plant defences can be summarized by the zigzag model (Jones and Dangl, 2006). Although the utility of this model has been challenged (Pritchard and Birch, 2014), it remains a powerful and widely used method for summarizing the basic principles of the function and evolution of the plant immune system. Plants are confronted with a constant barrage of potential pathogen threats. Most of these potential threats are defeated after activation of pattern triggered immunity (PTI) in response to detection of highly conserved pathogen molecules called pathogen associated molecular patterns (PAMPs). PAMPs are detected via cell surface pattern recognition receptors, which activate PTI. Successful biotrophic pathogens are able to suppress PTI using effectors that are secreted into the host. In order to counter this effector triggered susceptibility (ETS), plants use a second layer of immune receptors encoded by resistance genes that activate a strong localized defence response leading to effector triggered immunity (ETI). Many resistance genes (R) encode nucleotide-binding-leucine-rich repeat (NB-LRR) proteins. The strong selection pressure imposed by pathogens and plants on one another leads to constant coevolution of effectors and the cognate *NB-LRR* genes. The zigzag model is summarized in Fig. 4.2. Although the zigzag model was developed around microbial (i.e. bacterial, fungal and oomycete) pathogens, there is a great deal of evidence to support the idea that the principles behind it apply equally well to plant–nematode interactions. First, a nematode PAMP was recently identified (Manosalva *et al.*, 2015). This molecule, an ascaroside called ascr#18, induces expression of genes associated with PTI and activates defence responses in a wide range of plants that are effective against diverse pathogens. Second, effectors that suppress PTI have been identified from plant-parasitic nematodes (below). Finally, with the exception of soybean resistance genes to the soybean cyst nematode, *H. glycines* (Liu *et al.*, 2012; Cook *et al.*, 2012), many NB-LRRs have been identified against nematodes (reviewed in Caromel and Gebhardt, 2011; Kandoth and Mitchum, 2013) and the effector that is recognized by the *Gpa2* R gene has been determined (Sacco *et al.*, 2009). Each of these findings suggests that many of the processes governing the interactions between plants and plant-parasitic nematodes are similar to those that underpin plant–microbe interactions.

Assays have been developed that allow activation, and by extension suppression, of both PTI and ETI to be monitored. Defence responses associated with PTI can be assayed by measuring activation of specific marker genes or by physiological responses such as deposition of callose. Suppression of these responses was used as an assay for determining the role of a root-knot nematode calreticulin in suppression of PTI (Jaouannet *et al.*, 2013). Activation of ETI is frequently associated with a strong local cell death response. ETI can therefore be assayed by transiently co-expressing a resistance gene and its cognate avirulence gene and monitoring the

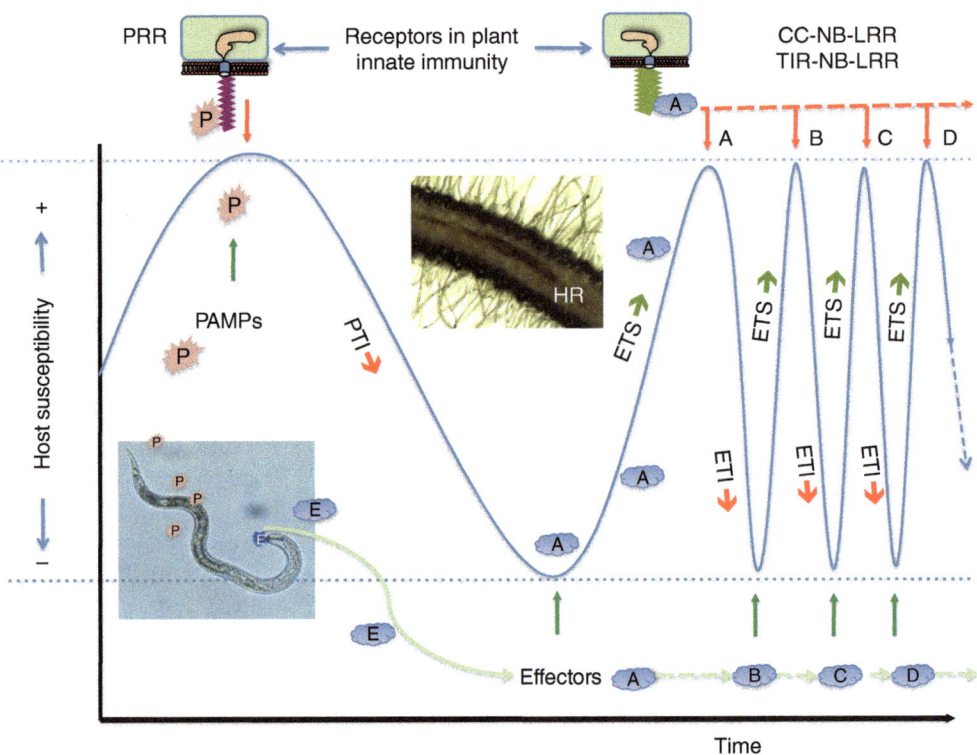

Fig. 4.2. The zigzag model (Jones and Dangl, 2006) as applied to plant–nematode interactions. PTI, pathogen (or PAMP) triggered immunity; ETI, effector triggered immunity; PRR, pattern recognition receptor; ETS, effector triggered susceptibility; HR, hypersensitive response. (From Smant and Jones, 2011.)

development of cell death within the tissues expressing these genes (Kamoun et al., 2003). Both PTI and ETI can be assayed by measurement of reactive oxygen species (ROS), which are associated with defence responses. The availability of these tools allows cyst nematode effectors to be screened in order to identify those that suppress these responses.

Effectors that suppress PTI and ETI have now been identified from cyst nematodes. Analysis of cyst nematode genome and transcriptome sequences has shown that a substantial gene family encoding SPRYSEC (secreted proteins with a SPRY domain) is present in these nematodes (e.g. Cotton et al., 2014), and several of these have been shown to suppress host defences. SPRYSEC 19 from *G. rostochiensis* suppresses cell death responses induced by the nematode resistance gene *Gpa2* as well as those induced by the related *Rx* resistance gene. The presence of SPRYSEC 19 also allowed an avirulent strain of *PVX* to infect plants in the presence of *Rx*, confirming the biological relevance of the suppression activity (Postma et al., 2012). Two SPRYSEC effectors from *G. pallida* suppressed *Gpa2*-induced defence responses but not those induced by other resistance genes (Mei et al., 2015). Although the mechanism of suppression remains unclear, SPRYSEC 19 interacts with the LRR domain of the *SW5* resistance gene (Postma et al., 2012). It is possible that SPRYSECs target resistance proteins directly, possibly explaining the specificity of their defence suppression activity.

In contrast to the highly specific effects of SPRYSECs, another cyst nematode effector (GrUBCEP12) has been identified that suppresses both PTI and ETI. This sequence is made up of a signal peptide, an ubiquitin domain and a short, but variable, C-terminal extension. The C-terminal extension may contain a nucleolar localization signal (Tytgat et al., 2004). This effector was first characterized from *H. schachtii* (Tytgat et al., 2004) and is also present in *G. pallida* (Thorpe et al., 2014), but the most detailed studies have been performed on the sequence from *G. rostochiensis*. The *G. rostochiensis* protein is cleaved *in planta*, and the released C-terminal extension peptide suppresses cell death induced by Gpa2 and Rx (Chronis et al., 2013). In addition, the peptide also suppresses flg22-induced defence responses including induction of marker genes and production of ROS, suggesting that it can also suppress PTI (Chen et al., 2013). Expression of the peptide in potato increases susceptibility to nematodes and to a bacterial pathogen, *Streptomyces scabies*, providing further evidence that the peptide suppresses general plant defence responses (Chronis et al., 2013). The broad activity of the GrUBCEP12 effector suggests that it may target the pathways that lead to generation of ROS.

A different effector that functions as a similarly broad-spectrum suppressor of plant defences has recently been identified from *H. glycines* (Noon et al., 2016). In heterologous immunosuppression assays, this GLAND18 effector suppressed the induction of both basal immunity and the hypersensitive cell death response (HR). Interestingly, the GLAND18 protein contains domains that are also present in the circumsporozoite protein (CSP), an immunosuppressant protein from malaria (*Plasmodium* spp.). These domains from the *Plasmodium* CSP were able to complement deletion mutants of the GLAND18 protein lacking these conserved regions, suggesting functional conservation. This effector is absent from other species of cyst nematodes, demonstrating that the ability to suppress host defences evolves rapidly.

All biotrophic pathogens need to suppress plant defences in order to infect their hosts and, although the precise molecular mechanisms by which they achieve this are varied, it has been suggested that diverse pathogens are likely to target a relatively small number of key host defence signalling proteins. Indeed, it has been shown that highly interconnected host proteins that interact with a large number of other plant proteins are more likely to be targets of effectors from bacterial and oomycete pathogens (Mukhtar et al., 2011). It can be hypothesized that nematodes are likely to target some of the same host proteins as other biotrophic pathogens in order to suppress host defences. In keeping with this, it has been shown that venom allergen proteins from *G. rostochiensis* suppress plant immunity mediated by cell surface receptors through targeting of the extracellular Rcr3 proteinase (Lozano-Torres et al., 2014). Rcr3 is also targeted by the Avr2 effector from the fungal pathogen *Cladosporium fulvum* and by two *Phytophthora infestans* secreted proteinase inhibitors EPIC 1 and 2 (Song et al., 2009). Although the Rcr3 protein is a virulence target for these pathogens it is guarded by the Cf-2 resistance receptor, which provides resistance against both *C. fulvum* and *G. rostochiensis* (Lozano-Torres et al., 2012). By contrast, Cf-2 does not provide resistance against *P. infestans*, suggesting that the effectors from this species are able to act without detection (Song et al., 2009). This is not the first example of an R gene that targets nematodes and other pathogens: *Mi* is active against aphids, whiteflies, tomato psyllids and root-knot nematodes (reviewed in Caromel and Gebhardt, 2011; Kandoth and Mitchum, 2013), suggesting that these diverse pathogens target a similar guarded virulence protein.

A variety of other effectors from cyst nematodes have been implicated in suppression of host defences on the basis of the plant proteins that they interact with in yeast 2 hybrid screens. The cyst nematode effector 4F01, which is similar to various members of the annexin family, interacts with an oxidoreductase protein implicated in plant defence responses (Patel et al., 2010). The 10A06 effector from *H. schachtii* interacts with a spermidine synthase. Overexpression of the effector or its target increases susceptibility to nematodes and causes upregulation of antioxidant proteins (Hewezi et al., 2010). The overexpression lines were also more susceptible to bacterial and viral pathogens and overexpression of the effector disrupted salicylic acid (SA) signalling. These lines of evidence strongly suggest a role in defence suppression for the 10A06 protein. Similarly, the 30C02 effector from *H. schachtii* interacts with a host β-1,3-endoglucanase and may suppress the normal role of this target protein as a pathogenesis-related protein (Hamamouch et al., 2012). As with other effectors, overexpression of the 30C02 effector in plants increases susceptibility to nematodes.

Chorismate mutase was one of the first effectors identified from any plant-parasitic nematode. Although the function of this protein remains uncertain, it is present in a wide range of nematodes including cyst (Jones et al., 2003) and root-knot (Lambert et al., 1999), as well as a variety of migratory endoparasitic nematodes (Bauters et al., 2014; Haegeman et al., 2011), suggesting a role essential to a variety of parasitic lifestyles. Chorismate mutase catalyses the conversion of chorismate to prephenate and therefore may reduce the production of SA from chorismate to suppress plant defence.

4.4 Manipulation of Host Biochemistry for Feeding Site Induction

Cyst nematodes induce the formation of a large, multinucleate feeding structure – the syncytium – in the roots of their host plants. A detailed description of the structure and development of the syncytium is provided by Sobczak and Golinowski (2011). During migration the J2 acts destructively, using the stylet mechanically to break host cells. However, once the nematode reaches root cells suitable for feeding site induction, its behaviour changes. Effectors are introduced into this cell, which induces a series of structural and biochemical changes. Cell wall openings are formed through widening of existing plasmodesmata followed by breakdown of the cell wall surrounding these areas. The cytoplasm of the initial cell becomes enriched with endoplasmic reticulum, mitochondria and other subcellular organelles, the central vacuole breaks down and the nucleus becomes enlarged. These are all changes indicative of a greatly increased metabolism. Cells surrounding the initial cell undergo similar changes and the protoplasts of these cells fuse with that of the initial syncytial cell in the region of the cell wall openings. This process is repeated until several hundred cells are incorporated into the syncytium.

It has become clear that cyst nematode effectors guide the unique formation of the syncytium by manipulating fundamental aspects of plant growth and development including mimicking meristem maintenance pathways and altering phytohormone transport and signalling. In plants, CLAVATA3/EMBRYO SURROUNDING REGION-like (CLE) peptides serve as ligands for LRR receptor-like proteins (RLPs) to regulate signalling pathways for shoot, root and vascular meristem maintenance by either promoting or inhibiting cell differentiation. The first animal CLE peptides were identified from *H. glycines* (Olsen and Skriver, 2003; Wang et al., 2005, 2010b) with similar sequences identified from other cyst nematodes (Lu et al., 2009; Wang et al., 2011). Secretion of CLE peptides is a unique feature of cyst nematode parasitism, granting these nematodes the ability to mimic endogenous plant peptides in a way that favours syncytium formation. *Globodera rostochiensis* CLEs were shown to be processed and post-translationally modified by plant cell machinery similar to plant CLE peptides (Chen et al., 2015). CLEs from *H. glycines*, *G. rostochiensis* and *H. schachtii* have been shown to interact with the receptor protein CLAVATA2 (CLV2) and the disruption of this gene in a variety of host plant species renders the plant more resistant to nematode infection (Guo et al., 2010; Replogle et al., 2011, 2013; Chen et al., 2015; Guo et al., 2015). Other receptors have also been implicated in cyst nematode CLE perception (Replogle et al., 2013) and simultaneous activation of parallel signalling pathways is not unexpected due to the complexity of this effector gene family in cyst nematodes. However, the molecular components of the downstream signalling pathways and their exact function in syncytium formation remain to be determined.

Auxin has long been known to play a key role in syncytium formation (Goverse et al., 2000; reviewed in Grunewald et al., 2009a). Local accumulation of this hormone occurs in response to differential regulation and relocalization of auxin transport proteins within developing feeding sites (Grunewald et al., 2009b; Lee et al., 2011). More recently, it was shown that cyst nematodes use their effectors to commandeer these changes in auxin levels and signalling. The 19C07 effector identified from *H. glycines* and *H. schachtii* was the first effector shown to interact with an auxin influx transporter protein (LAX3) to regulate auxin levels directly in feeding sites by promoting the activity of this transporter (Lee et al., 2011). This, in turn, mediates auxin-dependent expression of CWDEs required for feeding cell formation. Recent functional characterization of another novel effector unique to cyst nematodes confirmed that these nematodes are also capable of directly targeting auxin signalling pathways. The effector protein 10A07 is delivered to the cytoplasm of feeding cells where it undergoes host-mediated phosphorylation via its interaction with a plant protein kinase (Hewezi *et al.*, 2015). Phosphorylation of 10A07 is required for its subsequent trafficking to the nucleus, where it interferes with auxin signalling by interacting with AUX/IAA family member IAA16, a transcriptional repressor of auxin response factors (ARFs), to regulate expression of auxin-inducible genes. As a consequence of this interaction, it is proposed that binding of IAA16 to ARFs is inhibited, *ARF* expression is upregulated, and a subset of auxin-inducible genes is activated to promote syncytium formation (Hewezi, 2015). In support of this model, several IAA16-interacting ARFs shown to be upregulated in syncytia (Hewezi *et al.*, 2014) were also found to be upregulated in Arabidopsis plants overexpressing 10A07.

Recent studies have also confirmed an essential role for cytokinin in cyst nematode–plant interactions by demonstrating an early activation of cytokinin signalling pathways in nematode feeding cells and reduced nematode development on plants devoid of cytokinin receptors (Siddique *et al.*, 2015; Shanks *et al.*, 2016). Moreover, a nematode *tRNA-IPT* gene responsible for making cytokinin was found to be expressed predominantly in the pharyngeal gland cells of *H. schachtii* during the earliest stages of infection. Silencing of this gene prevented the nematode from activating cytokinin-dependent cell cycle changes required for proper feeding cell formation and provided the first evidence of a small molecule effector from a cyst nematode important for disease development (Siddique *et al.*, 2015).

Given that meristem maintenance pathways are tightly controlled and integrated with phytohormone signalling pathways, considerable interplay between nematode-secreted hormone and peptide signalling pathways probably exists. A better understanding of this cross-talk and its regulation of key downstream molecular components will significantly advance our understanding of syncytium formation.

4.5 Conclusions and Future Prospects

The field of nematode effector biology has made huge strides in the past decade, driven largely by increasing accessibility to tools for genome and transcriptome scale analysis. Researchers are no longer hampered by a lack of putative effector genes and are now able to focus efforts on studying gene function. High throughput sequencing will continue to have an impact on this field as new techniques that allow for the capture and subsequent high depth sequencing of genes of interest (e.g. Jupe *et al.*, 2013) are being applied to effectors. The ability to identify sequence differences in full effector complements of strains of a particular pathogen will assist in the characterization of the molecular basis of virulence in nematodes.

Each effector sequence that is functionally characterized acts as a probe for host biology, allowing important host processes to be described, sometimes for the first time. Therefore, given the nature of the changes induced by many sedentary endoparasites in the roots of their hosts, a better understanding of how nematode effectors are used to trigger these changes will lead to improved understanding of the host immune system in roots and plant development more generally.

4.6 Acknowledgements

The James Hutton Institute receives funding from the Rural and Environment Science and Analytical Services Division of the Scottish Government. This work benefited from interactions funded through COST Action FA1208. Research on cyst nematode effectors at the University of Missouri has been funded by the National Science Foundation, United States Department of Agriculture-National Institute of Food and Agriculture, United Soybean Board, Pioneer Hi-Bred, and North Central Soybean Research Program.

4.7 References

Ali, S., Magne, M., Chen, S., Côté, O., Stare, B.G., Obradovic, N., Jamshaid, L., Wang, X. Bélair, G. and Moffett, P. (2015) Analysis of putative apoplastic effectors from the nematode, *Globodera rostochiensis*, and identification of an expansin-like protein that can induce and suppress host defenses. *PLoS One* 10, e0115042. http://dx.doi.org/10.1371/journal.pone.0115042

Baldwin, J.G., Nadler, S.A. and Adams, B.J. (2004) Evolution of plant parasitism among nematodes. *Annual Review of Phytopathology* 42, 83–105. DOI: 10.1146/annurev.phyto.42.012204.130804

Bauters, L., Haegeman, A., Kyndt, T. and Gheysen, G. (2014) Analysis of the transcriptome of *Hirschmanniella oryzae* to explore potential survival strategies and host–nematode interactions. *Molecular Plant Pathology* 15, 352–363. DOI: 10.1111/mpp.12098

Boraston, A.B., Bolam, D.N., Gilbert, H.J. and Davies, G.J. (2004) Carbohydrate-binding modules: fine-tuning polysaccharide recognition. *Biochemical Journal* 382, 769–781. https://doi.org/10.1042/BJ20040892

Caromel, B. and Gebhardt, C. (2011) Breeding for nematode resistance: use of genomic information. In: Jones, J.T., Gheysen, G. and Fenoll, C (eds) *Genomics and Molecular Genetics of Plant-Nematode Interactions*. Springer Academic Publishers, Dordrecht, The Netherlands, pp. 492–465. DOI: 10.1007/978-94-007-0434-3_22

Chen, Q., Rehman, S., Smant, G. and Jones J.T. (2005) Functional analysis of pathogenicity proteins of the potato cyst nematode *Globodera rostochiensis* using RNAi. *Molecular Plant–Microbe Interactions* 18, 621–625. http://dx.doi.org/10.1094/MPMI-18-0621

Chen, S., Chronis, D. and Wang, X. (2013) The novel GrCEP12 peptide from the plant-parasitic nematode *Globodera rostochiensis* suppresses flg22-mediated PTI. *Plant Signaling and Behavior* 8, e25359. http://dx.doi.org/10.4161/psb.25359

Chen, S., Lang, P., Chronis, D., Zhang, S., De Jong, W.S., Mitchum, M.G. and Wang, X. (2015) In planta processing and glycosylation of a nematode CLE effector and its interaction with a host CLV2-like receptor to promote parasitism. *Plant Physiology* 167, 262–272. DOI:10.1104/pp.114.251637

Chronis, D., Chen, S., Lu, S., Hewezi, T., Carpenter, S.C.D., Loria, R., Baum, T.J. and Wang, X. (2013) A ubiquitin carboxyl extension protein secreted from a plant-parasitic nematode *Globodera rostochiensis* is cleaved *in planta* to promote plant parasitism. *The Plant Journal* 74, 185–196. DOI: 10.1111/tpj.12125

Cook, D.E., Lee, T.G., Guo, X. *et al*. (2012) Copy number variation of multiple genes at *Rhg1* mediates nematode resistance in soybean. *Science* 338, 1206–1209. DOI: 10.1126/science.1228746

Cotton, J.A., Lilley, C.J., Jones, L.M. *et al*. (2014) The genome and life-stage specific transcriptomes of *Globodera pallida* elucidate key aspects of plant parasitism by a cyst nematode. *Genome Biology* 15, R43. DOI: 10.1186/gb-2014-15-3-r43

Danchin, E.G., Rosso, M.N., Vieira, P., de Almeida-Engler, J., Coutinho, P.M., Henrissat, B. and Abad, P. (2010) Multiple lateral gene transfers and duplications have promoted plant parasitism ability in nematodes. *Proceedings of the National Academy of Sciences of the United States of America* 107, 17651–17656. DOI:10.1073/pnas.1008486107

Davis, E.L., Hussey, R.S. and Baum, T.J. (2004) Getting to the roots of parasitism by nematodes. *Trends in Parasitology* 20, 134–141. http://dx.doi.org/10.1016/j.pt.2004.01.005

Davis, E.L., Haegeman, A. and Kikuchi, T. (2011) Degradation of the plant cell wall by nematodes. In: Jones, J.T., Gheysen, G. and Fenoll, C (eds) *Genomics and Molecular Genetics of Plant–Nematode Interactions*. Springer Academic Publishers, Dordrecht, The Netherlands, pp. 255–272. DOI: 10.1007/978-94-007-0434-3_12

de Boer, J.M., McDermott, J.P., Davis, E.L., Hussey, R.S., Popeijus, H., Smant, G. and Baum, T.J. (2002) Cloning of a putative pectate lyase gene expressed in the subventral esophageal glands of *Heterodera glycines*. *Journal of Nematology* 34, 9–11.

Dropkin, V. (1963) Cellulase in phytoparasitic nematodes. *Nematologica* 9, 444–454. DOI: 10.1163/187529263X00980

Elling, A.A., Davis, E.L., Hussey, R.S. and Baum, T.J. (2007) Active uptake of cyst nematode parasitism proteins into the plant cell nucleus. *International Journal for Parasitology* 37, 1269–1279. http://dx.doi.org/10.1016/j.ijpara.2007.03.012

Eves-van den Akker, S., Lilley, C.J., Jones, J.T. and Urwin, P.E. (2014) Identification and characterisation of a hyper variable apoplastic effector gene family of *G. pallida*. *PLoS Pathogens* 10(9), e1004391. http://dx.doi.org/10.1371/journal.ppat.1004391

Eves-van den Akker, S., Laetsch, D.R., Thorpe, P. *et al.* (2016) The genome of the yellow potato cyst nematode, *Globodera rostochiensis*, reveals insights into the bases of parasitism and virulence. *Genome Biology* 17, 124. DOI: 10.1186/s13059-016-0985-1

Gao, B., Allen, R., Maier, T., Davis, E.L., Baum, T.J. and Hussey, R.S. (2003) The parasitome of the phytonematode *Heterodera glycines*. *Molecular Plant–Microbe Interactions* 16, 720–726. http://dx.doi.org/10.1094/MPMI.2003.16.8.720

Goellner, M., Wang, X. and Davis, E.L. (2001) Endo-β-1,4-glucanase expression in compatible plant-nematode interactions. *The Plant Cell* 13, 2241–2255. DOI: 10.1105/tpc.010219

Goverse, A., Overmars, H., Engelbertink, J., Schots, A., Bakker, J. and Helder, J. (2000) Both induction and morphogenesis of cyst nematode feeding cells are mediated by auxin. *Molecular Plant–Microbe Interactions* 13, 1121–1129. http://dx.doi.org/10.1094/MPMI.2000.13.10.1121

Grunewald, W., Van Noorden, G., Van Isterdael, G., Beeckman, T., Gheysen, G. and Mathesius, U. (2009a) Manipulation of auxin transport in plant roots during *Rhizobium* symbiosis and nematode parasitism. *The Plant Cell* 21, 2553–2562. doi:10.1105/tpc.109.069617

Grunewald, W., Cannoot, B., Friml, J. and Gheysen, G. (2009b) Parasitic nematodes modulate PIN-mediated auxin transport to facilitate infection. *PLoS Pathogens* 5, e1000266. http://dx.doi.org/10.1371/journal.ppat.1000266

Guo, X., Chronis, D., De La Torre, C.M., Smeda, J., Wang, X. and Mitchum, M.G. (2015) Enhanced resistance to soybean cyst nematode *Heterodera glycines* in transgenic soybean by silencing putative CLE receptors. *Plant Biotechnology Journal* 13, 801–810. DOI: 10.1111/pbi.12313

Guo, Y., Han, L., Hymes, M., Denver, R. and Clark, S.E. (2010) CLAVATA2 forms a distinct CLE-binding receptor complex regulating *Arabidopsis* stem cell specification. *The Plant Journal* 63, 889–900. DOI: 10.1111/j.1365-313X.2010.04295.x

Haegeman, A., Jones, J.T. and Danchin, E. (2011) Horizontal gene transfer in nematodes: a catalyst for plant parasitism? *Molecular Plant–Microbe Interactions* 24, 879–887. http://dx.doi.org/10.1094/MPMI-03-11-0055

Hamamouch, N., Li, C., Hewezi, T., Baum, T.J., Mitchum, M.G., Hussey, R.S., Vodkin, L.O. and Davis, E.L. (2012) The interaction of the novel 30C02 cyst nematode effector protein with a plant β-1,3-endoglucanase may suppress host defence to promote parasitism. *Journal of Experimental Botany* 63, 3683–3695. https://doi.org/10.1093/jxb/ers058

Hewezi, T. (2015) Cellular signaling pathways and posttranslational modifications mediated by nematode effector proteins. *Plant Physiology* 169, 1018–1026. DOI:10.1104/pp.15.00923

Hewezi, T. and Baum, T. (2013) Manipulation of plant cells by cyst and root-knot nematode effectors. *Molecular Plant–Microbe Interactions* 26, 9–16. http://dx.doi.org/10.1094/MPMI-05-12-0106-FI

Hewezi, T., Howe, P., Maier, T.R., Hussey, R.S., Goellner Mitchum, M., Davis, E.L. and Baum, T.J. (2008) Cellulose binding protein from the parasitic nematode *Heterodera schachtii* interacts with *Arabidopsis* pectin methylesterase: cooperative cell wall modification during parasitism. *The Plant Cell* 20, 3080–3093. DOI: 10.1105/tpc.108.063065

Hewezi, T., Howe, P.J., Maier, T.R., Hussey, R.S., Mitchum, M.G., Davis, E.L. and Baum, T.J. (2010) Arabidopsis spermidine synthase is targeted by an effector protein of the cyst nematode *Heterodera schachtii*. *Plant Physiology* 152, 968–984. doi:10.1104/pp.109.150557

Hewezi, T., Piya, S., Richard, G. and Rice, J.H. (2014) Spatial and temporal expression patterns of auxin response transcription factors in the syncytium induced by the beet cyst nematode *Heterodera schachtii* in Arabidopsis. *Molecular Plant Pathology* 15, 730–736. DOI: 10.1111/mpp.12121

Hewezi, T., Juvale, P., Piya, S., Maier, T.R., Rambani, A., Hollis Rice, J., Mitchum, M.G., Davis, E.L., Hussey, R.S. and Baum, T.J. (2015) The novel cyst nematode effector protein 10A07 targets and recruits host post-translational machinery to mediate its nuclear trafficking and promote parasitism. *The Plant Cell* 27, 891–907. doi:10.1105/tpc.114.135327

Hogenhout, S.A., Van der Hoorn, R.A.L., Terauchi, R. and Kamoun, S. (2009) Emerging concepts in effector biology of plant-associated organisms. *Molecular Plant–Microbe Interactions* 22, 115–122. http://dx.doi.org/10.1094/MPMI-22-2-0115

Huang, G., Dong, R., Allen, R., Davis, E.L., Baum, T.J. and Hussey, R.S. (2005) Developmental expression and molecular analysis of two *Meloidogyne incognita* pectate lyase genes. *International Journal for Parasitology* 35, 685–692. http://dx.doi.org/10.1016/j.ijpara.2005.01.006

Hussey, R.S. and Mimms, C.W. (1990) Ultrastructure of esophogeal glands and their secretory granules in the root-knot nematode *Meloidogyne incognita*. *Protoplasma* 162, 99–107. DOI: 10.1007/BF01666501

Jaouannet, M., Maglianao, M., Arguel, M.J., Gourges, M., Evangelisti, E., Abad, P. and Rosso, M.N. (2013) The root-knot nematode calreticulin Mi-CRT is a key effector in plant defense suppression. *Molecular Plant–Microbe Interactions* 26, 97–105. http://dx.doi.org/10.1094/MPMI-05-12-0130-R

Jones, J.D.G. and Dangl, J.L. (2006) The plant immune system. *Nature* 444, 323–329. DOI: 10.1038/nature05286

Jones, J.T., Furlanetto, C., Bakker, E., Banks, B., Blok, V.C., Chen, Q. and Prior, A. (2003) Characterisation of a chorismate mutase from the potato cyst nematode *Globodera pallida*. *Molecular Plant Pathology* 4, 43–50. DOI: 10.1046/j.1364-3703.2003.00140.x

Jones, J.T., Furlanetto, C. and Kikuchi, T. (2005) Horizontal gene transfer from bacteria and fungi as a driving force in the evolution of plant parasitism in nematodes. *Nematology* 7, 641–646. https://doi.org/10.1163/156854105775142919

Jones, J.T., Kumar, A., Pylypenko, L.A. *et al.* (2009) Identification and functional characterisation of effectors in expressed sequence tags from various life cycle stages of the potato cyst nematode *Globodera pallida*. *Molecular Plant Pathology* 10, 815–828. DOI: 10.1111/j.1364-3703.2009.00585.x

Jupe, F., Witek, K., Verweij, W. *et al.* (2013) Resistance gene enrichment sequencing (RenSeq) enables reannotation of the NB-LRR gene family from sequenced plant genomes and rapid mapping of resistance loci in segregating populations. *The Plant Journal* 76, 530–544. DOI: 10.1111/tpj.12307

Kandoth, P. and Mitchum, M.G. (2013) War of the worms: how plants fight underground attacks. *Current Opinion in Plant Biology* 16, 457–463. http://dx.doi.org/10.1016/j.pbi.2013.07.001

Kamoun, S., Hamada, W. and Huitema, E. (2003) Agrosuppression: a bioassay for the hypersensitive response suited to high-throughput screening. *Molecular Plant–Microbe Interactions* 16, 7–13. http://dx.doi.org/10.1094/MPMI.2003.16.1.7

Kikuchi, T., Shibuya, H., Aikawa, T. and Jones, J.T. (2006) Cloning and characterization of pectate lyases secreted by the pine wood nematode *Bursaphelenchus xylophilus*. *Molecular Plant–Microbe Interactions* 19, 280–287. http://dx.doi.org/10.1094/MPMI-19-0280

Kudla, U., Milac, A.L., Qin, L., Overmars, H., Roze, E., Holterman, M., Petrescu, A.J., Goverse, A., Bakker, J. and Helder, J. (2007) Structural and functional characterization of a novel, host penetration-related pectate lyase from the potato cyst nematode *Globodera rostochiensis*. *Molecular Plant Pathology* 8, 293–305. DOI: 10.1111/j.1364-3703.2007.00394.x

Lambert, K.N., Allen, K.D. and Sussex, I.M. (1999) Cloning and characterization of an esophageal gland specific chorismate mutase from the phytoparasitic nematode *Meloidogyne javanica*. *Molecular Plant–Microbe Interactions* 12, 328–336. http://dx.doi.org/10.1094/MPMI.1999.12.4.328

Lee, C., Chronis, D., Kenning, C., Peret, B., Hewezi, T., Davis, E.L., Baum, T.J., Hussey, R.S., Bennett, M. and Mitchum, M.G. (2011) The novel cyst nematode effector protein 19C07 interacts with the Arabidopsis auxin influx transporter LAX3 to control feeding site development. *Plant Physiology* 155, 866–880. DOI: 10.1104/pp.110.167197

Liu, S., Kandoth, P.K., Warren, S.D. *et al.* (2012) A soybean cyst nematode resistance gene points to a new mechanism of plant resistance to pathogens. *Nature* 492, 256–260. DOI: 10.1038/nature11651

Lozano-Torres, J.L., Wilbers, R.H.P., Gawronski, P. *et al.* (2012) Dual disease resistance mediated by the immune receptor Cf-2 in tomato requires a common virulence target of a fungus and a nematode. *Proceedings of the National Academy of Sciences of the United States of America* 109, 10119–10124. DOI: 10.1073/pnas.1202867109

Lozano-Torres, J.L., Wilbers, R.H.P., Warmerdam, S. *et al.* (2014) Apoplastic venom allergen-like proteins of cyst nematodes modulate the activation of basal plant innate immunity by cell surface receptors. *PLoS Pathogens* 10(12), e1004569. http://dx.doi.org/10.1371/journal.ppat.1004569

Lu, S.W., Chen, S., Wang, J., Yu, H., Chronis, D., Mitchum, M.G. and Wang, X. (2009) Structural and functional diversity of *CLAVATA3/ESR* (*CLE*)-like genes from the potato cyst nematode *Globodera rostochiensis*. *Molecular Plant–Microbe Interactions* 22, 1128–1142. http://dx.doi.org/10.1094/MPMI-22-9-1128

Maier, T.R., Hewezi, T., Peng, J.Q. and Baum, T.J. (2013) Isolation of whole esophageal gland cells from plant-parasitic nematodes for transcriptomic analyses and effector identification. *Molecular Plant–Microbe Interactions* 26, 31–35. http://dx.doi.org/10.1094/MPMI-05-12-0121-FI

Manosalva, P., Manohar, M., von Reuss, S.H. *et al.* (2015) Conserved nematode signalling molecules elicit plant defences and pathogen resistance. *Nature Communications* 6, 7795. DOI: 10.1038/ncomms8795

Mei, Y., Thorpe, P., Guzha, A., Haegeman, A., Blok, V.C., MacKenzie, K., Gheysen, G., Jones, J.T. and Mantelin, S. (2015) Only a small subset of the SPRY domain gene family in *Globodera pallida* is likely to encode effectors, two of which suppress host defences induced by the potato resistance gene *Gpa2*. *Nematology* 17, 409–424. https://doi.org/10.1163/15685411-00002875

Mitchum, M.G., Hussey, R.S., Baum, T.J., Wang, X., Elling, A.A., Wubben, M. and Davis, E.L. (2013) Nematode effector proteins: an emerging paradigm of parasitism. *New Phytologist* 199, 879–894. DOI: 10.1111/nph.12323

Mukhtar, M.S., Carvunis, A.R., Dreze, M. *et al.* (2011) Independently evolved virulence effectors converge onto hubs in a plant immune system network. *Science* 333, 596–601. DOI: 10.1126/science.1203659

Noon, J.B., Qi, M., Sill, D.N., Muppirala, U., Eves-van den Akker, S., Maier, T., Dobbs, D., Mitchum, M.G., Hewezi, T. and Baum, T.J. (2016) A *Plasmodium*-like virulence effector of the soybean cyst nematode suppresses plant immunity. *New Phytologist* 212, 444–460. DOI: 10.1111/nph.14047

Olsen, A.N. and Skriver, K. (2003) Ligand mimicry? Plant-parasitic nematode polypeptide with similarity to CLAVATA3. *Trends in Plant Science* 8(2), 55–57. http://dx.doi.org/10.1016/S1360-1385(03)00003-7

Patel, N., Hamamouch, N., Li, C., Hewezi, T., Hussey, R., Baum, T., Mitchum, M. and Davis, E. (2010) A nematode effector protein similar to annexins in host plants. *Journal of Experimental Botany* 61, 235–248. https://doi.org/10.1093/jxb/erp293

Peng, H., Gao, B.L., Kong, L.A., Yu, Q., Huang, W.K., He, X.F., Long, H.B. and Peng, D. (2013) Exploring the host parasitism of the migratory plant-parasitic nematode *Ditylenchus destructor* by expressed sequence tags analysis. *PLoS One* 8, e69579. http://dx.doi.org/10.1371/journal.pone.0069579

Popeijus, H., Blok, V.C., Cardle, L., Bakker, E., Phillips, M.S., Helder, J., Smant, G. and Jones, J.T. (2000a) Analysis of genes expressed in second stage juveniles of the potato cyst nematodes *Globodera rostochiensis* and *G. pallida* using the expressed sequence tag approach. *Nematology* 2, 567–574. https://doi.org/10.1163/156854100509358

Popeijus, H., Overmars, H., Jones, J., Blok, V., Goverse, A., Helder, J., Schots, A., Bakker, J. and Smant, G. (2000b) Degradation of plant cell walls by a nematode. *Nature* 406, 36–37. DOI: 10.1038/35017641

Postma, W.J., Slootweg, E.J., Rehman, S., Finkers-Tomczak, A., Tytgat, T.O.G., van Gelderen, K., Lozano-Torres, J.L., Roosien, J., Pomp, R., van Schaik, C., Bakker, J., Goverse, A. and Smant, G. (2012) The effector SPRYSEC-19 of *Globodera rostochiensis* suppresses CC-NB-LRR mediated disease resistance in plants. *Plant Physiology* 160, 944–954. DOI: 10.1104/pp.112.200188

Pritchard, L. and Birch, P.R.J. (2014) The zigzag model of plant–microbe interactions: is it time to move on? *Molecular Plant Pathology* 15, 865–870. DOI: 10.1111/mpp.12210

Qin, L., Kudla, U., Roze, E., Goverse, A., Popeijus, H., Nieuwland, J., Overmars, H., Jones, J.T., Schots, A., Smant, G., Bakker, J. and Helder, J. (2004) Identification of a functional expansin, a non-enzymatic, cell wall-loosening agent, from the plant parasitic nematode *Globodera rostochiensis*. *Nature* 427, 30. DOI: 10.1038/427030a

Replogle, A., Wang, J., Bleckmann, A., Hussey, R.S., Baum, T.J., Shinichiro, S., Davis, E.L., Wang, X., Simon, R. and Mitchum, M.G. (2011) Nematode CLE signaling in Arabidopsis requires CLAVATA2 and CORYNE. *The Plant Journal* 65, 430–440. DOI: 10.1111/j.1365-313X.2010.04433.x

Replogle, A., Wang, J., Paolillo, V., Smeda, J., Kinoshita, A., Durbak, A., Tax, F.E., Wang, X., Sawa, S. and Mitchum, M.G. (2013) Synergistic interaction of CLAVATA1, CLAVATA2, and RECEPTOR-LIKE PROTEIN KINASE 2 in cyst nematode parasitism of *Arabidopsis*. *Molecular Plant–Microbe Interactions* 26, 87–96. http://dx.doi.org/10.1094/MPMI-05-12-0118-FI

Sacco, M.A., Koropacka, K., Grenier, E., Jaubert, M.J., Blanchard, A., Goverse, A., Smant, G. and Moffett, P. (2009) The cyst nematode SPRYSEC protein RBP-1 elicits Gpa2- and RanGAP2-dependent plant cell death. *PLoS Pathogens* 5, e1000564. http://dx.doi.org/10.1371/journal.ppat.1000564

Shanks, C.M., Rice, J.H., Zubo, Y., Schaller, G.E., Hewezi, T. and Kieber, J.J. (2016) The role of cytokinin during infection of *Arabidopsis thaliana* by the cyst nematode *Heterodera schachtii*. *Molecular Plant–Microbe Interactions* 29, 57–68. http://dx.doi.org/10.1094/MPMI-07-15-0156-R

Siddique, S., Radakovic, Z.S., De La Torre, C.M. et al. (2015) A plant-parasitic nematode releases cytokinins that control cell division and orchestrate feeding-site formation in host plants. *Proceedings of the National Academy of Sciences of the United States of America* 112, 12669–12674. DOI:10.1073/pnas.1503657112

Smant, G. and Jones, J.T. (2011) Suppression of plant defences by nematodes. In: Jones, J.T., Gheysen, G. and Fenoll, C. (eds) *Genomics and Molecular Genetics of Plant–Nematode Interactions*. Springer Academic Publishers, Dordrecht, The Netherlands, pp. 273–286. DOI: 10.1007/978-94-007-0434-3_13

Smant, G., Stokkermans, J.P., Yan, Y. et al. (1998) Endogenous cellulases in animals: isolation of beta-1, 4-endoglucanase genes from two species of plant-parasitic cyst nematodes. *Proceedings of the National Academy of Sciences of the United States of America* 95, 4906–4911.

Sobczak, M. and Golinowski, W. (2011) Cyst nematodes and syncytia. In: Jones, J.T., Gheysen, G. and Fenoll, C. (eds) *Genomics and Molecular Genetics of Plant–Nematode Interactions*. Springer Academic Publishers, Dordrecht, The Netherlands, pp. 61–82. DOI: 10.1007/978-94-007-0434-3_4

Song, J., Win, J., Tian, M.Y., Schornak, S., Kaschani, F., Ilyas, M., van der Hoorn, R.A.L. and Kamoun, S. (2009) Kamoun unrelated eukaryotic plant pathogens target the tomato defense protein Rcr3. *Proceedings of the National Academy of Sciences of the United States of America* 106, 1654–1659. DOI: 10.1073/pnas.0809201106

Thorpe, P., Mantelin, S., Cock, P.J.A. *et al.* (2014) Characterisation of the full effector complement of the potato cyst nematode *Globodera pallida*. *BMC Genomics* 15, 923.

Truong, N.M., Nguyen, C.-N., Abad, P., Quentin, M. and Favery, B. (2015) Function of root-knot nematode effectors and their targets in plant parasitism. *Advances in Botanical Research* 73, 293–324. http://dx.doi.org/10.1016/bs.abr.2014.12.010

Tytgat, T., Vanholme, B., De Meutter, J. *et al.* (2004) A new class of ubiquitin extension proteins secreted by the dorsal pharyngeal gland in plant parasitic cyst nematodes. *Molecular Plant–Microbe Interactions* 17, 846–852. http://dx.doi.org/10.1094/MPMI.2004.17.8.846

van Megen, H., van den Elsen, S., Holterman, M., Karssen, G., Mooyman, P., Bongers, T., Holovachov, O., Bakker, J. and Helder, J. (2009) A phylogenetic tree of nematodes based on about 1200 full-length small subunit ribosomal DNA sequences. *Nematology* 11, 927–950. https://doi.org/10.1163/156854109X456862

Vanholme, B., Van Thuyne, W., Vanhouteghem, K., De Meutter, J., Cannoot, B. and Gheysen, G. (2007) Molecular characterization and functional importance of pectate lyase secreted by the cyst nematode *Heterodera schachtii*. *Molecular Plant Pathology* 8, 267–278. DOI: 10.1111/j.1364-3703.2007.00392.x

Vanholme, B., Haegeman, A., Jacob, J., Cannoot, B. and Gheysen, G. (2009) Arabinogalactan endo-1,4-beta-galactosidase: a putative plant cell wall-degrading enzyme of plant-parasitic nematodes. *Nematology* 11, 739–747. https://doi.org/10.1163/156854109X404599

Wang, J., Lee, C., Replogle, A., Joshi, S., Korkin, D., Hussey, R., Baum, T.J., Davis, E.L., Wang, X. and Mitchum, M.G. (2010a) Dual roles for the variable domain in protein trafficking and host-specific recognition of *Heterodera glycines* CLE effector proteins. *New Phytologist* 187, 1003–1017. DOI: 10.1111/j.1469-8137.2010.03300.x

Wang, J., Joshi, S., Korkin, D. and Mitchum, M.G. (2010b) Variable domain I of nematode CLEs directs post-translational targeting of CLE peptides to the extracellular space. *Plant Signaling and Behavior* 5, 1–3. http://dx.doi.org/10.4161/psb.5.12.13774

Wang, J., Replogle, A., Hussey, R., Baum, T., Wang, X., Davis, E.L. and Mitchum, M.G. (2011) Identification of potential host plant mimics of CLV3/ESR (CLE)-like peptides from the plant-parasitic nematode *Heterodera schachtii*. *Molecular Plant Pathology* 12, 177–186. DOI: 10.1111/j.1364-3703.2010.00660.x

Wang, X., Meyers, D., Yan, Y., Baum, T., Smant, G., Hussey, R. and Davis, E. (1999) In planta localization of a β-1,4-endoglucanase secreted by *Heterodera glycines*. *Molecular Plant–Microbe Interactions* 12, 64–67. http://dx.doi.org/10.1094/MPMI.1999.12.1.64

Wang, X., Mitchum, M.G., Gao, B., Li, C., Diab, H., Baum, T.J., Hussey, R.S. and Davis, E.L. (2005) A parasitism gene from a plant-parasitic nematode with function similar to *CLAVATA3/ESR* (*CLE*) of *Arabidopsis thaliana*. *Molecular Plant Pathology* 6, 187–191. DOI: 10.1111/j.1364-3703.2005.00270.x

Yan, Y., Smant, G., Stokkermans, J., Qin, L., Helder, J., Baum, T., Schots, A. and Davis, E. (1998) Genomic organization of four β-1,4-endoglucanase genes in plant-parasitic cyst nematodes and its evolutionary implications. *Gene* 220, 61–70. http://dx.doi.org/10.1016/S0378-1119(98)00413-2

5 Biochemistry

David J. Chitwood and Edward P. Masler
USDA-ARS, Beltsville, Maryland, USA

5.1 Introduction	89
5.2 Lipids	90
5.3 Carbohydrates	93
5.4 Proteins	93
5.5 Conclusions and Future Prospects	96
5.6 References	96

5.1 Introduction

Part of the framework for effective control or management of cyst nematodes is the detailed understanding of their biology. There is also the fascination in examining mechanisms and their interactions from a purely academic perspective to acquire new knowledge. Some of the information on individual components of the biological system is dated. There is a paucity of research directly examining aspects of biochemistry, perhaps because of the difficulties associated with the microscopic size of cyst nematodes as experimental animals, and perhaps because funding agencies have not viewed these areas of research as top priority. However, burgeoning genomic data now provide opportunities to make inferences about biochemical pathways through bioinformatic analyses of nematode gene content and expression. There are numerous genome projects, either completed or in progress, and the emphasis is on how the genes and the proteins they encode function in the nematodes. Part of this essential knowledge is information on the biochemical parameters that define behaviour and functionality. The metabolic pathways of nematodes are relatively unspecialized, which provides biochemical flexibility in responding to environmental challenges (Barrett, 2011).

In the post-genomic era, a somewhat daunting task awaits a reviewer of the biochemistry of any organism. For a group of animals the subject of intense genomic and proteomic investigation, the process is truly overwhelming. Readers interested in the principles and outcomes of dedicated genomic approaches to organismal biology, host–parasite interactions and plant resistance are referred to other chapters in this volume. Regardless, investigations targeting specific biochemical aspects of cyst nematodes can illuminate remarkable aspects of their biology and can provide a direction for the development of biorational control strategies.

5.2 Lipids

Although usually regarded as food reserves or structural components of membranes, lipids have several other roles in organisms. Because plant-parasitic nematodes contain and utilize abundant endogenous lipid and because ample evidence exists for fatty acid catabolic pathways in free-living nematodes, the most important food reserves in plant-parasitic nematodes are regarded as lipid (Barrett, 1976). The abundance of lipids in cyst nematodes, the diversity of lipid structure and the importance of lipids in survival when the nematode is not feeding have attracted much interest to these molecules. Unfortunately, much remains to be learned in cyst nematodes about some of the more interesting lipid structures found in other nematode species.

5.2.1 Content and utilization

An examination of several investigations of cyst nematode lipids in females, cysts and freshly hatched second-stage juveniles (J2) indicates that the mature female is the most lipid-rich on a dry weight basis and that the cyst contains the least. Specifically, lipid levels have been quantified in the following: females of *Globodera tabacum solanacearum*, 29.4% of dry weight (Orcutt *et al.*, 1978); and of *Heterodera zeae* 35.6% (Chitwood *et al.*, 1985); cysts of *G. rostochiensis*, 9.6–12.8% (Holz *et al.*, 1998a, b) and 17.1% (Holz *et al.*, 1998c); J2 of *G. rostochiensis* 29.2%, 24.3% and 26.1% (Holz *et al.*, 1997, 1998a, b); *G. pallida*, 27.2% (Holz *et al.*, 1997); and *H. oryzae*, 24.4% and 27.4% (Reversat, 1976, 1980). In addition to differences among these values being a function of species, some effect surely arises from variations in analytical or cultural procedures. For example, Oil Red O revealed that *G. rostochiensis* J2 from plants grown at cooler soil temperatures (12°C, 16°C) contained less lipid than that from plants grown at warmer soil temperatures (20°C, 24°C; Tiilikkala, 1992). As with most nematode eggs, the eggshells of *G. rostochiensis*, *H. schachtii* and *H. glycines* have a lipid-rich inner layer (Perry *et al.*, 1982; Perry and Trett, 1986; Burgwyn *et al.*, 2003).

The large quantities of lipids in cyst nematodes suggest a role in food storage, and direct confirmatory evidence of such a function exists. Oil Red O staining (Robinson *et al.*, 1987a, b) indicated that J2 of *G. pallida* and *G. rostochiensis* utilize lipid reserves during storage and that movement and infectivity decline as lipid levels decrease. Interestingly, *G. pallida* consumed lipid resources more slowly than *G. rostochiensis*. Similarly, Reversat (1980) observed lipid levels to decline by 70% in *H. oryzae* J2 stored for 5 weeks at 28°C. High lipid levels in cyst nematode J2 appear very important for successful host finding and penetration; for example, when *G. rostochiensis* J2 were stored at 20°C for 5–10 days, their movement and invasion of host roots was severely compromised when lipid fell to 65% of its original level (Robinson *et al.*, 1987a). The failure of lipid-low J2 of *G. rostochiensis* and *G. pallida* to infect plant roots and the decline of lipid levels in J2 of as much as 17% annually has led to suggestions that lipid quantification in field populations could be considered as partial determinants of potential economic damage in a field (Storey, 1984; Atkinson *et al.*, 2001). The importance of lipid levels to cyst nematodes was similarly capitalized upon by Riga *et al.* (2001), who evaluated the effects of many plant residues or root diffusates for ability to stimulate *H. glycines* hatch, reduce J2 lipid levels and thereby decrease subsequent infection of soybean plants; the ryegrass *Lolium multiflorum* was the most effective. In another interesting lipid study with management implications, *G. rostochiensis* J2 continuously exposed to a sublethal concentration of the nematicide oxamyl contained greater lipid levels than nematodes maintained in water, indicating that the infectivity of the former was at least equal to that of the latter (Wright *et al.*, 1989). One possibility is that the lack of movement in treated nematodes resulted in less utilization of lipid resources.

Lipid utilization in *G. rostochiensis* J2 is also indicated by treating *G. rostochiensis* J2 with 4-pentenoic acid, an inhibitor of fatty acid β-oxidation, which decreased oxygen consumption and heat production (Butterworth *et al.*, 1989). Unfortunately, investigation of the effects of other lipid-related inhibitors on cyst nematodes has not been extensive.

5.2.2 Lipid class composition and fatty acids

A painstaking series of investigations conducted nearly 20 years ago in *G. rostochiensis* and *G. pallida* and host cultivars by Holz *et al.* (1997, 1998a, b, c, 1999) remains the most comprehensive analysis of cyst nematode lipids and their host plants, especially with respect to analysis of specific lipid types. The major lipid classes were neutral lipids (73%), phospholipids (11–13%) and free fatty acids (14–16%). Although such large quantities of free fatty acids were considered as possible artefacts related to analytical issues often observed by lipidologists, the fact that lipids from yellow cysts of *G. rostochiensis* contained about the same level of triacylglycerol plus free fatty acid (75% + 2%) as the total in brown cysts (54% + 21%) could reflect a lack of artefactual results (Gibson *et al.*, 1995). In *G. rostochiensis* J2, 95% of the neutral lipids are triacylglycerols, which are typical storage lipids (Holz *et al.*, 1997). A similar role in cysts is indicated by the fact that the proportion of neutral lipids in *G. rostochiensis* egg lipids decreased by 16% when cysts stored for 3 and 7 years at 4°C were compared (Holz *et al.*, 1998a).

Approximately 20 fatty acids were identified by Holz *et al.* (1998a, c) in the total lipid from *G. rostochiensis* and *G. pallida*, ranging from C_{14}–C_{22} (i.e. chain length) and primarily being unsaturated (77–81%). The three most abundant fatty acids (20:4, 20:1 and 18:1, chain length: number of unsaturated bonds) constituted over 60% of the total fatty acids. Comparison of potato root fatty acids with *G. rostochiensis* cyst fatty acids indicated that cyst nematodes desaturate and/or elongate host root fatty acids. Indeed, the transcriptome of early parasitic *G. pallida* compared to the J2 transcriptome revealed upregulation of several fatty acid elongation proteins and two acyl coenzyme A desaturases (Cotton *et al.*, 2014). Neutral lipid of J2 of *G. rostochiensis* stored in potato root diffusate for 13 days at 18°C contained less 20:4, 20:1 and 20:3 fatty acid than that from freshly hatched J2. Although the fatty acids in stored J2 may change during storage, the neutral lipid fatty acid compositions in cysts were remarkably unaffected by a 13-year storage at 4°C. However, after *G. rostochiensis* cysts were soaked in potato root diffusate, freshly hatched J2 contained much larger proportions of polyunsaturated fatty acids, especially 20:4 (arachidonic) acid. Because the latter is a known elicitor of phytoalexins in potato roots, this fatty acid could be involved in potato cyst nematode infection (Gibson *et al.*, 1995; Holz *et al.* 1998c).

The major fatty acids of *G. tabacum solanacearum* females were 20:4, 20:1 and 18:1 acids (Orcutt *et al.*, 1978), as in potato cyst nematodes. By contrast, the fatty acid composition of *H. glycines* (Sekora *et al.*, 2009) is strikingly different from that of the few *Globodera* spp. analysed thus far, perhaps reflecting the effects of host fatty acids, intrinsic biochemical machinery and/or growth temperature. Indeed, the substantial unsaturation in *G. rostochiensis* fatty acids is often regarded as an adaptation to cold temperatures (Gibson *et al.*, 1995). With respect to host influence, although 20:4 and 18:1 acids were abundant in *H. glycines* females, cysts and J2, 20:1 acid was not detected in females, which contained 16:0 as the major acid. The major fatty acid in cysts and J2 were 18:1 acids; the similarity between cyst and juvenile fatty acids reflects the origin of the juveniles from eggs within the cyst. One might expect the fatty acid profile of females to be richer in the fatty acids common in soybean roots; indeed, the 16:0 acid abundant in *H. glycines* females is the most abundant fatty acid in soybean roots (Zenoff *et al.*, 1994).

5.2.3 Phospholipids

Phospholipids are the major lipid in cellular membranes; ethanolamine, choline and serine phosphoglycerides and sphingomyelin are major phospholipids of *G. rostochiensis* eggs and cysts (Atkinson and Taylor, 1983; Gibson *et al.*, 1995). During maturation of cysts from the white to the brown stage, the proportion of choline phosphoglycerides has been shown to decrease by half (Gibson *et al.*, 1995). Because treatment of eggs with phospholipase A_2 has been shown to reduce the amounts of ethanolamine phosphoglycerides and also decrease hatching, a membrane phospholipid may be involved in the control of hatching (Atkinson and Taylor, 1983; Chapter 3, this volume).

5.2.4 Hydrocarbons

Approximately 0.13% of the dry weight of *G. t. solanacearum* females is hydrocarbon, with 15 straight chain C_{15}–C_{29} compounds and numerous as yet unidentified others occurring (Orcutt et al., 1978).

5.2.5 Sterols

Although sterols are important fluidity-mediating components of membranes in many organisms, their investigation in plant-parasitic nematodes has been impeded by their low concentrations, for example, 0.01% of dry weight in *G. t. solanacearum* and 0.05% in *H. zeae* (Orcutt et al., 1978; Chitwood et al., 1985). Sterols in *G. t. solanacearum* females comprised 11.5% cholesterol, 25.6% typical plant sterols and 62.9% plant sterols in which the sterol nucleus was saturated, most likely by the nematode (Orcutt et al., 1978). Comparable values for the 14 sterols identified in *H. zeae* females were 7.2%, 81.8% and 10.0% (Chitwood et al., 1985), thereby indicating differences in the extent of saturation of plant sterols by the two species. The results strongly suggested that the species remove the C_{24} side chain substituent of host plant sterols to form cholesterol, as do microbivorous species such as *Caenorhabditis elegans*, although highly selective uptake of host cholesterol by either species cannot be excluded; for example, corn root sterols contained 0.5% cholesterol (Chitwood et al., 1985).

The role of sterols in cyst nematodes is poorly known; in the plant-parasitic nematodes examined thus far, the low concentration of sterol and the high proportion of unsaturated fatty acid probably preclude a role in modulation of fluidity, except perhaps in highly localized places. *Heterodera glycines* contains two homologues of 17β-hydroxysteroid dehydrogenase genes, though their sequences resemble mammalian genes that are involved in fatty acid elongation more than steroid hormone biosynthesis (Skantar et al., 2006). The expression in *H. glycines* of a gene with high homology to steroid 5α-reductase is inhibited by a resistant soybean cultivar, perhaps interfering with the metabolism of phytosterols (Klink et al., 2009), although the gene may also be a very-long-chain enoyl-CoA reductase involved in fatty acid elongation. A putative steroid dehydrogenase is also secreted by *H. schachtii* J2 (Vanholme et al., 2006). Although the only steroid hormone identified in nematodes thus far (dafachronic acid) has not yet been identified in cyst nematodes, it would not be surprising if a similar compound were detected, as dafachronic acid is a ligand for the nuclear hormone receptor DAF-12 in *C. elegans*, and *H. glycines* contains a homologue of *daf-12* (Toh et al., 2014).

5.2.6 Ascarosides

Ascarosides are a class of glycolipids with rather remarkable roles in nematodes. An entertaining description of the discovery of ascarosides, accompanied by a reference to flaming intestinal worms, is provided by Ludewig and Schroeder (2013). Ascarosides are glycosides incorporating the dideoxy sugar ascarylose; in *Ascaris*, the very long-chain C_{27}–C_{35} aglycones create an abundant lipid believed to be integral to the remarkable resistance of *Ascaris* eggs to environmental stresses. For this reason, Clarke and Hennessy (1977) examined *G. rostochiensis* cysts for the presence of ascaroside but were unsuccessful in detecting ascarylose or long chain alcohols in lipid hydrolysates. Nearly three decades later, nematode ascaroside research was radically invigorated when Jeong et al. (2005) discovered that a *C. elegans* dauer pheromone was an ascaroside with much shorter (C_7) aglycone. Since then, researchers have discovered in various nematode species dozens of ascarosides with a diverse array of small aglycones typically ranging from C_3–C_{11}. Perhaps the short length of the chains explains the failure of Clarke and Hennessy to detect them, or perhaps the fact that the aglycones are mainly hydroxylated fatty acids instead of alcohols. Regardless, the only cyst nematode to have been examined for presence of ascarosides thus far is *H. glycines*, where Manosalva et al. (2015) detected in excretory/secretory products small quantities of the ascaroside known as ascr#18 ((10R)-10-[(3,6-dideoxy-α-l-*arabino*-hexopyranosyl)oxy]undecanoic acid). Interestingly, treatment of *Arabidopsis thaliana* roots with ascr#18 enhanced resistance to *H. schachtii*, suggesting that ascarosides may be detected as pathogen-associated molecular patterns (PAMPs) by plants to indicate the presence of nematodes. Clearly, further investigation of cyst nematode ascarosides is warranted.

5.3 Carbohydrates

Like lipids, carbohydrates are major energy reserves for nematodes (Barrett, 1976). Analysis of mono- and disaccharides in *G. rostochiensis* cysts revealed the presence of glucose, fructose and trehalose (Clarke and Hennessy, 1976); interestingly, acid hydrolysis of the cyst wall of *G. rostochiensis* and *H. schachtii* released glucosamine as the major aminosugar in the former but galactosamine in the latter (Clarke, 1968, 1970). Because of its well studied role in nematodes as a facilitator of cryoprotection, trehalose has been a somewhat frequent subject of interest in cyst nematodes. In a 2-year monthly analysis of eggs of *H. glycines*, trehalose levels were correlated with lower temperatures, consistent with a cryoprotective role (Yen *et al.*, 1996). In *G. rostochiensis* eggshells, permeability changes are believed to result in a release of trehalose from eggs, thereby increasing the water content and turgor in the juveniles within to assist with the hatching process (Perry, 1989, 2002; Chapter 3, this volume). In recent years, trehalose contents in eggs of *G. rostochiensis* and *G. pallida* have been examined as a predictor of egg viability in field soils, with the accuracy and usefulness of the method increasing with time to be competitive with time-consuming visual assessment (van den Elsen *et al.*, 2012; Beniers *et al.*, 2014; Ebrahimi *et al.*, 2015). Transcriptomic analysis of *H. avenae* revealed several sequences encoding trehalose phosphate synthase (Kumar *et al.*, 2014; Yang *et al.*, 2017), the key enzyme involved in trehalose biosynthesis.

As in other nematodes, sugar residues are on the surfaces of cyst nematodes (Forrest and Robertson, 1986; Sharon and Spiegel, 1996; Duncan *et al.*, 1997), where they likely are involved in biotic interactions. The polysaccharides typical of other nematodes also occur in cyst nematodes, such as chitin (a structural component) in eggshells of *H. glycines* (Burgwyn *et al.*, 2003) and *G. rostochiensis* (Clarke *et al.*, 1967), and glycogen in *H. glycines* and *G. rostochiensis* cysts (Clarke and Hennessy, 1977; Yen *et al.*, 1996; Li *et al.*, 2009). Interestingly, the occurrence of higher levels of glycogen in *H. glycines* cysts when soil temperatures are cooler could indicate that this storage carbohydrate is preferentially utilized as a food reserve during cooler months (Yen *et al.*, 1996).

5.4 Proteins

Proteins certainly represent a substantial area, encompassing small bioactive peptides, through enzymes of complex developmental and metabolic pathways, to large structural complexes for maintaining cellular integrity and organism structure. Consequently, any individual study of them must be limited. In considering cyst nematode proteins, interest in selected life events (feeding, locomotion, metabolism, parasitism, reproduction, etc.), and their known or suspected involvement in such events, has informed the selection of specific proteins or protein families for examination. Fusion of biochemical and molecular genetic approaches to the analysis of nematode biology is now routinely employed in related protein studies. This is a productive strategy for exploring the fundamental biology of cyst nematodes, and for identifying new avenues for their control. The strategy is greatly enhanced by continual increases in bioinformation provided by genome, secretome and transcriptome resources.

Analyses of gene expression changes occurring across various life stages of *Globodera* (Cotton *et al.*, 2014; Eves-van den Akker *et al.*, 2016; Palomares-Rius *et al.*, 2016) and *Heterodera* (Kumar *et al.*, 2014; Fosu-Nyarko *et al.*, 2016; Yang *et al.*, 2017) cyst nematode species provide global illustrations useful for identifying shifts in protein type levels, and insights into changes in developmental biochemistry. Equally important, such studies can lead to identification of specific genes and protein products that are involved with individual developmental events. As the number of such identifications increases, the exploration and understanding of fundamental cyst nematode biology will accelerate. This section demonstrates, with some specific examples, the combination of biochemical studies with genetic and bioinformatic information.

5.4.1 Lipid-binding proteins

Fatty acid- and retinoid-binding (FAR) proteins are a family of lipid-transporting proteins unique to nematodes and suspected to be involved in lipid transport and host–parasite interactions (Jones et al., 2009; Kennedy et al., 2013). Several FAR proteins and the genes encoding them are present in G. pallida, H avenae and H. filipjevi (Prior et al., 2001; Vanholme et al., 2006; Jones et al., 2009; Le et al., 2016; Qiao et al., 2016). Far gene expression appears localized to the hypodermis in H avenae (Le et al., 2016; Qiao et al., 2016), reflecting the immunolocalization of G. pallida FAR on the J2 surface (Prior et al., 2001). Other lipid-conjugating genes in cyst nematodes include those in G. pallida and G. rostochiensis encoding protein prenyltransferases, which have been suggested as targets for evaluating the effects of known prenyltransferase inhibitors (Maurer-Stroh et al., 2003).

5.4.2 Feeding, hatching and parasitism

Some of the earliest success in molecular engineering of plant protection against cyst nematodes, by targeting specific protein type, involved manipulating the expression of transgenes encoding digestive protease inhibitors. Expression of oryzacystatin-1, a cysteine protease inhibitor from rice, in transformed tomato roots depressed G. pallida female production (Urwin et al., 1995). The expression of two inhibitors (oryzacystatin-1 and cowpea trypsin inhibitor), each targeting a different protease functional type, suppressed the growth of H. schachtii on transformed A. thaliana to a greater extent than did either transgene alone (Urwin et al., 1998). Lilley et al. (2004) demonstrated the benefits of targeted transgene expression by engineering cystatin expression at G. pallida feeding sites in potato. Urwin et al. (2003) were able to effect partial resistance of potato to G. pallida and G. rostochiensis by expressing cystatin in susceptible cultivars. Even more notable, expressing the same transgene in naturally resistant potato 'Sante' increased resistance to Globodera from partial to complete. The improvement of such engineering remains an important component of projected strategies for pest control (van Wyk et al., 2016).

Some of the most extensively examined features of cyst nematode biology are behaviours and biochemistry associated with infection and parasitism. The comprehensive analysis of nematode effectors, secretions from infective J2 that include enzymes and other proteins necessary for successful infection and the maintenance of the parasite in the plant, is particularly remarkable (Eves-van den Akker and Birch, 2016). This field is covered in detail in Chapter 4 of this volume and in recent reviews (Mitchum et al., 2013; Rehman et al., 2016), but is mentioned here as representing a benchmark for what can be achieved in parasitic cyst nematode research when genetic resources are exploited in exploring nematode biology and biochemistry.

Cotton et al. (2014) and Palomares-Rius et al. (2016) reported changes in G. pallida J2 gene expression associated with the shift from dormancy to hatch, and from free-living to infective stages. The great majority of detected G. pallida genes changing expression from dormancy to hatch (Cotton et al., 2014) were upregulated. Among them were those encoding enzymes and other proteins associated with cell membranes, secretion and ion transport, carbohydrate metabolism, and proteolysis (Cotton et al., 2014; Palomares-Rius et al., 2016). Carbohydrate hydrolases prominent in J2 are represented by cellulase, endoglucanase and pectate lyase, which aid in host invasion, and chitinase, which may be involved with hatch (Cotton et al., 2014; Kumar et al., 2014). Some metalloprotease genes upregulated (e.g. astacins; Cotton et al., 2014) may also be involved with hatch. Astacins are a large family of metalloproteases involved in multiple cellular processes including food processing and cuticle metabolism, as well as hatching (Park et al., 2010; Stepek et al., 2011).

The progression of newly infective J2 towards later parasitic stages includes significant upregulation of glutathione synthase genes. Glutathione is important for infective nematode development and reproduction (Cotton et al., 2014). Also upregulated are genes encoding enzymes involved with proteolysis and lipid metabolism. These changes suggest the beginning of feeding (Cotton et al., 2014) by the parasite. Among the protease genes detected are again astacins, and cysteine and serine proteases (Cotton et al., 2014), all of which may be involved with feeding and digestion, and provide potential control targets.

Stage changes in gene expression in *H. avenae* (Kumar *et al.*, 2014; Yang *et al.*, 2017) are globally similar to those in *G. pallida*. Pre-parasitic stages (egg, first-stage juvenile, free-living J2) strongly express genes involved with environmental information (sensory perception), nervous systems and signal transduction, whereas parasitic stages express genes encoding proteins involved with metabolism and cell growth. Female development includes upregulation of lipid metabolism, and chitin and strongly expressed vitellogenin genes (Cotton *et al.*, 2014; Kumar *et al.*, 2014), as egg formation and embryo nutritional demands take precedence.

5.4.3 Cell integrity

Some cytoskeletal proteins are differentially expressed in *H. avenae*, with spectraplakin upregulated in females *vs* J2, and dynein and radixin preferentially expressed in J2 (Kumar *et al.*, 2014). Since these are essential components of cellular structure and function (Labouesse, 2006; Oegema and Hyman, 2006; Suozzi *et al.*, 2012), they represent potential new targets for cyst nematode control. Tian *et al.* (2016) described the targeting of genes involved with cytoskeletal actin in *H. glycines*. Using the endogenous plant microRNA (miRNA) gene regulatory mechanism, *Glycine max* was transformed using vector constructs containing artificial miRNA (amiRNA) components targeting homologues of *C. elegans arx-1*, which encodes actin-related protein 3 (arp-3), and *C. elegans pfn-1*, which encodes profilin homologues. These actin-associated proteins are important at different stages of embryonic development.

Profilins bind to actin, affect actin dynamics and are widely expressed in both embryonic and adult stages, although distribution of three profilin homologues changes during development (Severson *et al.*, 2002; Polet *et al.*, 2006; Bernadskaya *et al.*, 2011). In the embryo, profilins are required for cytokinesis, and in the adult are associated with muscle and other tissues. Arp-3, as part of an arp-2/arp-3 complex, is essential for gastrulation and may be involved with cellular morphogenesis (Severson *et al.*, 2002; Polet *et al.*, 2006). Bernadskaya *et al.* (2011) demonstrated that Arp2/3 is essential for *C. elegans* adult tissue maintenance as well as embryonic development, and has a role in maintaining the epithelial junctional complex. RNAi feeding and depletion of Arp2/3 resulted in intestinal defects in adult *C. elegans*. In *H. glycines*, production of cysts and particularly eggs was significantly reduced on transformed *G. max* (Tian *et al.*, 2016), suggesting that silencing of *arx* and *pfn* genes may have affected J2 and adult stages and that the cytoskeleton and cellular motors are potential control targets.

5.4.4 Neuropeptides

Neuropeptides in nematodes are broadly classed into three families based upon structure and function: the FMRFamide-like proteins (FLPs), insulin-like proteins (ILPs) and neuropeptide-like proteins (NLPs). FLPs and ILPs are also discussed in Chapter 3, this volume. The most widely studied group comprises the FLPs, characterized by an amidated C-terminus with the structure -RFa. At least 32 *flp* genes are identified in *C. elegans*, with various subsets of these present in all parasitic nematodes, including cyst nematodes (Li and Kim, 2014; Peymen *et al.*, 2014). The 32 *C. elegans flp* genes encode 70 different peptides, with *flp* genes from plant-parasitic nematodes similarly prolific (McVeigh *et al.*, 2008; Li and Kim, 2014; Peymen *et al.*, 2014). FLPs are expressed throughout the nervous system, in sensory, motor and interneurons, affecting essentially all life processes including sensory perception, locomotion, mating, egg laying and others, and are ligands of G-protein coupled receptors (GPCRs). Consequently, nematode FLPs and associated GPCRs hold interest as control targets (McVeigh *et al.*, 2012), with FLP-GPCR discovery facilitated by genomic resources (McCoy *et al.*, 2014). Given the importance of FLPs to locomotion, it is not surprising that expression declines from motile to sedentary cyst nematode stages (Yang *et al.*, 2017), along with reduced expression of GPCRs (Cotton *et al.*, 2014; Kumar *et al.*, 2014).

Examinations of *G. pallida flp-32* expression (*Gp-flp-32*), its product peptide AMRNALVRFa and its related GPCR and gene (*Gp-flp-32R*) provide the first comprehensive functional analysis of FLP signalling in cyst nematodes (Atkinson et al., 2013, 2016). Silencing of *Gp-flp-32* causes increased locomotion in hatched J2, and the same effect is accomplished with *Gp-flp-32R* silencing (Atkinson et al., 2013). Another *flp* gene (*flp-11*) is found in *G. pallida*, *G. rostochiensis* and *H. glycines* (Peymen et al., 2014), and encodes multiple FLPs, one of which is a homologue of the FLP-32 sequence. The complex interaction of these peptides in *G. pallida* (Atkinson et al., 2016) reveals a striking biochemical functional complexity that had gone undetected and is only now being understood. This portends many more functional discoveries with regard to FLP control of cyst nematode behaviour and infectivity, as genes, gene products and nematode responses are examined in more complex studies. Expression of nine *flp* genes has been reported in *H. avenae*. Kumar et al. (2014) detected *flp* genes *-2*, *-3*, *-11* and *-18*, with *flp-18* particularly highly expressed in hatched J2. Yang et al. (2017) detected expression of *flp-1*, *-6*, *-12*, *-14*, *-16* and *-18*. Four of these five genes decreased in expression levels from pre- to post-parasitic stages by 4–6-fold; *flp-18* expression fell only 3-fold. Curiously, all of the *H. avenae flp* genes showed spikes in expression in the third-stage juvenile. The significance of this needs to be established and any possible association with early feeding explored.

Although ILPs have not been widely studied in cyst nematodes, several insulin genes have been found in *H. avenae* screens (Kumar et al., 2014). Expression levels change little between hatched J2 and female stages (Kumar et al., 2014). This is consistent with observations in *C. elegans* of dynamic and complex expression of *ins* genes throughout development (Ritter et al., 2013). As with all nematode neuropeptide genes, number coding follows the *C. elegans* system (Li and Kim, 2008). Of 40 genes encoding insulin-like peptides, one (*ins-17*) has been shown to be expressed in the *G. rostochiensis* secretome (Gahoi and Gautam, 2016). Involvement of insulin in *C. elegans* dauer formation and other functions (Li and Kim, 2008; Chapter 3, this volume), suggests that further elucidating expression of *ins* genes in cyst nematodes will uncover new molecular control targets.

NLPs do not fit into structural and functional characteristics that unify the FLP and ILP families. They are rather structurally diverse with comparatively little known about their functions in nematodes (Li and Kim, 2008; Warnock et al., 2017). However, through sequence mining coupled with bioassay, information from cyst nematodes is beginning to emerge. Four *nlp* genes *(-8, -14, -15, -21)* encoding 25 NLPs have been predicted from *G. pallida* genome BLASTp analysis (Warnock et al., 2017). Synthetic NLPs based upon the predicted sequences affected *G. pallida* chemotaxis, stylet thrusting and infectivity on tomato. Interestingly, infectivity was enhanced and reduced by different peptides that were encoded on the same gene (*nlp-21*). NLPs offer additional and functionally different alternatives to FLPs and NLPs in exploring these neuropeptide families for fundamental biology and cyst nematode control.

5.5 Conclusions and Future Prospects

A continuing theme in this chapter, and others in this volume, is that there is much value in exploring fundamental nematode biology by coupling molecular genetic information with well-defined and higher level biochemical pathways. There is a paucity of recent work involving classical biochemistry in cyst nematode studies compared with approaches applied to *C. elegans* and, for that matter, root-knot nematodes. In fact, research in *C. elegans* is frequently used to guide plant-parasitic nematode studies. Increased availability of cyst nematode expression data and transcriptome resources will make molecular-biochemical-basic biology couplings essential to any real progress in understanding cyst nematodes and acquiring insights for their control.

5.6 References

Atkinson, H.J. and Taylor, J.D. (1983) A calcium-binding sialoglycoprotein associated with an apparent eggshell membrane of *Globodera rostochiensis*. *Annals of Applied Biology* 102, 345–354. DOI: 10.1111/j.1744-7348.1983.tb02704.x

Atkinson, H.J., Holz, R.A., Riga, E., Main, G., Oros, R. and Franco, J. (2001) An algorithm for optimizing rotational control of *Globodera rostochiensis* on potato crops in Bolivia. *Journal of Nematology* 33, 121–125.

Atkinson, L.E., Stevenson, M., McCoy, C.J., Marks, N.J., Fleming, C., Zamanian, M., Day, T. A., Kimber, M.J., Maule, A.G. and Mousley, A. (2013) *flp-32* ligand/receptor silencing phenocopy faster plant pathogenic nematodes. *PLoS Pathogens* 9, e1003169. DOI: 10.1371/journal.ppat.1003169

Atkinson, L.E., Miskelly, I.R., Moffett, C.L. *et al.* (2016) Unraveling *flp-11/flp-32* dichotomy in nematodes. *International Journal for Parasitology* 46, 723–736. DOI: 10.1016/j.ijpara.2016.05.010

Barrett, J. (1976) Energy metabolism in nematodes. In: Croll, N.A. (ed.) *The Organization of Nematodes*. Academic Press, London, pp. 11–70.

Barrett, J. (2011) Biochemistry of survival. In: Perry, R.N. and Wharton, D.A. (eds) *Molecular and Physiological Basis of Nematode Survival*. CAB International, Wallingford, UK, pp. 282–310.

Beniers, J.E., Been, T.H., Mendes, O., van Gent-Pelzer, M.P.E. and van der Lee, T.A.J. (2014) Quantification of viable eggs of the potato cyst nematodes (*Globodera* spp.) using either trehalose or RNA-specific real-time PCR. *Nematology* 16, 1219–1232. DOI: 10.1163/15685411-00002848

Bernadskaya, Y.Y., Patel, F.B., Hsu, H.-T. and Soto, M. (2011) Arp3/3 promotes junction formation and maintenance in the *Caenorhabditis elegans* intestine by regulating membrane association of apical proteins. *Molecular Biology of the Cell* 22, 2886–2899. DOI: 10.1091/mbc.E10-10-0862

Burgwyn, B., Nagel, B., Ryerse, J. and Bolla, R.I. (2003) *Heterodera glycines*: eggshell ultrastructure and histochemical localization of chitinous components. *Experimental Parasitology* 104, 47–53. DOI: 10.1016/S0014-4894(03)00118-8

Butterworth, P.E., Perry, R.N. and Barrett, J. (1989) The effects of specific metabolic inhibitors on the energy metabolism of *Globodera rostochiensis* and *Panagrellus redivivus*. *Revue de Nématologie* 12, 63–67.

Chitwood, D.J., Hutzell, P.A. and Lusby, W.R. (1985) Sterol composition of the corn cyst nematode, *Heterodera zeae*, and corn roots. *Journal of Nematology* 17, 64–68.

Clarke, A.J. (1968) The chemical composition of the cyst wall of the potato cyst-nematode, *Heterodera rostochiensis*. *Biochemical Journal* 108, 221–224.

Clarke, A.J. (1970) The composition of the cyst wall of the beet cyst-nematode *Heterodera schachtii*. *Biochemical Journal* 118, 315–318.

Clarke, A.J. and Hennessy, J. (1976) The distribution of carbohydrates in cysts of *Heterodera rostochiensis*. *Nematologica* 22, 190–195. DOI: 10.1163/187529276X00283

Clarke, A.J. and Hennessy, J.A. (1977) Lipids of *Globodera rostochiensis*. *Rothamsted Experimental Station Report for 1976 Part 1*, p. 205.

Clarke, A.J., Cox, P.M. and Shepherd, A.M. (1967) The chemical composition of the egg shells of the potato cyst-nematode, *Heterodera rostochiensis* Woll. *Biochemical Journal* 104, 1056–1060.

Cotton, J.A., Lilley, C.J., Jones, L.M. *et al.* (2014) The genome and life-stage specific transcriptomes of *Globodera pallida* elucidate key aspects of plant parasitism by a cyst nematode. *Genome Biology* 15, R43. DOI: 10.1186/gb-2014-15-3-r43

Duncan, L.H., Robertson, L., Robertson, W.M. and Kusel, J.R. (1997) Isolation and characterization of secretions from the plant-parasitic nematode *Globodera pallida*. *Parasitology* 115, 429–438.

Ebrahimi, N., Viaene, N. and Moens, M. (2015) Optimizing trehalose-based quantification of live eggs in potato cyst nematodes (*Globodera rostochiensis* and *G. pallida*). *Plant Disease* 99, 947–953. DOI: 10.1094/PDIS-09-14-0940-RE

Eves-van den Akker, S. and Birch, P.R.J. (2016) Opening the effector protein toolbox for plant-parasitic cyst nematode interactions. *Molecular Plant* 9, 1451–1453. DOI: 10.1016/j.molp.2016.09.008

Eves-van den Akker, S., Laetsch, D.R., Thorpe, P. *et al.* (2016) The genome of the yellow potato cyst nematode, *Globodera rostochiensis*, reveals insights into the basis of parasitism and virulence. *Genome Biology* 17, 124. DOI: 10.1186/s13059-016-0985-1

Forrest, J.M.S. and Robertson, W.M. (1986) Characterization and localization of saccharides on the head region of four populations of the potato cyst nematode *Globodera rostochiensis* and *G. pallida*. *Journal of Nematology* 18, 23–25.

Fosu-Nyarko, J., Nicol, P., Naz, F. *et al.* (2016) Analysis of the transcriptome of the infective stage of the beet cyst nematode, *H. schachtii*. *PLoS ONE* 11, e0147511. DOI: 10/1371/journal.pone.0147511

Gahoi, S. and Gautam, B. (2016) Identification and analysis of insulin like peptides in nematode secretomes provide targets for parasite control. *Bioinformation* 12, 412–415. DOI: 10.6026/97320630012412

Gibson, D.M., Moreau, R.A., McNeil, G.P. and Brodie, B.B. (1995) Lipid composition of cyst stages of *Globodera rostochiensis*. *Journal of Nematology* 27, 304–311.

Holz, R.A., Wright, D.J. and Perry, R.N. (1997) The lipid content and fatty acid composition of hatched second-stage juveniles of *Globodera rostochiensis* and *G. pallida*. *Fundamental and Applied Nematology* 20, 291–298.

Holz, R.A., Wright, D.J. and Perry, R.N. (1998a) The effect of long-term storage on the lipid reserves and fatty acid composition of cysts and hatched juveniles of *Globodera rostochiensis* and *G. pallida*. *Journal of Helminthology* 72, 133–141. DOI: 10.1017/S0022149X0001631X

Holz, R.A., Wright, D.J. and Perry, R.N. (1998b) The influence of the host plant on lipid reserves of *Globodera rostochiensis*. *Nematologica* 44, 153–169. DOI: 10.1163/005325998X00045

Holz, R.A., Wright, D.J. and Perry, R.N. (1998c) Changes in the lipid content and fatty acid composition of 2nd-stage juveniles of *Globodera rostochiensis* after rehydration, exposure to the hatching stimulus and hatch. *Parasitology* 116, 183–190.

Holz, R.A., Troth, K. and Atkinson, H.J. (1999) The influence of potato cultivar on lipid content and fecundity of Bolivian and British populations of *Globodera rostochiensis*. *Journal of Nematology* 31, 357–366.

Jeong, P.Y., Jung, M., Yim, Y.H., Kim, H., Park, M., Hong, E., Lee, W., Kim, Y.H., Kim, K. and Paik, Y.K. (2005) Chemical structure and biological activity of the *Caenorhabditis elegans* dauer-inducing pheromone. *Nature* 433, 541–545. DOI: 10.1038/nature03201

Jones, J.T., Kumar, A., Pylypenko, L.A. *et al.* (2009) Identification and functional characterization of effectors in expressed sequence tags from various life cycle stages of the potato cyst nematode *Globodera pallida*. *Molecular Plant Pathology* 10, 815–823. DOI: 10.1111/j.1364-3703.2009.00585.x

Kennedy, M.W., Córsico, B., Cooper, A. and Smith, B.O. (2013) The unusual lipid-binding proteins of nematodes: NPAs, nemFABPs and FARs. In: Kennedy, M.W. and Harnett, W. (eds) *Parasitic Nematodes: Molecular Biology, Biochemistry and Immunology*, 2nd edn. CAB International, Wallingford, UK, pp. 397–412.

Klink, V.P., Hosseini, P., MacDonald, M.H., Alkharouf, N.W. and Matthews, B.F. (2009) Population-specific gene expression in the plant pathogenic nematode *Heterodera glycines* exists prior to infection and during the onset of a resistant or susceptible reaction in the roots of the *Glycine max* genotype Peking. *BMC Genomics* 10, 111. DOI: 10.1186/1471-2164-10-111

Kumar, M., Gantasala, N.P., Roychowdhury, T. *et al.* (2014) De novo transcriptome sequencing and analysis of the cereal cyst nematode, *Heterodera avenae*. *PLoS ONE* 9(5), e96311. DOI: 0.1371/journal.pone.0096311

Labouesse, M. (2006) Epithelial junctions and attachments. In: The *C. elegans* Research Community (eds) *WormBook*. WormBook, DOI: 10.1895/wormbook.1.56.1. Available at: www.wormbook.org (accessed 23 January 2018).

Le, X., Wang, X., Guan, T., Ju, Y. and Li, H. (2016) Isolation and characterization of a fatty acid- and retinoid-binding protein from the cereal cyst nematode *Heterodera avenae*. *Experimental Parasitology* 167, 94–102. DOI: 10.1016/j.exppara.2016.05.009

Li, C. and Kim, K. (2008) Neuropeptides. In: The *C. elegans* Research Community (eds) *WormBook*. WormBook, DOI: 10.1895/wormbook.1.142.1, www.wormbook.org.

Li, C. and Kim, K. (2014) Family of FLP peptides in *Caenorhabditis elegans* and related nematodes. *Frontiers in Endocrinology* 5, 150. DOI: 10.3389/fendo.2014.00150

Li, X., Wu, H., Shi, L., Wang, Z. and Liu, J. (2009) Comparative studies on some physiological and biochemical characters in white and brown cysts of *Heterodera glycines* race 4. *Nematology* 11, 465–470. DOI: 10.1163/156854109X447033

Lilley, C.J., Urwin, P.E., Johnston, K.A. and Atkinson, H.J. (2004) Preferential expression of a plant cystatin at nematode feeding sites confers resistance to *Meloidogyne incognita* and *Globodera pallida*. *Plant Biotechnology Journal* 2, 3–12. DOI: 10.1046/j.1467-7652.2003.00037.x

Ludewig, A.H. and Schroeder, F.C. (2013) Ascaroside signaling in *C. elegans*. In: The *C. elegans* Research Community (eds) *WormBook*. WormBook, DOI: 10.1895/wormbook.1.155.1.

Manosalva, P., Manohar, M., von Reuss, S.H. *et al.* (2015) Conserved nematode signalling molecules elicit plant defenses and pathogen resistance. *Nature Communications* 6, 7795. DOI: 10.1038/ncomms8795

Maurer-Stroh, S., Washietl, S. and Frank, E. (2003) Protein prenyltransferases: anchor size, pseudogenes and parasites. *Biological Chemistry* 384, 977–989. DOI: 10.1515/BC.2003.110

McCoy, C.J., Atkinson, L.E., Zamanian, M. *et al.* (2014) New insights into the FLPergic complements of parasitic nematodes: informing deorphanisation approaches. *EuPA Open Proteomics* 3, 262–272. DOI: 10.1016/j.euprot.2014.04.002

McVeigh, P., Alexander-Bowman, S., Veal, E. *et al.* (2008) Neuropeptide-like protein diversity in phylum Nematoda. *International Journal for Parasitology* 38, 1493–1503. DOI: 10.1016/j.ijpara.2008.05.006

McVeigh, P., Atkinson, L., Marks, N.J. *et al.* (2012) Parasite neuropeptide biology: seeding rational drug target design? *International Journal for Parasitology: Drugs and Drug Resistance* 2, 76–91. DOI: 10.1016/j.ijpara.2008.05.006

Mitchum, M., Hussey, R.S., Baum, T.J. *et al.* (2013) Nematode effector proteins: an emerging paradigm of parasitism. *New Phytologist* 199, 879–894. DOI: 10.1111/nph.12323

Oegema, K. and Hyman, A.A. (2006) Cell division. In: The *C. elegans* Research Community (eds) *WormBook*. WormBook, DOI: 10.1895/wormbook.1.72.1, www.wormbook.org.

Orcutt, D.M., Fox, J.A. and Jake, C.A. (1978) The sterol, fatty acid, and hydrocarbon composition of *Globodera solanacearum*. *Journal of Nematology* 10, 264–269.

Palomares-Rius, J.E., Hedley, P., Cock, P.J.A. *et al.* (2016) Gene expression changes in diapause or quiescent potato cyst nematode, *Globodera pallida*, eggs after hydration or exposure to tomato root diffusate. *PeerJ*. DOI: 10.7717/peerj.1654

Park, J.-O., Pan, J., Mohrlen, F. *et al.* (2010) Characterization of the astacin family of metalloproteases in *C. elegans*. *BMC Developmental Biology* 10, 14 biomedcentral.com/1471-213X/10/14

Perry, R.N. (1989) Dormancy and hatching of nematode eggs. *Parasitology Today* 5, 377–383. DOI: 10.1016/0169-4758(89)90299-8

Perry, R.N. (2002) Hatching. In: Lee, D.L. (ed.) *The Biology of Nematodes*. Taylor and Francis, London, pp. 147–169.

Perry, R.N. and Trett, M.W. (1986) Ultrastructure of the eggshell of *Heterodera schachtii* and *H. glycines* (Nematoda: Tylenchida). *Revue de Nématologie* 9, 399–403.

Perry, R.N., Wharton, D.A. and Clarke, A.J. (1982) The structure of the egg-shell of *Globodera rostochiensis* (Nematoda: Tylenchida). *International Journal for Parasitology* 12, 481–485. DOI: 10.1016/0020-7519(82)90080-7

Peymen, K., Watteyne, J., Frooninckx, L. *et al.* (2014) The FMRFamide-like peptide family in nematodes. *Frontiers in Endocrinology* 5, 90. DOI: 10.3389/fendo.2014.00090

Polet, D., Lambrechts, A., Ono, K., Mah, A., Peelman, F., Vandekerckhove, J., Baillie, D.L., Ampe, C. and Ono, S. (2006) *Caenorhabditis elegans* expresses three functional profilins in a tissue-specific manner. *Cell Motility and the Cytoskeleton* 63, 14–28. DOI: 10.1002/cm.20102

Prior, A., Jones, J.T., Blok, V.C., Beauchamp, J., McDermott, L., Cooper, A. and Kennedy, M.W. (2001) A surface-associated retinol- and fatty acid-binding protein (Gp-FAR-1) from the potato cyst nematode *Globodera pallida*: lipid binding activities, structural analysis and expression pattern. *Biochemical Journal* 356, 387–394. DOI: 10.1042/bj3560387

Qiao, F., Luo, L., Peng, H. *et al.* (2016) Characterization of three novel fatty acid- and retinoid-binding protein genes (*Ha-far-1*, *Ha-far-2* and *Hf-far-1*) from the cereal cyst nematodes *Heterodera avenae* and *H. filipjevi*. *PLoS ONE* 11, e016003. DOI: 0.1371/journal.pone.0160003

Rehman, S., Gupta, V.K. and Goyal, A.K. (2016) Identification and functional analysis of secreted effectors from phytoparasitic nematodes. *BMC Microbiology* 14, 48. DOI: 10.1186/s12866-016-0632-8

Reversat, G. (1976) Etude de la composition biochimique globale des juvéniles des nématodes *Meloidogyne javanica* et *Heterodera oryzae*. *Cahiers ORSTOM Série Biologie* 11, 225–234.

Reversat, G. (1980) Effect of *in vitro* storage time on the physiology of second-stage juveniles of *Heterodera oryzae*. *Revue de Nématologie* 3, 233–241.

Riga, E., Welacky, T., Potter, J., Anderson, T., Topp, E. and Tenuta, A. (2001) The impact of plant residues on the soybean cyst nematode, *Heterodera glycines*. *Canadian Journal of Plant Pathology* 23, 168–173. DOI: 10.1080/07060660109506926

Ritter, A.D., Shen, Y., Bass, J.F. *et al.* (2013) Complex expression dynamics and robustness in *C. elegans* insulin networks. *Genome Research* 23, 954–965. DOI: 10.1101/gr.150466

Robinson, M.P., Atkinson, H.J. and Perry, R.N. (1987a) The influence of soil moisture and storage time on the motility, infectivity and lipid utilization of second stage juveniles of the potato cyst nematodes *Globodera rostochiensis* and *G. pallida*. *Revue de Nématologie* 10, 343–348.

Robinson, M.P., Atkinson, H.J. and Perry, R.N. (1987b) The influence of temperature on the hatching, activity and lipid utilization of second stage juveniles of the potato cyst nematodes *Globodera rostochiensis* and *G. pallida*. *Revue de Nématologie* 10, 349–354.

Sekora, N.S., Lawrence, K.S., Agudelo, P., van Santen, E. and McInroy, J.A. (2009) Using FAME analysis to compare, differentiate, and identify multiple nematode species. *Journal of Nematology* 41, 163–173.

Severson, A.F., Baillie, D.L. and Bowerman, B. (2002) A formin homology protein and a profiling are required for cytokinesis and arp2/3-independent assembly of cortical microfilaments in *C. elegans*. *Current Biology* 12, 2066–2075. DOI: 10.1016/S0960-9822(02)01355-6

Sharon, E. and Spiegel, Y. (1996) Gold-conjugated reagents for the labelling of carbohydrate-recognition domains and glycoconjugates on nematode surfaces. *Journal of Nematology* 28, 124–127.

Skantar, A.M., Guimond, N.A. and Chitwood, D.J. (2006) Molecular characterisation of two novel 17β-hydroxysteroid dehydrogenase genes from the soybean cyst nematode *Heterodera glycines*. *Nematology* 8, 321–333. DOI: 10.1163/156854106778493439

Stepek, G., McCormack, G., Birnie, A.K. and Page, A.P. (2011) The astascin metalloprotease moulting enzyme NAS-36 is required for normal cuticle ecdysis in free-living and parasitic nematodes. *Parasitology* 138, 237–243. DOI: 10.1017/S0031182010001113

Storey, R.M.J. (1984) The relationship between neutral lipid reserves and infectivity for hatched and dormant juveniles of *Globodera* spp. *Annals of Applied Biology* 104, 511–520. DOI: 10.1111/j.1744-7348.1984.tb03034.x

Suozzi, K.C., Wu, X. and Fuchs, E. (2012) Spectraplakins: master orchestrators of cytoskeletal dynamics. *Journal of Cell Biology* 197, 465–475. DOI: 10.1083/jcb.201112034.

Tian, B., Li, J., Oakley, T.R. *et al.* (2016) Host-derived artificial microRNA as an alternative method to improve soybean resistance to soybean cyst nematode. *Genes* 7, 122. DOI: 10.3390/genes7120122

Tiilikkala, K.A. (1992) Influence of soil temperature on initial energy reserves of *Globodera rostochiensis* larvae. *Fundamental and Applied Nematology* 15, 49–54.

Toh, S., Holbrook-Smith, D., Stokes, M.E., Tsuchiya, Y. and McCourt, P. (2014) Detection of parasitic plant suicide germination compounds using a high-throughput *Arabidopsis* HTL/KAI2 strigolactone perception system. *Chemistry and Biology* 21, 988–998. DOI: 10.1016/j.chembiol.2014.07.005

Urwin, P.E., Atkinson, H.J., Waller, D.A. and McPherson, M.J. (1995) Engineered oryzacystatin-1 expressed in hairy roots confers resistance to *Globodera pallida*. *The Plant Journal* 8, 121–131. DOI: 10.1046/j.1365-313X.1995.08010121.x

Urwin, P.E., McPherson, M.J. and Atkinson, H.J. (1998) Enhanced transgenic plant resistance to nematodes by dual proteinase inhibitor constructs. *Planta* 204, 472–479. DOI: 10.1007/s004250050281

Urwin, P.E., Green, J. and Atkinson, H.J. (2003) Expression of a plant cystatin confers partial resistance to *Globodera*, full resistance is achieved by pyramiding a cystatin with natural resistance. *Molecular Breeding* 12, 263–269. DOI: 10.1023/A:1026352620308

van den Elsen, S., Ave, M., Schoenmakers, N., Landeweert, R., Bakker, J. and Helder, J. (2012) A rapid, sensitive, and cost-efficient assay to estimate viability of potato cyst nematodes. *Phytopathology* 102, 140–146. DOI: 10.1094/PHYTO-02-11-0051

van Wyk, S.G., Kunert, K.J., Cullis, C.A., Pillay, P., Makgopa, M.E., Schlüter, U. and Vorster, B.J. (2016) Review: the future of cystatin engineering. *Plant Science* 246, 119–127. DOI: 10.1016/j.plantsci.2016.02.016.

Vanholme, B., Mitreva, M., Van Criekinge, W., Logghe, M., Bird, D., McCarter, J. and Gheysen, G. (2006) Detection of putative secreted proteins in the plant-parasitic nematode *Heterodera schachtii*. *Parasitology Research* 98, 414–424. DOI: 10.1007/s00436-005-0029-3

Warnock, N.D., Wilson, L., Patten, C., Fleming, C.C., Maule, A.G. and Dalzell, J.J. (2017) Nematode neuropeptides as transgenic nematicides. *PLoS Pathogens* 13(2), e1006237. DOI: 10.1371/journal.ppat.1006237

Wright, D.J., Roberts, I.T.J. and Evans, S.G. (1989) Effect of the nematicide oxamyl on lipid utilization and infectivity in *Globodera rostochiensis*. *Parasitology* 98, 151–154. DOI: 10.1017/S0031182000059795

Yang, D., Chen, C., Liu, Q. and Jian, H. (2017) Comparative analysis of pre- and post-parasitic transcriptomes and mining pioneer effectors of *Heterodera avenae*. *Cell and Bioscience* 7, 11. DOI: 10.1186/s13578-017-0138-6

Yen, J.-H., Niblack, T.A., Karr, A.L. and Wiebold, W.J. (1996) Seasonal biochemical changes in eggs of *Heterodera glycines* in Missouri. *Journal of Nematology* 28, 442–450.

Zenoff, A.M., Hilal, M., Galo, M. and Moreno, H. (1994) Changes in roots lipid composition and inhibition of the extrusion of protons during salt stress in two genotypes of soybean resistant or susceptible to stress. Varietal differences. *Plant and Cell Physiology* 35, 729–735. DOI: 10.1093/oxfordjournals.pcp.a078650

6 Role of Population Dynamics and Damage Thresholds in Cyst Nematode Management

George W. Bird[1], Inga A. Zasada[2] and Gregory L. Tylka[3]

[1]Michigan State University, East Lancing, Michigan, USA; [2]USDA-ARS, Corvallis, Oregon, USA; [3]Iowa State University, Ames, Iowa, USA

6.1 Introduction	101
6.2 *Heterodera schachtii* (Sugar Beet Cyst Nematode)	103
6.3 *Globodera* spp. (Potato Cyst Nematodes)	107
6.4 *Heterodera glycines* (Soybean Cyst Nematode)	112
6.5 The Rest of the Heteroderinae	117
6.6 Conclusions and Future Prospects	118
6.7 References	119

6.1 Introduction

Nematode population dynamics and damage thresholds are applied aspects of population ecology. The domain of population ecology had its origins in the works of Lotka (1925) and Volterra (1926), leading to the contributions of Lack (1954) and Huffaker (1958). With the publication of Odum's *Fundamentals of Ecology* (1959), the basic principles and concepts pertaining to organization at the species population level were established. Currently, population ecology is a vibrant discipline, essential for developing a true understanding of how ecosystems work (Rockwood, 2015). It involves how a definable group of individuals of a single species interact with the biotic and abiotic attributes of their environment.

Nematode population dynamics and the concept of thresholds originated with the early 20th-century cyst nematode observations of Müller and Molz (1914) and Baunacke (1922). The concept of dynamics refers to the change that takes place in the size of the population (number of individuals) over a definable period of time, whereas a threshold represents the number of individuals required for a specific biological response to take place, such as a negative impact on crop yield or quality. Crop damage caused by cyst nematodes (Nematoda, Heteroderinae) was recognized by Herman Schacht in 1859 in Halle, Germany, and *Heterodera schachtii* was described by Schmidt (1871) more than 60 years before the science of nematology was catalysed by development of economically viable chemical nematicides (Bird, 1987). The chemotechnology era following World War II had positive attributes in regards to food production systems and also negative environmental and human health impacts. These resulted in the need for comprehensive nematode population and site-specific management systems (Bird, 2003). This includes the behavioural–biotic traits of nematode development, survivorship and reproduction associated with crop system resources and environmental resistance (Fig. 6.1). An equilibrium density is achieved when there is a relative balance between the biotic potential of the species and associated environmental resistance of the ecosystem. Key factors include the extent of population increase on host plants under various environmental conditions, population decline in the absence of host plants, crop damage thresholds and economic assessment. Research on cyst nematodes provides a significant portion of the knowledge base for application of population dynamics and damage thresholds for crop management decision-making related to nematodes. In the Heteroderinae, cyst-forming individuals must successfully complete more than ten processes, including hatch of second-stage juveniles (J2), emergence from the cyst, host detection, migration to the host, host penetration, identification of an appropriate feeding site and signalling for nurse cell (syncytial) establishment, in addition to completing four moults before becoming reproductively mature individuals. Prior to mating, males must emerge from root tissue and locate a female. All of these processes must be successfully completed for individuals to contribute to a positive net population growth rate per generation.

In 1970, Seinhorst summarized the contributions of 84 publications in his landmark review, *Dynamics of Populations of Plant-Parasitic Nematodes*. This was followed by *Computer Simulation and Population Models for Cyst-Nematodes (Heteroderidae: Nematoda)* by Jones *et al.* (1978). The first book on the *Ecology of Plant-parasitic Nematodes* was published by Norton in 1978. It reviewed the literature on population dynamics in relation to the soil environment, with a focus on strategies for survival. The second book, *Nematodes in Soil Ecosystems* (Freckman, 1982), expanded the concept of nematode population ecology to include interspecific interactions and model synthesis-validation, while including the roles of both bacterivores and fungivores. In 2010, Nehr described how population ecology relates to the overall science of the *Ecology of Plant and Free-Living Nematodes in Natural and Agricultural Soil*. In addition, it is imperative to recognize that populations comprise an ecosystem component, nested within complex non-linear systems having self-organization, bifurcation attributes and emergent properties (Wessels, 2013).

Field populations form functional units within a spatially defined area, but with potential for passive or active immigration and emigration. They have growth rates based on fecundity/fertility and mortality, life cycle stage distributions and spatial patterns. During their life histories, populations have the potential to undergo three phases: development, dynamic equilibrium and senescence (Fig. 6.2). Because of the difficulty associated with studying population intrinsic growth rates of plant-parasitic nematodes, they are most often expressed as density related to a unit of soil or plant tissue. Cyst nematode population assessment usually begins with an initial

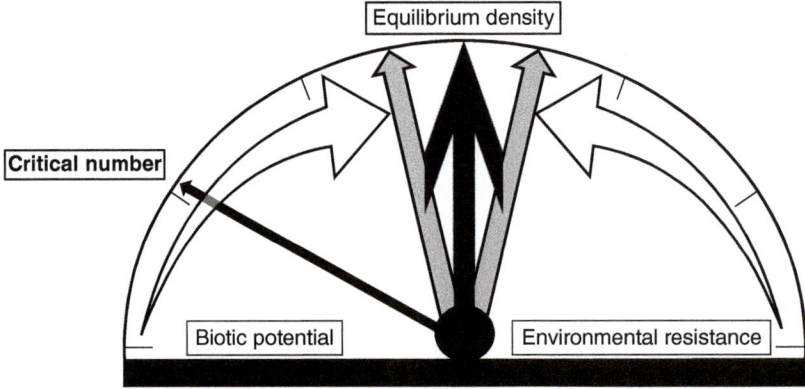

Fig. 6.1. Conceptual model of the dynamics of the biotic potential (e.g. reproductive rate, ability to survive adverse conditions, colonize new habitats) and environmental resistance (e.g. inadequate food, water, extreme temperatures, excessive parasitism/disease/competition) associated with the population dynamics of *Heterodera* spp. (Adapted from Wright, 2005.)

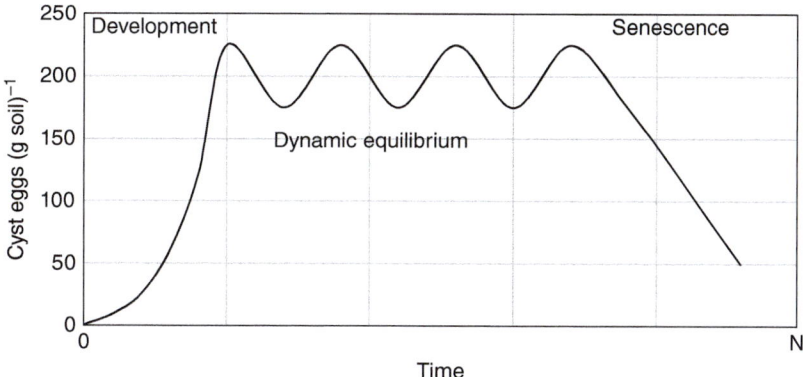

Fig. 6.2. Conceptual model of the three phases of population dynamics of *Heterodera schachtii*.

population density at a specific time (P_i) and projects to a future time (P_t). These latter are often designated as P_f (final population density at harvest or the end of a growing season) or P_m (the mid-season population density). An alternative is to describe the dynamics in relation to accumulated soil temperature degree days expressed at some biological base such as 8 or 10°C, depending on the nematode species, and very likely the individual local population (Griffin, 1988). Since most cyst nematode research is done using discrete units of time, differential equations have been used to estimate final population densities. Because cyst nematode immigration and emigration are not usually measured on a population-level basis, equations can be simplified, assuming that population growth is continuous with continuous population density monitoring.

Much of the early nematode population dynamics research was related to cyst nematodes and done by Jones (1956, 1959), Oostenbrink (1950, 1966) and Seinhorst (1964, 1965, 1967a, b). The focus was on sampling and extraction procedures, in addition to the quantitative relationships between the nematode–host plant interactions. The findings of these pioneering nematologists both created and validated the hypothesis that the rate of increase of a population density is based on its biological potential, with the upper limit forming an equilibrium density that is regulated by environmental resistant factors such as the quality and quantity of available food (Fig. 6.1). This formed the foundation for more recent cyst nematode population dynamics research that has become the basis for damage thresholds and information-based cyst nematode management systems (Fig. 6.3). This chapter describes the current state of population dynamics and damage threshold knowledge for *H. schachtii* (sugar beet cyst nematode), *Globodera* spp. (potato cyst nematodes), *H. glycines* (soybean cyst nematode) and selected information about more than 150 other species classified in the eight genera of cyst nematodes: *Betulodera*, *Cactodera*, *Dolichodera*, *Globodera*, *Heterodera*, *Paradolichodera*, *Punctodera* and *Vittatidera* (Turner and Subbotin, 2013; see Chapter 15, this volume).

6.2 *Heterodera schachtii* (Sugar Beet Cyst Nematode)

Successful sugar beet (*Beta vulgaris*) production systems develop, reach a state of dynamic equilibrium and have potential eventually to undergo the process of senescence. Symptoms of sugar beet crop damage caused by a cyst nematode were first characterized by Schacht in Germany in 1859 as Rübenmüdigkeit (beet weariness or tired soil). This nematode, a key global pest of sugar beet, was described as *H. schachtii* by Schmidt in 1871. It was first detected in the western hemisphere in Utah, USA in 1895. As early as 1914, crop rotation was recommended for control of *H. schachtii* (Müller and Molz, 1914; Baunacke, 1922), based on the hypotheses that: (i) crop yield loss is directly related to nematode population density; and (ii) egg mortality results from the lack of a suitable host. This was followed by comprehensive research designed to determine the extent of nematode

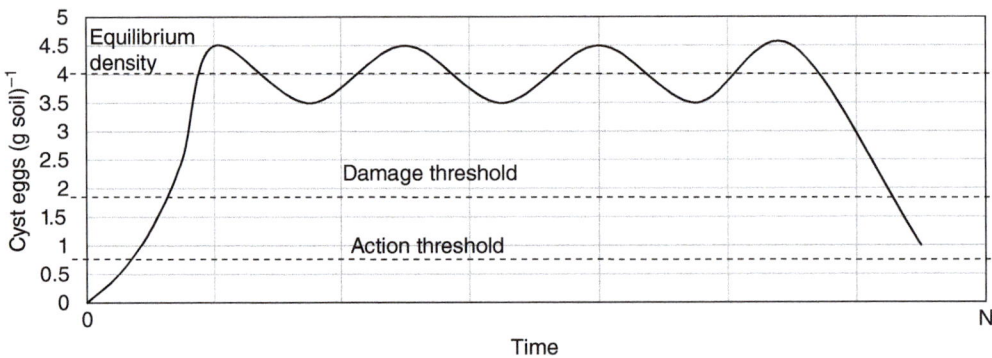

Fig. 6.3. Hypothetical relationship between *Heterodera schachtii* population density, equilibrium density, damage threshold and action threshold (time for implementation of a management tactic), without consideration of variation in key biotic and abiotic resistance and enhancement factors.

population increase on host plants under various environmental conditions, population decline in the absence of host plants, crop damage thresholds and economic assessment protocols.

6.2.1 Population dynamics

In the early 1930s, Jones initiated research designed to test the crop rotation hypotheses of Müller and Molz and Baunacke. Initially, this consisted of development and validation of nematode sampling, extraction and population density estimation protocols and their use in the evaluation of *H. schachtii* populations associated with multi-year crop rotations in 24 fields. In general, the results of the initial trials supported the hypotheses that population density was related to: (i) the frequency of growing susceptible crops; and (ii) the extent of sugar beet yield loss (Jones, 1945). This was followed by host range field studies conducted in 1945 and 1946, involving 36 plant families and 183 species. Sixty species representing eight families (Polygonaceae, Chenopodiaceae, Amaranthaceae, Aizoaceae, Caryophyllaceae, Crucifereae, Onagraceae and Labiatae) were identified as hosts of *H. schachtii* (Jones, 1950). Eighteen of the 20 Chenopodiaceae and 29 of the 45 Cruciferae species tested resulted in the production of mature cysts containing viable eggs (Jones, 1950). It was estimated that there were 2.5 to 3 generations of *H. schachtii* on sugar beets per growing season and between 1 and 4 on other host crops.

As long as soil moisture was not a limiting factor, temperature was the major driving variable in regards to cyst nematode development on sugar beets. An improved protocol for determining population densities of *Heterodera* spp. indicated that: 'Perhaps the greatest need with nematological techniques at present is careful study to see how time and effort may be saved without loss of accuracy' (Jones, 1955).

Under microplot conditions, population densities of *H. schachtii* increased more on Cruciferae species compared with Chenopodiaceae when the initial population densities were low (Jones, 1956). Greater population density decreases were associated with Cruciferae compared to Chenopodiaceae when the initial population densities were high. The concept of host status for population density increase, extent of shoot and root system symptoms, and equilibrium density were described in this work. It was determined that annual population density decline in the absence of a host was approximately 20, 40 and 50% for cysts, cysts containing eggs, and eggs, respectively. Wallace (1956) demonstrated the rate of juvenile emergence from cysts increased with soil aeration, which is related to soil moisture content, pore size, depth and oxygen consumption. This was followed by descriptions and models for good, intermediate, poor and non-hosts, with the maximum rate of reproduction of 20- to 50-fold and an equilibrium density (Fig. 6.2) of 200 eggs (g soil)$^{-1}$ for *H. schachtii* (Seinhorst, 1967a). Shepherd (1959) expanded the specificity of the concept of host suitability by finding that there is variation in the ability of *H. schachtii* populations to feed and reproduce on different cultivars of the same plant species. In 1959, Duggan confirmed the work of Jones (1955) by demonstrating that *H. schachtii* can complete more than two generations in a single growing season. This was expanded when Thomason and Fife (1962) reported that *H. schachtii* can have five generations per year under the sugar beet growing conditions in the Imperial Valley of California, USA. The impact of the quality of infected host tissue, underpopulation, overpopulation, sex ratio, interspecific competition and abiotic factors on population dynamics were described for other cyst species (Seinhorst, 1967b, 1968). Trudgill and Phillips (2007) found that *H. schachtii* completed a generation on a spring planted cover crop, but the first generation was not completed when it was planted in the autumn.

Raski and Johnson (1959) confirmed the observation of Baunacke (1922) and Nebel (1926) that there is no *H. schachtii* development on *B. vulgaris* at 6.0–9.5°C, slow development at 10–14°C and maximum development at 18–20°C, by demonstrating that infection by *H. schachtii* was significantly greater when soil temperatures were 16–28°C, compared to 10°C or below. Thomason and Fife (1962) found that the highest population density increase of *H. schachtii* occurred at 27.5°C, but eggs could remain viable for 3 months in cysts in surface field soil of 52°C. The optimum temperature for development on excised roots of *B. vulgaris* was 25°C (Johnson and Viglierchio, 1969a). Using an excised root system, Johnson found that nutrients, sucrose and vitamins impacted *H. schachtii* development and ability to penetrate root tissue (Johnson and Viglierchio, 1969a, b). Green *et al.* (1970) found that *H. schachtii* females can be inseminated by more than one male and that males inseminate multiple females.

6.2.2 Crop damage threshold

Seinhorst (1965) found that the *H. schachtii* microplot data of sugar beet yield of Jones (1956) provided a reasonably good fit of his model for the relationship between nematode population density and plant damage, providing a damage threshold of about 20 eggs (g soil)$^{-1}$. This was followed by development of a model determining the relationship between the initial and final population densities of sedentary nematodes based on the assumptions that the average individual nematode is the same at all densities. Additionally, it was assumed that nematodes do not attract or repel each other in the root penetration process and the size of the second generation is proportional to the feeding success of the initial generation (Seinhorst, 1967a). Under these conditions, the relationship between the initial and final population densities does not differ significantly from the common S-shaped (sigmoid) curve described in 1844–1845 by Pierre Francois Verhuist and referred to as the Verhuist or logistic curve. Under this model, the initial stage of population increase is close to exponential, followed by a decrease in growth rate and ending at maturity or equilibrium when the increase in population size stops (Figs 6.2, 6.3). Under glasshouse conditions, Cooke and Thomason (1979) found the damage threshold for *H. schachtii* to vary with soil temperature. The threshold was 65 eggs (100 g soil)$^{-1}$ at 23–27°C and 430 eggs (100 g soil)$^{-1}$ at 19°C, with no crop detectable damage taking place at 15°C. Jones *et al.* (1978) integrated the overall population dynamics concepts into a hypothetical nematode population dynamics–plant damage computer simulation.

Because of the complexity of the biotic and abiotic attributes of *H. schachtii* population dynamics, computer simulation models have been developed to predict nematode population dynamics and associated crop yield loss. In 1978, Jones published a general cyst nematode model. Caswell *et al.* (1978) published their single plant computer simulation model of *H. schachtii* infecting *B. vulgaris*. This model used five nematode development parameters (male development, female development, egg production, root penetration rate and feeding functions) and three sugar beet growth parameters (fibrous root growth, tap root growth and a tap root to fibrous root ratio) for the construction of 12 state variables. The initial population density of *H. schachtii* (eggs per g soil) and daily temperature were used as key driving variables for the beet growth and development, and nematode development subroutines. The specific simulation algorithms are recorded in the publication. The Schmidt et al. (1993) model for *H. schachtii* included juvenile stage development and population development associated with 11 potential rotation crops. In 2007, Moxnes and Hausken published a mathematical description of the population dynamics of potato cyst nematodes, providing a relation between initial potato cyst nematode (PCN) density, annual multiplication rate and harvesting strategy.

6.2.3 Information-based sugar beet cyst management

Information-based management of *H. schachtii* requires biological and abiotic environmental monitoring data, multi-year site-specific records and a decision-support system (Fig. 6.4). The biological monitoring component of the system must generate *H. schachtii* population density, and beet plant health/taproot yield, as well as information on other plant species such as weeds or cover crops, and other pathogen or beneficial soil-borne organisms. This must be done at appropriate intervals for incorporation into a multi-year permanent database. High resolution fixed-wing or drone-based remote sensing technologies are available for use in some systems. The abiotic environmental monitoring system should consist of soil texture, chemistry, temperature and moisture data, to be used by the system decision-maker in the selection of optimal system management strategies and tactics in a population ecology-based *H. schachtii* management programme (Fig. 6.5). Sugar companies keep detailed production system records and in many situations, a company agronomist is assigned to each individual grower to assist with beet production, including *H. schachtii* management practices. This is essential because both planting and marketing allotments are usually made on a contractual basis. The system controllers are responsible for the productivity and profitability of each site in relation to *H. schachtii* management (Fig. 6.5). System inputs include the use of resistant or tolerant cultivars, chemical or biological nematicides, seed treatments, inclusion of cover crops with trap crop capabilities and length of the rotation period. Use of electronic communications technology is available for enhancement of real-time decision-making. These data must be an integral component of the overall individual farm management plan, including the *H. schachtii* economic impact assessment information essential for interacting in a professional manner with the farm's enterprises financial institution and sugar company (Table 6.1).

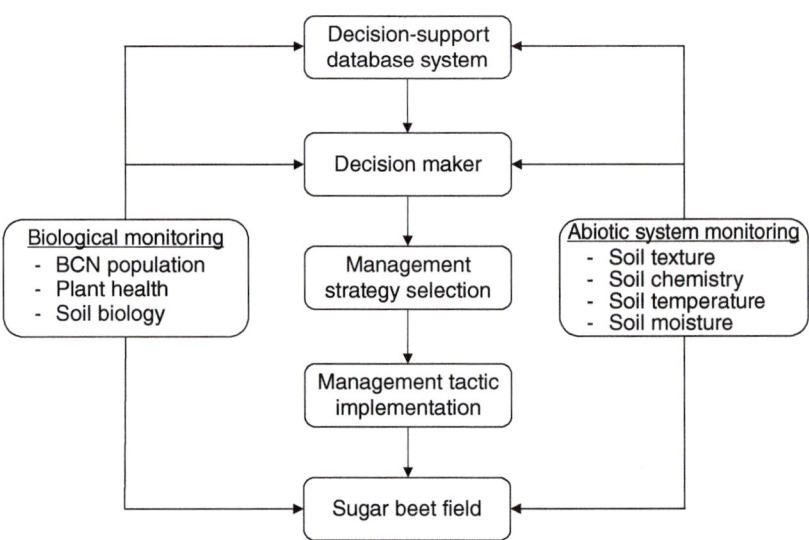

Fig. 6.4. Population ecology-based beet cyst nematode (BCN), *Heterodera schachtii*, management system.

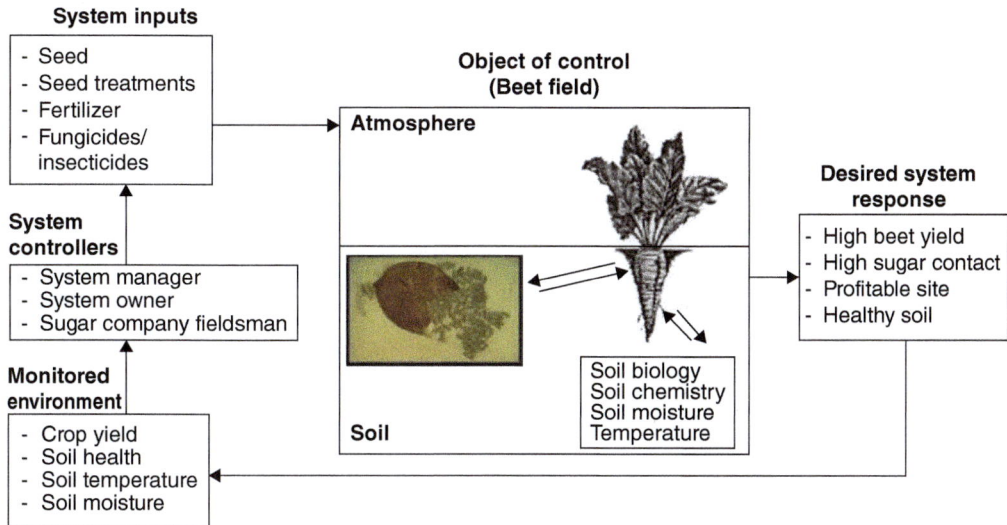

Fig. 6.5. Conceptual model of a sugar beet production management system in the presence of *H. schachtii*.

Table 6.1. Hypothetical sugar beet economic spreadsheet for *Heterodera schachtii* management decisions (after Michigan State University, Cropping Systems Economic Spreadsheet).

Field ledger category (per ha)	*H. schachtii* (< DT^a + SC^b)	*H. schachtii* (>DT^a +SC^b)	*H. schachtii* (>DT^a + TC^c)	*H. schachtii* (>DT^a + TC^c + ST^d)
Total CFC and CE ($ ha^{-1})^e	512	512	512	512
Total cash expenses ($ ha^{-1})	2279	2279	2299	2320
Total expenses ($ ha^{-1})	2791	2791	2811	2832
Beet yield (t ha^{-1})	64	45	79	88
Crop value ($ t^{-1})	124	124	124	124
Projected gross ($ ha^{-1})	3211	2223	3952	4150
Projected net ($ ha^{-1})	+420	−(568)	+1141	+1318

^{a}Damage threshold; ^{b}susceptible cultivar; ^{c}BCN tolerant cultivar; ^{d}seed treatment; ^{e}total cash fixed costs (CFC) and capital expenses (CE) in US$.

6.3 *Globodera* spp. (Potato Cyst Nematodes)

Research on two of the PCN, *G. rostochiensis* and *G. pallida*, has provided a significant portion of the knowledge base for the application of population dynamics and crop damage thresholds in cyst nematode management decisions. The population dynamics of PCN are relatively straightforward, as there is usually only one generation per year. Additionally, PCN have narrow host ranges, with tomato, potato and aubergine reported as crop hosts (Oostenbrink, 1950). Native to South America, the introduction(s) of PCN into Europe probably occurred in the 1850s. The first report of cyst-forming nematodes on potato was in 1881 by Kühn. However, PCN were not distinguished from the sugar beet cyst nematode until 1923 by Wollenweber. Because of the extreme economic impact of PCN on potato, there has been a major effort since the 1930s to understand the population dynamics and crop damage potential of PCN on potato. The population dynamics (population maintenance, increase and decline) of PCN driven by inherent nematode biology and environment-host interactions must be considered in the context of the PCN life cycle (see Chapter 1, this volume).

6.3.1 Population dynamics

J2 hatch from eggs in cysts, either after exposure to hatching cues (i.e. potato root diffusates; PRD) or spontaneously (see Chapter 3, this volume). Hatch of PCN varies among species and geographical locations. Spontaneous hatch in the absence of a host has been reported as low as 18% in Scotland (Grainger, 1964) to as high as 95% in Morocco (Schlüter, 1976). Across studies, spontaneous hatch averaged 30 to 33% (Turner and Evans, 1998). *Globodera pallida* appears to have lower spontaneous hatch than *G. rostochiensis* (Greet, 1974; Den Nijs and Lock, 1992). In the presence of PRD, hatch can also be variable, ranging from 40 to 80%. The hatching dynamics of *G. rostochiensis* and *G. pallida* have been reported to be markedly different, with *G. rostochiensis* hatching more readily than *G. pallida* (Clark and Perry, 1977; Ryan and Devine, 2005). The PCN, *G. ellingtonae*, has hatching dynamics similar to that of *G. rostochiensis* (Zasada et al., 2013). Hatch dynamics of PCN is influenced by soil texture and host plant. Maximum hatch of *G. rostochiensis* occurred in PRD collected 3 weeks after potato plant emergence, and declined after this time (Rawsthorne and Brodie, 1986). Although it has been demonstrated that hatch of PCN after exposure to PRD collected from different cultivars can vary (Evans, 1983), Rawsthorne and Brodie (1986) speculated that differences in length of the vegetative growth phase, extent of root growth and volume of roots, rather than the production of a more active cultivar-specific PRD, are important factors to consider.

A second factor in the hatching process is movement of chemical cues that stimulate hatch from host plant roots. Vertical and lateral distribution of PRD exhibited a concentration gradient. Hatching decreased with increasing distance from PRD (Rawsthorne and Brodie, 1986). Hatch of *G. rostochiensis* J2 in the UK was lower, 32% of cyst contents, in silt, clay and black fen soils compared to 60% hatch in a sandy soil (Cole and Howard, 1962a, b). Subsequent to hatch, survival of J2 is attained by the ability to disperse, migrate and locate a habitat in which its physiological characteristics can function best. This must include locating food and finding a mate (Wallace, 1968). Many factors come into play in regards to these processes (Fig. 6.6). These included particle size, gravity, pore size,

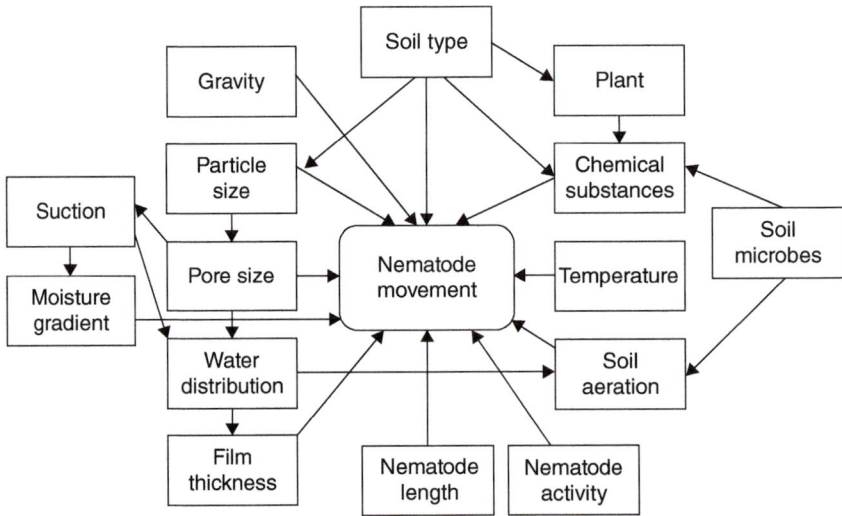

Fig. 6.6. Factors that influence the movement of *Globodera* spp. (Adapted from Wallace, 1960.)

water distribution, film thickness, moisture gradient, soil aeration, temperature, chemical substance, the plant and suction (Wallace, 1960). Whilst it is difficult to generalize because of the complexity of soil, G. rostochiensis mobility was the greatest in soil pore sizes ranging from 150 to 400 μm, regardless of soil type.

Once inside root tissue, population dynamics of PCN are influenced by host suitability and competition for resources. Multiplication rates of PCN are partially dependent on the initial population density (Seinhorst, 1965). Population increase is density-dependent, with the greatest multiplication observed at low initial population density (P_i) due to several factors. First, the size of the root system may be smaller due to nematode parasitism resulting in fewer available sites for nematode development. Second, if resources are limited, those individuals that are unable to induce successful syncytia become males and exit root tissue. When considering host suitability and its relationship to population dynamics, root biomass plays a major role in multiplication rates of PCN (Rawsthorne and Brodie, 1986). Additionally, if PCN is not able to establish a successful feeding site, such as in a resistant cultivar, multiplication rates will be reduced.

Seinhorst (1967a) proposed a population dynamics equation for sedentary nematodes with one generation based upon a competition curve (Nicholson, 1933):

$$P_f = \frac{a(1-q^{P_i})}{-\log q} \quad \text{(Eqn 6.1)}$$

Where P_f is the final population density, a is maximum multiplication rate and $1 - q$ is the proportion of available space that is exploited for food at a density of $P_i = 1$.

This equation was further modified to take into account the proportion of the PCN population that does not hatch, Cp (Seinhorst, 1986a). Recognizing that that sex determination is density-dependent in PCN (Trudgill, 1967), Jones and Kempton (1978) developed a logistic equation to incorporate this variable:

$$P_f = \frac{a(1-Cp)P_i}{1+(a-1)P_i / c(\frac{E}{C})^{P_i/E'}} + C_p P_i \quad \text{(Eqn 6.2)}$$

Where P_f is a proportion of the logistic equilibrium, a represents female development and fecundity, Cp is the proportion of the preplant population (P_i) that does not hatch, E' is the actual equilibrium density $(P_f = P_i)$ in eggs per g of soil, E is the proportional equilibrium density and c is a constant. To take into account female fecundity and incorporate the effects of potato cultivars on multiplication rates, the equation was further modified (Phillips et al., 1991). In most ecosystems there is only one generation of PCN per year. However, in some locations a partial second generation may occur. There was a second hatch of G. rostochiensis J2 and invasion of roots at 91 to 112 days after planting (DAP) in Slovakia (Renčo, 2007), at approximately 95 DAP in Italy (Greco et al., 1988) and at 75 to 100 DAP in Canada (Mimee et al., 2015). Globodera ellingtonae produced a second generation with J2 hatch in soil at approximately 80 DAP (Phillips, unpublished data). Greco and Moreno (1992) reported a lack of development of J2 of G. rostochiensis at the end of the spring potato season in Chile, suggesting a single generation in this ecosystem. In southern Italy and Venezuela, however, there are limited reports that G. rostochiensis is able to complete a second generation within a growing season (Greco et al., 1988; Jiménez-Pérez et al., 2009). The extent to which a partial or full second generation of PCN has on long-term population dynamics is not known.

PCN survives in soil as eggs in cysts, with up to 500 eggs per cyst. The length of time that eggs of PCN remain viable in soil depends on many factors, including initial density, dormancy, environmental conditions, neutral lipid reserves and the presence of biological control organisms (Lewis et al., 1960; Storey, 1984). The annual rate of decline in J2 viability varied with soil type in New Zealand, with annual attrition rates of 30–70% for G. pallida and 45–61% for G. rostochiensis (Marshall, 1998). In a study comparing G. pallida and G. rostochiensis hatch in absence of a host (Ryan and Devine, 2005), there was an initial difference in the rate of spontaneous hatch between the two species, with G. rostochiensis hatching in greater numbers than G. pallida. However, after 20 weeks spontaneous hatch was 37% for both species. In a 2-year study, the decline in egg numbers of G. rostochiensis was 57% in the first year and 40% in the second year in the absence of a host (Devine et al., 1999). In Bolivia, across 26 fields, the average rate of decline was 33% (Atkinson et al., 2001). The decline rate of G. ellingtonae is similar to its PCN relatives with approximately 50% reduction in viable unhatched J2 in a single season (Phillips et al., 2017). In cooler climates, egg decline rates have been reported to be lower than in warmer climates, with 10% annual decline in Ireland (Turner, 1996). In Scotland, PCN survived for up to 26 years in the absence of a host (Grainger, 1959). Juveniles present in Ireland in cysts more than 30 years old were not viable (Turner, 2008).

6.3.2 Crop damage thresholds

The development of population and crop loss models for PCN began in the 1950s and continued as a major focus through the 1980s. The history of modelling PCN population dynamics and yield loss has been extensively documented in numerous reviews (Jones and Perry, 1978; McSorley and Phillips, 1993; Phillips and Trudgill, 1998; Schmitt and Ferris, 1998; Trudgill et al., 2003). The basic requirements for PCN management models include P_i, yield loss functions (i.e. host tolerance to nematode damage), nematode dynamics, impact of control measures on tuber yield and economics of control measures (McSorley and Phillips, 1993).

Yield loss and PCN population increase rate are related to P_i. As P_i increases, yield decreases and multiplication rates of PCN decrease (Seinhorst, 1965). Yield loss is also affected by environmental factors such as soil type and differences between potato cultivars in their tolerance of damage by PCN (Trudgill, 1986). Seinhorst (1965) and Seinhorst and den Ouden (1971) developed models that are the basis for describing the relationship between nematode density and damage to plants. The Seinhorst (1965) equation describes a sigmoidal curve when P_i is plotted on a logarithmic scale. Proportional yield (Y) is given by:

$$Y = m + (1-m)z^{(P_i - T)} \text{ for } P_i \geq T \text{ and}$$
$$Y = 1.0 \text{ for } P_i < T \quad \text{(Eqn 6.3)}$$

Where P_i is the initial nematode density, Y is the ratio of the yield at a given P_i to the yield in the absence of PCN, T is the tolerance limit or threshold below which no damage occurs, m is the minimum yield at high P_i and z is a constant slightly less than 1.0, the proportion of the root system not damaged at $P_i = 1$. In fact, three parameters (m, T and z) are being used to describe differences in tolerance and overall rates of yield reduction. This equation, however, does not account for environmental factors and models have been developed to include these parameters.

Elston et al. (1991) introduced an inverse linear model for G. pallida that simplified the model to a single parameter c:

$$E(Y) = Y_{max} / [1 + (P_i / c)] \quad \text{(Eqn 6.4)}$$

Where $E(Y)$ is the expected value for a given P_i, Y equals yield in PCN-infested soil as a proportion of the nematode-free yield (Y_{max}) and P_i is the initial nematode population density. This equation was further modified (Phillips et al., 1998) to incorporate the influence of site (s) and cultivar (g) in the slope parameter:

$$E(Y) = Y_{max} [1 + (P_i / sg)] \quad \text{(Eqn 6.5)}$$

Initial population densities of PCN considered to result in economic damage to potato are usually fewer than 20 eggs (g soil)$^{-1}$. However, environmental conditions and interactions as well as different levels of host tolerance will result in a range of responses.

6.3.3 Application of PCN population dynamics and damage models

Online tools (NemaDecide for G. rostochiensis and G. pallida, and PCN Calculator Pallida for G. pallida) were developed to synthesize nematode population dynamics and damage function models into platforms for decision-making by potato growers in relation to PCN management. NemaDecide is an online decision tool that enables potato growers to estimate potential tuber yield loss, PCN population dynamics assessments and calculate management cost–benefit analyses (Been et al., 2006). PCN Calculator Pallida (AHDB potatoes) is an educational tool that provides information on G. pallida population and yield trends as well as predictions (Trudgill et al., 2014). The management models require specific system monitoring data (Table 6.2).

Table 6.2. Parameters included in online potato cyst nematode (PCN) resources (NemaDecide and PCN Calculator Pallida) applying PCN population dynamics and damage functions to management. (From Elliott et al., 2004; Been et al., 2006; Phillips, 2006.)

NemaDecide	PCN Calculator Pallida
Initial population density	Initial population density
Cultivar	Cultivar
	Soil type
Chemical control	Length of rotation
Financial information	Estimated maximum yield
Legislation	
Geographical data	Tolerance
	Resistance
	Treatment
	Percentage control of treatment[a]
	Percentage decline rate[b]

[a]Data suggest 80% control using a granular nematicide and 90% control using a fumigant nematicide.
[b]Assumed 20% decline per year.

NemaDecide (Been et al., 2006) consists of four components: (i) plant growth and tolerance; (ii) PCN population dynamics, plant resistance and nematode virulence; (iii) spatial distribution and sampling patterns; and (iv) the effect of control measures on the first and second components. The algorithms used are modified from Seinhorst (1986a, b), describing the relationship between P_i and plant weight and PCN population dynamics. Been et al. (2006) included a simulation example for G. pallida pathotype Pa3 at a P_i of 5 J2 (g soil)$^{-1}$, using a resistant cultivar in a 1:2 rotation that predicted a 4.4% yield loss the first year and a negligible yield loss if a resistant cultivar with the same degree of resistance is grown during the next rotation cycle. In this scenario, the population density of G. pallida declined to below the tolerance limit after 2 years.

PCN Calculator Pallida, known as The Model (Elliott et al., 2004; Phillips, 2006), utilizes the equations of Seinhorst and den Ouden (1971) and Elston et al. (1991) to predict population changes and crop damage over several rotations, involving different management strategies (Trudgill et al., 2014). To examine further the relationship between P_i of G. pallida and yield, research was conducted at two locations in Scotland and England chosen to represent a range of soil types (Phillips et al., 1998). Within these fields, potatoes were grown with the goal of creating large plots that encompassed a wide range of P_i. Established plots within these fields were planted with potato genotypes varying in tolerance and susceptibility to G. pallida. Models were fitted to the collected data (P_i, P_f and yield) from each site. The findings from this research included: (i) validation of the inverse linear model to describe the density-dependent effect of PCN on yield; (ii) support of the proportional nature of yield potentials to yield losses due to genotypic differences; (iii) confirmation that differences between potato genotypes in their tolerance to PCN were related to differences in the slopes of the regression between P_i and proportional to yield rather than to differences in the minimum yield; and (iv) yield losses due to PCN can be described in terms of genotypic and site differences (Phillips et al., 1998). To refine the model further, eight additional trials were conducted in nine commercial potato fields infested with either G. pallida or G. rostochiensis (Trudgill et al., 2014). In each of these sites, a wide range of P_i was included, in addition to paired nematicide-treated and non-treated plots. The goals were to determine nematicide efficacy and assess PCN decline rates. Data collected included P_i, P_f (at-harvest), yield and P_f 2–4 years after conclusion of the experiment. The Elston et al. (1991) model described the data from the nine fields equally as well as the Seinhorst (1965) model. The Seinhorst (1965) and Seinhorst and den Ouden (1971) population equation underestimated the maximum multiplication rate and overestimated the equilibrium density (Trudgill et al., 2014). This resulted from the underestimation of P_f at low P_i and overestimation of P_f at high P_i. The findings were integrated into The Model for enhancement of the PCN management strategy component. This data set allows for modelling of various integrated PCN management strategies, including cultivar resistance and tolerance, nematicide use and crop rotation length.

6.4 *Heterodera glycines* (Soybean Cyst Nematode)

The soybean cyst nematode *H. glycines* is a globally widespread and serious pathogen of soybean. This nematode is estimated to cause more yield loss than any other pathogen of soybean (Wrather *et al.*, 2001, 2010). The soybean cyst nematode was first reported in Japan by Hori in 1915 (cited in Ichinohe, 1959), and its potential for serious and persistent reduction in soybean yields resulted in *H. glycines* quarantines. It was designated as a quarantine pest in the USA (in 1957) and Canada, but deregulated in 1972 and 2013, respectively (Noel, 1992; Canadian Food Inspection Agency, 2013). As of 2017, *H. glycines* continues to be listed as a regulatory pest by the European Plant Protection Organization (EPPO/CABI, 1996). The population dynamics of *H. glycines* are governed by population densities on host crops, as well as by forces that impact survival over winter and during non-host crop periods and numerous other biotic and abiotic factors that increase and decrease population densities during the growing season. The interplay of these effects and factors determines the overall dynamics of *H. glycines* populations and their impacts on soybean yield.

6.4.1 Population dynamics

Key factors contributing to increases in *H. glycines* population densities include host status, population virulence on resistant cultivars, edaphic factors and pest interactions. Factors contributing to decreases in *H. glycines* population densities include non-host crops, avirulence on resistant soybeans, overwinter survival, use of nematode-protectant seed treatments, pest interactions and naturally occurring predators and pathogens.

Research in the 1980s in North Carolina, USA (Bonner and Schmitt, 1985) revealed that population dynamics of *H. glycines* differed depending on initial population densities. In general, population densities increased steadily for 3 months if initial densities were relatively low, but the increase in population densities occurred only for 1 month, presumably a single *H. glycines* generation, with high initial population densities, before decreasing over the next few months. Population densities were highest in both cases at harvest.

The most significant and direct factor affecting *H. glycines* population densities in commercial production systems is the cultivar of soybean grown. Increases in population densities of *H. glycines* on susceptible soybean cultivars can exceed 100-fold (Wheeler *et al.*, 1997), resulting in final population densities greater than 38,000 eggs $(100 \text{ cm}^3 \text{ soil})^{-1}$ (MacGuidwin *et al.*, 1995). Resistance and tolerance to *H. glycines* are characteristics or traits of soybean cultivars that are independent and well defined. Soybean cultivars resistant to *H. glycines* allow less than 10% reproduction of the nematode relative to the unchecked reproduction that occurs on a susceptible cultivar (Schmitt and Shannon, 1992). Soybean lines and cultivars supporting amounts of nematode reproduction greater than the 10% that occurs with resistance are described as moderately resistant, moderately susceptible, or susceptible, depending on the specific level of nematode reproduction (Schmitt and Shannon, 1992; see Chapter 9, this volume).

Within a few years of the nematode being first discovered in the USA, soybean cultivars with resistance to *H. glycines* were identified (Ross and Brim, 1957). They were already in existence and resistance to *H. glycines* was discovered through field screening of available soybean germplasm. Later, breeding lines with resistance genes were identified in glasshouse experiments (Miller, 1967) and purposeful crosses made in soybean breeding programmes to transfer resistance genes from known sources of resistance into soybean cultivars and germplasm lines with superior agronomic characteristics. Currently, there are many soybean breeding lines or sources of resistance known, with more than 100 in the USDA Soybean Germplasm Collection (Arelli *et al.*, 2000). Seven different plant genotypes are registered as sources of *H. glycines* resistance in soybean germplasm lines or cultivars (Brim and Ross, 1966; Hartwig and Epps, 1978; Hartwig and Young, 1990; Anand, 1992a, b; Nickell *et al.*, 1994a, b). In general, *H. glycines* population densities decrease, remain the same or increase only slightly during a growing season in which a *H. glycines*-resistant soybean cultivar is grown in a field infested by a population of the nematode that has not developed increased virulence on the resistant cultivar. For example, Todd *et al.* (1995) reported reproductive factors (RF = end-of-season or final population density divided by initial population density) from 0.7 to 1.3 on resistant cultivars and from 8.7 to 15.9 on susceptible soybean cultivars. In studies conducted throughout the north central USA, Rf of 0.3 to 2.6 occurred on resistant cultivars and *H. glycines* population densities increased more than 70-fold (Rf >70) on *H. glycines*-susceptible cultivars (Wang *et al.*, 2000). Similarly, Wheeler *et al.* (1997) observed Rf values of 0.1 to 5.5 on resistant and 0.4 to 112 on susceptible cultivars. Optimum control of *H. glycines* with resistant soybean cultivars occurs when the resistant cultivars are grown in fields infested with *H. glycines* populations that do not have increased virulence, or reproduction, on resistant cultivars. Such avirulent *H. glycines* populations are well controlled by resistant cultivars and are referred to as race 3 (Golden *et al.*, 1970) and, more recently, as HG Type 0 (Niblack *et al.*, 2002).

The availability of commercial soybean cultivars resistant to *H. glycines* has increased dramatically over the past four decades, but almost all of the resistant cultivars available in Argentina (Doucet *et al.*, 2008), Canada and the USA (Joos *et al.*, 2013; Tylka and Mullaney, 2017) have contained resistance genes from the soybean breeding line 'PI 88788'. A greater diversity of sources of resistance has been used to develop resistant cultivars in Brazil (Arias *et al.*, 2009). Prolonged use of the 'PI 88788' source of resistance in the USA has resulted in well-documented, widespread adaptation of *H. glycines* populations with increased reproduction or virulence on the formerly effective source of resistance (Niblack *et al.*, 2003, 2008; Mitchum *et al.*, 2007; Faghihi *et al.*, 2010). Populations with increased virulence on 'PI 88788' are classified as HG Type 2 (Niblack *et al.*, 2002). The continued selection for HG Type 2 populations through repeated use of resistant soybean cultivars with resistance genes from 'PI 88788' will likely lead to increased nematode reproduction, higher population densities and reduced bean yields of resistant soybean cultivars in the future.

Soybean cultivars also can be characterized as tolerant or intolerant to *H. glycines* based on the magnitude of decrease in plant growth or yield that occurs as a result of parasitism by the nematode (Boerma and Hussey, 1984). However, specific criteria for tolerance to *H. glycines* are not as well defined as for resistance, and it has not been possible to select purposefully for tolerant soybean cultivars for breeding purposes (Behm, 1997). Use of resistant soybean cultivars is more helpful in overall management of *H. glycines* than use of tolerant cultivars because resistant cultivars prevent significant increases in population densities and produce greater soybean yields than susceptible cultivars (MacGuidwin *et al.*, 1995; Wang *et al.*, 2000, 2003; Donald *et al.*, 2006; Tylka *et al.*, 2017). By contrast, tolerant soybean cultivars may be completely susceptible and allow for unchecked nematode reproduction resulting in large increases in population densities. In situations where susceptible soybean cultivars are grown in infested fields because of the unavailability of resistant cultivars or to slow the selection for virulent nematode populations, growing tolerant, susceptible cultivars allows for more profitable production of the crop than growing susceptible cultivars that are intolerant and suffer major yield reductions due to parasitism by the nematode.

Although soybean is the most economically important and widely grown host for *H. glycines*, numerous studies have been published revealing many additional hosts of the nematode. Riggs (1992) reviewed the host status of numerous plants, listing 63 species in 50 genera from 22 families of plants. From an economic perspective, *H. glycines* also can be a significant pest of many types of edible beans, *Phaseolus vulgaris*. Snap beans are well characterized as hosts for *H. glycines* (Noel *et al.*, 1982; Abawi and Jacobsen, 1984; Melton *et al.*, 1985). Poromarto and Nelson (2009, 2010) showed that *H. glycines* reproduced on most types of black, kidney, navy and pinto bean, with a range of reproduction levels including some greater than reproduction on susceptible soybean cultivars. Significant reductions in plant growth and yield of kidney, pinto and navy bean also have been documented in experiments conducted in *H. glycines*-infested soils (Poromarto *et al.*, 2010). Whilst yield reductions associated with *H. glycines* in edible bean production are of immediate economic significance, increases in *H. glycines* population densities as a consequence of growing bean crops are likely to have long-term economic effects on soybean production systems.

Weeds were determined as hosts for *H. glycines* within a few years of the nematode being discovered in the USA. These highly successful, but unwanted, plants can serve as inadvertent hosts for the nematode, resulting in increases in population densities, thereby negating management efforts implemented to maintain or reduce *H. glycines* population densities. The summer annual weed hemp sesbania (*Sesbania macrocarpa*) was reported by Epps and Chambers (1958) to serve as a host of *H. glycines*. Later, Smart (1964a, b) and Riggs and Hamblen (1966) reported numerous summer annual weeds as efficient hosts for *H. glycines*. Epps and Chambers (1958) also found henbit or common deadnettle (*Lamium amplexicaule*), a winter annual weed, to be a host for *H. glycines*. Additional winter annual weeds were discovered to be hosts of *H. glycines* in glasshouse experiments conducted by Venkatesh *et al.* (2000). The potential for winter annual weeds to affect population dynamics of *H. glycines* is determined not only by the host status of the weed, but also by soil temperatures. Development of *H. glycines* to adulthood and reproduction on winter annual weeds occurs only if a susceptible host weed is growing in an infested field and soil temperatures are above the basal threshold temperature of 5°C. In addition, hatching and penetration of roots by J2 of *H. glycines* require temperatures 14°C or greater (Alston and Schmitt 1988). Consequentially, soil temperatures must be sufficiently warm to support development and completion of the nematode life cycle. It is not clear whether juvenile stages of *H. glycines* can resume normal development following overwintering when soil temperatures rise above the basal temperature and weed growth resumes. Detailed bibliographies of weed hosts of *H. glycines* and other plant-parasitic nematodes have been published (Bendixen, 1988a, b).

Soybean cultivars are developed to be grown in specific geo-latitudes and are described as belonging to different maturity groups to reflect this characteristic. The length of the growing season needed for cultivars to mature increases as the maturity group number increases. Lower maturity group cultivars require less time (i.e. a shorter growing season) to mature than longer-season (higher maturity group) cultivars. Because of the obligatory parasitic nature of *H. glycines*, a longer growing season allows for additional generations in a single season. Riggs *et al.* (2000) found the lowest amount of *H. glycines* reproduction associated with a short-season (low maturity group) soybean cultivar, compared to four longer-season susceptible cultivars. Similarly, Wang *et al.* (2000) observed greater *H. glycines* reproduction in experiments conducted throughout the north central USA on longer-season susceptible cultivars ('MG III' and 'MG IV') than early season beans ('MG I' and 'MG II') and more consistent reductions in *H. glycines* population densities on earlier-season resistant cultivars than later-maturing cultivars. Hill and Schmitt (1989) found there was greater *H. glycines* reproduction associated with the latest-maturing cultivars, compared to early maturing cultivars. These results are consistent with the hypothesis that *H. glycines* population densities are affected by the length of time a susceptible host is grown at a site, hence determining the number of generations of the nematode that occur in a growing season.

Another management practice examined for possible beneficial effects on avoiding parasitism is delayed planting. The premise for using delayed planting as a management strategy is the starvation of *H. glycines* J2 that hatch several weeks before a delayed soybean crop is planted. The effects of delayed planting on *H. glycines* population densities have been variable. Some studies report no clear and consistent effect of delayed planting on *H. glycines* population densities (Hussey and Boerma, 1983), whilst others report inconsistent reductions in initial or final nematode population densities and yield loss (Koenning and Anand, 1991; Todd, 1993) and still others report consistently lower initial and often lower final population densities (Koenning *et al.*, 1996). A reasonable interpretation associated with delayed planting is that impacts are dependent on local growing conditions, aspects of the host soybean crop and the specific *H. glycines* population.

The other key factor affecting *H. glycines* population dynamics is the virulence of the nematode population. Shortly after *H. glycines* was discovered in the USA, resistant cultivars were available for use by soybean growers. Within a few years, researchers discovered *H. glycines* populations that reproduced well on resistant cultivars (Miller, 1967). Golden *et al.* (1970) developed and published guidelines for a standard method to assess reproduction of *H. glycines* populations on four known sources of resistance in soybeans. This *H. glycines* race test is a glasshouse bioassay to determine population virulence. It was revised and expanded in 2002 as the HG Type Test (Niblack *et al.*, 2002).

An additional biotic factor that affects *H. glycines* population dynamics is the interactions of this nematode with other pests and pathogens. Interactions with phytopathogenic fungi and plant-feeding insects impact *H. glycines* population densities. For example, feeding by soybean looper (*Chrysodeixis includens*) increased *H. glycines* juvenile and cyst population densities in roots and soil, respectively, in glasshouse studies (Russin *et al.*, 1989), whereas, in the same study, infection with the pathogenic fungus *Diaporthe phaseolorum* var. *caulivora*, causal agent of stem canker disease, decreased juvenile and cyst population densities. Concomitant infection of an *H. glycines*-susceptible soybean cultivar with *H. glycines* and *Cadophora gregata* (= *Phialophora gregata*) (the brown stem rot fungal pathogen) was found significantly to increase the incidence and severity of the brown stem rot disease, but the effects of the interaction of the nematode population densities was not reported (Sugawara *et al.*, 1997). In subsequent studies, infection of *H. glycines*-resistant and susceptible soybean cultivars by *H. glycines*, *C. gregata* and *Aphis glycines* (the soybean aphid) was found to decrease reproduction of the aphid and symptoms of the brown stem rot disease, but increased *H. glycines* population densities 5-fold compared to plants infected with *H. glycines* alone (McCarville *et al.*, 2012). Follow-up studies revealed that soybean aphid feeding increased *H. glycines* population densities only when overall pest pressure on the plants was low or moderate; that is, *H. glycines* population densities decreased at higher combined levels of pest pressure from both the nematode and the aphid (McCarville *et al.*, 2014). Interactions of *H. glycines* with the plant-pathogenic fungus *Fusarium virguliforme*, causal agent of soybean sudden death syndrome, results in earlier onset of the fungal disease and greater foliar symptom (McLean and Lawrence, 1993) and root-rot (Xing and Westphal, 2006) severity, but population densities of *H. glycines* juveniles and cysts on co-infected soybean plants are less than in the absence of the fungus (McLean and Lawrence, 1995; Gao *et al.*, 2006). Population dynamics of *H. glycines* are, therefore, affected by a myriad of biological factors and forces that result in the overall changes in population ecology.

A variety of abiotic factors impact *H. glycines* population dynamics. The clearest examples are the effects of soil moisture and soil pH. Francl (1993) found that *H. glycines* population densities were positively correlated with soil pH. The range of pH values in the experiments was from 5.0 to 6.4. In glasshouse experiments, Anand *et al.* (1995) observed greater reproduction of *H. glycines* on both susceptible and resistant soybean lines at pH 6.5 and 7.5 than at 5.5. On a larger spatial scale, Pedersen *et al.* (2010) documented a positive correlation between soil pH from 5.5 to 8.0 and *H. glycines* cyst and egg population densities in multiple states and across several sizes of field experiments. In the study by Francl (1993), population densities of the nematode were negatively correlated with fine texture soil. Fine soil texture has greater water-holding and cation exchange capacities, compared with coarse-textured soils. This relationship indicates greater *H. glycines* reproduction under drier soil conditions. Similarly, Koenning and Barker (1995) measured greater end-of-season *H. glycines* population densities in non-irrigated than in irrigated soils, supporting the observation of greater nematode reproduction in drier soils. Johnson *et al.* (1994) found no consistent trend in *H. glycines* reproduction under three different soil moisture regimes. As with the effects of planting date on *H. glycines* population dynamics, the effects of soil moisture may be dependent on specific soil conditions as well as specific aspects of the host soybean crop.

The length of time needed for *H. glycines* to complete a generation was shown to vary throughout a crop growing season in North Carolina, USA (Alston and Schmitt, 1988), varying from 28 days in early June to early July, 21 days in July–August and 28 days or more in September–November. The differences in generation times are generally attributed to the abiotic effects of soil temperature on the nematodes. Yet other factors may be involved. Yen *et al.* (1995) documented biochemical changes in levels of carbohydrates and proteins in *H. glycines* eggs at various times during the growing season. These dynamics, possibly related to changes in nutrition for nematodes from developing and senescing host soybean plants, may affect how various nematode life stages behave and respond to other biotic and abiotic factors and conditions throughout the growing season.

Whilst extensive research related to *H. glycines* population dynamics and densities have focused on in-season population dynamics, few studies have been conducted to characterize the survival of the nematodes during periods when non-host crops are grown and during overwinter periods. In Japan, Inagaki and Tsutsumi (1971) reported that encysted *H. glycines* eggs in infested field soil stored undried (26% soil moisture) in a room with no temperature control remained viable for 9 years, and Okada (1972) observed in laboratory experiments that cyst contents inhibited *H. glycines* hatching. Slack *et al.* (1972) reported that J2 of *H. glycines* survived longer in soil with no plants present in dry conditions (29–38 months) compared to soil that was flooded (up to 19 months). Greatest survival (90 months) occurred in soils that had moisture levels near field capacity. Francl and Dropkin (1986) found that overwintering survival of *H. glycines* in Missouri was density independent and ranged from 59 to 69% over three winters. Overwinter survival rates have been reported from as low as 27% (Bonner and Schmitt, 1985) to 67% in North Carolina (Ross, 1963). In a study conducted in five USA states Riggs *et al.* (2001) found that *H. glycines* populations had 50 to 100% survival over winter. In this same study, populations of *H. glycines* were transported to other states and overwintering survival compared to populations left in their resident state. In general, *H. glycines* populations survived best in their state of origin. Pérez-Hernández and Geisler (2014) reported that survival of *H. glycines* in soils following an annual crop of the non-host corn in Nebraska was affected by soil texture, with less survival in coarse-textured than in fine-textured soils.

6.4.2 Damage thresholds

Establishing consistent damage thresholds for *H. glycines* that apply for all situations is difficult, if not impossible. This is because of the key roles that environment and edaphic factors play in host–parasite relationships, nematode population dynamics and yield response of the host crop. Nonetheless, damage thresholds are frequently discussed. Population densities required to reduce soybean yields significantly are relatively low compared to the population densities encountered in infested soybean fields. For example, soybean yields were reduced at a population density of 3 cysts and 47 eggs (100 g soil)$^{-1}$ in field experiments conducted in Missouri (Francl and Dropkin, 1986). Schmitt *et al.* (1987) detected reductions in soybean yields at population densities of 63 eggs (100 g soil)$^{-1}$ or less, depending on soil type, with a damage threshold near zero in sandy soils. In the north central USA, Edwards (1988) reported that *H. glycines* population densities in soil greater than 20 eggs and J2 100 cm^{-3} on susceptible soybean cultivars were capable of reducing yields. Noel and Sikora (1990) measured significant reductions in susceptible soybean yields in Illinois at a *H. glycines* soil population density of 1 cyst 100 cm^{-3}. Niblack *et al.* (1992) artificially infested a non-infested field in Iowa with various inoculum densities of *H. glycines* J2 and eggs. The yield of a susceptible soybean cultivar was significantly reduced at inoculum densities of 10 and 50 eggs and J2 100 cm^{-3} during the 2 years of the study.

The damage threshold levels for *H. glycines* are below the detection level for some, if not all, cyst and egg extraction procedures. Consequently, considering the survival characteristics of *H. glycines* (Slack *et al.*, 1972; Riggs, 2004), coupled with its high reproductive potential (Wheeler *et al.*, 1997; Wang *et al.* 2000), a prudent recommendation is to consider any detected population of *H. glycines* to be at or exceeding the action threshold for implementation of an appropriate management tactic (Niblack *et al.*, 2006).

6.5 The Rest of the Heteroderinae

Reviews of nematode behaviour and survival strategies related to population dynamics were published by Lee (2002) and Wharton (2002), respectively. The behavioural characteristics of the more than 100 cyst-forming species classified in the Heteroderinae are well adapted for survival and parasitism in human-managed ecosystems. Some species classified in this taxon were described in the 19th century, such as the pea cyst nematode (*H. goettingiana*) and others in the 21st century (e.g. *Vittatidera zeaphila*). Of great significance, however, is that cyst nematodes are known to be the cause of major yield losses associated with the vast majority of the world's most important food, feed and fibre crops (Turner and Subbotin, 2013). Whilst most of the crop loss and nematode population dynamics research related to cyst nematodes has been conducted under temperate agriculture or laboratory conditions, more than 20 species are associated primarily with plants normally grown in hot tropic or dry arid biomes (Evans and Rowe, 1998). Some, like *H. avenae*, have relatively broad host ranges, whilst *H. carotae* has a narrow host range, feeding primarily on species of the Umbelliferae (Berney and Bird, 1992). Most cyst-forming species have been studied in regards to annual crops; however, a significant number are known to be associated with perennial crops such as banana, date palm, almond, cherry and fig. Very little is known about the population dynamics and host–parasite relationships of cyst-forming species associated with perennial crops and even less about species found in native (pristine) ecosystems. The relatively recent discovery of a significant number of cyst nematode species and current institutional, financial and personnel constraints has precluded development of comprehensive understandings of the population dynamics and damage thresholds of the majority of cyst nematode under local conditions. For example, 11 different cyst species have been confirmed in Michigan, USA. These include *Cactodera milleri*, *C. weissi*, *H. avenae*, *H. carotae*, *H. glycines*, *H. humuli*, *H. orientalis*, *H. schachtii*, *H. trifolii*, *H. ustinovi* and *Punctodera punctata* (Warner and Bird, unpublished data). During the time these identifications were made, *H. carotae* destroyed a vibrant organic soil carrot industry and forced it to move to mineral soil where there were more options for nematode management through extensive crop rotation.

Although there is a comprehensive population dynamics and damage threshold knowledge base for a few cyst nematodes, like the three described in this chapter, this is not true for most species. However, *H. avenae* is an exception. In 1982, Anderson published a review of the population dynamics of and control of *H. avenae*, indicating that in cereal crop production, the goal should be to keep population densities below 1 egg (g soil)$^{-1}$ at sowing time for oats and below 3 eggs (g soil)$^{-1}$ for barley, with additional information about differences in regards to autumn or spring planting. The influence of crop rotation, with non-hosts, resistant cultivars and fallow, on population dynamics of *H. avenae* has been extensively studied. The availability of *H. avenae* resistant wheat following evaluation and transfer of resistance from species such as *Triticum tauschii* and *Aegilops ventricosa* mandate the need for population density-based decisions based on damage thresholds (Eastwood *et al.*, 1991; Delibes *et al.*, 1993). In Australia (Fisher and Hancock, 1991), a density of 5 eggs (g soil)$^{-1}$ caused a 10% yield loss of susceptible wheat, while a resistant wheat cultivar maintained population densities of 1 egg (g soil)$^{-1}$, a level where crop loss was avoided. In the same study, percentage hatch within a growing season averaged 85% with approximately 7.5% of these individuals successfully invading roots. In Oregon, juvenile hatching dynamics in the spring present a greater level of risk to spring wheat than winter wheat with densities the highest in mid-April (Smiley *et al.*, 2007). In the same environment, at least two years between *H. avenae*-susceptible crops may be required to reduce nematode population densities and achieve management (Smiley, 1996). Population dynamics and damage threshold information for *H. avenae* have been used for evaluation of tolerant cultivars (Smiley *et al.*, 2013). In China, the dynamics of *H. avenae* infection of wheat was evaluated in Henan Province (Yuan *et al.*, 2014). J2 were found in wheat roots 14 days after planting (DAP) with a peak 150 DAP. Further evaluation of the population dynamics of *H. avenae* in China demonstrated that wheat in rotation with groundnut or maize were beneficial in reducing population densities while fallow was not (He *et al.*, 2016).

Information about J2 hatch and emergence, host finding, penetration, feeding, pathogenicity and interactions with other living organisms and abiotic factors is essential for future knowledge-based management systems. Cyst nematodes have the ability to regulate J2 hatch and emergence from cysts (Perry, 2002). This is species-specific and governed primarily by temperature and chemical stimulants/inhibitors (Hashmi and Krusberg, 1995; Sharma and Sharma, 1998; Chapter 3, this volume). Host root diffusates, other chemical stimuli and hatch inhibitors include a wide variety of inorganic and organic compounds, and their effects are usually dependent on dosage. In general, emergence temperature is also species-specific and ranges from ~10 to 30°C. *Heterodera avenae* is not dependent on root diffusates, but emergence can be chemically enhanced (Greco, 1981; Jones *et al.*, 1998). Local populations of *H. avenae* can exhibit summer dormancy in hot dry ecosystems, winter dormancy under cool wet conditions and adaptive behavioural characteristics due to changes in local or regional environmental conditions (Zancada and Sanchez, 1988). For example, J2 hatch and emergence needs to be studied from both a fundamental molecular basis and that of local population dynamics, for use in future knowledge-based cyst nematode management programmes. Without this level of detail, the concern of Jones *et al.* (1978) remains valid: 'Too much should not be expected of population models. They may provide quantitative information which is of value in understanding population dynamics of a species, but can not be used for serious prediction, particularly for individual fields, with crop cultivars, local farm practices and other factors affecting population development.'

6.6 Conclusions and Future Prospects

One thing is certain about complex non-linear systems, human-managed or natural: they are always changing. During the convergence of a cyst nematode population density to a steady-state or dynamic equilibrium, the population will rapidly encounter various unexpected bifurcation points (Wessels, 2013). The challenge associated with knowledge-based cyst nematode management is multi-dimensional and based on the overall dynamics of the ecosystem. With annual crops, for example, nematode host genetics may change as frequently as annually, and possibly more often with the inclusion of cover crops. Change impacting cyst nematodes also occurs within the soil microbial community, including bacteria, fungi, archaea, ciliates, flagellates, rotifers and possibly even some other currently unknown group or groups of soil-borne organisms. In addition, the production system management strategies and tactics change based on previous years' experiences of the managers, those of other progressive individuals and availability of various system inputs. This requires a comprehensive feedback system of biological and environmental monitoring information for appropriate decision-making (Fig. 6.6). The next generation of cyst nematode population dynamics, damage thresholds and management tactics need to include fundamental information on nematode behaviour, host finding, hatching, feeding and survival strategies such as that documented by Lee (2002), von Mende *et al.* (1998), Perry (2002), Wyss (2002) and Wharton (2002), respectively. The site specificity and human-management challenges may require at least one additional component. This could come from the social science domain of Action Research, originating with the works of Lewin at the Massachusetts Institute of Technology (Lewin, 1946). Action Research is a participatory problem-solving process that involves the applied research scientist working with actors (farm managers) and their actions (cyst nematode management practices) in as close to real-time mode as possible (Denscombe, 2010). In addition, it is highly probable that critical information could be learned from the investigation of cyst nematodes in natural ecosystems. Although only basic nematode population density and damage threshold information would be necessary in a true Anthropocene geologic epoch (Wilson, 2016), it is highly likely that it will be increasingly important to include individuals that are not professional nematologists, but progressive members of agricultural communities, in future applied cyst nematode population dynamics and information-based management research and implementation.

6.7 References

Abawi, G.S. and Jacobsen, B.J. (1984) Effect of initial inoculum densities of *Heterodera glycines* on growth of soybean and kidney bean and their efficiency as hosts under greenhouse conditions. *Phytopathology* 74, 1470–1474.

Alston, D.G. and Schmitt, D.P. (1988) Development of *Heterodera glycines* life stages as influenced by temperature. *Journal of Nematology* 20, 366–372.

Anand, S.C. (1992a) Registration of 'Hartwig' soybean. *Crop Science* 32, 1069–1070.

Anand, S.C. (1992b) Registration of 'Delsoy 4710' soybean. *Crop Science* 32, 1294.

Anand, S.C., Matson, K.W. and Sharma, S.B. (1995) Effect of soil temperature and pH on resistance of soybean to *Heterodera glycines*. *Journal of Nematology* 27, 478–482.

Anderson, S. (1982) Population dynamics and control of *Heterodera avenae* – a review with some original results. *Bulletin OPP EPPO Bulletin* 12, 463–475. DOI: 10-111/j.1365-2338.1982.tb01831.x

Arelli, A.P., Sleper, D.A., Yue, P. and Wilcox, J.A. (2000) Soybean reaction to races 1 and 2 of *Heterodera glycines*. *Crop Science* 40, 824–826. DOI: 10.2135/cropsci2000.403824x

Arias, C.A.A., Dias, W.P., Carneiro, G.E.S., Oliveira, M.F., de-Toledo, J.F.F., Carrão-Panizzi, M.C., Pipolo, A.E., Moreira, J.U.V., Kaster, M. and Bertagnolli, P. (2009) Resistance to soybean cyst nematode: genetics and breeding in Brazil. *Proceedings of the World Soybean Research Conference VIII*.

Atkinson, H.J., Holz, R.A., Riga, E., Main, G., Oros, R. and Franko, J. (2001) An algorithm for optimizing rotational control of *Globodera rostochiensis* on potato crops in Bolivia. *Journal of Nematology* 33, 121–125.

Baunacke, W. (1922) Ulnttersuchuugen zur Biologieund Bekämpfung der Rübennematoden *Heterodera schachtii* Schmidt. *Arbeit Biologie Reichsanstalt Land- und Forstwirtschaft Berlin-Dahlem* 11, 185–288.

Been, T.H., Schomaker, C.H. and Molendijk, L.P.G. (2006) NemaDecide: a decision support system for the management of potato cyst nematodes. Available at: https://www.researchgate.net/publication/40112909_NemaDecide_a_decision_support_system_for_the_management_of_potato_cyst_nematodes

Behm, J.E. (1997) Soybean tolerance to soybean cyst nematode (*Heterodera glycines* Ichinohe), and interactions between *H. glycines* and *Phialophora gregata*, the causal agent of brown stem rot of soybean. *Retrospective Theses and Dissertations*. Available at: http://lib.dr.iastate.edu/rtd/11774

Bendixen, L.E. (1988a) Weed hosts of *Heterodera*, the cyst, and *Pratylenchus*, the root-lesion, nematodes. Special Circular 117. The Ohio State University, Ohio Agricultural Research and Development Center, Wooster, Ohio, USA. Available at: http://hdl.handle.net/1811/71853

Bendixen, L.E. (1988b) A comparative summary of the weed hosts of *Heterodera*, *Meloidogyne*, and *Pratylenchus* nematodes. Special Circular 118. The Ohio State University, Ohio Agricultural Research and Development Center, Wooster, Ohio, USA. Available at: http://hdl.handle.net/1811/71854

Berney, M. and Bird, G. (1992) Distribution of *Heterodera carotae* and *Meloidogyne hapla* in Michigan carrot production. *Journal of Nematology* 24, 776–778.

Bird, G.W. (1987) Role of nematology in IPM, In: Veetch, J.A. and Dickson, D.W (eds) *Vistas in Nematology*. Society of Nematologists, Hyattville, Maryland, USA, pp. 114–121.

Bird, G.W. (2003) Role of integrated pest management and sustainable development. In: Maredia, K.M., Dakouo, D. and Mota-Sanchez, D. (eds) *Integrated Pest Management in the Global Arena*. CAB International, Wallingford, UK, pp. 73–85.

Boerma, H.R. and Hussey, R.S. (1984) Tolerance to *Heterodera glycines* in soybean. *Journal of Nematology* 16, 289–296.

Bonner, M.J. and Schmitt, D.P. (1985) Population dynamics of *Heterodera glycines* life stages on soybean. *Journal of Nematology* 17, 153–158.

Brim, C.A. and Ross, J.P. (1966) Registration of Pickett soybeans. *Crop Science* 6, 305.

Canadian Food Inspection Agency (2013) Pest risk management document for deregulation of *Heterodera glycines* Ichinohe (soybean cyst nematode). Risk Management Document RMD-11-02. Available at: www.inspection.gc.ca/plants/plant-pests-invasive-species/directives/risk-management/rmd-11-02/eng/1377523533087/1377523534384 (accessed 22 January 2018).

Caswell, E., MacGuidwin, A., Milne, K., Nelson, C., Thomason, I.J. and Bird, G. (1978) A simulation model of *Heterodera schachtii* infecting *Beta vulgaris*. *Journal of Nematology* 18, 512–519.

Clarke, A.J. and Perry, R.N. (1977) Hatching of cyst-nematodes. *Nematologica* 23, 350–368. DOI: 10.1163/187529277x00075

Cole, C.S. and Howard, H.W. (1962a) Further results from a field experiment on the effect of growing resistant potatoes on a potato root eelworm (*Heterodera rostochiensis*) population. *Nematologica* 7, 57–61. DOI: 10.1163/187529262x00747

Cole, C.S. and Howard, H.W. (1962b) The effect of growing resistant potatoes on a potato-root eelworm populations – a microplot experiment. *Annals of Applied Biology* 50, 121–127. DOI: 10.1111/j.1744-7348.1962.tb05993.x

Cooke, D.A. and Thomason, I.J. (1979) The relationship between population density of *Heterodera schachtii*, soil temperature and sugar beet yields. *Journal of Nematology* 11, 124–128.

Delibes, A., Romero, S., Auaded, S., Duce, A., Mena, M., Lopez-Brana, I., Andrés, M.-A.F., Martin-Sanchez, J.-A. and Garcia-Olmedo, T. (1993) Resistance to the cereal cyst nematode (*Heterodera avenae* Woll.) transferred from wild grass *Aegilops ventricosa* to hexaploid wheat by a 'stepping-stone' procedure. *Theoretical and Applied Genetics* 87, 402–408.

Den Nijs, L.J.M.F. and Lock, C.A.M. (1992) Differential hatching of the potato cyst nematodes *Globodera rostochiensis* and *G. pallida* in root diffusates and water of differing inonic composition. *Netherlands Journal of Plant Pathology* 98, 117–128. DOI: 10.1007/bf01996324

Denscombe, M. (2010) *The Good Research Guide for Small-Scale Social Research Projects*. McGraw-Hill, New York.

Devine, K.J., Dunne, C., O'Gara, F. and Jones, P.W. (1999) The influence of in-egg mortality and spontaneous hatching on the decline of *Globodera rostochiensis* during crop rotation in the absence of the host potato crop in the field. *Nematology* 1, 637–645. DOI: 10.1163/156854199508595

Donald, P.A., Pierson, P.E., St. Martin, S.K. *et al*. (2006) Assessing *Heterodera glycines*-resistant and susceptible cultivar yield response. *Journal of Nematology* 38, 76–82.

Doucet, M.E., Lax, P. and Coronel, N. (2008) The soybean cyst nematode *Heterodera glycines* Ichinohe, 1952 in Argentina. In: Ciancio, A. and Mukerji, K.G. (eds) *Integrated Management and Biocontrol of Vegetable and Grain Crops Nematodes*. Springer, Dordrecht, The Netherlands, pp. 127–148. DOI: 10.1007/978-1-4020-6063-2_7

Duggan, J. (1959) On the number of generations of beet eelworm, *Heterodera schachtii* Schmidt, produced in a year. *Nematologica* 4, 241–244. DOI: 10.1163/187629259X00435

Eastwood, R., Lagudah, E., Apples, R., Hannah, M. and Kollmargen, J. (1991) *Triticum tauschii*: a novel source of resistance to cereal cyst nematode. *Australian Journal of Agricultural Research* 42, 69–77.

Edwards, D.I. (1988) The soybean cyst nematode. In: Wyllie, T.D. and Scott, D.H (eds) *Soybean Diseases of the North Central Region*. APS Press, St. Paul, Minnesota, USA, pp. 81–86. DOI: 10.1071/app9910032a

Elliott, M.J., Trudgill, D.L., McNicol, J.W. and Phillips, M.S. (2004) Predicting PCN population changes and potato yields in infested soils. In: MacKerron, D.K.L. and Haverkort, A.J. (eds) *Decision Support Systems in Potato Production: Bring Models to Practice*. Wageningen Academic Publishers, Wageningen, The Netherlands, pp. 142–153. DOI: 10.3920/978-90-8686-527-7

Elston, D.A., Phillips, M.S. and Trudgill, D.L. (1991) The relationship between initial population density of potato cyst nematode *Globodera pallida* and the yield of partially resistant potatoes. *Revue de Nématologie* 14, 221–229. DOI: 10.2307/2404118

EPPO/CABI (1996) *Heterodera glycines*. In: Smith, I.M., McNamara, D.G., Scott, P.R. and Holderness, M. (eds) *Quarantine Pests for Europe*, 2nd edn. CAB International, Wallingford, UK, pp. 607–611.

Epps, J.M. and Chambers, A.Y. (1958) New host records for *Heterodera glycines* including one in Labiatae. *Plant Disease Reporter* 42, 194.

Evans, K. (1983) Hatching of potato cyst nematodes in root diffusates collected from twenty-five potato cultivars. *Crop Protection* 2, 97–103. DOI: 10.1016/0261-2194(83)90029-7

Evans, K. and Rowe J. (1998) Distribution and economic importance. In: Sharma, S. (ed.) *The Cyst Nematodes*. Kluwer Academic Publishers, Dordrecht, The Netherlands, pp. 1–30.

Faghihi, J., Donald, P.A., Noel, G., Welacky, T.W. and Ferris, V.R. (2010) Soybean resistance to field populations of *Heterodera glycines* in selected geographic areas. *Plant Health Progress* DOI: 10.1094/PHP-2010-0426-01-RS

Fisher, J.M., and Hancock, T.W. (1991) Population dynamics of *Heterodera avenae* Woll. In South Australia. *Australian Journal of Agriculture Research* 42, 53–68. DOI: 10.1071/AR9910053

Francl, L.J. (1993) Multivariate analysis of selected edaphic factors and their relationship to *Heterodera glycines* population density. *Journal of Nematology* 25, 270–276.

Francl, L.J. and Dropkin, V.H. (1986) *Heterodera glycines* population densities and relation of initial population to soybean yield. *Plant Disease* 70, 791–795. DOI: 10.1094/pd-70-791

Freckman, D. (1982) *Nematodes in Soil Ecosystems*. University of Texas Press, Austin, Texas, USA.

Gao, X., Jackson, T.A., Hartman, G.L. and Niblack, T.L. (2006) Interactions between the soybean cyst nematode and *Fusarium solani* f. sp. *glycines* based on greenhouse factorial experiments. *Phytopathology* 96, 1409–1415. DOI: 10.1094/phyto-96-1409

Golden, M.G., Epps, J.M., Riggs, R.D., Duclos, L.A., Fox, J.A. and Bernard, R.L. (1970) Terminology and identity of infraspecific forms of the soybean cyst nematode (*Heterodera glycines*). *Plant Disease Reporter* 54, 544–546.

Grainger, J. (1959) Population studies and successful control of the potato root eelworm. *European Potato Journal* 2, 184–198. DOI: 10.1007/bf02365564

Grainger, J. (1964) Factors affecting the control of eelworm disease. *Nematologica* 10, 5–20. DOI: 10.1163/187529264x00574

Greco N. (1981) Hatching of *Heterodera carotae* and *H. avenae*. *Nematologica* 27, 366–371. DOI: 10.1163/1875529281X0037

Greco, N., Inserra, R.N., Brandonisio, A., Tirro, A. and De Marinis, G. (1988) Life-cycle of *Globodera rostochiensis* on potato in Italy. *Nematologia Mediterranea* 16, 69–73.

Greco, N. and Moreno, L. (1992) Development of *Globodera rostochiensis* during three different growing seasons in Chile. *Nematropica* 22, 175–181.

Green, C.D., Greet, D.N. and Jones, F.G.W. (1970) The influence of multiple mating on the reproduction and genetics of *Heterodera rostochiensis* and *H. schachtii*. *Nematologica* 16, 511–519. DOI: 10.1163/187529270X00333

Greet, D.N. (1974) The response of five round-cyst nematodes (Heteroderidae) to five artificial hatching agents. *Nematologica* 20, 363–364. DOI: 10.1163/187529274X00410

Griffin, G.D. (1988) Factors affecting the biology and pathogenicity of *Heterodera schachtii* on sugarbeet. *Journal of Nematology* 20, 396–404.

Hartwig, E.E. and Epps, J.M. (1978) Registration of 'Bedford' soybeans. *Crop Science* 18, 915.

Hartwig, E.E. and Young, L.D. (1990) Registration of 'Cordell' soybean. *Crop Science* 30, 231.

Hashmi, S. and Krusberg L. (1995) Factors influencing emergence of juveniles from cysts of *Heterodera zeae*. *Journal of Nematology* 27, 362–269.

He, Q., Mo, A.S., Qiu, Z.Q., Zhou, X.B., and Wu, H.Y. (2016) Effect of rotation pattern on *Heterodera avenae* population in wheat field. *Journal of Animal Plant Science* 26, 211–216.

Hill, N.S., and Schmitt, D.P. (1989) Influence of temperature and soybean phenology on dormancy induction of *Heterodera glycines*. *Journal of Nematology* 21, 361–369.

Huffaker, C.B. (1958) Experimental studies on predation, dispersion factors and predator-prey oscillations. *Hilgardia* 27, 343–383. DOI: 10.3733/hilg.v27n14p343

Hussey, R.S. and Boerma, H.R. (1983) Influence of planting date on damage to soybean caused by *Heterodera glycines*. *Journal of Nematology* 15, 253–258.

Ichinohe, M. (1959) Studies on the soybean cyst nematode *Heterodera glycines* and its injury to soybean plants in Japan. *Plant Disease Reporter* 260, 239–248.

Inagaki, H. and Tsutsumi, M. (1971) Survival of the soybean cyst nematode, *Heterodera glycines* Ichinohe (Tylenchida : Heteroderidae) under certain storing conditions. *Applied Entomology and Zoology* 6, 156–162.

Jiménez-Pérez, N., Crozzoli, R. and Greco, N. (2009) The biology of *Globodera rostochiensis* in cultivated potato in Venezuela. *Nematologia Mediterranea* 37, 155–160.

Johnson, A.B., Scott, H.D. and Riggs, R.D. (1994) Response of soybean in cyst-infested soils at three soil-water regimes. *Journal of Nematology* 26, 329–335.

Johnson, R.N. and Viglierchio, D.R. (1969a) Sugar beet nematode (*Heterodera schachtii*) reared on axenic *Beta vulgaris* root explants. I. Selected environmental factors affecting penetration. *Nematologica* 15, 129–143. DOI: 10.1163/187529269X00164

Johnson, R.N. and Viglierchio, D.R. (1969b) Sugar beet nematode (*Heterodera schachtii*) reared on axenic *Beta vulgaris* root explants. II. Selected environmental and nutritional factors affecting development and sex ratio. *Nematologica* 15, 144–152. DOI: 10.1163/187529269X00173

Jones, F.G.W. (1945) Soil populations of beet eelworm (*Heterodera schachtii* Schm.) in relation to cropping. *Annals of Applied Biology* 32, 351–380. DOI: 10.1111/j.1744-7348.1945.tb06266.x

Jones, F.G.W. (1950) Observations on beet eelworm and other cyst-forming species of *Heterodera*. *Annals of Applied Biology* 37, 407–440. DOI: 10.1111/j.1744-7348.1950.tb00966.x

Jones, F.G.W. (1955) Quantitative methods in nematology. *Annals of Applied Biology* 42, 372–381. DOI: 10.1111/j.1744-7348.1955.tb02442.x

Jones, F.G.W. (1956) Soil populations of beet eelworm (*Heterodera schachtii* Schm.) in relation to cropping, II. Microplot and field plot results. *Annals of Applied Biology* 44, 25–56. DOI: 10.1111/j.1744-7348.1956.tb06845.x

Jones, F.G.W. (1959) Ecological relationships of nematodes. In: Holton, C., Fisher, G., Fulton, R., Hart, H. and McCallan, S. (eds) Plant Pathology. *Problems and Progress:* 1908–1958. University of Wisconsin Press, Madison, USA, pp. 399–411.

Jones, F.G.W. and Kempton, R.A. (1978) Population dynamics, population models and integrated control In: Southey, J. (ed.) *Plant Nematology*. MAFF ADAS Publication GD11, HMSO, London, pp. 333–361.

Jones, F.G.W. and Perry, J.N. (1978) Modelling populations of cyst nematodes (Nematoda: Heteroderidae). *Journal of Applied Ecology* 15, 349–371. DOI: 10.2307/2402596

Jones, F.G.W., Kempton, R.A. and Perry, J.N. (1978) Computer simulation and population models for cyst-nematodes (Heteroderidae: Nematoda). *Nematropica* 8, 36–56.

Jones, P.W., Tylka, G.L. and Perry, R.N. (1998) Hatching. In: Perry, R.N. and Wright, D.J. (eds) *The Physiology and Biochemistry of Free-living and Plant Parasitic Nematodes*. CAB International, Wallingford, UK, pp. 181–212.

Joos, D.K., Esgar, R.W., Henry, B.R. and Nafziger, E.D. (2013) Soybean variety test results in Illinois-2013. *Crop Sciences* Special Report 2013-04, University of Illinois, Urbana, Illinois, USA.

Koenning, S.R. and Anand, S.C. (1991) Effects of wheat and soybean planting date on *Heterodera glycines* population dynamics and soybean yield with conventional tillage. *Plant Disease* 75, 301–304.

Koenning, S.R. and Barker, K.R. (1995) Soybean photosynthesis and yield as influenced by *Heterodera glycines*, soil type and irrigation. *Journal of Nematology* 27, 51–62.

Koenning, S.R., Schmitt, D.P. and Barker, K.R. (1996) Soybean maturity group and planting date effects on seed yield and population densities of *Heterodera glycines*. *Fundamental and Applied Nematology* 19, 135–142.

Kühn, J. (1881) Die ergebnisse der Versuche zur Ermittelung der Ursache der Rübenmüdigkeit und zur Erforschung der Natur de Nematoden. *Berichte Physiologischen Laboratorium und der Versuchsanstalt des Landwirtschaftlichen Institutes der Universität Halle* 3, 1–153.

Lack, D. (1954) *The Natural Regulation of Animal Numbers*. Oxford University Press, Oxford, UK.

Lee, D. (2002) Behaviour. In: Lee, D.L. (ed.) *The Biology of Nematodes*. Taylor & Francis, Abingdon, UK, pp. 369–388.

Lewin, K. (1946) Action research and minority problems. *Journal of Social Issues* 2, 34–46. DOI: 10.1111/j.1540-4560.1946.tb02295.x

Lewis, F.J., Von, M. and Mai, W.F. (1960) Survival of encysted and free larvae of the golden nematode in relation to temperature and relative humidity. *Proceedings of the Helminthological Society of Washington* 2, 80–85.

Lotka, A.J. (1925) *Elements of Physical Biology*. Dover, New York, USA.

MacGuidwin, A.E., Grau, C.R. and Oplinger, E.S. (1995) Impact of planting 'Bell', a soybean cultivar resistant to *Heterodera glycines*, in Wisconsin. *Journal of Nematology* 27, 78–85.

Marshall, J.W. (1998) Potato cyst nematodes (*Globodera* species) in New Zealand and Australia. In: Marks, R.J. and Brodie, B.B. (eds) *Potato Cyst Nematodes: Biology, Distribution and Control*. CAB International, Wallingford, UK, pp. 353–394.

McCarville, M.T., O'Neal, M., Tylka, G.L., Kanobe C. and MacIntosh, G. (2012) A nematode, fungus, and aphid interact via a shared host plant: implications for soybean management. *Entomologia Experimentalis et Applicata* 143, 55–66. DOI: 10.1111/j.1570-7458.2012.01227.x

McCarville, M.T., Soh, D.H., Tylka, G.L. and O'Neal, M.E. (2014) Aboveground feeding by soybean aphid, *Aphis glycines*, affects soybean cyst nematode, *Heterodera glycines*, reproduction belowground. *PLoS ONE* 9(1), e86415. DOI: 10.1371/journal.pone.0086415

McLean, K.S. and Lawrence, G.W. (1993) Interrelationship of *Heterodera glycines* and *Fusarium solani* in sudden death syndrome of soybean. *Journal of Nematology* 25, 434–439.

McLean, K.S. and Lawrence, G.W. (1995) Development of *Heterodera glycines* as affected by *Fusarium solani*, the causal agent of sudden death syndrome of soybean. *Journal of Nematology* 27, 70–77.

McSorley, R. and Phillips, M.S. (1993) Modelling population dynamics and yield losses and their use in nematode management. In: Evans, K., Trudgill, D.L. and Webster, J.M. (eds) *Plant Parasitic Nematodes in Temperate Agriculture*. CAB International, Wallingford, UK, pp. 61–86.

Melton, T.A., Noel, G.R., Jacobsen, B.J. and Hagedorn, D.J. (1985) Comparative host suitability of snap beans to the soybean cyst nematode (*Heterodera glycines*). *Plant Disease* 69, 119–122. DOI: 10.1094/pd-69-119

Miller, L.I. (1967) Comparison of pathogenicity and development of Va.2 isolate of *Heterodera glycines* on Pickett and Lee soybeans. *Phytopathology* 57, 647.

Mimee, B., Dauphinais, N. and Bélair, G. (2015) Life cycle of the golden cyst nematode, *Globodera rostochiensis*, in Quebec, Canada. *Journal of Nematology* 47, 290–295.

Mitchum, M.G., Wrather, J.A., Heinz, R.D., Shannon, J.G. and Danekas, G. (2007) Variability in distribution and virulence phenotypes of *Heterodera glycines* in Missouri during 2005. *Plant Disease* 91, 1473–1476. DOI: 10.1094 / PDIS-91-11-1473

Moxnes, J.F. and Hausken, K. (2007) The population dynamics of potato cyst nematodes. *Ecological Modelling* 207, 339–348. DOI: 10.1016/j.ecolmodel.2007.06.020

Müller, H.C and Molz, E. (1914) Versuche zur Bekämpfung der Rübennematoden (*Heterodera schachtii*). *Zeitschrift des Vereins der Deutschen Zucker-Industrie* 64, 959–1050.

Nebel, B. (1926) Ein beieettag zur Physioloie des Rugennematoden *Heterodera schachtii* von Standpunkt der Bekämpfung. *Kuhn-Archiv* 12, 38–103.

Nehr, D.A. (2010) Ecology of plant and free-living nematodes in natural and agricultural soil. *Annual Review of Phytopathology* 48, 371–394. DOI: 10.1146/annurev-phyto-073009-114439

Niblack, T.L., Baker, N.K. and Norton, D.C. (1992) Soybean yield losses due to *Heterodera glycines* in Iowa. *Plant Disease* 76, 943–948.

Niblack, T.L., Arelli, P.R., Noel, G.R., Opperman, C.H., Orf, J.H., Schmitt, D.P., Shannon, J.G. and Tylka, G.L. (2002) A revised classification scheme for genetically diverse populations of *Heterodera glycines*. *Journal of Nematology* 34, 279–288.

Niblack, T.L., Wrather, J.A., Heinz, R.D. and Donald, P.A. (2003) Distribution and virulence phenotypes of *Heterodera glycines* in Missouri. *Plant Disease* 87, 929–932. DOI: 10.1094/PDIS.2003.87.8.929

Niblack, T.L., Lambert, K.N. and Tylka, G.L. (2006) A model plant pathogen from the kingdom Animalia: *Heterodera glycines*, the soybean cyst nematode. *Annual Review of Phytopathology* 44, 283–303. DOI: 10.1146/annurev.phyto.43.040204.140218

Niblack, T.L., Colgrove, A.L., Colgrove, K. and Bond, J.P. (2008) Shift in virulence of soybean cyst nematode is associated with use of resistance from PI 88788. *Plant Health Progress*. DOI: 10.1094/PHP-2008-0118-01-RS

Nicholson, A.J. (1933) Supplement: The balance of animal populations. *Journal of Animal Ecology* 2, 131–178. DOI: 10.2307/954

Nickell, C.D., Noel, G.R., Bernard, R.L., Thomas, D.J. and Frey, K. (1994a) Registration of soybean germplasm line 'LN89-5699' resistant to soybean cyst nematode. *Crop Science* 34, 1133–1134.

Nickell, C.D., Noel, G.R., Bernard, R.L., Thomas, D.J. and Pracht, J. (1994b) Registration of soybean germplasm line 'LN89-5612' moderately resistant to soybean cyst nematode. *Crop Science* 34, 1134.

Noel, G.R. (1992) History, distribution, and economics. In: Riggs, R.D. and Wrather, J.A. (eds) *Biology and Management of the Soybean Cyst Nematode*. APS Press, St. Paul, Minnesota, USA, pp. 1–14.

Noel, G.R. and Sikora, E.J. (1990) Evaluation of soybeans in maturity groups I-IV for resistance to *Heterodera glycines*. *Journal of Nematology* 22, 795–799.

Noel, G.R., Jacobsen, B.J. and Leeper, C.D. (1982) Soybean cyst nematode in commercial snap bean. *Plant Disease* 66, 520–522.

Norton, D. (1978) *Ecology of Plant-parasitic Nematodes*. John Wiley and Sons, New York.

Odum, E. (1953) *Fundamentals of Ecology*. W.B. Saunders Co., London.

Okada, T. (1972) Hatching inhibitory factor in the cyst contents of the soybean cyst nematode, *Heterodera glycines* Ichinohe (Tylenchida: Heteroderidae). *Applied Entomology and Zoology* 7, 99–102.

Oostenbrink, M. (1950) *Het aardappelaaltje (Heterodera rostochiensis Wollenweber), een gevaarlijke parasiet voor de eenzijdige aardappel-cultuur*. Verslagen en mededelingen van de Plantenzietktenkundige Dienst, Wageningen, The Netherlands.

Oostenbrink, M. (1966) *Major Characteristics of the Relation between Nematodes and Plants*. Mededelingen Landbouwhogeschool, Wageningen, The Netherlands.

Pedersen, P., Tylka, G.L., Mallarino, A.P., MacGuidwin, A.E., Koval, N.C. and Grau, C.R. (2010) Correlation between soil pH, *Heterodera glycines* population densities, and soybean yield. *Crop Science* 50, 1458–1464. DOI: 10.2135/cropsci2009.08.0432

Pérez-Hernández, O. and Geisler, L.J. (2014) Quantitative relationship of soil texture with the observed population density reduction of *Heterodera glycines* after annual corn rotation in Nebraska. *Journal of Nematology* 46, 90–100.

Perry, R. (2002) Hatching. In: Lee D.L. (ed.) *The Biology of Nematodes*. Taylor & Francis, Abingdon, UK, pp. 147–170.

Phillips, M.S. (2006) *Potato Council PCN Management Model*. Potato Council, Oxford, UK.

Phillips, M.S. and Trudgill, D.L. (1998) Population modelling and integrated control options for potato cyst nematodes. In: Marks, R.J. and Brodie, B.B. (eds) *Potato Cyst Nematodes: Biology, Distribution and Control*. CAB International, Wallingford, UK, pp. 153–164. DOI: 10.1046/j.1365-3059.1999.0378a.x

Phillips, M.S., Hackett, C.A. and Trudgill, D.L. (1991) The relationship between the initial and final population densities of the potato cyst nematode *Globodera pallida* for partially resistant potatoes. *Journal of Applied Ecology* 28, 109–119. DOI: 10.2307/2404118

Phillips, M.S., Trudgill, D.L., Hackett, C.A., Hancock, M., Holliday, J.M. and Spaull, A.M. (1998) A basis for predictive modelling of the relationship of potato yields to population densities of the potato cyst nematode, *Globodera pallida*. *Journal of Agricultural Science* 130, 45–51. DOI: 10.1017/s0021859697005054

Phillips, W.S., Kitner, M. and Zasada, I.A. (2017) Developmental dynamics of *Globodera ellingtonae* in field-grown potato. *Plant Disease* 101, 1182–1187. DOI: 10.1094/PDIS-10-16-1439-RE

Poromarto, S.H. and Nelson, B.D. (2009) Reproduction of soybean cyst nematode on dry bean cultivars adapted to North Dakota and northern Minnesota. *Plant Disease* 93, 507–511

Poromarto, S.H. and Nelson, B.D. (2010) Evaluation of northern-grown crops as host of soybean cyst nematode. *Plant Health Progress* DOI: 10.1094/PHP-2010-0315-02-RS

Poromarto, S.H., Nelson, B.D. and Goswami, R.S. (2010) Effect of soybean cyst nematode on growth of dry bean in the field. *Plant Disease* 94, 1299–1304. DOI: 10.1094 / PDIS-05-10-0326

Raski, D.J. and Johnson, R.T. (1959) Temperature and activity of the sugarbeet nematode as related to sugarbeet production. *Nematologica* 4, 136–141. DOI: 10.1163/187529259X00110

Rawsthorne, D. and Brodie, B.B. (1986) Relationship between root growth of potato, root diffusate production, and hatching of *Globodera rostochiensis*. *Journal of Nematology* 18, 379–384.

Renčo, M. (2007) Comparison of the life cycle of potato cyst nematode (*Globodera rostochiensis*) pathotype Ro1 on selected potato cultivars. *Biologie Bratislava* 62, 195–200. DOI: 10.2478/s11756-007-0029-0

Riggs, R.D. (1992) Host range. In: Riggs, R.D. and Wrather, J.A. (eds) *Biology and Management of the Soybean Cyst Nematode*. APS Press, St. Paul, Minnesota, USA, pp. 107–114.

Riggs, R.D. (2004) History and distribution. In: Riggs, R.D. and Wrather, J.A. (eds) *Biology and Management of the Soybean Cyst Nematode*, 2nd edn. APS Press, St. Paul, Minnesota, USA, pp. 9–40.

Riggs, R.D. and Hamblen, M.L. (1966) Additional weed hosts of *Heterodera glycines*. *Plant Disease Reporter* 50, 15–16.

Riggs, R.D., Wrather, J.A., Mauromoustakos, A. and Rakes, L. (2000) Planting date and soybean cultivar maturity group affect population dynamics of *Heterodera glycines*, and all affect yield of soybean. *Journal of Nematology* 32, 334–342.

Riggs, R.D., Niblack, T.L., Kinloch, R.A., MacGuidwin, A.E., Mauromoustakos, A. and Rakes, L. (2001) Overwinter population dynamics of *Heterodera glycines*. *Journal of Nematology* 33, 219–226.

Rockwood, L.L. (2015) *Introduction to Population Ecology*, 2nd edn. Wiley-Blackwell, Oxford, UK.

Ross, J.P. (1963) Seasonal variation of larval emergence from cysts of the soybean cyst nematode, *Heterodera glycines*. *Phytopathology* 53, 608–609.

Ross, J.P. and Brim, C.A. (1957) Resistance of soybeans to the soybean cyst nematode as determined by a double-row method. *Plant Disease Reporter* 41, 923–924

Russin, J.S., Layton, M.B., Boethel, D.J., McGawley, E.C., Snow, J.P. and Berggren, G.T. (1989) Development of *Heterodera glycines* on soybean damaged by soybean looper and stem canker. *Journal of Nematology* 21, 108–114.

Ryan, A. and Devine, K.J. (2005) Comparison of the in-soil hatching responses of *Globodera rostochiensis* and *G. pallida* in the presence and absence of the host potato crop cv. British Queen. *Nematology* 7, 587–597. DOI: 10.1163/156854105774384804

Schlüter, K. (1976) The potato cyst eelworm *Heterodera rostochiensis* Woll. In Morocco: its distribution and economic importance. *Zeitschrift für Pflanzenkrankheiten und Planzenschutz* 83, 401–406.

Schmidt, A. (1871) Ueber den Rügennemtoden. *Zeitschrift Ver. Rügenzuckerindustrie Zollver* 22, 67–75.

Schmidt, K., Sikora, R.A. and Richter, O. (1993) Modeling the population dynamics of the sugar beet cyst nematode *Heterodera schachtii*. *Crop Protection* 12, 490–496. DOI: 10.1016/0261-2194(93)90088-Z

Schmitt, D.P. and Ferris, H. (1998) Pathogenicity and damage levels. In: Sharma, S.B. (ed.) *The Cyst Nematodes*. Kluwer Academic Publishers, Dordrecht, The Netherlands, pp. 239–265. DOI: 10.1007/978-94-015-9018-1_10

Schmitt, D.P. and Shannon, G. (1992) Differentiating soybean responses to *Heterodera glycines*. *Crop Science* 32, 275–277. DOI: 10.2135/cropsci1992.0011183X003200010056x

Schmitt, D.P., Ferris, H. and Barker, K.R. (1987) Response of soybean to *Heterodera glycines* races 1 and 2 in different soil types. *Journal of Nematology* 19, 240–250.

Seinhorst, J.W. (1964) Population dynamics of some root infesting nematodes. *Nematologica* 10, 61 (abstract). DOI: 10.1163/187529264X00646

Seinhorst, J.W. (1965) The relationship between nematode density and damage to plants. *Nematologica* 11, 137–154. DOI: 10.1163/187529265x00582

Seinhorst, J.W. (1967a) The relationships between population increase and population density in plant-parasitic nematodes. II. Sedentary nematodes. *Nematologica* 13, 157–171. http://dx.doi.org/10.1163/187529267x01048

Seinhorst J.W. (1967b) The relationship between population increase and population density in plant parasitic nematodes. III Definition of the terms host, host status and resistance. IV The influence of external conditions on the regulation of population density. *Nematologica* 13, 429–442.

Seinhorst, J.W. (1968) Underpopulation in plant parasitic nematodes. *Nematologica* 14, 549–553. DOI: 10.1163/187529268X00246

Seinhorst, J.W. (1970) Dynamics of populations of plant parasitic nematode. *Annual Review of Phytopathology* 8, 131–156. DOI:10.1146/annurev.py.08.090170.001023

Seinhorst, J.W. (1986a) The development of individuals and populations of cyst nematodes on plants. In: Lamberti, F. and Taylor, C.E. (eds) *Cyst Nematodes*. Plenum Press, London, pp. 101–117. DOI: 10.1007/978-1-4613-2251-1_5

Seinhorst, J.W. (1986b) The effect of nematode attack on the growth and yield of crop plants. In: Lamberti, F. and Taylor, C.E. (eds) *Cyst Nematodes*. Plenum Press, New York, pp. 191–210. DOI: 10.1007/978-1-4613-2251-1_11

Seinhorst, J.W. and den Ouden, H (1971) The relation between density of *Heterodera rostochiensis* and growth and yield of two potato varieties. *Nematologica* 17, 347–369. DOI: 10.1163/187529271x00585

Sharma, S. and Sharma, R. (1998) Hatch and emergence. In: Sharma, S. (ed.) *The Cyst Nematodes*. Kluwer Academic Publishers, Dordrecht, The Netherlands, pp. 191–216.

Shepherd, A.M. (1959) The invasion and development of some species of *Heterodera* in plants of different host status. *Nematologica* 4, 253–267. DOI: 10.1163/187529259X00453

Slack, D.A., Riggs, R.D. and Hamblen, M.L. (1972) The effect of temperature and moisture on the survival of *Heterodera glycines* in the absence of a host. *Journal of Nematology* 4, 263–266.

Smart, G.C., Jr (1964a) Additional hosts of the soybean cyst nematode *Heterodera glycines*, including hosts in two additional plant families. *Plant Disease Reporter* 48, 388–390.

Smart, G.C., Jr (1964b) Physiological strains and one additional host of the soybean cyst nematode *Heterodera glycines*. *Plant Disease Reporter* 48, 542–543.

Smiley, R.W. (1996) Diseases of wheat and barley in conservation cropping systems of the semiarid Pacific Northwest. *American Journal of Alternative Agriculture* 11, 95–103. DOI: 10.1017/S0889189300006858

Smiley, R.W., Sheedy, J., Pinkerton, J., Easley, S., Thompson, A. and Yan, G. (2007) Cereal cyst nematode: distribution, yield reductions, and crop management strategies. *Dryland Agriculture Research Annual Report* 1074, 15–29.

Smiley, R.W., Marshall, J., Goulie, J. et al. (2013) Spring wheat tolerance and resistance to *Heterodera avenae* in the Pacific Northwest. *Plant Disease* 97, 590–600.

Storey, R.M.J. (1984) The relationship between neutral lipid reserves and infectivity for hatched and dormant juveniles of *Globodera* spp. *Annals of Applied Biology* 104, 511–520. DOI: 10.1111/j.1744-7348.1984.tb03034.x

Sugawara, K., Kobayashi, K. and Ogoshi, A. (1997) Influence of the soybean cyst nematode (*Heterodera glycines*) on the incidence of brown stem rot in soybean and adzuki bean. *Soil Biology and Biochemistry* 29, 1491–1498. DOI: 10.1016/S0038-0717(97)00033-3

Thomason, I.J. and Fife, D. (1962) The effect of temperature on development and survival of *Heterodera schachtii* Schm. *Nematologica* 7, 139–145. DOI: 10.1163/187529262X00873

Todd, T.C. (1993) Soybean planting date and maturity effects on *Heterodera glycines* and *Macrophomina phaseolina* in southeastern Kansas. *Journal of Nematology* 25, 731–737.

Todd, T.C., Schapaugh, W.T., Jr, Long, J.H. and Holmes, B. (1995) Field response of soybean in maturity groups III-V to *Heterodera glycines* in Kansas. *Journal of Nematology* 27, 628–633.

Trudgill, D.L. (1967) The effect of environment on sex determination in *Heterodera rostochiensis*. *Nematologica* 13, 263–272. DOI: 10.1163/187529267x00120

Trudgill, D.L. (1986) Yield losses caused by potato cyst nematodes: a review of the current position in Britain and prospects for improvement. *Annals of Applied Biology* 108, 181–198. DOI: 10.1111/j.1744-7348.1986.tb01979.x

Trudgill, D.L. and Phillips, M.S. (2007) Nematode population dynamics, threshold levels and estimation of crop losses. *FAO Corporate Document Repository*. Available at: www.fao.org/docrep/V9978E/v9978e07.htm

Trudgill, D.L., Elliott, M.J., Evans, K. and Phillips, M.S. (2003) The white potato cyst nematode (*Globodera pallida*) – a critical analysis of the threat in Britain. *Annals of Applied Biology* 143, 73–80. DOI: 10.1111/j.1744-7348.2003.00073.x

Trudgill, D.L., Phillips, M.S. and Elliott, M.J. (2014) Dynamics and management of the white potato cyst nematode *Globodera pallida* in commercial potato crops. *Annals of Applied Biology* 164, 18–34. DOI: 10.1111/aab.12085

Turner, S.J. (1996) Population decline of potato cyst-nematodes (*Globodera rostochiensis, G. pallida*) in field soils in Northern Ireland. *Annals of Applied Biology* 117, 385–397. DOI: 10.1111/j.1744-7348.1996.tb05754.x

Turner, S.J. (2008) Population decline of potato cyst nematodes (*Globodera rostochiensis, G. pallida*) in field soils in Northern Ireland. *Annals of Applied Biology* 129, 315–322. DOI: 10.1111/j.1744-7348.1996.tb05754.x

Turner, S.J. and Evans, K. (1998) The origins, global distribution and biology of potato cyst nematodes (*Globodera rostochiensis* (Woll.) and *G. pallida* (Stone)). In: Marks, R.J. and Brodie, B.B. (eds) *Potato Cyst Nematodes: Biology, Distribution, and Control*. CAB International, Wallingford, UK, pp. 7–26.

Turner, S.J. and Subbotin, S.A. (2013) Cyst nematodes. In: Perry, R.N. and Moens, M. (eds) *Plant Nematology*, 2nd edn. CAB International, Wallingford, UK, pp. 109–143.

Tylka, G.L. and Mullaney, M.P. (2017) Soybean cyst nematode-resistant soybeans for Iowa. Extension Publication IPM 1649. Iowa State University Extension, Ames, Iowa, USA.

Tylka, G.L., Gebhart, G.D., Marett, C.C. and Mullaney, M.P. (2017) Evaluation of soybean varieties resistant to soybean cyst nematode in Iowa – 2017. Extension Publication IPM 52. Iowa State University Extension, Ames, Iowa, USA.

Venkatesh, R., Harrison, S.K. and Riedel, R.M. (2000) Weed hosts of soybean cyst nematode (*Heterodera glycines*) in Ohio. *Weed Technology* 14, 156–160. DOI: 10.1614/0890-037X(2000)014[0156:WHOSCN]2.0.CO;2

Volterra, V. (1926) Fluctuations in the abundance of a species considered mathematically. *Nature* 118, 558–560. DOI: 10.1038/118558a0

von Mende, N., Gravato-Nobre, M. and Perry, R.N. (1998) Host finding invasion and feeding. In: Sharma, (ed.) *The Cyst Nematodes*. Kluwer Academic Publishers, Dordrecht, The Netherlands, pp. 217–238.

Wallace, H.R. (1956) Soil aeration and the emergence of larvae from cysts of the beet eelworm, *Heterodera schachtii* Schm. *Annals of Applied Biology* 44, 57–66.

Wallace, H.R. (1960) VI. The influence of soil type, moisture gradients and host plant roots on the migration of the potato-root eelworm *Heterodera rostochiensis* Wollenweber. *Annals of Applied Biology* 48, 107–120. DOI: 10.1111/j.1744-7348.1960.tb03509.x

Wallace, H.R. (1968) The dynamics of nematode movement. *Annual Review of Phytopathology* 6, 91–114. DOI: 10.1146/annurev.py.06.090168.000515

Wang, J., Donald, P.A., Niblack, T.L. *et al.* (2000) Soybean cyst nematode reproduction in the north central United States. *Plant Disease* 84, 77–82. DOI: 10.1094/PDIS.2000.84.1.77

Wang, J., Niblack, T.L., Tremaine, J.N., Wiebold, W.J, Tylka, G.L., Marett, C.C., Noel, G.R., Myers, O. and Schmidt, M. (2003) The soybean cyst nematode reduces soybean yield without causing obvious symptoms. *Plant Disease* 87, 623–628. DOI: 10.1094/PDIS.2003.87.6.623

Wessels, T. (2013) *The Myth of Progress: Towards a Sustainable Future*, 2nd edn. University of Vermont Press, Burlington, Vermont, USA.

Wharton, D. (2002) Nematode survival strategies In: Lee, D.L. (ed.) *The Biology of Nematodes*. Taylor & Francis, Abingdon, UK, pp. 389–412.

Wheeler, T.A., Pierson, P.E., Young, C.E., Riedel, R.M., Willson, H.R., Eisley, J.B., Schmitthenner, A.F. and Lipps, P.E. (1997) Effect of soybean cyst nematode (*Heterodera glycines*) on yield of resistant and susceptible soybean cultivars grown in Ohio. *Journal of Nematology* 29, 703–709.

Wilson, E.O. (2016) *Half-Earth: Our Plant's Fight for Life*. Liveright Publishing Co., New York, USA.

Wollenweber, H.W. (1923) *Krankheiten und Beschädigungen der Karooffel*. Arbeiten des Forschungsinstitutes für Kartoffelbau, Heft 7, Paul Parey, Berlin, Germany.

Wrather, J.A., Anderson, T.A., Arsyad, D.M., Tan, Y., Ploper, L.D., Porta-Puglia, A., Ram, H.H. and Yorinori, J.T. (2001) Soybean disease loss estimates for the top ten soybean-producing countries in 1998. *Canadian Journal of Plant Pathology* 23, 115–121. DOI: 10.1080/07060660109506918

Wrather, A., Shannon, G., Balardin, R. *et al.* (2010) Effect of diseases on soybean yield in the top eight producing countries in 2006. *Plant Health Progress* DOI: 10.1094/PHP-2010-0102-01-RS.

Wright, R.T. (2005) *Environmental Science: Towards a Sustainable Future*. Prentice Hall, Upper Saddle River, New Jersey, USA.

Wyss. U. (2002) Feeding behaviour of plant-parasitic nematodes. In: Lee, D.L. (ed.) *The Biology of Nematodes*. Taylor & Francis, Abingdon, UK, pp. 233–260.

Xing, L. and Westphal, A. (2006) Interaction of *Fusarium solani* f. sp. *glycines* and *Heterodera glycines* in sudden death syndrome of soybean. *Phytopathology* 96, 763–770. DOI: 10.1094/PHYTO-96-0763

Yen, J.H., Niblack, T.L. and Wiebold, W.J. (1995) Dormancy of *Heterodera glycines* in Missouri. *Journal of Nematology* 27, 153–163.

Yuan, H.X., Yan, H.T., Sun, B.J., Xing, X.P. and Li, H.L. (2014) Infection dynamics of two species of cereal cyst nematode in Zhengzhou, Henan Province. *Acta Phytopathology* 44, 74–79.

Zancada, M. and Sanchez A. (1988) Effect of temperature on juvenile emergence of *Heterodera avenae* Spanish pathotypes Ha81 and Ha22. *Nematologica* 34, 218–225. DOI: 10.1163/002825988X00314

Zasada, I.A., Peetz, A., Wade, N., Navarre, R.A. and Ingham, R.E. (2013) Host status of different potato (*Solanum tuberosum*) varieties and hatching in root diffusates of *Globodera ellingtonae*. *Journal of Nematology* 45, 195–201.

7 Quarantine, Distribution Patterns and Sampling

Jon Pickup[1], Adrian M.I. Roberts[2] and Loes J.M.F. den Nijs[3]

[1]*Science and Advice for Scottish Agriculture, Edinburgh UK;* [2]*Biomathematics and Statistics Scotland, Edinburgh, UK;* [3]*National Plant Protection Organisation/ Netherlands Food and Consumer Product Safety Authority, Wageningen, The Netherlands*

7.1 Introduction	128
7.2 Quarantine	129
7.3 Phytosanitary Status	132
7.4 Soil Sampling	135
7.5 Laboratory Diagnosis	144
7.6 Conclusions and Future Prospects	150
7.7 References	151

7.1 Introduction

Under the guidelines set out within the International Standard on Phytosanitary Measures (ISPM) 16 (FAO, 2002), quarantine pest status implies that specific regulations are implemented to ensure that phytosanitary measures address all transmission pathways to ensure that such pests are not introduced into or spread within the country, and that, if found, official control measures are implemented with the aim of eradication or containment. It is generally the responsibility of individual countries to make decisions of whether a pest is treated as a quarantine species.

Mature cysts of cyst nematodes contain and protect the eggs from pathogens and predators as well as from environmental factors that would compromise their long-term survival. Some cyst nematode genera also produce eggs outside the body, depending on environmental circumstances, e.g. *Heterodera glycines* (Turner and Subbotin, 2013). These eggs can hatch more easily and are protected by a gelatinous matrix forming a so-called eggmass (see Chapter 3, this volume). These species of cyst nematodes with external eggs frequently have more than one generation per year (Perry and Moens, 2011). Unhatched juveniles within cysts can survive in the soil over many years. Each growing season, a percentage of the second-stage juveniles (J2) hatch and emerge from the cyst. These J2 die unless they are successful in finding a host plant. For this reason, population assessments of cyst nematodes in soil assume that only eggs retained within cysts remain viable and, therefore, cysts are generally the entities that are used to determine population densities within field soils. Laboratory tests generally rely on the extraction of cysts from the soil and, since cysts will contain a varying number of eggs and many may contain no eggs, assessments of the actual cyst contents also need to be made.

Cyst nematodes can be introduced into a field in many ways. Initial introductions generally occur as a result of poor biosecurity, with the most common pathways believed to be with soil associated with plants for planting (or seed potatoes in the case of potato cyst nematodes), or the movement of agricultural machinery. Once cyst nematodes are established in an area, allowing high populations to develop on susceptible host crops increases the subsequent risk of spread via the movement of soil with wind, water, domestic and wild animals, as well as via the various methods of human transport. The means of introduction is likely to determine the initial population that will be bulked up each time a susceptible host crop is grown. As these nematodes are only capable of unassisted movement over short distances of 10–50 cm per year (Perry and Curtis, 2013), spread over greater distances in the field relies on assistance from external factors, for example, cultivation practices moving cysts both horizontally and vertically (Been and Schomaker, 2013). Infestations are presumed to start with one infected plant and, where there is a predominant direction of cultivation over the years, the resulting hotspot is usually skewed in that direction (Schomaker and Been, 1999). New foci can arise subsequently through further contamination within the field. Therefore, over time, the distribution within a field can vary from a highly aggregated initial focus of infestation to a more even and widespread distribution across an entire field, termed a 'full-field' distribution.

The distribution of cysts within a field has a critical influence on the effectiveness of any sampling scheme. Distributions that are more highly aggregated require more intensive sampling than less aggregated or random distributions. The spatial nature of the aggregation will determine the optimum approach to sampling, that is, determining the optimum frequency, distribution and size of sampling points (soil cores). Sampling procedures can also be developed for different purposes: primarily detection, that is, determining whether cysts are present within the field or, when known to be present, for estimating population levels to underpin management strategies.

7.2 Quarantine

A quarantine pest is defined within the Glossary of Phytosanitary Terms published in ISPM 5 by the International Plant Protection Convention (IPPC) as 'a pest of potential economic importance to the area endangered thereby and not yet present there, or present but not widely distributed and being officially controlled' (FAO, 2007). Quarantine pests are usually categorized by countries or by Regional Plant Protection Organizations as either A1, when the pest is not present within that area, or A2, when the pest is present in that area but not widely distributed there and being officially controlled (e.g. EPPO, 2016). Not all countries discriminate on this basis. A further definition is provided for pests that are not quarantine pests but which may be subject to phytosanitary measures because their presence on plants for planting results in economically unacceptable impacts (Table 7.1). These pests are defined in the IPPC as regulated non-quarantine pests (RNQPs). FAO (2002) differentiates between the two categories based on distribution, pathways for the transmission of the pest, economic impact and requirements for control. So the extent to which a pest is regulated by the plant protection authorities within a country or a region depends largely on the economic

Table 7.1. Comparison of quarantine pests and regulated non-quarantine pests (RNQPs) extracted from International Standard on Phytosanitary Measures 16. (From FAO, 2002.)

Defining criteria	Quarantine pest	Regulated non-quarantine pest
Pest status	Absent or of limited distribution	Present and may be widely distributed
Pathway	Phytosanitary measures for any pathway	Phytosanitary measures only on plants for planting
Economic impact	Impact is predicted	Impact is known
Official control	Under official control if present with the aim of eradication or containment	Under official control with respect to the specified plants for planting with the aim of suppression

importance of the host crop(s) and the current distribution of the pest within the country or region, that is, whether the pest is endemic or the extent to which it has already become established. Further details of the process of pest risk analysis required for determining quarantine status are covered in ISPM 11 (FAO, 2013).

The most highly regulated species of cyst nematodes are the potato cyst nematodes (PCN), *Globodera pallida* and *G. rostochiensis*, and the soybean cyst nematode, *Heterodera glycines*, and the beet cyst nematode *H. schachtii* (Table 7.2). The two species of *Globodera* are treated as quarantine species in most areas of the world because they cause extensive economic damage, are native only to relatively few South American countries and are readily transmitted with the trade in potatoes. However, *G. ellingtonae*, a recently described third species of potato cyst nematode (Handoo *et al.*, 2012), is not currently considered to be a quarantine pest, largely due to its lesser pathogenicity compared to *G. rostochiensis* and *G. pallida*, and hence much reduced economic impact (I. Zasada, 2017, pers. comm.). The two species of *Heterodera* have in general been less stringently regulated, partly because they probably originated from a wider geographical area, but most likely because their transmission to previously uninfested land is less likely to occur with the planting of their host crops, which are generally propagated from true seed. PCN are mostly spread with the planting of infested tubers and associated soil residues and populations are then amplified on the subsequent crop. Therefore, regulatory measures can be mostly targeted against this primary method of transmission (Hockland *et al.*, 2013). Therefore, the great wealth of work on quarantine, and particularly sampling for cyst nematodes, has been carried out with PCN as the subject, and they

Table 7.2. Countries and regional plant protection organizations (RPPOs) listing species of cyst nematodes as quarantine organisms. (From EPPO, 2017.)

Species	A1 Category	A2 Category	Quarantine pest
Globodera pallida	Argentina, Brazil, Uruguay, Bahrain, China, Kazakhstan, Uzbekistan, Azerbaijan, Moldova, Russia	Canada, Chile, Turkey, APPPC, COSAVE, EPPO, PPPO	USA, Israel, Jordan, Belarus, Norway
G. rostochiensis	Argentina, Brazil, Paraguay, Uruguay, Bahrain, China, Kazakhstan, Uzbekistan, Azerbaijan, Moldova	Canada, Chile, Russia, Turkey, Ukraine, APPPC, COSAVE, CPPC, EPPO, IAPSC, PPPO	USA, Israel, Jordan, Belarus, Norway
G. tabacum sensu *lato*			Jordan
Heterodera avenae	Brazil		Jordan
H. cajani	CPPC		
H. cruciferae			Jordan
H. fici	Argentina		Israel
H. glycines	Argentina, Chile, Paraguay, Uruguay, Russia, Turkey, Ukraine, CAN, CPPC, IAPSC	Canada, COSAVE, EPPO	Israel, Jordan
H. goettingiana	Brazil		
H. hordecalis			USA
H. latipons	Brazil		
H. oryzae	APPPC		
H. schachtii	Argentina, Brazil, IAPSC	China, APPPC	
H. zeae	Brazil		Jordan

The acronyms used for the RPPOs are: EPPO, European and Mediterranean Plant Protection Organization; APPPC, Asia and Pacific Plant Protection Commission; CAN, Comunidad Andina; COSAVE, Comite de Sanidad Vegetal del Cono Sur; CPPC, Caribbean Plant Protection Commission; IAPSC, Inter-African Phytosanitary Council; and PPPO, Pacific Plant Protection Organization.

EPPO also maintain an Alert List to draw the attention of member countries to pests that may become of quarantine concern. This list can also be used to select candidates which may be submitted to a pest risk analysis. One species of cyst nematode, *Heterodera elachista*, is included on this list (EPPO, 2017).

will be the prime example for the discussion throughout this chapter.

7.2.1 Soybean cyst nematode *Heterodera glycines*

Following rapid expansion in the area of soybean grown in the USA from the mid-1900s, federal quarantine was established for *H. glycines* in 1957 to help bring this pest under control. Infestations of *H. glycines* now occur in every soybean-producing state within the USA (see Chapter 6, this volume), with estimates of annual soybean yield losses of nearly US$1 billion (www.apsnet.org/edcenter/intropp/lessons/nematodes/pages/soycystnema.aspx). It appears that by the time quarantine measures were introduced, the extent of the spread of *H. glycines* into new soybean production areas through traditional cultivation practices resulting in the unintentional movement of infested soil was probably too widespread to make the introduced measures effective. Consequently, quarantine measures were lifted in 1972, with the pest having become widely distributed and the official control measures deemed to have had limited effect. *Heterodera glycines* remains a quarantine pest elsewhere where it is of much more limited distribution, for example, in Europe where it is listed as an A2 pest by the European and Mediterranean Plant Protection Organization (EPPO), having been reported from Italy in 2000 (Manachini, 2000); A2 is the classification given to quarantine pests that are locally present in the EPPO region. In North America *H. glycines* has recently spread to Canada where quarantine regulations have been imposed (Mimee *et al.*, 2014) and is spreading within China (Peng *et al.*, 2016) and within South America (Mendes and Dickson, 1993).

7.2.2 Potato cyst nematodes *Globodera pallida* and *G. rostochiensis*

The two very closely related species of PCN, *G. rostochiensis* and *G. pallida*, co-evolved with the potato in South America but have subsequently been introduced elsewhere with the movement and cultivation of potatoes. Outbreaks of these species have now occurred in most of the major potato growing areas of the world (Turner and Evans, 1998). China is probably the only country in the world growing a large acreage of potatoes that has yet to report an outbreak of PCN. Reports of PCN remain scarce from some countries with extensive potato acreages, most notably Australia, Canada, USA, India and, probably, some parts of the former USSR. Knowledge of the extent of PCN infestations varies with the amount of resources that have been invested in surveillance programmes, particularly in developing countries. However, PCN continues to be treated as a quarantine pest by regional plant protection organizations throughout the world.

Both *G. rostochiensis* and *G. pallida* are listed as A1 or A2 pests, throughout the world (Table 7.2). EPPO has also suggested that at some future time, consideration should be given to treating the individual pathotypes or virulence groups as quarantine organisms in their own right, rather than the species themselves. Some pathotypes are widely distributed in Europe and others less so, plus some are more economically important than others, for example, pathotypes of *G. rostochiensis* with the ability to overcome the resistance conferred by the *H1* gene, or the new virulence groups of *G. pallida* recently found in Germany (Niere *et al.*, 2014) and The Netherlands (den Nijs, 2016). Unfortunately, our inability to categorize pathotypes or virulence groupings robustly has hindered progress towards establishing detailed knowledge of the distribution of these groupings, although recent developments in molecular methods are opening up the potential for improved description of the genetic diversity within species (Eves-van den Akker *et al.*, 2015). Furthermore, we now know that a far greater diversity of virulence is present in the Andean region of South America, which is considered the genetic origin of *G. pallida* and *G. rostochiensis* (Plantard *et al.*, 2008). The risk to the continued use of resistant cultivars presented by the potential introduction into Europe of non-European populations of both species has strengthened the case for using plant health legislation to safeguard against the introduction of populations that would increase the genetic diversity of PCN in Europe (Hockland *et al.*, 2012).

7.3 Phytosanitary Status

Where a country participates in international trade then, under the standards of the IPPC, the country is required to provide evidence of the phytosanitary status of quarantine pests within that country. Any claims of pest absence must be supported by appropriate surveillance activities. Three phytosanitary approaches for any pest should be considered, based on the history of outbreaks of that pest and the long-term phytosanitary objective for the region or country under consideration: (i) the establishment and maintenance of pest free status with regards to the pest; (ii) the eradication of isolated outbreaks of the pest and the restoration of pest free status; and (iii) the management of known infestations of the pest with the aim of limiting further increase and spread. The principles that underpin these phytosanitary options are set out in ISPM 1 (FAO, 2006) and are discussed in detail below using PCN as the example.

7.3.1 Pest free areas

Based on the IPPC definition, pest free areas (PFA), or areas of freedom, are defined as areas where a specific pest does not occur as demonstrated by scientific evidence and in which, where appropriate, this condition is being officially maintained (FAO, 2007). In the case of *G. rostochiensis* and *G. pallida*, where detection is dependent upon soil sampling, the soil sampling strategy employed determines the detection level, and levels of probability of detection can be calculated for distributions of PCN that are known to be typical of field outbreaks. Random, stratified, risk-based, national surveys need to be conducted to demonstrate this.

From a scientific point of view, 'areas' should be defined by logical agriculture production systems and geographical barriers rather than political boundaries. ISPM 4 ('Requirements for the establishment of pest free areas') stresses the importance of delimiting a PFA with relevance to the biology of the pest concerned (FAO, 1995). For PCN, and for cyst nematodes in general, which are dispersed mainly by cultivation practices, the crop production system is very relevant to their epidemiology. Therefore, rather than using political boundaries, such as those between states and countries, PFAs should be delimited by a more detailed assessment of risk, for example, where production systems within a region are isolated, then these should be treated separately. Similarly, if production systems extend across political boundaries, then these should probably be assessed as belonging to the same PFA. The EPPO standard 'Pest-free areas and pest-free production and distribution systems for quarantine pests of potato' (EPPO, 2004) sets out the following phytosanitary measures relevant to the control of PCN in potato production systems:

- general surveillance or specific surveys to demonstrate that production systems of potatoes and other host plants are free from the pest (see ISPM 6; FAO, 1997);
- prohibition of the movement of potatoes into the PFA for storage, dressing or packing, except from other PFAs;
- prohibition of the planting of potatoes in the PFA, originating from areas which are not PFAs;
- establishment of a buffer zone around the PFA with other measures depending on the biology of the pest and the geographical area;
- precautions on the movement of machinery used in potato production into the PFA from areas where the pest occurs; and
- restriction of the movement and use, in the PFA, of commodities other than potatoes that may be infected by the pest.

Using PCN as an example, Table 7.3 sets out the implications of adopting four different levels of regulatory status in relation to several important criteria, including pest distribution, surveillance, risks and impact on crop production. For North America, PCN status is mainly *Pest free area* with some areas under *Eradication/Containment*. For Europe, some areas can be considered as under *Eradication/Containment* but most areas are currently at *Management* status, as reflected by the conditions laid out in the European Union (EU) PCN Directive 2007/33/EC (Anonymous, 2007).

7.3.2 PCN eradication and containment

When isolated outbreaks of PCN occur, the authorities with phytosanitary responsibilities

Table 7.3. Implications of regional pest status on the regulatory issues in relation to potato cyst nematodes (*Globodera pallida* and *G. rostochiensis*) and the production of host crops.

Pest status	Pest free area (PFA)	Eradication/ containment	Management	Unregulated
Pest distribution	Absent	Isolated outbreak(s)	Widely distributed, eradication no longer feasible	Widely distributed
Surveillance (soil testing)	Sufficient to justify PFA status Could use soil associated with harvested crops – e.g. tare dirt	Intensive localized soil testing in outbreak and associated fields, track back, track forward Field-based sampling	Lower intensity sampling Field-based sampling, emphasis on plants for planting	Grower's discretion
Impact of new introduction	Very severe	Severe	Moderate, except for high risk populations (South American or resistance breaking)	Low, except for high risk populations
Potato (host crop) production	Unrestricted	Tightly controlled in outbreak and associated areas	Seed potato production prohibited in infested fields	Unregulated
Biosecurity	PFA based General prohibition of planting of potatoes unless from a PFA Precautions on the movement of machinery into PFA Precautions on other commodities that could introduce PCN	Very strict controls on all pathways for movement of contaminated soil and produce from infested fields Precautions on soil movement for associated fields	Limiting the potential for new introductions by ensuring seed potatoes are produced on PCN-free land	Unregulated
PCN control measures	Pest is absent, so none required	Reduction of population levels within known outbreaks, e.g. use of soil sterilants, trap crops and biofumigants Potato and other host crop production prohibited in infested fields Highly resistant variety cultivation for fields that have been decontaminated and for associated fields	Use of soil sterilants, trap crops, biofumigants, inundation/ anaerobic disinfestations Use of nematicides to reduce initial population levels and protect yields Use of resistant varieties to protect land for future potato production Increased rotation periods	High dependency on: Use of nematicides to reduce initial population levels and protect yields Use of resistant varieties to protect land for future potato production Increased rotation periods Use of soil sterilants, trap crops, inundation and biofumigants

Continued

Table 7.3. Continued.

Pest status	Pest free area (PFA)	Eradication/containment	Management	Unregulated
Costs	Soil testing – low Full access to export markets	Soil testing – high Prohibition of potato production for a few infested fields Loss of production associated with trap and biofumigant crop use Variety choice restricted for associated fields Majority of growers in area unaffected Export markets unaffected for areas of production not associated with the outbreak	Soil testing – moderate – required for seed, growers choice for ware Environmental and financial cost of widespread nematicide use Loss of production associated with trap and biofumigant crop use Restricted choice of varieties Extended crop rotations Many/most growers affected Risk of selection for resistance breaking PCN populations. Export markets restricted	Soil testing – as required by the grower Dependency upon widespread nematicide use – high environmental and financial cost Loss of production associated with trap and biofumigant crop use Restricted choice of varieties Extended crop rotations Many/most growers affected Risk of selection for resistance breaking PCN populations Loss of export markets

have to make a decision on whether eradication is feasible, or whether to move into a management programme with the aim of limiting the potential of these pests to cause damage. Looking at the historical records of PCN outbreaks, eradication has only been claimed to have been achieved when the outbreak has occurred in a very discrete area, for example, on the Saanich peninsula on Vancouver Island, Canada (Rott et al., 2010) and in Western Australia (Collins et al., 2010).

Soil sampling is an essential component for monitoring during an eradication programme, although it has considerable limitations when population densities are extremely low. Infestations are very unlikely to be found when population densities are below the detection limits for any respective sampling programme. Once outbreaks are known to have occurred in an area, an increased level of surveillance can be justified, with more intensive soil sampling increasing the probability of detecting outbreaks at an earlier point in their development.

Eradication requires strict hygiene measures to ensure that all risks of spread of PCN from the infested site are addressed. To assist with this process in situations where PCN populations have reached levels that cause visual symptoms, a means of rapid reduction of the nematode population is highly advisable, either by soil fumigation, resistant varieties, trap crops, anaerobic soil disinfestation by means of flooding/inundation or a combination of these methods (see Chapter 13, this volume). At all points during an eradication programme, it is essential to have measures in place that will prevent any further introductions. If the practical operation of such measures is not feasible, then careful consideration should be given at the outset as to whether to commence with an eradication programme.

7.3.3 PCN management

In the event of PCN becoming established and endemic within an area, then a less stringent approach can be adopted. This can be considered similar to that taken across most member states of the EU, where eradication is now considered an unobtainable goal. Under these circumstances, pre-crop soil sampling may be applied routinely to all areas of potato production and effective PCN management procedures can be used. However, the costs of managing PCN can be considerable, when factors such as yield loss due to PCN, nematicide treatment, the cultivation of resistant varieties, extended rotation periods, the production of trap and/or biofumigant crops or inundation methods are considered.

Where regulatory authorities adopt more relaxed official control programmes and/or quarantine measures, the risk of spread of infestations is inevitably increased. In regions where infestations are very localized, looser regulation is likely to result in an increased proportion of growers having to manage infestations and hence incur the additional associated costs on a regular basis. By contrast, those growers in outbreak areas in which infestations have hitherto been highly regulated can be granted greater freedom to resume commercial production of host crops. Balancing the costs and benefits of regulatory control can be highly complex and challenging.

7.4 Soil Sampling

Each time a susceptible potato crop is grown in a field, the PCN population has the potential to increase by up to 100-fold (Evans and Kerry, 2007) and will, if not appropriately managed, lead to land becoming unsuitable for the commercial production of potatoes. To determine whether PCN are present in land and, if present, to assess the species and population level, field soils need to be sampled and tested for PCN. In each hectare of a field, the top 20 cm depth of soil equates to 2 million litres of soil, only a very small proportion of which can feasibly be tested. PCN are also unevenly distributed on a horizontal scale across the field, with an aggregated distribution that reduces the probability of detection. Therefore, the challenge when developing a strategy for soil sampling is how best to take a soil sample that represents the PCN status of a field. If the approach to sampling is not optimal, then the results from any subsequent laboratory testing will be compromised. The larger the sample, the more expensive the test will be, although solutions to this problem are being found (see section 7.4.4). Field sampling costs are largely dependent upon the time taken to draw the sample, and in turn, this will depend on how many cores are drawn and the pattern used to sample the field. Finding low-level and patchy cyst infestations in a field is particularly challenging. A single cyst in a 400 cm^3 sample of soil drawn from 1 ha will equate approximately to a population of 5 million cysts ha^{-1}, a level that is likely to have taken several field generations to reach a detectable level. Earlier detection would require soil sampling at a rate that is at least an order of magnitude more intensive and with a proportionate increase in laboratory costs. The official described sampling method of the EU, 1500 ml ha^{-1}, which is thought to be adequate to show a field free of PCN, translates to a detection level of 90% of finding one cyst, which is still equivalent to 1.5 million cysts ha^{-1}, based on a scenario where four foci of infestation are randomly distributed within a 1-ha field; three foci with a central population density of 50 cysts kg^{-1} and one with 150 cysts kg^{-1} (Schomaker and Been, 1999). Therefore, it is unrealistic to expect to be able reliably to detect PCN populations at very early stages of infestation. For this reason, the dissemination of cyst nematodes, particularly PCN, has proved very difficult to control since early detection requires soil sampling at extremely intensive rates. With seed potato production, following an initial infestation, it is highly likely that at least three seed crops will have been produced in a field and planted elsewhere before that initial infestation has been amplified to a level at which it can be detected by a realistic soil sampling programme. Once PCN cysts are known to be present in a field, national control programmes will typically prohibit the production of seed potatoes.

Fortunately, it is feasible to detect PCN by soil sampling before the pest can cause significant economic damage. Tools to manage PCN and provide commercial yields of ware potatoes are available, so monitoring and estimating population levels becomes critically important for the sustainable management of PCN, minimizing the economic impact on potato crops grown in land infested by these pests.

Therefore, there are two purposes for testing fields for cyst nematodes: *detection* and *quantification or population estimation*, that is, is PCN present and, if so, how much PCN is present? For detection, it is usually assumed that the distribution of cysts within the field will be typical of an early stage of infestation, although it is also possible that an older infestation may be present that has been managed so that population levels have reduced to levels below which cysts cannot be reliably detected. For the quantification of populations, it is more likely that infestations will be well established within the field and have been spread more widely around the field by cultivation practices. Therefore, different distribution models to describe PCN infestations typical of these differing scenarios can be developed and applied to the two approaches to sampling: detection and population density estimation.

When sampling for detection, the purpose is often to minimize the risk of transmission of cyst nematodes with infested planting material. When sampling for population density estimation in land unsuitable for the production of seed potatoes, the purpose can be either the protection of the crop that will be grown next in the field, or the protection of the land for future host crop production – or, hopefully, both.

The point at which cysts begin to cause damage within a crop is dependent upon the population level/viability of the cysts, the tolerance of the cultivar, the soil type and other environmental factors. Such variables have made the establishment of recommended threshold population levels for the protection of the crop very difficult. The theoretical tolerance level for potato plants to PCN is around 2 eggs g^{-1} soil, according to Seinhorst (1965). Typical threshold values range from 5 to 20 eggs g^{-1} soil, with higher values more acceptable in more moisture- and nutrient-retaining soils. When working with such threshold values, it is important to establish what the confidence limits associated with and population estimates are likely to be.

A single crop of a susceptible cultivar has the capacity to increase the PCN population by over 50-fold, that is, taking the population density from the limits of detection (using many commonly used sampling programmes) to highly damaging in one cropping season. Therefore, to protect land known or suspected to be infested, it is highly advisable not to grow cultivars susceptible to the species of PCN present without taking additional measures to mitigate the population increase. When considering thresholds for PCN management, most attention is generally paid to the protection of the crop to maintain yield, whereas the implications for the future use of the land are too frequently overlooked.

7.4.1 Detection of cysts/juveniles

PCN survive in the soil in eggs contained within cysts, which provide a physical barrier that protects the eggs within from pathogens and predators as well as from environmental factors that would compromise their long-term survival. Laboratory extraction methods are generally targeted at isolating cysts from soil, on the assumption that eggs within the cysts will contain viable J2 capable of invading a potato root. For species of cyst nematodes that produce eggs that are not contained within the cysts, extraction methods that recover eggs and J2, as well as cysts are required. However, for PCN it is safe to assume that eggs only exist in cysts; thus, cysts are generally the entities that are used for determining population densities within field soils.

Each growing season, a small percentage of J2 will hatch and emerge from the cyst; typically decline rates of 20–50% have been found, depending upon environmental conditions and species, although in The Netherlands declines of up to 70% have been recorded in the first year (Schomaker and Been, 2000). J2 emerging from cysts will die unless they are successful in finding a host plant (see Chapter 3, this volume). Unhatched J2 can survive within cysts in the soil over many years, and up to 30 years or more is typical. However, empty cysts can persist for a considerably longer time and records from fields in Edinburgh, Scotland, that have not grown potatoes since the 1940s indicate that non-viable empty PCN cysts can be extracted from the soil over 70 years after they were originally produced. Scottish data indicate that even in seed land, where findings of PCN have often been assumed to be a consequence of more recent infestations, empty cysts have been more regularly encountered than live cysts – of 8195 samples containing cysts, 970 (12%) had live but not empty cysts, whereas 3541 (43%) had empty but not live cysts (unpublished SASA, UK, records, 1975–2010).

Therefore, for quarantine purposes and production of planting material with an associated risk of transferring cysts with soil, for example, seed potato production, where knowing whether PCN are present or absent in a sample is the key question, the aim is to establish whether the sample contains a cyst with viable unhatched J2. If the sampling is for advisory or management purposes, for example, ware potato production, the aim is more likely to be the provision of a population density estimate. In this case, information on the number of cysts present is less useful than an assessment of the eggs that they contain. In summary, for detection, cysts with live content are the target, whereas for population density estimation, estimates of eggs per gram of soil are the target. However, as eggs are aggregated within cysts, this level of aggregation has an important bearing on the confidence limits associated with testing soil for PCN.

7.4.2 Distributions of PCN within fields

The distribution of cysts within a field has a critical influence on the effectiveness of any sampling scheme. A random pattern means that the chance of a cyst being found at any particular point in the field is dependent neither upon the positions of other cysts in the field nor the position within the field. There can be deviations from randomness in two directions: towards more aggregated distributions and towards more regularly dispersed patterns. In the first case, cysts are more likely to occur in the vicinity of other cysts, and in the latter case they are less likely to be near other cysts. Distributions that are more highly aggregated require more intensive sampling than less aggregated or random distributions. The spatial nature of the aggregation will determine the optimum approach to determining the frequency and distribution of sampling points (cores).

The nature of the life cycle of cyst nematodes means that unassisted movement of cysts is slow. These species are only capable of moving short distances, 10–50 cm per year, without assistance from external factors, for example, cultivation processes such as ploughing. Where there is a predominant direction to this movement over the years, the resulting hotspot will be skewed in the direction of cultivation. Modelling of PCN distributions and the development of infestation foci is well described by Schomaker and Been (1999). New foci can develop subsequently in the field through further contamination within, or between, fields.

The negative binomial distribution (McSorley and Parrado, 1982; Gilligan, 1988; Madden *et al.*, 2007) has commonly been used to describe aggregated distributions, such as found with cyst nematodes. To allow for the possibility that some parts of the field may be free from cysts, it may be necessary to allow for an excess of zero counts. A potential solution lies in the zero-inflated negative binomial distribution (e.g. Denwood *et al.*, 2008).

Whilst these statistical spatial methods are good for illustrating and understanding the pattern of PCN infestation in individual cases, they may be less useful in setting out generally applicable models that can, in turn, drive decisions on sampling strategies. Alternatively, Schomaker and Been (1999) proposed a more mechanistic model that has the benefit of using the mapping information whilst also forming the basis for understanding the benefits of different sampling strategies (Been and Schomaker, 2000). The Schomaker and Been model builds in an understanding of the nature of nematode infestations, that populations build from initial points of arrival and that cysts are moved relatively slowly and predominantly by cultivation. The authors reviewed infestations in many potato fields in The Netherlands, finding that infestations were 'approximately lozenge-shaped and cysts densities decreased exponentially away from the focus centre, but more slowly in the length than in the width direction'. They constructed a relatively simple model for such distributions. Two parameters, defining the rates of decline in the direction of and perpendicular to the direction of cultivation, were fitted in each field. Estimates for these parameters have been used for subsequent evaluation of sampling schemes. Thus, with the model a field infestation can be defined simply by the number of hotspots and density of PCN at their centres (the central population density, CPD). This method has been the core of the Dutch PCN decision-making system over recent years and is well described (e.g. Been and Schomaker, 2000). Note that this model describes the 'medium scale' of PCN. For the small-scale distribution (approximately 1 m^2), a negative binomial distribution is used to account for the high local variability in PCN counts. Again the authors fitted these models to the infested fields, leading to an estimate of the degree of aggregation that could then be used in sampling simulations.

Whilst the Schomaker and Been model provides a convenient basis to explore sampling options, it is not certain how well it describes established infestations that are more widely spread across fields, situations that may be commonly found, particularly where ware potatoes have been repeatedly grown on PCN-infested land. In a study commissioned by AHDB Potatoes, Pickup et al. (2016) collated available data from PCN-infested ware fields within the UK. These data sets comprised individual soil test results for different areas within fields, providing information on the degree of heterogeneity of cyst distributions within fields. These data sets were used to evaluate appropriate distributional models. A zero-inflated negative binomial model provided the best overall fit to the PCN count data from fields in England and Scotland, although there were data sets where there were no zeroes and the occasional case where the fit of the negative binomial provided a slightly better fit. However, neither the Poisson nor the zero-inflated Poisson distribution fitted the data as well as the negative binomial. This was due to the substantial heterogeneity in the data. Figure 7.1 shows the relationship between the egg density and the heterogeneity index, which is a key parameter for the zero-inflated negative binomial distribution. A spline fitted trend is included, which shows that heterogeneity tends to decline with higher levels of infestation, that is, the higher the field population, the less aggregated the distribution. In Fig. 7.1, the proportion of zeroes is also shown – this decreases with higher levels of infestation.

It may be that the Schomaker and Been model provides a description of early stage infestations and, therefore, is a more appropriate model to use for determining the most appropriate way to

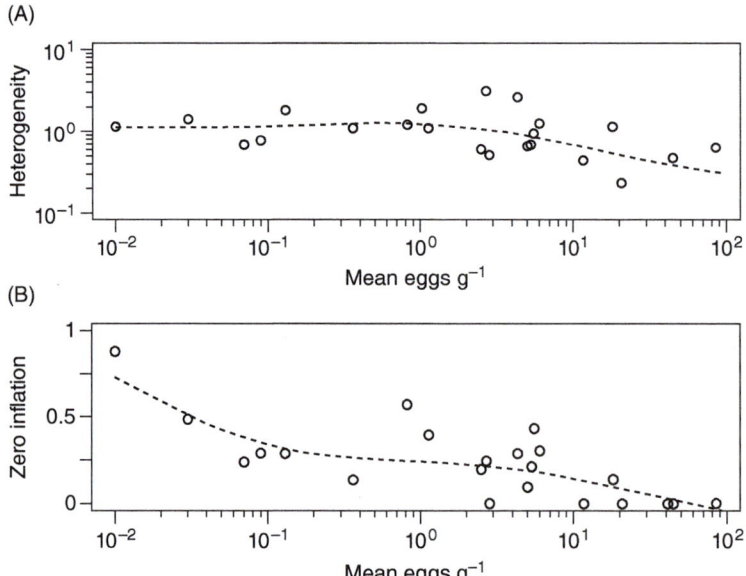

Fig. 7.1. Heterogeneity index (A) and zero-inflation proportion (B) for the zero-inflated negative binomial distribution against egg density of *Globodera rostochiensis* and *G. pallida* (larger values of the heterogeneity index indicate greater heterogeneity and zero means no heterogeneity).

sample a field to detect the presence or absence of PCN. The zero-inflated negative binomial model may represent more established infestations and be more appropriate for developing sampling programmes suited to quantifying field populations that are already known to be present. Whilst these models have been developed for distributions of PCN cysts, the development of infestations of other cyst nematodes, and their subsequent distribution patterns should make them applicable to the sampling for all cyst nematodes.

7.4.3 Sampling for detection

For detection of PCN for quarantine purposes, including requirements for seed potato classification schemes, the models of early infestations developed by Schomaker and Been provide the best available distributions on which to design optimum sampling strategies. Table 7.4 shows the results from simulations based on the Schomaker and Been model with a single focus in a 1 ha field area. Comparisons of numbers of cores, core size and subsample size are based on a square grid of sampling points. For populations of 500,000 or fewer cysts per hectare, the chances of detection are very low. In this situation, the likelihood of detecting a cyst using 100 cores and testing 400 cm^3 of soil is 8% (the lower sampling rate stipulated within EU PCN Directive 2007/33/EC; Anonymous, 2007), rising to 26% using the EU standard rate of 1500 cm^3 ha^{-1}. It is only once the number of cysts per hectare reaches 20 million that there is a 98% chance of detection at the EU standard rate (70% at the EU lower rate). In recognition that soil bulk density can vary considerably, and with the aim of establishing a harmonized sampling rate, the EU Directive stipulates a volume of soil to be tested rather than a weight of soil. However, most growers and agronomists are more familiar with assessments of populations as cysts or eggs per unit of weight (usually grams).

Varying the number of cores changes the probability of detection in relation to CPD. If all the soil is tested, based on a similar core size, the probability of detection increases as the number of cores increases, primarily because more soil is tested. If the increase in number of cores is controlled for by taking smaller volumes so as to reach a total sample of 1500 g for laboratory analysis (Fig. 7.2), the effect of increasing the number of cores is more moderate. However, over a range of CPDs from 20 to 1000 cysts kg^{-1} there is a clear increase in the probability of detection when sampling is based on more cores/ha. At rates of over 100 cores ha^{-1} the resulting improvement is relatively low. Therefore, these data

Table 7.4. Detection probability using the Schomaker and Been (1999) model with one focus.

			EU standard	EU lower
		Number of cores	100	100
		Volume tested (cm^3)	1500	400
Cysts ha^{-1}	Mean cysts l^{-1}	CPD (cysts kg^{-1})	Detection probability	
500,000	0.25	66.2	26%	8%
1,000,000	0.5	127.9	41%	14%
2,000,000	1.0	250.7	58%	23%
5,000,000	2.5	617.8	81%	40%
10,000,000	5.0	1228.8	92%	55%
20,000,000	10	2450.3	98%	70%
50,000,000	25	6113.4	100%[a]	86%
100,000,000	50	12217.7	100%[a]	93%

[a]Rounded up to 100%.
Focus with length (L) and width (W) parameters of 0.77 and 0.55, respectively, and a common coefficient of aggregation of 70. A soil bulk density of 1.6 kg l^{-1} has been assumed. CPD = central population density.

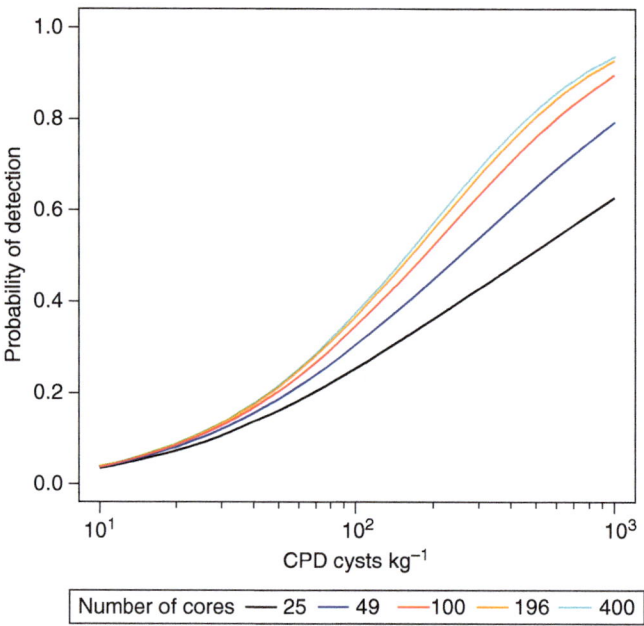

Fig. 7.2. Effect of number of cores on the probability of detection with a fixed sample volume of 1500 cm³. CPD = central population density.

support sampling based on a minimum of 100 cores ha^{-1} when sampling for detection.

The effect of increasing the amount of soil subsampled for laboratory testing from 100 g to 2000 g is shown in Fig. 7.3 (based on a 7 × 7 grid of 49 cores of 50 cm³ each). It is clear that it is highly beneficial to test more soil.

7.4.4 Sampling for population density estimation

Although the Schomaker and Been model is considered to provide the most appropriate description of early stage PCN infestations, it has been suggested that the zero-inflated negative binomial distribution may better describe well-established PCN infestations, particularly in fields that have been under long-term PCN management programmes and where 'full field' infestations can be considered to have developed. Using the available data from PCN-infested ware fields within the UK, Pickup et al. (2016) used the zero-inflated negative binomial distribution to fit spline curves to field data (Fig. 7.1) to estimate the levels of zero-inflation and heterogeneity. These parameters provide the basis for examining the effect on the coefficient of variation (CV) of varying factors such as the number of cores and the volume of the subsample of soil tested across a range of egg densities.

It is clear that a soil sample should consist of a minimum of 50 cores, and preferably 100 or more (Fig. 7.4), although the core size is almost immaterial as long as the core volume is in excess of 4 cm³ (data not shown). Assuming 1 egg cyst^{-1}, subsampling for laboratory testing, using subsamples of between 100 g and 2000 g has little effect once the egg density exceeds 2 eggs g^{-1} (Fig. 7.5). Below 1 egg g^{-1} there is a small but diminishing benefit in testing an increasing volume of soil. Therefore, assuming a similar distribution of eggs and cysts, that is, 1 egg cyst^{-1}, a laboratory subsample of 100 g of soil would be sufficient for PCN population estimation, providing that this subsample is representative of the whole sample and that the target egg density for management decisions is above 2 eggs g^{-1}. It is not possible to infer from these distribution models which sampling pattern works best; the models ignore spatial aspects of the distribution of PCN. However, in the absence of knowledge about the potential pattern of PCN in the field, sampling principles would imply that a square grid would be the most prudent choice.

For simplicity of modelling, the previous section to determine the most important factors influencing population estimation was based on

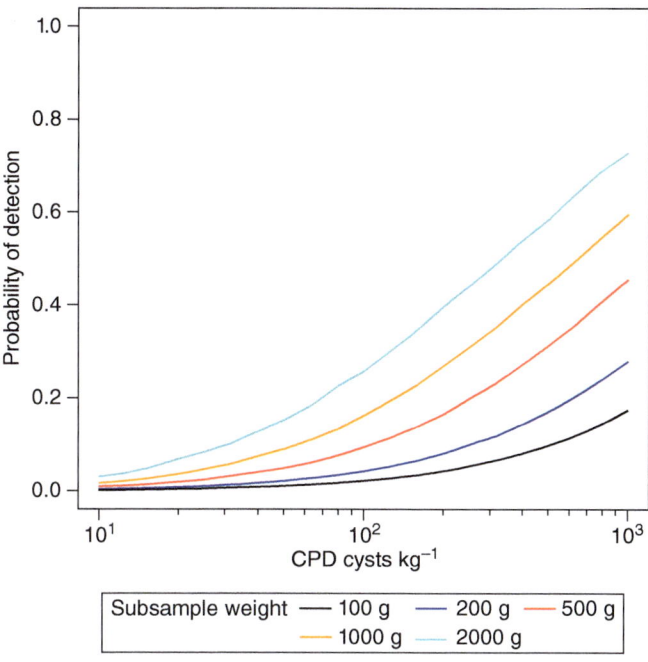

Fig. 7.3. Effect of soil subsample size on the probability of detection with 49 cores and cores of 50 cm³. CPD = central population density. In summary, for the detection of early infestations, the key factor for increasing the probability of detection is increasing the amount of soil tested at the laboratory. Sampling should also involve drawing a minimum of 100 cores ha^{-1}.

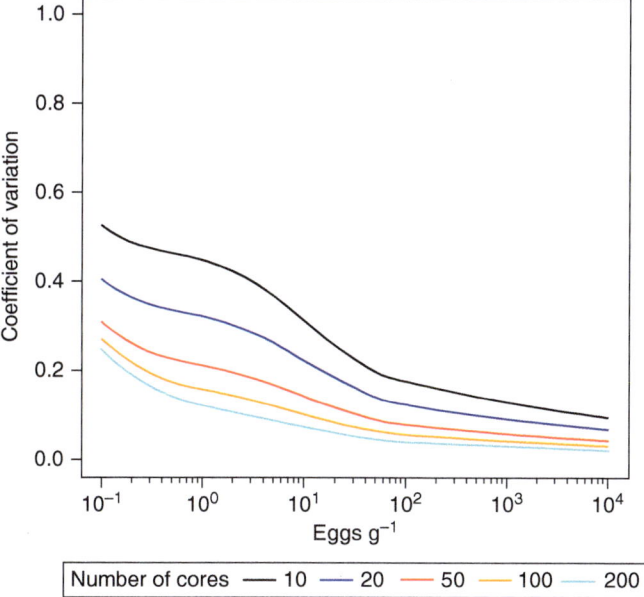

Fig. 7.4. Effect of number of cores (10, 20, 50, 100 and 200 ha^{-1}) on the coefficient of variation for the egg density with core size of 50 cm³ and a subsample of 200 g, assuming 1 egg cyst^{-1}. Sampled area of 1 ha.

having 1 egg cyst^{-1} throughout the sample. This is only possible if eggs were independent from cysts or were found at a rate of 1 egg cyst^{-1}. Figure 7.6 illustrates the effect on the CV when the aggregation of eggs within cysts is increased from 1 to 10 and 50 eggs cyst^{-1}.

The precision of egg density estimates is markedly reduced when the aggregation of eggs into cysts is taken into account. The higher the level of aggregation of eggs within cysts, the greater the CV. The increase in the CV is larger at lower egg densities and with smaller subsamples.

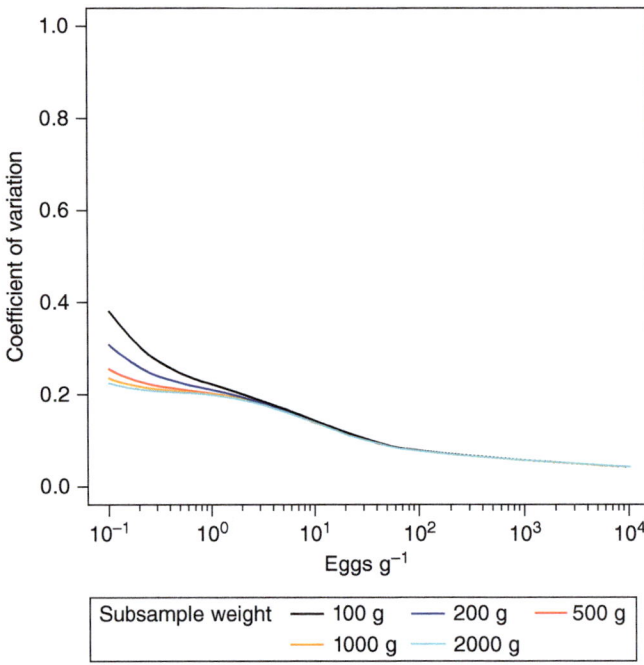

Fig. 7.5. Effect of amount of soil subsampled for laboratory analysis on the coefficient of variation (CV) over a range of egg densities using 50 cores of 50 cm³, assuming 1 egg cyst⁻¹.

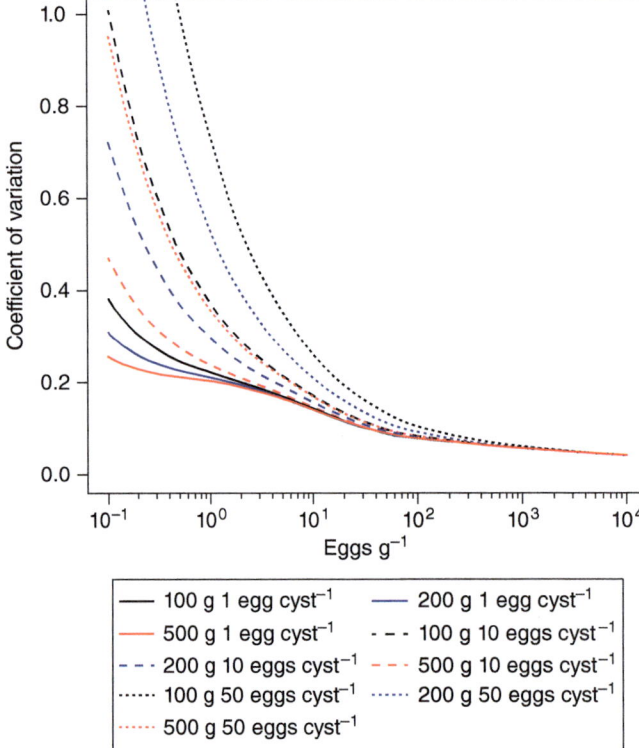

Fig. 7.6. Effect of aggregation into cysts combined with soil subsampling for testing on the coefficient of variation for the egg density with 50 cores of 50 cm³. Based on 100,000 simulations except when 1 egg cyst⁻¹ is assumed (direct calculation).

Without any aggregation of eggs into cysts, 100 g, 200 g and 400 g subsamples have similar CVs at population levels above 2 eggs g^{-1}. This convergence occurs at higher densities when aggregation is taken into account, for example, above 100 eggs g^{-1} when the average number of eggs per viable cyst exceeds 50. In these situations, however, the growing of a healthy potato crop is hardly feasible. For PCN, freshly produced cysts will frequently contain over 300 eggs.

Table 7.5 summarizes the relationship between the confidence interval associated with sampling as the PCN population increases, the aggregation of eggs into cysts increases and the size of the subsample taken increases. A confidence interval of less than the target population has arbitrarily been taken as acceptable (green shading); a confidence interval of between 1.0 and 1.5× the target population is considered marginally acceptable (yellow shading) and a confidence interval of greater than 1.5× the target population is considered unacceptable (red shading).

Table 7.5 clearly shows that smaller subsamples can be taken when the target populations are higher and there are fewer eggs per cysts. For example, with a target population of 20 eggs g^{-1}, the confidence interval associated with an aggregation of 200 eggs cyst^{-1} and a subsample of 200 g would be 10.3 to 29.7 eggs g^{-1}. With a target population of 5 eggs g^{-1}, the confidence interval associated with a similar level of aggregation and subsampling would be 0.4 to 9.6 eggs g^{-1}, which is viewed as unacceptable.

The aggregation into eggs per cyst will depend upon initial number of eggs per fresh cyst, typically 300–400, and the annual decline rate. Decline rates are typically higher for *G. rostochiensis* populations than for *G. pallida*. Low levels of eggs per cyst (below 100 eggs cyst^{-1}) are unlikely to be encountered in fields that have grown susceptible potato cultivars in the past

Table 7.5. Approximate 95% confidence limits (CL) using the zero-inflated negative binomial distribution based on soil subsampling for two target populations of A: 5 eggs (g soil)$^{-1}$, and B: 20 eggs (g soil)$^{-1}$.

	A: Target population = 5 eggs g^{-1}							
Subsample	100 g		200 g		400 g		1000 g	
Eggs cyst^{-1}	Lower CL	Upper CL	Lower CL	Upper CL	Lower CL	Upper CL	Lower CL	Upper CL
10	2.9	7.1	3.1	6.9	3.3	6.8	3.4	6.7
50	1.5	8.5	2.3	7.7	2.8	7.2	3.2	6.8
100	0.4	9.6	1.6	8.4	2.4	7.6	3.0	7.0
150	0.0	10.6	1.0	9.0	2.0	8.0	2.8	7.2
200	0.0	11.4	0.4	9.6	1.7	8.4	2.7	7.3
400	0.0	13.8	0.0	11.3	0.6	9.5	2.1	7.9
	B: Target population = 20 eggs g^{-1}							
Subsample	100 g		200 g		400 g		1000 g	
Eggs cyst^{-1}	lower CL	upper CL	lower CL	upper CL	lower CL	upper CL	lower CL	upper CL
10	14.7	25.3	15.1	24.9	15.3	24.7	15.4	24.6
50	12.4	27.6	13.8	26.2	14.6	25.4	15.2	24.8
100	10.3	29.7	12.4	27.6	13.8	26.2	14.9	25.1
150	8.5	31.5	11.4	28.6	13.2	26.8	14.6	25.4
200	7.0	33.0	10.3	29.7	12.6	27.5	14.3	25.7
400	2.2	37.9	7.1	33.0	10.5	29.5	13.4	26.6

Data are tabulated for four subsampling rates of 100, 200, 400 and 1000 g, and for six levels of aggregation of 10, 50, 100, 150, 200 and 400 eggs cyst^{-1}. A colour coding system is used: green if the difference between the upper and lower confidence limits is less than the target population, yellow if the difference is greater than the target population by a factor of <1.5 and red if the difference is >1.5× the target population. Where the calculated lower confidence limit is less than zero, a zero value has been recorded.

5 years (typical of UK rotation periods). Therefore, subsampling to as little as 100 g of soil per sample, is only likely to be acceptable under a scenario where the target population is 20 or more eggs g^{-1} of soil and the land has been out of potato production for at least 5 years. A subsample of 200 g is recommended for shorter rotations. For target populations of 10 eggs g^{-1}, a subsample of at least 400 g of soil is recommended.

Table 7.6 indicates the effect of increasing the number of cores used to collect a sample (25 to 400) when the target population is 10 eggs (g soil)$^{-1}$ and a subsample of 200 g is used for laboratory analysis. Two levels of aggregation are illustrated, 100 and 400 eggs cyst^{-1}. As with sampling to detect, taking more than 100 cores ha^{-1} provides minimal additional benefit; the differences between 50 and 100 cores appear to be less marked. This is in agreement with Figure 7.5, which presents the results of a similar analysis in graphical format, albeit without any aggregation of eggs within cysts.

7.4.5 Sampling for quarantine

Depending on the task at hand, the intensity of sampling for quarantine cyst nematodes can vary enormously, as can be seen in Table 7.7. This should be seen in view of the regulatory status in different countries (see Table 7.2) and the amount of effort required to achieve the regulatory aim, for example, eradication where the species is still within an initial outbreak phase, or control to limit further spread where the species is already well established.

7.5 Laboratory Diagnosis

7.5.1 Extraction of nematodes

Before cyst nematodes can be identified and enumerated, they generally need to be extracted from the soil or plant material. Diagnostic methods have been developed that do not require extraction from field collected samples such as soil (e.g. Ophel-Keller *et al.*, 2008) and a method in use for PCN in Scotland only requires partial extraction (Reid *et al.*, 2015; see section 7.5.4). Although the cyst nematode life cycle passes through various stages, the dormant cyst stage is generally the longest lived and will also be the life stage that the nematodes are most likely to be at when agronomic decisions need to be made about future crop production. When host plants are growing, cyst nematodes can be found within and outside the plant roots, and extracting cysts from the soil may not reveal a true picture of the infestation – in the case of import consignments soil might not be present at all. Therefore, additional to the cyst extraction techniques, other extraction techniques for plant-parasitic nematodes in general may be more appropriate when these species are present at stages of their life cycle other than the dormant cyst. For this reason we include these techniques.

Table 7.6. The effect on confidence limits (CL) of changing the number of cores based on the zero-inflated negative binomial model with aggregation of eggs into cysts.

	Target population = 10 eggs (g soil)$^{-1}$		Subsample size = 200 g	
Eggs cyst^{-1}	100		400	
No. of cores	Lower CL	Upper CL	Lower CL	Upper CL
25	4.3	15.7	0.8	19.2
50	4.9	15.1	1.0	19.0
100	5.3	14.7	1.1	18.9
200	5.4	14.6	1.2	18.8
400	5.5	14.5	1.2	18.8

Here the target population is 10 eggs (g soil)$^{-1}$ with 100 eggs cyst^{-1} assumed, core size of 50 cm^3 and subsample size of 200 g. A similar colour coding system to Table 7.5 has been used to indicate the size of the confidence interval in relation to the target population.

Table 7.7. Examples of sampling requirements for quarantine cyst nematodes.

	Country/area	Purpose	Method	Detection level	Comments
PCN	Europe	For propagation material including seed potatoes	1500 ml of soil from 1 ha taken by at least 100 cores in a grid pattern[a]	90% likelihood for 1 cyst found in field infested with 4 foci (1 × 100 and 3 × 50 cysts kg^{-1} soil)	Routine sampling, every field
	America and Canada	For plants for planting, surveillance of fields for seed production	5000 ml ha^{-1}, 1000 cores ha^{-1} (Method B)[b]	90% likelihood for 1 cyst found in field infested with 1 focus (75 cysts kg^{-1} soil)	75,000 samples for 20 infested fields. No routine sampling
		For regulated fields	20,000 ml of soil ha^{-1} (Method A)	90% likelihood for 1 cyst found in field infested with 1 focus (25 cysts kg^{-1} soil)	
	Canada	Delimiting surveys	20,000 ml of soil ha^{-1} (Method A)		
		Phytosanitary certification of seed potatoes	5000 ml ha^{-1}, 1000 cores ha^{-1} (Method B)		

[a]Anonymous (2007);
[b]Anonymous (2014).

7.5.1.1 Free-living stages from soil

The free-living stages of cyst nematodes, the infective J2 and males, can be found in the soil; they are only present for a very short time relative to the sedentary stage of the female and cysts. Sampling for these stages should coincide with the time in their life cycle when they can be found in the soil sample. Methods for processing soil samples for the extraction of motile nematodes are numerous, ranging from simple stirring, decanting and sieving methods, to the more elaborate methods such as the Seinhorst or Oostenbrink elutriator and the zonal centrifuge (Southey, 1986; EPPO, 2013a). Such methods are based on the principle that the nematodes are very light in weight and thus float easily in water in comparison with sand particles and heavier debris. After stirring, the nematodes float in the water, which can then be decanted and concentrated. The suspension can be cleared, using sieves or a cotton wool filter placed on a steel mesh. The latter can be used overnight, which gives the nematodes the opportunity to move through the cotton wool filter into the clear water just beneath the steel mesh. This technique creates a clean suspension of nematodes extracted from soil, ready to be analysed with the microscope. For more details see EPPO (2013a).

7.5.1.2 Endoparasitic and sedentary stages from roots

Juveniles, young males and females can be found in the roots. Techniques such as incubation and the use of a mist chamber, which depend on the motility of the nematodes, can be used to extract the motile stages such as males and second stage juveniles, from the roots. The roots are placed on sieves, with or without cotton wool filters, depending on the technique used, in a very moist environment. The nematodes leave the roots and, using their geotrophic movement, are then collected in a tray under the sieve. The sedentary young males and females, together with all juvenile stages, can be extracted from the roots using a technique involving maceration and centrifugation. Using a solution in which the nematodes float (e.g. sugar solution or magnesium sulphate) and adding a coagulate to precipitate the debris will produce a solution containing the nematodes. A clean-up step is required to remove the floatation solution. A suspension with clean nematodes is the end product for analysis under the microscope. For more details see EPPO (2013a).

7.5.1.3 Cysts from soil

Cysts are the life stage that can be found at all times in the soil and are used for the great majority of soil tests carried out for analysis for species of cyst nematodes. The cysts contain the viable eggs; the eggs can be in all stages of their development, either early embryogenesis to J2, ready to hatch when the circumstances are right (see Chapter 3, this volume). Cysts can range from being full of eggs (PCN cysts may contain over 400 eggs), just after formation on the roots, to empty ('dead cysts') because all J2 have hatched either in response to the presence of a host plant or simply due to age. Several methods are available to extract cysts from the soil, ranging from simple floatation in a bucket, via Wye Washer, Fenwick Can, Seinhorst and Kort elutriators to centrifuge methods, such as the Schuiling centrifuge, the Coolen and D'Herde method or the zonal centrifuge. A proficiency test running for 7 years involving 38 European national laboratories carrying out PCN diagnoses found the Schuiling centrifuge and the Fenwick Can to be the extraction methods most commonly in use (Ladeveze and Anthoine, 2010). The latter is only reliable when samples are greater than 100 ml and less than 500 ml. The reduced efficiency effect is stronger at lower cyst densities (Bellvert *et al.*, 2008). All relevant methods are mentioned in the EPPO diagnostic protocol PM7/40 (EPPO, 2013b) for *G. rostochiensis* and *G. pallida* cysts and protocol PM7/89 (EPPO, 2008) for *H. glycines* and are well described in the EPPO standard extraction methods for nematodes (EPPO, 2013a). These documents are under the supervision of EPPO and are kept up to date with new techniques and developments for these specific nematode groups. Documents are accessible via the EPPO website.

The basis of many cyst extraction methods is the characteristic that cysts can float due to a hydrophobic covering, a property that can be enhanced by drying the cysts. PCN are capable of withstanding this drying process, under ideal conditions of temperatures not higher than 35°C over a period of 3 days at low relative humidity (EPPO, 2013b), which creates a bubble of air inside the cyst. Methods such as simple floatation, the Wye Washer and Fenwick Can are based on this floating characteristic. Descriptions of these techniques can be found in Turner (1998) and EPPO (2013a). Unfortunately, not all species of cyst nematode can withstand this drying process, especially *Heterodera* spp., which should not be dried at temperatures higher than 25°C or at humidities lower than 40% RH. Therefore, methods to extract cysts from wet soil are also available and should be used when cyst nematodes other than PCN are involved. Seinhorst and Kort elutriators, the centrifuge methods, such as the Schuiling centrifuge, the Coolen and D'Herde method, or the zonal centrifuge can all be applied on wet soil samples.

In this first step of the extraction process, a fraction of small organic material containing cyst nematodes, the so-called 'float', is removed from the rest of the soil. Depending on the soil type, this float can be very small and cysts can be picked out very easily by hand using a brush. When the soil has a very high organic matter content, separation of cysts in this way is not feasible and a second separation technique should be applied, for example, extraction by acetone, ethanol or water or a second centrifugation step is necessary using a solution with a higher specific density (1.22–1.25) (EPPO, 2013a).

7.5.2 Identification and viability assessment

The purpose of identification is to assign the correct species name to a specimen. In most cases, there is an assumption that the specimens are in a Petri dish, counting tray or other device, depending on the method used. Identification by morphology requires skilled and experienced taxonomists and authoritative keys covering the relevant taxonomic groups. For morphological identification of most cyst nematodes, both juveniles and/or cysts with viable eggs are needed. When the required expertise is available, diagnosis is relatively quick and simple. However, when large numbers of samples require analysis and suitable qualified taxonomists are not available or too expensive, other methods are needed.

In recent years, molecular tools have been developed to identify most commercially important cyst nematodes, permitting higher throughput diagnoses. For more details on morphology and molecular methods for all Heteroderidae see Chapters 14, 15, 16 and 17, this volume. For PCN an updated protocol for identification is maintained by the EPPO (2013b).

The purpose of the diagnosis determines the next step in the process. When identification is necessary for import or incursions assessment (detection) it will differ from management information reasons (detection and quantity). Quantification usually requires the number of cysts and its contents to be counted. Identification of the cysts can be performed and, for the assessment of the contents, the numbers of J2 and eggs within the cysts should be estimated and the viability of the contents assessed. As PCN are regulated in many countries, with legislation based on the presence or absence of cysts with live contents, for example, countries within the EU, it is important to have good methods available to determine this. A proficiency test for the extraction technique for PCN and the identification of life contents of cysts running for 16 years showed that the variance between the cysts counts was stable but that the variance between the live juvenile counts was reduced by experience and instruction (van den Berg *et al.*, 2014). Different methods of assessing egg viability are available, such as hatching assays, microscopy, quantification of trehalose or an mRNA test on the cyst contents. These methods are extensively described in the EPPO protocol for *Globodera* (EPPO, 2013b); they can be in principle applied to *Heterodera* cysts as well. In cases when detection of a live infestation is the purpose, and enough time is available, it is also possible to use a bioassay that relies on viable nematodes reproducing on a host plant and checking for newly produced cysts after the growing period.

Molecular methods have been developed to screen soil samples and detect and identify possible cyst nematodes in the same process (see section 7.5.4). When based on DNA, one should be aware that there could be a considerable time lag between viability of the pathogen and the detectability of the pathogen (van den Elsen *et al.*, 2012). Methods based on RNA and trehalose measure viability of the eggs inside the cysts (Beniers *et al.*, 2014; Ebrahimi *et al.*, 2015).

The latter techniques are very useful when statutory sampling is the reason for this research; when cysts are found during import inspections the possibility of unknown species being present needs to be considered and thus polymerase chain reaction (PCR) using primers targeting a limited range of species may not be sufficient. Morphological analysis and/or sequencing of PCR products may be required in such cases.

7.5.3 Optimization for statutory or commercial purposes

Drawing soil samples for analysis for PCN occurs worldwide to meet quarantine requirements, facilitate trade in regulated products and provide information to underpin pest management programmes. Similar measures are required for other cyst nematodes in areas where they have considerable commercial impact. As there has always been an incentive both to minimize the cost of sampling and diagnosis, and to decrease the error associated with the sampling and testing process, equipment has been developed to improve efficiency.

Equipment for the automated soil sampling is available, such as four-wheel drive vehicles (e.g. quad bikes and compact tractors; Fig. 7.7) with automated soil sampling devices for PCN sampling and/or global positioning system (GPS) technology (www.agri-tech.co.uk/pcn.html), AMS nematode samplers (USA) (http://www.ams-samplers.com/industry/agriculture/nematode-samplers.html), or an automated precision soil sampler (Smrtnik *et al.*, 2016) for *H. glycines*. These automated samplers allow the number of cores taken to be related to field distributions of cysts and sampling occurs in a grid pattern. Precision mapping, facilitated by the use of GPS technology may also help improve the efficiency of soil sampling (Trudgill *et al.*, 1997). Equipment has also been developed for sampling the soil during the harvest (Goeminne *et al.*, 2015). However, relating the results from soil samples collected in this way to the field distribution can be very difficult, but for detection purposes this method is considered very efficient (Viaene, 2016).

Automation of the sieving and floatation processes widely used to extract cysts from the soil (Turner and Subbotin, 2013) has led to the

Fig. 7.7. Automated sampling using a quadbike. (Courtesy of NAK, The Netherlands.)

development of the so-called carousels, the first of which was introduced by a commercial laboratory (then called BLGG) in The Netherlands in the 1970s. This machine was adapted by NAK Emmeloord in 1980 and then by Landwirtschaftkammer Munster (Germany), with the further introduction of microprocessor controlled operation at the Science and Advice for Scottish Agriculture (SASA) (Scotland) in 2010 (by MEKU Erich Pollähne Gmbh, Germany; Fig. 7.8). In 2016, the 1980 carousel at the NAK was replaced with a 'rinsing' machine that extracts cysts from soil and also cysts from the float, and enables one person reliably to process 700 samples per day (Luimes, 2016; Fig. 7.9). Another automated technique is the zonal centrifuge, a method based on the centrifuge-floatation method of Coolen and D'Herde (EPPO, 2013a). A logical next development is to couple automated extraction devices with detection/identification. Recent developments in molecular diagnostics have been combined with automation of the extraction process. In the EU, recent developments using PCR diagnostics and robotics have enabled SASA to introduce a high throughput method to detect and identify the cysts of PCN directly from the float, eliminating the need for visual examination (Reid et al., 2015). In relation to quarantine requirements (identification of PCN species coupled with an assessment of viability) a method has been implemented by the NAK in The Netherlands, based on the RNA method of Beniers et al. (2014), where after the automated extraction of the cyst the amount of live nematodes are determined by real-time PCR. Most of these techniques have been developed for PCN, due to the worldwide demand for testing soil for the presence of these species. However, the extraction techniques and diagnostic methods can be applied equally well for other species of cyst nematodes, for example, the SASA process can be adapted for use in advisory testing for *H. schachtii* (A. Reid, UK, 2016, pers. comm.).

7.5.4 Quantitative assessments

For successful management, growers need a PCN population estimate that is indicative of the potential of the nematodes to invade the roots of the planted crop. Consequently, most laboratories provide an estimation of the number of eggs

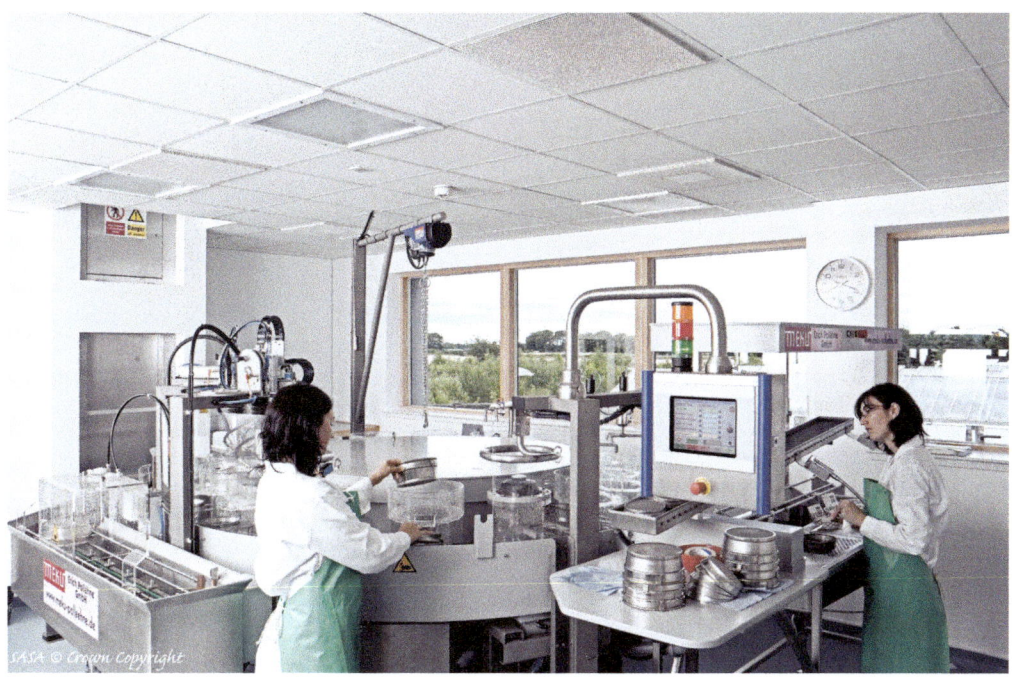

Fig. 7.8. Automated cyst nematode extraction facility (carousel) by MEKU Erich Pollähne Gmbh, Germany, in operation at SASA, Edinburgh, UK. (Courtesy of SASA, UK.)

Fig. 7.9. A carousel 'rinsing' machine from NAK (the Dutch General Inspection Service) that extracts cysts from soil and also cysts from the float, enabling one person to process 700 samples per day. (Courtesy of the NAK, The Netherlands.)

per gram of soil based on the soil sample provided. Traditionally this estimate has been made by crushing the cysts extracted from a volume of soil and enumerating the live eggs/juvenile nematodes. The development of quantitative molecular methods, for example, PCR, provides the potential for more efficient population assessments. Where more than one species of cyst nematode may be present, for example, for PCN infestations, such methods provide the potential for accurate enumeration of each species. This is rarely feasible using traditional microscopy methods. Both counting (van den Berg et al., 2014) and PCR are subject to error and this should be considered as part of the overall assessment of the accuracy of the overall sampling/testing procedure. The accuracy of the species identification should also be incorporated.

Sampling procedures can be compared under different scenarios related to the purpose of sampling. The level and nature of the PCN infestation may vary. Computer simulation is a valuable tool here, allowing random elements to be included for both the sampling and the distribution of PCN (Lin et al., 1979; McSorley, 1982; Been and Schomaker, 2000). Software has been produced both to carry out such simulations (SAMPLE, Been and Schomaker, 2000) and as part of a wider decision-making tool (PCN Calculator, Elliot et al., 2004, Phillips, 2006; NemaDecide, Been et al., 2005).

7.5.5 Laboratory diagnostics

Laboratory tests generally rely on the extraction of cysts from the soil and, since cysts will contain a varying number of eggs, or no eggs, assessments of the actual cyst contents need to be made as counts of cysts are of limited value. Many factors determine how many cysts within a field will be viable and how many will be dead. Regular cropping with potatoes on infested land, particularly using susceptible varieties, will produce large numbers of new, highly viable cysts, as well as leaving dead cysts from which J2 have emerged. Cultivation of varieties with high levels of PCN resistance will result in a much lower ratio of viable to dead cysts. For successful management, growers need a PCN population estimate that is indicative of the potential of the nematodes to invade the roots of the planted crop. Consequently, most laboratories provide an estimation of the number of eggs per gram of soil based on the soil sample provided.

7.6 Conclusions and Future Prospects

Cyst nematodes are important crop pests, causing major economic losses in many parts of the world. Depending upon their economic impact and their distribution within a region, decisions are made by regulatory authorities as to whether their pest status justifies national and international regulatory control. Official control measures, including classification as a quarantine pest, can be stringent, for example, when eradication is targeted, or more relaxed when the pest is accepted as having become established, but there is economic value in minimizing the further spread of the pest through the regulation of the most important pathways of transmission. Regulatory control also requires surveillance, to establish whether the pest is present, and for cyst nematodes, this usually requires field sampling and subsequent laboratory analysis of soil samples. Sampling strategies can only be optimized in the knowledge of the likely underlying field distributions.

Sampling strategies for cyst nematodes are reviewed above based on knowledge of PCN field distributions. When sampling for the purpose of detection viable cysts, sampling strategies should be based on the model of Schomaker and Been (1999), which describes early infestations developing as highly aggregated foci. The probability of detection is increased when at least 100 cores ha^{-1} are obtained and larger quantities of soil are analysed. Based on the sampling strategy adopted, estimates of the probability of detecting population levels of cyst nematode infestations can be made.

When a field is known to be infested, soil is primarily tested for cysts nematodes with the purpose of estimating population densities (eggs g^{-1} soil) to underpin management decisions. Under these circumstances, field infestations are likely to be more widespread and may be better described by a zero-inflated negative binomial distribution. Sampling at 50 cores ha^{-1}

is recommended and confidence intervals of population estimates can be calculated for the amount of soil that is subject to laboratory analysis. Reducing the volume of soil tested decreases the precision of the estimate, as does the aggregation of eggs into relatively fewer cysts. Provision of this information allows decisions on field sampling and laboratory analysis to be made in the light of the implications on the precision of field population estimates.

Extraction techniques and identification tools are available and a tendency to use automated sampling equipment with automated cyst extraction machines and when possible even automated species identification will make management of these nematodes easier in future.

7.7 References

Anonymous (2007) Council Directive 2007/33/EC of 11 June 2007 on the control of potato cyst nematodes and repealing Directive 69/465/EEC. *Official Journal of the European Union*, 156/12.

Anonymous (2014) Guidelines on Surveillance and Phytosanitary Actions for the Potato Cyst Nematodes *Globodera rostochiensis* and *Globodera pallida*, 7 May 2014. Available at: https://www.aphis.usda.gov/plant_health/plant_pest_info/nematode/downloads/potato_guidelines.pdf (accessed 19 January 2017).

Been, T.H. and Schomaker, C.H. (2000) Development and evaluation of sampling methods for fields with infestation foci of potato cyst nematodes (*Globodera rostochiensis* and *G. pallida*). *Phytopathology* 90, 647–656. DOI: 10.1094/PHYTO.2000.90.6.647

Been, T.H. and Schomaker, C.H. (2013) Distribution patterns and sampling. In: Perry, R.N. and Moens, M. (eds) *Plant Nematology*, 2nd edn. CAB International, Wallingford, UK, pp. 331–358.

Been, T.H., Schomaker, C.H. and Molendijk, L.P.G. (2005) NemaDecide: a decision support system for the management of potato cyst nematodes. In: Haverkort A.J. and Struik, P.C. (eds) *Potato in Progress: Science Meets Practice*. Wageningen Academic Publishers, Wageningen, The Netherlands, pp. 154–167.

Bellvert, J., Crombie, K. and Horgan, F.G. (2008) Effect of sample size on cyst recovery by flotation methods: recommendations for sample processing during EU monitoring of potato cyst nematodes (*Globodera* spp.). *Bulletin OEPP/EPPO Bulletin* 38, 205–210. DOI: 10.1111/j.1365-2338.2008.01204.x

Beniers, J.E., Been, T., Mendes, O., van Gent-Pelzer, M.P.E. and van der Lee, T.A.J. (2014) Quantification of viable eggs of the potato cyst nematodes (*Globodera* spp.) using either trehalose or RNA-specific Real-time PCR. *Nematology* 16, 1219–1232. DOI: 10.1163/15685411-00002848

Collins, S.A., Marshall, J.M., Zhang, X.H. and Vanstone, V.A. (2010) Area freedom from *Globodera rostochiensis* in Western Australia. *Aspects of Applied Biology* 103, 55–62.

den Nijs, L.J.M.F. (2016) New or old problem? The occurence of new virulent populations of potato cyst nematodes. 32nd Symposium European Society of nematologists. Braga, Portugal. Abstract p. 331.

Denwood, M.J., Stear, M.J., Mathews, L., Reid, S.W.J., Toft, N. and Innocent, G.T. (2008) The distribution of the pathogenic nematode *Nematodirus battus* in lambs is zero-inflated. *Parasitology* 135, 1225–1235. DOI: 10.1017/S0031182008004708

Ebrahimi, N., Viaene, N. and Moens, M. (2015) Optimizing trehalose-based quantification of live eggs in potato cyst nematodes (*Globodera rostochiensis* and *G. pallida*) *Plant Disease* 99, 947–953. DOI: org/10.1094/PDIS-09-14-0940-RE

Elliot, M.J., Trudgill, D.L., McNicol, J.W. and Phillips, M.S. (2004) Predicting PCN population changes and potato yields in infested soils. In: MacKerron, D.K.L. and Haverkort, A.J. (eds) *Decision Support Systems in Potato Production: Bringing Models to Practice*. Academic Publishers, Wageningen, The Netherlands, pp. 143–152.

EPPO (2004) Phytosanitary Procedure PM 3/61(1) Pest-free areas and pest-free production and distribution systems for quarantine pests of potato. *Bulletin OEPP/EPPO Bulletin* 34, 441–442.

EPPO (2008) Phytosanitary Procedure PM7/89(1) *Heterodera glycines*. *Bulletin OEPP/EPPO Bulletin* 38, 379–389.

EPPO (2013a) Diagnostic protocol PM 7/119(1) Nematode extraction. *Bulletin OEPP/EPPO Bulletin* 43, 471–495.

EPPO (2013b) Diagnostic protocol PM 7/40(3) *Globodera rostochiensis* and *Globodera pallida*. *Bulletin OEPP/EPPO Bulletin* 43, 119–138.

EPPO (2016) PM1/002(25) – EPPO A1 and A2 Lists of pests recommended for regulation as quarantine pests. Available at: https://gd.eppo.int/taxon/HETDPA/documents (accessed 4 May 2017).

EPPO (2017) EPPO Global database. Available at: https://gd.eppo.int (accessed 4 May 2017).

Evans, K. and Kerry, B. (2007) Changing priorities in the management of potato cyst nematodes. *Outlooks on Pest Management* 18, 265–269. DOI: 10.1564/18dec07

Eves-van den Akker, S., Lilley, C.J., Reid, A., Pickup, J., Anderson, E., Cock, P.J.A., Blaxter, M., Urwin, P.E., Jones, J.T. and Blok, V.C. (2015) A metagenetic approach to determine the diversity and distribution of cyst nematodes at the level of the country, the field and the individual. *Molecular Ecology* 24, 5842–5851. DOI: 10.1111/mec.13434

FAO (1995) *International Standards for Phytosanitary Measures. ISPM 4: Requirements for the Establishment of Pest Free Areas*. IPPC, FAO, Rome.

FAO (1997) *International Standards for Phytosanitary Measures. ISPM 6: Guidelines for Surveillance*. IPPC, FAO, Rome.

FAO (2002) *International Standards for Phytosanitary Measures. ISPM 16: Regulated Non-quarantine Pests: Concept and Application*. IPPC, FAO, Rome.

FAO (2006) *International Standards for Phytosanitary Measures. ISPM 1: Phytosanitary Principles for the Protection of Plants and the Application of Phytosanitary Measures in International Trade*. IPPC, FAO, Rome.

FAO (2007) *International Standards for Phytosanitary Measures. ISPM 5: Glossary of Phytosanitary Terms*. IPPC, FAO, Rome.

FAO (2013) *International Standards for Phytosanitary Measures. ISPM 11: Pest Risk Analysis for Quarantine Pests*, IPPC, FAO, Rome.

Gilligan, C.A. (1988) Analysis of the spatial pattern of soilborne pathogens. In: Kranz, J. and Rotem, J. (eds) *Experimental Techniques in Plant Disease Epidemiology*. Springer, Berlin, pp. 85–98.

Goeminne, M., Demeulemeester, K., Lanterbecq, D., de Proft, M. and Viaene, N. (2015) Detection of field infestations of potato cyst nematodes (PCN) by sampling soil from harvested potatoes. *Aspects of Applied Biology* 130, 105–110.

Handoo, Z., Carta, L., Skantar, A.M. and Chitwood, D.J. (2012) Description of *Globodera ellingtonae* n. sp. (Nematoda: Heteroderidae) from Oregon. *Journal of Nematology* 44, 40–57.

Hockland, S., Niere, B., Grenier, E., Blok, V., Phillips, M., den Nijs, L., Anthoine, G., Pickup, J. and Viaene, N. (2012) An evaluation of the implications of virulence in non-European populations of *Globodera pallida* and *G. rostochiensis* for potato cultivation in Europe. *Nematology* 14, 1–13. DOI: 10.1163/138855411X587112

Hockland, S., Inserra, R.N. and Kohl, L.M. (2013) International plant health – putting legislation into practice. In: Perry, R.N. and Moens, M. (eds) *Plant Nematology*, 2nd edn. CAB International, Wallingford, UK, pp. 359–382.

Ladeveze, L. and Anthoine, G. (2010) Outcome of a seven-year proficiency test for the detection and identification of potato cyst nematodes. *Aspects of Applied Biology* 103, 1–9.

Lin, C.S., Poushinsky, G. and Mauer, M. (1979) An examination of five sampling methods under random and clustered disease distributions using simulation. *Canadian Journal of Plant Science* 59, 121–130. DOI: 10.4141/cjps79-017

Luimes, J. (2016) Cyst extraction of >140,000 soil samples a year can be done by one person. *Proceedings of the 32nd Symposium of the European Society of Nematology*. European Society of Nematology, Braga, Portugal 28 Aug–1 Sept 2016, p. 226.

McSorley, R. (1982) Simulated sampling strategies for nematodes distributed according to a negative binomial model. *Journal of Nematology* 14, 517–522.

McSorley, R. and Parrado, J.L. (1982) Estimating relative error in nematode numbers from single soil samples composed of multiple cores. *Journal of Nematology* 14, 522–529.

Madden, L.V., Hughes, G. and van den Bosch, F. (2007) *The Study of Plant Disease Epidemics*. APS Press, St. Paul, Minnesota, USA.

Manachini, B. (2000) First report of *Heterodera glycines* Ichinohe on soybean in Italy. *Bolletino di Zoologia Agraria e di Bachicoltura, Serie II* 32, 261–267.

Mendes, M.L. and Dickson, D.W. (1993) Detection of *Heterodera glycines* on soybean in Brazil. *Plant Disease* 77, 499–500.

Mimee, B., Peng, H., Popovic, V., Yu, Q., Duceppe, M.O., Tétreault, M.P. and Bélair, G. (2014) First report of soybean cyst nematode (*Heterodera glycines* Ichinohe) on soybean in the province of Quebec, Canada. *Plant Disease* 98, 429. DOI: org/10.1094/PDIS-07-13-0782-PDN

Ophel-Keller, K., McKay, A., Hartley, D., Herdina and Curran, J. (2008) Development of a routine DNA-based testing service for soilborne diseases in Australia. *Australasian Plant Pathology* 37, 243–253. DOI: 10.1071/AP08029

Niere, B., Kruessel, S. and Osmer, K. (2014) Auftreten einer aussergewönlich virulenten Population der Kartoffelzystennematoden. *Journal für Kulturplflanzen* 66, 426–427.

Peng, D.L., Peng, H., Wu, D.Q., Huang, W.K., Ye, W.X. and Cui, J.K. (2016) First report of soybean cyst nematode (*Heterodera glycines*) on soybean from Gansu and Ningxia China. *Plant Disease* 100, 229. DOI: 10.1094/PDIS-04-15-0451-PDN

Perry, R.N. and Curtis, R.H.C. (2013) Behaviour and sensory perception. In: Perry, R.N. and Moens, M. (eds) *Plant Nematology*, 2nd edn. CAB International, Wallingford, UK, pp. 246–273.

Perry, R.N. and Moens, M. (2011) Survival of parasitic nematodes outside the host. In: Perry, R.N. and Wharton, D.A. (eds) *Molecular and Physiological Basis of Nematode Survival*. CAB International, Wallingford, UK, pp. 1–22.

Phillips, M.S. (2006) *Potato Council PCN Management Model*. Potato Council, Oxford, UK.

Pickup, J., Roberts, A. and Davie, K. (2016) *PCN Soil Sampling*. Reference, 1100018. Agriculture and Horticulture Development Board, Kenilworth, UK.

Plantard, O., Picard, D., Valette, S., Scurrah, M., Grenier, E. and Mugniéry, D. (2008) Origin and genetic diversity of Western European populations of the potato cyst nematode (*Globodera pallida*) inferred from mitochondrial sequences and microsatellite loci. *Molecular Ecology* 17, 2208–2218. DOI: 10.1111/j.1365-294X.2008.03718.x

Reid, A., Evans, F., Mulholland, V., Cole, Y. and Pickup, J. (2015) High-throughput diagnosis of potato cyst nematodes in soil samples. In: Lacomme, C. (ed.) *Plant Pathology: Techniques and Protocols. Methods in Molecular Biology Vol. 1302*, Springer Science+Business Media, New York, pp. 137–148.

Rott, M., Lawrence, T., Belton, M., Sun, F. and Kyle, D. (2010) Occurrence and detection of *Globodera rostochiensis* on Vancouver Island, British Columbia: an update. *Plant Disease* 94, 1367–1371. DOI: 10.1094/PDIS-03-10-0213

Schomaker, C.H. and Been, T.H. (1999) A model for infestation foci of potato cyst nematodes *Globodera rostochiensis* and *G.pallida*. *Phytopathology* 89, 583–590. DOI: 10.1094/PHYTO.1999.89.7.583

Schomaker, C.H. and Been, T.H. (2000) *Decrease of Potato Cyst Nematodes Populations in the Absence of Hosts*. Abstract in Proceedings of the 52nd International Symposium on Crop Protection, 9 May, 2000, Ghent, Belgium, p. 63.

Seinhorst, J.W. (1965) The relationship between nematode density and damage to plants. *Nematologica* 11, 137–154. DOI: 10.1163/187529265X00582

Smrtnik, E., Niblack, T., Dorrance, P.P., Dorrance, A. and Bruns, D. (2016) Comparison of two soil sampling methods for estimating population densities of *Heterodera glycines* cysts. *Plant Health Research* 17, 167–171. DOI: 10.1094/PHP-RS-16-0026

Southey, J.F. (1986) *Laboratory Methods for Work with Plant and Soil Nematodes*. Ministry of Agriculture, Fisheries and Food, HMSO, London.

Trudgill, D., Brown, D. and Phillips, M. (1997) GPS systems for mapping of nematodes. *Potato Review* 7, 44–48.

Turner, S.J. (1998) Sample preparation, soil extraction and laboratory facilities for the detection of potato cyst nematodes. In: Marks, R.J. and Brodie, B.B. (eds) *Potato Cyst Nematodes: Biology, Distribution and Control*. CAB International, Wallingford, UK, pp. 75–90.

Turner, S.J. and Evans, K. (1998) The origins, global distribution and biology of potato cyst nematodes (*Globodera rostochiensis* (Woll.) and *Globodera pallida* Stone). In: Marks, R.J. and Brodie, B.B. (eds) *Potato Cyst Nematodes: Biology, Distribution and Control*. CAB International, Wallingford, UK, pp. 7–26.

Turner, S.J. and Subbotin, S.A. (2013) Cyst nematodes. In: Perry, R.N. and Moens, M. (eds) *Plant Nematology*, 2nd edn. CAB International, Wallingford, UK, pp. 109–143.

van den Berg, W., Hartsema, O. and den Nijs, L.J.M.F. (2014) Statistical analysis of nematode counts from interlaboratory proficiency tests. *Nematology* 16, 229–243. DOI: 10.1163/15685411-00002761

van den Elsen, S., Ave, M., Schoenmakers, N., Landeweert, R., Bakker, J. and Helder, J. (2012) A rapid, sensitive and cost-efficient assay to estimate viability of potato cyst nematodes. *Phytopathology* 102, 140–146. DOI: org/10.1094/PHYTO-02-11-0051

Viaene, N. (2016) An alternative way of soil sampling for detection of field infestations with *Globodera* spp. *Proceedings of the 32nd Symposium of the European Society of Nematology*. European Society of Nematology, Braga, Portugal 28 August–1 September 2016, p. 155.

8 Mechanisms of Resistance to Cyst Nematodes

Aska Goverse and Geert Smant
Laboratory of Nematology, Wageningen University, The Netherlands

8.1 Introduction	154
8.2 Basal Immunity to Cyst Nematodes	155
8.3 Host-specific Immunity to Cyst Nematodes	156
8.4 Quantitative Aspects of Host-specific Immunity	161
8.5 Transcriptional Reprogramming and Defence Gene Expression	162
8.6 Hormone-mediated Defence Responses by Cyst Nematodes	164
8.7 Conclusions and Future Prospects	166
8.8 References	168

8.1 Introduction

Plants are constantly under attack by a wide range of pathogens and pests including bacteria, viruses, fungi, oomycetes, insects and nematodes. Fortunately, the majority of plant–pathogen interactions are incompatible due to the failure of the pathogen to localize a potential host plant or to recruit the appropriate battery of modifying enzymes needed for pathogenicity. The co-evolution between plants and pathogens resulted in the development of an immune system that, in contrast to animals that have both an adaptive and an innate immune system, is completely innate (Zipfel and Felix, 2005). This plant innate immune system is composed of overlapping layers, which allow the plant to respond effectively to invading microbial organisms (Cook *et al.*, 2015).

The first layer confers basal immunity to a broad range of microbes – both pathogenic and non-pathogenic – upon specific recognition of conserved microbial-derived components or danger signals by so-called pattern recognition receptors (PRR). However, pathogens have evolved strategies to overcome basal immunity. To counteract these virulent pathogens, plants have evolved a second layer of defence, which is based on the recognition of pathogen-specific components or host cell modifications by immune receptors (R proteins). Moreover, in response to the local activation of this multilayered immune system, systemic defence signalling is induced, which results in defence gene expression in distant plant parts that primes the plant for future attack (Pieterse *et al.*, 2012).

Like for other pathogens and pests the plant immune system is adapted to detect invading nematodes and to defend themselves by the activation of basal and host-specific immune responses (Goverse and Smant, 2014). However, compared to other pathosystems, knowledge on the mechanisms underlying plant immunity to plant-parasitic nematodes is still fragmentary. The development of molecular techniques has made it possible gradually to uncover the mechanisms underlying the different layers of plant defence. This has resulted in the identification of an increasing number of components involved in defence against nematodes, including host-specific immune receptors encoded by *R* genes as well as defence signalling and response networks, discussed in this chapter. In addition, research has focused on the identification of nematode-derived compounds responsible for the activation or suppression of plant defence responses (see Chapter 4, this volume). Recently, a novel defence mechanism involved in host-specific resistance to cyst nematodes was described based on natural variation in either copy number or genes essential for cyst nematode parasitism. This shows that plants have evolved additional layers of defence to block specific pathways required for cyst nematode parasitism.

In this review, we will focus on mechanisms involved in cyst nematode resistance by addressing: (i) molecular aspects involved in basal and host-specific immune responses to cyst nematodes (sections 8.2 and 8.3); (ii) evolutionary and quantitative aspects of host-specific resistance to cyst nematodes (section 8.4); (iii) the transcriptional reprogramming of defence gene expression upon cyst nematode infection (section 8.5); and (iv) the induction of local and systemic hormone-mediated defence responses by cyst nematodes (section 8.6). The aim of this review is not only to present a comprehensive overview of the recent advances in the molecular aspects involved in plant resistance responses to cyst nematodes, but also to place this in the context of what is currently known in the field of disease resistance in general, which provides directions for future research and resistance breeding.

8.2 Basal Immunity to Cyst Nematodes

8.2.1 Recognition of cyst nematodes during host invasion

Basal defence responses are activated by extracellular pattern recognition receptors (PRR) that are located on the cell surface to survey the apoplast for conserved immunogenic epitopes (Boller and Felix, 2009). These immunogenic epitopes can originate directly from invasive pathogens and parasites (pathogen/parasite-associated molecular patterns (PAMPs)) but can also be part of plant-derived compounds that are specifically released by damaged host cells (DAMPs). Invasion-triggered immunity therefore provides plants basal resistance against a broad spectrum of pathogen species. Based on the conserved mechanisms involved in basal immunity, it can be expected that it also plays a role in the interaction between host plants and parasitic cyst nematodes. Indeed, an increasing number of studies provide evidence for the activation of basal immune responses to cyst nematodes (reviewed by Holbein *et al.*, 2016). However, the underlying mechanisms are not yet well understood. Cyst nematodes produce the conserved ascaroside, Asc#18, which triggers both local and systemic immune responses in the plant and thus acts as a nematode-associated molecular pattern (NAMP; Manosalva *et al.*, 2015). Interestingly, application of Asc#18 resulted in increased resistance to most pathogen groups in both monocotyledons and dicotyledons. The induction of defence responses by Asc#18 in leaves after infiltration and spraying suggests that conserved PRRs are involved in the perception of this nematode-derived compound both in roots and shoots. Considering the fact that cyst nematodes use a whole battery of plant cell wall degrading enzymes to facilitate penetration and intracellular migration through the roots (Wieczorek and Seifert, 2012), it is likely that plant-derived components resulting from tissue damage and cell wall degradation play also an important role as elicitors in the activation of basal immunity to nematodes. How such DAMPs and NAMPs are perceived by pattern recognition receptors in the roots remains elusive and is currently under investigation by several research groups.

8.2.2 Activation of basal immune responses upon root invasion

Upon perception of conserved molecular patterns by PRRs, the rapid activation of intracellular downstream signalling pathways and defence gene expression is induced. Mitogen-activated protein kinases (MAPKs) are known to play a central role in the transduction of extracellular signals during the activation of basal immune responses (Meng and Zhang, 2013). However, the role of MAPKs in defence signalling during cyst nematode parasitism is largely unknown. The coordinated activity of MAPK signalling components in basal defence to the beet cyst nematode *Heterodera schachtii* was shown using the *Arabidopsis* mutants *mpk-3*, *mpk-6* and *ap2c1* (Sidonskaya *et al.*, 2016). Inoculation of *mpk-3* and *mpk-6* resulted in increased susceptibility, whereas a reduction was observed for *ap2c1*. MPK3 and MPK6 play a positive role in defence signalling, whereas AP2C1 is a negative regulator of basal immunity. A rapid induction of these genes was shown during invasion of the roots by cyst nematodes. Interestingly, mechanical wounding and cellulase treatment also results in a rapid induction, but these responses peaked earlier and the amplitude is higher, which could indicate the active suppression of immune responses by the invading nematodes.

The activation of basal immune responses results in the induction of a broad range of rapid chemical responses, including the release of reactive oxygen species (ROS), antimicrobial secondary metabolites, hydrolytic enzymes and protease inhibitors. Moreover, local fortifications of plant cell walls by lignification and callose deposits are induced (Schwessinger and Zipfel, 2008). Comparative analyses of *Arabidopsis thaliana* infected with the soybean cyst nematode *H. glycines* and the beet cyst nematode *H. schachtii* have shed more light on basal defences to nematodes in plants. *Heterodera glycines* is able to invade roots of *Arabidopsis*, but cannot form a proper syncytium due to non-host defence responses. Extensive callose depositions and necrosis in cells along the migratory tract of the nematodes was observed (Grundler *et al.*, 1997). This necrosis is not the result of mechanical damage caused by migrating cyst nematodes but involves the release of hydrogen peroxide (Waetzig *et al.*, 1999). Furthermore, overexpression of the ethylene response transcription factor *RAP2.6* increases the number of callose deposits in nematode-infected roots of *Arabidopsis* and reduces the development of *H. schachtii* (Ali *et al.*, 2013). Like many defence-related genes, *RAP2.6* is strongly downregulated inside syncytia of *H. schachtii* in susceptible *Arabidopsis* plants, which implies that basal defence responses can indeed reduce the development and reproduction of cyst nematodes.

8.2.3 Modulation of basal immune responses

To overcome basal immune responses upon host invasion, virulent pathogen species have evolved mechanisms to modulate the plant's immune system to cause disease. This involves pathogen effectors that are able to suppress basal immunity and result in effector-triggered susceptibility of the plant (Jones and Dangl, 2006). In cyst nematodes, apoplastic effectors are released to suppress invasion-triggered immune responses. For example, the secretion of venom allergen-like proteins (VAPs) in the extracellular space of host tissue modulates the activation of basal defence responses induced by cell surface immune receptors (Lozano-Torres *et al.*, 2014). This allows cyst nematodes to invade the roots of a host plant and the establishment of a compatible interaction.

8.3 Host-specific Immunity to Cyst Nematodes

8.3.1 *R* gene-mediated cyst nematode resistance

To defend themselves against virulent pathogens that are able to overcome basal immune responses, plants have evolved a second layer of defence based on host-specific immune receptors that detect pathogen-specific effectors (effector-triggered immunity). These immune receptors are encoded by so-called resistance (*R*) genes, which confer host-specific resistance to different pathogen species and strains carrying a corresponding avirulence (*Avr*) gene. Pathogens lacking this *Avr* gene are still virulent on such resistant plant genotypes and, *vice versa*, pathogens carrying this *Avr* gene are virulent on susceptible plant genotypes lacking the cognate *R* gene. This genetic interaction between host genotypes (*R/r*) and pathogen genotypes

(*Avr/avr*) is also known as the gene-for-gene concept (Flor, 1942). For the resistance gene *H1* in potato, Mendelian proof for a gene-for-gene interaction was demonstrated for two avirulent and virulent lines of the potato cyst nematode *Globodera rostochiensis* (Janssen *et al.*, 1991).

In case of a matching *R/Avr* pair, a local and effective resistance response is induced that blocks the invading pathogen from spreading further. This incompatible interaction between host and pathogen often involves a hypersensitive response (HR) that results in necrosis of the infected tissue. Highly specific nematode resistances can be divided roughly into three major types based on their timing in the ontogeny of feeding structures and their characteristic cytological features (reviewed by Goverse and Smant, 2014). The first type involves a rapid induction of cell death upon root invasion, but this response is not common for resistance to cyst nematodes. The second type and third type of cyst nematode resistance can be divided into early responses that result in a bias towards males or late responses that block female development. Examples of the early response type are the *G. rostochiensis* resistance genes *H1* from potato and the *Hero A* gene from tomato, which allow the initiation of a syncytium but restrict their expansion by the formation of a layer of necrotic cells around the feeding structure (Rice *et al.*, 1985; Sobczak *et al.*, 2005). This small syncytium sustains the development of males but not females, as the sex of cyst nematodes is epigenetically determined within the first week after the initiation of a feeding structure (Trudgill, 1967). An example of the late resistance response type is the potato cyst nematode resistance gene *Gpa2* (Koropacka, 2010), which allows normal initiation and expansion of the syncytium and the initial development of females. However, further development and reproduction of the nematode is arrested in a later stage due to starvation. The formation of a layer of dead cells disconnects the syncytium from the vascular bundle, thereby preventing nutrient uptake. This delayed resistance response type can occur with or without the characteristic cell death features of a HR (Das *et al.*, 2008; Kandoth *et al.*, 2011), suggesting that different mechanisms may contribute to the deterioration of the feeding structures.

8.3.2 Extra- and intracellular immune receptors encoded by *R* genes

The identification and characterization of *R* genes revealed that *R* genes encode both extra- and intracellular immune receptors (Fig. 8.1). Extracellular immune receptors are typically receptor-like proteins (RLP) and receptor-like kinases (RLK). These multidomain receptor proteins consist of an extracellular leucine-rich repeat (LRR) domain, which is anchored to the plasma membrane by a transmembrane domain. RLKs also harbour a cytoplasmic kinase domain, which is replaced by a short non-descriptive tail region in RLPs (Gish and Clark, 2011). They belong to the same group of proteins as PRRs involved in basal immunity, which underscores that basal immunity and host-specific resistance are overlapping layers in plant immunity. However, in the case of RLP/RLK encoded by *R* genes the proteins detect – directly or indirectly – apoplastic effectors from specific pathogen strains, which results in the activation of highly specific resistance responses. For cyst nematodes, two RLP immune receptors are known including the *R* genes *Hs1^{pro-1}* from sugar beet, which confers resistance to *H. schachtii* (Cai *et al.*, 1997), and *Cf-2* from tomato (Lozano-Torres *et al.*, 2012), which confers resistance to *G. rostochiensis* (Fig. 8.1).

The vast majority of *R* genes, however, encode for so-called intracellular NB-LRR immune receptors that detect – directly or indirectly – pathogen effectors inside the host cell. NB-LRR immune receptors all have in common a central NB domain attached to a carboxy-terminal LRR domain, but can be subdivided on the basis of the domain at their amino-terminus: either of a coiled-coil (CC) domain or a Toll-interleukin receptor (TIR)-like domain. Cyst nematode *R* genes encode for both types of intracellular immune receptors (Fig. 8.1). For instance, the potato gene *Gro1-4* encodes a TIR-NB-LRR protein (Paal *et al.*, 2004), whereas *Gpa2* from potato encodes a CC-NB-LRR protein (Van der Vossen *et al.*, 2000). They confer resistance against specific populations of the potato cyst nematodes *G. rostochiensis* and *G. pallida*, respectively. Moreover, several CC-NB-LRR immune receptors have an extended N-terminus in which the CC is fused to a so-called SD domain that is specific for Solanaceous plants. The resistance gene *Hero A* in tomato belongs to this subclass (Fig. 8.1) and confers

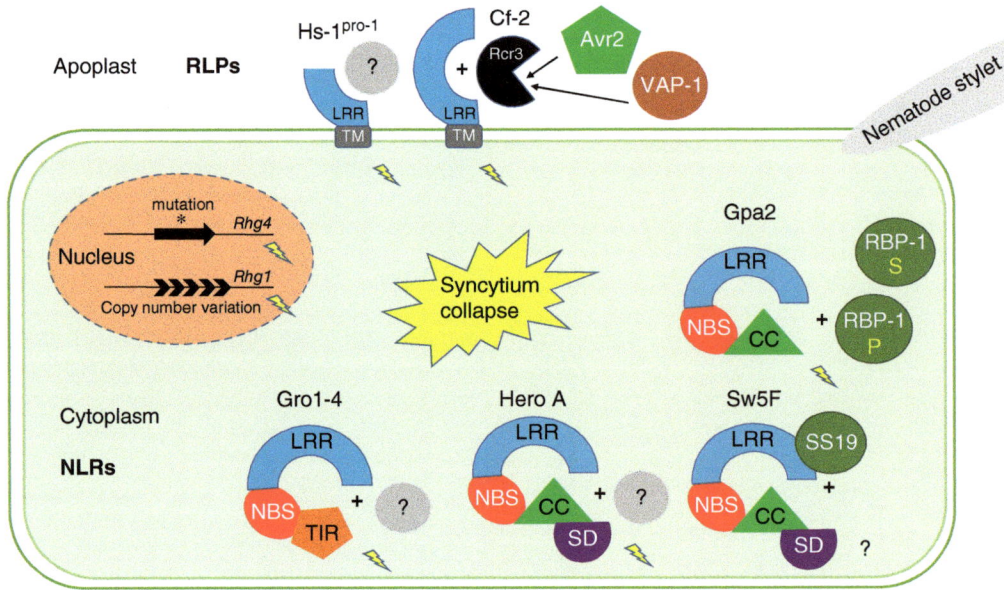

Fig. 8.1. Schematic overview of cyst nematode resistance genes currently identified in major crop species. Extracellular immune receptors: the resistance genes $Hs1^{pro-1}$ from sugar beet and *Cf-2* from tomato encode receptor-like proteins (RLPs), which consist of a variable extracellular LRR domain fused to a transmembrane domain (TM). The Cf-2 immune receptor confers host-specific resistance to the potato cyst nematode, *Globodera rostochiensis*, and the fungus, *Cladosporium fulvum*, by sensing the modifications of a common host factor Rcr3 upon the release of the nematode effector VAP-1 or the fungal effector Avr2 in the apoplast. The activation mechanism involved in $Hs1^{pro-1}$ mediated resistance is unknown. Intracellular immune receptors: the potato resistance gene *Gro1-4* encodes a TIR-NB-LRR immune receptor, whereas the *Gpa2* gene from potato belongs to the subclass encoding CC-NB-LRR immune receptors. The *Hero A* gene from tomato also belongs to this subclass but carries an extra SD domain fused to its N-terminus. They all confer host-specific resistance to potato cyst nematode species or populations. Only for the intracellular immune receptor Gpa2 the corresponding nematode effector is known, which results in the activation of a cell death response in agroinfiltration assays. Intracellular RBP-1 recognition by Gpa2 is highly specific and determined by a single amino acid polymorphism (P/S). However, whether RBP-1 detection is direct or indirect is unknown. RBP-1 belongs to the SPRYSECs, a highly diverse effector family present in potato cyst nematodes. Interestingly, SPRYSEC19 (SS19) interacts directly with the LRR domain of the tomato SD-CC-NB-LRR homologue Sw5F *in planta* but this does not activate a cell death response. Hence, the role of Sw5F in cyst nematode resistance and plant immunity remains elusive. Natural variation: the *Rhg1* and *Rhg4* loci in soybean confer host-specific resistance to *Heterodera glycines* populations by a novel mechanism based on copy number variation and mutations in genes essential for syncytium formation. In this case, syncytium collapse is not the result of effector-triggered immune responses but the result of metabolic stress either on the feeding cell or on the nematode.

broad-spectrum resistance to potato cyst nematode species and populations (Ernst *et al.*, 2002).

8.3.3 Effector recognition by extra- and intracellular R proteins

Plant immune receptors involved in nematode resistance show a remarkable difference in recognition specificities. However, for most *R* gene encoded nematode immune receptors the corresponding effector is unknown. Only for the extracellular immune receptor Cf-2 and the intracellular immune receptor Gpa2, the cognate nematode effector is known (Fig. 8.1), providing two model systems to study the molecular mechanisms underlying nematode detection and the activation of downstream defence responses to cyst nematodes.

Gpa2 is able to recognize specific variants of the secretory protein RBP-1 from *G. pallida*, which results in the activation of a HR in an agroinfiltration assay on leaves of *Nicotiana benthamiana*. Recognition of RBP-1 variants from several European and Latin American nematode populations by Gpa2 is highly specific and determined by just one single nucleotide polymorphism (Fig. 8.1; Sacco *et al.*, 2009; Carpentier *et al.*, 2012). Structure–function analysis revealed that RBP-1 recognition is determined by the C-terminus of the LRR domain of Gpa2 (Slootweg *et al.*, 2013). This corresponds to the same region in Sw5F, a homologue of the tomato SD-CC-NB-LRR immune receptor Sw5B conferring resistance to TSWV, which interacts directly with the homologous nematode effector SPRYSEC19 from *G. rostochiensis* (Fig. 8.1; Postma *et al.*, 2012). Unlike RBP-1 and Gpa2, this interaction does not activate an Sw5F-mediated cell death response and its role in nematode resistance remains elusive. For Gpa2, however, there is no evidence for a physical interaction between the distal end of the LRR domain and RBP-1. Therefore, an indirect recognition model may apply in which Gpa2 is able to sense the presence of RBP-1 by perturbations of its cellular host target involved in nematode parasitism or a decoy (guard or decoy model).

The guard model applies for Cf-2, which is able to detect perturbations of the apoplastic papaine-like cysteine protease Rcr3 upon targeting by the potato cyst nematode effector VAP-1 (Fig. 8.1). This results in the activation of a specific cell death response and nematode resistance in the roots. Cf-2 confers also effector-triggered immunity to the leaf mould fungus *Cladosporium fulvum* in tomato upon targeting *rcr3* by the effector *Avr2* (Rooney *et al.*, 2005; Lozano-Torres *et al.*, 2012). Apparently, effector repertoires of two different plant pathogens have independently converged on the same host target (Fig. 8.1). By guarding such common virulence targets, plants can efficiently expand the recognition spectrum of their limited innate immune receptor repertoire. Moreover, such a mechanism explains the dual recognition specificity of *R* genes like *Cf-2*.

8.3.4 Activation of effector-triggered immune responses

Effector recognition by immune receptors results in the rapid activation of downstream defence responses, which often includes a cell death response. For extracellular R proteins, effector recognition by the LRR domain of RLP/RLK immune receptors results in the complex formation with intracellular host components and the activation of downstream signalling pathways, resulting in host-specific resistance (Gust and Felix, 2014). The activation of Cf-2 upon rcr-3 targeting by the nematode effector VAP-1 probably involves the activation of the same defence pathways, although it cannot be excluded that defence signalling to nematodes involves root-specific components that differ from defence signalling to *C. fulvum* in leaves. Also for the RLP immune receptor Hs1^{pro-1}, it seems likely that it acts via the formation of a co-receptor complex to activate downstream resistance to beet cyst nematodes as it lacks a kinase signalling domain.

Intracellular NB-LRR immune receptors consist of a central nucleotide-binding domain, flanked by a N-terminal CC or TIR domain and a C-terminal LRR domain. These multidomain proteins combine the function of pathogen detection and the activation of downstream signalling responses. They act as a molecular switch and their activity is tightly controlled by inter- and intradomain interactions (Takken and Goverse, 2012). In the absence of a pathogen, NB-LRR proteins are present in the cell in a signalling competent resting state and upon pathogen detection, they adopt an activated state to trigger the activation of downstream defence responses. Biochemical and structure-informed analyses revealed novel insights in NLR functioning, including the cooperation between domains and the formation of pre- and post-activation complexes (Sukarta *et al.*, 2016).

The NB-LRR immune receptor Gpa2 is a close homologue of Rx1, which confers resistance to PVX in potato (Bendahmane *et al.*, 1999). Despite their high sequence similarity (88% identity), they confer resistance to two distinct pathogens, which make them a model to study mechanisms underlying pathogen recognition and NB-LRR immune activation. Extensive sequence exchange between the Gpa2 and Rx1 showed the close cooperation between the NB-ARC and N-terminal end of the LRR domain in controlling the activity of the immune receptor. Together, these domains form a switch module and incompatibility between domains due to subtle changes in the interface result either in autoactivation or loss of function of the immune receptor (Slootweg et al., 2013). So apparently, the LRR domain of Gpa2 plays a dual role in effector-triggered immunity in which the C-terminal end is involved in pathogen detection and the N-terminal end in controlling the resting state of the protein. However, the molecular mechanisms and pathways involved in the activation of Gpa2-mediated downstream defence responses remain elusive.

The CC domain of NB-LRR immune receptors is thought to be involved in the recruitment of signalling components and the induction of downstream signalling events. The N-terminal response domains of different NB-LRR immune receptors associate with a large diversity of host proteins involved in various plant processes. Like Rx1, the CC domain of Gpa2 interacts with the host protein RanGAP2 (Sacco et al., 2007; Tameling and Baulcombe, 2007) and the crystal structure of the CC domain of Rx1 was resolved in complex with the WPP domain of RanGAP2 (Hao et al., 2013). Gpa2 is predicted to reside in the cytoplasm of the cell as it lacks discernable nuclear localization signals. However, Gpa2 is present in both the cytoplasm and the nucleus as reported for Rx1 (Slootweg et al., 2010; Goverse, unpubl.). RanGAP2 functions as a cytoplasmic retention factor in the nucleocytoplasmic distribution of Rx1 and Gpa2 (Tameling et al., 2010; Slootweg, unpubl.). Moreover, RanGAP2 plays a role in Gpa2-mediated immunity (Sacco et al., 2009). Silencing of RanGAP2 resulted in reduced immune responses, whereas artificial tethering of RBP-1 and RanGap2 resulted in enhanced cell death. However, no patterns of positive selection were observed in allelic variants of RanGAP2 (Carpentier et al., 2013), which is a hallmark for co-evolution between host and pathogen. So, whether RanGAP2 plays a role as a host factor involved in cyst nematode recognition needs further analysis.

Unfortunately, the interacting host proteins for other cyst nematode NB-LRR proteins are unknown as they could provide novel insights in the specific activation mechanisms and pathways involved in effector-triggered immunity to cyst nematodes. De la Torre et al. (2013) showed that the calcium sensor system Cbl10/cipk6 is required for the activation of effector-triggered immunity mediated by different plant immune receptors, including Gpa2. This shows that signalling pathways involved in NB-LRR mediated plant immunity are conserved and, therefore, likely to play a role in host-specific immune responses to cyst nematodes as well.

8.3.5 Modulation of effector-triggered immune responses

To counteract host-specific immune responses, cyst nematodes have evolved a range of effector proteins to promote virulence on resistant plant genotypes (see Chapter 4, this volume). For example, a pharyngeal gland-specific ubiquitin carboxyl extension protein (GrUBCEP12) of G. rostochiensis suppresses host defence responses in plants (Chronis et al., 2013). Ectopically expressed GrUBCEP12 is cleaved in plants into ubiquitin and a small carboxyl extension protein of 12 amino acids (GrCEP12). GrCEP12 alone abrogates the cell death triggered by two CC-NB-LRR resistance proteins of potato and reduces the release of reactive oxygen species. Moreover, the suppression of defence-related genes in Nicotiana benthamiana treated with the immunogenic peptide flg22 was observed (Chen et al., 2013a), indicating that GrCEP12 acts on molecular mechanisms involved in both basal- and host-specific immunity. Another example of immunomodulation is the SPRYSEC effector family from potato cyst nematodes, which are named after SPRY domain-containing secretory proteins produced in the dorsal pharyngeal gland (reviewed by Diaz-Granados et al., 2016). They encode a highly expanded gene family present in the genome of G. rostochiensis and G. pallida, but only a small subset encodes for effectors that could play a role in plant parasitism (Mei et al., 2015). Several SPRYSEC family members of G. rostochiensis selectively suppress defence-related programmed cell death and disease resistance mediated by CC-NB-LRR type of plant immune receptors (Postma et al., 2012; Ali et al., 2015). Although the underlying mechanisms are unknown, it is hypothesized that SPRYSEC effectors enhance the degradation of immune receptors by targeting them to the 26S proteasome (Rehman et al., 2009).

8.4 Quantitative Aspects of Host-specific Immunity

8.4.1 NB-LRR gene clusters linked to quantitative cyst nematode resistance

Genome-wide sequence analysis and genetic mapping of *R* genes have shown that disease resistance genes are often located in clusters of homologous *NB-LRR* genes spread throughout the plant genome (e.g. Bakker *et al.*, 2011; Jupe *et al.*, 2012). *R* gene clusters from different genotypes and even related species are often located in the same chromosomal region. These regions are therefore called 'hot-spots of resistance'. With the exception of the beet cyst nematode resistance gene *Hs1^{pro-1}* (Cai *et al.*, 1997), all nematode resistance genes cloned to date reside in such complex loci harbouring tandemly repeated *R* gene homologues irrespective of whether they encode extra- or intracellular immune receptors. For example, the potato cyst nematode resistance gene *Hero A* (Ernst *et al.*, 2002) encoding a NB-LRR protein is located in a genomic region containing at least 14 homologous genes on chromosome IV of tomato and *Cf-2* encoding an extracellular RLP resides at a complex locus on chrVI of tomato. In addition, both CNLs and TNLs are located in clusters and the number of homologues per cluster can differ. For example, the gene *Gro1-4* (Paal *et al.*, 2004), which encodes a TNL, is a member of a large cluster containing 13 *R* gene homologues located on chromosome VII of potato, whereas the potato gene *Gpa2* (Van der Vossen *et al.*, 2000) encoding a CNL is present in a relatively small cluster of four highly homologous genes on chromosome XII. The specificities of the other members of these nematode *R* gene clusters are often unknown, except for the *Gpa2* locus in potato, which also harbours the resistance gene *Rx1* conferring extreme resistance to Potato Virus X.

These clusters of closely related genes are thought to be the results of diversifying selection. Unequal crossing-over between homologues at complex *R* gene loci followed by gene conversion play a role in the diversification of *R* genes and the evolution of novel recognition specificities. For example, *Rx1* and *Gpa2* are located in the same cluster and sequence exchange between homologues has contributed to polymorphic residues showing patterns of positive selection (Bendahmane *et al.*, 1999; Bakker *et al.*, 2011). Interestingly, mapping the sequence variability of Gpa2 and Rx1 homologues on the *in silico* 3D structure model of the LRR domain revealed that they show a polar distribution (Slootweg *et al.*, 2013). Polymorphic sites are often located at the surface of the protein consistent with a role of the LRR in pathogen recognition and the co-evolution between host and pathogen.

Several single dominant cyst nematode resistance traits map to regions harbouring NB-LRR clusters, providing candidate genes encoding intracellular immune receptors involved in nematode effector-triggered immunity. For example, four NB-LRR candidate genes could be identified for the *H1* gene conferring resistance to the potato cyst nematode *G. rostochiensis* (Finkers-Tomczak *et al.*, 2011). Similarly, co-localization of the *Cre3* gene with NB-LRR genes was observed, which confers resistance to the cereal cyst nematode *H. avenae* providing good candidates for further isolation and characterization (de Majnik *et al.*, 2003). Remarkably, quantitative trait loci (QTLs) conferring resistance to the potato cyst nematode also often co-localize with clusters of NB-LRR resistance genes, suggesting that they may contribute to quantitative resistance to nematodes. For example, NB-LRR clusters are located in the region of the *Grp1* locus on chrV in potato, conferring quantitative resistance to both *G. pallida* and *G. rostochiensis* (Finkers-Tomczak *et al.*, 2009). However, another explanation for this observation is that also in this case canonical NB-LRR *R* genes contribute to cyst nematode resistance, but the quantitative nature of the response is due to a mixture of virulent and avirulent genotypes in the potato cyst nematode field populations used to screen for resistance. Most of the cyst nematodes reproduce by obligate outcrossing, and there is generally great variation in host range and response to specific resistance genes between and within field populations (Bakker *et al.*, 1993). With genome sequences available for an increasing number of crops, it should be possible to pinpoint the genes responsible for both qualitative and quantitative cyst nematode resistance.

8.4.2 Resistance based on copy number variation and natural mutations

Quantitative resistance to cyst nematodes may also reflect the contribution of multiple or polymorphic genes present at different loci in the case of major and minor QTLs or even one particular locus. This could involve not only classical R genes encoding plant immune receptors as described above, but also essential plant genes required for the induction and maintenance of feeding structures. This latter was recently shown for the cyst nematode resistance loci *Rhg1* (Cook *et al.*, 2012) and *Rhg4* in soybean (Liu *et al.*, 2012). The *Rhg1* locus harbours three genes within a single 31-kb genomic segment, which encode three unrelated proteins: an α-soluble NSF attachment protein (α-SNAP), a wound-inducible domain protein (WI-12) and an amino acid transporter. The resistance mediated by the dominant *Rgh4* locus is linked to two nucleotide polymorphisms in a single copy gene encoding a serine hydroxymethyltransferase (Shmt), which has a key role in one-carbon folate metabolism (Liu *et al.*, 2012). These mutations affect the kinetic activity of Shmt, which could result in folate deficiency inside syncytia of soybean cyst nematodes. The metabolic stress associated with folate deficiency might trigger the activation of a cell death response in the syncytia. Alternatively, lower folate levels in syncytia can also indirectly cause folate deficiency in feeding nematodes, which are thought to rely on host plants for their folate requirements. The death of the nematodes by insufficient folate supplies might thus remove the physiological and molecular stimuli required for the maintenance of feeding structures, which could also result in their degeneration.

Although further studies have to resolve the exact biochemical mechanisms of *Rgh1*- and *Rgh4*-mediated resistances, these resistances demonstrate that plants defend themselves by disturbing developmental or metabolic processes that are either crucial for the feeding structure or nematode development (Bekal *et al.*, 2015). Interestingly, the *Rha2* locus in barley also confers resistance to the cereal cyst nematode, *H. avenae*, by influencing the viability of the nematodes during two phases of parasitism: feeding site establishment and subsequent cyst maturation. Histological studies and transcript profiling revealed that *Rha2*-mediated resistance involves the degradation of feeding sites, which coincides with specific changes in cell wall-related transcript abundance and changes in cell wall composition. Based on these observations it was concluded that (1,3;1,4)-beta-glucan may influence nematode feeding site development by limiting solute flow (Aditya *et al.*, 2015). This may point to another example in which cyst nematode resistance is based on qualitative differences in host plant genotypes that compromise normal feeding cell formation and functioning. It will be interesting to see whether similar mechanisms underlie other qualitative resistance traits to cyst nematodes in various crops.

8.5 Transcriptional Reprogramming and Defence Gene Expression

Basal and host-specific immunity are overlapping layers of the plant immune system that share common signalling components involved in the transcriptional reprogramming of the cell. However, the dynamics and amplitude of their activities are different. For NLRs involved in effector-triggered immunity (ETI), it is observed that they can interact directly with transcription factors to form a complex in the nucleus that regulates gene expression activity (reviewed by Tsuda and Somssich, 2015). Transcription factors are key components in the transcriptional reprogramming of the host cell, which is achieved by the specific binding to regulatory elements in the promoter region of defence-regulated genes. Moreover, the activity of TFs can be regulated by other immune signalling components like Ca^{2+} and MAPKs, thereby linking intracellular immune signalling pathways induced upon pathogen recognition and the activation of local and systemic defence responses.

8.5.1 Local induction of defence gene expression in susceptible host plants

Gene expression analyses of nematode-infected roots suggest that basal defence is activated in susceptible plants during the onset of parasitism, in particular the invasion stage and early feeding cell development. Several time course experiments show the transcriptional dynamics in gene expression during cyst nematode parasitism in infected whole roots or in syncytia-enriched tissues (reviewed by Li *et al.*, 2009). High-throughput gene expression analyses using various techniques revealed dramatic changes in gene expression of different functional categories, including genes involved in plant defence. Transcript profiling in susceptible tomato roots infected with *G. rostochiensis* revealed a first wave of defence-related gene expression that was followed by a second wave of developmental regulated gene expression in later stages of the infection (Swiecicka *et al.*, 2009). This is most likely the result of active suppression of basal immune responses by the invading juvenile nematode and the activation of developmental pathways involved in feeding cell formation. Comparison of the transcriptome of syncytia induced by *H. schachtii* in roots of *Arabidopsis* with other plant tissues and organs showed that cyst nematodes induce a transcriptional programme that is specific for metabolically active feeding cells (Szakasits *et al.*, 2009).

8.5.2 Local induction of defence gene expression in resistant host plants

In the incompatible interaction, similar dramatic changes in the transcriptional regulation of defence-related genes are observed. However, the dynamics of the responses differ. Compared to transcriptome analyses of quantitative resistance responses, surprisingly little information is available on changes in defence gene expression induced upon the activation of extra- or intracellular immune receptors encoded by single dominant *R* genes. Comparing defence gene expression patterns between susceptible and resistant tomato plants carrying the *Hero A* gene revealed increased transcript levels of genes encoding phenylalanine ammonia lyase (PAL), pyruvate decarboxylase (PDC), alcohol dehydrogenase (ADH) and Myb-related genes at 3 dpi in the incompatible interaction (Uehara *et al.*, 2010). This is consistent with the hypothesis that NB-LRR immune receptors encoded by *R* genes like *Hero A* result in the activation of a more rapid and stronger defence response, which blocks nematode parasitism. Furthermore, a strong increase in PR-1 transcript levels was observed at 3 dpi in resistant plants, which is a marker for SA-mediated defence responses. This was not observed in NahG tomato plants harbouring the *Hero A* gene. From this, it was concluded that PR-1 is a marker for the cultivar resistance conferred by *Hero A* against potato cyst nematodes. Interestingly, these five genes were downregulated in the compatible interaction at 3 dpi, supporting the hypothesis that basal immunity is suppressed by cyst nematodes in susceptible host plants to establish a feeding relationship. More detailed studies on the transcriptional reprogramming of cells upon the activation of plant immune receptors are needed to resolve the defence signalling networks underlying effector-triggered immunity to cyst nematodes.

For quantitative resistance mediated by the *Rhg1* and *Rhg4* loci, however, transcriptional changes in defence gene expression have been studied extensively providing novel insights into the underlying resistance mechanisms. Two different types of defence responses can be distinguished based on the soybean 'Peking' and the genotype 'PI 88788'. *Glycine max* 'Peking' confers a potent and rapid defence response including the formation of cell wall appositions (CWAs), whereas no such structures are formed during the more prolonged resistance response to soybean cyst nematode infection of *G. max* 'PI 88788'. These differences in resistance seem to correlate with differences in the amplitude of defence gene expression. Microarray analyses revealed differences in gene expression in the pericycle and surrounding cells of both genotypes even before infection, including higher relative expression levels of the arachidonic acid 1 gene (*Gm-DEA1*) and a protease inhibitor in *G. max* 'Peking' as compared to 'PI 88788'. This indicates that genotypic differences can also contribute to the observed differences in resistance response to cyst nematode infection (Klink *et al.*, 2007a, b, 2011).

In addition to common changes in defence gene expression, such as a strong increase in PAL and lipoxygenase-9 and lipoxygenase-4 (Klink *et al.*, 2009, 2010; Mazarei *et al.*, 2011), several genotype-specific transcriptome changes were observed during cyst nematode infection. For example, an increase in expression levels of other components of the phenylpropanoid pathway was observed specifically during the resistance response of *G. max* 'PI88788', which may correspond to the degeneration of syncytia during its slow resistance response. Moreover, the expression of an amino acid transporter and an α soluble NSF attachment protein (α-SNAP) was detected specifically in syncytia undergoing defence responses mediated by the *Rhg1* locus (Matsye *et al.*, 2011). Comparing gene expression profiles of developing syncytia in soybean near-isogenic lines differing at *Rhg1* revealed that several stress-related genes were upregulated in the resistant line, including those encoding homologues of enzymes that lead to increased levels of ROS and proteins associated with the unfolded protein response. These results indicate that syncytia induced in the resistant line are undergoing severe oxidative stress and imbalanced endoplasmic reticulum homeostasis, both of which probably contribute to the resistance reaction (Kandoth *et al.*, 2011).

The detection of shared differential expression levels of genes in both response types may indicate that a conserved transcriptional programme underlies soybean resistance to cyst nematodes, on which genotype-specific defence responses are built (Klink *et al.*, 2011). Recently, a whole-genome gene expression study confirmed these previous findings, but also revealed the overrepresentation of GC-rich elements (e.g. GCATGC) in the promoter regions of certain groups of genes that could point at new defence-responsive regulatory elements involved in cyst nematode–plant interactions (Wan *et al.*, 2015). In addition, comparative proteomics of infected susceptible and resistant soybean roots resulted in the identification of proteins that belong to the categories of disease/defence, but also metabolism, energy, cell growth and division, transcription, protein synthesis, protein destination and storage, signal transduction and secondary metabolism (Chen *et al.*, 2013b). Similar results were obtained when comparing incompatible and compatible wheat cultivars infected with *H. avenae*, which revealed that phospholipases associated with ROS production are important in several early defence responses pathways involved in quantitative cyst nematode resistance (Kong *et al.*, 2015).

8.6 Hormone-mediated Defence Responses by Cyst Nematodes

Activation of cellular immune responses in both susceptible and resistant plants as discussed in the previous sections include among others ROS production, MAPK activation and transcriptional reprogramming of the infected cell. These intracellular signalling pathways are interconnected to form a signalling network (Tsuda and Somssich, 2015), which ultimately results in the induction of local and systemic defence responses. Key components of the core defence system in plants are the plant hormones salicylic acid (SA) and jasmonic acid/ethylene (JA/ET), which play antagonistic roles in the activation of host defence responses depending on the pathogen feeding strategy (Pieterse *et al.*, 2012). Whilst the SA pathway is induced by biotrophic pathogens, the JA pathway – which consists of two antagonistic branches – provides resistance to necrotrophic pathogens (the ERF branch) and leaf chewing herbivores (the MYC2 branch). Information about hormone signalling in local and systemic defence responses induced by plant-parasitic nematodes in roots is still fragmentary. However, a picture emerges showing that root feeding by nematodes also activates an integrated signalling network, which is highly regulated by plant hormones (reviewed by Li *et al.*, 2015 and Kyndt *et al.*, 2014a). Below, studies on the role of the core defence system in local and systemic induced defence responses upon cyst nematode infection are discussed.

8.6.1 Local induced hormone-mediated defence responses

Root feeding by cyst nematodes results in the production and accumulation of SA, JA and ET at the site of infection (Kammerhofer *et al.*, 2015a), which triggers the core defence system. This ultimately results in the transcriptional reprogramming of host cells and the activation of SA- and JA/ET-mediated defence pathways. From several gene expression studies, mutant analysis and overexpression assays it is clear that local plant defence responses to cyst nematodes depends on SA biosynthesis, SA signalling and SA-regulated gene expression. For example, *Arabidopsis* SA-deficient mutants are more susceptible to cyst nematodes, whereas exogenous application of SA results in reduced susceptibility (Wubben *et al.*, 2008; Uehera *et al.*, 2010; Hamamouch *et al.*, 2011). SA signalling induced by cyst nematodes depends on the upstream components PAD4 (Wubben *et al.*, 2008; Youssef *et al.*, 2013) and EDS1 (Nguyen *et al.*, 2016). Moreover, downstream signalling of cyst nematode defence responses depends on the SA receptor NPR1. Overexpression of NPR1 and the transcription factor TGA2, which interacts in the nucleus with NPR1 to activate SA responsive genes, increases cyst nematode resistance (Matthews *et al.*, 2014). Similar results were obtained for the NPR1 suppressors SNC1 and SNI1, but also NHL genes upon cyst nematode infection of soybean roots (Maldonado *et al.*, 2014a, b). The induction of SA-mediated signalling results in the accumulation of various defence compounds like the pathogenicity related proteins, PR2 and PR5, which were shown to have a negative effect on cyst nematode parasitism of roots (papers cited by Li *et al.*, 2015). From this, it is clear that SA-mediated signalling plays a major role in host defence to cyst nematodes, similar to the role of SA in plant defence to other (foliar) pathogens with a biotrophic lifestyle.

For JA, however, opposite roles are reported in cyst nematode parasitism. A positive role for JA was shown by an increase in susceptibility of *Arabidopsis* mutants showing elevated JA levels in the roots (Ozalvo *et al.*, 2014). Interestingly, an increase in susceptibility was not only associated with changes in lysyl oxidase (LOX) gene expression, but also ERF4 expression levels, which could point to the activation of the ERF2 branch in JA signalling. On the other hand, the upregulation of VSP2, a marker for the MYC2 branch in JA defence signalling, was detected upon infection of soybean roots by *H. glycines* (Puthoff *et al.*, 2003), suggesting overlap in herbivore-induced and nematode-induced JA signalling. This is supported by the observation that the immune regulator SRFR1 reduces defence responses to both *H. schachtii* and the caterpillar *Spodoptera exigua* (Nguyen *et al.*, 2016). The tissue damage caused by intracellular migration by cyst nematodes may account for the activation of damaged triggered immune responses by the release of DAMPs during migration, comparable to leaf chewing herbivores or necrotrophic pathogens. Sidonskaya *et al.* (2016) showed that cyst nematode-induced JA-mediated defence responses depend on AP2C1, which is a negative regulator of MPK3 and MPK6, two positive regulators of plant defence responses controlling ET and JA biosynthesis. Infection of the *Arabidopsis* mutants *mpk3* and *mpk6* resulted in higher nematode infection rates, consistent with a reduced wound-induced defence phenotype of these plants. Together, these data show that JA/ET signalling pathways promote wound-induced defence responses during nematode invasion of the roots (necrotrophic phase) before the establishment of a permanent feeding site (biotrophic phase). This change in feeding behaviour may explain the observed dual role of JA in cyst nematode parasitism.

8.6.2 Systemic induced hormone-regulated defence responses

In addition to local cellular immune responses, systemic signalling is induced that results in defence responses in non-infected distant parts of the plant. Where local responses are effective in blocking the invader from further infection and prevent additional damage to plant tissues, systemic responses will prime the defence system of the plant to ward off a wide range of future invaders on the same or different plant parts (Pieterse *et al.*, 2012). This phenomenon is also known as systemic acquired resistance (SAR). Systemic induced defence requires long distance communication between the site of infection and distant plant tissues upon attack. However, the underlying principles and signals responsible for the systemic induction of the SA/JA/ET core defence system are not fully understood. For cyst nematodes, transcriptome analysis revealed that systemic changes in gene expression and defence responses are induced in the shoots of susceptible host plants upon local infection of the roots (Klink *et al.*, 2011). Moreover, comparing different systemic responses induced by cyst nematode root feeding showed species-specific changes in glucosinolate profiles in the shoots (Hofmann *et al.*, 2010; Hol *et al.*, 2013). Also other defence compounds like PR proteins accumulate in the shoots after cyst nematode root infection (Wubben *et al.*, 2008).

In resistant plants, systemic changes in gene expression are induced upon activation of a local *R* gene-mediated host-specific defence response. For example, in resistant potato plants carrying the *H1* gene the accumulation of novel gene products was observed encoding PR proteins (Hammond-Kosack *et al.*, 1989). Follow-up studies documented the changes in β-1,3-glucanase activity in leaves after infecting potato roots of cultivars carrying different resistance specificities with four potato cyst nematode populations. The resulting range of compatible and incompatible interactions elicited various classes of β-1,3-glucanases (Rahimi et al., 1996). One of them, β-D-glucosidase, was demonstrated to be significantly more active in the resistant cultivar carrying the *H1* gene infected with an avirulent potato cyst nematode population. The same potato cultivars were later tested for increased chitinase activity after *Globodera* spp. infection. The intercellular fluid of leaves obtained from all the nematode-infected plants showed a significant increase in exochitinase activity (Rahimi et al., 1998). Strong induction of PR-1 transcription and a slight induction of PR-5 expression was observed for tomato roots carrying the nematode resistance gene *Hero A* after infection with the potato cyst nematode *G. pallida* (Sobczak et al., 2005), but also in systemic leaves (Poch et al., 2006). PR protein expression is typically associated with the induction of SA-mediated defence signalling, thereby further supporting the key role of SA in host defence responses to cyst nematodes.

Induction of systemic defence responses by cyst nematodes may affect other pathogens and pests feeding on other plant parts, either by promoting or by inhibiting their performance. Indeed, root feeding by *H. schachtii* made *Arabidopsis* plants more attractive to the spider mite *Tetranychus urticae* and enhanced its development on the shoots (Kammerhofer et al., 2015b). In contrast, feeding by the thrips *Frankliniella occidentalis* on leaf epidermal cells increases root infection by *H. schachtii*, but feeding by the spider mite *T. urticae* does not (Kammerhofer et al., 2015b). Thrips feeding resulted in upregulation of the ET/JA marker gene *HEL* in systemic roots, whereas shoot feeding by mites resulted in downregulation of this marker in the roots. For both arthropods, no change in the SA marker *PR5* or the ET/JA marker *PDF1.2* was observed in the roots. Remarkably, the study by Kammerhofer et al. (2015b) shows that increased susceptibility is achieved by systemic induction of the ET/JA pathway, which promotes the attractiveness of the roots for following infective juveniles probably due to changes in root diffusates.

8.6.3 Modulation of local and systemic hormone-mediated defence responses

To counteract defence responses, nematodes have evolved mechanisms to suppress local defence responses during the onset of parasitism to establish a permanent feeding site in a hostile environment (reviewed by Goverse and Smant, 2014, and Mantelin et al., 2015). This may include effectors that target specific hormone-regulated defence pathways involved in host defence (Haegeman et al., 2012). For example, induction of PR1, PR2 and PR5 is suppressed in *Arabidopsis* upon overexpression of the SCN effector 10A06, which resulted in increased susceptibility to cyst nematodes (Hewezi et al., 2010). The production of chorismate mutase by both sedentary and migratory nematodes suggests that nematodes with different feeding strategies are indeed able to prevent the biosynthesis of SA, which will facilitate local invasion of the host (Mantelin et al., 2015). Given the importance of the SA pathway in cyst nematode defence responses as described above, it is not surprising that cyst nematodes have evolved strategies to manipulate this pathway. The modulation of immune responses is also reflected in the transcriptional dynamics observed during cyst nematode infection. For example, many defence-related genes upregulated within syncytia of *Rhg1* resistant plants, like salicylic acid-mediated defence signalling, were either partially suppressed or not induced to the same level by a virulent soybean cyst nematode population (Kandoth et al., 2011). Due to such local suppression of defence responses, the initiation of systemic induced defence responses may also be suppressed in the shoots upon cyst nematode infection, as observed in rice during root rot and root-knot nematode feeding (Kyndt et al., 2012, 2014b).

8.7 Conclusions and Future Prospects

Co-evolution between parasitic cyst nematodes and their host plants has resulted in the development of a gamut of resistances in the centre of origin of different crop plants. Resistance to cyst nematodes provides an important durable crop protection strategy and, as such, breeders have put much effort into the identification of resistance sources from various wild relatives. However, just a small portion of the genetic diversity has been explored to date. From this limited set of genetic resources, only a few genes underlying nematode resistance traits have been characterized. It is anticipated that this number will increase in the near future as the result of ongoing plant genome sequencing efforts and the development of high-throughput analysis methods. Meanwhile, sequence information related to cyst nematode resistance genes has become available and is being used to speed up the breeding process through marker-assisted selection.

The identification of cyst nematode resistance genes shows that they often encode canonical extra- and intracellular immune receptors, which share structural and functional similarities to other plant pathogen disease resistance genes. They are able to detect the presence of invading specific cyst nematode species or populations during different stages of parasitism, either in the apoplastic space or inside the cell. Upon the recognition of specific nematode effectors or effector-induced host cell modifications, a host-specific immune response is induced that blocks feeding cell development and nematode reproduction. Although a large repertoire of cyst nematode effectors is available, only a few are known to activate effector-triggered immunity in plants. The identification of RBP-1 as an activator of the intracellular immune receptor Gpa2 and VAP-1 as a modifier of the extracellular host protein Rcr-3, which is guarded by apoplastic immune receptor Cf-2, provides two model systems for further study of the molecular mechanisms underlying effector-triggered immunity to cyst nematodes. Such pairs of effectors and immune receptors allow the performance of functional studies using agroinfiltration assays on leaves of *N. benthamiana* instead of time consuming and highly variable nematode infection assays. This also opens the door for effector-based breeding approaches to identify novel sources of cyst nematode resistances in various crop species.

Interestingly, identification of genes underlying cyst nematode resistance mediated by the *Rhg1* and *Rhg4* loci in soybean revealed a novel mechanism of resistance based on natural variation in genes essential for cyst nematode parasitism. This implies that resistance in plants to cyst nematodes is determined by complementary layers of defence involving both receptor-mediated innate immunity and natural variation in plant genes that prevent the establishment of a successful feeding relationship. Considering the large number of genes required for proper feeding cell formation and its function as a metabolic sink for the uptake of nutrients by the nematode, it is likely that natural variation in such genes occurs among host plant genotypes that could provide additional sources of quantitative resistance to cyst nematodes. Resistance based on natural variation in parasitism genes may be more durable and difficult to overcome by nematodes compared to single dominant R genes. However, also for resistant genotypes carrying the *Rhg4* and *Rhg1* loci virulent *H. glycines* populations are reported. This indicates that cyst nematodes have also evolved strategies to circumvent the effects of impaired gene expression, but to understand the underlying mechanisms needs more research.

In contrast to host-specific resistance, research was less focused in the past on basal defence responses to cyst nematodes. However, it is evident that this first layer of immunity contributes to plant defence pre- and post-cyst nematode invasion of roots. Although the mechanistic insights are still limited it is clear that similar concepts and principles are involved in basal defence to cyst nematodes as well as other plant pathogens or pests. This may depend on their life strategies, as differences in invasion and feeding behaviour determine the activation of specific hormone-regulated defence signalling networks. The transcriptional dynamics in defence gene expression suggest that cyst nematodes activate defence signalling networks that resemble defence pathways induced by necrotrophic and biotrophic pathogens. This corresponds with the change in lifestyle from a cell-damaging migratory stage to a sedentary stage that includes feeding on living tissues. Therefore, it is important to distinguish between different parasitic stages when investigating the role of specific immune responses and signalling pathways in cyst nematode defence.

In addition, it is important to distinguish between local and systemic induced defence responses as several studies are based on syncytium-enriched plant material, whereas others are based on cyst nematode infected whole roots. The activation of systemic defence responses upon local root feeding by cyst nematodes is evident, but how this affects defence to other organisms in distant plant parts is still fragmentary and the molecular mechanisms obscure (Biere and Goverse, 2016). Like cyst nematodes, other plant pathogens and pests are able to manipulate their host by suppressing local and systemic defence responses or changing the food quality of the host, which could affect – either positively or negatively – the infection and performance of cyst nematodes as nicely shown by Kammerhofer et al. (2015b). Although this research area is still underexplored, knowledge on the cross talk between different plant pathogens and insects could provide novel insights on the susceptibility of host plants and the effectiveness of plant resistance to cyst nematodes under field conditions.

8.8 References

Aditya, J., Lewis, J., Shirley, N.J., Tan, H.T., Henderson, M., Fincher, G.B., Burton, R.A., Mather, D.E. and Tucker, M.R. (2015) The dynamics of cereal cyst nematode infection differ between susceptible and resistant barley cultivars and lead to changes in (1,3;1,4)-beta-glucan levels and HvCslF gene transcript abundance. *New Phytologist* 207, 135–147. DOI: 10.1111/nph.13349

Ali, M.A., Abbas, A., Kreil, D.P. and Bohlmann, H. (2013) Overexpression of the transcription factor RAP2.6 leads to enhanced callose deposition in syncytia and enhanced resistance against the beet cyst nematode *Heterodera schachtii* in Arabidopsis roots. *BMC Plant Biology* 13, 47. DOI: 10.1186/1471-2229-13-47

Ali, S., Magne, M., Chen, S.Y., Obradovic, N., Jamshaid, L., Wang, X.H., Belair, G. and Moffett, P. (2015) Analysis of *Globodera rostochiensis* effectors reveals conserved functions of SPRYSEC proteins in suppressing and eliciting plant immune responses. *Frontiers in Plant Science* 6. DOI: 10.3389/Fpls.2015.00623

Bakker, J., Folkertsma, R.T., Rouppe van der Voort, J.N.A.M., de Boer, J.M. and Gommers, F.J. (1993) Changing concepts and molecular approaches in the management of virulence genes in potato cyst nematodes. *Annual Review of Phytopathology* 31, 169–190. DOI: 10.1146/annurev.py.31.090193.001125

Bakker, E., Borm, T., Prins, P. et al. (2011) A genome-wide genetic map of NB-LRR disease resistance loci in potato. *Theoretical and Applied Genetics* 123, 493–508. DOI: 10.1007/s00122-011-1602-z

Bekal, S., Domier, L.L., Gonfa, B., Lakhssassi, N., Meksem, K. and Lambert, K.N. (2015) A SNARE-like protein and biotin are implicated in soybean cyst nematode virulence. *PLos ONE* 10 (12). DOI: 10.1371/journal.pone.0145601

Bendahmane, A., Kanyuka, K. and Baulcombe, D.C. (1999) The *Rx* gene from potato controls separate virus resistance and cell death responses. *The Plant Cell* 11, 781–792. DOI: doi.org/10.1105/tpc.11.5.781

Biere, A. and Goverse, A. (2016) Plant-mediated systemic interactions between pathogens, parasitic nematodes, and herbivores above- and belowground. *Annual Review of Phytopathology* 54, 499–527. DOI: 10.1146/annurev-phyto-080615-100245

Boller, T. and Felix, G. (2009) A renaissance of elicitors: perception of microbe-associated molecular patterns and danger signals by pattern-recognition receptors. *Annual Review of Plant Biology* 60, 379–407. DOI: 10.1146/annurev.arplant.57.032905.105346

Cai, D., Kleinem, M., Kifle, S. et al. (1997) Positional cloning of a gene for nematode resistance in sugar beet. *Science* 275, 832–834. DOI: 10.1126/science.275.5301.832

Carpentier, J., Esquibet, M., Fouville, D., Manzanares-Dauleux, M.J., Kerlan, M.C. and Grenier, E. (2012) The evolution of the *Gp-Rbp-1* gene in *Globodera pallida* includes multiple selective replacements. *Molecular Plant Pathology* 13, 546–555. DOI: 10.1111/j.1364-3703.2011.00769.x

Carpentier, J., Grenier, E., Esquibet, M., Hamel, L.P., Moffett, P., Manzanares-Dauleux, M.J. and Kerlan, M.C. (2013) Evolution and variability of Solanum RanGAP2, a cofactor in the incompatible interaction between the resistance protein GPA2 and the *Globodera pallida* effector Gp-RBP-1. *BMC Evolutionary Biology* 13. DOI: 10.1186/1471-2148-13-8

Chen, S., Chronis, D. and Wang, X. (2013a) The novel GrCEP12 peptide from the plant-parasitic nematode *Globodera rostochiensis* suppresses flg22-mediated PTI. *Plant Signaling and Behaviour* 8. DOI: 10.4161/psb.25359

Chen, X., MacDonald, M.H., Khan, F., Garrett, W.M., Matthews, B.F. and Natarajan, S.S. (2013b) Dynamic proteome analysis of soybean roots displaying compatible and incompatible interactions to different *Heterodera glycines* populations. *Current Proteomics* 10, 278–291. DOI: 10.2174/157016464113106660007

Chronis, D., Chen, S.Y., Lu, S.W., Hewezi, T., Carpenter, S.C.D., Loria, R., Baum, T. J. and Wang, X. H. (2013) A ubiquitin carboxyl extension protein secreted from a plant-parasitic nematode *Globodera rostochiensis* is cleaved in planta to promote plant parasitism. *The Plant Journal* 74, 185–196. DOI: 10.1111/tpj.12125

Cook, D.E., Geon Lee, T., Guo, X. *et al.* (2012) Copy number variation of multiple genes at Rhg1 mediates nematode resistance in soybean. *Science* 338, 1206–1209. DOI: 10.1126/science.1228746

Cook, D.E., Mesarich, C.H. and Thomma, B. (2015) Understanding plant immunity as a surveillance system to detect invasion. *Annual Review of Phytopathology* 53, 541–563. DOI: 10.1146/annurev-phyto-080614-120114

Das, S., DeMason, D.A., Ehlers, J.D., Close, T.J. and Roberts, P.A. (2008) Histological characterization of root-knot nematode resistance in cowpea and its relation to reactive oxygen species modulation. *Journal of Experimental Botany* 59, 1305–1313. DOI: 10.1093/jxb/ern036

de la Torre, F., Gutiérrez-Beltrán, E., Pareja-Jaime, Y., Chakravarthy, S., Martin, G.B. and del Pozo, O. (2013) The tomato calcium sensor Cbl10 and its interacting protein kinase cipk6 define a signaling pathway in plant immunity. *The Plant Cell* 25, 2748–2764. DOI: 10.1105/tpc.113.113530

de Majnik, J., Ogbonnaya, F.C., Moullet, O. and Lagudah, E.S. (2003) The cre1 and cre3 nematode resistance genes are located at homeologous loci in the wheat genome. *Molecular Plant Microbe Interactions* 16, 1129–1134. DOI: 10.3389/fpls.2016.01575

Diaz-Granados, A., Petrescu, A., Goverse, A. and Smant, G. (2016) SPRYSEC effectors: a versatile protein-binding platform to disrupt plant innate immunity. *Frontiers in Plant Sciences* 7, 1575. DOI: 10.3389/fpls.2016.01575

Ernst, K., Kumar, A., Kriseleit, D., Kloos, D.U., Phillips, M.S. and Ganal, M.W. (2002) The broad-spectrum potato cyst nematode resistance gene (*Hero*) from tomato is the only member of a large gene family of NBS-LRR genes with an unusual amino acid repeat in the LRR region. *The Plant Journal* 31, 127–136. DOI: 10.1046/j.1365-313X.2002.01341.x

Finkers-Tomczak, A., Danan, S., van Dijk, T., Beyene, A., Bouwman, L., Overmars, H., van Eck, H., Goverse, A., Bakker, J. and Bakker, E. (2009) A high-resolution map of the Grp1 locus on chromosome V of potato harbouring broad-spectrum resistance to the cyst nematode species *Globodera pallida* and *Globodera rostochiensis*. *Theoretical and Applied Genetics* 119, 165–173. DOI: 10.1007/s00122-009-1026-1

Finkers-Tomczak, A., Bakker, E., de Boer, J., van der Vossen, E., Achenbach, U., Golas, T., Suryaningrat, S., Smant, G., Bakker, J. and Goverse, A. (2011) Comparative sequence analysis of the potato cyst nematode resistance locus H1 reveals a major lack of co-linearity between three haplotypes in potato (*Solanum tuberosum* ssp.). *Theoretical and Applied Genetics* 122, 595–608. DOI: 10.1007/s00122-010-1472-9

Flor, H.H. (1942) Inheritance of pathogenicity in *Metampsora tini*. *Phytopathology* 32, 653–669.

Gish, L.A. and Clark, S.E. (2011) The RLK/Pelle family of kinases. *The Plant Journal* 66, 117–127. DOI: 10.1111/j.1365-313X.2011.04518.x

Goverse, A. and Smant, G. (2014) The activation and suppression of plant innate immunity by parasitic nematodes. *Annual Review of Phytopathology* 52, 243–265. DOI: 10.1146/annurev-phyto-102313-050118

Grundler, F.M.W., Sobczak, M. and Lange, S. (1997) Defence responses of *Arabidopsis thaliana* during invasion and feeding site induction by the plant-parasitic nematode *Heterodera glycines*. *Physiology and Molecular Plant Pathology* 50, 419–429. DOI: 10.1006/pmpp.1997.0100

Gust, A.A. and Felix, G. (2014) Receptor like proteins associate with SOBIR1-type of adaptors to form bimolecular receptor kinases. *Current Opinion in Plant Biology* 21, 104–111. DOI: 10.1016/j.pbi.2014.07.007

Haegeman, A., Mantelin, S., Jones, J.T. and Gheysen, G. (2012) Functional roles of effectors of plant-parasitic nematodes. *Gene* 492, 19–31. DOI: 10.1016/j.gene.2011.10.040

Hamamouch, N., Li, C.Y., Seo, P.J., Park, C.M. and Davis, E.L. (2011) Expression of *Arabidopsis* pathogenesis-related genes during nematode infection. *Molecular Plant Pathology* 12, 355–364. DOI: 10.1111/j.1364-3703.2010.00675.x

Hammond-Kosack, K.E., Atkinson, H.J. and Bowles, D.J. (1989) Systemic accumulation of novel proteins in the apoplast of the leaves of potato plants following root invasion by the cyst-nematode *Globodera rostochiensis*. *Physiology and Molecular Plant Pathology* 35, 495–506. DOI: 10.1016/0885-5765(89)90091-X

Hao, W., Collier, S.M., Moffett, P. and Chai, J. (2013) Structural basis for the interaction between the potato virus X resistance protein (Rx) and its cofactor Ran GTPase-activating protein 2 (RanGAP2). *Journal of Biological Chemistry* 288, 35868–35876. DOI: 10.1074/jbc.M113.517417

Hewezi, T., Howe, P.J., Maier, T.R., Hussey, R.S., Mitchum, M.G., Davis, E.L. and Baum, T.J. (2010) Arabidopsis spermidine synthase is targeted by an effector protein of the cyst nematode *Heterodera schachtii*. *Plant Physiology* 152, 968–984. DOI: 10.1104/pp.109.150557

Hofmann, J., El Ashry, A., Anwar, S., Erban, A., Kopka, J. and Grundler, F. (2010) Metabolic profiling reveals local and systemic responses of host plants to nematode parasitism. *The Plant Journal* 62, 1058–1071. DOI: 10.1111/j.1365-313X.2010.04217.x

Hol, W.H.G., De Boer, W., Termorshuizen, A.J., Meyer, K.M., Schneider, J.H.M., Van Der Putten, W.H. and Van Dam, N.M. (2013) *Heterodera schachtii* nematodes interfere with aphid-plant relations on *Brassica oleracea*. *Journal of Chemical Ecology* 39, 1193–1203. DOI: 10.1007/s10886-013-0338-4

Holbein, J., Grundler, F.M.W. and Siddique, S. (2016) Plant basal resistance to nematodes: an update. *Journal of Experimental Botany* 67, 2049–2061. DOI: 10.1093/jxb/erw005

Janssen, R., Bakker, J. and Gommers, F.J. (1991) Mendelian proof for a gene-for-gene relationship between virulence of *Globodera rostochiensis* and the *H1* resistance gene in *Solanum tuberosum* ssp. *andigena* CPC 1673. *Revue de Nématologie* 14, 213–219.

Jones, J.D.G. and Dangl, J.L. (2006) The plant immune system. *Nature* 444, 323–329. DOI: 10.1038/nature05286

Jupe, F., Pritchard, L., Etherington, G.J. *et al*. (2012) Identification and localisation of the NB-LRR gene family within the potato genome. *BMC Genomics* 13, 75. DOI: 10.1186/1471-2164-13-75

Kammerhofer, N., Radakovic, Z., Regis, J.M.A., Dobrev, P., Vankova, R., Grundler, F.M.W., Siddique, S., Hofmann, J. and Wieczorek, K. (2015a) Role of stress-related hormones in plant defence during early infection of the cyst nematode *Heterodera schachtii* in *Arabidopsis*. *New Phytologist* 207, 778–789. DOI: 10.1111/nph.13395

Kammerhofer, N., Egger, B., Dobrev, P., Vankova, R., Hofmann, J., Schausberger, P. and Wieczorek, K. (2015b) Systemic above- and belowground cross talk: hormone-based responses triggered by *Heterodera schachtii* and shoot herbivores in *Arabidopsis thaliana*. *Journal of Experimental Botany* 66, 7005–7017. DOI: 10.1093/jxb/erv398

Kandoth, P.K., Ithal, N., Recknor, J., Maier, T., Nettleton, D., Baum, T.J. and Mitchum, M.G. (2011) The soybean *Rhg1* locus for resistance to the soybean cyst nematode *Heterodera glycines* regulates the expression of a large number of stress- and defence-related genes in degenerating feeding cells. *Plant Physiology* 155, 1960–1975. DOI: 10.1104/pp.110.167536

Klink, V.P., Overall, C.C., Alkharouf, N.W., MacDonald, M.H. and Matthews, B.F. (2007a) Laser capture microdissection (LCM) and comparative microarray expression analysis of syncytial cells isolated from incompatible and compatible soybean (*Glycine max*) roots infected by the soybean cyst nematode (*Heterodera glycines*). *Planta* 226, 1389–1409. DOI: 10.1007/s00425-007-0578-z

Klink, V.P., Overall, C.C., Alkharouf, N.W., MacDonald, M.H. and Matthews, B.F. (2007b) A time-course comparative microarray analysis of an incompatible and compatible response by *Glycine max* (soybean) to *Heterodera glycines* (soybean cyst nematode) infection. *Planta* 226, 1423–1447. DOI: 10.1007/s00425-007-0581-4

Klink, V.P., Hosseini, P., Matsye, P., Alkharouf, N.W. and Matthews, B.F. (2009) A gene expression analysis of syncytia laser microdissected from the roots of the *Glycine max* (soybean) genotype PI 548402 (Peking) undergoing a resistant reaction after infection by *Heterodera glycines* (soybean cyst nematode). *Plant Molecular Biology* 71, 525–567. DOI: 10.1007/s11103-009-9539-1

Klink, V.P., Hosseini, P., Matsye, P., Alkharouf, N.W. and Matthews, B.F. (2010) Syncytium gene expression in *Glycine max* ([PI 88788]) roots undergoing a resistant reaction to the parasitic nematode *Heterodera glycines*. *Plant Physiology and Biochemistry* 48, 176–193. DOI: 10.1016/j.plaphy.2009.12.003

Klink, V.P., Hosseini, P., Matsye, P., Alkharouf, N.W. and Matthews, B.F. (2011) Differences in gene expression amplitude overlie a conserved transcriptomic program occurring between the rapid and potent localized resistant reaction at the syncytium of the *Glycine max* genotype Peking (PI 548402) as compared to the prolonged and potent resistant reaction of PI 88788. *Plant Molecular Biology* 75, 141–165. DOI: 10.1007/s11103-010-9715-3

Kong, L.A., Wu, D.Q., Huang, W.K., Peng, H., Wang, G.F., Cui, J.K., Liu, S.M., Li, Z.G., Yang, J. and Peng, D.L. (2015) Large-scale identification of wheat genes resistant to cereal cyst nematode *Heterodera avenae* using comparative transcriptomic analysis. *BMC Genomics* 16. DOI: 10.1186/s12864-015-2037-8

Koropacka, K. (2010) Molecular contest between potato and potato cyst nematode *Globodera pallida*: modulation of Gpa2-mediated resistance. Wageningen University, Wageningen. 132 pp. Available at: http://library.wur.nl/WebQuery/wurpubs/389198 (accessed 29 January 2018).

Kyndt, T., Nahar, K., Haegeman, A., De Vleesschauwer, D., Hofte, M. and Gheysen, G. (2012) Comparing systemic defence-related gene expression changes upon migratory and sedentary nematode attack in rice. *Plant Biology* 14, 73–82. DOI: 10.1111/j.1438-8677.2011.00524.x

Kyndt, T., Fernandez, D. and Gheysen, G. (2014a) Plant-parasitic nematode infections in rice: molecular and cellular insights. *Annual Review of Phytopathology* 52, 135–153. DOI: 10.1146/annurev-phyto-102313-050111

Kyndt, T., Denil, S., Bauters, L., Van Criekinge, W. and De Meyer, T. (2014b) Systemic suppression of the shoot metabolism upon rice root nematode infection. *PLoS ONE* 9. DOI: 10.1371/journal.pone.0106858

Li, R.J., Rashotte, A.M., Singh, N.K., Weaver, D.B., Lawrence, K.S. and Locy, R.D. (2015) Integrated signaling networks in plant responses to sedentary endoparasitic nematodes: a perspective. *Plant Cell Reports* 34, 5–22. DOI: 10.1007/s00299-014-1676-6

Li, Y., Fester, T. and Taylor, C.G. (2009) Transcriptome analysis of nematode infestation. In: Berg, R.H. and Taylor, C.G. (eds) *Cell Biology of Plant Nematode Parasitism*. Springer Plant Cell Monographs, Springer, Berlin, pp. 189–220. DOI: 10.1007/978-3-540-85215-5_7

Liu, S., Kandoth, P.K., Warren, S.D. *et al.* (2012) A soybean cyst nematode resistance gene points to a new mechanism of plant resistance to pathogens. *Nature* 492, 256–260. DOI: 10.1038/nature11651

Lozano-Torres, J.L., Wilbers, R.H.P., Gawronski, P. *et al.* (2012) Dual disease resistance mediated by the immune receptor Cf-2 in tomato requires a common virulence target of a fungus and a nematode. *Proceedings of the National Academy of Science of the United States of America* 109, 10119–10124. DOI: 10.1073/pnas.1202867109

Lozano-Torres, J.L., Wilbers, R.H.P., Warmerdam, S. *et al.* (2014) Apoplastic venom allergen-like proteins of cyst nematodes modulate the activation of basal plant innate immunity by cell surface receptors. *PLos PATHOGENS* 10. DOI: 10.1371/journal.ppat.1004569

Maldonado, A., Youssef, R., McDonald, M., Brewer, E., Beard, H. and Matthews, B. (2014a) Overexpression of four *Arabidopsis thaliana* NHL genes in soybean (*Glycine max*) roots and their effect on resistance to the soybean cyst nematode (*Heterodera glycines*). *Physiological and Molecular Plant Pathology* 86, 1–10. http://dx.DOI.org/10.1016/j.pmpp.2014.02.001

Maldonado, A., Youssef, R., McDonald, M., Brewer, E., Beard, H. and Matthews, B. (2014b) Modification of the expression of two NPR1 suppressors, SNC1 and SNI1, in soybean confers partial resistance to the soybean cyst nematode, *Heterodera glycines*. *Functional Plant Biology* 41, 714–726. DOI: 10.1071/FP13323

Manosalva, P., Manohar, M., von Reuss, S.H. *et al.* (2015) Conserved nematode signalling molecules elicit plant defences and pathogen resistance. *Nature Communications* 6, 7795. DOI: 10.1094/MPMI.2003.16.12.1129

Mantelin, S., Thorpe, P. and Jones, J.T. (2015) Suppression of plant defences by plant-parasitic nematodes. *Advances in Botanical Research* 73, 325–337. DOI: 10.1016/bs.abr.2014.12.011

Matsye, P.D., Kumar, R., Hosseini, P., Jones, C.M., Tremblay, A., Alkharouf, N.W., Matthews, B.F. and Klink, V.P. (2011) Mapping cell fate decisions that occur during soybean defense responses. *Plant Molecular Biology* 77, 513–528. DOI: 10.1007/s11103-011-9828-3

Matthews, B.F., Beard, H., Brewer, E., Kabir, S., MacDonald, M.H. and Youssef, R.M. (2014) *Arabidopsis* genes, AtNPR1, AtTGA2 and AtPR-5, confer partial resistance to soybean cyst nematode (*Heterodera glycines*) when overexpressed in transgenic soybean roots. *BMC Plant Biology* 14. DOI: 10.1186/1471-2229-14-96

Mazarei, M., Liu, W., Al-Ahmad, H., Arelli, P.R., Pantalone, P.R. and Stewart, C.N. Jr (2011) Gene expression profiling of resistant and susceptible soybean lines infected with soybean cyst nematode. *Theoretical and Applied Genetics* 123(7), 1193–1206. DOI 10.1007/s00122-011-1659-8

Mei, Y.Y., Thorpe, P., Guzha, A., Haegeman, A., Blok, V.C., MacKenzie, K., Gheysen, G., Jones, J.T. and Mantelin, S. (2015) Only a small subset of the SPRY domain gene family in *Globodera pallida* is likely to encode effectors, two of which suppress host defences induced by the potato resistance gene Gpa2. *Nematology* 17, 409–424. DOI: 10.1163/15685411-00002875

Meng, X. and Zhang, S. (2013) MAPK cascades in plant disease resistance signaling. *Annual Review of Phytopathology* 51, 245–266. 10.1146/annurev-phyto-082712-102314

Nguyen, P.D.T., Pike, S., Wang, J.Y., Poudel, A.N., Heinz, R., Schultz, J.C., Koo, A.J., Mitchum, M.G., Appel, H.M. and Gassmann, W. (2016) The Arabidopsis immune regulator SRFR1 dampens defences against herbivory by *Spodoptera exigua* and parasitism by *Heterodera schachtii*. *Molecular Plant Pathology* 17, 588–600. DOI: 10.1111/mpp.12304

Ozalvo, R., Cabrera, J., Escobar, C., Christensen, S.A., Borrego, E.J., Kolomiets, M.V., Castresana, C., Iberkleid, I. and Brown Horowitz, S. (2014) Two closely related members of Arabidopsis 13-lipoxygenases (13-LOXs), LOX3 and LOX4, reveal distinct functions in response to plant-parasitic nematode infection. *Molecular Plant Pathology* 15, 319–332. DOI: 10.1111/mpp.12094

Paal, J., Henselewski, H., Muth, J., Meksem, K., Menendez, C.M., Salamini, F., Ballvora, A. and Gebhardt, C. (2004) Molecular cloning of the potato *Gro1-4* gene conferring resistance to pathotype Ro1 of the root cyst nematode *Globodera rostochiensis*, based on a candidate gene approach. *The Plant Journal* 38, 285–297. DOI:10.1111/j.1365-313X.2004.02047.x

Pieterse, C.M., Van der Does, D., Zamioudis, C., Leon-Reyes, A. and Van Wees, S.C. (2012) Hormonal modulation of plant immunity. *Annual Review of Cell Developmental Biology* 28, 489–521. DOI: 10.1146/annurev-cellbio-092910-154055

Poch, H.L.C., Lopez, R.H.M. and Kanyuka, K. (2006) Functionality of resistance gene Hero, which controls plant root-infecting potato cyst nematodes, in leaves of tomato. *The Plant Cell and Environment* 29, 1372–1378. DOI: 10.1111/j.1365-3040.2006.01517.x

Postma, W.J., Slootweg, E.J., Rehman, S. *et al.* (2012) The effector SPRYSEC-19 of *Globodera rostochiensis* suppresses CC-NB-LRR-mediated disease resistance in plants. *Plant Physiology* 160, 944–954. DOI: 10.1104/pp.112.200188

Puthoff, D.P., Nettleton, D., Rodermel, S.R. and Baum, T.J. (2003) *Arabidopsis* gene expression changes during cyst nematode parasitism revealed by statistical analyses of microarray expression profiles. *The Plant Journal* 33, 911–921. DOI: 10.1046/j.1365-313X.2003.01677.x

Rahimi, S., Perry, R.N. and Wright, D.J. (1996) Identification of pathogenesis-related proteins induced in leaves of potato plants infected with potato cyst nematodes, *Globodera* species. *Physiology and Molecular Plant Pathology* 49, 49–59.

Rahimi, S., Perry, R.N. and Wright, D.J. (1998) Detection of chitinases in potato plants following infection with the potato cyst nematodes, *Globodera rostochiensis* and *G. pallida*. *Nematologica* 44, 181–193. DOI: 10.1163/005325998X00063

Rehman, S., Postma, W., Tytgat, T. *et al.* (2009) A secreted SPRY domain-containing protein (SPRYSEC) from the plant-parasitic nematode *Globodera rostochiensis* interacts with a CC-NB-LRR protein from a susceptible tomato. *Moecular Plant-Microbe Interactions.* 22, 330–340. DOI: 10.1094/MPMI-22-3-0330

Rice, S.L., Leadbeater, B.S.C. and Stone, A.R. (1985) Changes in cell structure in roots of resistant potatoes parasitzed by potato cyst-nematodes. I. Potatoes with resistance gene *H1* derived from *Solanum tuberosum* ssp. *andigena*. *Physiology and Molecular Plant Pathology* 27, 219–234. DOI: 10.1016/0048-4059(85)90069-4

Rooney, H.C., Van't Klooster, J.W., van der Hoorn, R.A., Joosten, M.H., Jones, J.D. and de Wit, P.J. (2005) Cladosporium Avr2 inhibits tomato Rcr3 protease required for Cf-2-dependent disease resistance. *Science* 308, 1783–1786. DOI: 10.1126/science.1111404

Sacco, M.A., Koropacka, K., Grenier, E., Jaubert, M.J., Blanchard, A., Goverse, A., Smant, G. and Moffett, P. (2009) The cyst nematode SPRYSEC protein RBP-1 elicits Gpa2- and RanGAP2-dependent plant cell death. *PLoS PATHOGENS* 5. DOI: 10.1371/journal.ppat.1000564

Sacco, M.A., Mansoor, S. and Moffett, P. (2007) A RanGAP protein physically interacts with the NB-LRR protein Rx, and is required for Rx-mediated viral resistance. *The Plant Journal* 52, 82–93. DOI: 10.1111/j.1365-313X.2007.03213.x

Schwessinger, B. and Zipfel, C. (2008) News from the frontline: recent insights into PAMP-triggered immunity in plants. *Current Opinion in Plant Biology* 11, 389–395. DOI: 10.1016/j.pbi.2008.06.001

Sidonskaya, E., Schweighofer, A., Shubchynskyy, V., Kammerhofer, N., Hofmann, J., Wieczorek, K. and Meskiene, I. (2016) Plant resistance against the parasitic nematode *Heterodera schachtii* is mediated by MPK3 and MPK6 kinases, which are controlled by the MAPK phosphatase AP2C1 in *Arabidopsis*. *Journal of Experimental Botany* 67, 107–118. DOI: 10.1093/jxb/erv440

Slootweg, E., Roosien, J., Spiridon, L.N. *et al.* (2010) Nucleocytoplasmic distribution is required for activation of resistance by the potato NB-LRR receptor Rx1 and is balanced by its functional domains. *The Plant Cell* 22, 4195–4215. DOI: 10.1105/tpc.110.077537

Slootweg, E.J., Spiridon, L.N., Roosien, J. *et al.* (2013) Structural determinants at the interface of the ARC2 and leucine-rich repeat domains control the activation of the plant immune receptors Rx1 and Gpa2. *Plant Physiology* 162, 1510–1528. DOI: 10.1104/pp.113.218842

Sobczak, M., Avrova, A., Jupowicz, J., Phillips, M.S., Ernst, K. and Kumar, A. (2005) Characterization of susceptibility and resistance responses to potato cyst nematode (*Globodera* spp.) infection of tomato lines in the absence and presence of the broad-spectrum nematode resistance *Hero* gene. *Molecular Plant Microbe Interactions* 18, 158–168. DOI: 10.1094/MPMI-18-0158

Sukarta, O.C., Slootweg, E. and Goverse, A. (2016) Structure-informed insights for NLR functioning in plant immunity. *Seminars in Cell and Developmental Biology* 56, 134–149. DOI: 10.1016/j.semcdb.2016.05.012

Swiecicka, M., Filipecki, M., Lont, D., Van Vliet, J., Qin, L., Goverse, A., Bakker, J. and Helder, J. (2009) Dynamics in the tomato root transcriptome on infection with the potato cyst nematode *Globodera rostochiensis*. *Molecular Plant Pathology* 10, 487–500. DOI: 10.1111/j.1364-3703.2009.00550.x

Szakasits, D., Heinen, P., Wieczorek, K., Hofmann, J., Wagner, F., Kreil, D.P., Sykacek, P., Grundler, F.M. and Bohlmann, H. (2009) The transcriptome of syncytia induced by the cyst nematode *Heterodera schachtii* in Arabidopsis roots. *The Plant Journal* 57, 771–784. DOI: 10.1111/j.1365-313X.2008.03727.x.

Takken, F.L.W. and Goverse, A. (2012) How to build a pathogen detector: structural basis of NB-LRR function. *Current Opinion in Plant Biology* 15, 375–384. DOI: 10.1016/j.pbi.2012.05.001

Tameling, W.I.L. and Baulcombe, D.C. (2007) Physical association of the NB-LRR resistance protein Rx with a Ran GTPase-activating protein is required for extreme resistance to potato virus X. *The Plant Cell* 19, 1682–1694. DOI: 10.1105/tpc.107.050880

Tameling, W.I., Nooijen, C., Ludwig, N., Boter, M., Slootweg, E., Goverse, A., Shirasu, K. and Joosten, M.H. (2010) RanGAP2 mediates nucleocytoplasmic partitioning of the NB-LRR immune receptor Rx in the Solanaceae, thereby dictating Rx function. *The Plant Cell* 22, 4176–4194. DOI: 10.1105/tpc.110.077461

Trudgill, D.L. (1967) The effect of environment on sex determination in *Heterodera rostochiensis*. *Nematologica* 13, 263–72. DOI: 10.1163/187529267X00120

Tsuda, K. and Somssich, I.E. (2015) Transcriptional networks in plant immunity. *New Phytologist* 206, 932–947. DOI: 10.1111/nph.13286

Uehara, T., Sugiyama, S., Matsuura, H., Arie, T. and Masuta, C. (2010) Resistant and susceptible responses in tomato to cyst nematode are differentially regulated by salicylic acid. *Plant and Cell Physiology* 51, 1524–1536. DOI: 10.1093/pcp/pcq109

Van der Vossen, E.A.G., Rouppe van der Voort, J.N.A.M., Kanyuka, K., Bendahmane, A., Sandbrink, H., Baulcombe, D.C., Bakker, J., Stiekema, W.J. and Klein-Lankhorst, R.M. (2000) Homologues of a single resistance-gene cluster in potato confer resistance to distinct pathogens: a virus and a nematode. *The Plant Journal* 23, 567–576. DOI: 10.1046/j.1365-313x.2000.00814.x

Waetzig, G.H., Sobczak, M. and Grundler, F.M.W. (1999) Localization of hydrogen peroxide during the defence response of *Arabidopsis thaliana* against the plant-parasitic nematode *Heterodera glycines*. *Nematology* 1, 681–686. DOI: 10.1163/156854199508702

Wan, J.R., Vuong, T., Jiao, Y.Q., Joshi, T., Zhang, H.X., Xu, D. and Nguyen, H.T. (2015) Whole-genome gene expression profiling revealed genes and pathways potentially involved in regulating interactions of soybean with cyst nematode (*Heterodera glycines* Ichinohe). *BMC Genomics* 16. DOI: 10.1186/s12864-015-1316-8.

Wieczorek, K. and Seifert, G.J. (2012) Plant cell wall signaling in the interaction with plant-parasitic nematodes. In: Witzany, G. and Baluska, F. (eds) *Biocommunication of Plants*. Springer-Verlag, Berlin, Heidelberg, Germany, pp. 139–155. DOI: 10.1007/978-3-642-23524-5_8

Wubben, M.J.E., Jin, J. and Baum, T.J. (2008) Cyst nematode parasitism of *Arabidopsis thaliana* is inhibited by salicylic acid (SA) and elicits uncoupled SA-independent pathogenesis-related gene expression in roots. *Molecular Plant-Microbe Interactions* 21, 424–432. DOI: 10.1094/MPMI-21-4-0424

Youssef, R.M., MacDonald, M.H., Brewer, E.P., Bauchan, G.R., Kim, K.H. and Matthews, B.F. (2013) Ectopic expression of AtPAD4 broadens resistance of soybean to soybean cyst and root-knot nematodes. *BMC Plant Biology* 13, 67. DOI: 10.1186/1471-2229-13-67

Zipfel, C. and Felix, G. (2005) Plants and animals: a different taste for microbes? *Current Opinion in Plant Biology* 8, 353–360. DOI: 10.1016/j.pbi.2005.05.004

9 Resistance Breeding

Vivian C. Blok[1], Gregory L. Tylka[2], Richard W. Smiley[3], Walter S. de Jong[4] and Matthias Daub[5]

[1]*The James Hutton Institute, Dundee, UK;* [2]*Iowa State University, Ames, Iowa, USA;* [3]*Oregon State University, Pendleton, Oregon, USA;* [4]*Cornell University, Ithaca, New York, USA;* [5]*Julius Kühn-Institut, Braunschweig, Germany*

9.1	Introduction	174
9.2	Resistance and Tolerance	176
9.3	Sources of Resistance and Genetics	178
9.4	Virulence, Pathotypes and Races	178
9.5	Screening (Phenotyping)	179
9.6	Cereal Cyst Nematodes	181
9.7	Potato Cyst Nematodes	187
9.8	Soybean Cyst Nematodes	191
9.9	Sugar Beet Cyst Nematodes	197
9.10	Other Cyst Nematodes	201
9.11	Conclusions and Future Prospects	202
9.12	References	203

9.1 Introduction

Cyst nematodes cause serious economic losses to the world's four most important food crops: cereals (wheat, maize and rice) and potatoes, as well as to important cash crops such as soybean and sugar beet. Due to a lack of obvious or distinct symptoms on the aerial parts of the host plants resulting from nematode parasitism of the roots, combined with the requirements for specific isolation processes and specialist skills to identify plant-parasitic nematodes, there is often a lack of awareness of their presence and the losses they can cause. This can be particularly problematic in subsistence agriculture where multiple contributing factors and a lack of nematological expertise can compromise attributing crop damage to plant-parasitic nematodes. Significant reductions in yields can eventually provide the motivation to prompt an 'in-depth' investigation.

Where plant-parasitic nematodes are implicated in yield losses, various management options may be available. Host resistance can protect yield and limit multiplication of plant-parasitic nematodes and can also provide a convenient, effective and environmentally benign method of controlling nematodes. Resistance is particularly appealing if the resistant cultivar has the same attributes as commercially successful susceptible cultivars already in use in the region, and if there are no additional costs required to purchase the resistant seed. Resistance is much preferred to the application of non-specific nematicides, which can have detrimental environmental impacts. Resistance not only provides a means of improving crop yields, but also it can have an impact on the targeted nematode populations, which are typically lower following a crop with a resistant cultivar. This can be beneficial where other crops within the rotation are hosts. Resistance can also help to ameliorate secondary damage to the host by reducing synergistic damage from other pathogens, such as soil-borne fungi (see Chapter 12, this volume). Resistance, however, is not usually a stand-alone solution to a plant-parasitic nematode problem but is often one component of an integrated management programme in which several strategies such as rotations, biocontrol, biofumigation and chemical control are used to minimize damage to the crop, control the nematode population(s) and enhance the durability of the resistance.

In this chapter general principles concerning resistance to cyst nematodes are presented together with more detailed information that relates specifically to soybean, cereal, potato or sugar beet cyst nematodes. Cyst nematodes tend to have restricted host ranges and this has the advantage that they are not likely to multiply on other crops within rotations. However, weeds that are hosts and ground keepers (i.e. volunteer potatoes) can sustain nematode populations between host crops and thus cause significant problems that require management. In temperate regions, the use of resistant crops within a rotation scheme combined with good weed control can provide an excellent management solution. In the tropics and developing countries, the use of resistance is often much more challenging due to the greater diversity of crops and varieties grown, more varied farming systems, a wider range of plant-parasitic nematodes species that are present and temperatures that promote faster nematode life cycles.

To achieve effective nematode population management, a high level of resistance in the crop is preferred. This is because cultivation of a partially resistant crop can in some circumstances result in a higher final nematode population than if a susceptible cultivar is grown, and the resulting population may be more virulent (Niblack *et al.*, 1986). Resistance is often highly specific and, thus, if there is diversity in the nematode population in virulence towards the resistance, the resistance will only be effective against the avirulent subset of the population and the virulent component will have an advantage. Partially resistant cultivars (see section 9.2) can be used successfully within a management programme if combined with other control measures that maintain population levels below the damaging threshold.

Sources of resistance exist against many of the agriculturally important plant-parasitic nematodes; however, incorporating resistance into commercially viable crops and successful adoption by the respective industry requires extended commitment to breeding programmes, sustained efforts of plant breeders and nematologists, as well as sources of resistance that are amenable to introgression into agronomically desirable germplasm. A major consideration in the breeding of nematode resistant crops is the potential economic benefits from increased yields *versus* the investment required in an extended breeding programme involving the introgression of a trait that may be specific to a particular species or subspecific race of nematode. Thus, an initial investigation to gauge the benefits that could be achieved from a resistant (or tolerant) crop and the effort required to achieve a commercially acceptable cultivar is needed. A preliminary evaluation of the performance of resistant and tolerant genotypes in the field to determine their agronomic qualities and performance in relation to nematode control would provide an indication of how much backcrossing might be necessary to generate commercially acceptable genotypes. Surveys of nematode distributions and damage assessments will provide background for evaluating the economic costs and benefits and whether future problems are likely to increase through spread and/or more favourable environmental conditions that may occur with, for example, climate change. Pathogenicity assessments of several plant-parasitic nematode populations should be performed to determine if there are differences in the damage and yield losses they incur (i.e. races or pathotypes) and whether the sources of resistance available are likely to be effective to some or all of them. It may also be worthwhile to review alternative control strategies that are available, such as trap and cover crops, where nematodes hatch but are unable to reproduce to see if these could be effective options.

Producing resistant cultivars follows the traditional plant breeding process: (i) make a cross between susceptible host and new resistance source to bring resistance into an acceptable genetic background; (ii) screen the progeny of the cross to select those that are resistant; (iii) make further crosses either to the susceptible parent or other desirable susceptible parents to reduce undesirable characters; and (iv) screen each new progeny for resistant individuals. The genetic structure of the resistance source and that of the crop will determine the appropriate breeding strategy to use and will influence the number of generations it will take to break linkages associated with characters that are agronomically undesirable. In practice, extensive breeding programmes against plant-parasitic nematodes have focused primarily on a few of the most economically important crops. Experience gained from these combined with recent advances in molecular technology should increase the speed of mapping and identification of resistance genes in the future for other crops. However, public and private funders need persuading that breeding for nematode resistance is a worthwhile investment and urgently needs support, given that nematicide use is less acceptable, access to genetically modified nematode resistant crops remains controversial and the demand for stable food production is increasing with the continuing rise in the human population.

9.2 Resistance and Tolerance

In plant-parasitic nematology, resistance is generally defined as the ability of a plant to inhibit or suppress the reproduction of a nematode species relative to the reproduction on a susceptible plant lacking such resistance (Table 9.1) (Cook and Evans, 1987; Roberts, 2002). Resistance can range from low or moderate (partial) resistance, permitting low levels of reproduction, or high level in which no or very little reproduction occurs. The reason for the small amount of reproduction that is sometimes observed on even highly resistant hosts is unclear. These nematodes may have escaped the host's resistant response or they could represent a very small number of virulent individuals in the population. If the latter, they represent a potential risk from a future breakdown of the resistance. Resistance is also described as vertical, which is race-specific, or horizontal, which is effective against all variants of the disease or parasite.

Table 9.1. Plant–nematode relationships as predicted from the gene-for-gene hypothesis: for each plant resistance gene (*R* gene) the nematode has a corresponding avirulence gene (*Avr* gene). If the host plant is resistant and the nematode avirulent the relationship is incompatible. However, if the host is susceptible and the nematode is avirulent or virulent, the relationship is compatible when these genes are dominant. Four possible genotypic combinations give rise to two phenotypic combinations.

	Avr/Avr	Avr/avr	avr/Avr	avr/avr
R/R	Ra	Ra	Ra	rV
R/r	Ra	Ra	Ra	rV
r/R	Ra	Ra	Ra	rV
r/r	rV	rV	rV	rV

Genotypic combinations: R, r = dominant, recessive alleles of plant *R* gene; Avr, avr = dominant, recessive alleles of nematode *Avr* gene.

	Avr	avr
R	Ra	rV
r	rV	rV

Phenotypic combinations: R, r = resistant, susceptible; a, V = avirulent, virulent nematode.

Resistance is differentiated from non-host status in which the host usually is not invaded or damaged by the nematode. Non-host status is not well understood but may provide a novel avenue for modifying host susceptibility if the underlying basis can be incorporated into crops. The expression of resistance can be affected by environmental conditions and susceptible hosts can differ in the amount of nematode reproduction they support, probably due to the influence of minor effect genes or the growth attributes of the particular genotype.

Tolerance refers to the ability of the host to tolerate the nematode parasitism without incurring an unacceptable yield penalty (Starr et al., 2013). Tolerance involves the sensitivity of the host and is also related to the population density of the nematode. Tolerance can be a combination of root responses and root growth characteristics, as well as general plant stress responses and, thus, is a complex trait and is generally more difficult to assess than resistance to nematodes. Tolerance and resistance are independent traits (Trudgill, 1991). Genotypes differ in their ability to withstand parasitism, some being highly sensitive even though they may be resistant to the nematode (resistant/intolerant), whereas others may be susceptible/tolerant and can support high nematode population multiplication (Fig. 9.1). Other combinations of resistance/tolerance and susceptible/intolerance are also possible, producing a complex set of interactions that, when combined with environmental factors such as moisture and temperature, may favour the host, nematode or both and have consequences for yield and nematode reproduction:

1. Tolerant and resistant – best combination for yield and controlling nematode populations.
2. Intolerant and resistant – suitable for low nematode population levels, otherwise the host is liable to be damaged.
3. Tolerant and susceptible – nematode population levels may increase above damage threshold level.
4. Intolerant and susceptible – not suitable for use with a nematode infestation.

Whilst tolerant varieties that are susceptible can be deleterious in a management programme as they can result in increases in nematode population levels, in a well-managed series of crop rotations and with relatively low nematode population levels, they can be incorporated successfully. The initial population density is the important determining factor. Damage is highly density dependent; thus, with some intolerant varieties if there are high nematode densities they can be severely damaged, whereas a tolerant susceptible variety may be able to produce a reasonable yield. Consequently, quite different initial nematode population levels can produce a similar final population density depending on the genotype. At high nematode densities even tolerant genotypes will incur damaged root systems leading to reduced efficiency of the host to support nematode reproduction and a final population that may be lower than the initial population. This generally occurs well above the damage threshold and, thus, is observed only at very high nematode population levels.

To determine the tolerance of a particular genotype, a comparison of the plant growth of an infected and non-infected resistant genotype *versus* the standard cultivar is usually performed in field conditions or plots, though a pot test can sometimes be sufficient. For plot or field tests this requires that in the infected areas the inoculum density and environmental conditions are uniform. Replicating trials in successive years

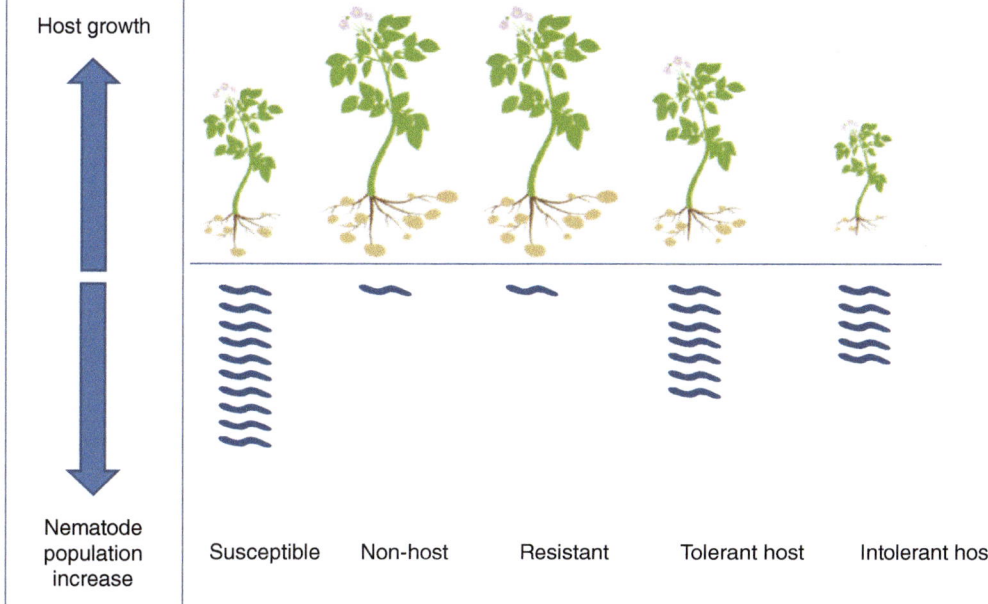

Fig. 9.1. Diagramatic representation of the effects of host growth and nematode population increase with susceptible, non-host, resistant, tolerant and intolerant hosts. (Adapted from McKenry and Roberts, 1985.)

may require moving to different field sites, adding further environmental variables to these assessments. Yields as well as nematode reproduction are determined in these assays.

9.3 Sources of Resistance and Genetics

Resistance is available to the main cyst nematode parasites of major crops. The genetic basis of these resistances can be monogenic (single gene), oligogenic (a few genes) or polygenic (many genes). Resistance genes are also referred to as major genes (having a large effect) or minor (having a small but detectable effect). Monogenic major gene resistance is more easily identified in resistance screening and genetically simpler to introgress and, thus, is most appealing in breeding programmes where resources must be distributed amongst a number of agronomically important traits. Major gene resistance typically results in a classic localized hypersensitive necrotic reaction in the host near to the nematode. For potato cyst nematodes (PCN), classic gene-for-gene interactions between a host resistance gene (R gene) and nematode avirulence gene (Avr gene) have been demonstrated. These are a consequence of direct or indirect interaction of the products of these genes that result in a cascade of host responses that can lead to localized cell death. By contrast, polygenic resistance can involve a number of genes dispersed over several chromosomes that collectively contribute differing amounts to the overall resistance. This sort of resistance greatly increases the difficulty in breeding. The host reaction can be a classic localized hypersensitive response or alternatively a less strong response that results in limited syncytium development and reduced nematode development and fecundity. In principle, monogenic resistance can be durable; however, polygenic quantitative resistance is considered to be more durable. A single resistance gene may fail due to intraspecific diversity within the nematode population and where there are multiple generations of the nematode within one growing season, rapid selection for population virulence can occur.

Resistance sources that are closely related to the crop of interest are usually preferred as these are likely to have fewer cytogenetic barriers to hybridization and fewer undesirable characters genetically linked to the resistance. However, wild relatives may be the only source of resistance available and this may extend the breeding time needed. National and international germplasm collections for different crops can provide a broad range of genetic resources that are relatively easy to access to start searching for a resistance source.

9.4 Virulence, Pathotypes and Races

Virulence is defined as the ability of the nematode parasite to reproduce on a host plant that possesses resistance. Avirulent nematodes are not able to reproduce, whereas virulent nematodes can, although the fecundity of the latter may be less than on a susceptible host. The use of resistance exerts a strong selection for virulent individuals within a population and where there are high levels of genetic variability within the parasite population this can result in selection for virulent individuals and a reduction of the resistance effect. Nematode populations often contain a mixture of virulent and avirulent nematodes and the proportion can change over time due to both host selection and vigour of the various nematode genotypes. Intraspecific genotypic and phenotypic diversity is typical, particularly with amphimictic species of plant-parasitic nematodes, so it is prudent to test a number of populations from the field against potential sources of resistance for their virulence before embarking on a breeding programme. The diversity of a nematode population can reflect the extent of the genetic diversity that has been introduced from a larger gene pool from the centre of origin, and this can also give an indication of the geographic restrictions on the effectiveness of a resistance. Plant-parasitic nematodes that have been introduced (i.e. are non-indigenous) are likely to have a smaller genotypic and phenotypic spectrum than in their centre of origin and, thus, also have fewer potential virulent individuals in the population(s) that could be selected when growing resistant cultivars. This helps to make breeding for resistance a strategy that can be worthwhile for the goal of producing an effective and durable control method. Reducing selection pressure by mixing resistant and susceptible cultivars within a rotation and pyramiding different resistances can also help to sustain the durability of resistances.

For some cyst nematodes, populations that show intraspecific variation in their responses to host resistances are referred to as host races or pathotypes. These pathotype and race schemes have been developed to assist with identifying which resistance sources are most effective. A set of host differentials that represent the spectrum of resistance sources is available and nematode populations that represent the different virulence groups are used to produce a matrix of interactions that can direct the choice of resistance sources that are most effective for use in breeding programmes. These schemes may need updating as new sources of resistance become available or new nematode populations with different virulence spectrums are identified.

9.5 Screening (Phenotyping)

For each crop, resistance tests (screens) have been developed to provide phenotypic information for their respective breeding programmes. Their aims include the identification of new sources of resistance and the identification of resistant progeny in segregating populations. Screening for nematode resistance is dependent on the availability of resistant germplasm and the progress of breeding programmes to generate new genotypes to test. The latter can reflect the complexity of the genetics of the host, the success of crossing, the life cycle of the crop and the resources that are available for screening. Resistance screens can be performed in a variety of ways and they aim to provide efficient, consistent and reproducible results so that there is repeatable ranking of host genotype susceptibility between years and in independent tests.

Resistance is evaluated by comparing reproduction to that on susceptible plants. A susceptible and (preferably) a resistant control are included to ensure the test has been successful and to allow comparisons between tests. Including a susceptible plant is important to determine relative susceptibility and to make comparisons between experiments. When screening wild species and genotypes with different genetic backgrounds from the susceptible control, which may contain various genes with small effects on the nematode multiplication, the multiplication rates on these plants may be significantly lower (or higher) than with the susceptible cultivar. Differences in the size of the root system and plant vigour can also have a very large impact on nematode multiplication and need to be taken into consideration when evaluating the results of the test. The assessment is based on an index of nematode reproduction; Reproduction Index (RI = 100 P_f/P_i) where P_i and P_f are the initial and final nematode numbers. At the end of the assay the nematodes counted may be females, cysts or eggs. Replication within the test is essential to improve accuracy. It is important for genetic studies, as well as for the industry that is using resistant cultivars in a nematode management programme, that the resistance information on which decisions are based is robust. Screens performed in controlled environments are usually preferable as resistance responses and nematode multiplication can be influenced by environmental conditions. Standardized procedures allow better control and greater efficiency; however, assessment of the performance in the field is also needed to evaluate other agronomically important characters and may be needed for tolerance assessments.

Generally there are no readily quantifiable host responses resulting from nematode parasitism and, thus, resistance in nematology generally is determined by the amount of nematode reproduction. The new generation of females is used to determine the reproduction index. The females visible on the root surface can be counted before they become cysts and have tanned, or if the roots are washed, the number of females that remain on the roots and those extracted by floatation and sieving the soil are counted. Cysts can also be extracted from the soil and counted, and then eggs within the cysts counted to obtain a measure of the fecundity on different host genotypes. However, this is very labour intensive and may not be warranted for routine screening of large progeny populations. When females or cysts are to be recovered from the soil, it is important that a high sand content is used to minimize the amount of organic material that contaminates the female/cyst sample and that the soil is dry.

9.5.1 Inoculum

Various nematode stages can be used for screening, including infested soil, cysts, eggs or active vermiform juveniles. The order of these corresponds to the increasing accuracy of the amount of the inoculum but is also the order of their increasing sensitivity to the environmental conditions of the test. Hatched juveniles allow a precise control of the number of individuals in the initial population (P_i) and more synchrony of development. Cysts can be enclosed in a mesh container so that juveniles can exit but these cysts can be recovered and differentiated from the final population (P_f). The stage used will also be determined by the life cycle of the particular cyst nematode. PCN require a period of diapause before they will hatch (Chapter 3, this volume) and, thus, a gap of several months is required between reproductive cycles; however, the cysts can be stored for many years and diapause can be circumvented in cultured cysts (Janssen et al., 1987).

The stage that is used will be determined by the nematode species, costs of producing inoculum, the amount of inoculum that is available, the scale of the assay and the outcomes required. Tolerance tests, for example, may require testing in plots or the field, whereas a routine screen to separate susceptible and highly resistant genotypes might be performed in small containers (Phillips et al., 1980). Usually the inoculum of cyst nematodes should be produced in clean conditions to minimize contamination with other organisms such as fungi, which can affect the quality of the inoculum, and contamination with other plant-parasitic nematodes should be prevented; however, production of inoculum in the field or microplots may also be suitable.

Selection of the nematode population(s) to test with is a critical component of the assay. It may be necessary to use several populations to represent the diversity found in the field or alternatively a mixture of the populations could be used. The scale of the screening assay may require that a representative population is chosen and then resistant genotypes selected from this initial screen are retested with more populations. Which approach is used will also depend on whether the screen is in relation to a specific resistance gene or aimed at identifying broad-based resistance. Prior assessment of the phenotypic diversity within the nematode populations in the geographic region(s) where the resistance is being developed is needed to make this judgement. To avoid applying selection pressure on the population being reared for use in the assays, a mixture of susceptible genotypes should be used and also multiplying on resistant genotypes can be helpful for maintaining species purity. Single cyst lines can also be produced but this can take several years to produce sufficient inoculum for use in screening assays.

The amount of inoculum will also depend on the assay type and the life stage of nematode used. There must be sufficient root volume to allow nematode invasion without causing root damage or there being detrimental effects through competition for feeding sites. The length of the life cycle needs to be considered so that sufficient time elapses to allow the nematodes to develop to the adult stage. More than one generation may occur during the assay with cyst nematodes that have short life cycles.

9.5.2 Marker assisted selection

Advances in sequencing technology and molecular methods have transformed the speed with which genomic regions that contain host resistance genes can be identified. These regions can provide genetic markers for use in the distinction of resistant and susceptible individuals. Marker assisted selection (MAS) can give equivalent information concerning the phenotype of segregating progeny as that from a nematode bioassay. Once markers have been developed, MAS can provide an alternative to nematode assays that is efficient and cost effective. These markers are generally associated with single, major resistance genes, although MAS with strong quantitative trait loci (QTLs) may also be useful, particularly where different resistances have been pyramided. Verification of the phenotype of genotypes selected with MAS is still required to ensure there is no loss of association between the marker and the resistance gene; however, there are likely to be many fewer individuals to test once a preliminary screen has been performed with MAS.

9.6 Cereal Cyst Nematodes

9.6.1 Scale of the problem

Wheat, barley, oats, rye and triticale are hosts for many species of cyst nematodes (Subbotin *et al.*, 2010) but only four species cause the greatest global damage: *H. avenae*, *H. australis* (= *H. avenae* Ha13), *H. filipjevi* and *H. latipons* (Rivoal and Cook, 1993; McDonald and Nicol, 2005; Smiley and Nicol, 2009). The most widely distributed of these species is *H. avenae*, followed by *H. filipjevi*, *H. latipons* and *H. australis*. Populations of *H. avenae* in China were recently redescribed as a new species, *H. sturhani* (Subbotin, 2015). Reports of economic damage in many countries were published during workshops of the International Cereal Nematodes Initiative (Riley *et al.*, 2009; Dababat *et al.*, 2015a). For example, cereal cyst nematodes were estimated to reduce wheat production by 20,933 metric tons (t) annually in three states of the USA. Only a small percentage of fields (<0.05%) were infested but losses in highly infested fields reached 39% for winter wheat and 64% for spring wheat (Smiley, 2009; Smiley *et al.*, 1994, 2013).

Populations of *H. avenae* and *H. filipjevi* from different locations are highly variable in their virulence and reproductive capacity (fitness) on the same host (Rivoal *et al.*, 2001; Mokabli *et al.*, 2002). They are also heterogeneous with respect to virulence to specific host genotypes. An International Test Assortment of barley, oat and wheat differentials was established to define pathotypes of *H. avenae* in northern Europe (Andersen and Andersen, 1982). Fewer than two decades later the Test Assortment was incapable of defining more than half the virulence phenotypes known to occur globally (Cook and Rivoal, 1998) and was declared incomplete and in need of updating (Cook and Noel, 2002; Smiley and Nicol, 2009). New virulence phenotypes continued to be identified and some were reported without proposing a new pathotype designation (Smiley *et al.*, 2011). Periodic revisions of the Test Assortment were suggested to account for the taxonomic distinction of *H. filipjevi* and *H. australis* from *H. avenae* (Turner and Subbotin, 2013).

Mixtures of *H. avenae* virulence phenotypes occur within individual fields (Cook and Williams, 1972) and may interbreed to produce new virulence phenotypes (Andersen, 1961; Person and Rivoal, 1979; Andersen and Andersen, 1982). Moreover, mixtures of *H. avenae* with *H. filipjevi* or *H. latipons* also occur (Abidou *et al.*, 2005; Holgado *et al.*, 2009; Yan and Smiley, 2010; Smiley and Yan, 2015). Further complexity occurs where mixtures of *H. avenae* virulence phenotypes occur in the same fields with mixtures of species (Holgado *et al.*, 2009). Cultivars that exhibit resistance to a specific population of *H. avenae* do not necessarily exhibit that same reaction to another population of *H. avenae* or another *Heterodera* species (Ireholm, 1994; Rivoal *et al.*, 2001; Smiley and Yan, 2015). These complexities, where they occur, present a strong challenge to the goal of managing crop losses by deploying resistant varieties. The spectrum of virulence of each regional population must be well defined before the expense of a targeted breeding programme can be justified.

9.6.2 Sources of resistance

Eleven *Cre* genes for resistance to *H. avenae* have been designated in wheat (Seah *et al.*, 1998; Nicol *et al.*, 2003; McDonald and Nicol, 2005; McIntosh *et al.*, 2008; Smiley and Nicol, 2009). Some of these genes are also effective against *H. australis*, *H. filipjevi*, *H. latipons* and/or *H. sturhani* (Rathjen *et al.*, 1998; Jahier *et al.*, 2001; Rivoal *et al.*, 2001; Lewis *et al.*, 2009; İmren *et al.*, 2013; Smiley and Yan, 2015; Wu *et al.*, 2016). In wheat, the *Cre*1 gene on chromosome 2BL and the *Cre*8 gene on chromosome 6BL were identified directly in *Triticum aestivum*. The *Cre*2 gene on chromosome N^v, *Cre*5 (= *CreX*) gene on chromosome 2AS, and *Cre*6 gene on chromosome $5N^v$ were derived from *Aegilops ventricosa*. The *Cre*3 and *Cre*4 genes are both on chromosome 2DL and were derived from *Ae. tauschii* (syn. *T. tauschii* and *Ae. squarrosa*). The *Cre*7 gene was derived from *Ae. triuncialis* and the *Cre*3S gene on chromosome 3S and the *CreY* gene (not yet located) were derived from *Ae. variabilis*. The *Cre*R gene on chromosome 6RL was identified in triticale (× *Triticosecale*).

Four major resistance genes are known in barley. The genes *Rha*1, *Rha*2 and *Rha*3 are located on chromosome 2HL and *Rha*4 is located on chromosome 5HL. The resistance expressed by *Rha*E in 'Morocco' and 'Emir' has not been mapped to a chromosome location. Resistance to *H. avenae* in the wild oat '*Avena sterilis* I376' consists of three unnamed resistance genes that collectively confer resistance to all known pathotypes of *H. avenae*. Other undefined sources of resistance in wheat, barley and oat are also present among entries of the Test Assortment (McDonald and Nicol, 2005; Smiley *et al.*, 2011) and in local collections of genotypes (Dababat *et al.*, 2015b; Marshall and Smiley, 2016; Smiley and Marshall, 2016; Wu *et al.*, 2016).

9.6.3 Selection of nematode populations for inoculum

9.6.3.1 Naturally infested soil

The simplest screenings are conducted with soil collected from naturally infested fields. Whilst field soil includes other root-invading nematodes and fungi, these other pathogens generally do not significantly complicate development of white females of cereal cyst nematodes. Moist soil containing cysts can be collected at the time when juveniles are expected to start emerging from cysts in the field. Alternatively, dry soil can be collected after an affected cereal crop has been harvested. Before brown cysts in soil can be used in assays they must be primed under low-temperature storage for at least 2 months to overcome diapause.

9.6.3.2 Cysts as inoculum

Cysts can be used to inoculate non-infested soil that has been sterilized, pasteurized, left untreated, or mixed with sand or other substrates. The cysts can be extracted from an infested soil or from plant cultures produced in the glasshouse or outdoors. The mean number of eggs plus juveniles in the cyst inoculum must be determined by extracting cysts from soil, selecting cysts of various sizes and breaking the cyst walls to release eggs plus juveniles for counting (Hooper and Evans, 1993). Inoculation with cysts rather than juveniles is often preferred because the collection of large numbers of juveniles over a short period of time can be unpredictable.

9.6.3.3 Juveniles as inoculum

Inoculation of soil with juveniles provides the greatest precision for pot tests. For studies in which other species of nematodes and soil microbes are not worrisome, naturally infested soil or elutriated cysts plus debris from soil that has been kept cold during the winter (or has been incubated in a refrigerator at 4°C for at least 4 months) can be used. The soil (or cysts plus debris) are placed onto racks in a shallow tub in which water to wet but not submerge the soil layer has been added. Emerging juveniles will settle on the bottom of the tub from where they can be collected at frequent intervals and immediately transferred to the refrigerator. Juveniles should be used as inoculum within 2 weeks because they quickly decline in aggressiveness. For studies requiring a greater purity of inoculum, the cysts can be surface sterilized before being used to collect juveniles (Pariyar *et al.*, 2016).

Different numbers of juveniles are required to screen different small grain cereals; more are required for barley and wheat than oats. The goal is to supply sufficient inoculum for the formation of at least 20 white females per root system, the minimal number for reliably distinguishing resistant and susceptible phenotypes (Andersen and Andersen, 1982).

9.6.3.4 Preparation of a new inoculum

To establish a new source of inoculum, only newly produced mature white females from field-grown plants or from pot cultures should be used. They should be stored at 3°C on a 250-µm mesh nylon screen suspended in tap water. If fungi become evident, cysts are transferred to a filter paper and then rolled with an artist's brush to remove the fungi. They are then returned to the screen in cold water at 3°C or moved to 7°C for hatching. Juveniles in eggs in white females will moult readily to second-stage juveniles capable of hatching without the cold treatment that is required to break the diapause exhibited by juveniles in brown cysts. If newly produced brown cysts are used they should be stored in an air-dry state at 3°C for a minimum of 4 months in sealed glass tubes to overcome diapause. Cysts will retain near-original egg viability for up to 5 years using this system. To induce hatching, unbroken white females or brown cysts are incubated at

7°C for 2 weeks on a screen in a tap water-filled watch glass inside a covered but unsealed Petri plate.

9.6.4 Screening assays

9.6.4.1 Field trials

Based upon extensive experience with *H. australis* (= *H. avenae* Ha13), Fisher (1982a) concluded that 'an accurate assessment of resistance in the field is not possible'. Screening cultivars for resistance is therefore often initially conducted almost entirely under controlled conditions (Lewis *et al.*, 2009). Nevertheless, field trials using high numbers of replicates are reported in many countries. These trials are almost always conducted in naturally infested soils. Before establishing a trial the investigator must determine the spatial heterogeneity of nematode density within the desired field and, ideally, also predetermine from multiple sites across the proposed area whether the population consists of more than one species or virulence group. Fields should be sampled on a grid pattern to identify high-density patches large enough to accommodate an experiment.

The 'microsquare' is a method for screening cultivars for resistance in fields where the nematode density is heterogeneous (Andersen, 1961). The microsquare (1 m × 1 m) is placed within an area of high nematode density. The soil is mixed well to a depth of 15–20 cm before sowing. Ten cultivars are sown as single seeds at each of ten points separated by a distance of 10 cm. Andersen (1961) determined that when the susceptible control contained about 100 white females per plant the standard error could be reduced to ±25 females and a satisfactory distinction could be made of resistant cultivars.

Estimates of both resistance and tolerance traits have been performed on adjacent blocks of land that have a higher versus a lower density of cereal cyst nematodes. These blocks are usually prepared in advance by growing for two or more years a highly susceptible cereal on blocks that are to be used as highly infested blocks. Adjacent low-density 'control' blocks can be continuously fallowed or planted to either a non-host or a resistant cereal. Differences in density of pathogenic fungi and other parasitic nematodes can complicate such comparisons.

Field trials that compare cultivars under conditions of high or low nematode 'pressure' can also be performed by planting cultivars side-by-side in replicated strip plots. Alternating drill strips are either treated with a nematicide or left untreated and the cultivars are changed at intervals within each drill strip to allow side-by-side comparisons (Smiley *et al.*, 2013; Marshall and Smiley, 2016). The nematicide used for this purpose is often aldicarb because it minimizes effects on other members of the microflora and microfauna, and minimizes potential growth stimulatory effects on plants (Smiley *et al.*, 2013).

9.6.4.2 Pot trials

Cultivars can be screened more efficiently and accurately for resistance under controlled conditions of a glasshouse or outdoor nursery. Large numbers of replicates are required due to high variability (Smiley *et al.*, 2011; Pariyar *et al.*, 2016).

Screening assays require approximately 9 weeks from planting until inspection of roots and/or extraction of swollen white females from soil and roots. In general, sampling occurs at the time plants are in anthesis. Heat units calculated as growing-degree days dictate the rates of juvenile emergence from cysts, invasion of roots, development of each *Heterodera* stage, and the total time required between the date of juvenile invasion and development of mature white females or brown cysts. Recent incubation schemes including lighting and humidity were described by Al-Hazmi *et al.* (2001), Cook and Noel (2002), Şahin *et al.* (2010), Hajihasani *et al.* (2010), Dababat *et al.* (2014), Smiley and Yan (2015), Pariyar *et al.* (2016), Wu *et al.* (2016) and others. Watering should be monitored closely to avoid saturating the soil, which may be allowed to become dry at the surface so long as the plants do not become severely wilted.

Pot studies are often very successful when performed outdoors at a time period when juvenile densities are greatest in naturally infested fields. Pots are prepared as for glasshouse studies and then the bottoms are buried outdoors into a layer of non-infested sand or soil (Smiley *et al.*, 2011). The pots are watered initially from the surface to facilitate seedling establishment. After roots have emerged from the drainage holes, the plants may be watered from 'leaky' hoses buried in the sand or soil.

Larger outdoor pot studies, called microplots, consist of drainage pipes or other tubing being buried into soil and either watered from the surface or by exposing the plants only to natural rainfall events (Andersen, 1961; Hajihasani et al., 2010, 2013). Various procedures have been used to fill the pipes, including different layering techniques and the use of naturally infested or inoculated soils. Initial amounts of inoculum must be recalibrated for different methods of filling microplots with soil (Andersen, 1961).

Identification of genotypic resistance generally requires counting white females, either those immediately visible on the surface of intact root balls pulled from pots (Lewis et al., 2009), or the total number after roots are washed (Andersen, 1961). Counting females on the surface of root balls greatly increases the number of plants that can be examined each day.

9.6.4.3 Miniaturized trials

Distinctions among phenotypic reactions are most precise in miniaturized screening systems in which plants are grown using either a Petri dish method or a test tube method. Published descriptions (Rivoal et al., 1978, 1991; Bekal et al., 1998) lack important details critical to the repeatability of results. Rivoal (2002, pers. comm.) provided the following, more comprehensive descriptions of the two methods. Both methods are challenging but produce precise assessments of cultivar resistance and very clean cysts for use in subsequent experiments. Results from these methods are well correlated (Şahin et al., 2006; Yavuzaslanoğlu et al., 2016).

In the Petri plate method the assessment is done in sterile conditions on agar. Sterilized seeds are germinated on water-agar and then single fungus-free wheat plants with 0.5-cm-long roots are placed on four to ten replicate water-agar Petri plates and slices of agar from another plate are used to 'tack down' each root. The reproductive efficiency (fitness) differs for each cyst nematode population. Preliminary experiments need to be performed to determine the number of juveniles required to produce a given number of white females. The goal is to obtain about ten females per plate. The number of females will be fewer than half the juveniles transferred to each root because not all juveniles are aggressive and the population will develop into approximately half males and half females. If the fitness estimate is unknown, place at least 16 juveniles onto each of three roots.

To inoculate the roots, juveniles in water are prepared as described (section 9.6.3) and placed into two watch glasses from where they are transferred to root tips (where they invade) by using a 60× dissecting microscope and a 'needle' constructed of a short eyebrow hair glued to the end of a wood doweling. A black mark is made on the plate under the root tips to indicate the point of inoculation. Juveniles will start penetrating the root within a few hours after inoculation; plates are checked and any 'wanderers' moved back to the root tip. An extra plate placed on top of the stack will prevent the seedlings from pushing up the lid of the uppermost assay plate. An average of one or more swollen white females on the entire plant indicates that it is susceptible. It is also possible to make quantitative assessments and to collect pure white females for future experiments or DNA analysis.

In the test tube method plastic tubes are used with a 1-mm hole drilled through the base to allow water and/or nutrients to wick up. The tubes are filled with a mixture of sand (pure white, naturally rounded, very fine sand of <250 μm diameter) and kaolin (4:1, by volume) and then wetted with a complete horticultural nutrient solution. Next, the surface of wheat seeds are sterilized with 5% sodium hypochlorite for 7 min and then rinsed with tap water. The wheat seeds are pre-germinated on a water-agar plate (see above). One plant with 0.5 cm long roots is placed onto the growth substrate in each tube and covered with additional sand–kaolin mixture (eight to ten replicates of each genotype).

To inoculate the wheat plants, it is critical to understand that juveniles only penetrate the meristematic region of the root tip. A glass pipette is used to inoculate sufficient numbers of juveniles from a well-agitated suspension to get ~10 cysts per plant. Afterwards, do not add water or nutrient solution to the tube from the top; top watering greatly diminishes reliability of the method by creating thick water films that impede or prevent root invasion, and by washing juveniles deeper into the substrate.

The plant cultures are watered and fed as necessary during an incubation period of 12–16 weeks until white females mature. The plants are then allowed to dry out naturally for at least a month after watering is stopped. The cysts will turn brown and be clearly visible against the white growth medium; all cysts will occur in the top centimetre of the tube. The debris-free cysts can be extracted by washing the sand and roots through a 1-mm sieve nested over a 250-μm mesh sieve. An average of ≤1 cyst per plant indicates that the plant is resistant to the nematode population being tested. Jahier et al. (1996) modified the test tube method by inoculating each tube with either two thermally primed cysts that contained about 220 juveniles, or with 220 motile juveniles. Yavuzaslanoğlu et al. (2016) filled the tubes with a mixture of sand, non-infested field soil and organic matter (70:29:1) and inoculated each tube with 100 juveniles on each of two successive days.

9.6.5 Interpretation of 'resistance'

Many issues complicate the distinction of resistant and susceptible genotypes. First, for large populations of host genotypes, some of which are highly resistant, there exists a continuum of phenotypic responses relative to numbers of white females produced on roots. There is seldom a definite borderline to distinguish between resistance and susceptibility, particularly in wheat. It remains unknown as to why one or a few white females will form on genotypes generally considered highly resistant to a specific nematode population, or why cultivars carrying the same single dominant resistance gene can vary considerably in phenotypic response. Numbers of eggs may vary greatly within cysts that may otherwise appear similar. Non-gravid but otherwise normal-appearing white females may also develop parthenogenetically after a root has been inoculated with a single second-stage juvenile. Whilst these and other issues challenge a definitive definition of resistance, rating systems can distinguish phenotypic reactions. The systems are of two basic types depending upon either an absolute number of white females per root when averaged across replicates, or the percentage of white females relative to a pre-designated control genotype known to be highly susceptible to the nematode population being evaluated.

Using the 'absolute number' system, Mathur et al. (1974) and Ireholm (1994) defined genotypes in glasshouse screenings as resistant if the average was ≤3 females per root system, and susceptible if the average was above 3.0. Variations of this system were used to screen wheat in miniaturized tests (Bekal et al., 1998; Mokabli et al., 2002) and in field trials (Smiley et al., 2013). O'Brien and Fisher (1974) described four reaction types for field trials, with a range from resistant (no females) to very susceptible (>50 females per plant). Variants of the grouping system have been used by Kaur et al. (2008), Nicol et al. (2009), Marshall and Smiley (2016), Pariyar et al. (2016), and Smiley and Marshall (2016).

Brown (1969) reported that European workers using the 'percentage system' agreed to consider a cultivar resistant if the total number of newly formed white females did not exceed 5% of the number formed on a comparative susceptible variety tested under the same glasshouse conditions. The inoculation procedure and number of replicates must be sufficient to produce a minimum of 200 newly formed white females on the pre-designated susceptible genotype control. This rating scale continues to be used (Holgado et al., 2004). However, Lücke (1976) described a graduated scale with six groups, from highly resistant (≤1% of white females produced on the susceptible control) to susceptible (>60%). Variants of this scale have been used by Valocká et al. (1994), Smiley et al. (2011) and Dababat et al. (2014).

9.6.6 Development of resistant cultivars

Recent screenings of cultivar collections for phenotypic resistance reactions have been summarized in workshops (Riley et al., 2009; Dababat et al., 2015a) and elsewhere (Dababat et al., 2014; Pariyar et al., 2016). Locations where resistance genes have been mapped on chromosomes have been summarized in reviews (McDonald and Nicol, 2005; Smiley and Nicol, 2009). MAS has been reported for detection of resistance genes such as Cre1, Cre3, and Rha1, Rha2 and Rha4 (Eagles et al., 2001; Chełkowski et al., 2003; Çalişkan et al., 2011).

Very few papers summarize breeding programmes in which cultivars were developed for resistance to cereal cyst nematodes. Two papers are considered essential reading: Rathjen *et al.* (1998) and Fisher (1982b). The review by Rathjen *et al.* (1998) examined breeding strategies that failed to produce effective cultivars in Australia, and the breeding approach that succeeded in greatly reducing the density of *H. australis* (= *H. avenae* Ha13) across that continent. They described the folly of breeding for resistance without a dedicated and equally strong consideration of cultivar tolerance, a concept that was previously advocated vigorously by Fisher (1982b). Briefly, resistance is of little practical agricultural value when it resides in intolerant cultivars, because the yields of highly sensitive plants can be greatly reduced even though reproduction of the nematode will be prevented. Rathjen *et al.* (1998) described the release of resistant but intolerant cultivars that were rejected by farmers because of partial crop failures when those cultivars were planted into highly infested fields.

Research on genetic improvement in many countries has been focused initially on screening potential sources of resistance already existing within adapted genotypes of wheat, barley and oat collections. When resistance is identified and characterized, almost all efforts to introgress resistance into susceptible, agronomically adapted cultivars have been through crosses based on the use of single dominant resistance genes. Crosses between susceptible and resistant genotypes are typically followed by up to five additional backcrosses with the susceptible genotype (Andersen, 1961). The susceptible genotype must be selected with care to assure that it is phenotypically tolerant to invasion by the nematode, because introgression of resistance into an intolerant/susceptible genotype is likely to produce intolerant/resistant genotypes that are unlikely to be accepted by farmers. Rathjen *et al.* (1998) concluded that tolerance is highly heritable but has received little attention by breeders and has not been effectively evaluated with respect to mode of action or genetics. Simultaneous screenings for resistance plus tolerance of wheat and barley collections under field conditions were reported by Marshall and Smiley (2016) and Smiley and Marshall (2016).

Development of resistant cultivars has been practised for more than five decades in northern Europe (Andersen, 1961), where it is common for multiple pathotypes and/or species of cereal cyst nematodes to occur within individual fields. Cook and Noel (2002) concluded that continuous production of cultivars with resistance to *H. avenae* has not been particularly successful in Europe due to a gradual selection for either another pathotype of *H. avenae* or another species, such as *H. filipjevi*. In those situations, management of cereal cyst nematodes depends more heavily on strategies other than, or in addition to, genetic resistance (Holgado *et al.*, 2009). Cook and Noel (2002) pointed out that the exception appears to be in Australia, where only a single pathotype of *H. australis* (= *H. avenae* Ha13) is reported to occur (Riley and McKay, 2009). More than half the cultivars grown in infested regions of Australia during the past three decades have possessed moderate to high resistance as well as tolerance to this population (Rathjen *et al.*, 1998; Lewis *et al.*, 2009). Densities of *H. australis* have therefore declined markedly (Riley and McKay, 2009). Resistance breeding efforts in Australia include high-throughput phenotyping methods (Lewis *et al.*, 2009) and MAS (Eagles *et al.*, 2001; Williams *et al.*, 2003).

While only a single pathotype of *H. avenae* is known to occur in the Pacific Northwest region of the USA (Smiley *et al.*, 2011) and in India (Bishnoi, 2009), infestations of *H. filipjevi* also occur in each country (Smiley *et al.*, 2008; Bishnoi, 2009). In India, Bishnoi (2009) reported field and glasshouse trials that identified sources of resistance to *H. avenae*. The resistances were then successfully introgressed into agronomically adapted cultivars. However, the resistance was ineffective against the Indian population of *H. filipjevi*.

9.6.7 Future challenges

Breeding for resistance to cereal cyst nematodes will be benefited by improvements of practical methods to identify spatial distributions of pathotypes and species. Current methods for identifying species are based upon soil extractions and morphological and/or molecular identifications, which are tedious and difficult for workers other than nematode taxonomists. Current methods to identify virulence types (pathotypes) require multiple 3-month bioassays, half of which may fail to provide reliable results. Direct evaluation of DNA extracted from soil is now capable of identifying and quantifying cereal cyst nematode species but not pathotypes. Greater uniformity of screening methods to assess resistance and tolerance is required to facilitate more reliable comparisons among regional and international assays. Widespread use of MAS will be required to improve the capacity of wheat breeders to pyramid resistance genes capable of encompassing the diversity of species and/or virulence types occurring in many regions of the world. A much greater understanding of tolerance mechanisms and of rapid methods to identify tolerance traits will be required before breeders can more effectively manage programmes to introgress resistance and tolerance into individual cultivars. Lastly, advances in breeding efficiency are anticipated with the development of novel approaches to influence host–parasite interactions, such as gene editing.

9.7 Potato Cyst Nematodes

9.7.1 Scale of the problem

Potatoes, and the PCN, *Globodera rostochiensis* and *G. pallida*, originate from South America (Stone, 1979). Potatoes were first brought to Europe in the late 1500s, and subsequently disseminated to the rest of the world from there. PCN appear to have been introduced via a similar route, probably in soil associated with unwashed potatoes, and are now found on all seven continents, with *G. rostochiensis* reported in at least 65 countries and *G. pallida* in 41 (Turner and Evans, 1998; CABI Invasive Species Compendium, www.cabi.org/isc). Their total global economic impact has not been fully quantified; however, even a small introduction can incur considerable costs to put in place surveys and ongoing monitoring systems, in addition to the costs to the growers (Hodda and Cook, 2009). Discovery of an infestation can have an impact on trade of potatoes and other commodities between countries. Protection of seed land from infestations using pre-plant soil sampling is a vital aspect of preventing spread into non-infested land. The persistence of viable cysts in the soil for many years and the difficulty in detecting new infestations below detection thresholds, makes preventing the spread of PCN problematic. In addition to *G. rostochiensis* and *G. pallida*, a potential third PCN, was reported by Skantar *et al.* (2011) and has subsequently been described as *G. ellingtonae* (Handoo *et al.*, 2012). Potato is a host for this species but its pathogenicity on potato is still under investigation (Phillips and Zasada, 2016, pers. comm.) and this species has not been able to reproduce on potato genotypes that have *H1* resistance (Zasada, 2016, pers. comm.). This species has been reported in Oregon and Idaho in the USA and in the Andean region (Lax *et al.*, 2014).

The degree to which cyst nematodes reduce potato tuber yield is a function of a number of factors including the nematode population density in the soil (Seinhorst and den Ouden, 1971) and the susceptibility and tolerance of the host. The higher the inoculum level, the more yield is reduced until a point is reached well beyond the damage threshold where competition between the juvenile nematodes disrupts this relationship between population level and yield loss. At high nematode population densities the yield of harvested potatoes can be less than the amount of seed planted (Mai, 1977). Management schemes aim to keep nematode population levels below damage thresholds through PCN eradication programmes; prohibiting the cultivation of potato in infested land is another strategy that has been employed.

PCN are genetically much more diverse in their centre of origin in South America than elsewhere (Plantard *et al.*, 2008; Grenier *et al.*, 2010), as are their hosts. This relative lack of diversity of PCN in potato growing regions outside South America makes it possible to consider controlling these cyst nematodes through potato breeding: the nematodes have less pre-existing variation to adapt to the resistance genes that breeders aim to deploy. For example, most of the *G. rostochiensis* found outside South America is pathotype Ro1, which is sensitive to the *H1* resistance gene (Turner and Evans, 1998). Similarly, most of the *G. pallida* outside South America fall within pathotypes Pa2 and Pa3 (Pa2/3) (Kort *et al.*, 1977; Canto Saenz and Mayer de Scurrah, 1977; Phillips and Trudgill, 1998).

9.7.2 Sources of resistance

Screens of potato germplasm (Ellenby, 1948, 1952, 1954; van Soest et al., 1983; Dellaert and Hoekstra, 1987; Turner, 1989; Rousselle-Bourgeois and Mugniéry, 1995; Ruiz de Galarreta et al., 1998; Castelli et al., 2003) have identified accessions of many wild and cultivated species that are resistant to G. rostochiensis and G. pallida or both. These include *Solanum tuberosum* ssp. *andigena*, *S. vernei*, *S. spegazzinii*, *S. sparsipilum*, *S. multidissectum* and others (Castelli et al., 2003). Of greatest practical importance to date have been resistance genes sourced from *S. tuberosum* ssp. *andigena* and *S. vernei* (Ross, 1986).

Excellent resistance against G. rostochienis pathotype Ro1 was first identified in the Commonwealth Potato Collection (CPC), specifically in *S. tuberosum* ssp. *andigena* accession CPC 1673 (Ellenby, 1952). Subsequent genetic analyses revealed that resistance was conferred by the dominant *H1* gene, which is located on chromosome 5 (Gebhardt et al., 1993; Pineda et al., 1993). The *H1* gene has since been transferred, through conventional crossing and selection, into many cultivars around the world. As pathotype Ro1 is the predominant pathotype of G. rostochiensis, and *H1* provides such effective control, the discovery and subsequent deployment of *H1* stands as one of the most important examples of the value of crop genebanks in helping breeders address new production problems as they arise. Molecular markers tightly linked to *H1* have recently been developed (De Koeyer et al., 2010; Galek et al., 2011; Schultz et al., 2012), facilitating the further dissemination of this gene with MAS.

The *H2* gene from *S. multidissectum* confers strong resistance against G. pallida pathotype Pa1 (Dunnett, 1961) and partial resistance to G. pallida Pa2/3 (Blok and Phillips, 2012). As pathotype Pa1 is not widely distributed, this gene has limited value for breeding on its own; however, it may have utility when pyramided with other G. pallida resistances. The *H3* gene from *S. tuberosum* ssp. *andigena* CPC accession 2802 is, in fact, at least two genes, and confers moderate resistance against G. pallida pathotype Pa2/3 (Bryan et al., 2004). One component of *H3*, now known as $GpaIV^s_{adg}$, maps to chromosome 4, whilst another maps to chromosome 11 (Bryan et al., 2004). A diagnostic marker linked to $GpaIV^s_{adg}$ has been developed (Moloney et al., 2010).

The *Gpa2* gene from *S. tuberosum* ssp. *andigena* accession CPC 1673 is highly effective against a specific population of G. pallida pathotype Pa2, but is ineffective against most other populations (Rouppe van der Voort et al., 1997). *Gpa2* maps to chromosome 12 and has been cloned (van der Vossen et al., 2000). As the particular population for which this gene confers resistance is of very limited distribution, *Gpa2* has limited value for breeding.

The *Gpa5* gene, from *S. vernei*, is located on chromosome 5 (Rouppe van der Voort et al., 2000). *Gpa5* is the single most effective resistance gene against pathotype Pa2/3 currently deployed in cultivated potato. Although *Gpa5* resistance against Pa2/3 is not as effective as *H1* is against Ro1, the resistance conferred is recorded as having the highest score of 9 in 'Innovator' using the European Plant Protection Organization (EPPO) method to assess resistance (OEPP/EPPO, 2006). A diagnostic marker tightly linked to *Gpa5* has been developed (Sattarzadeh et al., 2006). *Gpa5* does not confer resistance to G. rostochiensis.

The *Gpa6* gene, also from *S. vernei*, maps to chromosome 9, but confers a lower level of resistance than *Gpa5* against Pa2/3 (Rouppe van der Voort et al., 2000). Together, *Gpa5* and *Gpa6* confer a higher level of resistance than either gene alone (Rouppe van der Voort et al., 2000). The *Grp1* gene, from *S. vernei*, confers broad spectrum resistance to both G. pallida and G. rostochiensis. *Grp1* maps to the same region of chromosome 5 as *Gpa5* (Rouppe van der Voort et al., 1998). The resistance conferred by *Grp1* against G. pallida is not as effective as that provided by *Gpa5*, but is still useful. The *Gro1* locus, probably from *S. spegazzinii*, confers broad spectrum resistance to all pathotypes of G. rostochiensis (Barone et al., 1990). *Gro1* maps to chromosome 7 and one member of this gene family, *Gro1-4*, has been cloned (Paal et al., 2004).

9.7.3 Selection of nematode populations

For purposes of resistance breeding, it is important to inoculate candidate potato varieties with cyst nematode populations that mirror, as closely as possible, the genetic diversity of the production areas where the varieties will be grown. An obvious starting point is to collect nematode populations directly from infested production areas. If the intended resistance gene(s) are effective against all populations sampled, resistance breeding is straightforward, and likely to be worthwhile. If the resistance gene(s) are only effective against some populations, additional resistance genes may be needed, and care taken to ensure that resistant gene(s) are only deployed in fields where they are known to be effective.

EPPO has produced a protocol for testing of potato varieties for resistance to *G. rostochiensis* and *G. pallida* (OEPP/EPPO, 2006). This test involves a pot test performed in standard conditions and with specific test populations of *Ro*1, *Ro*5, *Pa*1 and *Pa*3, which cover the main pathotypes found in Europe. Additional populations can be used but they should be assessed with standard resistant and susceptible reference potato genotypes that allow a pathotype classification to be made, usually with reference to the scheme of Kort *et al.* (1977). Whilst there have been criticisms of this scheme because it does not account for environmental sensitivity of the test, it remains the scheme that is of most relevance to the resistances used in current potato breeding programmes.

9.7.3.1 *Inoculum*

Cysts, eggs or juveniles can be used for inoculum. For all three types, cysts need to have experienced a period of cold treatment at 4°C for >3 months. Dry cysts are stored in a refrigerator until needed. A hatching test should be performed with cysts to determine their viability before using in an assay. Eggs are produced by crushing cysts in water in a homogenizer and then adjusting the number of eggs to the appropriate amount with water. They should be used within 24 h. Juveniles are hatched in root diffusate (tomato or potato) and collected over several days and stored in the refrigerator for no more than a few days.

9.7.3.2 *Screening protocols*

There are three issues of considerable practical importance for any potato breeder who wishes to develop resistant varieties: (i) what assay(s) will be used to identify resistant individuals; (ii) to what stage of the breeding process will these assay(s) be applied; and (iii) who will conduct the screening? A typical potato breeding programme generates, through crossing, 20,000 to 100,000 new potato seedlings annually, each with a unique genotype. That cohort is reduced to approximately 100 to 2000 individuals after the first year of field production, and further whittled down to approximately 10 to 50 clones after 4–5 years of evaluation in the field. The huge numbers of seedlings produced each year make it unrealistic to test for resistance at the seedling stage. However, testing is possible after a cohort has been reduced to a few hundred. The easiest tests are done with molecular markers. Although current markers are not 100% accurate, as recombination can occur between markers and the resistance genes they are linked to, accuracies greater than 90% have been observed, for example, with markers linked to the *H1* gene (Galek *et al.*, 2011; Schultz *et al.*, 2012). Current marker assays cost a few dollars per sample and can be completed in a day or two. After isolating DNA from each sample, the polymerase chain reaction (PCR) is used to amplify a diagnostic DNA fragment, whose presence or absence can then be visualized.

Even in programmes where molecular markers are used, bioassays ultimately need to be employed to confirm resistance. Bioassays are slow, and considerably more expensive due to the labour involved. In many countries PCN are quarantine pests, and only facilities with appropriate biosafety controls are permitted to work with them. In practice, this means that cyst nematologists, not potato breeding staff, perform bioassays. In a typical bioassay, potato seed pieces are placed in soil that has been inoculated with cysts or eggs, and the plant is allowed to grow for 7–9 weeks. As the plants develop, juveniles hatch and invade the roots, leading to the eventual formation of numerous cysts on susceptible clones, and no or fewer cysts on resistant individuals.

Although bioassays are typically conducted in pots, with the plants grown in a glasshouse, they can also be run in transparent 'canisters', margarine tub-sized containers with lids, where roots are allowed to develop in the dark (Phillips *et al.*, 1980). Canister tests are especially useful for screening large numbers of samples due to the method by which cysts are enumerated – namely, by counting any that are visible with the use of a magnifying lens through the transparent container walls. For pot assays, quick-and-dirty cyst counts can be obtained by gently removing the plant from the pot, and counting the number of cysts visible on the root ball or by using a Rootrainer system where the roots can be seen after opening the 'book' and females on the root surface are counted (Fig. 9.2). The pot test is the method currently used at Cornell University, USA, for inoculations with *G. rostochiensis* race Ro1, where plants with cyst counts of 0–4 are considered resistant, and plants with cyst counts of 5 or more are considered susceptible. Potatoes susceptible to Ro1 typically exhibit 30 or more cysts in these tests. Unfortunately, cysts of pathotype Ro2 easily detach from roots, and so assessment of resistance here requires that cysts first be extracted from soil before they can be counted, a far more time consuming procedure. For any bioassay it is important to perform multiple replicates, as 'escapes' (lack of cyst production for reasons other than resistance) are common.

Weighing the costs of the assays, and the importance of resistance to the programme as a whole, the Cornell University breeding programme currently begins testing for resistance to Ro1 after the second field generation, employing a pot bioassay, when the initial cohort of 20,000 has been reduced to about 200 individuals by selection for agronomic criteria, primarily yield, tuber appearance and fry colour. Any potato genotype that survives the third and subsequent years of selection is tested for resistance again, so that by the time a resistant variety has been released (typically 12–15 years after the cross was first made), it has been tested over ten times, and there is no doubt whatsoever that it is resistant. Because testing for resistance to Ro2 is more costly, these tests begin after 3–4 years of field selection, when the cohort has been reduced to 50 or fewer clones.

9.7.4 Development of resistant cultivars

The art of breeding for cyst nematode resistance consists of creating, and then identifying, individual potato clones that carry one or more genes for cyst nematode resistance in a commercially acceptable genetic background. In practice, creating resistant clones by crossing is relatively easy. Identifying resistant individuals that are also commercially acceptable is considerably more difficult. The difficulty arises because potato is highly heterozygous, so that in the offspring of any cross, thousands of genes will segregate, and very few of the offspring will possess the right combination of alleles to succeed in the marketplace. For a complete discussion of how breeders identify potatoes that are commercially acceptable, see Tiemens-Hulscher *et al.* (2013). Once resistant varieties are

Fig. 9.2. The four-cell Rootrainer 'book' used to assess resistance of potato genotypes to potato cyst nematodes. (A) The Rootrainer (Tildenet Ltd, Bristol, UK) into which soil, cysts and tubers are planted. (B) An opened Rootrainer showing the root systems of four potato plants. (C) Females of *Globodera pallida* visible on the root surface, which are counted to determine relative susceptibility of different potato genotypes.

developed it is important to consider how they will be deployed. In New York State, USA, where a few farms have Ro1, and some growers want to grow potatoes as many years as possible, studies revealed that a 4-year rotation consisting of 2 years of an Ro1-resistant variety, 1 year of a non-host and 1 year of a susceptible variety were sufficient to keep (already low) *G. rostochiensis* levels from increasing (Brodie, 1996).

When Cornell University first began releasing Ro1-resistant varieties, their agronomic properties were less than ideal, and affected growers did not want to grow them any more than New York State required (2 out of every 3 years of potato production). Over time, as the agronomic features of newer varieties improved to become competitive with, and occasionally better than, susceptible alternatives, some growers began to grow Ro1-resistant varieties many years in succession. This appears, in retrospect, to have led to the emergence of pathotype Ro2, a resistance-breaking strain (Brodie, 1995). New York State now requires growers with Ro1-infested land to grow a susceptible variety 1 out of every 3 years that potatoes are grown; in theory, this should slow the development of strains that can overcome *H1* but time will tell if it actually works. Over the past 50 years Cornell University has released 21 Ro1-resistant potato cultivars, seven of which are still widely grown. In conjunction with regulations that mandate the washing of all farm equipment as it leaves infested fields, the deployment of resistant varieties on Ro1-infested fields has resulted in negligible spread of *G. rostochiensis* within New York State for the past few decades.

9.7.5 Future challenges

It is one thing to control PCN by developing a suite of varieties with a highly effective, single dominant resistance gene, for example, deploying *H1* to control Ro1. It is quite another to develop potato varieties with multiple resistance genes, for example, combining resistance genes of smaller effect. Potato breeding is largely a numbers game, making crosses and hoping to create and find the rare offspring with a desirable combination of traits. The more desirable alleles that an offspring must have, the lower the probability of creating it in the first place.

Molecular markers should make it easier to pyramid multiple genes, for several reasons. The first is that, with some types of markers, it is possible to identify parents that harbour more than one copy of a resistance gene. Parents with one copy of a resistance gene (Rrrr) pass it on to half the progeny; parents with two copies (RRrr) pass it on to 5/6 of progeny; parents with three copies (RRRr) pass it on to almost all progeny (double reduction occasionally results in the formation of rr gametes, which do not pass on the trait); and parents with four copies (RRRR) pass it on to all offspring. The second is that markers make it possible to identify parents, as well as offspring, that harbour complementary genes. Indeed, Dalton *et al.* (2013) have shown that pyramiding $GpaIV^s_{adg}$ and *Gpa5* with the aid of diagnostic markers produces offspring more resistant to Pa2/3 than those containing either gene alone. Finally, markers make it possible, at least in principle, for a breeder to screen larger populations than can be handled in the field. By screening seedlings with markers, susceptible offspring can be discarded early, reducing a too-large population to something more manageable for agronomic evaluation.

From a technical perspective, it is easier to develop a commercially acceptable resistant potato variety by genetic engineering than it is through traditional crossing and selection. However, current societal attitudes towards genetic engineering will have to change before this becomes routine.

9.8 Soybean Cyst Nematodes

9.8.1 Scale of the problem

The soybean cyst nematode, *Heterodera glycines*, is a major yield-reducing pathogen of soybean, *Glycine max*, throughout the world (Wrather *et al.*, 2010). The nematode was probably discovered, or at least known to exist, as early as 239 to 235 BC in China (Noel, 1992; Riggs, 2004). There is no well-defined first report of the pathogen in China in the scientific literature, but damage to soybeans by *H. glycines* was reported from China in 1899 (Liu *et al.*, 1997). Hori generally is credited as first reporting *H. glycines* in the scientific literature in Japan in 1915 (Ichinohe, 1959). The nematode was first discovered in the USA in 1954 (Winstead *et al.*, 1955), in Canada in 1987 (Anderson and Welacky, 1988), in Argentina in 1991–1992 (Doucet *et al.*, 2008) and in Brazil in 1993 (Mendes and Dickson, 1993).

The soybean cyst nematode is considered the most damaging pathogen of soybean in the USA and many other countries. Annual yield losses are estimated to exceed US$1 billion in the USA and Canada (Koenning and Wrather, 2010) and more than US$120 million in China (Wang *et al.*, 2015). The nematode was recently reported to be found in four additional areas of China (Peng *et al.*, 2016).

A new species of cyst nematode parasitizing soybeans, *Heterodera sojae*, was recently reported in Korea (Kang *et al.*, 2016). No details have been reported yet on the damage and yield loss caused by this nematode.

9.8.2 Sources of resistance

Soon after *H. glycines* was initially found in the USA, it was discovered that some existing soybean cultivars, including 'Peking', possessed resistance to *H. glycines* (Ross and Brim, 1957) and the first soybean breeding lines with specific resistance genes were identified, including 'PI 90763' (Ross and Brim, 1957) and 'PI 88788' (Ross, 1962; Fig. 9.3). Crosses were made to transfer resistance genes into soybean cultivars and germplasm lines with superior agronomic characteristics for use in the USA in the 1960s. Additional sources of resistance were identified in the 1970s (Epps and Hartwig, 1972), and the first cultivar with 'PI 88788' resistance, named 'Bedford', was released (Hartwig and Epps, 1978). Currently, there are many soybean breeding lines with resistance to

Fig. 9.3. Soybean field trial, Waseca County, Minnesota, USA, 2004, with high densities of soybean cyst nematode (SCN), *Heterodera glycines*. Left plot: SCN-susceptible soybean variety (soil population density 17,400 eggs 100 cm^{-3}). Right plot: moderately resistant soybean variety (soil population density 10,500 eggs 100 cm^{-3}) with the PI 88788 source of resistance. (Courtesy of Senyu Chen, University of Minnesota, USA.)

H. glycines, with more than 100 in the United States Department of Agriculture's Soybean Germplasm Collection alone (Arelli *et al.*, 2000). Seven different plant genotypes are registered as sources of *H. glycines* resistance in soybean germplasm lines or cultivars (Brim and Ross, 1966; Hartwig and Epps, 1978; Hartwig and Young, 1990; Anand, 1992a, b; Nickell *et al.*, 1994a, b). Developing soybean germplasm and cultivars with resistance to *H. glycines* has been a major area of focus for soybean breeders in the USA and Canada. Some of the efforts of these breeding programmes have been summarized recently (Rincker *et al.*, 2017). Resistant soybean cultivars also have been developed in Argentina (Doucet *et al.*, 2008), Brazil (Arantes *et al.*, 1998; Mauro *et al.*, 1999) and China (Li *et al.*, 2011; Liu and Peng, 2016).

In the early 1990s, specific criteria were published for rating soybean cultivars for resistance-susceptibility to *H. glycines* in the USA and Canada based on values of an 'index of parasitism' (Schmitt and Shannon, 1992). This index now is commonly called the 'female index' and serves as the basis of the HG type test (Niblack *et al.*, 2002) for *H. glycines* populations as well. The *H. glycines* index of parasitism, or female index, is the mean number of females that form on a plant or cultivar when grown in soil infested with the nematode in a 30-day glasshouse experiment divided by the mean number of females that form on a standard, susceptible soybean cultivar grown under the same conditions, multiplied by 100. Resistant, moderately resistant, moderately susceptible and susceptible soybean genotypes or cultivars are those that allow *H. glycines* female indices of 0–9, 10–30, 31–60 and >60, respectively (Schmitt and Shannon, 1992). For the most part, these criteria and categories have been used by industry and university soybean breeders, nematologists and plant pathologists in characterizing the reaction of soybean cultivars to *H. glycines* in North America.

In the USA, there is no legal definition of, or criteria for, soybean cultivars to be described as resistant to *H. glycines* and there is no government oversight or testing of soybean cultivars labelled as resistant to the nematode. Many soybean cultivars designated as resistant to *H. glycines* were found to allow moderate to high levels of nematode reproduction in 30-day glasshouse evaluations, with female indices ranging from 40 to 60 or more (Niblack, 2005). Also, some cultivars described as resistant to *H. glycines* allowed large increases in soil egg population densities in field experiments (Tylka *et al.*, 2017). In Ontario, Canada, the performance of *H. glycines*-resistant soybean cultivars is evaluated annually in the field by the Ontario Soybean and Canola Committee (www.GoSoy.ca).

Knowledge of the genetic basis of soybean resistance to *H. glycines* has increased steadily over the past 50 years and has changed considerably in recent years. Resistance to *H. glycines* in soybean was reported to be a quantitative trait just a few years after the discovery of the nematode in the USA (Caldwell *et al.*, 1960). Up to five *rhg* genes (resistance to Heterodera glycines) were believed to confer resistance to the nematode in soybean. However, it was shown that the number of copies of a repeated, multi-gene segment affected the degree of resistance to *H. glycines* in soybean (Cook *et al.*, 2012). Also, it was reported that *H. glycines* resistance gene *Rhg4* coded for a serine hydroxymethyltransferase enzyme and that alleles of the gene had genetic polymorphisms that affected functioning of the enzyme (Liu *et al.*, 2012). The current state of understanding of the genetics of soybean resistance to *H. glycines* was recently reviewed (Mitchum, 2016) and was neatly summarized by the author in the following statement (in which *H. glycines* is abbreviated as SCN): 'Though the complexity of the SCN-soybean interaction was evident, no one expected soybean resistance to SCN to be controlled by a group of dissimilar genes at multiple locations and copy number variation. This surprising finding has confounded researchers and will require a level of investigation that goes well beyond the one gene model.' Recent, new findings about the genetic nature of soybean resistance to *H. glycines* underscores the importance of glasshouse phenotyping and field testing soybean genotypes in order to verify reduction or suppression of *H. glycines* reproduction when developing new resistant soybean cultivars.

9.8.3 Selection of nematode populations

Choosing which nematode population to use for research with *H. glycines* and for development of resistant soybean cultivars must be done carefully, with consideration of the genetic diversity present in nematode populations, not just the soybean host.

Field populations of *H. glycines* are genetically diverse and very appropriate and effective for assessing the host resistance-susceptibility of soybean cultivars for possible use by farmers in managing the nematode. However, using *H. glycines* field populations to attempt to map, or otherwise discern, specific genes responsible for resistance-susceptibility to *H. glycines* in a soybean genotype is unreasoned and counterproductive. The genetic heterogeneity within the *H. glycines* field population would confound results of genetic analyses related to the reaction of the soybean genotypes to nematode parasitism. Inbred *H. glycines* populations, such as those described by Niblack *et al.* (2002), with relatively homogeneous and stable genotypes and virulence phenotypes are much better suited for use in studies of the genetics of soybean reactions to *H. glycines* than field populations.

For resistance screening purposes in plant breeding studies, *H. glycines*-infested soil may be obtained from farmers' fields for direct use in glasshouse evaluations of host reaction to the nematode (resistance-susceptibility). Alternatively, nematode eggs and juveniles could be extracted from *H. glycines*-infested field soil and used to establish glasshouse cultures of field populations of the nematode on susceptible soybean cultivars, which allow unchecked reproduction of the nematode. These natural nematode populations are a diverse mixture of individual *H. glycines* nematodes with varying genotypes and virulence phenotypes capable of parasitizing and reproducing on soybean lines with resistance. The *H. glycines* race test (Golden *et al.*, 1970) and HG type test (Niblack *et al.*, 2002) were developed to describe the variation in virulence phenotypes occurring among natural populations of the nematode.

Much of the reported field experimentation and soybean breeding efforts of the past refer to or consider *H. glycines* races and HG types as pure strains of the nematode, treating them as genetically and phenotypically homogeneous. For example, soybean cultivars are described as resistant to a specific race or races of *H. glycines* and laboratory studies of nematode behaviour report using certain races of *H. glycines*. Conceptualizing *H. glycines* field populations in this way does not account for the heterogeneous nature of the naturally occurring nematode populations, as revealed in the results of *H. glycines* race (Golden *et al.*, 1970) and HG type (Niblack *et al.*, 2002) tests. Races and HG types are genetically and phenotypically heterogeneous mixtures of individual *H. glycines* nematodes occurring at varying frequencies among natural populations of the nematode, and the race and HG type concepts do not apply to individual nematodes (Beeman *et al.*, 2016; Tylka, 2016).

Despite imprecisely considering *H. glycines* races and HG types to be pure strains of the nematode, efforts to develop resistant soybean cultivars that limit *H. glycines* reproduction and produce profitable soybean yields have been successful, and such breeding efforts probably will successfully continue without regard to the true nature of *H. glycines* races and HG types. However, future research efforts to understand the molecular basis of the *H. glycines*–soybean host–parasite relationship and to elucidate the genetic and molecular bases of *H. glycines* resistance in soybean will be confounded and impeded if races and HG types of the nematode are conceptually viewed as pure populations or even mixtures of pure strains of the nematode.

9.8.4 Screening protocols

Thirty-day glasshouse bioassays, originally designed to determine the race of *H. glycines* populations (Golden *et al.*, 1970), have been used very successfully in soybean breeding to develop many soybean cultivars with resistance to *H. glycines*. The bioassay involves growing plant genotypes in soil infested with *H. glycines* for 30 days and then removing and counting newly formed adult *H. glycines* females from the roots and calculating a mean number of females formed for each soybean genotype included in the experiment. Additional details of the bioassay and a precise definition of female index were specified by Niblack et al. (2002) for conducting the *H. glycines* HG type test. A standardized method for conducting glasshouse bioassays, named SCE08, to determine *H. glycines* resistance phenotype was developed (Niblack et al., 2009). These methods recommend use of multiple replications per plant, control of environmental conditions and inclusion of soybean cultivars known to be susceptible to the nematode as positive controls.

The 30-day glasshouse bioassay had been used countless times for many decades to screen soybean breeding populations and determine the *H. glycines* resistance phenotype of soybean genotypes despite being relatively time, labour and space intensive. In the early 1990s, specific sequences in the soybean genome were discovered that were closely associated with resistance to the nematode (Concibido et al., 1994). This discovery led to development of methods to utilize these sequences as molecular markers to check for the presence of *H. glycines* resistance genes in populations of soybean plants, such as those developed for breeding and cultivar development purposes (Concibido et al., 2004). Use of MAS increased the efficiency with which a population of soybeans could be assessed for possible resistance to *H. glycines*. It is unclear how often commercial and academic soybean breeding programmes currently use MAS solely as the means to characterize the reaction of soybean genotypes to *H. glycines* and how often breeding programmes confirm the resistance or susceptibility of new soybean genotypes with glasshouse phenotyping using the 30-day bioassays.

Using glasshouse bioassays and/or MAS has led to success in developing *H. glycines*-resistant soybean cultivars for farmers. However, the true reactions of soybean cultivars to the nematode as well as yield may not be fully known until the cultivars are studied in the field. When more than 350 soybean cultivars were grown in experiments conducted in *H. glycines*-infested fields and in 30-day glasshouse bioassays using *H. glycines*-infested soil brought from the fields in which the experiments were conducted, the correlation between female indices in the glasshouse and changes in *H. glycines* egg population densities in the field on the exact same set of 350 varieties was poor (Tylka et al., 2013). Some resistant cultivars with relatively low female indices of 12 to 25% in the glasshouse bioassays allowed 6- to 15-fold increases in *H. glycines* population densities in the field, and some cultivars that had female indices of 55 to 67% in the glasshouse resulted in little or no increase in *H. glycines* egg population densities in the field. The lack of correlation of glasshouse and field results presumably was due, at least in part, to effects of environment on soybean resistance to the nematode in the field. Perhaps other, undefined factors affected expression of nematode resistance in the soybean cultivars over the course of the 4-month-long growing season. Expression of *H. glycines* resistance in soybean over an entire growing season has not been investigated and, therefore, temporal effects on resistance expression are not understood.

One of the greatest challenges in accurately evaluating soybean cultivars for resistance to *H. glycines* in the field is the aggregated distribution of the nematode. In naturally infested fields, incidence of the nematode is uneven and there is great spatial variability in population densities (Gavassoni et al., 2001). Ross and Brim (1957) overcame this difficulty by planting rows of susceptible soybeans 15 cm away from rows of experimental soybean genotypes and then comparing numbers of *H. glycines* females on the roots of plants in adjacent rows to determine if soybean cultivars had resistance to the nematode. In Brazil, susceptible soybeans were planted in infested fields, then the plants were removed after 45 days and soybean lines were planted in the specific areas where large numbers of *H. glycines* females and cysts were observed on roots of the removed plants (Arantes et al., 1998). Split plots of the same soybean genotype with or without nematicide were used by Boerma and Hussey (1984) to study the yield response of soybean cultivars being evaluated for tolerance to *H. glycines*. More recently, four-row-wide plots and collection of soil samples from the plots at the time of planting and harvest to determine initial and final nematode population densities, respectively, has been used to assess the effects of soybean cultivars on nematode population dynamics (Noel and Sikora, 1990; Todd *et al*., 1995; Wheeler *et al*., 1997; Tylka *et al*., 2017).

9.8.5 Developing resistant cultivars

Yield potential and resistance to *H. glycines* in soybean cultivars are independent traits. Developing resistant soybean cultivars to enable farmers to grow soybeans profitably in *H. glycines*-infested fields requires striking a balance between selecting soybean genotypes that produce maximum yield and selecting genotypes that offer maximum suppression of nematode reproduction. Ultimately, farmers grow soybeans to produce grain, which is sold for income, so innate yield potential is of great importance in a soybean cultivar. However, in situations where *H. glycines* parasitism reduces growth and development of the soybean plants, nematode resistance directly affects whether the soybean crop can achieve its full yield potential in the field.

Soybean germplasm lines being developed for *H. glycines* resistance often are evaluated in public breeding programmes in the Midwestern USA based only on yield data (Rincker *et al*., 2017). This approach is appealing from a practical perspective because it requires much less effort than assessing *H. glycines* population densities in individual field plots. However, the highest-yielding soybean lines may not be providing maximum, or even effective, resistance against *H. glycines*. The high reproductive potential and effective long-term survival of *H. glycines* requires sustained efforts to keep nematode population densities in check. Growing soybean cultivars in *H. glycines*-infested fields in an attempt to maximize soybean yields in the short term without considering effects of the cultivars on *H. glycines* population densities in the field will reduce the long-term soybean productivity of the land (Tylka *et al*., 2017).

The innate yield potential of soybean cultivars is most important in fields where *H. glycines* population densities are moderately low, low or undetectable. However, the effectiveness of nematode control provided by resistant soybean cultivars becomes more important and more influential on yield as *H. glycines* populations increase in density and in ability to reproduce on resistant soybean cultivars (virulence). Also, growing soybean cultivars with effective resistance to *H. glycines* is an important means of slowing the spread of *H. glycines* populations into non-infested fields or areas of fields by limiting the availability of susceptible host plants to support increases in densities of newly introduced *H. glycines* populations.

The benefits of keeping *H. glycines* population densities in check mentioned immediately above underscore the critical need to verify the effects of soybean cultivars on *H. glycines* reproduction and population densities. Doing so requires glasshouse phenotyping and evaluating soybean cultivars in field experiments rather than relying on the presence of markers or the use of yield data from non-infested fields as a proxy for gauging the utility of soybean cultivars for management of *H. glycines*.

9.8.6 Future challenges

Soybean cultivars that are resistant to *H. glycines* are critical management tools. These cultivars produce greater yields and allow less nematode reproduction than susceptible cultivars when grown in infested fields (Tylka *et al*., 2017). Although many sources of *H. glycines* resistance have been identified and reported for use in soybean cultivar development, almost all commercially available cultivars in the USA and Canada possess resistance from a single source of resistance, named 'PI 88788' (Tylka and Mullaney, 2017). *Heterodera glycines* populations with increased reproduction (or virulence) on the 'PI 88788' resistance have been reported to be widespread throughout the Midwestern USA (Kim *et al*., 1994; Niblack *et al*., 2003, 2008; Mitchum *et al*., 2007; Hershman *et al*., 2008; Chen *et al*., 2010; Faghihi *et al*., 2010; Zheng *et al*., 2006; Acharya *et al*., 2016). The increase in frequency of occurrence and virulence of *H. glycines* populations on 'PI 88788' resistance presumably is due to prolonged and widespread use of the resistance in soybean cultivars and the resultant selection in the nematode populations for individuals able to reproduce on these resistant cultivars. The increase in virulence on 'PI 88788' in field experiments in Iowa has been correlated with increases in end-of-season nematode population densities in fields and decreases in yields of *H. glycines*-resistant soybean cultivars with 'PI 88788' resistance (McCarville *et al.*, 2017). Increased reproduction of *H. glycines* populations from Brazil on 'PI 88788' as well as on 'PI 90763' and 'PI 548402' ('Peking') has been reported (Asmus *et al.*, 2009) but some surveys of *H. glycines* populations in the country have not found increased virulence of the nematode on these sources of resistance (Matsuo *et al.*, 2012; Santana *et al.*, 2009). It is not certain that resistant soybean cultivars will continue to be effective for managing *H. glycines* and it is clear that *H. glycines*-resistant soybean cultivars with resistance genes from a source other than 'PI 88788' are urgently needed.

Scientists have searched for new, useful *H. glycines* resistance genes in *Glycine* species other than *G. max*, such as *G. soja* (Wang *et al.*, 2001; Kim *et al.*, 2011) and *G. tomatella* (Bauer *et al.*, 2007), possibly to incorporate novel resistance genes into *G. max*. Whilst such efforts have identified some promising leads (Kabelka *et al.*, 2006), currently there are no soybean cultivars available to farmers with *H. glycines* resistance genes from any other *Glycine* species.

It is possible that scientists will develop or engineer soybean cultivars with novel *H. glycines* resistance, based on understanding the molecular mechanisms of the host–parasite relationship. For example, Siddique *et al.* (2016) found that plant hormone signalling was involved in feeding site development for *H. schachtii* in Arabidopsis and that cytokinin-deficient Arabidopsis mutants were less susceptible to cyst nematode parasitism. Such feeding site biochemical mechanisms are probably conserved among cyst nematode species and their hosts, and alteration of the cytokinin metabolism in root cells may lead to development of less susceptible (more resistant) plants in the future. Also, activation of a tracheary element differentiation inhibitory factor was recently shown to reduce infection by *H. schachtii* and also to reduce nematode feeding site size in Arabidopsis mutants (Guo *et al.*, 2017), indicating that this pathway may be a fertile target for further manipulation to engineer resistance to *H. schachtii* in sugar beet and possibly to *H. glycines* in soybean.

Technologies such as plant transformation and RNA interference are being used to alter soybean genotypes to decrease susceptibility to *H. glycines* in soybean. For example, soybeans transformed with three different artificial microRNA molecules allowed less *H. glycines* reproduction than soybeans transformed with the empty vector (Tian *et al.*, 2016). Soybeans transformed to overexpress the enzyme salicylic acid methyl transferase allowed less reproduction of *H. glycines* than untransformed control plants (Lin *et al.*, 2016). Should these disruptions in the molecular biology of the host–parasite relationship of *H. glycines* with soybeans prove to be highly effective at reducing reproduction of the nematode and to be robust in and among varied field environments, they may become the basis of new *H. glycines*-resistant soybean cultivars in the future.

9.9 Sugar Beet Cyst Nematodes

9.9.1 Scale of the problem

The sugar beet cyst nematode (BCN), *Heterodera schachtii*, was first discovered in Germany in 1859 by the botanist Schacht and described in 1871 by Schmidt, who established the genus *Heterodera* (Subbotin *et al.*, 2010). The BCN was the first cyst nematode species described and identified as a major pest of beets belonging to the genus *Beta*. They are distributed worldwide (Europe, North and South America, Australia, Africa and Russia) and occur in all major sugar beet production areas, favouring temperate regions but tolerating a broad range of climates (Subbotin *et al.*, 2010). In Europe, where awareness of BCN problems have a long history, estimation of the percentage of infested sugar beet-growing area largely exists on the basis of the personal experiences of experts and rarely on the basis of systematic surveys conducted on a national scale. Through combining both of these information sources, about 10–20% of the European sugar beet-growing area representing more than 150,000 ha (WVZ and VDZ, 2016) can be assumed to be BCN infested. In the core production areas of Europe (France, Germany,

The Netherlands and Belgium), where there is intensive cultivation of sugar beets, local BCN infestations have a long history and might exceed 50% of the regional sugar beet areas. During the past decade, sugar beet production experienced a significant change and high yielding tolerant cultivars became available to growers and, thus, older approximations on economic impact of BCN are probably not up to date, although in 1999 the economic damage of BCN in Europe was estimated at up to €90 million annually (Müller, 1999). In the USA, BCN was first reported in 1895 and is now found in 17 states, although sugar beet cultivation has terminated due to heavy infestations in Utah and Washington (Hafez and Seyedbagheri, 1997). Integrated management tools including rotation with non-hosts, good weed control and the use of tolerant cultivars is recommended.

Depending on the growing climate, environmental conditions and cropping intensity, three to five generations of BCN can develop in each growing season of a host plant. At high BCN pressures, population densities over 2000 eggs and juveniles $(100 \text{ ml soil})^{-1}$, the yield of susceptible sugar beet cultivars can decrease by 30–40% and by 10–20% in nematode tolerant cultivars under temperate conditions. Natural population decline during growing periods of non-hosts can vary between 40 and 70% over 3 years; hence, BCN management is essential to maintain high productivity rates in intensive production systems.

9.9.2 Sources of resistance

Resistance has been selected in sugar beets to *H. schachtii*, and even earlier in white mustard and oilseed radish. The use of resistant catch crops in European sugar beet cropping systems is a standard management tool to reduce BCN population density prior to planting the sugar beet crop. High resistance levels in oilseed radish and mustard against BCN has been known for more than 30 years. Root penetration in resistant and susceptible plants apparently does not differ, but induction of syncytia in procambial tissues in resistant plants is suppressed, whilst syncytia in the pericyle are not affected (Soliman *et al.*, 2005) and, thus, the development of females, which preferably derive from feeding sites in the procambium, is apparently hampered and the male to female ratio increases resulting in a reduction of further population development (Müller and Steudel, 1982; Wyss *et al.*, 1984).

Resistance in oilseed radish was studied by inter- and intraspecific crossings between resistant and non-resistant oilseed radish lines. The existence of the single dominant gene $HS1^{RPH}$ was confirmed by QTL mapping (Budahn *et al.*, 2009). In white mustard, genetic expression has not been investigated during the past decades. Apparently resistance in mustard also affects the male to female ratio as in oilseed radish but, in addition, a delay in the development of second- to fourth-stage juvenile post root penetration was observed (Soliman *et al.*, 2005). Resistance has been successfully transferred from mustard and oilseed radish lines into resistant hybrids with oilseed rape having a high grade of resistance (Lange *et al.*, 1989; Lelivelt *et al.*, 1993; Peterka *et al.*, 2010). As natural sources of resistance have so far not been found for oilseed rape, this is an agronomic feature of high importance taking into consideration that the very high susceptibility of oilseed rape to BCN is a vital restriction for integration of this crop in rotation systems including sugar beets. Agronomic value (yield potential and quality) of such oilseed rape hybrids with resistance to *Sclerotinia sclerotiorum* from *Sinapis alba* lines could be improved by further hybridization to produce lines with BCN resistance (Li *et al.*, 2009).

Except for one accession in the wild species *Beta vulgaris* ssp. *maritima*, no genetic variability for resistance has been found in *B. vulgaris* (Heller *et al.*, 1996; Müller and Klinke, 1996). In addition to the overall necrosis of syncytia, large cavities in surrounding tissues were observed in roots after penetration of BCN juveniles and consequently this reaction was distinguished as a hypersensitive reaction in resistant sugar beets (Yu and Steele, 1981). In fact this reaction type implies a high level of intolerance towards penetration of BCN juveniles and therefore the management value of this resistance is possibly limited to the reduction of BCN populations during crop rotations (Roberts, 1992). Within the genus *Beta*, complete resistance is only known to exist in the wild beet species *B. procumbens*,

B. webbiana and *B. patellaris*, which is inherited in a dominant mode. From previous work (Loptien, 1984; Lange *et al.*, 1993) some major genes are known to occur within these wild species that are located on chromosome 1 ($HS1^{pro-1}$, $HS1^{web-1}$), chromosome 7 ($HS2^{pro-7}$, $HS2^{web-7}$) and chromosome 8 ($HS3^{pro-8}$, $HS3^{web-8}$), and only one gene was identified in *B. patellaris* on chromosome 1 ($HS1^{pat-1}$), but Müller (1992) also found evidence for the existence of a second resistance encoding factor in *B. pattelaris*, which to date remains unidentified. There is some evidence that resistance carrying chromosomes from *B. procumbens* and *B. webbiana* are homologous (Heller *et al.*, 1996; Müller and Klinke, 1996). This might also be true for the *HS1* gene from *B. pattelaris* (Salentijn *et al.*, 1992).

9.9.3 Selection of nematode populations

Screening of 113 field populations of *H. schachtii* from Germany and other locations throughout Europe using the nearly 100% resistant sugar beet translocation line ($HS1^{pro-1}$) showed that there was widespread occurrence of virulent pathotypes. This indicates that virulence in field populations could readily increase after exposure to resistance genes introgressed from wild sugar beets and they could spread into other sugar beet regions (Müller, 1992). Selection studies using a virulent $HS1^{pro-1}$ breaking pathotype and an avirulent BCN population from the same origin in combination with monosomic additions originating from the three wild beet species and carrying resistance genes from chromosomes 1 and 7 clearly confirmed that there is a gene-for-gene relationship in the BCN pathosystem (Klinke *et al.*, 1996). Further, resistance-breaking pathotypes could be identified for HS1 but that were avirulent to HS2 accessions and *vice versa*. One pathotype was virulent to HS2 and another one was virulent to both genes. Following the schemes used for characterization of other cyst nematodes, Müller and Klinke (1996) suggested defining BCN pathotypes in accordance with the definition of their concurrent resistance gene. Accordingly pathotype schach1 (virulent to HS1), schach 2 (virulent to HS2), schach 12 (virulent to HS1 and HS2) and pathotype schach 0 (normal population avirulent to HS1 to HS2) are confirmed so far (Müller, 1999).

9.9.4 Screening protocols

Glasshouse testing protocols for the evaluation of BCN resistance in cultivated plants mostly have been performed with a simple pot experiment (Toxopeus and Lubberts, 1980) with standard soil using a defined BCN population that is inoculated with juveniles after appearance of plants, or prior to sowing with cysts mixed into the soil. In Germany, registration of mustard or oilseed radish cultivars certified as resistant against BCN requires validation and evaluation of a defined resistance level according to an official protocol (Müller and Rumpenhorst, 2000). Three resistance levels are predefined on the basis of the reproduction indices P_f/P_i, where P_i is the population density inoculated as cysts and P_f is the population density detected after terminating plant growth at 400 degree-days (cumulated daily temperature in pots above 10°C) plus 2 weeks at room temperature to allow maturation of eggs. Raw data from test lines are set against data from resistance reference cultivars that are also tested, and a long-term field constant function. Moderate resistance levels of catch crops in this test achieve P_f/P_i 0.1–0.3. Nevertheless, field resistance usually reaches clearly higher multiplication rates between 0.6 and 0.9, depending on environmental factors and cultivation practices. The nematode inoculum used for this protocol is the pathotype schach 0, the standard population of *H. schachtii* consistently maintained by using susceptible oilseed rape as the host that is inoculated with viable BCN juveniles.

9.9.5 Development of resistant cultivars

Resistance breeding relating to BCN and accompanying research has focused primarily on monogenic resistance genes due to the fact that target features, like cyst density per plant, was a simple and accessible dataset for evaluating resistant sources and their hybrids with *B. vulgaris*. Despite early reports of partial resistance in *B. maritima*, this resistance source was rarely followed up by breeding programmes and research. Heijbroek (1977) studied partial resistance of *B. maritima* accessions obtained from the US Department of Agriculture (USDA) collection, several botanical gardens and directly from seed bearing plants along the Atlantic and Mediterranean coasts of, particularly, England, France, Ireland and North Africa, and found that resistance was lost after backcrossing into *B. vulgaris*. This indicated a recessive polygenic character of resistance gene(s) in *B. maritima* and, therefore, was supposed not to provide a good basis for classical resistance screening techniques (Müller and Klinke, 1996). Similarly, in resistant catch crops, partial resistant sugar beet genotypes impair the development of females, thus producing a wider male to female ratio (Heijbroek, 1977).

Paradoxically, partial resistance with a *B. maritima* background probably is due to be rediscovered as modern tolerant sugar beet cultivars have been established since 2005 in the main sugar beet-growing areas with BCN history. In contrast to low levels of the four resistant cultivars, a significantly higher number (16) of tolerant cultivars have been registered in Germany, which reflects the situation in all major sugar beet areas throughout Europe. In sugar beet-growing areas where BCN is known to be widespread, the proportion of tolerant cultivars sometimes reaches 80–90% of all cropped sugar beet cultivars. Since first registration of tolerant cultivars in Germany the total white sugar yield in northern and western German sugar beet crops increased by approximately 25–30% in just 10 years. Resistant cultivars do not reach the yield potential of tolerant cultivars, thus carrying an economic short-term risk for growers. Consequently, classical resistant cultivars with a long breeding history do not play a considerable role any longer. Therefore, almost all sugar beet breeders have cancelled their breeding programmes with monogenic sources from the wild beets and have shifted their current activity entirely on to breeding programmes with tolerant material supposedly with a *B. maritima* background. The actual motivation of breeding was to provide tolerance but a certain degree of resistance was a welcomed but unintentional side effect.

By applying bulk segregant analysis (BSA) resistance in a tolerant *B. vulgaris* ssp. *maritima* source WB242 was localized and defined as Hs^{Bvm-1} (Stevanato et al., 2014). This finding probably will open up a new perspective to identify further resistance sources as a powerful pathway to the classical screening methods. According to the gene-for-gene hypothesis, selection for virulent pathotypes as described for monogenic resistance sources will unlikely occur under permanent cultivation of sugar beet cultivars carrying the Hs^{Bvm-1} genes. But due to the lower reproduction, some extent of selection pressure on the BCN population will occur, which probably could increase virulence in the long term. Experimentally this virulence shifting has not been confirmed for BCN but was demonstrated for *G. pallida* in wild polygenic resistant *S. verneii* hybrids after six reproduction cycles (Turner and Fleming, 2002). Evaluating partial resistance is a challenge as resistance parameters like BCN population density (cysts, eggs and juveniles) are highly variable and therefore difficult to distinguish between susceptible and partial resistant genotypes.

9.9.6 Future challenges

In the final breeding step of achieving a resistant sugar beet cultivar the resistance carrying donor is crossed with a *B. vulgaris* line. The resistance gene usually is not fully transferred into the resistant cultivar and minor parts of the plant population therefore show susceptibility. The proportion of susceptible plants in a defined number of plants of a resistant cultivar is defined as transmission rate. The protocol to test resistance in sugar beets in Germany originally focused on the detection of transmission rates and density of cysts per root. Today resistant sugar beet lines are tested in accordance to the test described for catch crops on the basis of P_f/P_i values, except that the evaluation of resistance primarily requires validated reproduction rates of below 1. Accordingly the same schach 0 standard population is used. This sort of test delivers sufficient data to evaluate a complete resistance type that most likely derives from a monogenic resistance source. The very restricted number of registered resistant cultivars (four cultivars in 16 years) justifies this simple evaluation.

In the near future, partial resistance of sugar beets with a B. maritima background will increasingly be the priority as a part of combined strategy to sustain high productivity in sugar beet crops by using tolerance and simultaneously maintaining BCN population density below a damage threshold by using partial resistance in the same cultivar. The challenge to evaluate partial resistance in sugar beets currently is to find a suitable, representative and reproducible reference to which partial resistance could be related. One evaluation concept, which is known from resistance testing of potatoes against PCN, is relative susceptibility that uses susceptible standard cultivars as the reference base (Anonymous, 2007). Current test protocols following approved nematological methods based on field experiments or pot experiments are able to distinguish three different susceptibility levels: (i) resistant cultivars with B. procumbens background; (ii) 'partial resistant' cultivars with B. maritima background; and (iii) susceptible cultivars without a resistance background (Westphal, 2013; Hauer et al., 2016). A classification of different resistance levels within the assortment of cultivars with a B. maritima background requires a high accuracy of test methods to minimize variability but this definitely is a future objective of resistance testing in sugar beets.

Another future challenge for resistance breeding as well as resistance testing concerns the lack of knowledge about differences in virulence within BCN populations, which, until now, could only be recognized and confirmed by selection studies using resistant monogenic hybrids described earlier in this chapter. Virulence of BCN shows natural differences between local populations even when comparing susceptible sugar beet genotypes (Griffin, 1981; Lange et al., 1993). This fact is of crucial importance concerning field resistance of partial resistant cultivars in particular. Whereas resistant cultivars show very low multiplication rates below 1 over a wide P_i range, partial resistant cultivars show a P_i multiplication rate that is P_i dependent, which also might be related to the local virulence type of the specific locality. In fact a number of biotic and abiotic environmental factors determine multiplication rates of BCN in the field (Schmidt et al., 1993) and usually BCN follows a heterogeneous horizontal distribution (Hbirkou et al., 2011); thus, high variability of P_i values have to be expected between plots in field experiments, which is even increased at lower P_i densities. Therefore resistance testing, which primarily focuses on detecting the genetic potential of a cultivar, should be conducted preferably under controlled conditions in the glasshouse. Test conditions should be maintained under the most favourable plant growth conditions to avoid damage effects of BCN on test plants that would affect the reproduction rate. Nevertheless, field testing is an essential amendment to glasshouse tests as field resistance of specific site conditions (soil, microclimate, nutrient and water availability and management) shows variation. Only by combining information concerning genetic and field resistance will there be a realistic approach for delivering a robust basis for management decisions.

9.10 Other Cyst Nematodes

9.10.1 Rice cyst nematodes

Both upland and irrigated rice can be infected by rice cyst nematodes: *Heterodera oryzicola* is widespread in the states in India where it has been recorded, *H. elachista* has been recorded from Japan and China and on corn in Italy, *H. sacchari* is found throughout West Africa (Bridge et al., 1990) and is highly pathogenic on susceptible upland rice cultivars in suitable conditions (Babatola, 1983; Coyne and Plowright, 1998). This species is also found in India, Trinidad, Pakistan and Thailand (Reversat and Destombes, 1998) and is reported to cause damage to sugar cane in Africa (Luc and Merny, 1963; Cadet and Merney, 1978) and many wild Cyperaceae and Graminae in West Africa are hosts. *Heterodera oryzae* is reported from West Africa, Bangladesh, India and Pakistan and also parasitizes banana.

No resistance to *Heterodera* spp. has been reported in rice, *Oryza sativa*; however, several lines of *O. glaberrima* and *O. breviligulata* are reported to have resistance to *H. sacchari* (Reversat and Destombes, 1998). One major gene, $Hsa\text{-}1^{Og}$, originating from *O. glaberrima* conferred resistance to *H. sacchari* and has been mapped to chromosome 11 (Lorieux *et al.*, 2003).

9.10.2 Pea cyst nematode

Heterodera goettingiana is widespread. It has been found in Asia, Europe and Mediterranean regions, and has been found in the USA and Algeria in Africa. In glasshouse trials *Pisum abyssinicum* accessions MG101791, MG101793, MG101788, MG101789 and MG101790, *P. elatius* accession MG100956 and *P. sativum* var. *arvense* accession MG101877 appeared to show some resistance to *H. goettingiana* (Di Vito and Perrino, 1978). A cross, *Pisum sativum* × *P. abyssinicum*, has yielded some moderately resistant genotypes to *H. goettingiana* (Di Vito, 1991).

9.10.3 Pigeon pea cyst nematode

Heterodera cajanii is widely distributed in India and Pakistan and has also been found in Egypt. In India, *H. cajanii* is considered to be one of the four most important cyst nematodes. It is an economic pest of cowpea and pigeon pea and also suppresses grain yield in mung bean (*Vigna radiata*) (Saxena and Reddy, 1987). A glasshouse technique to screen pigeon pea for resistance to *H. cajanii* is described by Sharma *et al.* (1991) and screening programmes to identify resistance have also been carried out in a wild relative of pigeon pea, *Cajanus platycarpus* (Sharma *et al.*, 1993; Sharma, 1995; Singh and Singh, 1995; Elyas and Sharma, 1997; Siddiqui *et al.*, 1998), mung bean (*Vigna radiata*) and black gram (*V. mungo*) (Devi and Gupta, 1987; Siddiqui *et al.*, 1999), as well as with cowpea, in which generally low levels of resistance have been reported (Sharma and Sethi, 1978; Devi and Gupta, 1991; Balasubramanian *et al.*, 1996) and resistance in sesame has been reported (Balasubramanian and Vadivelu, 2004).

9.10.4 Clover cyst nematode

Resistance to *H. trifolii*, the clover cyst nematode, has been found in some white clovers, *Trifolium repens* (Kuiper, 1960) and soybeans (Mankau and Linford, 1956; Grandison, 1963; Mulvey and Anderson, 1974). A New Zealand pedigree strain of white clover has a high degree of resistance (Grant *et al.*, 1996).

9.10.5 Carrot cyst nematode and corn cyst nematodes

Heterodera carotae, the carrot cyst nematode, is found in carrot-growing regions in Europe and has been reported from Cyprus, India, Russia and Michigan, USA. No resistance to this species has been found in carrot. Due to its narrow host range it is restricted to cultivated and wild carrots (Jones, 1950) and other *Daucus* spp. within the Apiaceae.

No source of resistance to the corn cyst nematode, *Heterodera zeae*, has been found in *Zea mays*.

9.11 Conclusions and Future Prospects

Breeding nematode resistance into economically important crops parasitized by cyst nematodes is an important strategy for crop improvement. However, despite the success in producing cultivars with high levels of nematode resistance, the uptake by industry has sometimes been disappointing. Undesirable agronomic characteristics due to linkage drag resulting in, for example, reduced yield can be the impediment; lack of market for new cultivars or resistance that is not sufficiently broad spectrum to control virulence types within the nematode populations are all potential constraints. Further backcrossing, education of the market and more resistance sources may be needed to achieve success. In the future, the use of MAS to aid the process of selection will greatly enhance the process and will facilitate the pyramiding of resistance genes. Also, direct transfer of resistance genes into commercially successful genotypes will further expedite the exploitation of resistance genes for nematode control. Technically these challenges are increasingly achievable within the complex process of crop improvement (Fig. 9.4); however, investment and acceptance are still constraints on their uptake.

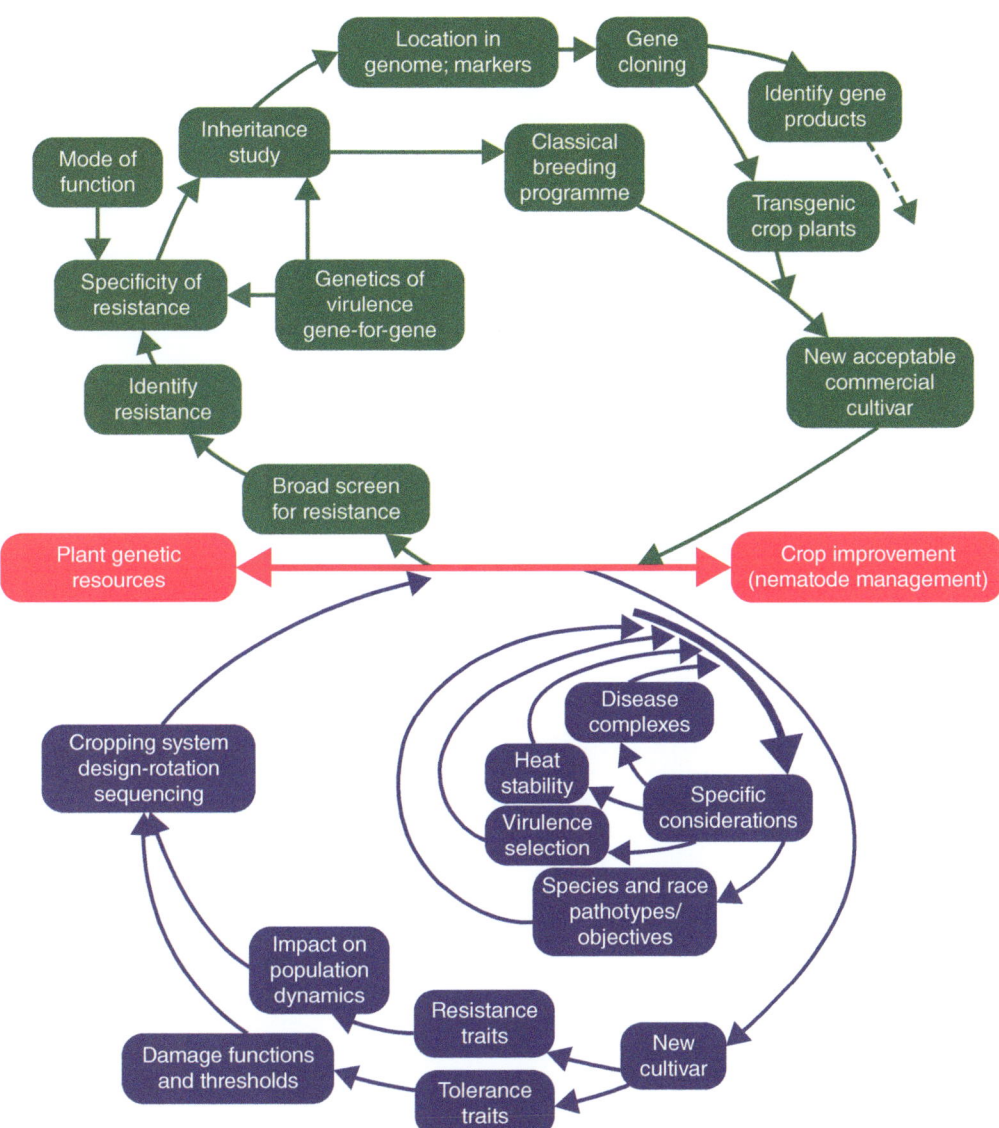

Fig. 9.4. A schematic representation of the complex factors that are involved in crop improvement for nematode control. The use of plant genetic resources to achieve this requires an understanding of the specific nematological aspects of each host–nematode relationship. (Adapted from Roberts, 1992.)

9.12 References

Abidou, H., Valette, S., Gauthier, J.P., Rivoal, R. El-Ahmed, A. and Yahyaoui, A. (2005) Molecular polymorphism and morphometrics of species of the *Heterodera avenae* group in Syria and Turkey. *Journal of Nematology* 37, 146–154.

Acharya, K., Tande, C. and Byamukama, E. (2016) Determination of *Heterodera glycines* virulence phenotypes occurring in South Dakota. *Plant Disease* 100, 2281–2286. DOI: dx.doi.org/10.1094/PDIS-04-16-0572-RE

Al-Hazmi, A.S., Cook, R. and Ibrahim, A.A.M. (2001) Pathotype characterisation of the cereal cyst nematode, *Heterodera avenae*, in Saudi Arabia. *Nematology* 3, 379–382. DOI: 10.1163/156854101317020312

Anand, S.C. (1992a) Registration of 'Hartwig' soybean. *Crop Science* 32, 1069–1070.

Anand, S.C. (1992b) Registration of 'Delsoy 4710' soybean. *Crop Science* 32, 1294.

Andersen, K. and Andersen, S. (1982) Classification of plants resistant to *Heterodera avenae*. *EPPO Bulletin* 12, 435–437. DOI: 10.1111/j.1365-2338.1982.tb01826

Andersen, S. (1961) Resistens mod Havreål *Heterodera avenae*. Copenhagen, Denmark, Meddelelse fra Den kgl. Veterinærog Landbohøskoles afdeling for landbrugets plantekultur, No. 68.

Anderson, T.R. and Welacky, T.W. (1988) First report of *Heterodera glycines* on soybeans in Ontario, Canada. *Plant Disease* 72, 453.

Anonymous (2007) *Council Directive 2007/33/EC of 11 June 2007 on the control of potato cyst nematodes and repealing Directive 69/465/EEC*. Council of the European Union, 11.

Arantes, N.E., Mauro, A.O. and Tihohood, D. (1998) An alternative field method for screening soybean genotypes for resistance to *Heterodera glycines*. *Journal of Nematology* 30, 542–546.

Arelli, A.P., Sleper, D.A., Yue, P. and Wilcox, J.A. (2000) Soybean reaction to races 1 and 2 of *Heterodera glycines*. *Crop Science* 40, 824–826. DOI: 10.2135/cropsci2000.403824x

Asmus, G.L. Teles, T.L., Anselmo, J. and Rosso, G.T. (2012) Races of *Heterodera glycines* in the northeast of Mato Grosso do Sul, Brazil. *Tropical Plant Pathology* 37, 146–148.

Babatola, J.O. (1983) Rice cultivars and *Heterodera sacchari*. *Nematologia Mediterranea* 11, 103–105.

Balasubramanian, P. and Vadivelu, S. (2004) Screening in *Sesame* L. genotypes for resistance to pigeonpea cyst nematode, *Heterodera cajani*, Koshy and Swarup, 1967. *Indian Journal of Agricultural Research* 38, 1–7.

Balasubramanian, P., Vadivelu, S. and Vijayakumar, J. (1996) Reaction of cowpea varieties against *Heterodera cajani* Koshy. *Indian Journal of Pulses Research* 9, 99–100.

Barone, A., Ritter, E., Schachtschabel, U., Debener, T., Salamini, F. and Gebhardt, C. (1990) Localization by restriction fragment length polymorphism mapping in potato of a major dominant gene conferring resistance to the potato cyst nematode *Globodera rostochienesis*. *Molecular and General Genetics* 224, 177–182. DOI: 10.1007/BF00271550

Bauer, S., Hymowitz, T. and Noel, G.R. (2007) Soybean cyst nematode resistance derived from *Glycine tomentella* in amphiploid (*G. max* × *G. tomentella*) hybrid lines. *Nematropica* 37, 277–285.

Beeman, A.Q., Harbach, C.J., Marett, C.C. and Tylka, G.L. (2016) Soybean cyst nematode HG type test results differ among multiple samples from the same field but the management implications are the same. *Plant Health Progress* 17, 160–162. DOI:10.1094/PHP-BR-16-0033

Bekal, S., Jahier, J. and Rivoal, R. (1998) Host responses of different Triticeae to species of the cereal cyst nematode complex in relation to breeding resistant durum wheat. *Fundamental and Applied Nematology* 21, 359–370.

Bishnoi, S.P. (2009) Importance of cereal cyst nematode in Rahasthan, India and its control through breeding for resistance. In: Riley, I.T., Nicol, J.M. and Dababat, A.A. (eds) *Cereal Cyst Nematodes: Status, Research and Outlook*. CIMMYT, Ankara, Turkey, pp. 143–148.

Blok, V.C. and Phillips, M.S. (2012) Biological characterisation of *Globodera pallida* from Idaho. *Nematology* 14, 817–826. DOI: 10.1163/156854112X627336

Boerma, H.R. and Hussey, R.S. (1984) Tolerance to *Heterodera glycines* in soybean. *Journal of Nematology* 16, 289–296.

Bridge, J., Luc, M. and Plowright, R.A. (1990) Nematode parasites of rice. In: Luc, M., Sikora, R. and Bridge, J. (eds) *Plant Parasitic Nematodes in Subtropical and Tropical Agriculture*. CAB International, Wallingford, UK, pp. 69–108.

Brim, C.A. and Ross, J.P. (1966) Registration of Pickett soybeans. *Crop Science* 6, 305.

Brodie, B.B. (1995) The occurrence of a second pathotype of potato cyst nematode in New York. *Journal of Nematology* 127, 493–494.

Brodie, B.B. (1996) Effect of initial nematode density on managing *Globodera rostochiensis* with resistant cultivars and nonhosts. *Journal of Nematology* 28, 510–519.

Brown, R.H. (1969) The occurrence of biotypes of the cereal cyst nematode (*Heterodera avenae* Woll.) in Victoria. *Australian Journal of Experimental Agriculture and Animal Husbandry* 9, 453–456.

Bryan, J., McLean, K., Pande, B., Purvis, A., Bradshaw, J.E., Waugh, R. and Hackett, C. (2004) Genetical dissection of H3-mediated polygenic PCN resistance in a heterozygous autotetraploid potato population. *Molecular Breeding* 14, 105–116. DOI: 10.1023/B:MOLB.0000037999.13581.9c

Budahn, H., Peterka, H., Mousa, M.A.A., Ding, Y., Zhang, S. and Li, J. (2009) Molecular mapping in oil radish (*Raphanus sativus* L.) and QTL analysis of resistance against beet cyst nematode (*Heterodera schachtii*). *Theoretical and Applied Genetics* 118, 775–782. DOI: 10.1007/s00122-008-0937-6

Cadet, P. and Merney, G. (1978) Influence of some factors on sex-ratio in *Heterodera oryzae* and *H. sacchari* (Nematode: Heteroderidae). *Revue de Nématologie* 1, 143–149.

Caldwell, B.E., Brim, C.A. and Ross, J.P. (1960) Inheritance of resistance of soybeans to the cyst nematode *Heterodera glycines*. *Agronomy Journal* 52, 635–636.

Çalişkan, M., Uranbey, S., Nicol, J., Akar, J., Elekçioğlu, H. and Kaya, G. (2011) Indirect selection of *Cre*1 gene in winter wheat populations. *Archives of Biological Sciences Belgrade* 63, 49–53. DOI: 10.2298/ABS1101049C

Canto Saenz, M. and Mayer de Scurrah, M. (1977) Races of the potato cyst nematode in the Andean region and a new system of classification. *Nematologica* 23, 340–349. DOI: 10.1163/187529277X00066

Castelli, L., Ramsay, G., Bryan, G., Neilson, S.J. and Phillips, M.S. (2003) New sources of resistance to the potato cyst nematodes *Globodera pallida* and *G. rostochiensis* in the Commonwealth Potato Collection. *Euphytica* 129, 377–386. DOI: 10.1023/A:1022264111096

Chełkowski, J., Tyrka, M. and Sobkiewicz, A. (2003) Resistance genes in barley (*Hordeum vulgare* L.) and their identification with molecular markers. *Journal of Applied Genetics* 44, 291–309.

Chen, S., Potter, B. and Orf, J. (2010) Virulence of the soybean cyst nematode has increased over years in Minnesota. *Journal of Nematology* 42, 238.

Concibido, V.C., Denny, R.L., Boutin, S.R., Hautea, R., Orf, J.H. and Young, N.D. (1994) DNA marker analysis of loci underlying resistance to soybean cyst nematode (*Heterodera glycines* Ichinohe). *Crop Science* 34, 240–246. DOI: 10.2135/cropsci1994.0011183X003400010044x

Concibido, V.C., Diers, B.W. and Arelli, P.R. (2004) A decade of QTL mapping for cyst nematode resistance in soybean. *Crop Science* 44, 1121–1131. DOI: 10.2135/cropsci2004.1121

Cook, D.E., Lee, T.G., Guio, X. et al. (2012) Copy number variation of multiple genes at Rhg1 mediates nematode resistance in soybean. *Science* 338, 1206–1209. DOI: 10.1126/science.1228746

Cook, R. and Evans, K. (1987) Resistance and tolerance. In: Brown, R.H. and Kerry, B.R. (eds) *Principles and Practice of Nematode Control in Crops*. Academic Press, Sydney, Australia, pp. 179–231.

Cook, R. and Noel, G.R. (2002) Cyst nematodes: *Globodera* and *Heterodera* species. In: Starr, J.L., Cook, R. and Bridge, J. (eds) *Plant Resistance to Parasitic Nematodes*. CAB International, Wallingford, UK, pp. 71–105.

Cook, R. and Rivoal, R. (1998) Genetics of resistance and parasitism. In: Sharma, S.B. (ed.) *The Cyst Nematodes*. Chapman and Hall, London, pp. 322–352.

Cook, R. and Williams, T.D. (1972) Pathotypes of *Heterodera avenae*. *Annals of Applied Biology* 71, 267–271.

Coyne, D.L. and Plowright, R.A. (1998) Use of solarisation to control *Heterodera sacchari* and other plant parasitic nematodes in the field: a modified technique for experimental purposes. *International Journal of Nematology* 8, 81–84.

Dababat, A.A., Erginbaş-Orakçi, G., Toktay, H., İmren, M., Akin, B., Braun, H.J., Dreisigacker, S., Elekçioğlu, İ.H. and Morgounov, A. (2014) Resistance of winter wheat to *Heterodera filipjevi* in Turkey. *Turkish Journal of Agriculture and Forestry* 38, 180–186.

Dababat, A.A., İmren, M. Erginbaş-Orakçi, G., Ashrafi, S., Yavuzaslanoğlu, E., Toktay, H., Pariyar, S.R., Elekçioğlu, İ.H., Morgounov, A. and Mekete, T. (2015a) The importance and management strategies of cereal cyst nematode, *Heterodera* spp., in Turkey. *Euphytica* 202, 173–188. DOI: 10.1007/s10681-014-1269-z

Dababat, A.A., Muminjanov, H. and Smiley, R.W. (eds) (2015b) *Nematodes of Small Grain Cereals: Current Status and Research*. FAO, Ankara, Turkey, 384 pp. Available at: www.fao.org/3/a-i4934e.pdf (accessed 22 January 2018).

Dalton, E., Griffin, D., Gallagher, T.F., de Vetten, N. and Milbourne, D. (2013) The effect of pyramiding two potato cyst nematode resistance loci to *Globodera pallida* Pa2/3 in potato. *Molecular Breeding* 31, 921–930. DOI: 10.1007/s11032-013-9845-9

De Koeyer, D., Douglass, K., Murphy, A., Whitney, S., Nolan, L., Song, Y. and De Jong, W. (2010) Application of high-resolution DNA melting for genotyping and variant scanning of diploid and autotetraploid potato. *Molecular Breeding* 25, 67–90. DOI: 10.1007/s11032-009-9309-4

Dellaert, L.M.W. and Hoekstra, R. (1987) Resistance to potato cyst nematodes, *Globodera* spp., in wild and primitive *Solanum* species. *Potato Research* 30, 579–587. DOI: 10.1007/BF02367639

Devi, L.S. and Gupta, P. (1987) Evaluation of greengram and blackgram varieties against cyst nematode – *Heterodera cajani*. *Indian Journal of Nematology* 17, 341.

Devi, L.S. and Gupta, P. (1991) Evaluation of cowpea varieties against cyst nematode. *Current Nematology* 2, 59–60.

Di Vito, M. (1991) The pea cyst nematode, *Heterodera goettingiana*. Florida Department of Agriculture and Consumer Services, *Nematology Circular* No. 188.

Di Vito, M. and Perrino, P. (1978) Reazione de *Pisum* spp. agli attacchi di *Heterodera goettingiana*. *Nematologia Mediterranea* 6, 113–118.

Doucet, M.E., Lax, P. and Coronel, N. (2008) The soybean cyst nematode *Heterodera glycines* Ichinohe, 1952 in Argentina. In: Ciancio, A. and Mukerji, K.G. (eds) *Integrated Management and Biocontrol of Vegetable and Grain Crops Nematodes*. Springer, New York, pp. 127–148. DOI: 10.1007/978-1-4020-6063-2_7

Dunnett, J.M. (1961) Inheritance of resistance to potato root eelworm in breeding lines stemming from *Solanum multidissectum* Hawkes. *Annual Report Scottish Plant Breeding Station 1961*, 39–46.

Eagles, H.A., Bariana, H.S., Ogbonnaya, F.C., Rebetzke, G.J., Hollamby, G.J., Henry, R.J., Henschke, P.H. and Carter, M. (2001) Implementation of markers in Australian wheat breeding. *Australian Journal of Agricultural Research* 52, 1349–1356. DOI: org/10.1071/AR01067

Ellenby, C. (1948) Resistance to the potato-root eelworm. *Nature* 162, 704. DOI: 10.1038/1701016a0

Ellenby, C. (1952) Resistance to the potato-root eelworm *Heterodera rostochiensis* Woll. *Nature* 170, 1016. DOI: 10.1038/1701016a0

Ellenby, C. (1954) Tuber forming species and varieties of the genus *Solanum* tested for resistance to the potato root eelworm *Heterodera rostochiensis* Wollenweber. *Euphytica* 3, 195–202. DOI: 10.1007/BF00055593

Elyas, Z. and Sharma, S.B. (1997) Mechanism of resistance to *Heterodera cajani* in *Cajanus platycarpus* accessions. *International Journal of Nematology* 7, 119–121.

Epps, J.M. and Hartwig, E.E. (1972) Reaction of soybean varieties and strains to race 4 of the soybean cyst nematode. *Journal of Nematology* 4, 222.

Faghihi, J., Donald, P.A., Noel, G., Welacky, T.W. and Ferris, V.R. (2010) Soybean resistance to field populations of *Heterodera glycines* in selected geographic areas. *Plant Health Progress* DOI: 10.1094/PHP-2010-0426-01-RS

Fisher, J.M. (1982a) Problems with the use of resistance in wheat to the Australian pathotype of *Heterodera avenae*. *EPPO Bulletin* 12, 417–421. DOI: 10.1111/j.1365-2338.1982.tb01824.x

Fisher, J.M. (1982b) Towards a consistent laboratory assay for resistance to *Heterodera avenae*. *EPPO Bulletin* 12, 445–449.

Galek, R., Rurek, M., De Jong, W.S., Pietkiewicz, G., Augustyniak, H. and Sawicka-Sienkiewicz, E. (2011) Application of DNA markers linked to the potato H1 gene conferring resistance to pathotype Ro1 of *Globodera rostochiensis*. *Journal of Applied Genetics* 52, 407–411. DOI: 10.1007/s13353-011-0056-y

Gavassoni, W.L., Tylka, G.L. and Munkvold, G.P. (2001) Relationship between tillage and spatial patterns of *Heterodera glycines*. *Phytopathology* 91, 534–545. DOI: dx.doi.org/10.1094/PHYTO.2001.91.6.534

Gebhardt, C., Mugniéry, D., Ritter, E., Salamini, F. and Bonnel, E. (1993) Identification of RFLP markers closely linked to the H1 gene conferring resistance to *Globodera rostochiensis* in potato. *Theoretical and Applied Genetics* 85, 541–544. DOI: 10.1007/BF00220911

Golden, M.G., Epps, J.M., Riggs, R.D., Duclos, L.A., Fox, J.A. and Bernard, R.L. (1970) Terminology and identity of infraspecific forms of the soybean cyst nematode (*Heterodera glycines*). *Plant Disease Reporter* 54, 544–546.

Grandison, G.S. (1963) The clover cyst nematode (*Heterodera trifolii* Goffart) in New Zealand. *New Zealand Journal of Agricultural Research* 6, 460–462. DOI: 10.1080/00288233.1963.10423290

Grant, J.L., Mercer, C.F. and Stewart, T.M. (1996) Effects of resistance in white clover (*Trifolium repens*) on *Heterodera trifoli*. *Journal of Nematology* 28, 492–500.

Grenier, E., Fournet, S., Petoe, E. and Anthoine G. (2010) A cyst nematode 'species factory' called the Andes. *Nematology* 12, 163–169. DOI: 0.1163/138855409X12573393054942

Griffin, G.D. (1981) Pathological differences in *Heterodera schachtii* populations. *Journal of Nematology* 13, 191–195.

Guo, X., Wang, J., Gardner, M., Fukada, H., Kondo, Y., Etchells, J.P., Wang, X. and Mitchum, M.G. (2017) Identification of cyst nematode B-type CLE peptides and modulation of the vascular stem cell pathway for feeding cell formation. *PLoS Pathogens* 13, e1006142. DOI: 10.1371/journal.ppat.1006142

Hafez, S.L. and Seyedbagheri, M.-M. (1997) *Sugarbeet Cyst Nematode: Impact of Sugarbeet Production in Idaho and Eastern Oregon. Bulletin no. 792.* University of Idaho, College of Agriculture, Cooperative Extension System, Agricultural Experiment Station, Moscow, Idaho.

Hajihasani, A., Tanha Maafi, Z. and Hajihasani, M. (2010) The life cycle of *Heterodera filipjevi* in winter wheat under microplot conditions in Iran. *Nematologia Mediterranea* 38, 53–57.
Hajihasani, A., Smiley, R.W. and Afshar, F.J. (2013) Effects of co-inoculations with *Pratylenchus thornei* and *Fusarium culmorum* on growth and yield of winter wheat. *Plant Disease* 97, 1470–1477. DOI: 10.1094/PDIS-02-13-0168-RE
Handoo, Z.A., Carta, L.K., Skantar, A.M. and Chitwood, D.J. (2012) Description of *Globodera ellingtonae* n. sp. (Nematodea: Heteroderidae) from Oregon. *Journal of Nematology* 44, 40–57.
Hartwig, E.E. and Epps, J.M. (1978) Registration of 'Bedford' soybeans. *Crop Science* 18, 915.
Hartwig, E.E. and Young, L.D. (1990) Registration of 'Cordell' soybean. *Crop Science* 30, 231.
Hauer, M., Koch, H.-J., Krüssel, S., Mittler, S. and Märländer, B. (2016) Integrated control of *Heterodera schachtii* Schmidt in Central Europe by trap crop cultivation, sugar beet variety choice and nematicide application. *Applied Soil Ecology* 99, 62–69. DOI: 10.1016/j.apsoil.2015.11.017
Hbirkou, C., Welp, G., Rehbein, K., Hillnhütter, C., Daub, M., Oliver, M.A. and Pätzold, S. (2011) The effect of soil heterogeneity on the spatial distribution of *Heterodera schachtii* within sugar beet fields. *Applied Soil Ecology* 51, 25–34. DOI: 10.1016/j.apsoil.2011.08.008
Heijbroek, W. (1977) Partial resistance of sugarbeet to beet cyst eelworm (*Heterodera schachtii* Schm.). *Euphytica* 26, 257–262. DOI: 10.1007/BF00026986
Heller, R., Schondelmaier, J., Steinrücken, G. and Jung, C. (1996) Genetic localization of four genes for nematode (*Heterodera schachtii* Schm.) resistance in sugar beet (*Beta vulgaris* L.). *Theoretical and Applied Genetics* 92, 991–997. DOI: 10.1007/BF00224039
Hershman, D.E., Heinz, R.D. and Kennedy, B.S. (2008) Soybean cyst nematode, *Heterodera glycines*, populations adapting to resistant soybean varieties in Kentucky. *Plant Disease* 92, 1475. DOI: dx.doi.org/10.1094/PDIS-92-10-1475B
Hodda, M. and Cook, D.C. (2009) Economic impact from unrestricted spread of potato cyst nematodes in Australia. *Phytopathology* 99, 1387–1393. DOI: 10.1094/PHYTO-99-12-1387
Holgado, R., Rowe, J., Andersson, S. and Magnusson, C. (2004) Electrophoresis and biotest studies on some populations of cereal cyst nematode, *Heterodera* spp. (Tylenchida: Heteroderidae). *Nematology* 6, 857–865. DOI: 10.1163/1568541044038551
Holgado, R., Andersson, S., Rowe, J., Clark, I. and Magnusson, C. (2009) Management strategies for cereal cyst nematodes *Heterodera* spp. in Norway. In: Riley, I.T., Nicol, J.M. and Dababat, A.A. (eds) *Cereal Cyst Nematodes: Status, Research and Outlook*. CIMMYT, Ankara, Turkey, pp. 154–159.
Hooper, D.J. and Evans, K. (1993) Extraction, identification and control of plant parasitic nematodes. In: Evans, K., Trudgill, D.L. and Webster, J.M. (eds) *Plant Parasitic Nematodes in Temperate Agriculture*. CAB International, Wallingford, UK, pp. 1–60.
Ichinohe, M. (1959) Studies on the soybean cyst nematode *Heterodera glycines* and its injury to soybean plants in Japan. *Plant Disease Reporter Supplement* 260.
İmren, M., Toktay, H., Bozbuğa, R., Erginbaş Orakçi, G., Dababat, A. and Elekçioğlu, İ.H. (2013) Identification of genetic resistance to cereal cyst nematodes; *Heterodera avenae* (Wollenweber, 1924), *Heterodera filipjevi* (Madzhidov, 1981) Stetler and *Heterodera latipons* (Franklin, 1969) in some international bread wheat germplasms. *Turkish Journal of Entomology* 37, 277–282.
Ireholm, A. (1994) Characterization of pathotypes of cereal cyst nematodes, *Heterodera* spp., in Sweden. *Nematologica* 4, 399–411. DOI: 10.1163/003525994X00283
Jahier, J., Tanguy, A.M., Abelard, P. and Rivoal, R. (1996) Utilization of deletions to localize a gene for resistance to the cereal cyst nematode, *Heterodera avenae*, on an *Aegilops ventricosa* chromosome. *Plant Breeding* 115, 282–284. DOI: 10.1111/j.1439-0523.1996.tb00919.x
Jahier, J., Abelard, P., Tanguy, A.M., Dedryver, F., Rivoal, R., Khatkar, S. and Bariana, H.S. (2001) The *Aegilops ventricosa* segment on chromosome 2AS of the wheat cultivar 'VPM1' carries the cereal cyst nematode resistance gene *Cre5*. *Plant Breeding* 120, 125–128. DOI: 10.1046/j.1439-0523.2001.00585.x
Janssen, R., Bakker, J. and Gommers, F.J. (1987) Circumventing the diapause of potato cyst nematodes. *Netherlands Journal of Plant Pathology* 93, 107–113. DOI: 10.1007/BF02000561
Jones, F.G.W. (1950) A new species of root eelworm attacking carrots. *Nature*, London 165, 81. DOI: 10.1038/165081a0
Kabelka, E.A., Carlson, S.R. and Diers, B.W. (2006) *Glycine soja* PI 468916 SCN resistance loci's associated effects on soybean seed yield and other agronomic traits. *Crop Science* 46, 622–629. DOI: 10.2135/cropsci2005.06-0131
Kang, H., Eun, G., Ha, J., Kim, Y., Park, N., Kim, D. and Choi, I. (2016) New cyst nematode, *Heterodera sojae* n. sp. (Nematoda: Heteroderidae) from soybean in Korea. *Journal of Nematology* 48, 280–289.

Kaur, D.J., Sharma, V.S., Sohu, V.S. and Bains, N.S. (2008) Reaction of wheat genotypes to a population of *Heterodera avenae* from Punjab, India. *Nematologia Mediterranea* 36, 157–160.

Kim, D.G., Riggs, R.D., Robbins, R.T. and Rakes, L. (1994) Distribution of races of *Heterodera glycines* in the United States. *Journal of Nematology* 29, 173–179.

Kim, M., Hyten, D.L., Niblack, T.L. and Diers, B.W. (2011) Stacking resistance alleles from wild and domestic soybean sources improves soybean cyst nematode resistance. *Crop Science* 51, 934–943. DOI: 10.2135/cropsci2010.08.0459.

Klinke, A., Müller., J. and Wricke, G. (1996) Characterization of nematode resistance genes in the section Procumbentes genus *Beta*: Response to two populations of *Heterodera schachtii*. *Theoretical and Applied Genetics* 93, 773–779. DOI: 10.1007/BF00224075

Koenning, S.R. and Wrather, J.A. (2010) Suppression of soybean yield potential in the continental United States by plant diseases from 2006 to 2009. *Plant Health Progress* DOI: 10.1094/PHP-2010-1122-01-RS

Kort, J., Jaspers, C.P. and Dijkstra, D.L. (1972) Testing for resistance to pathotype C of *Heterodera rostochiensis* and the practical application of *Solanum vernei* – hybrids in the Netherlands. *Annals of Applied Biology* 71, 289–294. DOI: 10.1111/j.1744-7348.1972.tb05098.x

Kort, J., Ross, H., Rumpenhorst, H.J. and Stone, A.R. (1977) An international scheme for identifying and classifying pathotypes of potato cyst-nematodes *Globodera rostochiensis* and *G. pallida*. *Nematologica* 23, 333–339. DOI: 10.1163/187529277X00057

Kuiper, K. (1960) Resistance of white clover varieties to the clover cyst eelworm *Heterodera trifolii* Goffarth. *Nematologica Supplement II*, 95.

Lange, W., Toxopeus, H., Lubberts, J.H., Dolstra, O. and Harrewijn, J.L. (1989) The development of Raparadish (× *Brassicoraphanus*, 2n=38), a new crop in agriculture. *Euphytica* 40, 1–14. DOI: 10.1007/BF00023291

Lange, W., Müller, J. and De Bock, T.S.M. (1993) Virulence in the beet cyst nematode (*Heterodera schachtii*) versus some alien genes for resistance in beet. *Fundamental and Applied Nematology* 16, 447–454.

Lax, P., Rondan Dueñas, J.C., Franco-Ponce, J., Gardenal, C.N. and Doucet, M.E. (2014) Morphology and DNA sequence data reveal the presence of *Globodera ellingtonae* in the Andean region. *Contributions to Zoology* 83, 227–243.

Lelivelt, C.L.C., Leunissen, E.H., Frederiks, H.J., Helsper, J.P. and Krens, F.A. (1993) Transfer of resistance to the beet cyst nematode (*Heterodera schachtii* Schm.) from *Sinapis alba* L. (white mustard) to the *Brassica napus* L. gene pool by means of sexual and somatic hybridization. *Theoretical and Applied Genetics* 85, 688–696. DOI: 10.1007/BF00225006

Lewis, J.G., Matic, M. and McKay, A.C. (2009) Success of cereal cyst nematode resistance in Australia: History and status of resistance screening systems. In: Riley, I.T., Nicol, J.M. and Dababat, A.A. (eds) *Cereal Cyst Nematodes: Status, Research and Outlook*. CIMMYT, Ankara, Turkey, pp. 137–142.

Li, A, Wei, C., Jiang, J., Zhang, Y., Snowdon, R.J. and Wang, Y. (2009) Phenotypic variation in progenies from somatic hybrids between *Brassica napus* and *Sinapis alba*. *Euphytica* 170, 289–296. DOI: 10.1007/s10681-009-9979-3

Li, Y.H., Qi, X.T., Chang, R.Z. and Qiu, L.J. (2011) Evaluation and utilization of soybean germplasm for resistance to cyst nematode in China. In: Sudarić, A. (ed.) *Soybean – Molecular Aspects of Breeding*. InTech, Rijeka, Croatia, pp. 373–396. DOI: 10.5772/14379

Lin, J., Mazarei, M., Hatcher, C.N. *et al.* (2016) Transgenic soybean overexpressing GmSAMT1 exhibits resistance to multiple-HG types of soybean cyst nematode *Heterodera glycines*. *Plant Biotechnology Journal* 14, 2100–2109. DOI: 10.1111/pbi.12566

Liu, S. and Peng, D. (2016) Recent progresses on soybean resistance to soybean cyst nematode. *Scientia Sinica Vitae* 46, 535–547. DOI: 10.1360/N052016-00162

Liu, S., Kandoth, P.K., Warren, S.D. *et al.* (2012) A soybean cyst nematode resistance gene points to a new mechanism of plant resistance to pathogens. *Nature* 492, 256–260. DOI: 10.1038/nature11651

Liu, X., Li, J. and Zhang, D. (1997) History and status of soybean cyst nematode in China. *International Journal of Nematology* 7, 18–25.

Loptien, H. (1984) Breeding nematode-resistant beets. II. Investigations into the inheritance of resistance to *Heterodera schachtii* Schm. in wild species of the section Patellares. *Zeitschrift für Pflanzenzuchtung* 93, 237–245.

Lorieux, M., Reversat, G., Garcia Diaz, S.X., Denance, C., Jouvenet, N., Orieux, Y., Bourger, N., Pando-Bahuon, A. and Ghesquière, A. (2003) Linkage mapping of Hsa-1^{Og}, a resistance gene of African rice to the cyst nematode, *Heterodera sacchari*. *Theoretical and Applied Genetics* 107, 691–696. DOI: 10.1007/s00122-003-1285-1

Luc, M. and Merny, G. (1963) *Heterodera sacchari* n. sp. (Nematoda: Tylenchoidae) parasite de la canne a sucre au Congo-Brazzaville. *Nematologica* 9, 31. DOI: 10.1163/187529263X00089

Lücke, E. (1976) Pathotypen-Untersuchungen mit *Heterodera avenae*-populationen (1966–1975). *Zeitschrift für Pflanzenkrankheiten und Pflanzenschutz* 83, 647–656.

Mai, W.F. (1977) Worldwide distribution of potato-cyst nematodes and their importance in crop production. *Journal of Nematology* 9, 30–34.

Mankau, G.R. and Linford, M.B. (1956) Soybean varieties tested as hosts of clover cyst nematode. *Plant Disease Reports* 40, 39–42.

Marshall, J.M. and Smiley, R.W. (2016) Resistance and tolerance of spring barley to *Heterodera avenae*. *Plant Disease* 100, 396–407. DOI: org/10.1094/PDIS-09-15-1055-RE

Mathur, B.N., Arya, H.C., Mathur, R.L. and Handa, D.K. (1974) The occurrence of biotypes of the cereal cyst nematode (*Heterodera avenae*) in the light soils of Rajasthan and Haryana, India. *Nematologica* 20, 19–26. DOI: 10.1163/187529274X00537

Matsuo, E., Sediyama, T., Oliveira, R.D.D., Cruz, C.D. and Oliveira, R.D.T. (2012) Characterization of type and genetic diversity among soybean cyst nematode differentiators. *Scientia Agricola* 69, 147–151. DOI: http://dx.doi.org/10.1590/S0103-90162012000200010

Mauro, A.O., de Oliveira, A.L. and Mauro, S.M.Z. (1999) Genetics of resistance to soybean cyst nematode, *Heterodera glycines* Ichinohe (race 3), in a Brazilian soybean population. *Genetics and Molecular Biology* 22, 257–260. dx.DOI.org/10.1590/S1415-47571999000200021

McCarville, M.C., Marett, C.C., Mullaney, M.P., Gebhart, G.D. and Tylka, G.L. (2017) Increase in soybean cyst nematode virulence and reproduction on resistant soybean varieties in Iowa from 2001 to 2015 and the effects on soybean yields. *Plant Health Progress* 18, 146–155. DOI: 10.1094/PHP-RS-16-0062.

McDonald, A.H. and Nicol. J.M. (2005) Nematode parasites of cereals. In: Luc, M., Sikora, R.A. and Bridge, J. (eds) *Plant Parasitic Nematodes in Subtropical and Tropical Agriculture*. CAB International, Wallingford, UK, pp. 131–191.

McIntosh, R.A., Yamazaki, Y., Dubcovsky, J., Rogers, J., Morris, C., Somers, D.J., Appels, R. and Devos, K.M. (2008) *Catalogue of Gene Symbols for Wheat*. 11th International Wheat Genetics Symposium, 24–29 August 2008, Brisbane, Queensland, Australia. Available at: http://wheat.pw.usda.gov/GG2/Triticum/wgc/2008/

McKenry, M.V. and Roberts, P.A. (1985) *Phytonematolgy Study Guide*. Publication 4405. University of California Press, Oakland, California.

Mendes, M.L. and Dickson, D.W. (1993) *Heterodera glycines* found on soybean in Brazil. *Plant Disease* 77, 499–500.

Mitchum, M.G. (2016) Soybean resistance to the soybean cyst nematode *Heterodera glycines*: an update. *Phytopathology* 106, 1444–1450. DOI: dx.doi.org/10.1094/PHYTO-06-16-0227-RVW.

Mitchum, M.G., Wrather, J.A., Heinz, R.D., Shannon, J.G. and Danekas, G. (2007) Variability in distribution and virulence phenotypes of *Heterodera glycines* in Missouri during 2005. *Plant Disease* 91, 1473–1476. DOI: dx.doi.org/10.1094/PDIS-91-11-1473

Mokabli, A., Valette, S., Gauthier, J.-P. and Rivoal, R. (2002) Variation in virulence of cereal cyst nematode populations from North Africa and Asia. *Nematology* 4, 521–525. DOI: 10.1163/156854102760290491

Moloney, C., Griffin, D., Jones, P.W., Bryan, G.J., McLean, K., Bradshaw, J.E. and Milbourne, D. (2010) Development of diagnostic markers for use in breeding potatoes resistant to *Globodera pallida* pathotype Pa2/3 using germplasm derived from *Solanum tuberosum* ssp. *andigena* CPC 2802. *Theoretical and Applied Genetics* 120, 679–689. DOI: 10.1007/s00122-009-1185-0

Müller, J. (1992) Detection of pathotypes by assessing the virulence of *Heterodera schachtii* populations. *Nematologica* 38, 50–64. DOI: 10.1163/187529292X00045

Müller, J. (1999) The economic importance of *Heterodera schachtii* in Europe. *Helminthologia* 36, 205–213.

Müller, J. and Klinke, A. (1996) Selektion virulenter Populationen von *Heterodera schachtii* und ihre Nutzung zur Charakterisierung von Resistenzgenen in Beta-Rüben. *50 Jahre Forschung am Standort MünsterBiologische Bundesanstalt für Land- und Forstwirtschaft, Institut für Nematologie und Wirbeltierkunde, Münster, Germany*, 317, 102–116.

Müller, J. and Rumpenhorst, H.J. (2000) *Testing of Crop Cultivars for Resistance to Noxious Organisms at the Federal Biological Research Centre*. Parey, Berlin, Germany.

Müller, J. and Steudel, W. (1982) Changes in the population of *Heterodera schachtii* on chines radish (*Raphanus sativus* L) under different environmental conditions. *Zuckerindustrie* 107, 1120–1123.

Mulvey, R.H. and Anderson, R.V. (1974) *Heterodera trifolii*. Commonwealth Institute of Helminthology Descriptions of Plant-parasitic Nematodes, Set 4, No. 46, 4 pp.

Niblack, T.L. (2005) Soybean cyst nematode management reconsidered. *Plant Disease* 89, 1020–1026. DOI: 10.1094/PD-89-1020

Niblack, T.L., Arelli, P.R. Noel, G.R., Opperman, C.H., Orf, J.H., Schmitt, D.P., Shannon, J.G. and Tylka, G.L. (2002) A revised classification scheme for genetically diverse populations of *Heterodera glycines*. *Journal of Nematology* 34, 279–288.

Niblack, T.L., Wrather, J.A., Heinz, R.D. and Donald, P.A. (2003) Distribution and virulence phenotypes of *Heterodera glycines* in Missouri. *Plant Disease* 87, 929–932.

Niblack, T.L., Colgrove, A.L., Colgrove, K. and Bond, J.P. (2008) Shift in virulence of soybean cyst nematode is associated with use of resistance from PI 88788. *Plant Health Progress*. DOI: 10.1094/PHP-2008-0118-01-RS

Niblack, T.L., Hussey, R.S. and Boerma, H.R. (1986) Effects of interactions among *Heterodera glycines*, *Meloidogyne incognita*, and host genotypes on soybean yield and nematode population densities. *Journal of Nematology* 18, 436–443.

Niblack, T., Tylka, G.L., Arelli, P. *et al.* (2009) A standard greenhouse method for assessing soybean cyst nematode resistance in soybean: SCE08 (standardized cyst evaluation 2008). *Plant Health Progress*. DOI: 10.1094/PHP-2009-0513-01-RV

Nickell, C.D., Noel, G.R., Bernard, R.L., Thomas, D.J. and Frey, K. (1994a) Registration of soybean germplasm line 'LN89-5699' resistant to soybean cyst nematode. *Crop Science* 34, 1133–1134.

Nickell, C.D., Noel, G.R., Bernard, R.L., Thomas, D.J. and Pracht, J. (1994b) Registration of soybean germplasm line 'LN89-5612' moderately resistant to soybean cyst nematode. *Crop Science* 34, 1134.

Nicol, J., Rivoal, R., Taylor, S. and Zaharieva, M. (2003) Global importance of cyst (*Heterodera* spp.) and lesion nematodes (*Pratylenchus* spp.) on cereals: distribution, yield loss, use of host resistance and integration of molecular tools. In: Cook, R. and Hunt, D.J. (eds) *Proceedings of the Fourth International Congress of Nematology, 8–13 June 2002, Tenerife, Spain. Nematology Monographs and Perspectives 2*. Brill, Leiden, The Netherlands, pp. 1–19.

Nicol, J.M., Ogbonnaya, F., Singh, A.K. *et al.* (2009) Current global knowledge of the usability of cereal cyst nematode resistant bread wheat germplasm through international germplasm exchange and evaluation. In: Riley, I.T., Nicol, J.M. and Dababat, A.A. (eds) *Cereal Cyst Nematodes: Status, Research and Outlook*. CIMMYT, Ankara, Turkey, pp. 149–153.

Noel, G.R. (1992) History, distribution, and economics. In: Riggs, R.D. and Wrather, J.A. (eds) *Biology and Management of the Soybean Cyst Nematode*. APS Press, St. Paul, Minnesota, pp. 1–14.

Noel, G.R. and Sikora, E.J. (1990) Evaluation of soybeans in maturity groups I–IV for resistance to *Heterodera glycines*. *Journal of Nematology* 22, 795–799.

O'Brien, P.C. and Fisher, J.M. (1974) Resistance within wheat, barley and oat cultivars to *Heterodera avenae* in South Australia. *Australian Journal of Experimental Agriculture and Animal Husbandry* 14, 399–404.

OEPP/EPPO (2006) Testing of potato varieties to assess resistance to *Globodera rostochiensis* and *Globodera pallida*. *OEPP/EPPO Bulletin* 36, 419–420.

Paal, J., Henselewski, H., Muth, J., Meksem, K., Menéndez, C.M., Salamini, F., Ballvora, A. and Gebhardt, C. (2004) Molecular cloning of the potato *Gro1-4* gene conferring resistance to pathotype Ro1 of the root cyst nematode *Globodera rostochiensis*, based on a candidate gene approach. *Plant Journal* 38, 285–297. DOI: 10.1111/j.1365-313X.2004.02047

Pariyar, S.R., Dababat, A.A., Siddique, S., Erginbaş-Orakçi, G., Elashry, A., Morgounov, A. and Grundler, F.M.W. (2016) Identification and characterization of resistance to the cereal cyst nematode *Heterodera filipjevi* in winter wheat. *Nematology* 18, 377–402. DOI: 10.1163/15685411-00002964

Peng, D.L., Pend, H., Wu, D.Q., Huang, W.K., Ye, W.X. and Cui, J.K. (2016) First report of soybean cyst nematode (*Heterodera glycines*) on soybean from Gansu and Ningxia China. *Plant Disease* 100, 229. DOI: dx.doi.org/10.1094/PDIS-04-15-0451-PDN

Person, F. and Rivoal, R. (1979) Hybridation entre les races Fr_1 et Fr_4 d'*Heterodera avenae* Wollenweber en France et étude du comportement d'aggressivité des descendants F_1. *Revue de Nématology* 2, 177–183.

Peterka, H., Budahn, H., Zhang, S.S. and Li, J.B. (2010) Nematode resistance of rape-radish chromosome addition lines. *Nematology* 12, 269–275. DOI: 10.1163/138855409X12506855979677

Phillips, M.S. and Trudgill, D.L. (1998) Variation of virulence, in terms of quantitative reproduction of *Globodera pallida* populations, from Europe and South America, in relation to resistance from *Solanum vernei* and *S. tuberosum* spp. *andigena* CPC 2802. *Nematologica* 44, 409–423. DOI: 10.1163/005525998X00070

Phillips, M.S., Forrest, J.M.S. and Wilson, L.A. (1980) Screening for resistance to potato cyst nematode using closed containers. *Annal of Applied Biology* 96, 317–322. DOI: 10.1111/j.1744-7348.1980.tb04782.x

Pineda, O., Bonierbale, M.W., Plaisted, R.L., Brodie, B.B. and Tanksley, S.D. (1993) Identification of RFLP markers linked to the H1 gene conferring resistance to the potato cyst nematode *Globodera rostochiensis*. *Genome* 36, 152–156. DOI: 10.1139/g93-019

Plantard, O., Picard, D., Valette, S., Scurrah, M. Grenier, E. and Mugniéry, D. (2008) Origin and genetic diversity of Western European populations of the potato cyst nematode *Globodera pallida* inferred from mitochondrial sequences and microsatellite loci. *Molecular Ecology* 17, 2208–2218. DOI: 10.1111/j.1365-294X.2008.03718

Rathjen, A.J., Eastwood, R.F., Lewis, J.G. and Dube, A.J. (1998) Breeding wheat for resistance to *Heterodera avenae* in southeastern Australia. *Euphytica* 100, 55–62. DOI: 10.1023/A:1018347704735

Reversat, G. and Destombes, D. (1998) Screening for resistance to *Heterodera sacchari* in the two cultivated rice species, *Oryza sativa* and *O. glaberrima*. *Fundamental and Applied Nematology* 21, 307–317.

Riggs, R.D. (2004) History and distribution In: Riggs, R.D. and Wrather, J.A. (eds) *Biology and Management of the Soybean Cyst Nematode*, 2nd edn. APS Press, St. Paul, Minnesota, pp. 9–40.

Riley, I.T. and McKay, A.C. (2009) Cereal cyst nematode in Australia: biography of a biological invader. In: Riley, I.T., Nicol, J.M. and Dababat, A.A. (eds) *Cereal Cyst Nematodes: Status, Research and Outlook*. CIMMYT, Ankara, Turkey, pp. 23–28.

Riley, I.T., Nicol, J.M. and Dababat, A.A. (eds) (2009) *Cereal Cyst Nematodes: Status, Research and Outlook*. CIMMYT, Ankara, Turkey, 244 pp. Available at: www.spipm.cgiar.org/reports (accessed 22 January 2018).

Rincker, K. Cary, T. and Diers, B.W. (2017) Impact of soybean cyst nematode resistance on soybean yield. *Crop Science* DOI: 10.2135/cropsci2016.07.0628

Rivoal, R. and Cook, R. (1993) Nematode pests of cereals. In: Evans, K., Trudgill, D.L. and Webster, J.M. (eds) *Plant Parasitic Nematodes in Temperate Agriculture*. CAB International, Wallingford, UK, pp. 259–303.

Rivoal, R., Person, F., Caubel, G. and Scotto La Massèse, C. (1978) Méthodes d'évaluation de la résistance des céréales au développement des nématodes: *Ditylenchus dipsaci*, *Heterodera avenae* et *Pratylenchus*. *Annales d'Amélioration des Plantes* 28, 371–394.

Rivoal, R., Lasserre, F., Hullé, M. and Doussinault, G. (1991) Evaluation de la résistance et de la tolérance à *Heterodera avenae* chez le blé par des tests miniaturisés. *Mededelingen van den Faculteit Landbouwwettenschappen Rijksuniv Gent* 56, 1281–1292.

Rivoal, R., Bekal, S., Valette, S., Gauthier, J.-P., Bel Hadj Fradj, M., Mokabli, A., Jahier, J., Nicol, J.M. and Yahyaoui, A. (2001) Variation in reproductive capacity and virulence on different genotypes and resistance genes of Triticeae, in the cereal cyst nematode species complex. *Nematology* 3, 581–592. DOI: 10.1163/156854101753389194

Roberts, P.A. (1992) Current status of the availability, development, and use of host plant-resistance to nematodes. *Journal of Nematology* 24, 213–227.

Roberts, P.A. (2002) Concepts and consequences of resistance. In: Starr, J.L., Cook, R. and Bridge, J. (eds) *Plant Resistance to Parasitic Nematodes*. CAB International, Wallingford, UK, pp. 23–41.

Ross, H. (1986) Potato breeding – problems and perspectives. *Journal of Plant Breeding* (Supplement 13), 132 pp.

Ross, J.P. (1962) Physiological strains of *Heterodera glycines*. *Plant Disease* 46, 766–769.

Ross, J.P. and Brim, C.A. (1957) Resistance of soybeans to the soybean cyst nematode as determined by a double-row method. *Plant Disease Reporter* 41, 923–924.

Rouppe van der Voort, J., Wolters, P., Folkertsma, R. *et al.* (1997) Mapping of the cyst nematode resistance locus Gpa2 in potato using a strategy based on comigrating AFLP markers. *Theoretical and Applied Genetics* 95, 874–880. DOI: 10.1007/s001220050638

Rouppe van der Voort, J., Lindeman, W., Folkertsma, R., Hutten, R., Overmars, H., Van Der Vossen, E., Jacobsen, E. and Bakker, J. (1998) A QTL for broad-spectrum resistance to cyst nematode species (*Globodera spp.*) maps to a resistance to gene cluster in potato. *Theoretical and Applied Genetics* 96, 654–661. DOI: 10.1007/s001220050785

Rouppe van der Voort, J., van der Vossen, E., Bakker, E., Overmars, H., van Zandvoort, P., Hutten, R., Klein Lankhorst, R.K. and Bakker, J. (2000) Two additive QTLs conferring broad-spectrum resistance in potato to *Globodera pallida* are localized on resistance gene clusters. *Theoretical and Applied Genetics* 101, 1122–1130. DOI: 10.1007/s001220051588

Rousselle-Bourgeois, F. and Muigniéry, D. (1995) Screening tuber bearing *Solanum* spp. for resistance to *Globodera rostochiensis* Ro1 Woll. and *G. pallida* Pa2/3 Stone. *Potato Research* 38, 241–249. DOI: 10.1007/BF02359906

Ruiz de Galarreta, J.I., Carrasco, A., Salazar, A., Barrena, I., Iturritxa, E., Marquinez, R., Legorburu, F.J. and Ritter, E. (1998) Wild *Solanum* species as resistant sources against different pathogens of potato. *Potato Research* 41, 57–68. DOI: 10.1007/BF02360262

Şahin, E., Nicol, J.M., Bolat, N., Yorgancilar, A. and Yildirim, A.F. (2006) *In vitro* resistance reaction of published *Cre* genes against the Turkish isolate of cereal cyst nematode *Heterodera filipjevi* (abstract). In: Falloon, R.E., Cromey, M.G., Stewart, A. and Jones, E.E. (eds) *Proceedings of the 4th Australasian Soilborne Diseases Symposium. Queenstown, New Zealand*, p. 97.

Şahin, E., Nicol, J.M., Elekçioğlu, H. and Rivoal, R. (2010) Hatching of *Heterodera filipjevi* in controlled and natural temperature conditions in Turkey. *Nematology* 12, 277–287. DOI: 10.1163/156854109X463738

Salentijn, E.M.J., Sandal, N.N., Lange, W., De Bock, Th. S.M., Krens, F.A., Marcker, K.S. and Stiekema, W.J. (1992) Isolation of DNA markers linked to a beet cyst nematode resistance locus in *Beta patellaris* and *Beta procumbens. Molecular and General Genetics* 235, 432–440. DOI: 10.1007/BF00279390

Santana, H., Pires, E., Comerlato, A.P., Nasu, E.G.C., and Furlanetto, C. (2009) Genetic variability in field populations of the soybean cyst nematode from the states of Parana and Rio Grande do Sul, Brazil. *Tropical Plant Pathology* 34, 261–264.

Sattarzadeh, A., Achenbach, U., Lubeck, J., Strahwald, J., Tacke, E., Hofferbert, H.-R., Rothsteyn, T. and Gebhardt, C. (2006) Single nucleotide polymorphism (SNP) genotyping as basis for developing a PCR-based marker highly diagnostic for potato varieties with high resistance to *Globodera pallida* pathotype Pa2/3. *Molecular Breeding* 18, 301–312. DOI: 10.1007/s11032-006-9026-1

Saxena, R. and Reddy, D.D.R. (1987) Crop losses in pigeonpea and mungbean by pigeonpea cyst nematode, *Heterodera cajani. Indian Journal of Nematology* 17, 91–94.

Schmidt, K., Sikora, R.A. and Richter, O. (1993) Modelling the population dynamics of the sugar beet cyst nematode *Heterodera schachtii. Crop Protection* 12, 490–496. DOI: 10.1016/0261-2194(93)90088-Z

Schmitt, D.P. and Shannon, G. (1992) Differentiating soybean responses to *Heterodera glycines* races. *Crop Science* 32, 275–277.

Schultz, L., Cogan, N.O.I., Mclean, K., Dale, F.B., Bryan, G.J., Forster, J.W. and Slater, A.T. (2012) Evaluation and implementation of a potential diagnostic molecular marker for H1-conferred potato cyst nematode resistance in potato (*Solanum tuberosum* L.). *Plant Breeding* 131, 315–321. DOI: 10.1111/j.1439-0523.2012.01949

Seah, S., Sivasithamparam, K., Karakousis, A. and Laguday, E.S. (1998) Cloning and characterisation of a family of disease resistance gene analogs from wheat and barley. *Theoretical and Applied Genetics* 97, 937–945. DOI: 10.1007/s001220050974

Seinhorst, J.W. and Den Ouden, H. (1971) The relation between density of *Heterodera rostochiensis* and growth and yield of two potato varieties. *Nematologica* 17, 347–369. DOI: 10.1163/187529271X00585

Sharma, N.K. and Sethi, C.L. (1978) Interaction between *Meloidogyne incognita* and *Heterodera cajani* on cowpea. *Indian Journal of Nematology* 6, 1–12.

Sharma, S.B. (1995) Resistance to *Rotylenchulus reniformis*, *Heterodera cajani*, and *Meloidogyne javanica* in accessions of *Cajanus platycarpus*. *Plant Disease* 79, 1033–1035. DOI: 10.1094/PD-79-1033

Sharma, S.B., Kumar, A. and McDonald, D. (1991) A greenhouse technique to screen pigeonpea for resistance to *Heterodera cajani*. *Annals of Applied Biology* 118, 315–356. DOI: 10.1111/j.1744-7348.1991.tb05635

Sharma, S.B., Remanandan, P.R. and Jain, K.C. (1993) Resistance to cyst nematodes (*Heterodera cajani*) in pigeonpea cultivars and in wild relatives of *Cajanus*. *Annals of Applied Biology* 123, 75–81. DOI: 10.1111/j.1744-7348.1993.tb04074

Siddique, S., Radakovic, Z.S., De La Torre, C.M. *et al.* (2016) A parasitic nematode releases cytokinin that controls cell division and orchestrates feeding site formation in host plants. *Proceedings of the National Academy of Sciences of the United States of America* 112, 12669–12674. DOI: 10.1073/pnas.1503657112

Siddiqui, M.R., Gupta, P. and Ali, H. (1998) Screening of pigeonpea varieties for resistance to the pigeonpea cyst nematode, *Heterodera cajani*. *Annals of Plant Protection Sciences* 6, 193–196.

Siddiqui, M.R., Ali, H. and Gupta, P. (1999) Evaluation of greengram and blackgram germplasm against pigeonpea cyst nematode, *Heterodera cajani*. *Journal of Mycology and Plant Pathology* 29, 135–136.

Singh, V.K. and Singh, K.P. (1995) Development and pathogenicity of *Heterodera cajani* on a susceptible and moderately susceptible cultivars of pigeonpea. *International Journal of Tropical Plant Disease* 13, 237–244.

Skantar, A.M., Handoo, Z.A., Zasada, I.A., Ingham, R.E., Carta, L.K. and Chitwood, D.J. (2011) Morphological and molecular characterization of *Globodera* populations from Oregon and Idaho. *Phytopathology* 101, 480–491. DOI: 10.1094/PHYTO-01-10-0010

Smiley, R.W. (2009) Occurrence, distribution and control of *Heterodera avenae* and *H. filipjevi* in the western USA. In: Riley, I.T., Nicol, J.M. and Dababat, A.A. (eds) *Cereal Cyst Nematodes: Status, Research and Outlook*. CIMMYT, Ankara, Turkey, pp. 35–40.

Smiley, R.W. and Marshall, J.M. (2016) Detection of dual *Heterodera avenae* resistance plus tolerance traits in spring wheat. *Plant Disease* 100, 1677–1685.

Smiley, R.W. and Nicol, J.M. (2009) Nematodes which challenge global wheat production. In: Carver, B.F. (ed.) *Wheat Science and Trade*. Wiley-Blackwell, Ames, Iowa, pp. 171–187.

Smiley, R.W. and Yan, G.P. (2015) Discovery of *Heterodera filipjevi* in Washington and comparative virulence with *H. avenae* on wheat. *Plant Disease* 99, 376–386.

Smiley, R.W., Ingham, R.E., Uddin, W. and Cook, G.H. (1994) Crop sequences for managing cereal cyst nematode and fungal populations of winter wheat. *Plant Disease* 78, 1142–1149. DOI: 10.1094/PD-78-1142

Smiley, R.W., Yan, G.P. and Handoo, Z.A. (2008) First record of the cyst nematode *Heterodera filipjevi* on wheat in Oregon. *Plant Disease* 92, 1136. DOI: 10.1094/PDIS-92-7-1136B

Smiley, R.W., Yan, G.P. and Pinkerton, J.N. (2011) Resistance of wheat, barley and oat to *Heterodera avenae* in the Pacific Northwest, USA. *Nematology* 13, 539–552. DOI: 10.1163/138855410X531862

Smiley, R.W., Marshall, J.M., Gourlie, J.A. et al. (2013) Spring wheat tolerance and resistance to *Heterodera avenae* in the Pacific Northwest. *Plant Disease* 97, 590–600.

Soliman, A.H., Sobazak, M. and Golinowski, W. (2005) Defence responses of white mustard, *Sinapis alba*, to infection with the cyst nematode *Heterodera schachtii*. *Nematology* 7, 881–889. DOI: 10.1163/156854105776186389

Starr, J.L., Mc Donald, A.M. and Claudius-Cole, A. (2013) Nematode resistance in crops. In: Perry, R.N. and Moens, M. (eds) *Plant Nematology*, 2nd edn. CAB International, Wallingford, UK, pp. 411–436.

Stevanato, P., Trebbi, D., Panella, L., Richardson, K., Broccanello, C., Pakish, L., Fenwick, A.L. and Saccomani, M. (2014) Identification and validation of a SNP marker linked to the gene HsBvm-1 for nematode resistance in sugar beet. *Plant Molecular Biology Reporter* 33, 474–479. DOI: 10.1007/s11105-014-0763-8

Stone, A.R. (1979) Co-evolution of nematodes and plants. *Symbolae Botanicae Uppsala* 22, 46–61.

Subbotin, S.A. (2015) *Heterodera sturhani* sp. n. from China, a new species of the *Heterodera avenae* species complex (Tylenchida: Heteroderidae). *Russian Journal of Nematology* 23, 145–152.

Subbotin, S.A., Mundo-Ocampo, M. and Baldwin, J.G. (2010) *Systematics of cyst nematodes (Nematoda: Heteroderinae)*. Nematology Monographs and Perspectives 8A (Series Editors: Hunt, D.J. and Perry, R.N.). Brill, Leiden, The Netherlands.

Tian, B., Li, J., Oakley, T.R., Todd, T.C. and Trick, H.N. (2016) Host-derived artificial microRNA as an alternative method to improve soybean resistance to soybean cyst nematode. *Genes* 7, 122. DOI: 10.3390/genes7120122

Tiemens-Hulscher, M., Delleman, J. Eisinger, E. and Lammerts Van Bueren, E.T. (2013) *Potato breeding: a practical manual for the potato chain*. Aardappelwereld BV, The Hague, The Netherlands.

Todd, T.C., Schapaugh, Jr, W.T., Long, J.H. and Holmes, B. (1995) Field response of soybean in maturity groups III-V to *Heterodera glycines* in Kansas. *Journal of Nematology* 27, 628–633.

Toxopeus, H. and Lubberts, H. (1980) Breeding for resistance to the sugarbeet nematode (*Heterodera schachtii* Schm.) in cruciferous crops. Eucarpia 'Cruciferae 1979' Conference, 1, 2, 3 October 1979. Wageningen, The Netherlands, p. 151.

Trudgill, D.L. (1991) Resistance to and tolerance of plant parasitic nematodes in plants. *Annual Review of Phytopathology* 29, 167–192. DOI: 10.1146/annurev.py.29.090191.001123

Turner, S.J. (1989) New sources of resistance to potato cyst nematodes in the Commonwealth Potato Collection. *Euphytica* 42, 145–153.

Turner, S.J. and Evans, K. (1998) The origins, global distribution and biology of potato cyst nematodes (*Globodera rostochiensis* (Woll.) and *Globodera pallida* Stone). In: Marks, R.J. and Brodie, B.B. (eds) *Potato Cyst Nematodes: Biology, Distribution and Control*. CAB International, Wallingford, UK, pp. 7–26.

Turner, S.J. and Fleming, C.C. (2002) Multiple selection of potato cyst nematode *Globodera pallida* virulence on a range of potato species. I. Serial selection on *Solanum*-hybrids. *European Journal of Plant Pathology* 108, 461–467.

Turner, S.J. and Subbotin, S.A. (2013) Cyst nematodes. In: Perry, R.N. and Moens, M. (eds) *Plant Nematology*, 2nd edn. CAB International, Wallingford, UK, pp. 109–143.

Tylka, G.L. (2016) Understanding soybean cyst nematode HG types and races. *Plant Health Progress* 17, 149–151. DOI: 10.1094/PHP-PS-16-0615

Tylka, G.L. and Mullaney, M.P. (2017) Soybean cyst nematode-resistant soybean varieties for Iowa. *Iowa State University Extension and Outreach Publications*. http://lib.dr.iastate.edu/extension_pubs/100

Tylka, G.L., McCarville, M.T., Marett, C.C., Gebhart, G.D., Soh, D.H., Mullaney, M.P. and O'Neal, M.E. (2013) Direct comparison of soybean cyst nematode reproduction on resistant soybean varieties in greenhouse and field experiments. *Journal of Nematology* 45, 322–323.

Tylka, G.L., Gebhart, G.D., Marett, C.C. and Mullaney, M.P. (2017) Evaluation of soybean varieties resistant to soybean cyst nematode in Iowa – 2017. *Iowa State University Extension and Outreach Publications* 99. http://lib.dr.iastate.edu/extension_pubs/99

Valocká, B., Sabová, M., and Lišková, M. (1994) Response of some winter wheat and spring barley cultivars to *Heterodera avenae* pathotype Ha 12. *Helminthologia* 31, 155–158.

van der Vossen, E.A., van der Voort, J.N., Kanyuka, K., Bendahmane, A., Sandbrink, H., Baulcombe, D.C., Bakker, J., Stiekema, W.J. and Klein-Lankhorst, R.M. (2000) Homologues of a single resistance-gene cluster in potato confer resistance to distinct pathogens: a virus and a nematode. *Plant Journal* 23, 567–576.

van Soest, L.J.M., Rumpenhorst, H.J. and Huijsman, C.A. (1983) Resistance to potato cyst-nematodes in tuber bearing *Solanum* species and its geographical distribution. *Euphytica* 32, 65–74.

Wang, D., Arelli, P.R., Shoemaker, R.C. and Diers, B.W. (2001) Loci underlying resistance to Race 3 of soybean cyst nematode in *Glycine soja* plant introduction 468916. *Theoretical and Applied Genetics* 103, 561–566. DOI: 10.1007/PL00002910

Wang, H.-M., Zhao, H.-H. and Chu, D. (2015) Genetic structure analysis of populations of the soybean cyst nematode, *Heterodera glycines*, from north China. *Nematology* 17, 591–600. DOI: 10.1163/15685411-00002893

Westphal, A. (2013) Vertical distribution of *Heterodera schachtii* under susceptible, resistant, or tolerant sugar beet cultivars. *Plant Disease* 97, 101–106.

Wheeler, T.A., Pierson, P.E., Young, C.E., Riedel, R.M., Willson, H.R., Eisley, J.B., Schmitthenner, A.F. and Lipps, P.E. (1997) Effect of soybean cyst nematode (*Heterodera glycines*) on yield of resistant and susceptible soybean cultivars grown in Ohio. *Journal of Nematology* 29, 703–709.

Williams, K.J., Lewis, J.G., Bogacki, P., Pallotta, M.A., Willsmore, K.L., Kuchel, H. and Wallwork, H. (2003) Mapping of a QTL contributing to cereal cyst nematode tolerance and resistance in wheat. *Australian Journal of Agricultural Research* 54, 731–737.

Winstead, N.N., Skotland, C.B. and Sasser, J.N. (1955) Soybean cyst nematode in North Carolina. *Plant Disease Reporter* 39, 9–11.

Wrather, A., Shannon, G., Balardin, R., Carregal, L., Escobar, R., Gupta, G.K., Ma, Z., Morel, W., Ploper, D. and Tenuta, A. (2010) Effect of diseases on soybean yield in the top eight producing countries in 2006. *Plant Health Progress* DOI: 10.1094/PHP-2010-0125-01-RS

Wu, L., Cui, L., Li, H., Sun, L., Gao, X., Qiu, D., Sun, Y., Wang, X. and Li, H. (2016) Characterization of resistance to the cereal cyst nematode in the soft white winter wheat 'Madsen'. *Plant Disease* 100, 679–685.

WVZ and VDZ (2016, 2015) Rübenanbaufläche in der EU 2015/16. Available at: www.zuckerverbaende.de/zuckermarkt/zahlen-und-fakten/eu-zuckermarkt/ruebenanbau.html (accessed 14 February 2018).

Wyss, U., Stender, C. and Lehmann, H. (1984) Ultrastructure of feeding sites of the cyst nematode *Heterodera schachtii* Schmidt in roots of susceptible and resistant *Raphanus sativus* I var *oleiformis* pers cultivars. *Physiological Plant Pathology* 25, 21–37. http://dx.doi.org/10.1016/0048-4059(84)90015-8

Yan, G.P. and Smiley, R.W. (2010) Distinguishing *Heterodera filipjevi* and *H. avenae* using polymerase chain reaction-restriction fragment length polymorphism and cyst morphology. *Phytopathology* 100, 216–224. DOI: 10.1094/PHYTO-100-3-0216

Yavuzaslanoğlu, E., Elekçioğlu, İ.H., Nicol, J.M. and Sheedy, J.G. (2016) Resistance of Iranian landrace wheat to the cereal cyst nematode, *Heterodera avenae*. *Australasian Plant Pathology* 45, 411–414.

Yu, M.H. and Steele, A.E. (1981) Host-parasite interaction of resistant sugarbeet and *Heterodera schachtii*. *Journal of Nematology* 13, 206–212.

Zheng, J., Li, Y. and Chen, S. (2006) Characterization of the virulence phenotypes of *Heterodera glycines* in Minnesota. *Journal of Nematology* 38, 383–390.

10 Plant Biotechnology Approaches: From Breeding to Genome Editing

Godelieve Gheysen[1] and Catherine J. Lilley[2]

[1]*Department of Biotechnology, Ghent University, Belgium;* [2]*Centre for Plant Sciences, University of Leeds, UK*

10.1	Introduction	215
10.2	PCR and Next Generation Sequencing	216
10.3	Molecular Research into Plant–Nematode Interactions: Using Molecular Biology to Understand the Parasite	217
10.4	Genetic Engineering: The Basic Principles	217
10.5	Breeding or Genetic Engineering with Natural Resistance Genes	218
10.6	Genetic Engineering Using Genes that Interfere with Nematode Feeding and Digestion	219
10.7	Secretion of Peptide Repellents from Root Tips Inhibits Nematode Invasion	220
10.8	RNAi Acting across Kingdom Borders	221
10.9	Susceptibility Genes as an Alternative Target for Nematode Resistance	222
10.10	Genome Editing Can Adapt Susceptibility Genes with Surgical Precision	224
10.11	Strategic Development of Resistance Technology	225
10.12	Conclusions and Future Prospects	229
10.13	References	230

10.1 Introduction

Molecular biotechnology and information technology and the combination of these (e.g. in bioinformatics) have dramatically changed scientific research and its applications in the past decades and hold the promise of continuing to bring novel insights and benefits to society. In this chapter we view biotechnology not only as genetic engineering, but also as molecular technology being used on living organisms whether it is for analysis or for modification. Indeed, DNA-based analysis has transformed the breeding process, whilst RNA and protein analysis have revolutionized our insight into living organisms, including cyst nematodes and their interactions with the host plants.

Genetic engineering has also helped to transform nematology research from a more descriptive science to one that uses experimental strategies that allow a better fundamental understanding of nematode biology and the plant–nematode interaction in particular. Much of this improved understanding is driven by the desire to develop new, effective and sustainable control strategies (Mantelin *et al.*, 2017). The applications of these recent advances are starting to be realized as nematode-resistant plants based on genetic engineering have reached the stage of field trials. When our ever-deepening insight into the molecular intricacies of the parasitic interaction is combined with the rapid advances in new genome editing technology, this raises exciting possibilities for future developments. We do not aim to provide a comprehensive review of all biotechnology approaches in relation to nematode control; instead this chapter focuses on the principles with some inspiring examples as illustration.

10.2 PCR and Next Generation Sequencing

PCR and next generation sequencing have turned difficult approaches into easy ones and made the impossible possible. Amplification of small amounts of DNA has been possible since the 1970s, but at that time it was a rather cumbersome process as the desired DNA needed to be inserted in a DNA vector to be transformed into bacteria (DNA cloning), after which a culture had to be grown from which the DNA could then be extracted. If the desired DNA was not already isolated in a pure form, the procedure involved more work as the resulting mass of bacteria had to be screened to find the one with the specific DNA of interest, a process that could take weeks or even months. Everything changed with the advent of the polymerase chain reaction (PCR) that allowed amplification of a specific piece of DNA from a small amount of a complex DNA mixture by simply adding reaction components and a polymerase enzyme and incubating the reaction mix through a cycling temperature programme for a couple of hours (Garibyan and Avashia, 2013). Similarly, DNA sequence analysis could initially only be done in laboratories that had experienced technicians, who were happy if they could generate 10 kb of sequence in a month. Currently, any piece of DNA can be sent for sequencing and the results are available the next day. Furthermore, massively parallel sequencing on dedicated machines now allows the generation of Gb of sequences per day (Goodwin *et al.*, 2016). Software has been developed to assemble a whole genome from smaller DNA sequences and to compare massive amounts of data.

What does this technical progress bring for plant nematology and in particular how can it assist efforts to improve nematode management and control? We refer to Waeyenberge (Chapter 17, this volume) and Blok *et al.* (Chapter 9, this volume) for more details but give here a brief summary.

Rapid and reliable identification of nematode species is a prerequisite for designing effective management strategies. However, the identification of nematodes down to the species level using morphological characteristics is time consuming and requires expert skills. Molecular identification provides an alternative, especially in cases where morphological characters may lead to ambiguous interpretation. Methods using PCR amplification of diagnostic regions (i.e. rapidly evolving regions of ribosomal DNA sequences and mitochondrial genes) are now available for easily distinguishing closely related species such as those within the genus *Heterodera* or between *Globodera rostochiensis* and *G. pallida*. High-throughput, simultaneous analysis of hundreds of samples is being facilitated by use of next generation sequencing techniques that also can allow the discrimination of discrete genotypes within a species. In this way, the distribution of different mitotypes of *G. pallida* with particular sequence variants in the *cytochrome B* gene was analysed both countrywide and within individual fields (Eves-van den Akker *et al.*, 2015a). Other work has focused on using new genomic technologies for rapid identification of markers for different potato cyst nematode (PCN) populations that may be linked to virulence characteristics and can therefore ultimately be used to assist in cultivar choice (Mimee *et al.*, 2015).

One of the most effective and environmentally friendly management strategies is the use of resistant crop varieties. Next generation sequencing has facilitated whole genome analysis to identify the genes underlying resistance. The identified DNA polymorphisms can then be used for tracking those genes in breeding populations. Furthermore, sequence comparison and functional analysis of different resistance genes and allelic variants have resulted in new insights coupled to exciting possibilities for engineering resistance. For example, the gene underlying the

Rhg4 resistance to *Heterodera glycines* in soybean has been compared with its alleles. This analysis revealed that only two nucleotide polymorphisms are responsible for the difference between resistance and susceptibility (Liu *et al.*, 2012). This opens new possibilities for rapid conversion of elite but susceptible soybean cultivars into resistant ones (see section 10.10).

10.3 Molecular Research into Plant–Nematode Interactions: Using Molecular Biology to Understand the Parasite

PCR in combination with DNA cloning has also facilitated detailed studies of nematode biology and parasitism-related genes. Individual genes important for nematode survival and parasitism could be isolated and studied. Paramount examples that have provided important information to assist the development of biotechnological control measures are the study of proteinase enzymes, neuropeptides and effectors from cyst nematodes covered in Chitwood and Masler (Chapter 5, this volume) and Jones and Mitchum (Chapter 4, this volume), respectively.

Being animals, plant-parasitic nematodes need to digest protein for their dietary needs and, correspondingly, proteinase activity and genes have been characterized in a range of species, with cyst nematodes the focus of much work (Lilley *et al.*, 1996, 1997; Urwin *et al.*, 1997a; Silva *et al.*, 2004; Prasad *et al.*, 2013). A cysteine proteinase was found to be the predominant protein degrading enzyme in feeding females of *G. pallida* (Koritsas and Atkinson, 1994) and this activity was subsequently localized to the intestine (Lilley *et al.*, 1996), corroborating the presumed role in digestion. Together with cloning of the corresponding genes (Urwin *et al.*, 1997a), this underpinned the rational design of appropriate proteinase inhibitors (PIs) that block nematode cysteine proteinase activity.

Plants produce a range of PIs that can act against pests in addition to carrying out endogenous roles. However, plants do not necessarily produce the most appropriate PIs against a certain pest. In addition, many PIs accumulate largely in seeds, whereas a high level of expression in roots is required for efficacy against plant-parasitic nematodes. Overexpression of a PI in the plant roots (see section 10.6) can therefore interfere with the nematode's digestion of their dietary protein intake and consequently with nematode growth and reproduction. With digestion of protein being a common requirement of nematodes, PI-based control could have efficacy against not just cyst nematodes, but a wide range of species, irrespective of their parasitic strategy. This would have particular utility in those field situations where a number of different nematode pests occur concurrently.

Extensive molecular studies have indicated that plant-parasitic nematodes secrete numerous proteins into their host during infection (Haegeman *et al.*, 2012). These proteins, called effectors, can have very diverse roles, such as facilitating migration of the nematode by degradation of plant cell walls. Some effectors have been shown to be involved in suppression of plant defences, while others can interact specifically with plant hormone pathways to stimulate the formation of nematode feeding sites. Functional analysis of these effectors has demonstrated that many of those proteins are essential for a successful infection, implying that any strategy (such as RNA interference (RNAi); see section 10.8) to interfere with the effectors may hamper plant parasitism.

10.4 Genetic Engineering: The Basic Principles

Genetic engineering is defined as the introduction (also called transformation) of a specific piece of DNA into an organism with the aim of changing its genetic characteristics. The first genetically engineered or modified organisms (GMOs) were bacteria in 1972, but a decade later the first description of a genetically engineered plant was published (Herrera-Estrella *et al.*, 1983). The application of genetic transformation to plants is mainly based on the use of the plant pathogen *Agrobacterium tumefaciens*. In nature this bacterium causes crown gall tumours, predominantly on dicotyledonous plants. The finding that this bacterium transfers a piece of its own DNA into the plant, where it is stably integrated in the plant DNA, opened the way for manipulation of this natural transformation system. For plants that are not susceptible to *Agrobacterium* infection, direct gene transfer methods were established. Later it was found that many monocotyledonous plants could also be transformed by *Agrobacterium* with appropriate modifications of the transformation protocol. Nowadays *Agrobacterium*-mediated transformation is the method of choice for transforming a wide range of plants because of its efficiency, simplicity and low cost (Gheysen *et al.*, 1998).

The use of *Agrobacterium* as a transformation vector is based on several important findings: *Agrobacterium* transfers part of its Ti-plasmid, namely the T-DNA, to the plant nucleus where it is integrated into the plant DNA by plant repair enzymes. The only T-DNA sequences essential for T-DNA transfer are the T-DNA border sequences: a 24-bp direct (non-identical) repeat at the T-DNA ends. The genes that are located on the T-DNA are necessary for tumour formation (via elevated biosynthesis of auxins and cytokinins) but they are not needed for transfer of the T-DNA. Therefore these tumour-inducing genes can be removed and replaced by any other gene that is of interest to enhance the plant's agronomical or food qualities; for example, a gene that confers cyst nematode resistance. The plant cells that have received the gene of interest are then regenerated into shoots that contain the new DNA as an integral part of their genome and the progeny plants inherit the new DNA according to the laws of Mendel.

Plant genetic engineering can basically be used for similar purposes as breeding; namely, to introduce genes for desirable crop characteristics with the following main differences:

1. The genes do not have to be from species with which the plant can be crossed (cisgenesis), but can also be from other plants or completely different organisms such as bacteria (transgenesis).
2. It is known exactly which genes are introduced, whilst during breeding hundreds of unknown genes 'hitch-hike' with the gene of interest.
3. The process is much faster.
4. The original cultivar characteristics can be retained.
5. Specific mutations can be introduced (genome editing).
6. The resulting plants need to be thoroughly tested before they can get an authorization for commercialization.

Specific examples of how transgenesis has been used in the context of nematode control are given in section 10.6.

10.5 Breeding or Genetic Engineering with Natural Resistance Genes

Many sources of resistance to cyst nematodes have been identified and several of the corresponding genes have been cloned (See Goverse and Smant, Chapter 8, this volume). The availability of a resistant plant allows introgression of the resistance by breeding into a crossable crop plant. For example, the resistance gene *Gpa2* from *Solanum tuberosum* ssp. *andigena* that confers resistance to *G. pallida* has been introgressed into potato cultivars. Even genes from different but related *Solanum* species, such as the *Gro1* gene (resistance to *G. rostochiensis*) from *S. spegazzinii*, have been introgressed into potato. The drawback is that introduction of these genes often necessitates a long breeding process to segregate out 'wild' plant genes (e.g. those involved in producing toxins or compromising desirable agronomic qualities) that are linked to the resistance. Development of resistant crops can potentially be accelerated by using genetic modification (GM) techniques to introduce an isolated *R* gene to a favoured crop variety. Cisgenesis is a genetic engineering concept that allows for the rapid transfer of resistance genes from wild relatives into domesticated crops avoiding co-transfer of genes with undesirable effects (linkage drag) and conserving the characteristics of the commercial variety. This could avoid the yield penalties such as that incurred upon the transfer of *H. schachtii* resistance from *Beta procumbens* to cultivated sugar beet (Panella and Lewellen, 2007). Cisgenesis is very similar to breeding in that in both cases only genes (with their own regulatory sequences) from the sexually compatible gene pool of the recipient plant are introduced.

Some interesting *R* genes, such as the broad-spectrum resistance *Hero* from the wild tomato *S. pimpinellifolium* that confers resistance to potato cyst nematode, cannot be introduced into the desired crop plant by breeding. As the *Hero* gene has been cloned, it can be introduced into potato by genetic engineering.

This process is called transgenesis as tomato cannot be crossed with potato. However, attempts to introduce nematode *R* genes directly into less closely related crops have so far met with variable success, probably due to differences in downstream signalling. Indeed, although direct transfer of *Hero A* into a susceptible tomato cultivar conferred comparable levels of resistance to *Globodera* species to that seen in an introgressed *Hero* line, no significant resistance was achieved in transgenic potato expressing the same construct (Sobczak *et al.*, 2005). This line of research is not without possibilities though. Some *R* genes effective against root-knot nematode species have been successfully introduced by transgenesis into more distantly related species. Expression of the *CaMi* gene from pepper (*Capsicum annuum*) in susceptible tomato cultivars conferred resistance to the root-knot nematode *Meloidogyne incognita* (Chen *et al.*, 2007), and the tomato *Mi1* gene has been used to generate transgenic lines of lettuce (*Lactuca sativa*) resistant to the same nematode (Zhang *et al.*, 2010).

The identification, cloning and direct transfer of new nematode *R* genes could be revolutionized by exploiting the latest developments in next generation sequencing. Finding *R* gene candidates has up to now been a slow, expensive process but a recent refinement of the *R* gene sequence capture (RenSeq) technique combined with single-molecule real-time (SMRT) sequencing (SMRT RenSeq) allowed the rapid cloning of a broad-spectrum resistance gene against *Phytophthora infestans* from the non-tuber forming *S. americanum* (Witek *et al.*, 2016). A subset of DNA sequences with similarity to typical *R* genes was selectively captured from a previously unsequenced plant accession using a method that selected for long DNA molecules. These DNA molecules were then sequenced using the novel long-read SMRT technology. The identified *Rpi-amr3i* gene provides an example of successful *R* gene transgenesis by conferring resistance to *P. infestans* when introduced into potato. Rapid cloning of multiple *R* genes from uncharacterized germplasm using this technology could provide candidates for engineering crops resistant to a range of pathogens, including cyst nematodes.

Typical *R* genes encode nucleotide-binding, leucine-rich repeat (NLR) proteins, which directly or indirectly recognize pathogen effectors, then activate appropriate defence mechanisms. For these genes to function correctly in a transgenic context requires all downstream components of the response cascade to be present in the susceptible recipient plant. A novel resistance mechanism in soybean, towards the cyst nematode *H. glycines*, may be more amenable to cross-species transfer. It is associated with the *Rhg1* quantitative trait resistance locus and involves copy number variation of a section of DNA harbouring three separate genes encoding an amino acid transporter, an α-SNAP protein and a WI12 (wound-inducible domain) protein (Cook *et al.*, 2012). One copy of the genes per haploid genome is present in susceptible soybean varieties but ten tandem copies, associated with higher expression, are found in the *rhg1-b* resistant haplotype. Overexpression of all three genes was necessary to recapitulate *Rhg1* resistance in a susceptible cultivar, posing the intriguing possibility that similar transgenic overexpression may provide a level of resistance in other host–cyst nematode interactions. More recently, amino acid polymorphisms in one of the three genes at the *rhg1* locus have been implicated in resistance to *H. glycines* (Cook *et al.*, 2014). The α-SNAP protein, with a role in vesicle-mediated transport, displayed amino acid polymorphisms likely to affect activity when present in resistant cultivars of soybean that possess only three tandem repeats of the *rhg1* locus. This suggests another possibility for engineering cyst nematode resistance, either by transgenic expression of the novel alleles or genome editing of endogenous genes.

10.6 Genetic Engineering Using Genes that Interfere with Nematode Feeding and Digestion

Natural nematode resistance genes have the limitation that they are typically effective against only one or a limited range of species – or even only against specific pathotypes – whilst crops are often exposed to several parasitic nematodes. Therefore, it is worthwhile to explore other strategies for nematode resistance such as approaches that interfere with the digestive system of plant-parasitic nematodes.

Plant PIs can be found that bind to nematode proteinases and inhibit their activity and therefore hamper nematode food digestion. For example, a cystatin from rice (oryzacystatin) is a good inhibitor of the major cysteine proteinases of cyst nematodes. Mutant cystatins were tested for improved inhibition and this identified variant OcIΔD86 as having the best activity (Urwin et al., 1995). The cystatin has proven effective in controlling a range of plant-parasitic nematodes, including the cyst nematodes G. pallida (Urwin et al., 1995) and H. schachtii (Urwin et al., 1997b). Its efficacy was initially tested in potato hairy root cultures or Arabidopsis, but subsequent work led to field trials of transgenic potatoes (see section 10.11.1). To have sufficient production of this PI in potato roots, the cystatin gene was fused to either the constitutive plant viral promoter CaMV35S (Urwin et al., 2003) or to a root-specific promoter with upregulated expression in the nematode feeding site (Lilley et al., 2004). These constructs were used for transformation of susceptible potato 'Desiree' and the best transgenic lines were shown to have commercially useful resistance to G. pallida (Urwin et al., 2003).

Serine proteinase inhibitors have been less developed for biotechnology approaches, but also could have potential for cyst nematode control. Transgenic expression of the sweet potato serine PI sporamin, in sugar beet hairy roots, inhibited growth and development of female H. schachtii (Cai et al., 2003).

Insect-resistant Bt crops expressing crystal proteins (Cry) from Bacillus thuringiensis are grown on over 80 million ha worldwide and have dramatically reduced the need for chemical insecticides (Brookes and Barfoot, 2016). Bt formulations are also being used as insecticidal sprays in organic farming. The success of Bt proteins is that they are orally toxic to a specific class of insects; for example, Cry1Aa is toxic to caterpillars and Cry3A is toxic to beetles (Schnepf et al., 1998). Beneficial insects that do not feed on the plant (such as ladybirds) or that belong to a different class (such as bees) are not affected by Bt transgenic plants. The potential of Bt toxins to control plant-parasitic nematodes was investigated using Caenorhabditis elegans as a model nematode and this revealed Cry5B and Cry6A to be nematotoxic (Marroquin et al., 2000). Both have been expressed in tomato hairy roots and this resulted in reduced reproduction of M. incognita (Li et al., 2007, 2008).

Recently more Cry proteins have demonstrated efficacy against root-knot nematodes, but information on the toxicity of Bt to cyst nematodes is lacking. The size of Bt toxins may potentially be a major constraint to their efficacy against cyst nematodes. Meloidogyne incognita can ingest the 54-kDa Cry6A protein and the efficacy of the 79-kDa Cry5B suggests that the larger protein can also be ingested from hairy roots by this nematode. However, H. schachtii appears not to ingest proteins of 23 kDa and larger, most likely due to the different structure of its feeding tube compared with root-knot nematodes. The feeding tubes of Heterodera and Globodera species appear to have very similar structures (Eves-van den Akker et al., 2014a, 2015b), so this size constraint may also exist for other cyst nematodes.

This concern led scientists at Monsanto specifically to seek smaller toxins or truncated Cry proteins that retained activity in assays against C. elegans. Bt proteins delivered from transgenic soybean plants were more effective against H. glycines when truncated at the C- and/or N-terminus. These smaller versions of Cry proteins typically ranged from 14–30 kDa, and so their improved bioactivity was likely due to their more efficient ingestion (Bowen et al., 2016).

10.7 Secretion of Peptide Repellents from Root Tips Inhibits Nematode Invasion

The first steps in nematode infection include attraction to and invasion of the plant roots. Blocking this step not only prevents nematode reproduction but also avoids the root damage associated with the destructive migration of cyst nematodes. This damage can reduce the rate of root growth and decrease rates of uptake for water and nutrients (Trudgill, 1991), probably contributing to reduced tolerance even amongst some resistant potato cultivars. Screening of a library of peptides has identified two peptides that bind to either acetylcholinesterase or nicotinic acetylcholine receptors (nAChRs) in the cyst nematode nervous system and disrupt chemosensory-mediated behaviour of cyst nematodes at very low concentrations (Winter et al., 2002). The route of uptake of the peptide by the nematodes has been explored using fluorescent-labelling studies. It was found to be transported efficiently from the exposed sensilla in the amphids along the chemosensory neurons of second-stage juveniles (J2) of *G. pallida* and *H. schachtii* to the neuronal cell bodies and beyond the nerve ring to a limited number of interneurons where it is presumed to exert its effect (Wang et al., 2011). Transgenic plants were subsequently developed that secrete the peptides from their roots (Liu et al., 2005; Lilley et al., 2011; Green et al., 2012). The suppressive effect of these peptides therefore probably occurs in the soil, at or near the rhizoplane, prior to invasion. They may also be taken up by nematodes following root penetration and thus impair the orientation within the root that is essential for selection of an appropriate feeding site location.

The acetylcholinesterase-inhibiting peptide suppressed the number of female *H. schachtii* that developed on Arabidopsis by more than 80%, whilst expression in the root tips of potato plants driven by the Arabidopsis *MDK4-20* promoter resulted in almost 95% resistance to *G. pallida* (Lilley et al., 2011). Potato plants secreting the nAChR-binding peptide from their root tips similarly provided effective resistance against potato cyst nematode (Green et al., 2012).

The cyst nematode nervous system is an attractive target for engineered resistance strategies as it is of proven value through the known mode of action of a number of effective chemical nematicides. Cholinergic signalling is one component of the nematode neurotransmitter repertoire. The biogenic amine serotonin (5-HT) has a known role in mediating stylet activity (Masler, 2007) that is required at many stages of the parasitic process so this signalling pathway could be an alternative target. Cyst nematodes also utilize an abundance of small neuropeptides, belonging to distinct families, which modulate neuronal responses and behaviour in concert with classical neurotransmitters (for reviews, see e.g. Holden-Dye and Walker, 2011; Mousley et al., 2013).

10.8 RNAi Acting across Kingdom Borders

All previous examples illustrate the use of genetic engineering to express one or more additional proteins to achieve nematode resistance in plants. However, genetic engineering can also be used to inactivate genes and thus prevent the production of specific proteins, either in the plant or in the parasitic nematode, by RNA silencing or RNAi. RNAi is triggered by double-stranded (ds) RNA and results in the rapid degradation of that dsRNA into small interfering RNAs (siRNAs). These siRNAs are recruited by a protein complex called RISC that will then bind to all single stranded RNAs (mainly mRNAs) that have sufficient sequence identity with the siRNA. Finally, depending on several factors, this binding results in the cleavage and subsequent degradation of the RNA or in the inhibition of mRNA translation into protein.

RNAi in plants has evolved as a defence mechanism against viruses. Many plant viruses generate dsRNAs in their replication cycle and these are targets for RNAi-mediated destruction. As the RNAi mechanism is also active in many other organisms, such as fungi, insects and nematodes, the question was if dsRNA or siRNA could be delivered from the host plant into these pathogens and pests to control them. In 2007, Baum et al. reported the efficacy of expressing dsRNA in corn to control the corn rootworm, a beetle that damages corn roots. Already in 2002, Urwin et al. had demonstrated that soaking of infective J2 of *H. glycines* or *G. pallida* in dsRNA for specific genes resulted in significantly lower mRNA levels of those genes. These results evoked the opportunity of engineering nematode resistance by the production of dsRNA in the plant to target essential nematode genes. The feeding nematode is then continuously exposed to the dsRNA or the siRNAs by ingestion of the plant cell cytoplasm. This technology, now often referred to as host-induced gene silencing (HIGS), can be effective against a range of insect pests, bacterial and fungal plant pathogens (Koch and Kogel, 2014). In practice, a plant that produces dsRNA to a nematode gene can be generated by transformation of a so-called hairpin construct. This type of construct contains a plant promoter driving the transcription of a nematode gene fragment present twice, once in sense and once in antisense orientation, and separated by an intron or spacer region (Fig. 10.1). The transcribed RNA forms a self-complementary hairpin structure with either the spacer region forming a loop or, more commonly, the intron sequence being

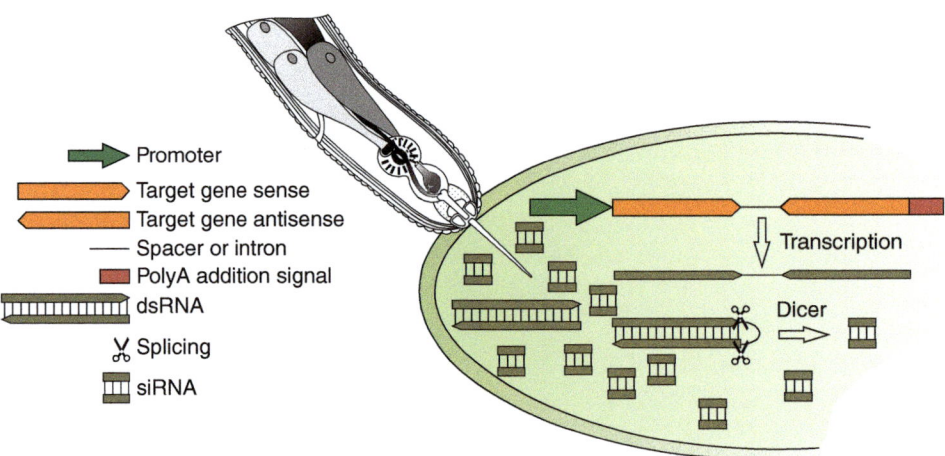

Fig. 10.1. Schematic diagram describing the process of host-induced gene silencing (HIGS) for cyst nematode control. A transcribed region of the target nematode gene is cloned in both forward and reverse orientations under the control of a promoter active in the nematode feeding site. A spacer sequence or intron separates the two nematode sequences. After the RNA is transcribed, the intron is removed by splicing or the spacer region forms a hairpin loop and the complementary strands representing the nematode sequence anneal and form a double-stranded (ds)RNA molecule. The dsRNA may be directly ingested by the feeding nematode or it may be processed within the syncytium to short interfering (si) RNAs that are similarly ingested to trigger RNAi gene silencing.

removed by splicing to leave a linear dsRNA molecule. It is not clear if the target transcript suppression observed arises from ingestion of this dsRNA that is subsequently processed by the nematode or from uptake of plant-derived siRNAs. Both molecules have been detected in the delivering host plant. siRNAs, for example, have been produced from constructs in Arabidopsis (Huang *et al.*, 2006; Sindhu *et al.*, 2009), potato (Dinh *et al.*, 2014), tobacco (Papolu *et al.*, 2013) and soybean (Li *et al.*, 2010a), whilst intact dsRNA has also been detected in some cases (Sindhu *et al.*, 2009; Papolu *et al.*, 2013).

Since the first reports of *in planta* RNAi against *Meloidogyne* in 2006 (Huang *et al.*, 2006; Yadav *et al.*, 2006), numerous other papers have followed, with root-knot nematodes dominating the list (Dutta *et al.*, 2015), but only a few successful experiments have been reported with cyst nematode species. Arabidopsis plants expressing dsRNA against a range of *H. schachtii* parasitism genes were significantly less susceptible, but the difference compared to the control plants (only a 23–64% reduction in females) was disappointingly lower than in the original publications on *Meloidogyne* (Patel *et al.*, 2008; Sindhu *et al.*, 2009; Kyndt *et al.*, 2013). The efficacy of RNAi against *H. glycines* in soybean varies between constructs (Steeves *et al.*, 2006; Klink *et al.*, 2009, Li *et al.*, 2010a, b; Youssef *et al.*, 2013). On the positive side, up to 93% fewer females of *H. glycines* were observed to develop on soybean roots expressing dsRNA derived from a region of the nematode synaptobrevin, a house-keeping gene involved in vesicle fusion (Klink *et al.*, 2009). Eves-van den Akker *et al.* (2014b) on the other hand used RNAi in potato to target an effector gene family of *G. pallida*. Exploiting a conserved region for the otherwise hyper-variable apoplastic effector gene family, the RNAi constructs in potato roots resulted in a 50–60% reduction in nematode numbers at 2 weeks after infection.

10.9 Susceptibility Genes as an Alternative Target for Nematode Resistance

A compatible interaction between a plant and a pathogen occurs if no resistance is present to ward off the attacker. However, a successful infection also requires the activity of specific plant genes that enable the pathogen to complete its infection cycle. These genes have been termed susceptibility (S) genes, and allelic variants or mutations can render a plant (more) resistant due to loss of susceptibility. The prototype example of an inactivated susceptibility gene is the homozygous *mlo* mutation that confers powdery mildew resistance in barley (Jørgensen, 1992). The *mlo* gene is a negative regulator of plant defence (Acevedo-Garcia *et al.*, 2014) and its mutation confers a broad-spectrum durable protection against powdery mildew that has been used since mid last century in barley breeding. The study of effector targets is currently enabling the identification of many more susceptibility genes as illustrated by the research on *Phytophthora infestans* (Boevink *et al.*, 2016).

Cyst nematodes are obligate biotrophic parasites – they completely depend on the suitability of the plant host and its receptive reaction to be able to penetrate the root and establish a metabolically active syncytium as a feeding site (see Chapter 1, this volume). The cyst nematode takes advantage of existing plant cell differentiation pathways to transform normal root cells into a syncytium and at the same time suppresses the plant defence. The complex mechanism of syncytium initiation and development remains largely enigmatic, but there are already some examples of nematode effectors that target plant proteins to facilitate this process. Obviously plant susceptibility genes must play a major role in the different steps of the nematode infection process.

Warmerdam *et al.* (2014) have taken a genomic approach to identify *Arabidopsis thaliana* genes involved in susceptibility to *M. incognita*. A large-scale infection experiment was done to compare the susceptibility of 364 different ecotypes of *A. thaliana*. The most and the least infected ecotypes differed nine-fold in their susceptibility to *M. incognita*. As the whole genome sequence is known for all these ecotypes, a genome wide association analysis could be performed to determine the loci underlying the differential susceptibility. This analysis looks for nucleotide polymorphisms (allelic variations) that can be linked with differences in susceptibility. Genes identified in this way were further studied by the infection of mutants in those genes to confirm their contribution to susceptibility. Genes involved in mitosis and endoreduplication, which are known to be important in feeding cell formation (de Almeida-Engler *et al.*, 1999; de Almeida-Engler and Gheysen, 2013) were identified in this analysis. Nevertheless, the presence of alleles that confer reduced susceptibility to nematodes in different ecotypes suggests that these allelic variants are not necessarily linked to a generally lower fitness of the plant, and that they may therefore have applications in plant breeding.

Although this analysis has been done for root-knot nematode susceptibility, similar results would be expected for cyst nematodes. Indeed, recently ArslanAnwer *et al.* (2018) described the discovery of a QTL on chromosome 4 influencing susceptibility to cyst nematode infection in Arabidopsis. Several candidate genes were identified in that region and further analyzed. *AtS40-3*, encoding a senescence-associated transcription factor, showed much lower expression in poorly susceptible Arabidopsis ecotypes than in highly susceptible ones. Furthermore, mutating the gene in a susceptible ecotype resulted in a significant decrease in the number of females and a corresponding increase in the number of males. The authors hypothesize that *AtS40-3* delays senescence of the syncytium allowing females to develop.

From molecular and cellular analysis of cyst nematode infection, several other plant genes are known to be important. Besides genes involved in the cell cycle (de Almeida-Engler and Gheysen, 2013) and auxin transport (Grunewald *et al.*, 2009; Lee *et al.*, 2011), the transcription factor AtWRKY23 is required for efficient cyst nematode infection of *A. thaliana* (Grunewald *et al.*, 2008). Another cellular change during syncytium development is the reorganization of the cell wall, which involves many different enzymes, and some of these have been identified as targets of nematode effectors. Hewezi *et al.* (2008) identified a methylesterase protein 3 (PME3) interacting with a cellulose-binding protein secreted by *H. schachtii*. Plants with higher expression of this *PME3* were more susceptible, lower *PME3* transcript resulted in lower susceptibility to this cyst nematode.

A different approach to identification of cyst nematode susceptibility genes that could be manipulated to provide resistance has been taken by Matthews et al. (2013). In this case, microarray data for both compatible and incompatible interactions between soybean and H. glycines were mined to identify 100 candidate genes for overexpression in a high-throughput soybean composite plant system. Nine of the overexpressed genes reduced the number of mature females by more than 50% with an ascorbate peroxidase providing the highest level of resistance at 74%. Overexpression of other genes, notably an oligopeptide transporter and UDP-glucuronate 4-epimerase enhanced susceptibility more than two-fold. Functional genomics approaches such as this can provide leads to candidate plant genes for either overexpression or RNAi/silencing approaches.

10.10 Genome Editing Can Adapt Susceptibility Genes with Surgical Precision

Plants obviously do not have susceptibility genes to benefit nematodes or other pathogens; these genes have other functions in the plant. Cell cycle genes and cell wall modifying genes evidently play important roles in a healthy plant. Mutation of those genes may lead to a less susceptible but probably also a less fit plant. Each susceptibility gene and its alleles need to be evaluated on a case-by-case basis.

Nonetheless, recent developments that allow very precise changes to be made in the genome of an organism have resulted in remarkable successes in engineering disease resistance in plants. The first examples were based on the use of TALEN to engineer powdery mildew resistance in wheat (Wang et al., 2014) and bacterial blight resistance caused by *Xanthomonas oryzae* in rice (Li et al., 2012). TALEN is a combination of a protein that binds a specific DNA sequence (TAL = transcription activator like, an effector of *X. oryzae*) and an enzyme that cuts DNA (EN = endonuclease). The engineered TALEN can be designed to recognize and cut any target sequence of choice. If a TALEN is expressed in a plant cell, it will cut the chosen genomic sequence, after which the plant DNA repair system tries to undo the damage, but nearly always introduces some small errors during this process. The result is a slightly changed or mutant sequence. In wheat this process was used to mutate the three *MLO* alleles into the recessive *mlo* alleles (Wang et al., 2014). The process can even be turned into real editing, changing a DNA sequence into a precisely defined variant by adding the desired end sequence as a template for repair.

Remarkably, this genome editing can be used to change susceptibility genes in a way that the plant becomes resistant without the drawback of a lower fitness as illustrated by Li et al. (2012). During infection of rice *X. oryzae* activates susceptibility genes such as the sucrose transporter or SWEET gene *Os11N3*. This is done by a TAL effector that binds to a specific sequence in the promoter of *Os11N3*. Using genome editing, the recognition site of the *Xanthomonas* effector was mutated, leaving the rest of the gene intact. This edited gene is still active during plant development but is no longer activated by the pathogen *X. oryzae* (Li et al., 2012), rendering the plant less susceptible to the disease.

Several alternative genome editing techniques have been developed recently with CRISPR-Cas9 being the most promising due to its simplicity, efficiency and relatively low cost. The first description of CRISPR was in 2012 (Jinek et al., 2012) but by 2014 there were already over 500 publications using CRISPR in a variety of organisms. The technique has now been demonstrated to work in a number of crop plants including soybean (e.g. Li et al., 2015), potato (Wang et al., 2015) and wheat (e.g. Shan et al., 2014) that are hosts to major cyst nematode species. CRISPR has been called the biggest game changer to hit biology since PCR (Ledford, 2015). The first examples of the utility of CRISPR for engineering pathogen resistance in plants are beginning to emerge. Targeted mutagenesis of an ethylene response factor in rice enhanced resistance against the rice blast pathogen, *Magnaporthe oryzae* (Wang et al., 2016). Tomato plants have been engineered with frameshift deletions in the *DMR6* gene, which is associated with salicylic acid homeostasis. The resulting plants accumulated higher levels of salicylic acid and showed broad-spectrum resistance against the bacterial pathogens *Xanthomonas gardneri*, *X. perforans* and *Pseudomonas syringae* and the oomycete *Phytophthora capsici* (Thomazella et al., 2016). Importantly, this resistance phenotype was not associated with any significant deleterious effect on plant growth and development.

It is without doubt that a combination of better understanding of nematode susceptibility and resistance genes in plants together with CRISPR-gene editing will allow strategies to render plants more resistant to nematodes to be designed and produced.

10.11 Strategic Development of Resistance Technology

So far this chapter has described the theory behind the various biotechnology approaches that are being advanced for cyst nematodes. How are those methods developing towards utility in the field and what are the factors that underpin their deployment and uptake? Much of the initial research and proof-of-concept demonstrations of engineered cyst nematode resistance have exploited model systems such as Arabidopsis, hairy root cultures or composite soybean plants with transgenic root systems. These more rapid, higher-throughput experiments are invaluable for providing early indications of efficacy that can then be followed up in stably transformed crop host plants. Many instances of increased resistance have been reported in these model systems, using the approaches described earlier. Examples of *in planta* RNAi to target nematode genes with effects on nematode development and reproductive success continue to accumulate, both for cyst and other plant-parasitic nematodes. Similarly, there are numerous cases where alteration of host gene expression or use of specific Arabidopsis mutants reduces susceptibility of this plant to *H. schachtii*. However, it is proving a considerable challenge to translate these results to useful levels of resistance in the field (see section 10.11.4).

Although Arabidopsis is able to support development of the cyst nematode *H. schachtii* it is unlikely to be a natural host. Small, and frequently variable, numbers of nematodes complete development on each plant in the assay systems typically used. Transgenic hairy root cultures can suffer from similar problems with inherent variability of nematode infection (Plovie *et al.*, 2003). Therefore, an important first step following promising results in these systems is to test the resistance strategy in a host crop of interest under containment glasshouse conditions. If the resistance is replicated and there are no deleterious effects on the plant under these more natural growth conditions, then field testing of the technology should be the ultimate goal.

10.11.1 Field trials of GM nematode-resistant crops

Despite almost two decades of research, only a few of the nematode resistance technologies described earlier have progressed to field trials as yet and none has reached the stage of commercialization. Field trials involving cyst nematodes have centred on genetically modified potato plants and their resistance to potato cyst nematodes. Engineered resistance based on cysteine proteinase inhibitors has been trialled successfully in potato plants on a number of occasions, confirming the efficacy of different cystatins. In the first field trial, constitutive expression of chicken egg white cystatin in the otherwise susceptible potato 'Desiree' conferred up to 70% resistance to a natural field population of potato cyst nematode that was dominated by *G. pallida* (Urwin *et al.*, 2001). When the same cystatin defence was subsequently stacked with the natural partial resistance of the two potato cultivars 'Sante' and 'Maria Huanca', the best transgenic lines of each were enhanced to full resistance (Urwin *et al.*, 2003). Future acceptability of transgenic traits was a consideration in later field trials, which utilized transgenes from widely consumed plant species. Both the modified rice cystatin (OcIΔD86) described earlier and a sunflower cystatin could provide levels of nematode resistance similar to that from chicken egg white when expressed in 'Desiree' potato plants (Urwin *et al.*, 2003). Enhanced biosafety was the driver behind a subsequent trial of plants in which production of the OcIΔD86 cystatin was limited mainly to the roots and, in particular, to the nematode feeding cells by use of promoters with restricted spatial expression (Fig. 10.2B). The levels of resistance displayed by these potato plants were similar to those achieved with constitutive expression. The Arabidopsis *ARSK1* promoter proved most effective in delivering resistance against potato cyst nematode (Lilley *et al.*, 2004).

The potential of the secreted peptide repellent technology to control PCN has also been assessed in field trials of transgenic potato plants

Fig. 10.2. Field trials of GM potato, Yorkshire, UK. (A) A small-scale field trial of GM potatoes expressing the secreted peptide nAChRbp, which reduced root invasion and establishment of the potato cyst nematode, *Globodera pallida*, providing up to 77% resistance in this trial. (B) The Arabidopsis *ARSK1* promoter is active in potato roots to drive localized gene expression in the syncytium of *G. pallida*. (C) The Arabidopsis *MDK4-20* promoter provides specific expression in the root tips of potato and was used to deliver the nAChRbp-repellent peptide to the rhizoplane.

(Fig. 10.2A). Three lines of potato that expressed a disulphide constrained 7-mer active peptide termed nAChRbp, under control of the *AtMDK4-20* root-cap-specific promoter (Fig. 10.2C), were first validated in a glasshouse trial before being advanced to field trial. The best performing line displayed 77% resistance to potato cyst nematode in the field (Green *et al.*, 2012). It was notable that the glasshouse trial was a good predictor of field performance, as had previously been reported for the cystatin technology. The resistance of each line in the field was not significantly different from the level it displayed in the glasshouse. Nevertheless, the inherent variation in nematode density across the site in field conditions reduced the significance of the resistance effect.

The only field tests of these two technologies against cyst nematodes have involved PCN; however, the efficacy of both in providing nematode resistance has also been proven in confined field trials of transgenic plantain in Uganda (Tripathi *et al.*, 2015). Plants expressing a maize cystatin, the nAChRbp-repellent peptide, or both defences in combination exhibited a level of resistance to both *Radopholus similis* and *Helicotylenchus multicinctus*. The highest resistance, of around 99% at harvest, was provided by a peptide-expressing transgenic line. Importantly, this line also showed improved agronomic performance relative to wild type plants under the same nematode challenge, with significantly greater yield due to larger bunches and reduced loss of plants from toppling. However, plant vigour and yield were not enhanced for all transgenic lines with increased nematode resistance, with some showing reduced performance. This highlights the importance of field trials that can assess the agronomic performance of transgenic crop plants subjected to typical environmental conditions. Not all deleterious phenotypes may be apparent under the optimum conditions normally used for glasshouse trials.

To our knowledge, the only other published report of field evaluation for GM nematode-resistant plants involved soybeans engineered to overexpress an endogenous salicylic acid methyl transferase (*GmSAMT1*) (Lin *et al.*, 2016). Transgene expression correlated with a reduction in the number of female *H. glycines* developing on the plants in a growth chamber assay, although the precise mechanism accounting for the observed resistance was unclear. A subsequent 1-year field test in Tennessee, USA, found no significant differences for a range of agronomic traits between four transgenic soybean lines and non-transgenic control plants. However, there was a negligible level of soybean cyst nematode present at the trial site and resistance of the plants was not measured.

Globally, nematode resistance traits account for only a very small proportion of the documented GM confined field trials (CFTs). A comprehensive survey in 2014 that gathered information for almost 41,000 CFTs conducted or applied for across 55 countries in the period 2009–2013 reported nematode resistance as a trait in only 1.3% of all CFTs (Rudelsheim and Smets, 2014). The relevant crops were potato in Europe, cotton, soybean, ornamentals and plum trees in the USA and banana in Uganda.

10.11.2 Biosafety considerations

An additional important role of field trials is to allow relevant biosafety assessments of the GM plants, especially in relation to any potential adverse effects on the environment or on non-target organisms. One concern for nematode resistance technologies is that, through their mode of action, they could conceivably impinge upon shared biological processes of non-plant-parasitic nematodes in the soil. It is therefore essential that biosafety of any transgenic approach is considered and only fully biosafe GM plants for nematode resistance are developed beyond the proof-of-concept stage.

Environmental biosafety aspects of the cystatin proteinase inhibitor technology were investigated in a range of studies over the course of multiple field trials. The transgenic potato plants did not harm insect aerial feeders such as aphids (Cowgill et al., 2002a) and leafhoppers (Cowgill and Atkinson, 2003) or their natural parasitoid associates (Cowgill et al., 2004). This was in stark contrast to the significant reductions in these insects observed when a standard nematicide treatment was used on the wild type plants (Cowgill et al., 2002a; Cowgill et al., 2004). Similarly, no deleterious effects were detected on the soil microbial community (Cowgill et al., 2002b). These studies highlight the importance of comparing the impact of GM plants on non-target organisms with that of the current management practices they could replace.

Soil nematodes are not only potential unintended targets of any nematode resistance technology. As a consequence of their predictable response to land management changes that reflect changes in soil microenvironments (Ferris et al., 2012) they also provide a reliable and rapid means of defining the wider environmental impact of GM plants on the soil environment. Their fluctuating numbers and feeding types in response to changes in the microbes they consume, the soil environment and disturbance make them organisms of choice as biological indicators of soil health. Nematodes participate at many different levels of the soil food web and their faunal analysis allows the enrichment and structure of the food web to be inferred, based on those genera that respond rapidly to environmental change (enrichment index) and those that prefer undisturbed habitats (structural index, Ferris et al., 2001). Neither the enrichment nor the structural indices of the soil around transgenic potato plants expressing two different nematode resistance technologies (cystatin and/or repellent peptide) changed more over the course of a field trial than for non-transgenic plants (Green et al., 2012). This suggested that there was no detrimental effect on the soil nematodes or the soil food web from growing the GM crop. Additional tests have been undertaken for the nAChRbp-repellent peptide, which was shown to have no acute toxicity against a range of standard test organisms selected to represent the main groups of organisms present in soil water, including macrofauna, meiofauna and microfauna (Wang, 2009).

Host-delivered RNAi as a strategy for nematode resistance is seemingly an attractive option from a biosafety point of view, as there is no necessity to produce a foreign protein in the GM plant (Atkinson et al., 2012). However, deployment of RNAi-based transgenic plants could potentially have off-target effects if there is sufficient homology between the produced siRNAs and either host plant sequences or genes in other, non-target organisms. Both of these considerations could be compounded if either dsRNA or small RNA molecules are shown to have prolonged environmental persistence, although the evidence to date suggests that dsRNA is degraded rapidly in soil (Dubelman et al., 2014). Bioinformatic analyses will be key in identifying any likely risks; however, there are still uncertainties surrounding, for example, the level of sequence mismatch between siRNAs and non-target genes that still allows silencing (Ramon et al., 2014). Prediction of unintended effects on some host crops and non-target organisms is currently hampered by a lack of complete genome data. Careful selection of nematode target genes can go some way to overcoming this limitation; for example, many effector genes are novel and lack known sequence homologues in other genera. A bioinformatics pipeline developed to deliver a short-list of novel RNAi targets in root-knot nematodes, with the main aim being to reduce the likelihood of non-target effects in future control strategies, contained primarily effector-like genes (Danchin et al., 2013). The more recent availability of comprehensive genome and transcriptome data for cyst nematodes, in particular G. pallida and G. rostochiensis (see Eves-van den Akker and Jones, Chapter 2, this volume) will allow similar approaches to be used for these species.

There is also an imperative to ensure any novel proteins expressed transgenically are safe for both human and livestock consumption. A prima facie case has been established for the food safety of GM cystatin-expressing nematode-resistant plants. The engineered rice cystatin that was expressed in potato plants is not toxic and is rapidly digested in simulated gastric fluid, providing a margin of exposure in excess of 2000-fold (Atkinson et al., 2004). The repellent peptide is destroyed by typical cooking temperatures and by simulated intestinal fluid, suggesting that it is unlikely to be a toxin (Roderick et al., 2012). Transgenically expressed proteins, particularly those that are novel to the diet, should also lack allergenicity. GM crops are currently assessed for potential food allergy risks on the basis of an international consensus guideline outlined by the Codex Alimentarius Commission (FAO, 2008). This focuses primarily on evaluation of the potential allergenicity of the newly expressed protein(s) but also recommends evaluation of potential increases in any known allergens already present in a food crop. In addition to direct experimentation using IgE tests, a number of bioinformatic tools are recommended for prediction of likely allergens. Using these, the repellent peptide was found to present no allergenic risk (Roderick et al., 2012).

10.11.3 Promoter choice can target transgene expression for enhanced biosafety

Expression of transgenes in GM crops is frequently driven by a strong constitutive promoter such as CaMV35S. This will provide a high level of expression in all parts of the plant, but this is unnecessary and even undesirable for a nematode control trait. Effective, tissue-specific promoters can reduce any hazard to non-target organisms and enhance food safety when transgene expression is minimized in the harvested product. The most appropriate expression pattern will correspond with the required timing and location of the particular nematode defence. Wound-inducible promoters that are activated by the destructive root migration of cyst nematode J2 have been characterized (Niebel et al., 1993; Hansen et al., 1996) and could potentially drive expression of a defence that targets the nematode invasion process. The repellent peptide that acts against invasion of J2 before feeding occurs has been expressed in potatoes using the MDK4-20 promoter of A. thaliana, which is active in root tips and also the root border cells that detach from the roots. This provided an additional benefit as the promoter delivered 94.9 ± 0.8% resistance to G. pallida in a glasshouse trial rather than the 34.4 ± 8.4% resistance achieved using CaMV35S (Lilley et al., 2011).

Those defence strategies, such as proteinase inhibitors, Bt endotoxins or RNAi, that rely on ingestion of transgenically expressed molecules, require promoters that are active in the nematode feeding site, the syncytium. The identification of feeding site-specific promoters has been the focus of research for more than two decades, with biotechnology applications being a main driver of such work. The model plant Arabidopsis has been used extensively in these studies; promoter-trap reporter lines (e.g. Barthels et al., 1997) and subsequently microarray analysis of roots infected with H. schachtii (Szakasits et al., 2009) have provided promoters with increased and relatively specific activity in syncytia. Whilst many genes are upregulated in syncytia, most are also expressed in several other plant tissues. Exceptions are two Arabidopsis myo-inositol oxygenase genes, MIOX4 and MIOX5 (Siddique et al., 2009) and the defensin gene Pdf2.1 (Siddique et al., 2011). These are all expressed strongly in the syncytia of H. schachtii with limited expression elsewhere (floral tissue for the MIOX genes and siliques for Pdf2.1). Promoters from genes such as these could have potential to drive preferential expression in syncytia for transgenic nematode defences, although there is the caveat that tissue-specific expression patterns are not always faithfully reproduced upon transfer to other plant species. Three Arabidopsis promoters (*TUB-1*, *RPL16A* and *ARSK-1*) that did direct feeding site expression in potato roots were utilized to deliver cystatins to cyst and root-knot nematodes (Lilley et al., 2004). Only the ARSK promoter was highly active in syncytia as well as giant cells and those plants displayed partial resistance to *Globodera* in the field. In another example a tobacco cellulase promoter was used for localized expression of RNAi constructs in the syncytia of *H. schachtii* (Patel et al., 2010).

Gene expression is regulated by transcription factors, which bind to defined sequence motifs (*cis*-regulatory elements) in promoter regions. Identification of elements associated specifically with syncytial expression offers the prospect of engineering synthetic promoters with the desired activity. Analysis of soybean microarray and transcriptome data in combination with computational discovery of shared promoter motifs led to the discovery of soybean cyst nematode-inducible regulatory elements (Liu et al., 2014). A number of synthetic promoters consisting of tetramer cis-elements fused upstream of a minimal CaMV35S promoter were shown to respond to the early stages of nematode parasitism. This strategy could be used to identify similar motifs in other crop plants, although it remains to be determined if these synthetic promoters can provide sufficiently strong expression to deliver effective transgenic resistance.

10.11.4 The route to uptake of future GM crops for nematode control

As yet, no GM crops with nematode resistance traits have developed beyond the small-scale field trial stage. Uptake of GM crops requires clear recognition of grower needs and either significant funding from public sources or a sufficiently high commercial value of the trait to warrant the investment from biotechnology companies. The uptake of Bt cotton establishes that the success of any deployed biopesticide trait depends on the extent of losses prevented, availability in locally adapted crop varieties and the level of other investment needed to reduce the yield gap (Hillocks, 2009). The public sector will have an important role, particularly in the developing world, when companies consider a limited market opportunity does not justify product development. In addition, there needs to be political and public support, together with implementation of effective regulations. Recent public surveys indicate that citizens are more positive about GMOs than before, certainly if these GMOs reduce pesticide use, such as apples that need less fungicides (Eurobarometer, 2010).

A major investment of both time and resources is required to bring any new biotech crop to market. A 2011 survey of six of the largest crop biotechnology companies found that the average time from start of a project to commercial launch was around 13 years, with regulatory science, registration and regulatory affairs accounting for the longest phase (Phillips McDougall, 2011). The mean length of this phase increased from 44.5 months to 65.5 months in the period from 2002 to 2011 and it incurred a quarter of the total average cost of US$136 million to bring a trait to market. However, a recent study of the regulatory cost for a public organization to release a GM potato to the farmers was estimated to be a feasible 1 to a few million US$ (Schiek et al., 2016). Furthermore, there is a general trend towards a decline in the length of time between project initiation and commercial launch, as procedures no doubt become more streamlined. Interestingly, there appeared to have been an increase in the time spent on optimization of transformation constructs and selection of the best lines to advance for commercialization.

The impetus for developing a nematode-resistant GM crop to utility in the field occurs when global financial losses to nematodes are high for a particular crop or high percentage yield losses occur in more specialist crops. A trait that offers broad-spectrum resistance to multiple nematode species rather than just cyst nematodes is likely to be a more attractive prospect.

10.12 Conclusions and Future Prospects

Modern biotechnology has revolutionized our insight into cyst nematode biology and the plant–nematode interaction. High-throughput DNA analysis methods have transformed the breeding process. This new knowledge and methodology enables the development of novel, effective and sustainable nematode control strategies. This may be in the shape of improvements in marker assisted selection to exploit natural resistance traits. However, when progress of conventional breeding is confounded by a lack of suitable resistance genes, nematode virulence or breeding difficulties, transgenic approaches can offer an alternative option. It is perhaps surprising, therefore, that no GM crops for nematode control have yet appeared on the market, given that 2015 marked the 20th anniversary of commercialization of biotech crops with 179.7 million ha planted (James, 2015). Indeed, the authors of a 2005 report for the Food and Agriculture Organization of the United Nations (FAO) expressed surprise at the lack of transgenic approaches for nematode resistance given the 'high frequency and ubiquity of nematode pest species' and the wide range of affected host plants (Dhlamini et al., 2005). The situation has not changed substantially in the subsequent 10 years. This is in stark contrast to the success of insect-resistant GM crops that produce Bt endotoxins. One factor could be the relative lack of investment – in 2005 only 0.2% of the value of crop loss due to nematodes was invested in their research – but there are also currently few available transgenes that alone provide a consistently high level of nematode resistance. One possible way forward is for a transgenic resistance trait to be introduced into the best available partially resistant crop varieties that are favoured by growers. This would maximize the resistance phenotype as exemplified by the expression of a cystatin in the potato cultivars Sante and Maria Huanca (Urwin et al., 2003).

Inclusion of transgenes for nematode control in a gene stacking strategy may enhance their marketability and hence their appeal to biotechnology companies. Gene stacking is the incorporation of more than one transgenic event in a single plant. This may be to address issues surrounding trait durability by, for example, expressing several different insecticidal Bt proteins, or to combine disparate traits (e.g. insect resistance and herbicide tolerance). The global cultivation of stacked trait GM crops has risen rapidly in recent years as they are favoured by farmers and in 2015 they occupied one third of the total planted area (James, 2015). Crops in the field are challenged by multiple pests and diseases; nematode control could be combined with disease resistance to deliver an attractive product for growers. One example, already in the research pipeline, is the generation of GM potatoes with resistance both to potato cyst nematode and late blight disease (*Phytophthora infestans*), two of the most important biotic constraints for potato production. These plants will combine the cystatin and repellent peptide technologies for control of nematodes with multiple *R* genes for control of late blight (P. Urwin, 2016, pers. comm.). The commercial prospects of the plants will be further enhanced by the addition of proven transgenic traits conferring reduced bruising and lower acrylamide content. The project is an example of a public–private partnership between the Sainsbury Laboratory Norwich and the University of Leeds in the UK and two commercial partners, JR Simplot and BioPotatoes Ltd.

Generation of the large DNA constructs necessary to deliver multiple traits in a single transgenic event is facilitated by new techniques such as Gibson assembly (Gibson et al., 2009) and Golden Gate cloning (Engler et al., 2014). Such advances, together with recent technologies such as CRISPR, signal the beginning of a new phase of biotech crops from which nematode control should benefit in the future.

10.13 References

Acevedo-Garcia, J., Kusch, S. and Panstruga, R. (2014) Magical mystery tour: MLO proteins in plant immunity and beyond. *New Phytologist* 204, 273–281. DOI: 10.1111/nph.12889

ArslanAnwer, M., Anjam M.S., Shah, S.J., Hasan, M.S., Naz, A.A., Grundler, F.M.W. and Siddique, S. (2018) Genome-wide association study uncovers a novel QTL allele of AtS40-3 that controls susceptibility to cyst nematode infection in Arabidopsis. *Journal of Experimental Botany* ery019. DOI: 10.1093/jxb/ery019

Atkinson, H.J., Johnston, K.A. and Robbins, M. (2004) Prima facie evidence that a phytocystatin for transgenic plant resistance to nematodes is not a toxic risk in the human diet. *Journal of Nutrition* 134, 431–434.

Atkinson, H.J., Lilley, C.J. and Urwin, P.E. (2012) Strategies for transgenic nematode control in developed and developing world crops. *Current Opinion in Biotechnology* 23, 251–256. DOI: 10.1016/j.copbio.2011.09.004

Barthels, N., van der Lee, F.M., Klap, J. *et al*. (1997) Regulatory sequences of Arabidopsis drive reporter gene expression in nematode feeding structures. *Plant Cell* 9, 2119–2134.

Baum, J.A., Bogaert, T., Clinton, W. *et al*. (2007) Control of coleopteran insect pests through RNA interference. *Nature Biotechnology* 25, 1322–1326. DOI: 10.1038/nbt1359

Boevink, P.C., McLellan, H., Gilroy, E.M. *et al*. (2016) Oomycetes seek help from the plant: *Phytophthora infestans* effectors target host susceptibility factors. *Molecular Plant* 9, 636–638. DOI: 10.1016/j.molp.2016.04.005

Bowen, D.J., Bunkers, G.J., Chay, C. *et al*. (2016) Pesticidal nucleic acids and proteins and uses thereof. Patent Publication No. US9328356 B2.

Brookes, G. and Barfoot, P. (2016) Environmental impacts of genetically modified (GM) crop use 1996–2014: impacts on pesticide use and carbon emissions. *GM Crops and Food* 7, 84–116. DOI: 10.1080/21645698.2016.1192754

Cai, D., Thurau, T., Tian, Y., Lange, T., Yeh, K.-W. and Jung, C. (2003) Sporamin-mediated resistance to beet cyst nematodes (*Heterodera schachtii* Schm.) is dependent on trypsin inhibitory activity in sugar beet (*Beta vulgaris* L.) hairy roots. *Plant Molecular Biology* 51, 839–849. DOI: 10.1023/A:1023089017906

Chen, R.G., Li, H.X., Zhang, L.Y., Zhang, J.H., Xiao, J.H. and Ye, Z. (2007) *CaMi*, a root-knot nematode resistance gene from hot pepper (*Capsicum annuum* L.) confers nematode resistance in tomato. *Plant Cell Reports* 26, 895–905. DOI: 10.1007/s00299-007-0304-0

Cook, D.E., Lee, T.G., Guo, X. *et al*. (2012) Copy number variation of multiple genes at Rhg1 mediates nematode resistance in soybean. *Science* 338, 1206–1209. DOI: 10.1126/science.1228746

Cook, D.E., Bayless, A.M., Wang, K. *et al*. (2014) Distinct copy number, coding sequence, and locus methylation patterns underlie *Rhg1*-mediated soybean resistance to soybean cyst nematode. *Plant Physiology* 165, 630–647. DOI: 10.1104/pp.114.235952

Cowgill, S.E. and Atkinson, H.J. (2003) A sequential approach to risk assessment of transgenic plants expressing protease inhibitors: effects on nontarget herbivorous insects. *Transgenic Research* 12, 439–449. DOI: 10.1023/A:1024215922148

Cowgill, S.E., Wright, C. and Atkinson, H.J. (2002a) Transgenic potatoes with enhanced levels of nematode resistance do not have altered susceptibility to nontarget aphids. *Molecular Ecology* 11, 821–827. DOI: 10.1046/j.1365-294X.2002.01482.x

Cowgill, S.E., Bardgett, R.D., Kiezebrink, D.T. and Atkinson, H.J. (2002b) The effect of transgenic nematode resistance on non-target organisms in the potato rhizosphere. *Journal of Applied Ecology* 39, 915–923. DOI: 10.1046/j.1365-2664.2002.00774.x

Cowgill, S.E., Danks, C. and Atkinson, H.J. (2004) Multitrophic interactions involving genetically modified potatoes, nontarget aphids, natural enemies and hyperparasitoids. *Molecular Ecology* 13, 639–647. DOI: 10.1046/j.1365-294X.2004.02078.x

Danchin, E.G.J., Arguel, M.J., Campan-Fournier, A. *et al*. (2013) Identification of novel target genes for safer and more specific control of root-knot nematodes from a pan-genome mining. *PLoS Pathogens* 9, e1003745. DOI: 10.1371/journal.ppat.1003745

de Almeida-Engler, J. and Gheysen, G. (2013) Nematode induced endoreduplication in plant host cells: why and how? *Molecular Plant-Microbe Interactions* 26, 17–24. DOI: 10.1094/MPMI-05-12-0128-CR

de Almeida-Engler, J., De Vleesschauwer, V., Burssens, S., Celenza, J.L. Jr, Inzé, D. Van Montagu, M., Engler, G. and Gheysen, G. (1999) The use of molecular markers and cell cycle inhibitors to analyze cell cycle progression in nematode induced galls and syncytia. *Plant Cell* 11, 793–807. DOI: 10.1105/tpc.11.5.793

Dhlamini, Z., Spillane, C., Moss, J.P., Ruane, J., Urquia, N. and Sonnino, A. (2005) *Status of Research and Application of Crop Biotechnologies in Developing Countries*. FAO, Rome.

Dinh, P.T.Y., Zhang, L., Brown, C.R. and Elling, A.A. (2014) Plant-mediated RNA interference of effector gene *Mc16D10L* confers resistance against *Meloidogyne chitwoodi* in diverse genetic backgrounds of potato and reduces pathogenicity of nematode offspring. *Nematology* 16, 669–682. DOI: 10.1163/15685411-00002796

Dubelman, S., Fischer, J., Zapata, F., Huizinga, K., Jiang, C., Uffman, J., Levine, S. and Carson, D. (2014) Environmental fate of double-stranded RNA in agricultural soils. *PLoS ONE*, 9, e93155. DOI: 10.1371/journal.pone.0093155

Dutta, T.K., Banakar, P. and Rao, U. (2015) The status of RNAi-based transgenic research in plant nematology. *Frontiers in Microbiology* 5, Article 760. DOI: 10.3389/fmicb.2014.00760

Engler, C., Youles, M., Gruetzner, R., Ehnert, T.-M., Werner, S., Jones, J.G.D., Patron, N.J. and Marillonnet, S. (2014) A Golden Gate modular cloning toolbox for plants. *ACS Synthetic Biology* 3, 839–843. DOI: 10.1021/sb4001504

Eurobarometer (2010) *Europeans and Biotechnology in 2010; Winds of Change?* Available at: http://ec.europa.eu/research/science-society/document_library/pdf_06/europeans-biotechnology-in-2010_en.pdf (accessed 1 February 2018).

Eves-van den Akker, S., Lilley C.J., Ault, J.R., Ashcroft, A.E., Jones, J.T. and Urwin, P.E. (2014a) The feeding tube of cyst nematodes: characterisation of protein exclusion. *PLoS ONE* 9(1), e87289. DOI: 10.1371/journal.pone.0087289

Eves-van den Akker, S., Lilley, C.J., Jones, J.T. and Urwin, P.E. (2014b) Identification and characterisation of a hyper-variable apoplastic effector gene family of the potato cyst nematodes. *Plos Pathogens* 10, e1004391. DOI: 10.1371/journal.ppat.1004391

Eves-van den Akker, S., Lilley, C.J., Reid, A., Pickup, J., Anderson, E., Cock, P.J., Blaxter, M., Urwin, P.E., Jones, J.T. and Blok, V.C. (2015a) A metagenetic approach to determine the diversity and distribution of cyst nematodes at the level of the country, the field and the individual. *Molecular Ecology* 24, 5842–5851. DOI: 10.1111/mec.13434

Eves-van den Akker, S., Lilley, C.J., Jones, J.T. and Urwin, P.E. (2015b) Plant-parasitic nematode feeding tubes and plugs: new perspectives on function. *Nematology* 17, 1–9. DOI: 10.1163/15685411-00002832

FAO (2008) Guideline for the conduct of food safety assessment of foods derived from recombinant-DNA plants. Available at: www.fao.org/fao-who-codexalimentarius/sh-proxy/en/?lnk=1&url=https%253A%252F%252Fworkspace.fao.org%252Fsites%252Fcodex%252FStandards%252FCAC%2B-GL%2B45-2003%252FCXG_045e.pdf (accessed 1 October 2016).

Ferris, H., Bongers, T. and de Goede, R.G.M. (2001) A framework for soil food web diagnostics: extension of the nematode faunal analysis concept. *Applied Soil Ecology* 18, 13–29. DOI: 10.1016/S0929-1393(01)00152-4

Ferris, H., Griffiths, B.S., Porazinska, D.L., Powers, T.O., Wang, K.-H. and Tenuta, M. (2012) Reflections on plant and soil nematode ecology: past, present and future. *Journal of Nematology* 44, 115–126.

Garibyan, L. and Avashia, N. (2013) Research techniques made simple: polymerase chain reaction (PCR). *Journal of Investigative Dermatology* 133, e6. DOI: 10.1038/jid.2013.1

Gheysen, G., Angenon, G. and Van Montagu, M. (1998) *Agrobacterium*-mediated plant transformation: a scientifically intriguing story with significant applications. In: Lindsey, K. (ed.) *Transgenic Plant Research*. Harwood Academic Publishers, Amsterdam, The Netherlands, pp. 1–33. Available at: http://hdl.handle.net/1854/LU-278758 (accessed 1 February 2018).

Gibson, D.G., Young, L., Chuang, R.-Y., Venter, J.C., Hutchison, C.A. and Smith, H.O. (2009) Enzymatic assembly of DNA molecules up to several hundred kilobases. *Nature Methods* 6, 343–345. DOI: 10.1038/nmeth.1318

Goodwin, S., McPherson, J.D. and McCombie, W.R. (2016) Coming of age: ten years of next-generation sequencing technologies. *Nature Reviews Genetics* 17, 333–351. DOI: 10.1038/nrg.2016.49

Green, J., Wang, D., Lilley, C.J., Urwin, P.E. and Atkinson, H.J. (2012) Transgenic potatoes for potato cyst nematode control can replace pesticide use without impact on soil quality. *Plos One* 7, e30973. DOI: 10.1371/journal.pone.0030973

Grunewald, W., Karimi, M., Wieczorek, K., Van de Cappelle, E., Grundler, F., Inzé, D., Beeckman, T. and Gheysen, G. (2008) A role for AtWRKY23 in feeding site establishment of plant-parasitic nematodes. *Plant Physiology* 148, 358–368. DOI: 10.1104/pp.108.119131

Grunewald, W., Cannoot, B., Friml, J. and Gheysen, G. (2009) Parasitic nematodes modulate PIN-mediated auxin transport to facilitate infection. *Plos Pathogens* 5(1), e1000266, 1–7. DOI: 10.1371/journal.ppat.1000266

Haegeman, A., Mantelin, S., Jones, J.T. and Gheysen, G. (2012) Functional roles of effectors of plant-parasitic nematodes: a molecular update. *Gene* 492, 19–31. DOI: 10.1016/j.gene.2011.10.040

Hansen, E., Harper, G. and McPherson, M.J. (1996) Differential expression patterns of the wound-inducible transgene *wun-1-uidA* in potato roots following infection with either cyst or root knot nematodes. *Physiological and Molecular Plant Pathology* 48, 161–170. DOI: 10.1006/pmpp.1996.0014

Herrera-Estrella, L., Depicker, A., Van Montagu, M. and Schell, J. (1983) Expression of chimaeric genes transferred into plant-cells using a Ti-plasmid-derived vector. *Nature* 303, 209–213. DOI: 10.1038/303209a0

Hewezi, T., Howe, P., Maier, T.R., Hussey, R.S., Mitchum, M.G., Davis, E.L., and Baum, T.J. (2008) Cellulose binding protein from the parasitic nematode *Heterodera schachtii* interacts with Arabidopsis pectin methylesterase: cooperative cell wall modification during parasitism. *Plant Cell* 20, 3080–3093. DOI: 10.1105/tpc.108.063065

Hillocks, R.J. (2009) GM cotton for Africa. *Outlook on Agriculture* 38, 311–316. DOI: 10.5367/000000009790422142

Holden-Dye, L. and Walker, R.J. (2011) Neurobiology of plant parasitic nematodes. *Invertebrate Neuroscience* 11, 9–19. DOI: 10.1007/s10158-011-0117-2

Huang, G.Z., Allen, R., Davis, E.L., Baum, T.J. and Hussey, R.S. (2006) Engineering broad root-knot resistance in transgenic plants by RNAi silencing of a conserved and essential root-knot nematode parasitism gene. *Proceedings of the National Academy of Sciences of the United States of America* 103, 14302–14306. DOI: 10.1073/pnas.0604698103

James, C. (2015) *20th Anniversary (1996 to 2015) of the Global Commercialisation of Biotech Crops and Biotech Crop Highlights in 2015*. ISAAA Brief No. 51. ISAAA, Ithaca, NY.

Jinek, M., Chylinski, K., Fonfara, I., Hauer, M., Doudna, J.A. and Charpentier, E. (2012) A programmable dual-RNA-guided DNA endonuclease in adaptive bacterial immunity. *Science* 337, 816–821. DOI: 10.1126/science.1225829

Jørgensen, I.H. (1992) Discovery, characterization and exploitation of Mlo powdery mildew resistance in barley. *Euphytica* 63, 141–152. DOI: 10.1007/BF00023919

Klink, V.P., Kim, K.H., Martins, V., MacDonald, M.H., Beard, H.S., Alkharouf, N.W., Lee, S.K., Park, S.C. and Matthews, B.F. (2009) A correlation between host-mediated expression of parasite genes as tandem inverted repeats and abrogation of development of female *Heterodera glycines* cyst formation during infection of *Glycine max*. *Planta* 230, 53–71. DOI: 10.1007/s00425-009-0926-2

Koch, A. and Kogel, K.-H. (2014) New wind in the sails: improving the agronomic value of crop plants through RNAi-mediated gene silencing. *Plant Biotechnology Journal* 12, 821–831. DOI: 10.1111/pbi.12226

Koritsas, V.M. and Atkinson, H.J. (1994) Proteinases of females of the phytoparasite *Globodera pallida* (potato cyst-nematode). *Parasitology* 109, 357–365. DOI: 10.1017/S0031182000078392

Kyndt, K., Ji, H., Vanholme, B. and Gheysen, G. (2013) Transcriptional silencing of RNAi constructs against nematode genes in *Arabidopsis*. *Nematology* 15, 519–528. DOI: 10.1163/15685411-00002698

Ledford, H. (2015) CRISPR, the disruptor. *Nature* 522, 20–24. DOI: 10.1038/522020a

Lee, C., Chronis, D., Kenning, C., Peret, B., Hewezi, T., Davis, E.L., Baum, T.J., Hussey, R., Bennett, M. and Mitchum, M.G. (2011) The novel cyst nematode effector protein 19C07 interacts with the Arabidopsis auxin influx transporter LAX3 to control feeding site development. *Plant Physiology* 155, 866–880. DOI: 10.1104/pp.110.167197

Li, J.R., Todd, T.C., Oakley, T.R., Lee, J. and Trick, H.N. (2010a) Host-derived suppression of nematode reproductive and fitness genes decreases fecundity of *Heterodera glycines* Ichinohe. *Planta* 232, 775–785. DOI: 10.1007/s00425-010-1209-7

Li, J.R., Todd, T.C. and Trick, H.N. (2010b) Rapid *in planta* evaluation of root expressed transgenes in chimeric soybean plants. *Plant Cell Reports* 29, 113–123. DOI: 10.1007/s00299-009-0803-2

Li, T., Liu, B., Spalding, M.H., Weeks, D.P. and Yang, B. (2012) High-efficiency TALEN-based gene editing produces disease-resistant rice. *Nature Biotechnology* 30, 390–392. DOI: 10.1038/nbt.2199

Li, X.Q., Wei, J.Z., Tan, A. and Aroian, R.V. (2007) Resistance to root-knot nematode in tomato roots expressing a nematicidal *Bacillus thuringiensis* crystal protein. *Plant Biotechnology Journal* 5, 455–464. DOI: 10.1111/j.1467-7652.2007.00257.x

Li, X.Q., Tan, A., Voegtline, M., Bekele, S., Chen, C.S. and Aroian, R.V. (2008) Expression of Cry5B protein from *Bacillus thuringiensis* in plant roots confers resistance to root-knot nematode. *Biological Control* 47, 97–102. DOI: 10.1016/j.biocontrol.2008.06.007

Li, Z., Liu, Z.B., Xing, A., Moon, B.P., Koellhoffer, J.P., Huang, L., Ward, R.T., Clifton, E., Falco, S.C. and Cigan, A.M. (2015) Cas9-guide RNA directed genome editing in soybean. *Plant Physiology* 169, 960–970. DOI: 10.1104/pp.15.00783

Lilley, C.J., Urwin, P.E., McPherson, M.J. and Atkinson, H.J. (1996) Characterisation of intestinally active proteases of cyst-nematodes. *Parasitology* 113, 415–424. DOI: 10.1017/S0031182000066555

Lilley, C.J., Urwin, P.E., Atkinson, H.J. and McPherson, M.J. (1997) Characterisation of cDNAs encoding serine proteases from the soybean cyst nematode *Heterodera glycines*. *Molecular and Biochemical Parasitology* 89, 195–207. DOI: 10.1016/S0166-6851(97)00116-3

Lilley, C.J., Urwin, P.E., Johnston, K.A. and Atkinson, H.J. (2004) Preferential expression of a plant cystatin at nematode feeding sites confers resistance to *Meloidogyne incognita* and *Globodera pallida*. *Plant Biotechnology Journal* 2, 3–12. DOI: 10.1046/j.1467-7652.2003.00037.x

Lilley, C.J., Wang, D., Atkinson, H.J. and Urwin, P.E. (2011) Effective delivery of a nematode-repellent peptide using a root-cap-specific promoter. *Plant Biotechnology Journal* 9, 151–161. DOI: 10.1111/j.1467-7652.2010.00542.x

Lin, J., Mazarei, M., Zhao, N. et al. (2016) Transgenic soybean overexpressing *GmSAMT1* exhibits resistance to multiple-HG types of soybean cyst nematode *Heterodera glycines*. *Plant Biotechnology Journal* 14, 2100–2109. DOI: 10.1111/pbi.12566

Liu, B., Hibbard, J.K., Urwin, P.E. and Atkinson, H.J. (2005) The production of synthetic chemodisruptive peptides *in planta* disrupts the establishment of cyst nematodes. *Plant Biotechnology Journal* 3, 487–496. DOI: 10.1111/j.1467-7652.2005.00139.x

Liu, S.M., Kandoth, P.K., Warren, S.D. et al. (2012) A soybean cyst nematode resistance gene points to a new mechanism of plant resistance to pathogens. *Nature* 492, 256–260. DOI: 10.1038/nature11651

Liu, W., Mazarei, M., Peng, Y., Fethe, M.H., Rudis, M.R., Lin, J., Millwood, R.J., Arelli, P.R. and Stewart, Jr, C.N. (2014) Computational discovery of soybean promoter *cis*-regulatory elements for the construction of soybean cyst nematode-inducible synthetic promoters. *Plant Biotechnology Journal* 12, 1015–1026. DOI: 10.1111/pbi.12206

Mantelin, S., Thorpe, P. and Jones, J.T. (2017) Translational biology of nematode effectors. Or, to put it another way, functional analysis of effectors – what's the point? *Nematology* 19, 251–261. DOI: 10.1163/15685411-00003048

Marroquin, L.D., Elyassnia, D., Griffitts, J.S., Feitelson, J.S. and Aroian, R.V. (2000) *Bacillus thuringiensis* (Bt) toxin susceptibility and isolation of resistance mutants in the nematode *Caenorhabditis elegans*. *Genetics* 155, 1693–1699.

Masler, E.P. (2007) Responses of *Heterodera glycines* and *Meloidogyne incognita* to exogenously applied neuromodulators. *Journal of Helminthology* 81, 421–427. DOI: 10.1017/S0022149X07850243

Matthews, B.F., Beard, H., MacDonald, M.H., Kabir, S., Youssef, R.M., Hosseini, P. and Brewer, E. (2013) Engineered resistance and hypersusceptibility through functional metabolic studies of 100 genes in soybean to its major pathogen, the soybean cyst nematode. *Planta* 237, 1337–1357. DOI: 10.1007/s00425-013-1840-1

Mimee, B., Duceppe, M.O., Veronneau, P.Y., Lafond-Lapalme, J., Jean, M., Belzile, F. and Bélair, G. (2015) A new method for studying population genetics of cyst nematodes based on Pool-Seq and genomewide allele frequency analysis. *Molecular Ecology Resources* 15, 1356–1365. DOI: 10.1111/1755-0998.12412

Mousley, A., McVeigh, P., Dalzell, J.J. and Maule, A.G. (2013) Nematode neuropeptide communication systems. In: Kennedy, M.W. and Harnett, W. (eds) *Parasitic Nematodes: Molecular Biology, Biochemistry and Immunology*, 2nd edn. CAB International, Wallingford, UK, pp. 279–307.

Niebel, A., de Almeida-Engler, J., Tire, C., Engler, G., Van Montagu, M. and Gheysen, G. (1993) Induction patterns of an extensin gene in tobacco upon nematode infection. *Plant Cell* 5, 1697–1710. DOI: 10.1105/tpc.5.12.1697

Panella, L. and Lewellen, R.T. (2007) Broadening the genetic base of sugar beet: introgression from wild relatives. *Euphytica* 154, 383–400. DOI: 10.1007/s10681-006-9209-1

Papolu, P.K., Gantasala, N.P., Kamaraju, D., Banakar, P., Sreevathsa, R. and Rao, U. (2013) Utility of host delivered RNAi of two FMRF amide like peptides, *flp-14* and *flp-18*, for the management of root-knot nematode, *Meloidogyne incognita*. *PLoS One* 8, e80603. DOI: 10.1371/journal.pone.0080603

Patel, N., Hamamouch, N., Li, C.Y., Hussey, R.S., Mitchum, M., Baum, T.J., Wang, X. and Davis, E.L. (2008) Similarity and functional analyses of expressed parasitism genes in *Heterodera schachtii* and *Heterodera glycines*. *Journal of Nematology* 40, 299–310.

Patel, N., Hamamouch, N., Li, C.Y., Hewezi, T., Hussey, R.S., Baum, T.J., Mitcham, M. and Davis, E.L. (2010) A nematode effector protein similar to annexins in host plants. *Journal of Experimental Botany* 61, 235–248. DOI: 10.1093/jxb/erp293

Phillips McDougall (2011) The cost and time involved in the discovery, development and authorisation of a new plant biotechnology trait. Available at: http://croplife.org/wp-content/uploads/2014/04/Getting-a-Biotech-Crop-to-Market-Phillips-McDougall-Study.pdf (accessed 22 January 2018).

Plovie, E., De Buck, S., Goeleven, E., Tanghe, M., Vercauteren, I. and Gheysen, G. (2003) Hairy roots to test for transgenic nematode resistance: think twice. *Nematology* 5, 831–841. DOI: 10.1163/156854103773040736

Prasad, C.V.S.S., Gupta, S., Gaponenko, A. and Tiwari, M. (2013) Molecular dynamic and docking interaction study of *Heterodera glycines* serine proteinase with *Vigna mungo* proteinase inhibitor. *Applied Biochemistry and Biotechnology* 170, 1996–2008. DOI: 10.1007/s12010-013-0342-8

Ramon, M., Devos, Y., Lanzoni, A., Liu, Y., Gomes, A., Gennaro, A. and Waigmann, E. (2014) RNAi-based GM plants: food for thought for risk assessors. *Plant Biotechnology Journal* 12, 1271–1273. DOI: 10.1111/pbi.12305

Roderick, H., Tripathi, L., Babirye, A., Wang, D., Tripathi, J., Urwin, P.E. and Atkinson, H.J. (2012) Generation of transgenic plantain (*Musa* spp.) with resistance to plant parasitic nematodes. *Molecular Plant Pathology* 13, 842–851. DOI: 10.1111/J.1364-3703.2012.00792.X

Rudelsheim, P.L.J. and Smets, G. (2014) Survey of field trials with genetically modified plants. COGEM Report No CGM 2014-04.

Schiek, B., Hareau, G., Baguma, Y., Medakker, A., Douches, D., Shotkoski, F. and Ghislain, M. (2016) Demystification of GM crop costs: releasing late blight resistant potato varieties as public goods in developing countries. *International Journal of Biotechnology* 14, 112–131. DOI: 10.1504/IJBT.2016.077942

Schnepf, E., Crickmore, N., Van Rie, J., Lereclus, D., Baum, J., Feitelson, J., Zeigler, D.R. and Dean, D.H. (1998) *Bacillus thuringiensis* and its pesticidal crystal proteins. *Microbiology and Molecular Biology Reviews* 62, 775–806.

Shan, Q., Wang, Y., Li, J. and Gao, C. (2014) Genome editing in rice and wheat using the CRISPR/Cas system. *Nature Protocols* 9, 2395–2410. DOI: 10.1038/nprot.2014.157

Siddique, S., Endres, S., Atkins, J.M. *et al.* (2009) Myo-inositol oxygenase genes are involved in the development of syncytia induced by *Heterodera schachtii* in Arabidopsis roots. *New Phytologist* 184, 457–472. DOI: 10.1111/j.1469-8137.2009.02981.x

Siddique, S., Wieczorek, K., Szakasits, D., Kreil, D.P. and Bohlmann, H. (2011) The promoter of a plant defensin gene directs specific expression in nematode-induced syncytia in Arabidopsis roots. *Plant Physiology and Biochemistry* 49, 1100–1107. DOI: 10.1016/j.plaphy.2011.07.005

Silva, F.B., Batista, J.A., Marra, B.M., Fragoso, R.R., Monteiro, A.C., Figueira, E.L. and Grossi-de-Sá, M.F. (2004) Prodomain peptide of HGCP-Iv cysteine proteinase inhibits nematode cysteine proteinases. *Genetics and Molecular Research* 3, 342–355.

Sindhu, A.S., Maier, T.R., Mitchum, M.G., Hussey, R.S., Davis, E.L. and Baum, T.J. (2009) Effective and specific *in planta* RNAi in cyst nematodes: expression interference of four parasitism genes reduces parasitic success. *Journal of Experimental Botany* 60, 315–324. DOI: 10.1093/jxb/ern289

Sobczak, M., Avrova, A., Jupowicz, J., Phillips, M.S., Ernst, K. and Kumar, A. (2005) Characterization of susceptibility and resistance responses to potato cyst nematode (*Globodera* spp.) infection of tomato lines in the absence and presence of the broad-spectrum nematode resistance *Hero* gene. *Molecular Plant-Microbe Interactions* 18, 158–168. DOI: 10.1094/MPMI-18-0158

Steeves, R.M., Todd, T.C., Essig, J.S. and Trick, H.N. (2006) Transgenic soybeans expressing siRNAs specific to a major sperm protein gene suppress *Heterodera glycines* reproduction. *Functional Plant Biology* 33, 991–999. DOI: 10.1071/FP06130

Szakasits, D., Heinen, P., Wieczorek, K., Hofmann, J., Wagner, F., Kreil, D.P., Sykacek, P., Grundler, F.M.W. and Bohlmann, H. (2009) The transcriptome of syncytia induced by the cyst nematode *Heterodera schachtii* in Arabidopsis roots. *The Plant Journal* 57, 771–784. DOI: 10.1111/j.1365-313X.2008.03727.x

Thomazella, D.P., Brail, Q., Dahlbeck, D. and Staskawicz, B.J. (2016) CRISPR-Cas9 mediated mutagenesis of a *DMR6* ortholog in tomato confers broad-spectrum disease resistance. *bioRxiv* 064824. DOI: 10.1101/064824

Tripathi, L., Babirye, A., Roderick, H., Tripathi, J.N., Changa, C., Urwin, P.E., Tushemereirwe, W.K., Coyne, D. and Atkinson, H.J. (2015) Field resistance of transgenic plantain to nematodes has potential for future African food security. *Scientific Reports* 5, 8127. DOI: 10.1038/srep08127

Trudgill, D.L. (1991) Resistance to and tolerance of plant parasitic nematodes in plants. *Annual Review of Phytopathology* 29, 167–192. DOI: 10.1146/annurev.py.29.090191.001123

Urwin, P.E., Atkinson, H.J., Waller, D.A. and McPherson, M.J. (1995) Engineered oryzacystatin-I expressed in transgenic hairy roots confers resistance to *Globodera pallida*. *The Plant Journal* 8, 121–131. DOI: 10.1046/j.1365-313X.1995.08010121.x

Urwin, P.E., Lilley, C.J., McPherson, M.J. and Atkinson, H.J. (1997a) Characterisation of two cDNAs encoding cysteine proteases from the soybean cyst nematode *Heterodera glycines*. *Parasitology* 114, 605–613.

Urwin, P.E., Lilley, C.J., McPherson, M.J. and Atkinson, H.J. (1997b) Resistance to both cyst- and root-knot nematodes conferred by transgenic *Arabidopsis* expressing a modified plant cystatin. *The Plant Journal* 12, 455–461. DOI: 10.1046/j.1365-313X.1997.12020455.x

Urwin, P.E., Troth, K.M., Zubko, E.I. and Atkinson, H.J. (2001) Effective transgenic resistance to *Globodera pallida* in potato field trials. *Molecular Breeding* 8, 95–101. DOI: 10.1023/A:1011942003994

Urwin, P.E., Lilley, C.J. and Atkinson, H.J. (2002) Ingestion of double-stranded RNA by preparasitic juvenile cyst nematodes leads to RNA interference. *Molecular Plant-Microbe Interactions* 15, 747–752. DOI: 10.1094/MPMI.2002.15.8.747

Urwin, P.E., Green, J. and Atkinson, H.J. (2003) Expression of a plant cystatin confers partial resistance to *Globodera*, full resistance is achieved by pyramiding a cystatin with natural resistance. *Molecular Breeding* 12, 263–269. DOI: 10.1023/A:1026352620308

Wang, D. (2009) Reducing the environmental risks and hazards of crop production by biosafe use of transgenic crops. PhD Thesis, University of Leeds, UK.

Wang, D., Jones, L.M., Urwin, P.E. and Atkinson, H.J. (2011) A synthetic peptide shows retro- and anterograde neuronal transport before disrupting the chemosensation of plant-pathogenic nematodes. *PLoS ONE* 6, e17475. DOI: 10.1371/journal.pone.0017475

Wang, F., Wang, C., Liu, P., Lei, C., Hao, W., Gao, Y., Liu, Y.-G. and Zhao. K. (2016) Enhanced rice blast resistance by CRISPR/Cas9-targeted mutagenesis of the ERF transcription factor gene *OsERF922*. *PLoS ONE* 11(4), e0154027. DOI: 10.1371/journal.pone.0154027

Wang, S., Zhang, S., Wang, W., Xiong, X., Meng, F. and Cui, X. (2015) Efficient targeted mutagenesis in potato by the CRISPR/Cas9 system. *Plant Cell Reports* 34, 1473–1476. DOI: 10.1007/s00299-015-1816-7

Wang, Y.P., Cheng, X., Shan, Q.W., Zhang, Y., Liu, J.X., Gao, C. and Qiu, J.-L. (2014) Simultaneous editing of three homoeoalleles in hexaploid bread wheat confers heritable resistance to powdery mildew. *Nature Biotechnology* 32, 947–951. DOI: 10.1038/nbt.2969

Warmerdam, S., van Schaik, C., Lozano-Torres, J.L., Finkers-Tomczak, A.M., Bakker, J., Goverse, A. and Smant, G. (2014). Exploiting natural variation in susceptibility of *Arabidopsis thaliana* to *Meloidogyne incognita* to breed broad-spectrum resistance to root knot nematodes *M. incognita*. XVI International Congress on Molecular Plant-Microbe Interactions, 2014, 6–10 July, Rhodes, Greece, poster 453.

Witek, K., Jupe, F., Witek, A.I., Baker, D., Clark, M.D. and Jones, J.D.G. (2016) Accelerated cloning of a potato late blight-resistance gene using RenSeq and SMRT sequencing. *Nature Biotechnology* 34, 656–60. DOI: 10.1038/nbt.3540

Winter, M.D., McPherson, M.J. and Atkinson, H.J. (2002) Neuronal uptake of pesticides disrupts chemosensory cells of nematodes. *Parasitology* 125, 561–565. DOI: 10.1017/S0031182002002482

Yadav, B.C., Veluthambi, K. and Subramaniam, K. (2006) Host-generated double-stranded RNA induces RNAi in plant-parasitic nematodes and protects the host from infection. *Molecular and Biochemical Parasitology* 148, 219–222. DOI: 10.1016/j.molbiopara.2006.03.013

Youssef, R.M., Kim, K.-H., Haroon, S.A. and Matthews, B.F. (2013) Post-transcriptional gene silencing of the gene encoding aldolase from soybean cyst nematode by transformed soybean roots. *Experimental Parasitology* 134, 266–274. DOI: 10.1016/j.exppara.2013.03.009

Zhang, L.Y., Zhang, Y.Y., Chen, R.G., Zhang, J.H., Wang, T.T., Li, H.-X. and Ye, Z.-B. (2010) Ectopic expression of the tomato *Mi-1* gene confers resistance to root knot nematodes in lettuce (*Lactuca sativa*). *Plant Molecular Biology Reporter* 28, 204–211. DOI: 10.1007/s11105-009-0143-y

11 Biological Control of Cyst Nematodes through Microbial Pathogens, Endophytes and Antagonists

Keith G. Davies[1,2], Sharad Mohan[3] and Johannes Hallmann[4]

[1]*University of Hertfordshire, Hatfield, UK;* [2]*Norwegian Institute of Bioeconomy Research, Ås, Norway;* [3]*Indian Agricultural Research Institute, New Delhi, India;* [4]*Julius Kühn-Institut Federal Research Centre for Cultivated Plants, Münster, Germany*

11.1 Introduction	237
11.2 Nematophagous and Antagonistic Fungi	238
11.3 Rhizobacteria and Root Endophytic Bacteria	245
11.4 Microbial–Nematode Molecular Interactions	249
11.5 Biological Control Products	253
11.6 Managing Biological Control	254
11.7 Conclusions and Future Prospects	257
11.8 References	258

11.1 Introduction

The biological control of nematode pests has a long history dating back at least to the 1930s when Linford and his colleagues showed that the natural enemies of plant-parasitic nematodes restricted their population growth (Linford, 1937; Linford *et al.*, 1938; Linford and Yap, 1939). However, since the middle of the last century synthetic chemicals, as byproducts of the petrochemical industry, became the dominant method to control plant-parasitic nematodes as they were both effective and easily applied. Following publication of the book *Silent Spring* (Carson, 1962) that highlighted some of the problems associated with the application of synthetic chemicals and concern over the environmental and human toxicity of these products, there has been an increasing investment in research for alternatives. The use of natural enemies to control pests and diseases as an alternative to chemical pesticides has grown since the middle of the 1970s and built on foundational concepts defined by Baker and Cook (1974). Stirling (2014) characterizes biological control methods into three broad groups: (i) 'inundative' approaches whereby a particular natural enemy (fungus, bacteria, etc.) is identified, mass produced and then reapplied into the soil at high densities to control the pest nematode; this approach, according to Stirling (2014), is simply to replace a synthetic chemical with a biological control agent; (ii) 'inoculative' methods that basically follow the approach in (i) above but leave out the mass production step and the biological control agent is added at a low density, insufficient to control the pest but it is expected to establish itself in the soil and build up naturally to a level that controls the pest nematode; and (iii) 'conservation' biological control in which the community of natural enemies is managed in such a manner to control the pest nematodes. This latter approach is fundamentally an ecologically based strategy taking advantage of ecosystem services in order to suppress the pest nematodes.

The idea that soils can become suppressive to nematodes is well established and nematode suppressive soils can develop over prolonged periods of time (6–8 years) and were first recognized in the decline of cereal cyst nematode, *Heterodera avenae* (Gair *et al.*, 1969). Research showed that the application of formalin as a drench increased the density of *H. avenae* (Williams, 1969; Kerry *et al.*, 1980, 1982a, b) and it was suggested that the formalin was destroying the natural microbial enemies that were suppressing the nematode populations; this allowed the nematodes to escape control and their populations to increase. Subsequent research showed that the fungal parasites *Pochonia chlamydosporia* (= *Verticillium chlamydosporium*) and *Nematophthora gynophila* were the major microbial enemies prohibiting nematode reproduction. Further studies at one of the experimental sites showed that a bacterium similar to *Pasteuria penetrans* was adhering to and infecting over 50% of the second-stage juveniles (J2) of *H. avenae* and preventing them from reaching the root system (Davies *et al.*, 1990). More recently, Wei *et al.* (2015) induced suppressiveness towards the soybean cyst nematode, *H. glycines*, by continuously growing soybean for more than 15 years. However, suppressiveness did not occur when soybean was rotated with cereals, such as wheat or maize. The suppressive soil was associated with higher levels of antagonistic fungi, such as *Trichoderma harzianum*, *P. chlamydosporia* and *Purpureocillium lilacinum* (= *Paecilomyces lilacinus*). Thus, different microbial enemies operate in concert to reduce the population of *H. glycines*.

Nematode suppression is a general concept that can be thought of as being a continuum between two extremes; general suppression is usually non-specific, affecting all plant-parasitic nematodes irrespective of species or biotype, whereas specific suppression is usually where one natural enemy is responsible for controlling one particular nematode species or biotype.

The objectives of this chapter are to describe the main microbial antagonists of cyst nematodes, their mode of action and how such antagonists can be used in a sustainable manner to suppress cyst nematodes and improve plant yield.

11.2 Nematophagous and Antagonistic Fungi

When Julius Kühn studied the beet cyst nematode, *Heterodera schachtii*, in 1877 he observed several cysts containing eggs infected by a fungus he described as *Tarichium auxiliarum*, currently named *Catenaria auxiliaris*. He immediately recognized the enormous potential of such antagonistic fungi and declared them the biggest helper of farmers in controlling *H. schachtii* (Kühn, 1877). Since then numerous fungi have been identified that interfere with the lifecycle of cyst nematodes resulting in reduced nematode propagation. According to their lifestyle and mode of action they can be classified into nematophagous fungi and antagonistic fungi (Table 11.1).

11.2.1 Nematophagous fungi

So far, more than 700 species of nematophagous fungi have been described. Nematophagous fungi are known from all four major fungal clades, Ascomycota, Basidiomycota, Zygomycota and Chytridiomycota, indicating that the nematophagous behaviour has evolved independently multiple times through convergent evolution (Li *et al.*, 2015). According to their mode of action nematophagous fungi can be categorized in four groups: nematode-trapping fungi, endoparasitic fungi, egg and female parasitic fungi, and toxin-producing fungi.

11.2.1.1 Nematode-trapping fungi

This group of soil-borne fungi forms mycelial traps that capture and kill their host nematode. Different fungal taxa produce different trapping structures. The simplest structures are fungal hyphae (*Stylophage* spp.) or short erect tapering branches (*Monacrosporium cionopagum*), both

Table 11.1. Fungal groups and their mode of action indicating potential biocontrol of cyst nematodes.

Fungal group	Mode of action	Examples
Nematophagous fungi	Carnivorous fungi specialized in trapping and complete digestion of nematodes	
Nematode-trapping fungi (predatory fungi)	Vermiform nematode stages are captured by mycelial trapping devices, e.g. adhesive branches, adhesive knobs, adhesive three-dimensional nets, non-adhesive constricting rings; low host specificity and parasitic ability	*Arthrobotrys oligospora*, *Drechslera dactyloides*, *Nematroctonus robustus*
Endoparasitic fungi	Vermiform nematode stages are infected by fungal spores (conidia or zoospores); the fungus spends its entire vegetative life inside the infected nematode; obligate parasites, high host specificity	*Catenaria auxiliaris*, *Dactylella oviparasitica*, *Hirsutella rhossiliensis*, *H. minnesotensis*
Egg and female parasitic fungi	Nematode eggs or females are infected by fungal appressoria or zoospores	*Pochonia chlamydosporida*, *Purpureocillium lilacinum*
Toxin-producing fungi	Vermiform nematode stages are first immobilized by fungal toxins, prior to hyphal penetration through the nematode cuticle	*Coprinus comatus*, *Stropharia rugosoannulata*, *Pleurotus* spp.
Antagonistic fungi	Plant-associated fungi that suppress plant-parasitic nematodes and protect the plant against nematode damage	
Endophytic fungi		*Trichoderma* spp., *Piriformospora indica*
Arbuscular mycorrhiza fungi		*Glomus fasciculatum*

covered with adhesive material (Stirling, 2014). In some fungi, lateral branches grow from the vegetative mycelium, curve around and form an adhesive loop. Usually several loops are formed that result in a two- or three-dimensional network. Those adhesive networks are the most common trapping mechanism (e.g. *Arthrobotrys oligospora*, *A. superba*, *Dactylella pseudoclavata*). The most sophisticated trapping structures are constricting rings formed by fungi such as *A. dactyloides* and *Monacrosporium doedycoides*. On lateral branches of vegetative hyphae a three-celled ring is formed, about 20 μm in diameter. If a nematode enters the ring the three ring cells inflate instantly and hold the prey tightly.

In most cases trapping structures are induced in response to signals from the environment, including peptides and other compounds secreted by the host nematode (Dijksterhuis *et al.*, 1994). However, traps can also be initiated spontaneously. The mechanisms have been reviewed in detail by Tunlid and Ahrén (2011). In brief, once a nematode becomes attached to the trap surface, the fungus forms a penetration tube to pierce the nematode's cuticle. This process involves mechanical pressure plus the activity of hydrolytic enzymes that solubilize components of the cuticle (Tunlid and Ahrén, 2011). Once penetration has succeeded the nematode becomes paralysed. The penetration tube then forms an infection bulb that produces assimilative hyphae that invade the body and consume its content. The entire infection process is usually completed within 48–60 h (Dijksterhuis *et al.*, 1994). Abundance and diversity of nematode-trapping fungi varies considerably. Population densities are generally found to be highest in the upper soil layer (0–30 cm) and towards the end of the season (Persmark *et al.*, 1996).

Nematode-trapping fungi are not specific in their prey and trap free-living as well as plant-parasitic nematodes. This is an important aspect to be considered when it comes to the development of a biocontrol product and its effect against non-target organisms. Furthermore, susceptibility of plant-parasitic nematodes towards nematode-trapping fungi can also vary. As shown by Jaffee and Muldoon (1997), *Meloidogyne javanica*, *M. incognita* and *M. chitwoodi* were strongly suppressed by the nematode-trapping fungi

Monacrosporium ellipsosporum and *M. cionopagum*, but *H. schachtii* was not. Similarly, den Belder and Jansen (1994) observed efficient attachment of *A. oligospora* to root-knot nematodes but not to *Globodera pallida* and *G. rostochiensis*. That cyst nematodes are less susceptible to nematode-trapping fungi than root-knot nematodes cannot be concluded from these data as there is a lack of comparative studies. Nonetheless, numerous nematode-trapping fungi have been described as being associated with cyst nematodes (Table 11.1). For a few of these fungi their potential to control cyst nematodes has been demonstrated under laboratory and glasshouse conditions. *Monacrosporium lysipagum*, which produces adhesive knobs, infects mobile stages of *H. avenae* (Khan *et al.*, 2006a, b) and *Arthrobotrys cladodes* effectively controls *G. rostochiensis* (Davide and Zorilla, 1995).

11.2.1.2 Endoparasitic fungi

Endoparasitic fungi use conidia or zoospores that attach to the nematode cuticle where they germinate rapidly and penetrate the nematode body using assimilative hyphae. This group of nematophagous fungi includes mostly obligate parasites with no or only a limited saprophytic phase in the soil. Although they are highly pathogenic and specific, the low competiveness of endoparasitic fungi in the soil ecosystem usually hinders their use in biocontrol applications. However, where naturally present, they can play a significant role in soil suppressiveness towards plant-parasitic nematodes. The best-studied endoparasitic fungi are *Catenaria auxiliaris*, *Dactylella oviparasitica*, *Hirsutella rhossiliensis* and *H. minnesotensis*.

The obligate parasitic fungus *C. auxiliaris*, attacks saccate females of *Heterodera* spp. Using a single flagellum, zoospores of the fungus swim towards their female prey where they enter through natural openings or penetrate the cuticle directly. If infection occurs at an early stage, females will be completely destroyed by the fungus (Tribe, 1977). Fungal infection at a later stage may result in death of the female but eggs remain unharmed. *Catenaria auxiliaris* is widespread on *H. schachtii* in Europe and also occurs on *H. avenae* in England and Australia and *H. glycines* in the USA (Tribe, 1977; Crump *et al.*, 1983; Stirling and Kerry, 1983). However, the level of parasitism is usually low. As *C. auxiliaris* requires water for zoospore movement its distribution is restricted to moist, fine-textured soils (Stirling and Kerry, 1983).

The ascomycete *D. oviparasitica* was originally detected in peach orchards of the San Joaquin Valley of California parasitizing eggs of *Meloidogyne* (Stirling and Mankau, 1978). However, specificity seems to be low as the fungus successfully parasitizes eggs of plant-parasitic nematodes, *M. incognita*, *M. arenaria*, *M. javanica*, *M. hapla*, *H. schachtii* and *Tylenchulus semipenetrans*, as well as bacteria-feeding nematodes, *Diplenteron* sp. and *Acrobeloides* sp. (Stirling and Mankau, 1979). *Dactylella oviparasitica* was found to be associated with cysts of *H. schachtii* and identified as the key component of nematode suppressiveness occurring at a 1-ha field site at the University of California, Riverside's Agricultural Experiment Station (Westphal and Becker, 2001; Borneman and Becker, 2007). The mode of parasitism is not yet fully understood and seems to vary between root-knot and cyst nematodes. In the first case, hyphae of *D. oviparasitica* rapidly colonize egg masses of *Meloidogyne* and form appressoria after coming in contact with eggs (Stirling and Mankau, 1979). The fungus then penetrates the eggshell by mechanical force, but possibly also uses chitinases to facilitate this process (Stirling and Mankau, 1979). Eggs containing J2 usually escape parasitism. With *H. schachtii*, *D. oviparasitica* parasitizes the eggs in eggsacs outside the female but also invades the white females and cysts (Stirling and Mankau, 1979). More eggs are parasitized in white females than in cysts as females contain a higher ratio of eggs in the embryonic stage, which are preferred by *D. oviparasitica*, whereas eggs that contain J2 generally are not parasitized. However, studies by Becker *et al.* (2013) indicate that parasitism of developing juveniles is the main mode of action; these authors studied the mode of parasitism of *H. schachtii* by *D. oviparasitica* on *Arabidopsis thaliana* and cabbage and found that the fungus mainly parasitized and killed sedentary juveniles, whereas viable eggs seemed to be resistant to parasitism. However, parasitism of J2 only occurred outside the root after they had emerged from the root. By contrast, in *Arabidopsis* with its very fine roots, both male and developing females ruptured the root and were infected by *D. oviparasitica*; in the slightly thicker roots of cabbage only J2 within adult females that have broken through the root surface were infected by the fungus. Male juveniles remained enclosed within the cabbage root tissue avoiding infection. When developing females were infested, fecundity was reduced (Becker et al., 2013).

Hirsutella is an ascomycete genus. To date, two Hirsutella species, H. rhossiliensis and H. minnesotensis, have been identified as parasitizing cyst nematodes. However, overall host specificity is low and both species infect a broad spectrum of nematodes including plant-parasitic, bacteriovorous and entomopathogenic species (Timper et al., 1991; Tedford et al., 1994; Liu and Chen, 2001). According to Tedford et al. (1994), all 25 isolates of H. rhossiliensis obtained from various sources infected H. schachtii at a similar level, suggesting low host specificity between isolates. By contrast, Liu and Chen (2001) showed that isolates obtained from bacteriovorous nematodes did not infect H. glycines, indicating host specificity between isolates. Both species of Hirsutella produce adhesive spores on the tips of bottle-shaped phialides. Those spores attach to passing vermiform nematodes and are detached from the phialides (Stirling, 2014). The spores then germinate, produce an infection peg and hyphae penetrate the nematode's cuticle, filling its body completely with mycelium. Death of the juvenile usually occurs within 20 to 30 h after infection (Sun et al., 2015). Finally, hyphae start growing out of the nematode carcass about 3 days after infection to initiate a new infection cycle. The two Hirsutella species differ in their occurrence and host spectrum: H. rhossiliensis is widespread (Jaffee et al., 1994; Tedford et al., 1994) and infects a wide range of hosts, including nematodes of the genera Globodera, Heterodera, Meloidogyne, Xiphinema and Rotylenchus (Velvis and Kamp, 1995; Sun et al., 2015); H. minnesotensis has been reported from the USA, Germany, Poland and China infecting nematodes of the genera Aphelenchoides, Heterodera, Mesocriconema, Belonolaimus, Hoplolaimus, Steinernema and Heterorhabditis (Liu and Chen, 2001; Sun et al., 2015). In China, H. minnesotensis is the dominant parasite of H. glycines and is considered a main contributor to the suppression of H. glycines in the field (Xiang et al., 2010). Field trials have demonstrated its efficiency for the biological control of H. glycines (Liu and Chen, 2005). Although Hirsutella is highly pathogenic, slow growth characteristics and weak competiveness under field conditions limit its biocontrol potential. Several attempts have been made to develop a suitable formulation that protects the fungus from unfavourable soil conditions and supports its growth at the same time, but so far have not resulted in a biocontrol product (Jaffee and Muldoon, 1997; Patel et al., 2011).

11.2.1.3 Fungi parasitizing eggs and females

The group of nematophagous fungi that parasitize eggs and females use appressoria, a specialized penetration peg or lateral mycelia branches to invade and finally completely colonize eggs or females. Numerous parasites of eggs and females of cyst nematodes have been described (reviewed in Stirling, 2014) of which the best-studied are P. chlamydosporia and P. lilacinum.

Pochonia chlamydosporia has been isolated from eggs and cysts of numerous cyst nematode species (Zare and Gams, 2001). However, the host spectrum of P. chlamydosporia is very broad infecting eggs of several other plant-parasitic nematode taxa, and eggs of the intestinal roundworm, Ascaris lumbricoides, as well as eggs of slugs and snails (Zare and Gams, 2001). The role of P. chlamydosporia in reducing H. avenae in cereal monoculture was shown by Kerry (1975). During a survey on 30 cereal fields in southern England P. chlamydosporia was found in 76% of the sites and considered to be the most important egg parasite; P. chlamydosporia does not prevent J2 from invading roots and, therefore, is most effective when plants are tolerant of nematode damage (Kerry, 2000).

Purpureocillium lilacinum has been found to be associated with cyst nematode species from around the world (Khan et al., 2006a). Due to its broad host spectrum and high control potential the fungus was developed as a biological nematicide that is now marketed under different trade names worldwide (Table 11.2). The various products differ in the fungal isolate being used and the formulation but they are all registered for the control of cyst nematodes, among others. Purpureocillium lilacinum works by parasitizing nematode eggs and to a lesser extent juveniles and immature cysts (Khan et al., 2006b). In addition, fungal filtrates inhibit hatching and cause juvenile mortality, as shown on H. glycines by Manhong et al. (2002). The antagonistic potential of P. lilacinum has been demonstrated for *H. avenae* on barley (Khan *et al.*, 2006a), *H. schachtii* on sugar beet (Westphal and Becker, 2001) and *Globodera* spp. on potato (Franco *et al.*, 1981; Davide and Zorilla, 1995).

11.2.1.4 Toxin-producing fungi

Some nematophagous fungi produce toxins to immobilize nematodes before fungal hyphae penetrate through the cuticle and colonize the nematode body (Li *et al.*, 2015). In addition to toxins, a few species such as *Coprinus comatus* and *Stropharia rugosoannulata* use sharp structures to damage the nematode cuticle causing leakage of the body fluid. The toxins produced by these fungi include diverse chemical groups such as alkaloids, peptides, terpenoids, sterols and aromatic compounds. Some of those compounds show promising features for further development as biological nematicides.

11.2.2 Antagonistic fungi

Antagonistic fungi usually do not directly attack their nematode host but do protect the plant against nematode damage by mechanisms such as induced systemic resistance, competition for nutrients and infection sites, niche exclusion and improved plant growth.

11.2.2.1 Endophytic fungi

Here we describe two groups of antagonistic fungi, first *Trichoderma*, a soil and rhizosphere colonizing fungus that can also colonize the root internally and, second, 'true' fungal endophytes of plant roots.

Trichoderma is an opportunistic symbiont and various strains have been developed into products that are used to promote plant growth and control a range of soil-borne fungal pathogens (Stirling, 2014). Some *Trichoderma* strains have shown promising potential to reduce nematode populations and the level of plant damage. The majority of this work has been done on root-knot nematodes (Sharon *et al.*, 2011). However, several *Trichoderma* species have also been reported to be antagonistic towards cyst nematodes, including *T. harzianum*, *T. longibrachiatum* and *T. virens* (Saifullah and Khan, 2014; Zhang *et al.*, 2014). For example, *T. longibrachiatum* has been shown in glasshouse trials to reduce *H. avenae* on wheat (Zhang *et al.*, 2014). Following recognition of the cysts, spores of *T. longibrachiatum* germinated into hyphae that rapidly colonized the surface of the cysts, which were dissolved by fungal metabolites and the contents destroyed. Extracellular chitinases were the main mechanism used by *T. longibrachiatum* to kill *H. avenae* (Zhang *et al.*, 2014). Studying the parasitism of *G. rostochiensis* by *T. harzianum*, Saifullah and Khan (2014) used low temperature scanning electron microscopy to demonstrate that *T. harzianum* infected mature cysts of *G. rostochiensis* by directly penetrating the cyst wall or by entering via the mouth opening. In the first case, following penetration of the cyst wall, fungal mycelium penetrated the eggshell by chemical or mechanical means and then colonized the entire egg. From the work with root-knot nematodes, it is known that *Trichoderma* species are rhizosphere-competent, colonizing the root surface where they provide a protective shield against nematode attack (Sharon *et al.*, 2011). The main mechanisms involved in root-knot nematode suppression include antibiosis, parasitism, inducing host plant resistance and competition. However, it still needs to be determined if these mechanisms also explain the antagonistic effect against cyst nematodes. Future research should probably not focus on developing cyst nematode-specific *Trichoderma* strains into a product, but instead aim to add nematode suppression to the plant growth promotion and disease-control properties already available (Stirling, 2014).

Endophytic fungi are a very promising source for nematode control as they increase both plant tolerance and resistance to various plant-parasitic nematodes. The definition of an endophyte can be rather controversial depending on whether plant pathogens are included or only plant symbionts (Schouten, 2016). Within the context of this chapter the term endophytic fungi will be used exclusively for plant-colonizing fungi forming a mutualistic or synergistic interaction leading to reduced damage by plant-parasitic nematodes. Endophytic fungi form a rather heterogenic group and include genera of the phylum Ascomycota, such as *Fusarium*, *Trichoderma* and *Acremonium* and, to a lesser extent, Basidiomycota, such as *Piriformospora indica* (Schouten, 2016). A specific form of endophytes are arbuscular mycorrhizal fungi belonging to the phylum Glomeromycota. Furthermore, several nematode-trapping and egg-parasitic fungi spend a significant part of their life cycle as root endophytes (Bordallo *et al.*, 2002). Depending on the species, endophytic fungi can affect plant-parasitic nematodes by direct repellency, attack or killing through toxic compounds or toxins, or by inducing plant defence mechanisms. A thorough review on the mechanisms involved in nematode control by endophytic fungi and the molecular mechanisms behind it is given by Schouten (2016).

Although the beneficial effect of fungal endophytes controlling plant-parasitic nematodes is very well documented for the genera *Meloidogyne*, *Pratylenchus* and *Radopholus* (Sikora, 1992), studies on cyst-forming nematodes are very limited. Hallmann and Sikora (1996) showed that fungal metabolites of the endophyte *Fusarium oxysporum* Fo162 completely inactivated J2 of *H. schachtii* within 24 h of incubation, but not of the mycophagous species *Aphelenchoides composticola* and the myophagous species *Panagrellus redivivus*. As an explanation for this selective effect, the authors discussed differences in surface coat or detoxification systems, which have been developed in free-living nematodes to cope with fungal metabolites within the soil environment, whilst endoparasitic nematodes do not possess those traits due to their protected life within the root environment. An interesting finding from various genome sequences of plant-parasitic nematodes is that several different endoparasitic nematodes have reduced numbers of genes associated with immune responses (Gravato-Nobre and Hodgkin, 2011). The only other examples demonstrating the antagonistic potential of endophytic fungi towards cyst-forming nematodes are those by *P. indica* and arbuscular mycorrhizal fungi.

Piriformospora indica, originally isolated from the Thar Desert of Rajasthan, India, has been shown to enhance plant growth (Varma *et al.*, 1999) and protect a wide range of plants against various biotic and abiotic stresses (Deshmukh and Kogel, 2007). When used as a soil amendment, *P. indica* significantly reduced soil populations of *H. glycines* on soybean by over 30% compared to the control (Bajaja *et al.*, 2015). At the same time, shoot biomass increased by 30% and flowering by 75%. Daneshkhah *et al.* (2013) studied the potential of this fungus to induce resistance against *H. schachtii* on *A. thaliana* as model plant. The fungus successfully colonized the roots of *Arabidopsis* and significantly reduced reproduction of *H. schachtii*. Besides this direct beneficial interaction, application of fungal exudates as well as cell wall extracts also reduced nematode multiplication indicating that *P. indica* may induce systemic resistance in the plant (Daneshkhah *et al.*, 2013). Although fungal hyphae never colonized the vascular cylinder in the absence of *H. schachtii*, in the presence of *H. schachtii* the fungus penetrated into the vascular cylinder colonizing the cells previously destroyed by *H. schachtii* during their intracellular migration through the root system. Unfortunately, the hyphae of *P. indica* were never found inside nematode-induced syncytia. Daneshkhah *et al.* (2013) suggested that fungal proliferation around syncytia may affect the developing nematode by a hyphal network mechanically restricting syncytium expansion or chemically by the exudation of fungal compounds that may affect enzymatic processes of plant cells. Attraction of *H. schachtii* J2 towards *Arabidopsis* roots was significantly reduced by root diffusates collected from plants treated with *P. indica* 3 days before nematode inoculation compared to diffusates from plants treated with *P. indica* 7 days before nematode inoculation (Daneshkhah *et al.*, 2013). Thus, inoculation of *Arabidopsis* with *H. schachtii* during the biotrophic colonization stage of *P. indica* led to a significant reduction in the number of nematode infection sites and disturbed nematode development. Potential mechanical barriers formed by the fungal hyphae hindering J2 migration or syncytium expansion may play a secondary and rather minor role. Hence, the data suggest that *P. indica*-derived chemicals, exudates and cell wall compounds cause the major inhibitory effects on the development of *H. schachtii* in *Arabidopsis* roots.

11.2.2.2 Arbuscular mycorrhiza fungi

Arbuscular mycorrhiza fungi (AMF) have been reported from the vast majority of wild and cultivated plants, where they form a symbiotic relationship resulting in improved plant growth and health (Hol and Cook, 2005). AMF are obligate symbionts that penetrate the plant root tissue intercellularly and eventually form extracellular structures such as arbuscules and vesicles. The potential of AMF to reduce plant infection by plant-parasitic nematodes has been demonstrated for cyst nematodes (Hol and Cook, 2005). For example, reproduction of *H. cajani* on cowpea and pigeon pea was reduced in the presence of *Glomus fasciculatum* (Jain and Sethi, 1987; Siddiqui and Mahmood, 1995). Soybean tolerance towards *H. glycines* was increased following application of an AMF inoculum mix (Tylka *et al.*, 1991). The main factors affecting this tritrophic interaction are time and intensity of AMF colonization, efficacy of the AMF strain and identity and density of the nematode. In general, the earlier and more intensive the root colonization by AMF the better the nematode suppression. AMF introduction 15 days prior to inoculation of *H. cajani* adversely affected nematode root penetration to a greater extent than at simultaneous inoculation (Jain and Sethi, 1988). Nematode root penetration was considerably hampered when AMF colonization was above 60% of the root system but development or female fecundity were unaffected if nematodes were able to settle in the root (Jain and Sethi, 1988).

However, plant colonization by AMF does not always result in mutual benefits and sometimes even adverse effects are reported. For example, cowpea colonized by *Glomus epigaeus* stimulated reproduction of *H. cajani* and high nematode densities can even reduce AMF root infection and spore production (Jain and Sethi, 1987). Tylka *et al.* (1991), studying the effect of AMF on *H. glycines*, found that soil population densities of *H. glycines* were greater in microplots infested with AMF than in control microplots. The overall effect of AMF varied with time and nematode density. *Heterodera glycines* population densities were reduced by AMF by as much as 73% at the highest nematode inoculum level 7 weeks after planting, but were later comparable on AMF-treated and non-treated plants (Tylka *et al.*, 1991). Early nematode infestation can even reduce AMF colonization. Winkler *et al.* (1994) demonstrated that colonization of soybean roots by AMF was negatively correlated with *H. glycines* population densities and concluded that this effect was due to nematode antagonism to the mycorrhizal fungi rather than suppression of nematode populations.

The effect of the nematode on the outcome of the AMF × cyst nematode interaction was further studied by Ryan *et al.* (2000, 2003) on potato. If potato plants 'Golden Wonder' were inoculated with Vaminoc, a mycorrhizal inoculum consisting of three *Glomus* species, hatch of *G. pallida* was stimulated by almost 25% compared with non-mycorrhizal potato plants (Ryan *et al.*, 2000). No such effect was observed for *G. rostochiensis*. This raises the question of how two closely related species react so differently to AMF inoculation. The differences in hatching behaviour between the *Globodera* species were observed in soil and *in vitro* using independent batches of potato plants, indicating that this effect is not an artefact and was most likely caused by AMF-altered production of *G. pallida* selective hatching chemicals from potato roots (Ryan *et al.*, 2000). On strawberry, a non-host for *G. pallida*, AMF was not able to induce plant-derived hatching factors, further supporting the hypothesis that those hatching factors are host plant specific. Analysis of potato root diffusates by anion exchange liquid chromatography detected several potato cyst nematode species-specific hatching factors in the diffusate of mycorrhizal plants but not in non-mycorrhizal plants (Ryan and Jones, 2004). The composition of the potato root diffusate also differed in its main components with AMF colonization resulting in a 20% increase in carbon but a 48% decrease in nitrogen concentration compared with non-mycorrhizal diffusate. In addition to affecting hatch, AMF also affected nematode development and reproduction. Twelve weeks after inoculation, the multiplication rate of *G. rostochiensis* was significantly higher on AMF-treated potato plants than non-treated plants, whereas no such difference was observed for *G. pallida*. Whilst the different reaction of the two potato cyst nematodes species still remains unsolved, the increase in multiplication rate can be attributed to the larger root system of AMF-treated plants compared with non-treated plants. In addition, the period of cyst production in mycorrhizal plants was longer than in non-mycorrhizal plants, which could be due to a greater food supply for the nematodes resulting in a higher proportion of females (Ryan *et al.*, 2003). In contrast to *G. rostochiensis*, Deliopoulos *et al.* (2007) showed for *G. pallida* that the interaction with AMF was mutually inhibitory, each reducing the population of the other, with AMF reducing *G. pallida* reproduction more than *G. pallida* reduced AMF root colonization. In conclusion, the most frequently reported effect of AMF × cyst nematode interactions are enhanced plant tolerance to nematode attack in response to mycorrhization (Tylka et al., 1991; Hol and Cook, 2005). However, generalizations of the effects of AMF on cyst nematode development and plant growth are difficult as the interactions between AMF and cyst nematode are highly specific and depend on the particular host plant and nematode combination (Deliopoulos et al., 2008).

11.3 Rhizobacteria and Root Endophytic Bacteria

Rhizobacteria colonize the root rhizosphere and form a barrier to protect the root against pathogen attack (Weller, 1988). Some rhizobacteria also have the ability to colonize plant roots (Schroth and Hancock, 1982). Depending on their effects on plant health and plant growth, these bacteria were termed as 'plant health promoting rhizobacteria' (PHPR) by Sikora (1988) and 'plant growth promoting rhizobacteria' (PGPR) by Kloepper et al. (1991). They are widely studied for their potential to control plant-parasitic nematodes. Rhizobacteria regulate nematode behaviour by interfering with plant–nematode host signals (Oostendorp and Sikora, 1990; Sikora and Hoffmann-Hergarten, 1993), promote plant growth (El-Nagdi and Youssef, 2004), induce systemic resistance (Hasky-Günther et al., 1998) or directly antagonize nematodes by means of the production of toxins, enzymes and other metabolic products (Siddiqui and Mahmood, 1999).

Root endophytic bacteria grow internally in the root tissue, without causing any detrimental effect on the plants (McInory and Kloepper, 1995; Hallmann et al., 1997, 1999). Although endophytic bacteria occupy a different habitat from rhizobacteria, they both share similar mechanisms for controlling nematodes. The effects of endophytic bacteria may be manifested through the production of inhibitory chemicals and through inducing systemic resistance in plants to reduce nematodes and increase plant health (Hallmann et al., 1998; Hallmann, 2001; Compant et al., 2005). Thus, endophytes have three roles in nematode control: (i) to inhibit hatching (Kluepfel et al., 1993; Westcott and Kluepfel, 1993); (ii) to degrade the hatching factor(s) (Oostendorp and Sikora, 1989); and (iii) to reduce mobility or kill the infective stages of nematode, thus reducing nematode population even before the onset of infection (Ali et al., 2002). In addition, the colonization of endophytic bacteria in the same root tissue where sedentary nematodes, such as cyst nematodes, establish their feeding sites, reduces sites available to the nematodes and thus creates direct competition between females, reducing their individual fecundity, and thus enhancing the potential of endophytic bacteria as biocontrol agents (Siddiqui and Ehteshamul-Haque, 2000; Siddiqui et al., 2003).

11.3.1 *Pseudomonas* species

Pseudomonas spp. are among the most dominant bacterial populations in the rhizosphere that are able to antagonize plant-parasitic nematodes (Rovira and Sands, 1977; Krebs et al., 1998). Many strains of *Pseudomonas fluorescens* are known to produce an antibiotic compound 2,4-diacetylphloroglucinol (DAPG), which is toxic to many organisms, including plants, fungi, viruses, bacteria and nematodes (McSpadden Gardener, 2007; Weller, 2007). Application to soil of strains of *P. fluorescens* capable of producing DAPG will further induce plant resistance against pathogens (Iavicoli et al., 2003; Siddiqui and Shaukat, 2003, 2004; Weller et al., 2004; Van Loon and Bakker, 2005; Bakker et al., 2007) and increase crop yields (McSpadden Gardener et al., 2006a, b).

Fenton et al. (1992) compared the DAPG-producing *P. fluorescens* F113 strain with the DAPG-negative biosynthetic mutant F113G22 under *in vitro* and soil microcosm conditions and found that strain F113 increased the hatch of *G. rostochiensis* but reduced J2 mobility. Similar results were obtained by exposing *G. pallida* cysts and J2 to synthetic DAPG (Cronin et al., 1997). A more recent study (Meyer et al., 2009) indicated that DAPG is not universally toxic and its nematicidal efficacy can vary between different taxonomic groups of nematodes and different life stages of a species. For example, neither hatch nor viability of J2 or adults of *H. glycines*, *Pratylenchus scribneri*, *P. pacificus* and *Rhabditis rainai* were affected by DAPG treatment. Thus, for field application it is important to know the control spectrum of DAPG-producing *P. fluorescens* strains. A fluorescent *Pseudomonas* strain (FPs6) isolated from soil infested with *H. cruciferae* was able to inhibit hatching *in vitro* and the unhatched nematodes and the hatched J2 were infected by the bacteria (Aksoy and Mennan, 2004).

The efficacy of pseudomonads as seed treatment against cyst nematodes has been successfully demonstrated. Oostendorp and Sikora (1989) recorded a 68% reduction in root invasion by *H. schachtii* following coating of the beet seeds with *P. fluorescens* P523. In glasshouse tests, eight bacteria isolates including *P. fluorescens*, isolated from sugar beet rhizosphere and applied as a seed treatment, suppressed early root infection by *H. schachtii*. In the first year of the trial, one isolate reduced penetration significantly by 75% and three isolates caused significant increases in yield, although in the second year one isolate significantly reduced nematode penetration but none of the bacterial treatments affected plant growth or yield (Oostendorp and Sikora, 1989).

Andreoglou *et al.* (2003) reported that temperature and soil moisture are the key factors determining the control potential of *P. oryzihabitans* against *G. rostochiensis*. Invasion of J2 in potato roots was reduced more at 25°C and 21°C than at 17°C. The bacterial motility *in vitro* was inhibited at temperatures below 18°C, while in soil its movement and survival were suppressed at 16°C. At both temperatures the biocontrol agent moved faster in the wetter (−0.03 MPa) than in the drier soil (−0.1 MPa). In a pot trial with *P. putida* strain 3(2) and *P. aurantiacea* strain 13(2), the application of 35 ml aqueous cell suspension (10^8 cells ml^{-1}) in 600 g soil at the time of planting significantly reduced a population of *G. rostochiensis* by 40.7% and 42.2%, respectively, compared to the control (Trifonova *et al.*, 2014).

11.3.2 *Bacillus* species

Rhizobacteria belonging to the *Bacillus* group have been intensively studied for their nematicidal activity. Lian *et al.* (2007) reported that a cuticle-degrading protease enzyme, which is released by several *Bacillus* strains, possesses strong nematicidal activity. They identified an extracellular cuticle-degrading protease, Apr219, from *Bacillus* sp. strain RH219, which shared high similarity with cuticle-degrading proteases from *Brevibacillus laterosporus* strain G4 and *B. nematocida* strain B16. Huang *et al.* (2005) reported strong nematicidal activity of *B. nematocida* (B16) against *Panagrellus redivivus*. However, the effects of the cuticle-degrading protease enzyme against cyst nematodes have yet to be ascertained. In glasshouse and laboratory experiments, treatment of soybean roots with *B. subtilis* inhibited the migration of *H. glycines* J2 towards the roots. Following the treatment of both soil and seeds with either wettable powder or a solution containing *B. subtilis*, a further reduction in the number of females was observed in the soybean root (de Araújo *et al.*, 2002).

In vitro studies showed that the culture filtrates of *B. cereus* (09B18) and *Achromobacter xylosoxidans* (09X01), isolated from cysts of *H. filipjevi* in Henan, China, reduced hatch of *H. filipjevi* by 83.1% and 58.9% after a month, and caused 100% and 99.5% mortality of the J2 within 72 h, respectively (Zhang *et al.*, 2016b). The application of bacterial suspensions in glasshouse trials significantly reduced the numbers of females in wheat roots by 75.9% for 09B18 and 70.2% for 09X01, whilst in the field the reduction was 43.5% by 09B18 and 51.1% by 09X01. There was an associated increase in the wheat yields by 15.9% (09B18) and 13.2% (09X01) compared to the untreated control.

Bacillus firmus produces chemicals that inhibit hatching and cause paralysis of J2 of various different species including *H. glycines* (Mendoza *et al.*, 2008; Terefe *et al.*, 2009; Schrimsher *et al.*, 2011). Geng *et al.* (2016) reported that the fermentation supernatant of *B. firmus* strain DS-1 from 18-h to 36-h-old cultures caused 92.5% and 97.8%, mortality, respectively, of *H. glycines*, which was much higher than for *M. incognita*. A peptidase S8 super-family protein, designated as Sep1, exhibiting serine protease activity, was found to be responsible for degradation of the intestinal tissues of *Caenorhabditis elegans* and *M. incognita*, and may also affect cyst nematodes in a similar manner.

Bacillus thuringiensis (Bt) is known to produce proteinaceous protoxin crystals (Cry protein) during sporulation (Schnepf *et al.*, 1998) that show toxin activity on caterpillars, beetles and nematodes. Since the first report by Prasad *et al.* (1972), showing significant reduction of *M. incognita* by *B. thuringiensis* var. *thuringiensis*, several studies have followed. Cry5B is the most extensively studied Bt protein against nematodes; following application to juveniles of *M. incognita* and *C. elegans*, Cry5B interacted with specific receptors located on the membrane of gut epithelial cells causing lytic pores in the intestine (Griffitts *et al.*, 2005; Vachon *et al.*, 2012). Another protein, Cry6Aa2, was found to cause growth inhibition, reduced brood size and abnormal motility in *C. elegans* (Luo *et al.*, 2013; Zhang *et al.*, 2016a). A combination of the two crystal proteins can enhance the nematicidal toxicity of Bt and provide an alternative to overcome the potential resistance of plant-parasitic nematodes (Yu *et al.*, 2014). Besides Cry proteins, Bt also produces additional virulence factors with nematicidal activity (Nielsen-LeRoux *et al.*, 2012). Unfortunately, none of the Bt toxins has yet been targeted against cyst nematodes, although Bt may be a candidate for control of cyst nematodes in the future.

11.3.3 *Rhizobium etli*

The rhizobacterium *Rhizobium etli* G12, initially identified as *Agrobacterium radiobacter*, was originally isolated from the rhizosphere of potatoes (Racke and Sikora, 1992). *Rhizobium etli* G12 is known to reduce early root infection by *G. pallida* (Hasky-Günther *et al.*, 1998). Split-root experiments with live and heat-killed cells of *R. etli* G12 indicated that these are responsible for induced systemic resistance in potato roots against *G. pallida* infection (Hoffmann-Hergarten *et al.*, 1997; Hasky-Günther *et al.*, 1998). *Rhizobium etli* has two heat-stable surface carbohydrates, exopolysaccharides (EPS), which form a sticky layer around the bacterial cell as an additional capsule, and lipopolysaccharides (LPS), which are an integral part of the outer membrane of the cell. These together mediate the recognition process in the symbiotic interaction between *Rhizobium* and legumes (Leigh and Coplin, 1992; Denny, 1995). Reitz *et al.* (2000) found that, unlike EPS, the application of LPS to potato roots induced systemic resistance to *G. pallida* infection at concentrations as low as 1.0 and 0.1 mg ml^{-1}. In split-root experiments, soil treatment of one half of the root system with LPS resulted in up to 37% lower infection rates of *G. pallida* in the untreated half of the root system. Reitz *et al.* (2002) demonstrated that the core region of the LPS was the main trigger of the resistance response, whereas the O-antigen region only had a minor effect. In a split-root system both the entire LPS and its core region systemically reduced *G. pallida* infection significantly by 45% of the untreated control. By contrast, the lipid A fraction provided an insignificant 20% reduction in nematode infection. Mahdy *et al.* (2001) reported that *R. etli* G12 was effective in inducing resistance to *G. pallida* in potato. Hackenberg and Sikora (1994) observed that seed potato treated with *R. etli* G12 exhibited reduced root penetration of potato by *G. pallida* J2 in glasshouse studies.

11.3.4 Chitinase-producing bacteria

Chitin is present in the outer layer of eggshells of *H. schachtii* and *H. glycines* (Perry and Trett, 1986), which could make them susceptible to chitinase-producing bacteria. A screening of 3200 bacterial isolates for chitinase production showed 137 to be chitinase-positive. Out of these, *Stenotrophomonas maltophilia* and *Chromobacterium* sp. at 10^6 CFU ml^{-1} cell density or greater and an incubation time of 2 weeks reduced hatch of *G. rostochiensis*, both in *in vitro* and soil microcosm assays (Cronin *et al.*, 1997).

The amendment of chitin in soil stimulates chitin-producing bacteria that degrade chitin, first of the provided chitin source, and second from other sources such as nematode eggs. If the chitin in eggshells is degraded, premature hatch will occur with J2 less viable than under normal conditions. The depolymerization of chitin by chitinase in soil releases ammonia, which is nematicidal to J2 at certain concentrations. In a glasshouse study, Tian *et al.* (2000) reported suppression of *H. glycines* by chitinolytic bacterial isolates in heat-treated silt loam soil amended with 0.6% (w/w) chitin. A synergistic effect between the isolates and the chitin substrate reduced nematode numbers. Further details of the molecular aspects of chitinases will be developed in section 11.4.2.2.

11.3.5 *Pasteuria* species

The *Pasteuria* group of bacteria are members of the *Bacillus–Clostridium* clade that produce endospores and are hyperparasites of plant-parasitic nematodes and water fleas (Cladocerca: Daphnidae) (Chen and Dickson, 1998; Charles *et al.*, 2005). A major attribute of *Pasteuria* is their host specificity. Some isolates display cross-generic parasitism of nematodes. Among the five documented species of *Pasteuria* that can parasitize plant-parasitic nematodes, *P. nishizawae* was reported to infect cyst nematodes (Sayre *et al.*, 1991). Attachment of the endospores (Fig. 11.1) has been observed on J2 of *G. rostochiensis*, *H. lespedezae*, *H. schachtii* and *H. trifolii*, but the completion of the life cycle of *P. nishizawae* has been confirmed only on *H. glycines* (Atibalentja *et al.*, 2004; Noel *et al.*, 2005). Mohan *et al.* (2011) reported that a *Pasteuria* population (HcP), originally isolated from *H. cajani* in India, could also parasitize and successfully complete its life cycle on *G. pallida* from the UK. Characterization of a 1430-bp contig constructed from the 16S rRNA gene of the HcP isolate showed it to be most closely related to *P. nishizawae* with 98.6% identity. As such, it is not clear whether *P. nishizawae* and the HcP isolate are the same species, or are different biotypes of the same species. In 2013, Syngenta launched Clariva®, which is a seed treatment product carrying *P. nishizawae* for the control of *H. glycines*. A considerable body of research has accumulated on the biology of *Pasteuria*, including information on attachment and molecular aspects (see section 11.4).

11.3.6 *Brevibacillus laterosporus*

Brevibacillus laterosporus is an opportunistic parasitic bacterium and an emerging entomopathogen with broad-spectrum biocontrol activity reported against several insects, nematodes and molluscs; antimicrobial activity of certain strains are of pharmaceutical interest as they produce an array of antibiotics (Ruiu, 2013). Unlike *Pasteuria*, which is an obligate nematode parasite, *B. laterosporus* lives a saprophytic life but qualifies as an opportunistic parasitic bacterium, as it can puncture through the cuticle to infect and kill a nematode host and digest the host tissue for nutrients. Pathogenicity of *B. laterosporus* strain G4 has been reported against *H. glycines*, *Trichostrongylus colubriformis*, *Bursaphelenchus xylophilus* and *P. redivivus* (Singer, 1996; Oliveira *et al.*, 2004). Huang *et al.* (2005) and Tian *et al.* (2006) demonstrated that BLG4, an extracellular alkaline serine protease, contributed to the pathogenic activity and speculated that some other extracellular enzymes or toxins are probably also important for the nematicidal activities.

11.3.7 Others

Acinetobacter sp. SB13, *Arthrobacter* sp. SB14 and *Bacillus* sp. SB15, isolated from the rhizosphere of sugar beet plants, significantly increased hatch of *G. pallida in vitro* and *in vivo* when applied as a seed coat application by colonizing the

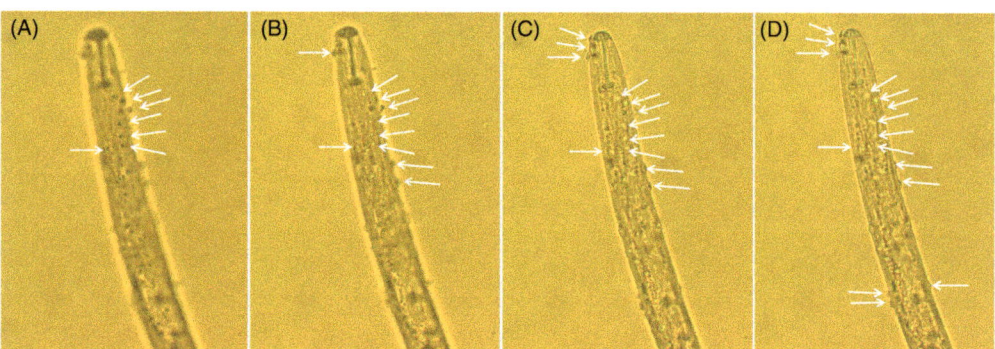

Fig. 11.1. Anterior of a second-stage juvenile of a cyst nematode (*Heterodera* sp.) at increasingly deeper focal planes. A, B, C and D show endospores of a *Pasteuria* species (arrows) adhering to the nematode cuticle. (Courtesy of K.G. Davies.)

rhizosphere (Lettice and Jones, 2015). In a screening programme, 16 bacterial isolates out of 179 isolated from root and cysts caused more than 25% reduction in *G. pallida* penetration of potato roots. In glasshouse and field experiments, *Bacillus sphaericus* (B43) reduced the penetration of *Globodera* J2 by 30–40% (Racke and Sikora, 1992; Hackenberg and Sikora, 1994). Comparisons were made by applying three different bioagents including *B. subtilis*, the N-fixing bacterium and the vesicular-arbuscular mycorrhiza (VAM) fungi *Bradyrhizobium japonicum* and the AMF *Glomus fasciculatum* alone or in combination for the management of wilt disease complex of pigeon pea caused by *H. cajani* and *Fusarium udum*. Application of all the three bioagents not only reduced the nematode multiplication and wilting index, but also increased the plant shoot dry weight, nodulation and phosphorus uptake (Siddiqui and Mahmood, 1995).

11.4 Microbial–Nematode Molecular Interactions

As is clear from the previous sections, suppression of plant-parasitic nematodes can either be general or specific – arguably the distinction is currently exceedingly difficult to determine as insufficient examples have been investigated to ascertain the modes of action and mechanisms by which any given nematode suppressive soil is manifest. In relation to assessing if a potential biological control agent is to be successful, it is interesting to consider the nematicide aldicarb as a 'thought' experiment. Aldicarb is a cholinesterase inhibitor, a mode of action that is effective across a broad range of animals, including mammals; therefore, it can be regarded as 'generalist' in application. However, investigations looking at mutations in the synaptic gap using *C. elegans*, shows that sensitivity to aldicarb can be altered both positively and negatively (Mahoney *et al.*, 2006). These results indicate that different types of resistant subpopulations can arise that are the result of highly specific genetic changes. Indeed, what appears to have a general mode of action can lead to populations with highly specific resistance that, given time, would result in a co-evolutionary arms races. Therefore, *a priori*, and applying the aldicarb example to the concept of 'general' soil suppressiveness, understanding the molecular interactions between a particular natural enemy and the host nematode it is controlling, becomes a crucial step in understanding the nematode suppressiveness of a soil and how to manage it. Using natural enemies as biological control agents in the field requires an understanding of the molecular interactions between microbes and nematodes at the population level if the deployment of a potential biological control agent is to be successful.

Understanding molecular interactions between microbes and nematodes in this detail is in its infancy, and knowledge of molecular interactions between microbes and cysts nematodes is sparse. From a generic biological perspective this interaction is fundamentally a host–parasite interaction where the host is the nematode and the parasite the microbe, and our knowledge is focused at this important interface. Building on section 11.3.2, in this section we will use *C. elegans* as a model nematode–microbial pathogen system and then examine some key examples to gain insights into using microbial enemies as potential control agents.

11.4.1 Insights from *Caenorhabditis elegans*

One of the major issues confronting the use of microbes to control nematodes is their inability robustly to control plant-parasitic nematodes in the field. Different isolates of the same species of microorganism do not all behave with the same efficacy. Each isolate maintains itself and is adapted to a particular environmental niche, which in the rhizosphere is always in continual biotic and abiotic flux. Therefore, each isolate has its own set of molecular and biochemical characteristics that make it more or less amenable to use as a control agent. Understanding the molecular and biochemical interactions between microbial enemies of plant-parasitic nematodes and their hosts is currently rudimentary and focused on those nematode–microbial interactions where the microorganism may have potential as a commercial biological control agent. Interestingly, with increasing knowledge in the common themes that run across the diseases of the plant and animal kingdoms under the broad title of innate immunity, substantial insights have been made using *C. elegans* as a model of microbial infection processes (Gravato-Nobre and Hodgkin, 2011). Knowledge gained from these model systems may help in the selection of the most suitable organisms necessary for their deployment in the field and, as we will see, perhaps be useful in developing novel biological control strategies.

Caenorhabditis elegans has to combat multiple encounters with microbes and it has evolved a defence system and strategies that lead to the production of effectors with antimicrobial properties (Ausubel, 2005; Kim and Ausubel, 2005; Schulenburg and Ewbank, 2007; Gravato-Nobre and Hodgkin, 2011). As many of these defence mechanisms are shared with higher organisms it is likely that these mechanisms will also be used by plant-parasitic nematodes. Any microbial enemy being used as a control agent will therefore have to avoid such defence responses.

Initiation of infection is key and one obvious difference between *C. elegans* and plant-parasitic nematodes is that internal ingestion of microbes by plant-parasitic nematodes is unlikely. Nematodes are protected by a tough multi-layered cuticle that serves as an external skeleton and the surface of which acts as the first line of defence against microbial attack (Curtis *et al.*, 2011; Davies and Curtis, 2011). The surface coat of the cuticle is labile and there is a continual turnover of proteins, carbohydrates and lipids, the rate of which can change in response to environmental cues (Grenache *et al.*, 1996; Olsen *et al.*, 2007). The cuticle surface coat can be seen as part of the nematode's immune defence system that any microbial parasite must evade.

Mutational studies of the surface coat of *C. elegans* have shown altered responses to fungal (Mendoza de Gives *et al.*, 1999; Fekete *et al.*, 2008) and bacterial (Hodgkin *et al.*, 2000; Ewbank, 2002; Gravato-Nobre and Hodgkin, 2005) infections. The most commonly studied mutants fall into three groups: (i) the *srf* mutants that were originally identified by altered binding of lectins to the cuticle surface coat (Politz and Phillip, 1992; Silverman *et al.*, 1997); (ii) the *bus* (bacterially unswollen) mutants that conferred resistance to the bacterium *Microbacterium nematophilum*, which produced nematodes with a swollen anal region (Gravato-Nobre and Hodgkin, 2005); and (iii) the *bah* (bacteria absent from the head) mutants that were identified by the absence of biofilm development around the head region in cultures grown in the presence of *Yersinia pestis* (Darby *et al.*, 2002; Joshua *et al.*, 2003). These mutational studies have proved useful in identifying the types of genes involved in surface coat production and have revealed that there is a close link between microbial attachment and glycosylation and/or post-translational modification pathways that are involved in building up large complex molecules on the nematode surface (Parsons *et al.*, 2014). The surface coat is present in all known nematodes and currently where mutational analysis has led to their genetic characterization, orthologues and/or homologues have been found in both animal- and plant-parasitic nematodes (Gravato-Nobre and Hodgkin, 2011); *C. elegans*, therefore, provides a starting point to investigate this fundamental interaction between plant-parasitic nematodes and potential microbiological control agents.

11.4.2 Microbial diversity and molecular interactions

As can be seen from Table 11.1 nematophagous fungi are a phylogenetically very diverse group. With the exception of the obligate parasites, most of the groups listed in Table 11.1 can be cultured saprophytically on artificial media and under the control of environmental cues can switch from vegetative growth into parasitic growth. Upon switching to a parasitic life cycle strategy the fungus must breach the protective layer of the nematode cuticle or eggshell. In essence, these protective structures are multilayered, consisting of proteins and lipids embedded in some form of micro-fibrous structure; chitin in eggshells and collagen in nematode cuticles, on the surface of which is a glycocalyx with a large carbohydrate component (Bird and Bird, 1991; Davies and Curtis, 2011). One of the key characteristics of the nematophagous fungi is that they produce an array of extracellular enzymes that in concert are capable of weakening the protective layer of the nematode cuticle or eggshell so that hyphae can penetrate and set up an infection. The most important groups of enzymes thought to be involved in weakening this protective barrier are proteases and chitinases (Marciá-Vicente et al., 2011).

11.4.2.1 Microbial inter- and intraspecific extracellular enzyme diversity

Nematophagous fungi are a cosmopolitan group of fungi that can be found in diverse habitats wherever nematodes are present. For example, *P. chlamydosporia* has been isolated from economically important plant-parasitic nematodes from many countries and continents (Kerry and Hirsch, 2011). Although characterized as an egg parasite, the associated molecular mechanism remains poorly understood. Hyphae of the fungus, usually at the tip, develop appressoria that bind tightly to the eggshell and then penetrate using an infection peg, which subsequently produces an infection bulb that then differentiates further leading to mycelial growth that destroys the contents of the egg (Segers et al., 1996). On synthetic growth media the fungus has been shown to produce a cocktail of enzymes (Dackman et al., 1989; Dupont et al., 1999; Esteves et al., 2009); however, their role as determinants of infection and host range is currently unclear. Two groups of enzymes that are generally thought to be of key importance are the chitinases and proteinases.

11.4.2.2 Chitinases

Chitin is a polymer of N-acetylglucosamine and is the major component of the exoskeleton of crustacea and nematode eggs. Applications of chitin, usually in the form of ground up crab shells, to soil are thought to stimulate chitinolytic microbes and enhance nematode control through chitinases that can degrade the eggshell (Godoy et al., 1983; Rodríguez-Kábana, 1986; Rodríguez-Kábana et al., 1987). This can be from bacteria that produce ammonia (section 11.3.4) or fungi. One of the most intensively studied groups of fungal chitinases are those occurring in *Trichoderma* spp. (Sharon et al., 2011). Using media in which chitin is the major carbon source, the various chitinases produced can be characterized into three major types: (i) the N-acetylglucosaminidases, a large group of enzymes that are increasingly expressed in the presence of glucosamine and are also thought to be important in triggering the expression of other chitinolytic enzymes (Haran et al., 1996; Viterbo et al., 2002); (ii) exochitinases, which by hydrolysis cleave off two subunits from either the reducing or non-reducing ends of the chitin chain and have been shown to be active when *Trichoderma* has been grown in the presence of crabshell chitin (Harman et al., 2006); and (iii) endochitinases, which comprise a large family and can cut randomly within a chitin chain and are thought to be important in a number of mycoparasitic interactions (Carsolio et al., 1999; Zeilinger et al., 1999; De Marco et al., 2000). Although many of these chitinases have been characterized (Tikhonov et al., 2002; Dong et al., 2007; Gan et al., 2007), the nature of their inter- and intraspecific diversity and the detailed role they may play in nematode–microbial interactions remains largely unexplored.

11.4.2.3 Proteinases

The expression of the gene encoding a 31-kDa serine proteinase (Prb1) produced by *T. atroviride* (Geremia et al., 1993) was upregulated by chitin, and transgenic fungal lines carrying multiple copies of the gene improved biological control activity of *Rhizoctonia solani* (Flores et al., 1997); one line of *T. harzianum* (P-2) exhibited increased control of root-knot nematodes in in vitro bioassays (Sharon et al., 2001). Further characterization of proteinases from a number of different *Trichoderma* spp. revealed a range of proteolytic capacities, perhaps explaining differences in the aggressiveness of the different strains against different life stages of the nematode (Sharon et al., 2011).

Proteinases from other nematophagous fungi that can potentially be developed into biological control agents have been investigated. A serine proteinase from *Pochonia rubescens* (= *Verticillium suchlasporium*) and designated P32 was implicated in the infection of *H. schachtii* (López-Llorca, 1990) and another related serine proteinase (designated VCP1) was capable of digesting specifically the vitelline membrane of nematode eggs (Segers et al., 1996). Interestingly, amino acid substitutions at particular points, thought to be involved in an active site of the protein sequence of this enzyme across different isolates of this fungus, altered its specificity and changed its ability to digest different vitelline membranes of cyst and root-knot nematodes (Morton et al., 2003, 2004).

The proportion of chitin to protein differs between different species and genera; for example, chitin takes up a greater proportion of eggshell in root-knot nematodes compared to cyst nematodes, and the eggshells of *Globodera* spp. tend to be thicker than in species of *Heterodera* (Marciá-Vicente *et al.*, 2011). The implications of this suggest that co-evolutionary arms races would therefore optimize the proportions and types of chitinases and proteinases produced to suit the host and lead to adaptation that would favour specialization. One could hypothesize that similar arguments could be applied to the nematode cuticle.

11.4.3 Attachment and surface recognition

The secretion of extracellular enzymes by potential microbial enemies of plant-parasitic nematodes is clearly an important step in any infection process. However, prior to infection and enzyme secretion, binding of the microbe to a potential host surface is fundamental. The investigation of binding of microbes to nematode cuticle surfaces is poorly understood with most detailed genetic studies focusing on *C. elegans* (Gravato-Nobre and Hodgkin, 2011).

Extensive and comprehensive investigations on microbial recognition and attachment have been on endospores of the Gram-positive obligate bacterial parasite *Pasteuria* spp. and root-knot nematodes. Early work showed that different populations of endospores did not attach equally well to all populations of root-knot nematodes (Stirling, 1985; Davies *et al.*, 1988; Espanol *et al.*, 1997) and monoclonal antibodies revealed that both the surface of the endospore and the surface of the infective J2 cuticle were biochemically highly heterogeneous (Davies and Danks, 1993; Davies *et al.*, 1994, Davies and Redden, 1997), and some form of protein–carbohydrate interaction was occurring. Interestingly, at the level of endospore attachment, Fahrenholtz's Rule, which simply states that the phylogeny of hosts and parasites mirror each other, was not upheld (Davies *et al.*, 2001) and suggests that endospore binding to cuticle is in some form of reiterative dynamic process that is generationally determined. The most recent research from *C. elegans* would suggest that glycosylation of cuticle components of the surface coat may be playing a role (Parsons *et al.*, 2014) on the nematode side of the host–parasite interaction and hypothetically could involve mucin-like molecules (Davies and Curtis, 2011).

Co-evolutionary arms races between the nematode surface coat and a population of attaching *Pasteuria* endospores would necessarily produce surface heterogeneity on the part of the endospore populations. *Pasteuria* spp. are phylogenetically closely related to the endospore-forming *Bacillus* spp. of bacteria (Charles *et al.*, 2005). This group of bacteria can broadly be divided into animal pathogens (e.g. *B. cereus* and *B. thuringiensis*) and non-pathogens (e.g. *B. subtilis*). Fibronectin is an integrin that occurs in the extracellular matrix across the animal kingdom and has been found to be responsible for the adhesion of microbes, and in particular the Gram-positive *Staphylococcus aureus*, to host cells through **M**icrobial **S**urface **C**omponents **R**ecognizing **A**dhesive **M**atrix **M**olecules (MSCRAMM) mediated adhesion processes (Patti *et al.*, 1994; Menzies, 2003). Fibronectin (Fn) is a heterodimer (~440 kDa) and is made up of several functional domains consisting of: (i) repeat amino acids of RGD that is a cell-binding domain; (ii) two fibrin binding domains, one at the N-terminal end and another at the COOH-terminal; and (iii) a collagen or gelatin-binding domain (GBD) (Fig. 11.2; Mohan *et al.*, 2001). Antibodies that recognized the different domains of fibronectin also recognized *M. javanica* cuticle extracts and in *Pasteuria* endospore inhibition assays the GBD fragment of Fn blocked endospore attachment more effectively than any of the other domains (Mohan *et al.*, 2001).

Fibronectin is not known to occur in the extracellular matrix of *C. elegans* cuticle (Kramer, 1997) and, therefore, it was surprising that the antibodies to Fn recognized aspects of the cuticle extracts. Interestingly, it was only after animal-parasitic *Bacillus* spp. had been sequenced (Read *et al.*, 2003; Ivanov *et al.*, 2003) and shown to contain genes encoding for collagen-like sequences with G-x-y repeats (Sylvester *et al.*, 2002; Steichen *et al.*, 2003) that it became obvious that collagen-like sequences would have been present both in the nematode cuticle and also on the surface of *Pasteuria*, potentially accounting for the cross reactivity of the antibodies. Sylvester *et al.* (2002) showed that different strains of

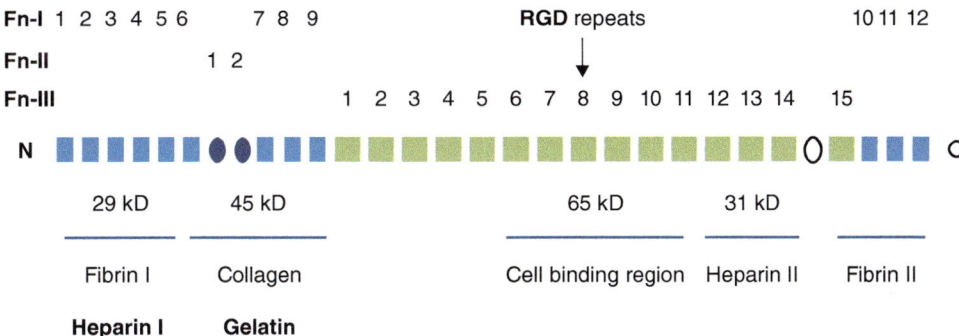

Fig. 11.2. Schematic representation of monomers of fibronectin (Fn-I, Fn-II and Fn-III). The regions of homology are represented by rectangles, ovals and squares. The Heparin 1 and gelatin-binding domains inhibited *Pasteuria* endospore attachment. Fn-III domain labelled RGD repeats are triplets of arginine, glycine and aspartic acid. (Adapted from Mohan *et al.*, 2001.)

B. anthracis had different lengths of G-x-y repeat sequences, and the number of repeats related to the length of the hair-like nap that were expressed and shown to occur on the endospore surface. Genome survey sequences of *Pasteuria* showed that both *P. penetrans* and *P. ramosa* also contained collagen-like G-x-y repeat sequences (Davies and Opperman, 2006; Mouton *et al.*, 2009; McElroy *et al.*, 2011) and it became clear that the GBD of Fn was interfering with the endospore surface collagen-like proteins involved in endospore binding to the nematode cuticle surface and inhibiting attachment.

The assembly of the multilayered endospore of *Bacillus* species is an active area of research (McKenney *et al.*, 2013) and it is clear that there are major differences between *Bacillus* spp. that are pathogens (*B. anthracis*, *B. cereus*, *B. thuringiensis*), which contain a rhamnose cluster operon that includes the gene BclA, producing the hair-like nap on the surface of the endospore, when compared to the non-pathogen (*B. subtilis*) that lack these genes (Todd *et al.*, 2003). Interestingly, a phylogenetic analysis using genes from the genome survey sequence (Bird *et al.*, 2003; Charles *et al.*, 2005), *Pasteuria* aligns most closely to the pathogen-related *Bacillus* spp. and subsequent analysis using several key exosporium genes, including *bclA*, *exsJ* and *vrrB*, displayed a degree of micro-synteny across all the groups (Srivastava *et al.*, 2014). To date most of this work has concentrated on the interactions between *Pasteuria* and *Meloidogyne* spp., which are highly host-specific; attachment studies indicate that interactions between *Pasteuria* and cyst nematodes appear more promiscuous (Sharma and Davies, 1996; Mohan *et al.*, 2011) suggesting that the nature of the J2 cuticle of cyst nematodes is very different from root-knot nematodes and requires further exploration.

11.5 Biological Control Products

After the enormous potential of fungal antagonists to control plant-parasitic nematodes was recognized by Kühn (1877) it took 60 years until Linford (1937) and Linford and Yap (1939) started first attempts using nematode-trapping fungi reared on sterilized sugarcane bagasse to control root-knot nematodes on pineapple. However, it was another 40 years before research on biological control was to develop with rapidly increasing numbers of publications in the 1980s. Since then numerous antagonistic bacteria and fungi have been studied for the control potential on various plant-parasitic nematodes (Stirling, 2014; Askary and Martinelli, 2015). Most of them performed well *in vitro* or in pot experiments, but when it came to field application control efficacy and consistency often failed or was unsatisfactory. Some of the main reasons explaining such poor performance are inconsistent efficacy, variable quality of the biocontrol product, unsatisfactory cost–benefit ratio, difficulties in scale-up production, formulation, delivery system and short shelf life (Moosavi and Askary, 2015). In addition, overall costs for registration, distribution and marketing were often underestimated. However, those drawbacks also stimulated research to improve understanding of the underlying biocontrol mechanisms and to increase biocontrol efficacy using modern molecular tools. In parallel, the development of new formulation and application technologies led to increased performance under field conditions. Two of the first biocontrol products on the market were Royal 300 and Royal 350 based on *Arthrobotrys robusta* and *A. irregularis* (Cayrol *et al.*, 1978; Cayrol, 1983). Those nematode-trapping fungi were produced on barley as substrate and applied as fresh mycelium, a method routinely used for the production of mushroom inoculum.

The main challenges for the development of a biocontrol product are its compatibility with commercial development and production methods, and its suitability for field implementation (Lumsden and Vaughn, 1995). Table 11.2 lists some of the currently available 'inundative' biocontol products for control of cyst nematodes.

11.6 Managing Biological Control

Managing biological control refers to the 'conservation' approach in which the community of natural antagonists is managed in such a manner that plant-parasitic nematode densities are maintained below the economic threshold level (Stirling, 2014). Although being challenging, this is probably the most sustainable approach in nematode management. For this approach it is mandatory to protect and nurture the microbial antagonists in the soil responsible for nematode suppression. There is no single method to achieve this goal, but there are numerous measures the farmer can apply to contribute to nematode suppression and thus soil sustainability. The key factor is to properly feed the soil biota by increasing the organic matter content of the soil. As a consequence, microbial activity is enhanced and thus is the antagonistic potential in the soil. Besides, microbivorous nematodes feeding on bacteria and fungi are stimulated which results in increased nutrient mineralization (Ferris *et al.*, 2004). Those conditions not only suppress plant-parasitic nematodes but at the same time increase the plant capacity to better tolerate damage by plant-parasitic nematodes. Methods that increase soil organic content are among others elimination of bare fallow, minimum tillage, retaining crop residues on the field, planting green manure crops and applying organic amendments. Besides, pesticides should be used judiciously and compaction and other suboptimal soil constraints need to be avoided. Below are discussed some of the major aspects in managing biological control. For a more thorough coverage on this topic see Sikora (1992), Kerry (2000), Hildalgo-Diaz and Kerry (2008) and Stirling (2014).

11.6.1 Effect of tillage

As a consequence of increasing problems associated with soil erosion reliance on 'conventional' tillage (mouldboard plough and disc harrow), incorporating residue of the preceding crop into the soil, shifted since the 1980s towards 'conservation' tillage, combining agricultural practices less disturbing to the soil with crop residue left on the soil surface. Those latter methods ranging from no-till to various forms of minimum tillage reduce soil erosion, increase water infiltration and retain at least 30% residue cover on the soil surface after planting. The type of tillage practice adopted for a cropping system influences the population density of nematodes (Kimpinski and Kunelius, 1988; Govaerts *et al.*, 2007; Ito *et al.*, 2015a, b) and crop damage severity (Rovira, 1982) and these may be the result of indirect effects through alterations in the populations of nematode antagonists in the soil (Stirling, 2014).

Minimum tillage generates large amounts of organic matter in the soil, which enhances the activity of specific groups of antagonists leading to general nematode suppression (Sikora, 1992; Stirling, 2011). Pathogenic activity of conidia of *Hirsutella* spp. is largely dependent on the fact that they remain attached to the conidiophores and, therefore, are sensitive to soil disturbance during tillage (McInnis and Jaffee, 1989). Parasitism of *H. glycines* eggs was found to be >10% at different tillage regimes for a period of 2 years with comparatively greater parasitism seen in disc ploughed compared to mouldboard ploughed or the no-till treatments (Bernard *et al.*, 1996). However, more recent studies are contradictory; a comparison between a 'simulated soil disturbance', by passing soil through a sieve (aperture 5 mm), and no-disturbance (non-sieve) showed that *H. glycines*

Table 11.2. Examples of commercially available biological control products by name, active microbe, formulation, application and crop and company for the control of cyst nematodes.

Product name	Active antagonist	Formulation	Application form	Crop	Company, country
BioAct WG, MeloCon WG	*Purpureocillium lilacinum*	Water-dispersible granulate	Drench, drip irrigation	Vegetables, tuber and row crops	Bayer CropScience, Germany
MyTech	*P. lilacinum*	Wettable powder	Drench	Agricultural and horticultural crops	Dudutech, Kenya
Yorker	*P. lilacinum*	Water soluble powder	Drench	Agricultural and horticultural crops	Agriland Biotech, India
Nema	*P. lilacinum*	Wettable powder	Drench		Tropical Agro, India
Hocusia	*P. lilacinum*	Water-dispersible granulate	Drench, drip irrigation	Potato, brinjal	Bayer, India
Paecilo	*P. lilacinum*	Wettable powder	Drench	Agricultural and horticultural crops	Agri Life, India
Nemoend-PL	*P. lilacinum*	Wettable powder	Drench	Vegetables, turf	Exotic Naturals, India
KlamiC	*Pochonia chlamydosporia*	Granulate	Soil incorporation	Vegetables	CENSA, Cuba
BioShield	*P. chlamydosporia*	Water soluble powder	Drench	Agricultural and horticultural crops	Agriland Biotech, India
Rizotec	*P. chlamydosporia*	Granulate	Soil application	Vegetables	Stoller, Brazil
DiTera	*Myrothecium verrucaria*	Granulate	Soil application	Agricultural and horticultural crops	Valent Bioscience, USA
VOTiVo	*Bacillus firmus*	Liquid	Seed treatment	Soybean, cotton, maize	Bayer CropScience, USA
Nortica	*B. firmus*	Wettable powder	Drench	Turf, sports greens	Bayer CropScience, USA
Nemoend-BF	*B. firmus*	Wettable powder	Drench	Vegetables, turf	Exotic Naturals, India
BioNemaGon	*B. firmus*	Wettable powder	Soil applied, drip irrigation	Agricultural and horticultural crops	Agri Life, India
Clariva	*Pasteuria nishizawae*	Liquid	Seed treatment	Soybean	Syngenta, USA

egg densities increased and J2 parasitization by *H. rhossiliensis* decreased due to soil disturbance (Bao et al., 2011). In contrast, Chen and Liu (2005) did not observe differences in parasitism of *H. glycines* J2 by *H. rhossiliensis* and *H. minnesotensis* between conventional tillage and no-tillage at three different sites growing soybean.

Noel et al. (2010) observed that attachment of *P. nishizawae* endospores to *H. glycines* J2 remained the same in conventional and no-tillage plots growing soybean, although Talavera et al. (2002) observed a higher percentage of *P. penetrans* endospore attachment to *M. incognita* J2 under no-tillage than deep rotary tillage. In conclusion, minimum tillage secures the soil integrity by conserving soil structure, organic matter content, moisture retention and assisting in the establishment of nematode antagonists.

11.6.2 Organic matter content

As indicated above, soil organic matter plays a key role in nematode suppression. Supplementing soil with organic matter provides food for the soil organisms, thus increasing the diversity and activity of soil microbes, many of which are antagonistic to plant-parasitic nematodes (Chen et al., 1999; Viaene and Abawi, 2000). For example, exposure of eggs of H. schachtii to aqueous extract of poultry manure compost and neem cake for 3 min made them more than twofold more susceptible to parasitism by P. chlamydosporia in comparison to the control (Pandey and Sikora, 2000). Renčo et al. (2011) evaluated the nematicidal potential of four composts made from horse, poultry, pig and cattle manure, sugar beet pomace, grapevine and medical plant wastes, against different pathotypes of G. rostochiensis (Ro1) and G. pallida (Pa2 and Pa3) and observed significant reduction in number of cysts, eggs and J2 per cyst and eggs and J2 per gram soil over unamended soil. A similar reduction of H. schachtii population was observed in soils amended with compost (Schlang, 1993) and it is likely that some of these effects are indirect through effects on the nematode antagonists.

Incorporating the cover crop into the soil as green manure positively enhances soil quality by increasing organic matter and effectively controlling diseases and plant-parasitic nematodes through either chemical or biological means. The chemical mechanism involves the release of compounds that are toxic or antagonistic to the nematodes and the biological mechanisms provide a suitable environment for the increase of nematode antagonistic microorganisms. Gray (1987) suggested that the composition of the organic matter may affect the activity of nematophagous fungi, especially trap-forming fungi. In this regard, the influence of rape and mustard as green manure in comparison to barley straw was tested for the antagonistic efficacy of five strains of Arthrobotrys spp. As a result, early root penetration of sugar beet by H. schachtii was reduced significantly by 30–35% in fungal-treated mustard-amended soil and 29% in straw-amended soil, whilst no effect was observed in the rape-amended treatment when compared to non-inoculated control. This clearly indicated that the antagonistic capacity of the nematode-trapping fungi varied depending on the source of organic matter applied (Hoffmann-Hergarten and Sikora, 1993). Liu and Chen (2009) reported that the effectiveness of H. rhossiliensis and H. minnesotensis against H. glycines differed regarding soil pH, texture and organic matter content. The relationship between parasitism of J2 by H. rhossiliensis and organic matter content changed from negative at days 0 and 10 to positive at day 30; however, there was no significant correlation between egg reduction and soil organic matter content. It was further shown that infection of H. glycines J2 by H. rhossiliensis was faster than by H. minnesotensis, most likely because H. rhossiliensis established under the given conditions more rapidly than H. minnesotensis (Chen and Liu, 2005; Chen, 2007).

11.6.3 C:N ratio

The efficacy of organic matter in controlling plant-parasitic nematodes not only depends on the composition of the organic matter but also on its C:N ratio. According to Rodríguez-Kábana (1986) the amount of 'protein' N in organic amendments is directly related to nematode suppression, whereas ammonical nitrogen is more damaging to the nematode than nitrate nitrogen. In general, the nematicidal potential of organic matter increases with C:N ratio decreasing. Organic soil amendments are considered nematicidal with a C:N ratio less than 20. If the C:N ratio drops below 11 it might become phytotoxic requiring a time gap between application of the organic amendment and planting of the crop (Rodríguez-Kábana et al., 1987). A low C:N ratio or high content of ammonia results either in plasmolysis of nematodes, or proliferation of nematophagous fungi due to the release of ammonium (Rodríguez-Kábana, 1986). Nematode antagonists that benefit under those conditions are fungal egg parasites, trapping fungi, endoparasitic fungi, fungal parasites of females, endomycorrhizal fungi, plant health-promoting rhizobacteria and obligate bacterial parasites.

One of the better studied nitrogenous amendments is chitin containing about 7% nitrogen. The amendment of chitin in soil stimulates chitin-producing bacteria that degrade chitin (see section 11.3.4). The depolymerization of chitin by chitinase in soil releases ammonia, which is also nematicidal to J2. In a glasshouse study Tian et al. (2000) reported suppression of *H. glycines* by chitinolytic bacterial isolates in heat-treated silt loam soil amended with 0.6% (w/w) chitin. A synergistic effect between the isolates and the chitin substrate reduced the nematode numbers.

In summary, the balance between the plant and its plant-parasitic nematode population is complex and affected by any changes to crop management practices such as tillage and soil amendments. Such changes, as reviewed above, may have beneficial effects and reduce the phytonematode population, but this clearly cannot just be taken for granted. There are clear examples where nematode antagonists acting, either directly or indirectly, will not necessarily lead to phytonematode control; each soil and each management activity will be soil-specific and understanding the key factors to maintain phytonematode populations below economic thresholds is the key challenge for managing biological control.

11.7 Conclusions and Future Prospects

At the beginning of this chapter three approaches to the management of plant-parasitic nematodes using biological control were introduced: (i) inundation; (ii) inoculation; and (iii) 'conservation'. All three approaches aim to use microbial enemies to suppress plant-parasitic nematodes to a point where they no longer cause yield losses. The three approaches can be seen as a continuum from managing the soil rhizosphere microflora in such a way that no new microbial enemies are added and control is built based on the indigenous microflora already present, through to where a natural enemy is added in low amounts and allowed to build up of its own accord using the pest nematode as its means of *in situ* mass production, to where the mass production of the natural enemy is external and where large quantities are added. One, if not the most, important factor underpinning any of these three approaches is the compatibility of the natural enemy/ies for their host plant-parasitic nematode/s. Notice here the use of the singular and plural forms of host and enemy because, until very recently, the selection of biological control agents all rested on empirical approaches as the tools were not available to understand the biodiversity of hosts and enemies. Molecular biology and the genomics era is now providing new tools that allow the subtle interactions and biological diversity necessary to dissect the complexity of these nematode suppressive soils (Mendes et al., 2011).

The application of molecular biology to plant nematodes has generally focused on looking at the interactions between the nematode and the plant and this has been fruitful. More recently, however, new insights have been gained from potential biological control agents. As far back as 1959 it was recognized that nematodes produced a substance, nemin, that elicited certain groups of fungi to produce traps (Pramer and Stoll, 1959). Nemin, it turns out, is not a single substance but belongs to a family of compounds known as ascarosides (Hsueh et al., 2013) and these appear to have multiple roles involved in signalling and regulating aspects of nematode behaviour such as mate finding and dauer formation (Choe et al., 2012). They are a diverse family of compounds and, interestingly, not only do they appear to elicit trap formation in fungi, but also play a role in microbe-associated molecular pattern (MAMP)-triggered immunity in plants and elicit plant defence responses against nematodes (Manosalva et al., 2015). MAMPs are recognized by specific pattern recognition receptors in the plant host and those identified to date include a range of different compounds as diverse as flagellin, lipopolysaccharide and peptidoglycan (Bittel and Robatzek, 2007; Boller and Felix, 2009; Zipfel, 2014). Similar to Nemin, it was recently shown that FAR proteins have a dual role; they modulate *Pasteuria* endospore attachment to nematode cuticle and also have a role in nematode plant interactions (Phani et al., 2017).

The biochemical and molecular studies discussed in this chapter appear on a continuum from being highly specific, for example the specificities of endospore attachment in *P. penetrans* and the proteinase activity of enzymes produced by *P. chlamydosporia*, on the one hand, to conserved molecular signatures of MAMPs, for example, the broad plant defence responses elicited by endoparasitic nematodes, on the other. What we therefore see today as a conserved molecular signature eliciting a broad defence response may indeed turn out to be more specific when scrutinized sufficiently. Indeed, the ZigZag Model of effector co-evolutionary arms races would suggest increasing molecular diversity and specificity (Jones and Dangl, 2006). However, if some of these compounds are indeed involved in some other aspects of the organism's biology that constrains their diversity, for example, ascarocides are involved in mate finding and/or dauer formation, these broadly conserved signalling compounds can possibly be exploited to develop new and broad-spectrum nematode control strategies.

How might these strategies be developed and applied? It is widely acknowledged that different biological control agents, or even isolates of a particular microbe, operate through differing modes of action and it has been suggested that these distinct modes of action can be combined by applying a cocktail of organisms (Sharon *et al.*, 2011). Such an approach would require multiple fermentation processes as the different organisms would require different conditions. However, research focusing on the design of minimal genomes (Hutchison *et al.*, 2016) and the rapid development of new molecular tools for gene editing (Ledford, 2016), underpins the suggestion that synthetic biology and the production of designer biological control organisms (Davies and Spiegel, 2011) are increasingly real possibilities.

The idea of designer biological control agents can be evaluated through an adapted version of Koch's postulates. The application of Koch's postulates for the production of transgenically engineered biologically based nematicides for use in proof of concept experiments would:

1. Identify and clone a particular gene from an organism that causes a disease or affects behaviour in the host nematode thought to be involved in infection.
2. Remove the gene from the microorganism such that the mode of action of the gene is lost and the microbe can no longer infect the host nematode or at least there is a quantitative reduction in the amount of infection.
3. Replace the gene in the microorganism to restore the amount of infection.

This transgenic bionematicidal approach can be applied from the identification and mode of action of a particular gene to the situation where two or more genes are combined in a single microbe. Indeed, synthetic biology is increasingly commonplace within the microbiological community. Already research is being undertaken to construct synthetic bacterial endospores to produce versatile display platforms for drugs and vaccines for medical applications, and for enzymes that neutralize pollutants for environmental remediation (Wu *et al.*, 2015). It is therefore not a huge leap of imagination to start using this technology for the development of increasingly effective biological control agents. Recently, such a transgenic approach has been demonstrated to work where neuropeptide-like proteins have been cloned into *B. subtilis* and *Chlamydomonas reinhardtii* and have been shown to reduce plant-parasitic nematode infection by 90% (Warnock *et al.*, 2017). Such an approach clearly heralds a way forward for biological control in the 21st century.

11.8 References

Aksoy, H.M. and Mennan, S.S. (2004) Biological control of *Heterodera cruciferae* (Tylenchida: Heteroderidae) Franklin 1945 with fluorescent *Pseudomonas* spp. *Journal of Phytopathology* 152, 514–518. DOI: 10.1111/j.1439-0434.2004.00890.x

Ali, N.I., Shaukat, S.S. and Zaki, M.J. (2002) Nematicidal activity of some strains of *Pseudomonas* spp. *Soil Biology and Biochemistry* 34, 1051–1058. DOI: 10.1016/S0038-0717(02)00029-9

Andreoglou, F.I., Vagelas, I.K., Wood, M., Samaliev, H.Y. and Gowen, S.R. (2003) Influence of temperature on the motility of *Pseudomonas oryzihabitans* and control of *Globodera rostochiensis*. *Soil Biology and Biochemistry* 35, 1095–1101. DOI: 10.1016/S0038-0717(03)00157-3

Askary, T.H. and Martinelli, P.R.P. (eds) (2015) *Biocontrol Agents of Phytonematodes*. CAB International, Wallingford, UK. DOI: 10.1079/9781780643755.0081

Atibalentja, N., Jakstys, B.P. and Noel, G.R. (2004) Life cycle, ultrastructure, and host specificity of the North American isolate of *Pasteuria* that parasitizes the soybean cyst nematode, *Heterodera glycines*. *Journal of Nematology* 36, 171–180.

Ausubel, F.M. (2005) Are innate immune signaling pathways in plants and animals conserved? *Nature Immunology* 6, 973–979. DOI: 10.1038/ni1253

Bajaja, R., Hu, W., Huang, Y.Y., Chen, S., Prasad, R., Varma, A. and Bushley, K.E. (2015) The beneficial root endophyte *Piriformospora indica* reduces egg density of the soybean cyst nematode. *Biological Control* 90, 193–199. DOI: 10.1016/j.biocontrol.2015.05.021

Baker, K.F. and Cook, R.J. (1974) *Biological Control of Plant Pathogens*. APS, St. Paul, MN.

Bakker, P.A.H.M., Pieterse, C.M.J. and van Loon, L.C. (2007) Induced systemic resistance by fluorescent *Pseudomonas* spp. *Phytopathology* 97, 239–243. DOI: 10.1094/PHYTO-97-2-0239.

Bao, Y., Neher, D.A. and Chen, S.Y. (2011) Effect of soil disturbance and biocides on nematode communities and extracellular enzyme activity in soybean cyst nematode suppressive soil. *Nematology* 13, 687–699. DOI: 10.1163/138855410X541230

Becker, J.S., Borneman, J. and Becker, J.O. (2013) *Dactylella oviparasitica* parasitism of the sugarbeet cyst nematode observed in trixenic culture plates. *Biological Control* 64, 51–56. DOI: 10.1016/j.biocontrol.2012.10.007

Bernard, E.C., Self, L.H. and Tyler, D.D. (1996) Fungal parasitism of soybean cyst nematode, *Heterodera glycines* (Nemata: Heteroderidae), in differing cropping-tillage regimes. *Applied Soil Ecology* 5, 57–70. DOI: 10.1016/S0929-1393(96)00125-4

Bird, A.F. and Bird, J. (1991) *The Structure of Nematodes*. Academic Press, San Diego, CA.

Bird, D., Opperman, C.H. and Davies, K.G. (2003) Interactions between bacteria and plant-parasitic nematodes: now and then. *International Journal of Parasitology* 33, 1269–1276. DOI: 10.1016/S0020-7519(03)00160-7

Bittel, P. and Robatzek, S. (2007) Microbe-associated molecular patterns (MAMPs) probe plant immunity. *Current Opinion in Plant Biology* 10, 335–341. DOI: 10.1016/j.pbi.2007.04.021

Boller, T. and Felix, G.A. (2009) Renaissance of elicitors: perception of microbe-associated molecular patterns and danger signals by pattern-recognition receptors. *Annual Review of Plant Biology* 60, 379–406. DOI: 10.1146/annurev.arplant.57.032905.105346

Bordallo, J.J., López-Llorca, L.V., Jansson, H.-B., Salinas, J., Persmark, L. and Asensio, L. (2002) Colonization of plant roots by egg-parasitic and nematode-trapping fungi. *New Phytologist* 154, 491–499. DOI: 10.1046/j.1469-8137.2002.00399.x

Borneman, J. and Becker, J.O. (2007) Identifying microorganisms involved in specific pathogen suppression in soil. *Annual Reviews of Phytopathology* 45, 153–172. DOI: 10.1146/annurev.phyto.45.062806.094354

Carsolio, C., Benhamou, N., Haran, S., Cortés, C., Gutiérrez, A., Chet, I. and Herrera-Estrella, A. (1999) Role of the *Trichoderma harzianum* endochitinase gene, *ech42*, in mycoparasitism. *Applied and Environmental Microbiology* 65, 929–935. PMCID: PMC91125

Carson, R. (1962) *Silent Spring*. Houghton Mifflin, Boston, MA.

Cayrol, J.-C. (1983) Lutte biologique contre les *Meloidogyne* au moyen d'*Arthrobotrys irregularis*. *Revue de Nématologie* 6, 265–273.

Cayrol, J.-C., Frankowski, J.P., Laniece, A., D'Hardemare, G. and Talon, J.P. (1978) Contre les nématodes en champignonniere. Mise au point d'une methode de lutte biologique a l'aide d'un Hyphomycete predateur: *Arthrobotrys robusta* souche *Antipolis* (Royal 300). *Revue Horticole* 184, 23–30.

Charles, L., Carbone, I., Davis, K.G., Bird, D., Burke, M., Kerry, B.R. and Opperman, C.H. (2005) Phylogenetic analysis of *Pasteuria penetrans* by use of multiple genetic loci. *Journal of Bacteriology* 187, 5700–5708. DOI: 10.1128/JB.187.16.5700-5708.2005

Chen, S. (2007) Suppression of *Heterodera glycines* in soils from fields with long-term soybean monoculture. *Biocontrol Science and Technology* 17, 125–134. DOI: 10.1080/09583150600937121

Chen, Z.X. and Dickson, D.M. (1998) Review of *Pasteuria penetrans*: biology, ecology, and biological control potential. *Journal of Nematology* 30, 313–340.

Chen, S. and Liu, S. (2007) Effects of tillage and crop sequence on parasitism of *Heterodera glycines* juveniles by *Hirsutella* spp. and on juvenile population density. *Nematropica* 37, 93–106.

Chen, S.Y. and Liu, X.Z. (2005) Control of the soybean cyst nematode by the fungi *Hirsutella rhossiliensis* and *Hirsutella minnesotensis* in greenhouse studies. *Biological Control* 32, 208–219. DOI: 10.1016/j.biocontrol.2004.09.013

Chen, J., Abawi, G.S. and Zuckerman, B.M. (1999) Suppression of *Meloidogyne hapla* and its damage to lettuce grown in a mineral soil amended with chitin and biocontrol organisms. *Supplement to the Journal of Nematology* 31, 719–725.

Choe, L., von Reuss, S.H., Kogan, D., Gasser, R.B., Platzer, E.G., Schroeder, F.C and Sternberg, P.W. (2012) Ascarocide signalling is widely conserved among nematodes. *Current Biology* 22, 772–780. DOI: 10.1016/j.cub.2012.03.024.

Compant, S., Duffy, B., Nowak, J., Clement, C. and Barka, E.A. (2005) Use of plant growth-promoting bacteria for biocontrol of plant diseases: principles, mechanisms of action, and future prospects. *Applied Environmental Microbiology* 71, 4951–4959. DOI: 10.1128/aem.71.9.4951-4959.2005

Cronin, D., Moenne-Loccoz, Y., Dunne, C. and O'Gara, F. (1997) Inhibition of egg hatch of the potato cyst nematode *Globodera rostochiensis* by chitinase-producing bacteria. *European Journal of Plant Pathology* 103, 433–440. DOI: 10.1023/a:1008662729757

Crump, D.H., Sayre, R.M. and Young, L.D. (1983) Occurence of nematophagous fungi in cyst nematodes. *Plant Disease* 67, 63–64.

Curtis, R.H.C., Jones, J.T., Davies, K.G., Sharon, E. and Spiegel, Y. (2011) Plant nematode surfaces. In: Davies, K.G. and Spiegel, Y. (eds) *Biological Control of Plant-Parasitic Nematodes: Building Coherence between Microbial Ecology and Molecular Mechanisms*. Springer, Dordrecht, The Netherlands, pp. 115–144. DOI: 10.1007/978-1-4020-9648-8_5.

Dackman, C., Chet, I. and Nordbring-Hertz, B. (1989) Fungal parasitism of the cyst nematode *Heterodera schachtii*: infection and enzymic activity. *FEMS Microbial Ecology* 62, 201–208. DOI: 10.1111/j.1574-6968.1989.tb03694.x

Daneshkhah, R., Cabello, S., Rozanska, E, Sobczak, M., Grundler, F.M.W., Wieczorek, K. and Hofmann, J. (2013) *Piriformospora indica* antagonizes cyst nematode infection and development in *Arabidopsis* roots. *Journal of Experimental Botany* 64, 3763–3374. DOI: 10.1093/jxb/ert213

Darby, C., Hsu, J.W., Ghori, N. and Falkow, S. (2002) *Caenorhabditis elegans* – Plague bacteria biofilm blocks food intake. *Nature* 417, 243–244. DOI: 10.1038/417243a

Davide, R.G. and Zorilla, R.A. (1995) Evaluation of three nematophagous fungi and some nematicides for the control of potato cyst nematode *Globodera rostochiensis*. *BioControl* 3, 45–55.

Davies, K.G. and Curtis, R.H.C. (2011) Cuticle surface coat of plant-parasitic nematodes. *Annual Review of Phytopathology* 49, 135–156. DOI: 10.1146/annurev-phyto-121310-111406

Davies, K.G. and Danks, C. (1993) Carbohydrate/protein interactions between the cuticle of infective juveniles of *Meloidogyne incognita* and spores of the obligate hyperparasite *Pasteuria penetrans*. *Nematologica* 39, 54–64. DOI: 10.1163/187529293x00033

Davies, K.G. and Opperman, C.H. (2006) A potential role for collagen in the attachment of *Pasteuria penetrans* to nematode cuticle. In: Raaijmakers, J.M. and Sikora, R.A. (eds) *Multitrophic Interactions in the Soil and Integrated Control. IOBC WPRS Bulletin* 29, 11–15.

Davies, K.G. and Redden, M. (1997) Diversity and partial characterisation of putative virulence determinants in *Pasteuria penetrans*, the hyperparasite of root-knot nematodes. *Journal of Applied Microbiology*, 83, 227–235. DOI: 10.1046/j.1365-2672.1997.00223.x

Davies, K.G. and Spiegel, Y. (2011) Root patho-systems nematology. In: Davies, K. and Spiegel, Y. (eds) *Biological Control of Plant-Parasitic Nematodes: Building Coherence between Microbial Ecology and Molecular Mechanisms*. Springer, Dordrecht, The Netherlands, pp. 291–303.

Davies, K.G., Kerry, B.R. and Flynn, C.A. (1988) Observations on the pathogenicity of *Pasteuria penetrans*, a parasite of root-knot nematodes. *Annals of Applied Biology* 112, 1491–1501. DOI: 10.1111/j.1744-7348.1988.tb02086.x

Davies, K.G., Flynn, C.A., Laird, V. and Kerry, B.R. (1990) The life-cycle, population dynamics and host specificity of a parasite of *Heterodera avenae*, similar to *Pasteuria penetrans*. *Revue de Nématologie* 13, 303–309.

Davies, K.G., Redden, M. and Pearson, T.K. (1994) Endospore heterogeneity in *Pasteuria penetrans* related to attachment to plant-parasitic nematodes. *Letters in Applied Microbiology* 19, 370–373. DOI: 10.1111/j.1472-765X.1994.tb00478.x

Davies, K.G., Fargette, M., Balla, G. et al. (2001) Cuticle heterogeneity as exhibited by *Pasteuria* spore attachment is not linked to the phylogeny of parthenogenetic root-knot nematodes (*Meloidogyne* spp.). *Parasitology* 122, 111–120. DOI: 10.1017/s0031182000006958

de Araújo, F.F., Silva, J.F.V. and de Araújo, F.A.S. (2002) Influence of *Bacillus subtilis* on the *Heterodera glycines* eclosion, orientation and infection in soybean. *Ciência Rural* 32, 197–203. http://dx.doi.org/10.1590/S0103-84782002000200003

Deliopoulos, T., Devine, K.J., Haydock, P.P.J. and Jones, P.W. (2007) Studies on the effect of mycorrhization of potato roots on the hatching activity of potato root leachate towards the potato cyst nematode, *Globodera pallida* and *G. rostochiensis*. *Nematology* 9, 719–729. DOI: 10.1163/156854107782024758

Deliopoulos, T., Haydock, P.P. and Jones, P.W. (2008) Interaction between arbuscular mycorrhizal fungi and the nematicide aldicarb on hatch and development of the potato cyst nematode, *Globodera pallida*, and yield of potatoes. *Nematology* 10, 783–799. DOI: 10.1163/156854108786161427

De Marco, J.L., Lima, L.H.C., valle de Sousa, M. *et al.* (2000) A *Trichoderma harzianum* chitinase destroys the cell wall of the phytopathogen *Crinipellis perniciosa*, the casual agent of witches' broom disease of cocoa. *World Journal of Microbiology and Biotechnology* 16, 383–386. DOI: 10.1023/A:1008964324425

den Belder, E. and Jansen, E. (1994) Capture of plant-parasitic nematodes by an adhesive hyphae forming isolate of *Arthrobotrys oligospora* and some other nematode-trapping fungi. *Nematologica* 40, 423–437. DOI: 10.1163/003525994x00300

Denny, T.P. (1995) Involvement of bacterial polysaccharides in plant pathogenesis. *Annual Review of Phytopathology* 33, 173–195. DOI: 10.1146/annurev.py.33.090195.001133

Deshmukh, S.D. and Kogel, K.H. (2007) *Piriformospora indica* protects barley from root rot caused by *Fusarium graminearum*. *Journal of Plant Diseases and Protection* 114, 263–268. DOI: 10.1007/bf03356227

Dijksterhuis, J., Veenhuis, M., Harder, W. and Nordbring-Hertz, B. (1994) Nematophagous fungi: physiological aspects and structure–function relationships. *Advances in Microbial Physiology* 36, 111–143. DOI: 10.1016/S0065-2911(08)60178-2

Dong, L.Q., Yang, J.K. and Zhang, K.Q. (2007) Cloning and phylogenetic analysis of the chitinase gene from the facultative pathogen *Paecilomyces lilacinus*. *Journal of Applied Microbiology* 103, 2476–2488. DOI: 10.1111/j.1365-2672.2007.03514.x

Dupont, A., Segers, R. and Coosemans, J. (1999) The effect of chitinase from *Verticillium chlamydosporium* on the egg of *Meloidogyne incognita*. *Communications in Agricultural and Applied Biological Sciences* 64, 383–389.

El-Nagdi, W.M.A. and Youssef, M.M.A. (2004) Soaking faba bean seed in some bio-agent as prophylactic treatment for controlling *Meloidogyne incognita* root-knot nematode infection. *Journal of Pest Science* 77, 75–78. DOI: 10.1007/s10340-003-0029-y

Espanol, M., Verdejo-Lucas, S., Davies, K.G. and Kerry, B.R. (1997) Compatibility between *Pasteuria penetrans* and *Meloidogyne* populations from Spain. *Biocontrol Science and Technology* 7, 219–230. DOI: 10.1080/09583159730910

Esteves, I., Peteira, B., Atkins, S., Magan, N. and Kerry, B.R. (2009) Production of extracellular enzymes by different isolates of *Pochonia chlamydosporia*. *Mycological Research* 113, 867–876. DOI: 10.1016/j.mycres.2009.04.005

Ewbank, J.J. (2002) Tackling both sides of the host pathogen equation with *Caenorhabditis elegans*. *Microbes and Infection* 4, 247–256. DOI: 10.1016/s1286-4579(01)01531-3

Fekete, C.M., Tholander, B., Rajasheker, D., Ahrén, D., Friman, E., Johansson, T. and Tunlid, A. (2008) Paralysis of nematodes: shifts in the transcriptome of the nematode trapping fungus *Monacrosporium haptoxylum* during infection of *Caenorhabditis elegans*. *Environmental Microbiology* 10, 364–375. DOI: 10.1111/j.1462-2920.2007.01457.x

Fenton, A.M., Stephens, P.M., Crowley, J., O'Callaghan, M. and O'Gara, F. (1992) Exploitation of gene(s) involved in 2,4-diacetylphloroglucinol biosynthesis to confer a new biocontrol capability to a *Pseudomonas* strain. *Applied Environmental Microbiology* 58, 3873–3878. PMID1476431

Ferris, H., Venette, R.C. and Scow, K.M. (2004) Soil management to enhance bacterivore and fungivore nematode populations and their nitrogen mineralisation function. *Applied Soil Ecology* 25, 19–35. DOI: 10.1016/j.apsoil.2003.07.001

Flores, A., Chet, I. and Herrera-Estrella, A. (1997) Improved biocontrol activity of *Trichoderma harzianum* by over-expression of the proteinase-encoding gene *prb*1. *Current Genetics* 31, 30–37. DOI: 10.1007/s002940050173

Franco, J., Jatala, P. and Bocangel, M. (1981) Efficiency of *Paecilomyces lilacinus* as a biocontrol agent of *Globodera pallida*. *Journal of Nematology* 13, 438–439.

Gair, R., Mathias, P.L. and Harvey, P.N. (1969) Studies of cereal cyst nematode populations and cereal yield under continuous or intensive culture. *Annals of Applied Biology* 63, 503–512. DOI: 10.1111/j.1744-7348.1969.tb02846.x

Gan, Z., Yang, J., Tao, N., Liang, L., Mi, Q., Li, J. and Zhang, K.-Q. (2007) Cloning of the gene *Lecanicillium psalliotae* chitinase Lpchi 1 and identification of its potential role in the biocontrol of root-knot nematode

Meloidogyne incognita. Applied Microbiology and Biotechnology 76, 1309–1317. DOI: 10.1007/s00253-007-1111-9

Geng, C., Nie, X., Tang, Z., Zhang, Y., Lin, J., Sun, M. and Penga, D. (2016) A novel serine protease, Sep1, from *Bacillus firmus* DS-1 has nematicidal activity and degrades multiple intestinal-associated nematode proteins. *Scientific Reports* 6, 25012. DOI: 10.1038/srep25012.

Geremia, R., Goldman, G., Jacobs, D., Ardiles, W., Vila, S.B., Van Montagu, M. and Herrera-Estrella, A. (1993) Molecular characterization of the proteinase-encoding gene *prb1* related to mycoparasitism by *Trichoderma harzianum. Molecular Microbiology* 8, 603–613. DOI: 10.1111/j.1365-2958.1993.tb01604.x

Godoy, G., Rodríguez-Kábana, R., Shelby, R.A. and Morgan-Jones, G. (1983) Chitin amendments for the control of *Meloidogyne arenaria* in infested soil II. Effects on microbial population. *Nematropica* 13, 63–74.

Govaerts, B., Fuentes, M., Mezzalama, M., Nicol, J.M., Deckers, J., Etchevers, J.D., Figueroa-Sandoval, B. and Sayre, K.D. (2007) Infiltration, soil moisture, root rot and nematode populations after 12 years of different tillage, residue and crop rotation managements. *Soil Tillage Research* 94, 209–219. DOI: 10.1016/j.still.2006.07.013

Gravato-Nobre, M.J. and Hodgkin, J. (2005) *Caenorhabditis elegans* as a model of innate immunity to pathogens. *Cell Microbiology* 7, 741–751. DOI: 10.1111/j.1462-5822.2005.00523.x

Gravato-Nobre, M.J. and Hodgkin, J. (2011) Microbial interactions with *Caenorhabditis elegans*: lessons from a model organism. In: Davies, K.G. and Spiegel, Y. (eds) *Biological Control of Plant-Parasitic Nematodes: Building Coherence between Microbial Ecology and Molecular Mechanisms*. Springer, Dordrecht, The Netherlands, pp. 65–90. DOI: 10.1007/978-1-4020-9648-8_3

Gray, N.F. (1987) Nematophagous fungi with particular reference to their ecology. *Biological Revue* 62, 245–304. DOI: 10.1111/j.1469-185x.1987.tb00665.x

Grenache, D.G., Caldicott, I., Albert, P.S., Riddle, D.L. and Politz, S.M. (1996) Environmental induction and genetic control of surface antigen switching in the nematode *Caenorhabditis elegans. Proceedings of the National Academy of Sciences of the United States of America* 93, 12388–12393. DOI: 10.2307/40627

Griffitts, J.S., Haslam, S.M., Yang, T., Garczynski, S.F., Mulloy, B., Morris, H., Cremer, P.S., Dell, A., Adang, M.J. and Aroian, R.V. (2005) Glycolipids as receptors for *Bacillus thuringiensis* crystal toxin. *Science* 307, 922–925. DOI: 10.1126/science.1104444

Hackenberg, C. and Sikora, R.A. (1994) Influence of temperature and soil moisture on the biological control of the potato-cyst nematode *Globodera pallida* using the plant-health promoting rhizobacterium *Agrobacterium radiobacter. Journal of Phytopathology* 142, 338–344. DOI: 10.1111/j.1439-0434.1994.tb00031.x

Hallmann, J. (2001) Plant interactions with endophytic bacteria. In: Jeger, M.J. and Spence, N.J. (eds) *Biotic Interactions in Plant-Pathogen Interactions*. CAB International, Wallingford, UK, pp. 87–119.

Hallmann, J. and Sikora, R.A. (1996) Toxicity of fungal endophyte secondary metabolites to plant parasitic nematodes and soil-borne plant pathogenic fungi. *European Journal of Plant Pathology* 102, 155–162. DOI: 10.1007/bf01877102

Hallmann, J., Quadt-Hallmann, A., Mahaffee, W.F. and Kloepper, J.W. (1997) Bacterial endophytes in agricultural crops. *Canadian Journal of Microbiology* 43, 895–914. DOI: 10.1139/m97-131

Hallmann, J., Quadt-Hallmann., A., Rodríguez-Kábana, A. and Kloepper, J.W. (1998) Interactions between *Meloidogyne incognita* and endophytic bacteria in cotton and cucumber. *Soil Biology and Biochemistry* 30, 925–937. DOI: 10.1016/s0038-0717(97)00183-1

Hallmann, J., Rodríguez-Kábana, R. and Kloepper, J.W. (1999) Chitin-mediated changes in bacterial communities of the soil, rhizosphere and within roots of cotton in relation to nematode control. *Soil Biology and Biochemistry* 31, 551–560. DOI: 10.1016/s0038-0717(98)00146-1

Haran, S., Schickler, H. and Chet, I. (1996) Molecular mechanisms of lytic enzymes involved in the biocontrol activity of *Trichoderma harzianum. Microbiology* 142, 2321–2331. DOI: 10.1099/00221287-142-9-2321

Harman, G.E. (2006) Overview of mechanisms and uses of *Trichoderma* spp. *Phytopathology* 96, 190–194. DOI: 10.1094/PHYTO-96-0190

Hasky-Günther, K., Hoffmann-Hergarten, S. and Sikora, R.A. (1998) Resistance against the potato cyst nematode *Globodera pallida* systemically induced by the rhizobacteria *Agrobacterium radiobacter* (G12) and *Bacillus sphaericus* (B43). *Fundamental and Applied Nematology* 21, 511–517.

Hildalgo-Diaz, L. and Kerry, B.R. (2008) Integration of biological control with other methods of nematode management. In: Ciancio, K. and Mukerji, K.G. (eds) *Integrated Management and Biocontrol of Vegetable and Grain Crops Nematodes*. Springer, Dordrecht, The Netherlands, pp. 29–49. DOI: 10.1007/978-1-4020-6063-2_2

Hodgkin, J., Kuwabara, E. and Corneliussen, B. (2000) A novel bacterial pathogen, *Microbacterium nematophilum*, induces morphological changes in the nematode *C. elegans*. *Current Microbiology* 10, 1615–1618. DOI: 10.1016/S0960-9822(00)00867-8

Hoffmann-Hergarten, S. and Sikora, R.A. (1993) Enhancing the biological control efficacy of nematode-trapping fungi towards *Heterodera schachtii* with green manure. *Journal of Plant Diseases and Protection* 100, 170–175.

Hoffmann-Hergarten, S., Hasky-Günther, K., Reitz, M. and Sikora, R.A. (1997) Induced systemic resistance by rhizobacteria toward the cyst nematode *Globodera pallida* on potato. In: Ogoshi, A., Kobayashi, K., Homma, Y., Kodama, F., Kondo, N. and Akino, S. (eds) *Plant Growth-Promoting Rhizobacteria – Present Status and Future Prospects*. Nakanishi Printing, Sapporo, Japan, pp. 292–295.

Hol, G.W.H. and Cook, R. (2005) An overview of arbuscular mycorrhizal fungi – nematode interactions. *Basic and Applied Ecology* 6, 489–503. DOI: 10.1016/j.baae.2005.04.001

Hsueh, Y.P., Mahanti, P., Schroeder, F.C. and Sternberg, P.W. (2013) Nematode-trapping fungi eavesdrop on nematode pheromones. *Current Biology* 23, 83–86. DOI: 10.1016/j.cub.2012.11.035.

Huang, X.-W., Niu, Q.-H., Zhou, W. and Zhang, K.-Q. (2005) *Bacillus nematocida* sp. nov., a novel bacterial strain with nematotoxic activity isolated from soil in Yunnan, China. *Systemic and Applied Microbiology* 28, 323–327. DOI: 10.1016/j.syapm.2005.01.008

Hutchison, C.A., Chuang, R.-Y., Noskov, V.N. *et al*. (2016) Design and synthesis of a minimal bacterial genome. *Science* 351, 6280. DOI: 10.1126/science.aad6253

Iavicoli, A., Boutet, E., Buchala, A. and Métraux, J.P. (2003) Induced systemic resistance in *Arabidopsis thaliana* in response to root inoculation with *Pseudomonas fluorescens* CHA0. *Molecular Plant-Microbe Interactions* 16, 851–858. DOI: 10.1094/MPMI.2003.16.10.851

Ito, T.M., Higashi, T., Komatsuzaki, M., Araki, M., Kaneko, N. and Ohta, H. (2015a) Responses of soil nematode community structure to soil carbon changes due to different tillage and cover crop management practices over a nine-year period in Kanto, Japan. *Applied Soil Ecology* 89, 50–58. DOI: 10.1016/j.apsoil.2014.12.010

Ito, T.M., Komatsuzaki, M., Araki M., Kaneko, N. and Ohta, H. (2015b) Soil nematode community structure affected by tillage systems and cover crop managements in organic soybean production. *Applied Soil Ecology* 86, 137–147. DOI: 10.1016/j.apsoil.2014.10.003

Ivanov, V., Tay, S.T.L. and Tay, J.-H. (2003) Monitoring of microbial diversity by fluorescence *in-situ* hybridization and fluorescence spectrometry. *Water Science and Technology* 45, 133–138.

Jaffee, B.A. and Muldoon, A.E. (1997) Suppression of the root-knot nematode *Meloidogyne javanica* by alginate pellets containing the nematophagous fungi *Hirsutella rhossiliensis*, *Monacrosporium cionopagum* and *M. ellipsosporum*. *Biocontrol Science and Technology* 7, 203–218. DOI: 10.1080/09583159730901

Jaffee, B.A., Ferris, H., Stapleton, J.J., Norton, M.V.K. and Muldoon, A.E. (1994) Parasitism of nematodes by the fungus *Hirsutella rhossiliensis* as affected by certain organic amendments. *Journal of Nematology* 26, 152–161.

Jain, R.K and Sethi, C.L. (1987) Pathogenicity of *Heterodera cajani* on cowpea as influenced by the presence of VAM fungi *Glomus fasciculatum* or *G. epigaeus*. *Indian Journal of Nematology* 17, 165–170.

Jain, R.K. and Sethi, C.L. (1988) Interaction between vesicular-arbuscular mycorrhiza, *Meloidogyne incognita* and *Heterodera cajani* on cowpea as influenced by time of inoculation. *Indian Journal of Nematology* 18, 264–268.

Jones, J.D.G. and Dangl, J.L. (2006) The plant immune system. *Nature* 444, 323–329. DOI: 10.1002/cbic.201000694

Joshua, G.W., Karlyshev, A.V., Smith, M.P., Isherwood, K.E., Titball, R.W. and Wren, B.W. (2003) A *Caenorhabditis elegans* model for *Yersinia* infection: biofilm formation on a biotic surface. *Microbiology* 149, 3221–3229. DOI: 10.1099/mic.0.26475-0

Kerry, B.R. (1975) Fungi and the decrease of cereal cyst-nematode populations in cereal monoculture. *European Plant Protection Organisation Bulletin* 5, 353–361. DOI: 10.1111/j.1365-2338.1975.tb02485.x

Kerry, B.R. (2000) Rhizosphere interactions and the exploitation of microbial agents for the biological control of plant-parasitic nematodes. *Annual Review of Phytopathology* 38, 423–441. DOI: 10.1146/annurev.phyto.38.1.423

Kerry, B.R. and Hirsch, P.R. (2011) Ecology of *Pochonia chlamydosporia* in the rhizosphere at the population, whole organism and molecular scales. In: Davies, K.G. and Spiegel, Y. (eds) *Biological Control of Plant-Parasitic Nematodes: Building Coherence between Microbial Ecology and Molecular Mechanisms*. Springer, Dordrecht, The Netherlands, pp. 171–182. DOI: 10.1007/978-1-4020-9648-8_7

Kerry, B.R., Crump, D.H. and Mullen, L.A. (1980) Parasitic fungi, soil moisture and multiplication of the cereal cyst nematode, Heterodera avenae. *Nematologica* 26, 57–68.

Kerry, B.R., Crump, D.H. and Mullen, L.A. (1982a) Studies of the cereal cyst nematode, Heterodera avenae under continuous cereals, 1975–1978. II. Fungal parasitism of nematode eggs and females. *Annals of Applied Biology* 100, 489–499. DOI: 10.1111/j.1744-7348.1982.tb01415.x

Kerry, B.R., Crump, D.H. and Mullen, L.A. (1982b) Natural control of cereal cyst nematode, Heterodera avenae Woll. by soil fungi at three sites. *Crop Protection* 1, 99–109. DOI: 10.1016/0261-2194(82)90061-8

Khan, A., Williams, K.L. and Nevalainen, H.K.M. (2006a) Infection of plant-parasitic nematodes by *Paecilomyces lilacinus* and *Monacrosporium lysipagum*. *BioControl* 51, 659–678. DOI: 10.1007/s10526-005-4242.

Khan, A., Williams, K.L. and Nevalainen, H.K.M. (2006b) Control of plant-parasitic nematodes by *Paecilomyces lilacinus* and *Monacrosporium lysipagum* in pot trials. *BioControl* 51, 643–658. DOI: 10.1007/s10526-005-4241.

Kim, D.H. and Ausubel, F.M. (2005) Evolutionary perspectives on innate immunity of *Caenorhabditis elegans*. *Current Opinion in Immunology* 17, 4–10. DOI: 10.1016/j.coi.2004.11.007

Kimpinski, J. and Kunelius, H.T. (1988) Effects of conventional and minimum tillage on nematode populations in red clover and timothy. *Forage Notes* 32, 14–16.

Kloepper, J.W., Rodríguez-Kábana, R., McInroy, J.A. and Collins, D.J. (1991) Analysis of populations and physiological characterization of microorganisms in rhizospheres of plants with antagonistic properties to phytopathogenic nematodes. *Plant and Soil* 136, 95–102. DOI: 10.1007/BF02465224

Kluepfel, D.A., McInnis, T.M. and Zehr, E.I. (1993) Involvement of root colonizing bacteria in peach orchid soils suppressive of the nematode *Criconemella xenoplax*. *Phytopathology* 3, 1240–1245. DOI: 10.1094/Phyto-83-1240.

Kramer, J.M. (1997) Extracellular matrix. In: Riddle, D.L., Blumenthal, T., Meyer, B.J. and Priess, J.R. (eds) *C. elegans* II. Cold Spring Harbor Laboratory Press, Cold Spring Harbor, New York, pp. 471–500.

Krebs, B., Höding, B., Kübart, S., Workie, M.A., Junge, H., Schmiedeknecht, G., Grosch, R., Bochow, H. and Hevesi, M. (1998) Use of *Bacillus subtilis* as biocontrol agent. I. Activities and characterization of *Bacillus subtilis* strains. *Zeitschrift für Pflanzenkrankheiten und Pflanzenschutz* 105, 181–197. Available at: www.jstor.org/stable/43215232 (accessed 22 January 2018).

Kühn, J. (1877) Vorläufiger Bericht über die bisherigen Ergebnisse der seit dem Jahre 1875 im Auftrage des Vereins für Rübenzucker-Industrie ausgeführten Versuche zur Ermittlung der Ursache der Rübenmüdigkeit des Bodens uns zur Erforschung der Natur der Nematoden. *Zeitschrift des Vereins für die Rübenzucker-Industrie im Zollverein* 27, 452–457.

Ledford, H. (2016) CRISPR: gene editing is just at the beginning. *Nature* 531, 156–159. DOI: 10.1038/531156a

Leigh, J.A. and Coplin, D.L. (1992) Exopolysaccharides in plant-bacterial interactions. *Annual Review of Microbiology* 46, 307–346. DOI: 10.1146/annurev.mi.46.100192.001515

Lettice, E.P. and Jones, P.W. (2015) Evaluation of rhizobacterial colonisation and the ability to induce *Globodera pallida* hatch. *Nematology* 17, 203–212. DOI: 10.1163/15685411-00002863

Li, J., Zou, C., Xu, J., Ji, X., Niu, X., Yang, J., Huang, X. and Zhang, K.Q. (2015) Molecular mechanisms of nematode-nematophagous microbe interactions: basis for biological control of plant-parasitic nematodes. *Annual Review of Phytopathology* 53, 67–95. DOI: 10.1146/annurev-phyto-080614-120336

Lian, L.H., Tian, B.Y., Xiong, R., Zhu, M.Z., Xu, J. and Zhang, K.Q. (2007) Proteases from *Bacillus*: a new insight into the mechanism of action for rhizobacterial suppression of nematode populations. *Letters in Applied Microbiology* 45, 262–269. DOI: 10.1111/j.1472-765x.2007.02184.x

Linford, M.B. (1937) Stimulated activity of natural enemies of nematodes. *Science* 85, 123–124. DOI: 10.1126/science.85.2196.123

Linford, M.B. and Yap, F. (1939) Root-knot nematode injury restricted by a fungus. *Phytopathology* 29, 596–608. DOI: 10.2307/42563476

Linford, M.B., Yap, F. and Oliveira, J.M. (1938) Reduction in soil populations of root-knot nematode during the decomposition of organic matter. *Soil Science* 45, 127–142. DOI: 10.1097/00010694-193802000-00004

Liu, S.F. and Chen, S.Y. (2001) Nematode hosts of the fungus *Hirsutella minnesotensis*. *Phytopathology* 91, S138 (Abstr.).

Liu, S.F. and Chen, S.Y. (2005) Efficacy of the fungi *Hirsutella minnesotensis* and *H. rhossiliensis* from liquid culture for control of the soybean cyst nematode *Heterodera glycines*. *Nematology* 7, 149–157. DOI: 10.1163/1568541054192153

Liu, X.Z. and Chen, S.Y. (2009) Effectiveness of *Hirsutella minnesotensis* and *H. rhossiliensis* in control of the soybean cyst nematode in four soils with various pH, texture, and organic matter. *Biocontrol Science and Technology* 19, 595–612. DOI: 10.1080/09583150902960979

López-Llorca, L.V. (1990) Purification and properties of extracellular proteases produced by the nematophagous fungus *Verticillium suchlasporium*. *Canadian Journal of Microbiology* 36, 530–537. DOI: 10.1139/m90-093

Lumsden, R.D. and Vaughn, J.L. (1995) *Pest Management: Biotechnology-based Technologies*. American Chemical Society, Washington, DC.

Luo, H., Xiong, J., Zhou, Q., Xia, L. and Yu, Z. (2013) The effects of *Bacillus thuringiensis* Cry6A on the survival, growth, reproduction, locomotion, and behavioral response of *Caenorhabditis elegans*. *Applied Microbiology and Biotechnology* 97, 10135–10142. DOI: 10.1007/s00253-013-5249-3

Mahdy, M., Hallmann, J. and Sikora, R.A. (2001) Influence of plant species on the biological control activity of the antagonistic rhizobacterium *Rhizobium etli* strain G12 toward the root knot nematode *Meloidogyne incognita*. *Mededelingen van de Rijksuniversiteit Gent Fakulteit Landbouwkundige en Toegepaste Biologische Wetenschappen* 66, 655–662. PMID: 12425090

Mahoney, T.R., Luo, S. and Nonet, M.L. (2006) Analysis of synaptic transmission in *Caenorhabditis elegans* using and aldicarb-sensitivity assay. *Nature Protocols* 1, 1772–1777. DOI: 10.1038/nprot.2006.281

Manhong, S., Xingzhong, L. and Zhibo, J. (2002) Effects of *Paecilomyces lilacinus* on egg hatching and juvenile mortality of *Heterodera glycines*. *Acta Phytophylacica Sinica* 29, 57–61.

Manosalva, P., Manohar, M., Von Reuss, S.H. et al. (2015) Conserved nematode signalling molecules elicit plant defenses and pathogen resistance. *Nature Communications* 6, 7759. DOI: 10.1038/ncomms8795

Marciá-Vicente, J.G., Palma-Guerrero, J., Gómez-Vidal, S. and López-Llorca, L.V. (2011) New insights on the mode of action of fungal pathogens of invertebrates for improving their biological performance. In: Davies, K.G. and Spiegel, Y. (eds) *Biological Control of Plant-Parasitic Nematodes: Building Coherence between Microbial Ecology and Molecular Mechanisms*. Springer, Dordrecht, The Netherlands, pp. 203–225. DOI: 10.1007/978-1-4020-9648-8_9.

McElroy, K., Mouton, L., Du Pasquier, L., Qi, W. and Ebert, D. (2011) Characterisation of a large family of polymorphic collagen-like proteins in the endospore-forming bacterium *Pasteuria ramosa*. *Research in Microbiology* 162, 701–714. DOI: 10.1016/j.resmic.2011.06.009

McInnis, T.M. and Jaffee, B.A. (1989) An assay for *Hirsutella rhossiliensis* spores and the importance of phialides for nematode inoculation. *Journal of Nematology* 21, 229–234.

McInory, J.A. and Kloepper, J.W. (1995) Survey of indigenous bacterial endophytes from cotton and sweet corn. *Plant and Soil* 173, 337–342. DOI: 10.1007/bf00011472

McKenney, P.T., Driks, A. and Eichenberger, P. (2013) The *Bacillus subtilis* endospore: assembly and functions of the multilayered coat. *Nature Reviews Microbiology* 11, 33–44. DOI: 10.1038/nrmicro2921

McSpadden Gardener, B.B. (2007) Diversity and ecology of biocontrol *Pseudomonas* spp. in agricultural systems. *Phytopathology* 97, 221–226. DOI: 10.1094/PHYTO-97-2-0221

McSpadden Gardener, B.B., Benitez, M.S., Camp, A. and Zumpetta, C. (2006a) Evaluation of a seed treatment containing aphlD+ strain of *Pseudomonas fluorescens* on organic soybeans, 2005. *Biological and Cultural Tests for Control of Plant Diseases Report* 21, FC046.

McSpadden Gardener, B.B., Kroon van Diest, C. and Beuerlein, J. (2006b) Evaluation of biological seed treatments containing phlD+ strains of *Pseudomonas fluorescens* on soybeans grown in Ohio, 2005. *Biological and Cultural Tests for Control of Plant Diseases Report* 21, FC045.

Mendes, R., Kruijt, M., Bruijn, I. et al. (2011) Deciphering the rhizosphere microbiome for disease-suppressive bacteria. *Science* 332, 1097–1100. DOI: 10.1126/science.1203980

Mendoza, A.R., Kiewnick, S. and Sikora, R.A. (2008) In vitro activity of *Bacillus firmus* against the burrowing nematode *Radopholus similis*, the root-knot nematode *Meloidogyne incognita* and the stem nematode *Ditylenchus dipsaci*. *Biocontrol Science and Technology* 18, 377–389. DOI: 10.1080/09583150801952143

Mendoza de Gives, P., Davies, K.G., Clark, S.J. and Behnke, J.M. (1999) Predatory behaviour of trapping fungi against srf mutants of *Caenorhabditis elegans* and different plant and animal parasitic nematodes. *Parasitology* 119, 95–104. DOI: 10.1017/S0031182099004424

Menzies, B.E. (2003) The role of fibronectin binding proteins in the pathogenesis of *Staphylococcus aureus* infections. *Current Opinion in Infectious Diseases* 16, 225–229. DOI: 10.1097/01.qco.0000073771.11390.75

Meyer, S.L.F., Halbrendt, J.M., Carta, L.K., Skantar, A.M., Liu, T., Abdelnabby, H.M.E. and Vinyard, R.T. (2009) Toxicity of 2,4-diacetylphloroglucinol (DAPG) to plant-parasitic and bacterial-feeding nematodes. *Journal of Nematology* 41, 274–280.

Mohan, S., Fould, S. and Davies, K.G. (2001) The interaction between the gelatine binding domain of fibronectin and the attachment of *Pasteuria penetrans* endospores to nematode cuticle. *Parasitology* 123, 271–276.

Mohan, S., Mauchline, T.M., Rowe, J., Hirsch, P.R. and Davies, K.G. (2011) *Pasteuria* endospores from *Heterodera cajani* (Nematoda: Heteroderidae) exhibit inverted attachment and altered germination in cross-infection studies with *Globodera pallida* (Nematoda: Heteroderidae). *FEMS Microbiology Ecology* 79, 675–684. DOI: 10.1111/j.1574-6941.2011.01249.x

Moosavi, M.R. and Askary T.H. (2015) Nematophagous fungi: commercialization. In: Askary, T.H. (ed.) *Biocontrol Agents of Phytonematodes*. CAB International, Wallingford, UK, pp. 187–202.

Morton, C.O., Hirsch, P.R., Peberdy, J.P. and Kerry, B.R. (2003) Cloning of and genetic variation in the protease VCP1 from the nematophagous fungus *Pochonia chlamydosporia*. *Mycological Research* 107, 38–46. DOI: 10.1017/s0953756202007050

Morton, C.O., Hirsch, P.R. and Kerry, B.R. (2004) Infection of plant-parasitic nematodes by nematophagous fungi – a review of the application of molecular biology to understand infection processes and improve biological control. *Nematology* 6, 161–170. DOI: 10.1163/1568541041218004

Mouton, L., Traunecker, E., McElroy, K., Du Pasquier, L. and Ebert, D. (2009) Identification of a polymorphic collagen-like protein in the crustacean bacteria *Pasteuria ramosa*. *Research in Microbiology* 160, 792–799. DOI: 10.1016/j.resmic.2009.08.016.

Nielsen-LeRoux, C., Gaudriault, S., Ramarao, N., Lereclus, D. and Givaudan, A. (2012) How the insect pathogen bacteria *Bacillus thuringiensis* and *Xenorhabdus/Photorhabdus* occupy their hosts. *Current Opinion in Microbiology* 15, 220–231. DOI: 10.1016/j.mib.2012.04.006

Noel, G.R., Atibalentja, N. and Domier, L.L. (2005) Emended description of *Pasteuria nishizawae*. *International Journal of Systematic and Evolutionary Microbiology* 55, 1681–1685. DOI: 10.1099/ijs.0.63174-0

Noel, G.R., Atibalentja, N. and Bauer, S.J. (2010) Suppression of *Heterodera glycines* in a soybean field artificially infested with *Pasteuria nishizawae*. *Nematropica* 40, 41–52.

Oliveira, E.J., Rabinovitch, L., Monnerat, R.G., Passos, L.K.J. and Zahner, V. (2004) Molecular characterization of *Brevibacillus laterosporus* and its potential use in biological control. *Applied and Environmental Microbiology* 70, 6657–6664. DOI: 10.1128/AEM.70.11.6657-6664.2004

Olsen, D.P., Phu, D., Libby, L.J.M., Cormier, J.A., Montez, K.M., Ryder.E.F. and Politz, S.M. (2007) Chemosensory control of surface antigen switching in the nematode *Caenorhabdits elegans*. *Genes Brain Behaviour* 6, 240–252. DOI: 10.1111/j.1601-183X.2006.00252.x

Oostendorp, M. and Sikora, R.A. (1989) Seed treatment with antagonistic rhizobacteria for the suppression of *Heterodera schachtii* early root infection of sugar beet. *Revue de Nématologie* 12, 77–83.

Oostendorp, M. and Sikora, R.A. (1990) *In-vitro* interrelationships between rhizosphere bacteria and *Heterodera schachtii*. *Revue de Nématologie* 13, 269–274.

Pandey, R. and Sikora, R.A. (2000) Influence of aqueous extracts of organic matter on the sensitivity of *Heterodera schachtii* (Schmidt) and *Meloidogyne incognita* (Kofoid and White) Chitwood eggs to *Verticillium chlamydosporium* Goddard infection. *Journal of Plant Diseases and Protection*, 107, 494–497. Available at: www.jstor.org/stable/43215352 (accessed 22 January 2018).

Parsons, L.M., Mizanur, R.M., Jankowska, E., Hodgkin, J., O'Rourke, D., Stroud, D., Ghosh, S. and Cipollo, J.F. (2014) *Caenorhabditis elegans* bacterial pathogen resistant *bus-4* mutants produce altered mucins. *PLoS ONE* 9, e107250. DOI: 10.1371/journal.pone.0107250

Patel, A.V., Jakobs-Schönwandt, D., Rose, T. and Vorlop, K.D. (2011) Fermentation and micro encapsulation of the nematophagous fungus *Hirsutella rhossiliensis* in a novel type of hollow beads. *Applied Microbiology and Biotechnology* 89, 1751–1760. DOI: 10.1007/s00253-010-3046-9

Patti, J.M., Allen, B.L., McGavin, M.J. and Hook, M. (1994) MSCRAMM-mediated adherence of microorganisms to host tissues. *Annual Review of Microbiology* 48, 585–617. DOI: 10.1146/annurev.mi.48.100194.003101

Perry, R.N. and Trett, M.W. (1986) Ultrastructure of the eggshell of *Heterodera schachtii* and *H. glycines* (Nematoda: Tylenchida). *Revue de Nématologie* 9, 399–403.

Persmark, L., Banck, A. and Jansson, H.B. (1996) Population dynamics of nematophagous fungi and nematodes in an arable soil: vertical and seasonal fluctuations. *Soil Biology and Biochemistry* 28, 1005–1014. DOI: 10.1016/0038-0717(96)00060-0

Phani, V., Shivakumara, T.N., Davies, K.G. and Rao, U. (2017) *Meloidogyne incognita* fatty acid- and retinol- binding protein (Mi-FAR-1) affects nematode infection of plant roots and the attachment of *Pasteuria penetrans* endospores. *Frontiers in Microbiology* 8, 2122. DOI: 10.3389/fmicb.2017.02122

Politz, S.M. and Phillip, M. (1992) *Caenorhabditis elegans* as a model for parasitic nematodes: a focus on the cuticle. *Parasitology Today* 8, 6–12. DOI: 10.1016/0169-4758(92)90302-i

Pramer, D. and Stoll, N.R. (1959) Nemin: a morphogenic substance causing trap formation by predaceous fungi. *Science* 129, 966–967. DOI: 10.1126/science.129.3354.966

Prasad, S.S.V., Tilak, K. and Gollakota, K. (1972) Role of *Bacillus thuringiensis* var. *thuringiensis* on the larval survivability and egg hatching of *Meloidogyne* spp., the causative agent of root knot disease. *Journal of Invertebrate Pathology* 20, 377–378. DOI: 10.1016/0022-2011(72)90177-2

Racke, J. and Sikora, R.A. (1992) Isolation, formulation and antagonistic activity of rhizobacteria toward the potato cyst nematode *Globodera pallida*. *Soil Biology and Biochemistry* 24, 521–526. DOI: 10.1016/0038-0717(92)90075-9

Read, T.D., Peterson, S.N., Tourasse, N. *et al.* (2003) The genome sequence of *Bacillus anthracis* Ames and comparison to closely related bacteria. *Nature* 423, 81–86. DOI: 10.1038/nature01586

Reitz, M., Rudolph, K., Schröder, I., Hoffmann-Hergarten, S., Hallmann, J. and Sikora, R.A. (2000) Lipopolysaccharides of *Rhizobium etli* strain G12 act in potato roots as an inducing agent of systemic resistance to infection by the cyst nematode *Globodera pallida*. *Applied and Environmental Microbiology* 66, 3515–3518. DOI: 10.1128/aem.66.8.3515-3518.2000

Reitz, M., Oger, P., Meyer, A., Niehaus, K., Farrand, S.K., Hallmann, J. and Sikora, R.A. (2002) Importance of the O-antigen, core-region and lipid A of rhizobial lipopolysaccharides for the induction of systemic resistance in potato to *Globodera pallida*. *Nematology* 4, 73–79. DOI: 10.1163/156854102760082221

Renčo, M., Sasanelli, N. and Kováčik, P. (2011) The effect of soil compost treatments on potato cyst nematodes *Globodera rostochiensis* and *Globodera pallida*. *Helminthologia* 48, 184–194. DOI: 10.2478/s11687-011-0027-1

Rodríguez-Kábana, R. (1986) Organic and inorganic nitrogen amendments to soil as nematode suppressants. *Journal of Nematology* 18, 129–135.

Rodríguez-Kábana, R., Morgan-Jones, G. and Chet, I. (1987) Biological control of nematodes: soil amendments and microbial antagonists. *Plant and Soil* 100, 237–247. DOI: 10.1007/bf02370944

Rovira, A.D. (1982) *Management Strategies for Controlling Cereal Cyst Nematode*. CSIRO Division of Soils, Adelaide, Australia.

Rovira, A.D. and Sands, D.C. (1977) Fluorescent pseudomonas – a residual component in the soil microflora. *Journal of Applied Bacteriology* 34, 253–259. DOI: 10.1111/j.1365-2672.1971.tb02284.x

Ruiu, L. (2013) *Brevibacillus laterosporus*, a pathogen of invertebrates and a broad-spectrum antimicrobial species. *Insects* 4, 476–492. DOI: 10.3390/insects4030476

Ryan, A. and Jones, P. (2004) The effect of mycorrhization of potato roots on the hatching chemicals active towards the potato cyst nematodes, *Globodera pallida* and *G. rostochiensis*. *Nematology* 6, 335–342. DOI: 10.1163/1568541042360456

Ryan, N.A., Duffy, E.M., Cassells, A.C. and Jones, P.W. (2000) The effect of mycorrhizal fungi on the hatch of potato cyst nematodes. *Applied Soil Ecology* 15, 233–240. DOI: 10.1016/s0929-1393(00)00099-8

Ryan, N.A., Deliopoulos, T., Jones, P. and Haydock, P.P.J. (2003) Effects of mixed-isolate mycorrhizal inoculum on the potato – potato cyst nematode interaction. *Annals of Applied Biology* 143, 111–119. DOI: 10.1111/j.1744-7348.2003.tb00275.x

Saifullah, A. and Khan, N.U. (2014) Low temperature scanning electron microscopic studies on the interaction of *Globodera rostochiensis* Woll. and *Trichoderma harzianum* Rifai. *Pakistan Journal of Botany* 46, 357–361.

Sayre, R.M., Wergin, W.P., Schmidt, J.M. and Starr, M.P. (1991) *Pasteuria nishizawae* sp. nov., a mycelial and endospore-forming bacterium parasitic on cyst nematodes of genera *Heterodera* and *Globodera*. *Research in Microbiology* 142, 551–564. DOI: 10.1016/0923-2508(91)90188-G

Schlang, J. (1993) Mit Kompostierung Nematoden bekämpfen? *DLG-Mitteilungen* 9, 30–31.

Schnepf, E., Crickmore, N., Van Rie, J., Lereclus, D., Baum, J., Feitelson, J., Zeigler, D.R. and Dean, D.H. (1998) *Bacillus thuringiensis* and its pesticidal crystal proteins. *Microbiology and Molecular Biology Reviews* 62, 775–806. PMID: 9729609

Schouten, A. (2016) Mechanisms involved in nematode control by endophytic fungi. *Annual Review of Phytopathology* 54, 121–142. DOI: 10.1146/ANNUREV-PHYTO-080615-100114

Schrimsher, D.W., Lawrence, K.S., Castillo, J., Moore, S.R. and Kloepper, J.W. (2011) Effects of *Bacillus firmus* GB-126 on the soybean cyst nematode mobility *in vitro*. *Phytopathology* 101, S161 (Abstr.)

Schroth, M.N. and Hancock, J.G. (1982) Disease-suppressive soil and root colonizing bacteria. *Science* 216, 1376–1381. DOI: 10.1126/science.216.4553.1376

Schulenburg, H. and Ewbank, J.J. (2007) The genetics of pathogen avoidance in *Caenorhabditis elegans*. *Molecular Biology* 66, 563–570. DOI: 10.1111/j.1365-2958.2007.05946.x

Segers, R., Butt, T.M., Kerry, B.R., Beckett, A. and Peberdy, J.F. (1996) The role of the proteinase VCP1 produced by the nematophagous *Verticillium chlamydosporium* in the infection process of nematode eggs. *Mycological Research* 100, 421–428. DOI: 10.1016/s0953-7562(96)80138-9

Sharma, S.B. and Davies, K.G. (1996) Characterisation of *Pasteuria* isolated from *Heterodera cajani* using morphology, pathology and serology of endospores. *Systematic and Applied Microbiology* 19, 106–112. DOI: 10.1016/s0723-2020(96)80017-8

Sharon, E., Bar-Eyal, M. and Chet, I. (2001) Biocontrol of the root-knot nematode *Meloidogyne javanica* by *Trichoderma harzianum*. *Phytopathology* 91, 687–693. DOI: 10.1094/PHYTO.2001.91.7.687

Sharon, E., Chet, I. and Spiegel, Y. (2011) *Trichoderma* as a biological control agent. In: Davies, K.G. and Spiegel, Y. (eds) *Biological Control of Plant-Parasitic Nematodes: Building Coherence between Microbial Ecology and Molecular Mechanisms*. Springer, Dordrecht, The Netherlands, pp. 183–201.

Siddiqui, I.A. and Ehteshamul-Haque, S. (2000) Use of *Pseudomonas aeruginosa* for the control of root rot-root knot disease complex in tomato. *Nematologia Mediterranea* 28, 189–192.

Siddiqui, I.A. and Shaukat, S.S. (2003) Suppression of root-knot disease by *Pseudomonas fluorescens* CHA0 in tomato: Importance of bacterial secondary metabolite, 2,4-diacetylphloroglucinol. *Soil Biology and Biochemistry* 35, 1615–1623. DOI: 10.1016/j.soilbio.2003.08.006

Siddiqui, I.A. and Shaukat, S.S. (2004) Systemic resistance in tomato induced by biocontrol bacteria against the root-knot nematode, *Meloidogyne javanica* is independent of salicylic acid production. *Journal of Phytopathology* 152, 48–54. DOI: 10.1046/j.1439-0434.2003.00800.x

Siddiqui, I.A., Shaukat, S.S. and Hamid, M. (2003) Suppression of *Meloidogyne incognita* by *Pseudomonas fluorescens* strain HAO and its genetically modified derivatives: the influence of oxygen. *Nematologia Mediterranea* 31, 105–109.

Siddiqui, Z.A and Mahmood, I. (1995) Biological control of *Heterodera cajani* and *Fusarium udum* by *Bacillus subtilis*, *Bradyrhizobium japonicum* and *Glomus fasciculatum* on pigeon pea. *Fundamental and Applied Nematology* 18, 559–566.

Siddiqui, Z.A. and Mahmood, I. (1999) Role of bacteria in the management of plant parasitic nematodes. A review. *Bioresource Technology* 69, 167–179. DOI: 10.1016/s0960-8524(98)00122-9

Sikora, R.A. (1988) Interrelationship between plant health-promoting rhizobacteria, plant parasitic nematodes and soil microorganisms. *Mededelingen Faculteit Landbouwkundige Rijksuniversiteit Gent* 53, 867–878.

Sikora, R.A. (1992) Management of the antagonistic potential in agricultural ecosystems for the biological control of plant parasitic nematodes. *Annual Review of Phytopathology* 30, 245–270. DOI: 10.1146/annurev.py.30.090192.001333

Sikora, R.A. and Hoffmann-Hergarten, S. (1993) Biological control of plant parasitic nematodes with plant health-promoting rhizobacteria. In: Lumsden, R.D. and Vaughn, J.L. (eds) *Pest Management: Biotechnology-Based Technologies*. American Chemical Society, Washington DC, pp. 166–172.

Silverman, M.A., Blaxter, M.L. and Link, C.D. (1997) Biochemical analysis of *Caenorhabditis elegans* surface mutants. *Journal of Nematology* 29, 296–305.

Singer, S. (1996) The utility of morphological group II *Bacillus*. *Advances in Applied Microbiology* 42, 219–261. DOI: 10.1016/s0065-2164(08)70374-5

Srivastava, A., Hall, A.M., Graeme-Cook, K. and Davies, K.G. (2014) Exploiting genomics to improve the biological control potential of *Pasteuria* spp. an organism with potential to control plant-parasitic nematodes. *Aspects of Applied Biology* 127, 9–14. DOI: 10.13140/RG.2.1.4578.3127

Steichen, C., Chen, P., Kearney, J.F. and Turnbough, C.L., Jr (2003) Identification of the immunodominant protein and other proteins of the *Bacillus anthracis* exosporium. *Journal of Bacteriology* 185, 1903–1910. DOI: 10.1111/j.1365-2958.2007.05658.x

Stirling, G.R. (1985) Host specificity of *Pasteuria penetrans* within the genus *Meloidogyne*. *Nematologica* 31, 203–209. DOI: 10.1163/187529285X00265

Stirling, G.R. (2011) Biological control of plant parasitic nematodes: an ecological perspective, a review of progress and opportunities for further research. In: Davies, K.G. and Spiegel, Y. (eds) *Biological Control of Plant-Parasitic Nematodes: Building Coherence between Microbial Ecology and Molecular Mechanisms*. Springer, Dordrecht, The Netherlands, pp. 1–38.

Stirling, G.R. (2014) *Biological Control of Plant-parasitic Nematodes*. CAB International, Wallingford, UK.

Stirling, G.R. and Kerry, B.R. (1983) Antagonists of the cereal cyst nematode *Heterodera avenae* Woll. in Australian soils. *Animal Production Science* 23, 318–324. DOI: 10.1071/EA9830318

Stirling, G.R. and Mankau, R. (1978) *Dactylella oviparasitica*, a new fungal parasite of *Meloidogyne* eggs. *Mycologia* 70, 774–783. DOI: 10.2307/3759357

Stirling, G.R. and Mankau, R. (1979) Mode of parasitism of *Meloidogyne* and other nematode eggs by *Dactylella oviparasitica*. *Journal of Nematology* 11, 282–288.

Sun, J., Park, S.-Y., Kang, S., Liu, X., Qiu, J. and Xiang, M. (2015) Development of a transformation system for *Hirsutella* spp. and visualization of the mode of nematode infection by GFP-labeled *H. minnesotensis*. *Scientific Reports* 5, 10477. DOI: 10.1038/srep10477

Sylvester, P., Couture-Tosi, E. and Mock, M. (2002) A collagen-like surface glycoprotein is a structural component of the *Bacillus anthracis* exosporium. *Molecular Microbiology* 45, 169–178. DOI: 10.1046/j.1365-2958.2000.03000.x

Talavera, M., Mizukubo, T., Ito, K. and Aiba, S. (2002) Effect of spore inoculum and agricultural practices on the vertical distribution of the biocontrol plant-growth-promoting bacterium *Pasteuria penetrans* and growth of *Meloidogyne incognita* infected tomato. *Biology and Fertility of Soils* 35, 435–440. DOI: 10.1007/s00374-002-0491-3

Tedford, E.C., Jaffee, B.A. and Muldoon, A.E. (1994) Variability among isolates of the nematophagous fungus *Hirsutella rhossiliensis*. *Mycological Research* 98, 1127–1136. DOI: 10.1016/S0953-7562(09)80198-6

Terefe, M., Tefera, T. and Sakhuja, P.K. (2009) Effect of a formulation of *Bacillus firmus* on root-knot nematode *Meloidogyne incognita* infestation and the growth of tomato plants in the greenhouse and nursery. *Journal of Invertebrate Pathology* 100, 94–99. DOI: 10.1016/j.jip.2008.11.004

Tian, B.Y., Li, N., Lian, L.H., Liu, J.W., Yang, J.K. and Zhang, K.Q. (2006) Cloning, expression and deletion of the cuticle-degrading protease BLG4 from nematophagous bacterium *Brevibacillus laterosporus* G4. *Archives of Microbiology* 186, 297–305. DOI: 10.1007/s00203-006-0145-1

Tian, H., Riggs, R.D. and Crippen, D.L. (2000) Control of soybean cyst nematode by chitinolytic bacteria with chitin substrate. *Journal of Nematology* 32, 370–376.

Tikhonov, V.E., López-Llorca, L.V., Salinas, J. and Janssen, H.-B. (2002) Purification and characterisation of chitinases from nematophagous fungi *Verticillium chlamydosporium* and *V. suchlasporium*. *Fungal Genetics and Biology* 35, 67–78. DOI: 10.1006/fgbi.2001.1312

Timper, P., Kaya, H.K. and Jaffee, B.A. (1991) Survival of entomogenous nematodes in soil infested with the nematode-parasitic fungus *Hirsutella rhossiliensis* (Deuteromycotina: Hyphomycetes). *Biological Control* 1, 42–50. DOI: 10.1016/1049-9644(91)90100-E

Todd, S.J., Moir, S.J.G., Johnson, M.J. and Moir, A. (2003) Genes of *Bacillus cereus* and *Bacillus anthracis* encoding proteins of the exosporium. *Journal of Bacteriology* 185, 3373–3378. DOI: 10.1128/JB.185.11.3373-3378.2003

Tribe, H.T. (1977) Pathology of cyst-nematodes. *Biological Reviews* 52, 477–507. DOI: 10.1111/j.1469-185X.1977.tb00857.x

Trifonova, Z., Tsvetkov, I., Bogatzevska, N. and Batchvarova, R. (2014) Efficiency of *Pseudomonas* spp. for biocontrol of the potato cyst nematode *Globodera rostochiensis* (Woll.). *Bulgarian Journal of Agricultural Science* 20, 666–669.

Tunlid, A. and Ahrén, D. (2011) Molecular mechanisms of the interaction between nematode-trapping fungi and nematodes: lessons from genomics. In: Davies, K. and Spiegel, Y. (eds) *Biological Control of Plant-Parasitic Nematodes: Building Coherence between Microbial Ecology and Molecular Mechanisms*. Springer, Dordrecht, The Netherlands, pp. 145–169.

Tylka, G.L., Hussey, R.S. and Roncadori, R.W. (1991) Interactions of vesicular-arbuscular mycorrhizal fungi, phosphorus, and *Heterodera glycines* on soybean. *Journal of Nematology* 23, 122–133.

Vachon, V., Laprade, R. and Schwartz, J.-L. (2012) Current models of the mode of action of *Bacillus thuringiensis* insecticidal crystal proteins: a critical review. *Journal of Invertebrate Pathology* 111, 1–12. DOI: 10.1016/j.jip.2012.05.001

Van Loon, L.C. and Bakker, P.A.H.M. (2005) Induced systemic resistance as a mechanism of disease suppression by rhizobacteria. In: Siddiqui, Z.A. (ed.) *PGPR: Biocontrol and Biofertilization*. Springer, Dordrecht, The Netherlands, pp. 39–66.

Varma, A., Verma, S., Sudha, Sahay, N., Bütehorn, B. and Franken, P. (1999) *Piriformospora indica*, a cultivable plant-growth-promoting root endophyte. *Applied and Environmental Microbiology* 65, 2741–2744. PMCID: PMC91405

Velvis, H. and Kamp, P. (1995) Infection of second stage juveniles of potato cyst nematodes by the nematophagous fungus *Hirsutella rhossiliensis* in Dutch potato fields. *Nematologica* 41, 617–627. DOI: 10.1163/003925995X00558

Viaene, N.M. and Abawi, G.S. (2000) *Hirsutella rhossiliensis* and *Verticillium chlamydosporium* as biological control agents of *Meloidogyne hapla* on lettuce. *Journal of Nematology* 32, 85–100.

Viterbo, A., Montero, M., Ramot, O., Friesem, D., Monte, E., Llobell, A. and Chet, I. (2002) Expression regulation of an endochitinase gene (chit36) from *Trichoderma asperellum* (*T. harzianum* T-203). *Current Genetics* 42, 114–122. DOI: 10.1007/s00294-002-0345-4

Warnock, N.D., Wilson, L., Patten, C., Fleming, C.C., Maule, A.G. and Dalzell, J.J. (2017) Nematode neuropeptides as transgenic nematicides. *PLOS Pathogens* DOI: 10.1371/journal.ppat.1006237

Wei, W., Xua, Y., Lib, S., Zhuc, L. and Songa, J. (2015) Developing suppressive soil for root diseases of soybean with continuous long-term cropping of soybean in black soil of Northeast China. *Acta Agriculturae Scandinavica, Section B – Soil and Plant Science* 65, 279–285. DOI: 10.1080/09064710.2014.992941

Weller, D.M. (1988) Biological control of soil-borne plant pathogens in the rhizosphere with bacteria. *Annual Review of Phytopathology* 26, 379–407. DOI: 10.1146/annurev.py.26.090188.002115

Weller, D.M. (2007) *Pseudomonas* biocontrol agents of soil borne pathogens: looking back over 30 years. *Phytopathology* 97, 250–256. DOI: 10.1094/PHYTO-97-2-0250

Weller, D.M., van Pelt, J.A., Mavrodi, D.V., Pieterse, C.M.J., Bakker, P.A.H.M. and van Loon, L.C. (2004) Induced systemic resistance (ISR) in *Arabidopsis* against *Pseudomonas syringae* pv. tomato by 2,4-diacetylphloroglucinol (DAPG)-producing *Pseudomonas fluorescens*. *Phytopathology* 94, S108 (Abstr.).

Westcott, S.W. and Kluepfel, D.A. (1993) Inhibition of *Criconemella xenoplax* egg hatch by *Pseudomonas aurefaciens*. *Phytopathology* 83, 1245–1249. DOI: 10.1094/Phyto-83-1245

Westphal, A. and Becker, J.O. (2001) Components of soil suppressiveness against *Heterodera schachtii*. *Soil Biology and Biochemistry* 33, 9–16. DOI: 10.1016/s0038-0717(00)00108-5

Williams, T.D. (1969) The effects of formalin, nabam, irrigation and nitrogen on *Heterodera avenae* Woll., *Ophiobolus graminis* Sacc. and the growth of spring wheat. *Annals of Applied Biology* 64, 325–334. DOI: 10.1111/j.1744-7348.1969.tb02882.x

Winkler, H.E., Hetrick, B.A.D. and Todd, T.C. (1994) Interactions of *Heterodera glycines*, *Macrophomina phaseolina*, and mycorrhizal fungi on soybean in Kansas. *Journal of Nematology* 26, 675–682.

Wu, I.-L., Narayan, K., Castaing, J.-P., Tian, F., Subramaniam, S. and Ramamurthi, K.S. (2015) A versatile nano display platform from bacterial spore coat protein. *Nature Communications* 6, 6777. DOI: 10.1038/ncomms7777

Xiang, M.C., Xiang, P.A., Jiang, X.Z., Duan, W.J. and Liu, X.Z. (2010) Detection and quantification of the nematophagous fungus *Hirsutella minnesotensis* in soil with real-time PCR. *Applied Soil Ecology* 44, 170–175. DOI: 10.1016/j.apsoil.2009.12.002

Yu, Z., Luo, H., Xiong, J., Zhou, Q., Xia, L., Sun, M., Li, L. and Yu, Z. (2014) *Bacillus thuringiensis* Cry6A exhibits nematicidal activity to *Caenorhabditis elegans* bre mutants and synergistic activity with Cry5B to *C. elegans*. *Letters in Applied Microbiology* 58, 511–519. DOI: 10.1111/lam.12219

Zare, R. and Gams, W. (2001) A revision of *Verticillium* section Prostrata. VI. The genus *Haptocillium*. *Nova Hedwigia* 73, 271–292. DOI: 10.1127/nova.hedwigia/73/2001/271

Zeilinger, S., Galhaup, C., Payer, K., Woo, S.L., Mach, R.L., Fekete, C., Lorito, M. and Kubicek, C.P. (1999) Chitinase gene expression during mycoparasitic interaction of *Trichoderma harzianum*. *Fungal Genetics and Biology* 26, 131–140. DOI: 10.1006/fgbi.1998.1111

Zhang, F., Peng, D., Cheng, C. et al. (2016a) *Bacillus thuringiensis* crystal protein Cry6Aa triggers *Caenorhabditis elegans* necrosis pathway mediated by aspartic protease (ASP-1). *PLoS Pathogens* 12, p.e1005389. DOI: 10.1371/journal.ppat.1005389

Zhang, J., Li, Y., Yuan, H., Sun, B. and Li, H. (2016b) Biological control of the cereal cyst nematode (*Heterodera filipjevi*) by *Achromobacter xylosoxidans* isolate 09X01 and *Bacillus cereus* isolate 09B18. *Biological Control* 92, 1–6. DOI: 10.1016/j.biocontrol.2015.08.004

Zhang, S.W., Gan, Y.T., Xu, B.L. and Xue, Y.Y. (2014) The parasitic and lethal effects of *Trichoderma longibrachiatum* against *Heterodera avenae*. *Biological Control* 72, 1–8. DOI: 10.1016/j.biocontrol.2014.01.009

Zipfel, C. (2014) Plant pattern-recognition receptors. *Trends in Immunology* 35, 345–351. DOI: 10.1016/j.it.2014.05.004

12 Interactions with Other Pathogens

Horacio D. Lopez-Nicora and Terry L. Niblack
Department of Plant Pathology, The Ohio State University, Columbus, Ohio, USA

12.1 Introduction	271
12.2 Defining Interactions	272
12.3 Methodologies for Investigation of Interactions between Nematodes and Other Organisms	272
12.4 Biological and Statistical Evidence of Interactions	274
12.5 Mechanisms of Interactions	276
12.6 Interactions between Cyst Nematodes and Other Organisms	281
12.7 Conclusions and Future Prospects	294
12.8 References	295

12.1 Introduction

Interactions between nematodes and other organisms are one component of the vast ecological network of biotic and abiotic interactions with plants. Quantifying the effect of each component alone is not easy and multiple components ('determinants') present larger challenges (Wallace, 1978, 1989). Complex interactions between plant-parasitic nematodes and other pathogenic organisms generate uncertainties in our abilities to predict host damage. The lack of comprehension of the mechanisms of these interactions and under- or overestimation of damage and economic thresholds impede the development and implementation of management strategies.

More than a century ago, Atkinson (1892) reported that more prevalent and severe fungal disease took place in cotton fields infested with root-knot nematode. This report was the starting point in a chain reaction of interaction studies in nematology. Fawcett (1931) encouraged the community of plant pathologists to focus on interaction studies, because under field conditions plants are not solely exposed to one pathogen at a time. Several excellent reviews have been published on the study of nematode interactions with other organisms and the complexity of understanding them (Powell, 1971; Bergeson, 1972; Wallace, 1978, 1983; Sikora and Carter, 1987; Taylor, 1990; Khan, 1993a; Abawi and Chen, 1998; Barker and McGawley, 1998; Back *et al.*, 2002; Bond and Wrather, 2004). As will become clear from the present review, studies on the same interactions often present contradictory evidence and conclusions, primarily because measurements are taken on the indirect effects of interactions (i.e. disease symptoms, plant growth parameters, pathogen development, etc.) rather than direct effects.

After more than a century of studies conducted to evaluate disease complexes involving plant-parasitic nematodes, we still do not understand the mechanisms of the interactions reported in the literature. We organize the information presented in this chapter into three main sections: the first part aims to provide the reader with the terminologies and methodologies used in interaction studies between nematodes and other organisms; the second part is a detailed literature review of studies on interactions between cyst nematodes and other organisms; finally, we provide a summary of the chapter with future directions for this area of research.

12.2 Defining Interactions

Throughout the chapter, we discuss results from several studies on interaction between cyst nematodes and other organisms. The authors of these studies reported on additive, synergistic and/or antagonistic interactions. The terminology used by these authors must be clearly defined.

The term 'interaction' must be used after biological and statistical evidence has been obtained (Wallace, 1983; Sikora and Carter, 1987; Khan and Dasgupta, 1993). A significant statistical interaction does not explain the mechanism by which the two pathogens interact (Wallace, 1983). In typical usage, a synergistic interaction between a nematode and another organism is one in which the combined effect of both organisms is greater than the additive effects caused by each alone. At the opposite end of the spectrum of possible interactions, antagonism describes a combined effect that is less than the additive effect that each causes alone.

12.3 Methodologies for Investigation of Interactions between Nematodes and Other Organisms

Most of the experiments conducted to investigate the relationship between a nematode and another organism (as it pertains to disease development and/or yield loss) have involved four treatments to a plant: nematode alone, 'other' organism alone, both organisms together and a control with neither organism (Wallace, 1983; Abawi and Chen, 1998). In most of the cases, the 'other' organisms were also plant pathogens. Data recorded from these experiments included plant growth parameters (height, weight, development, yield, etc.), nematode and interacting organism damage and/or reproduction, and plant disease incidence and severity. Therefore, through the plant (or pathogen) response the interaction between the nematode and the associated plant pathogen is measured indirectly.

In decreasing order of controlled environmental conditions, experiments that evaluated interactions between nematodes and other plant pathogens have been conducted in growth chambers (Tabor *et al.*, 2003, 2006), glasshouses (Gao *et al.*, 2006; Bhattarai *et al.*, 2010; Hillnhütter *et al.*, 2011a), microplots (McLean and Lawrence, 1993a; Xing and Westphal, 2006; Westphal *et al.*, 2014) and in fields (Back *et al.*, 2006; Hillnhütter *et al.*, 2011b; Lopez-Nicora *et al.*, 2013, 2014; Marburger *et al.*, 2014). Each method has important advantages and disadvantages that must be considered when designing interaction experiments.

Experiments conducted in growth chambers or glasshouses involved plants grown in pots with pasteurized, sterilized or fumigated soil (Khan, 1993a; Abawi and Chen, 1998; Barker and McGawley, 1998). Abiotic factors such as temperature, irrigation and fertilization are also under the control of the researchers. Such experiments were criticized because they do not closely reflect field conditions (Sikora and Carter, 1987; Evans and Haydock, 1993; Abawi and Chen, 1998). In addition, the use of unrealistic pathogen inoculum levels may result in a significant interaction between the two organisms that might not occur under field environmental conditions.

Nonetheless, evaluation of nematode–pathogen interactions in growth chambers and glasshouse experiments can be conducted in shorter periods of time and/or beyond the local growing season. Controlled conditions help researchers identify and reduce sources of variation during data acquisition. The application of varying levels of each pathogen is relatively simple, and allows evaluation of the interaction effect for each combination. Moreover, infestation at different plant phenological stages is easily done in controlled environments. The information these experiments generate can complement the design of field studies; for example, the cultivars that will be used based on identified resistance levels, the optimum inoculum levels of both organisms, and the plant and pathogen parameters that should be measured in the field. The combinations of inoculum levels, timing of infestations, conditions before and after infestations, and age of plants, should all be considered and determined using epidemiological knowledge of the pathogens that are involved in the interaction (Sikora and Carter, 1987; Evans and Haydock, 1993; Khan and Dasgupta, 1993; Abawi and Chen, 1998).

The timing of infestation of one pathogen with respect to the other can easily be manipulated in controlled environments. Pathogens can be introduced to plants at the same time (i.e. simultaneous, concomitant or co-infestation). Simultaneous infestation most likely represents what typically takes place under field conditions (Sikora and Carter, 1987). Sequential infestations, however, have been conducted in which the nematode is introduced 2–4 weeks before the introduction of the other pathogen. The objective is to 'predispose' plants to physiological changes caused by the nematode and to evaluate whether these changes have an effect on the interaction between the two organisms. Sedentary nematodes produce feeding sites that peak in metabolic activity at about 2–4 weeks after root invasion by the nematodes. Studies have shown that nematode infection before the introduction of the other pathogen resulted in more disease damage (Taylor, 1990; Khan and Dasgupta, 1993; Back et al., 2002). The former sequential infestation (nematode introduced before other pathogen) mainly evaluates interaction effect on the host, the latter sequential infestation (pathogen before nematode) could reveal whether the previous establishment of the interacting pathogen affects the nematode damage and/or reproduction.

Other techniques used to evaluate interactions between nematodes and other pathogens were borrowed from plant propagation techniques in horticulture. These techniques include bridging, layering, grafting and double root or split root (Khan, 1993a; Sugawara et al., 1997; Abawi and Chen, 1998; Barker and McGawley, 1998; Sikora et al., 2007). The most common technique is the split root, wherein a plant is forced to produce two root systems, each half growing in a different pot but connected to the same shoot on which observations and measurements on the effect of the treatments can be made (Sugawara et al., 1997; Sikora et al., 2007). The nematode and the interacting pathogen can be introduced to the same root half, and the effect measured on the plant could be localized (Khan, 1993a; McLean and Lawrence, 1993b). By contrast, a systemic effect could be detected when the nematode is introduced in one half of the root system and the interacting pathogen in the other half. The idea is that translocatable substances will travel systemically and produce the observed effect on the host (Ko et al., 1984; Khan, 1993b).

Microplots were used in several studies to evaluate interactions between nematodes and other pathogens with a wide range of inoculum levels under field climatic conditions (McLean and Lawrence, 1993a; Melgar et al., 1994; Abawi and Chen, 1998; Barker and McGawley, 1998; Xing and Westphal, 2006; Westphal et al., 2014). The ability to decide which plots will receive a specific treatment permits proper factorial designs to be adopted in microplot studies. In most cases, soil used in microplots is treated to eliminate natural soil organisms. This caveat can be addressed by using natural field soil in which pathogens of interest have not been detected.

Finally, field studies have also been conducted to evaluate interaction between nematodes and other organisms. Small plots are commonly used to account for the spatial aggregation of soil-borne pathogens. Previously, in several interaction studies, soil fumigants (e.g. methyl bromide), nematicides (e.g. aldicarb) and fungicides (e.g. benomyl) were applied to generate control plots, nematode-free and fungus-free plots, respectively (Minton et al., 1985; Khan and Dasgupta, 1993; Winkler et al., 1994). Disadvantages of the use of these chemicals in field studies were their elevated costs of application, toxicity and scarcity after many were removed from the market. As an alternative, the use of plant breeding lines with homogeneous genetic backgrounds differing only in genes for resistance (e.g. isogenic lines susceptible and resistant to one of the pathogens) can be used to account for the damage caused by that pathogen just as chemical applications did (Lopez-Nicora et al., 2014).

In fields naturally infested with a specific nematode, the initial population density of the nematode can be surveyed before or at planting. Small plots that fall in approximately the same nematode initial population can be paired and an interacting pathogen (e.g. a fungal pathogen) could be introduced in one plot but not on the adjacent one (Back et al., 2006; Lopez-Nicora et al., 2014).

12.4 Biological and Statistical Evidence of Interactions

Across multiple disciplines, defining interactions is a complex topic of debate, including the statistical analysis of data collected in interaction studies (Powell, 1971; Bergeson, 1972; Wallace, 1978, 1983; Berenbaum, 1989; Taylor, 1990; Foucquier and Guedj, 2015; Piggott et al., 2015). Nematologists have argued that the term 'interaction' ought to be used only in a mathematical or statistical sense (Wallace, 1983; Sikora and Carter, 1987; Kahn and Dasgupta, 1993).

Interaction studies have been designed and conducted to generate data measured on the response variable (i.e. plant growth, disease severity, pathogen development, etc.) as the result of the combined effect of a nematode and another organism (e.g. a pathogen) and the effect of each pathogen alone, compared with the growth of the host with no pathogens (Fig. 12.1). In interaction studies with cyst nematodes and other organisms, data have been analysed almost without exception with additive effect models, also referred to as the linear interaction effect (Wallace, 1983; Khan and Dasgupta, 1993; Slinker, 1998; Foucquier and Guedj, 2015; Piggott et al., 2015). Additive effect models include two-way (multivariate) analysis of variance (ANOVA or MANOVA) and multiple regression analysis.

Under the lack of significant interaction (combined effect of both pathogens) the effect of each one alone (main effects) is described as additive (Fig. 12.2A). Interpretations of the main effects on the response variable can take place independently. Significant interaction, by contrast, results in non-additive effects. Interpretations of the main effects, under significant interaction, can be misleading because one is a function of the other. Evaluation of each main effect must be done relative to a hypothetical additive effect or comparing one main effect at different levels of the other. It is in doing so that a significant interaction effect can be understood as synergistic (Fig. 12.2B) or antagonistic (Fig. 12.2C).

Interaction effects can be illustrated in bar graphs (Fig. 12.1), regression lines (Fig. 12.2) or response surfaces for regression model with interaction effect (Fig. 12.3). Wallace (1983), however, recommended that with statistical evidence, the researcher should conclude quantitatively on positive, negative or lack of interaction effect rather than using qualitative terms such as synergistic, antagonistic or additive effect. Wallace (1983) also noted that a significant interaction will not reveal the mechanism of interaction but underscores the need for further physiological experiments.

One of the main constraints of the additive effect model is that it must meet several assumptions such as normal and homogeneous distribution of the residuals, and independence of the observations (Kutner et al., 2004); furthermore, it assumes the pathogens operate independently one from the other. This will not be the case when both organisms have the same mechanism of pathogenicity (e.g. nematode–nematode interactions). In addition, the ability of the plant to react systemically to pathogen infection may contribute to a lack of independence. Selecting the appropriate model and meeting its assumptions is therefore of utmost importance to avoid misleading interpretation of results.

Slinker (1998) reviewed the importance of using two-way ANOVA in interaction studies as opposed to one-way ANOVA with *post hoc* multiple comparisons, which generates misleading conclusions. In addition, power analysis to estimate the sample size for an experiment is absent from the majority of interaction studies. Replications for each treatment are chosen, it seems, at the convenience of the researcher. Slinker's (1998) review highlights potential problems resulting from improper analysis of interaction study data. Moreover, statistical analysis for interaction effects with additive effect models is becoming more detailed and advanced in other research areas such as ecology. Piggott et al. (2015) reviewed and revisited concepts on synergism and antagonism in ecology, expanding on the additive effect models to classify further the different types of synergistic and antagonistic effects that can take place.

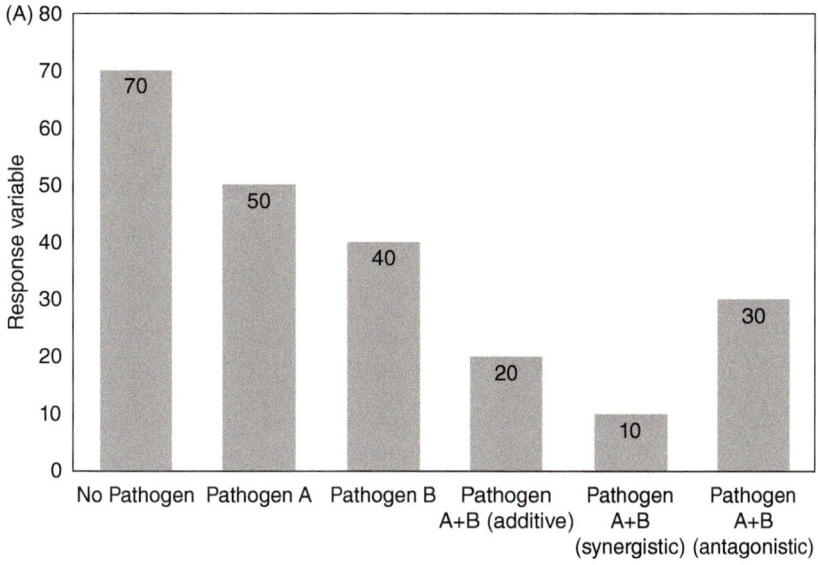

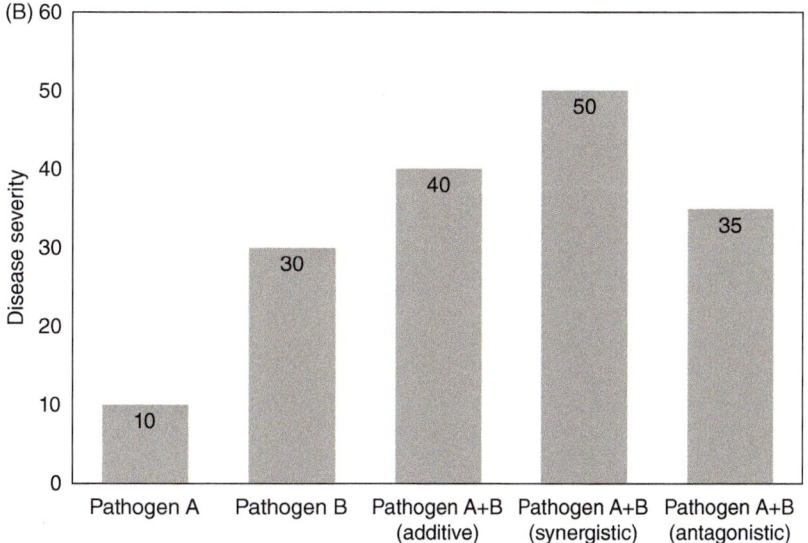

Fig. 12.1. Different hypothetical effects from pathogen interaction studies on plant growth (A) and on disease severity (B).

Other types of analyses have been suggested to evaluate interaction study results. Stetina et al. (1997) assessed the usefulness of replacement series to evaluate nematode–nematode interactions. Jolliffe (2000) reviewed the advantages and disadvantages of this method and identified and explained potential factors of using this technique that may introduce bias. A dose effect model approach was used in interaction studies (Burrows, 1987; Khan and Dasgupta, 1993; Xu et al., 2011; Foucquier and Guedj, 2015). The mathematical framework most commonly used is Loewe additivity. This model assumes that the interacting factors act through a similar or common mechanism as opposed to the additive effect models, which assume independence between the factors. Also, Loewe additivity enables complementation of the analysis with graphical tools such as an isobologram analysis. The use of Loewe additivity and isobologram analyses is described in detail and commonly used in biocontrol

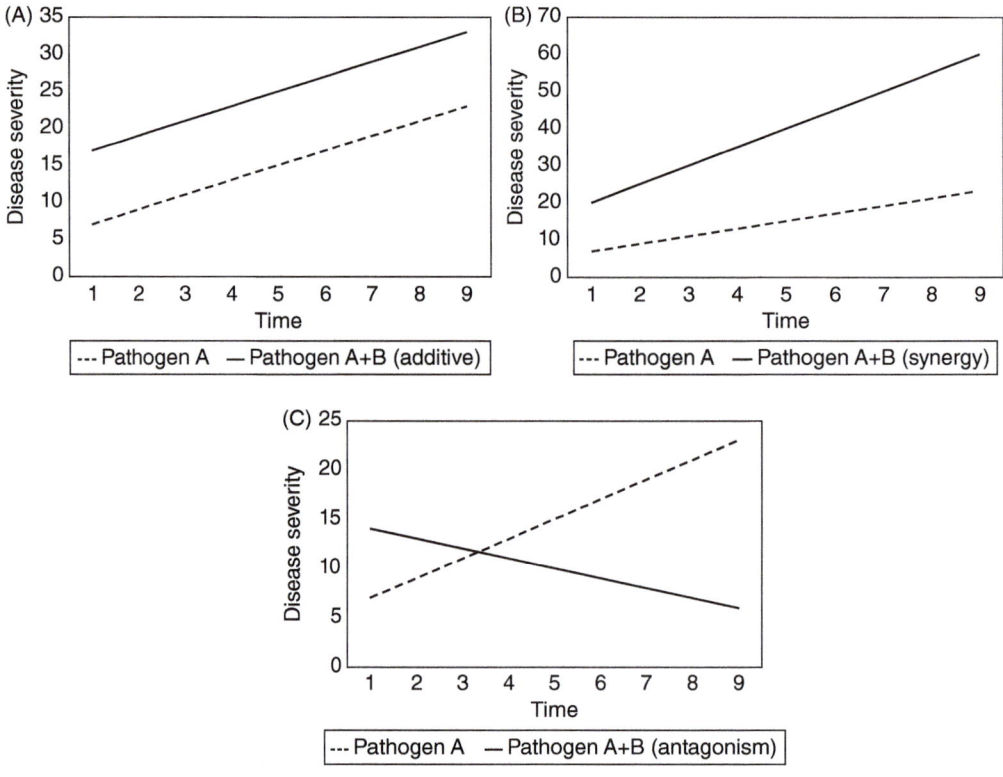

Fig. 12.2. Additive (A), synergistic (B) and antagonistic (C) effect on disease severity when combinations of pathogens are compared to one alone over time.

(interaction) studies (Xu *et al.*, 2011) and pharmacological drug interaction studies (Foucquier and Guedj, 2015). Extrapolating statistical analyses from other fields in science, in which rapid advancements and developments in interaction studies analyses are occurring, may help us to understand the mechanisms of interaction between nematodes and other organisms.

12.5 Mechanisms of Interactions

A statistically significant interaction effect in an experiment will rarely explain the mechanisms of interaction. Most authors speculate on the mechanisms of interaction when they identify a significant one. Supporting data are rarely presented to provide further evidence on how these interactions operate. Wallace (1983) suggested that, after identifying the existence of a significant interaction in a study, further hypotheses and different experiments should be designed to address how the interaction operates.

Different mechanisms of interactions between plant-parasitic nematodes and other pathogens have been proposed (Powell, 1971; Bergeson, 1972; Taylor, 1990; Evans and Haydock, 1993; Francl and Wheeler, 1993; Khan, 1993b; Barker and McGawley, 1998; Back *et al.*, 2002). The mechanisms that apply to cyst nematodes and their interacting pathogens will depend on whether the interaction is synergistic or antagonistic. Proposed mechanisms of synergistic interactions are: nematodes act as a wounding agent; they physiologically modify the host; they modify the rhizosphere; and/or they act as resistance-breakers (Bergeson, 1972; Taylor, 1990; Back *et al.*, 2002). When the significant interaction is antagonistic, however, the mechanisms are quite different (Evans and Haydock, 1993). Proposed mechanisms are: both the interacting pathogen partners compete for root space; metabolites that are toxic to the nematode may be produced by the interacting

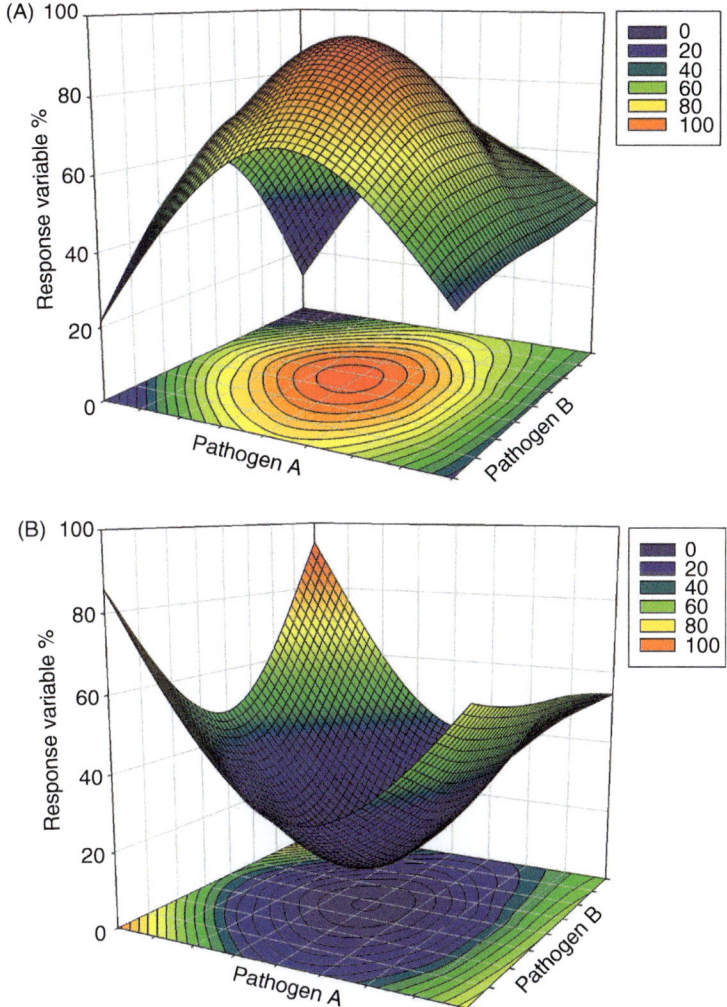

Fig. 12.3. Response surfaces to evaluate interaction effect. The effect is synergistic when the response variable measured is disease severity (A) and plant growth (B). The effect is antagonistic when the response variable measured is disease severity (B) and plant growth (A). A plane, rather than a concave (or convex) surface will represent additive effect.

organism; and/or the nematode could be infected by the antagonistic organism.

12.5.1 Mechanisms of synergistic interactions

12.5.1.1 The 'wounding agent' effect

Cyst nematodes generate physical damage to their host during three stages of their cycle (Lauritis et al., 1983; Koenning, 2004; see Chapter 6, this volume): during the infective second-stage juvenile (J2), when J2 penetrate roots; after the fourth moult, when males, having regained their vermiform shape, emerge from the fourth-stage juvenile (J4) cuticle and leave the roots; and when the maturing female ruptures the epidermis, protruding from the root to mate and produce eggs. Nematode penetration sites may provide and facilitate the entry to soil-borne pathogens, and this was thought to be the main mechanism responsible for synergistic interactions (Khan, 1993a; Back et al., 2002). Cyst nematode J2 create penetration tracks in the root cortical region as they migrate to the vascular cylinder to establish feeding sites (syncytia). Some researchers attempted to validate the wounding mechanism hypothesis by conducting histological studies in plants infested with a cyst nematode and a soil-borne fungal pathogen concomitantly (Polychronopoulos et al., 1969; Storey and Evans, 1987; McLean and Lawrence, 1995). These histological studies provided evidence that the fungal pathogen used the invasion tracks created in the root cortex, both physically and enzymatically, by the cyst nematode on its way to the vascular cylinder.

Very few studies, however, have a histological component. It would be useful to include histological analyses at different points in the life cycle of the nematodes under co-infestation with another soil-borne pathogen, and to evaluate the effect of the mechanical or enzymatic damage caused by the male leaving the roots and the mature female rupturing the epidermis.

In future research, the combination of microscopy, rhizosphere chambers and fluorescent tags as described by Dinh et al. (2014) could be used to study nematodes and soil-borne pathogens in co-infected roots. For example, several studies have used transformed soil-borne pathogens in confocal microscopy studies (Lorang et al., 2001). Non-destructive imaging techniques such as the one described by Dinh et al. (2014) could be used to replace more time-consuming histological techniques and provide evidence of the role of each pathogen in the infection court, as well as subsequent development of each organism and their interaction.

12.5.1.2 *The 'physiological changes' effect*

Some authors consider mechanical wounding to be a less important mechanism in synergistic interactions between plant-parasitic nematodes and other organisms (Bergeson, 1972; Taylor, 1990). Evidence showed that more severe disease developed when the nematodes infected plants 2–4 weeks before the introduction of the interacting pathogen, in most cases a fungal pathogen. The predisposition effect is believed to be linked to syncytial development (Bergeson, 1972; Taylor, 1990; Evans and Haydock, 1993; Francl and Wheeler, 1993; Khan, 1993b; Back et al., 2002). Under conducive environmental conditions, cyst nematode feeding sites (syncytia) reach their peak metabolic activity and development at approximately 2–4 weeks after root infection (Lauritis et al., 1983; Koenning, 2004). Several researchers, therefore, included two different infestation time points in interaction studies: sequential and simultaneous infestation.

Cyst nematode syncytia are nutrient-rich and metabolically active cells that serve as a food source for soil-borne pathogens and could also generate translocatable metabolites that enhance synergistic interaction with other pathogens (Khan, 1993a). Therefore, the effect of the physiological changes produced when the syncytia are formed can be localized or systemic. Several studies have used the split-root technique to evaluate the synergistic interaction between cyst nematodes and other organisms (Polychronopoulos et al., 1969; Corbett and Hide, 1971; McLean and Lawrence, 1993b; Sugawara et al., 1997). Polychronopoulos et al. (1969), for example, showed that a fungal pathogen was attracted to and developed better in syncytial cells, from which it later colonized neighbouring healthy cells. Other studies have also used split roots to find evidence that the physiological changes generated by the cyst nematode had a localized rather than a systemic effect in the interaction (Corbett and Hide, 1971; Tylka et al., 1991; McLean and Lawrence, 1993b; Sugawara et al., 1997).

The physiological changes induced by the cyst nematodes, however, can be systemic. Effect of cyst nematodes in one half of a split-root system will systemically affect the development of the interacting organisms on the other half of the root, contributing to the synergistic interaction effect. For example, evaluating the interaction between cyst nematodes and nodule-forming rhizobia in soybean plants, Ko et al. (1984) presented results from a split-root experiment showing that cyst nematodes (*Heterodera glycines*) on one half of the root system (where infection took place) reduced the nodulation on the other half (where the nematode was not present).

Furthermore, results from recent studies demonstrate that plant-parasitic nematodes can suppress host defences (Smant and Jones, 2011; Haegeman et al., 2012; Goverse and Smant, 2014; Mantelin et al., 2015). Cyst nematodes, particularly, can suppress plant defence responses during the infection process, feeding cell formation and throughout the maintenance of the syncytium (Goverse and Smant, 2014; Gardner et al., 2015; Mantelin et al., 2015). The suppression of the host innate immunity by cyst nematodes may render the plant more susceptible to subsequent infections from other pathogens.

12.5.1.3 The 'modified rhizosphere' effect

The rhizosphere is the region wherein the roots, soil and soil organisms dynamically interact (Bais et al., 2006). In this region, root diffusates play a significant role in several interactions between the plant and the soil biota. Quantitative and qualitative modifications of root diffusates by plant-parasitic nematodes can affect the response of soil-borne pathogens present in the rhizosphere and are considered a mechanism of interaction (Bergeson, 1972; Evans and Haydock, 1993; Kerry, 2000; Back et al., 2002).

Quantitative modifications of root diffusates by nematode infection can be either direct or indirect (Bergeson, 1972; Back et al., 2002). Direct modification by cyst nematodes involves physical damage to roots during the infection process. Root penetration can increase root diffusate leakage to the rhizosphere attracting or stimulating soil-borne pathogens (Abawi and Chen, 1998; Back et al., 2002). The physiological changes cyst nematodes generate in their hosts can serve as an indirect way to increase root diffusates in the rhizosphere. For example, infection by both the sugar beet (Cooke, 1993) and the potato (Evans and Stone, 1977) cyst nematodes stimulate the production of secondary roots, resulting in increased root diffusate release in the rhizosphere.

Qualitative modifications involve alteration in the biochemical and nutritional composition of root diffusates due to nematode infection. The modified root diffusates can make the host more attractive to soil-borne pathogens (Bergeson, 1972; Back et al., 2002). Back et al. (2010) determined that root diffusates from potato plants infected with *Globodera rostochiensis* had more sucrose than those from non-infected plants. Subsequently, they amended media with this modified root diffusates and showed that *Rhizoctonia solani* grew faster in this medium compared with one amended with root diffusates from non-infected plants. In addition, modified root diffusates can reduce the population densities of organisms antagonistic to plant pathogens, such as actinomycetes (Bergeson, 1972).

Nematodes generate physiological changes that may stimulate a synergistic interaction with other soil-borne pathogens, and the inverse situation is also possible (Evans and Haydock, 1993; Back et al., 2002). Although few reports are found in the literature, plants infected by fungi can render the plant more attractive and susceptible to nematodes (Nordmeyer and Sikora, 1983a). In a different study, a fungal enzyme involved in root cell wall degradation was thought to be facilitating cyst nematode penetration (Nordmeyer and Sikora, 1983b). Another effect of modified root diffusates was stimulation of cyst nematode hatching (Ryan et al., 2000, 2003; Ryan and Jones, 2004).

Ample evidence that rhizosphere modifications play a significant role in cyst nematode–soil-borne pathogen interactions has been reported. Bergeson (1972) suggested that modifications within the rhizosphere could be the first step to synergistic interactions. Metabolite profile analyses can contribute to understanding of compounds released in the presence of potential interacting organisms. Recent contributions have come from work on the ecology and evolution of nematode chemotaxis (Rasmann et al., 2012; Ali and Davidson-Lowe, 2015; Hiltpold et al., 2015), which provides improved techniques to evaluate root leachates. In addition, proteomic and metabolomic approaches will allow identification of compounds that may be enriched during a synergistic or antagonistic interaction.

12.5.1.4 The 'resistance breaker' effect

Another hypothesis on the mechanism of synergistic interactions is that nematode infection can modify the resistance levels of plants to other soil-borne pathogens (Bergeson, 1972; Khan, 1993a). Cultivars resistant to a specific pathogen may lose expression of a defence response following nematode infection; in this way, the nematode could reduce levels of resistance to the interacting plant pathogen. Several authors agree that polygenic (quantitative) resistance seems to be more easily altered by nematodes than monogenic (qualitative) resistance (Evans and Haydock, 1993; Francl and Wheeler, 1993;

Khan, 1993b; Back et al., 2002). Quantitative resistance may be affected by physiological changes generated by the nematode, whereas monogenic resistance to the interacting pathogen may operate independently from the changes caused by nematode infection. In order to test a 'resistance-breaking' effect, experiments should be conducted with cultivars of homogeneous genetic backgrounds. Isogenic lines differing in resistance to either or each pathogen should be used in interaction studies to avoid error introduced by comparisons of cultivars with heterogeneous backgrounds.

Khan (1993a) suggested that the 'break down' of resistance is due to physiological changes caused by nematodes with no effect on gene expression; however, advances in molecular techniques have revealed that physiological changes in the plant are accompanied by significant changes in host gene expression (Gheysen and Mitchum, 2009; Li et al., 2009; Escobar et al., 2011). Furthermore, nematodes can manipulate hormone signalling pathways, which are responsible for host defences (Goverse and Smant, 2014; Mantelin et al., 2015) and development of functional syncytia (Goverse and Bird, 2011; Haegeman et al., 2012; Cabrera et al., 2015; Gardner et al., 2015). For example, cyst nematodes can manipulate the salicylic acid (SA)- and the jasmonic acid (JA)-mediated signalling pathways (Prior et al., 2001; Uehara et al., 2010) suppressing the host defence response. The SA- and JA-mediated signalling pathways are involved in defence responses not only to nematodes but also to a wide range of plant pathogens and pests (Smith et al., 2009; Robert-Seilaniantz et al., 2011; Pieterse et al., 2012). Hormone signalling pathway manipulation by the nematode may lead to an increase in host susceptibility and consequently favour infection and colonization by other pathogens.

12.5.2 Mechanisms of antagonistic interactions

The interaction between the nematode and other pathogens can be detrimental for either or both organisms. In close proximity in the root or rhizosphere, the interacting pathogens can antagonistically affect each other directly or indirectly (Evans and Haydock, 1993; Back et al., 2002). Direct antagonisms are possible when soil-borne pathogens have the ability to infect the nematodes (Kerry, 2000; Back et al., 2002). For example, the fungal causal agent of sudden death syndrome (SDS) of soybean infects different life stages (eggs, juveniles and females) of the soybean cyst nematode (SCN), *H. glycines*, resulting in reduced viability and reproduction of the nematode (McLean and Lawrence, 1993a). Indirect antagonistic effects in disease interactions are less understood – reduction in the development and reproduction of the interacting pathogens could be the result of competition for root space and substrate (Jorgenson, 1970; Eisenback, 1993; Evans and Haydock, 1993; Back et al., 2002). Fungal pathogens may invade roots and colonize cyst nematodes' feeding sites. Feeding sites colonized by the fungus *Rhizoctonia solani* were disrupted and consequently generated delay or suppression of cyst nematode development (Polychronopoulos et al., 1969). Another hypothesis is that the interacting pathogen could produce metabolites that reduce nematode development and reproduction (Evans and Haydock, 1993). Studies have been conducted on the effect of fungal culture filtrates on nematode eggs and juveniles; preventing hatch or killing juveniles suggested a toxic effect produced by the interacting pathogen (Evans and Haydock, 1993). The reverse could also take place: root invasion by nematodes triggers and activates the plant defence responses. These responses are either hypersensitive-like responses or systemic acquired resistance (SAR), which ultimately restrict further attack by other soil-borne pathogens (Back et al., 2002). Sequential infestation techniques, introducing the nematode before the interacting pathogen may reveal that previous nematode infection suppresses or delays the infection by the other pathogen.

Identifying pathogen- or plant-derived molecules that could be used in crop protection by activating host defence responses in the absence of the pathogen and its detrimental effects would be beneficial to the improvement of integrated disease management strategies. Transcriptomics, proteomics and metabolomics approaches are improving our molecular understanding of the potential applications of these phenomena in agriculture (Wiesel et al., 2014).

12.6 Interactions between Cyst Nematodes and Other Organisms

The majority of the studies conducted to evaluate cyst nematode interaction with other organisms involve a fungal pathogen; thus, most of the examples presented in this section are interactions between cyst nematodes and fungal pathogens. We grouped these studies within major cyst genera. We also include the interaction of cyst nematodes with bacteria, specifically nitrogen-fixing bacteria, since this is one area in which more research has been done. Nematode–nematode interactions, interactions with aboveground pests and pathogens, and multiple pathogen/pest interactions are also included.

12.6.1 Cyst nematodes – oomycete and fungal pathogen interactions

12.6.1.1 Interactions between Heterodera glycines *and fungal pathogens*

Heterodera glycines causes significant annual losses to the soybean industry (Wrather *et al.*, 2001; Niblack *et al.*, 2006; Allen *et al.*, 2017) and can affect a variety of other host species (Poromarto and Nelson, 2010); it reduces soybean yield by up to 30% without any visible symptoms in the host plant (Wang *et al.*, 2003; Niblack, 2005). Yield loss due to SCN is exacerbated by specific abiotic and biotic factors. Biotic interaction studies between *H. glycines* and soilborne fungal and oomycete pathogens include many on *Fusarium* spp., *Macrophomina phaseolina*, *Phialophora gregata*, *R. solani*, *Diaporthe phaseolorum*, *Phytophthora sojae* and *Calonectria ilicicola*. The interaction studies between *H. glycines* and these pathogens described the effect on the host as well as the effect on the reproduction of either pathogen involved in the disease complex.

HETERODERA GLYCINES AND FUSARIUM OXYSPORUM. *Fusarium oxysporum*, an ascomycete fungus, causes soybean to wilt and produce fewer seeds. Research suggests that these symptoms are intensified through interaction with *H. glycines*. Ross (1965) suggested that wounds caused by infective juveniles intensified the interactions between *F. oxysporum* and *H. glycines* but not between *F. oxysporum* and *Meloidogyne incognita* (another nematode known to infect soybean). Whilst *M. incognita* infects soybean through the elongation zone of roots and migrates intercellularly to the vascular system, *H. glycines* J2 penetrate any part of the root and migrate both inter- and intracellularly, producing more physical damage in the cortical region of the root (Endo, 1965; Ross, 1965). In addition, the *H. glycines* population density increased more than two-fold in plots infested with *F. oxysporum*, compared with plots infested with *H. glycines* alone. Ross (1965) suggested that the oat–*Fusarium* inoculum used to infest plots may have enhanced soybean root growth, also supporting infection of the root by *H. glycines*. Stiles *et al.* (1993) also found that *H. glycines* increased its population in pots co-infested with *F. oxysporum* after one generation, supporting the results of Ross (1965). Stiles *et al.* (1993) did not use oat or any substrate in their inoculum but used a suspension of fungal mycelia and spores.

HETERODERA GLYCINES AND FUSARIUM VIRGULIFORME. Synergistic interactions between *H. glycines* and the fungus that causes SDS in soybean have been studied for over 30 years. Severe symptoms of SDS in fields infested with SCN were noted early on and suggested an association between the nematode and the unknown culprit (Hirrel, 1983, 1987). Roy *et al.* (1989) concluded, after completion of Koch's postulates, that *F. solani* Form-A (FS-A) was the causal agent of SDS. They also evaluated the effect of co-infesting plants with *H. glycines* and the fungus: more severe foliar symptoms were observed when plants were infested with both pathogens compared with plants infested with the fungus alone. In 1997 the nomenclature of the fungus changed to *F. solani* f. sp. *glycines* (Roy, 1997) and once again in 2003 to *F. virguliforme* (Aoki *et al.*, 2003). Three other species of *Fusarium* are also associated with SDS, including *F. tucumaniae*, *F. brasiliense* and *F. crassistipitatum* (Aoki *et al.*, 2003, 2005, 2012). All four species are responsible for causing SDS in South America; however, only *F. virguliforme* has been associated with SDS in North America (Aoki *et al.*, 2005, 2012). All interaction studies between SCN and SDS have been conducted with *F. virguliforme*.

Interactions between *F. virguliforme* and *H. glycines* have been evaluated in glasshouse (Roy *et al.*, 1989; McLean and Lawrence, 1993b; Gao *et al.*, 2006), microplot (McLean and Lawrence, 1993a; Melgar *et al.*, 1994; Xing and Westphal, 2006; Westphal *et al.*, 2014) and field studies (Hartman *et al.*, 1995; Rupe *et al.*, 1997, 1999; Brzostowski *et al.*, 2014; Marburger *et al.*, 2014). Even though the fungus can infect soybean plants and produce characteristic SDS symptoms, in the presence of SCN plant damage occurs earlier and is more severe (Fig. 12.4; McLean and Lawrence, 1993a; Westphal *et al.*, 2014). Conversely, the population density of the fungus is inversely related to SCN population density.

HETERODERA GLYCINES AND PHIALOPHORA GREGATA. *Phialophora gregata* (= *Cadophora gregata*) is another yield-reducing soybean pathogen and the causal agent of brown stem rot (BSR). Symptoms of BSR include interveinal chlorosis, necrosis and abscission of leaves, and discoloration of vascular and pith tissue (Allington and Chamberlain, 1948; Impullitti and Malvick, 2014). *Phialophora gregata*, similar to *F. virguliforme*, infects soybean plants early in the growing season, but symptoms appear at the onset of the reproductive stage.

Carris *et al.* (1989) isolated *P. gregata* from *H. glycines* cysts and demonstrated pathogenicity of the isolates in soybean plants, but did not test for interactions. Researchers later

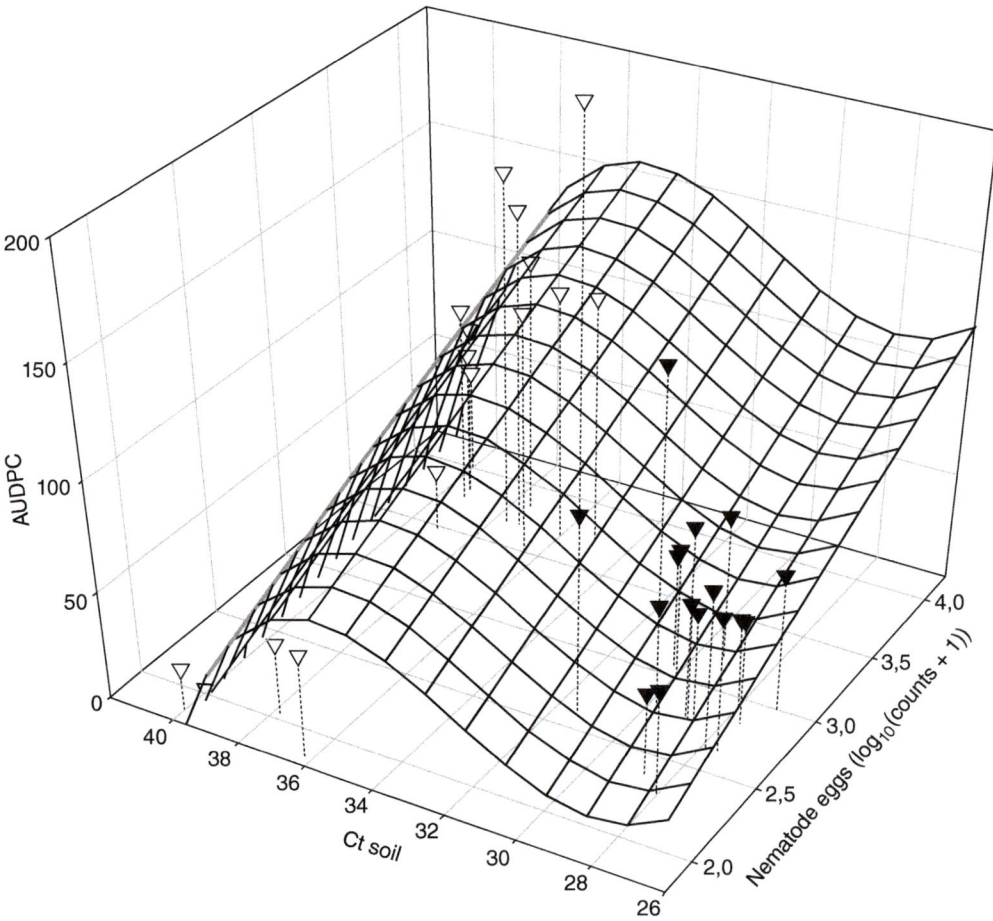

Fig. 12.4. Value of the area under the disease progress curve (AUDPC) in relation to soil DNA of *Fusarium virguliforme* (Ct soil) and population densities of *Heterodera glycines* (LogPi$_{H.\ glycines}$) at planting. (From Westphal *et al.*, 2014.)

noticed that BSR symptoms developed more often in fields infested with SCN (Niblack et al., 1992; Tubajika et al., 1994; Tabor et al., 2003), suggesting an interaction between pathogens was likely. Sugawara et al. (1997), using glasshouse and growth chamber studies, found increased BSR incidence and severity in the presence of *H. glycines*. Additionally, the authors suggested that the interaction likely occurred due to wounding rather than the disease being systemic, since they found no evidence of interaction in co-infested pots using split-root assays. Tabor et al. (2003, 2006), using a complete factorial experimental design in a glasshouse and growth chamber, found that cultivars that were susceptible to SCN and susceptible and resistant to BSR revealed a synergistic effect on BSR disease development in the presence of both pathogens. The authors also reported a reduction in disease severity for cultivars with resistance to SCN. The mechanisms of interaction were not evaluated, but the authors suggested that wounding and physiological changes in the SCN-susceptible cultivars may play an important role in the disease-enhancing results.

HETERODERA GLYCINES AND MACROPHOMINA PHASEOLINA. *Macrophomina phaseolina* causes charcoal rot in hundreds of plant species worldwide, and has resulted in significant reductions of soybean yield in the southern USA (Koenning and Wrather, 2010; Allen et al., 2017). The potential interaction effects between *M. phaseolina* and *H. glycines* is less clear. Todd et al. (1987) evaluated *H. glycines*-resistant and -susceptible soybean cultivars in fields infested with both soil-borne pathogens and reported a positive correlation between colonization of root tissue by the fungus and number of *H. glycines* in roots. Greater yield loss was observed when soybean plants were concomitantly infected by *H. glycines* and *M. phaseolina* than by either one alone, suggesting a synergistic interaction. Winkler et al. (1994) found that *H. glycines*-susceptible soybean cultivars had higher levels of *M. phaseolina* in roots when soybean plants were infected with *H. glycines*. In a similar study, however, an interaction effect on soybean yield was not observed when *H. glycines* and *M. phaseolina* were quantified from soybean roots (Todd, 1993). Population densities of both pathogens were measured in soil and no evidence of interaction was found between *H. glycines* and *M. phaseolina* on soybean production (Francl et al., 1988). Lopez-Nicora et al. (2014) observed that *H. glycines* and *M. phaseolina* reduced soybean performance through early infection. Additionally, chlorosis and premature senescence of soybean plants were observed in plots that have elevated population densities of both *H. glycines* and *M. phaseolina* (Fig. 12.5).

HETERODERA GLYCINES AND RHIZOCTONIA SOLANI. Dave (1975) and Frohning (2013) evaluated the effect of *R. solani* and *H. glycines* in soybean. They both concluded that the reproduction of the nematode was significantly reduced in presence of the fungus. Frohning (2013) conducted

Fig. 12.5. Premature chlorosis (A) and senescence (B) of soybean plants in field naturally infested with *Heterodera glycines* and artificially infested with *Macrophomina phaseolina*.

glasshouse and field studies where seed treatments were applied to SCN-resistant and -susceptible soybean cultivars. In the glasshouse studies, *H. glycines* reproduction factor (ratio of final to initial population, RF = Pf/Pi) was greater in pots where *R. solani* was reduced by the fungicide. An interaction effect, however, was not observed when yield, dry weight and root growth parameters were measured.

HETERODERA GLYCINES AND DIAPORTHE PHASEOLORUM VAR. CAULIVORA. *Diaporthe phaseolorum* var. *caulivora* is a fungus, and the causal agent of soybean stem canker. Russin *et al.* (1989) evaluated the potential interactions among *D. phaseolorum* var. *caulivora*, *Pseudoplusia includes* (a lepidopterous defoliator) and *H. glycines* in soybean plants in glasshouse studies. The researchers concluded that the damage to soybean plants was additive rather than synergistic. Russin *et al.* (1989) suggested that the infection and development of the nematode in soybean roots triggers physiological changes that may be responsible for the reduction in severity of stem canker lesions. Moreover, the nematode population was reduced when plants were infested with both the fungus and the nematode.

Pacumbaba (1992) conducted a three-way interaction study between *H. glycines*, *D. phaseolorum* var. *caulivora* and *Pseudomonas syringae* pv. *glycinea*, a bacterial pathogen. Based on the disease symptomology, there was no synergistic interaction detected, but only an additive effect on yield and agronomic characteristics measured.

HETERODERA GLYCINES AND CALONECTRIA ILICICOLA. *Calonectria ilicicola* (= *Calonectria crotalariae*) is a fungus that causes crown rot in soybean and groundnut. Overstreet and McGawley (1988) and Overstreet *et al.* (1990) conducted several glasshouse experiments to evaluate the interaction between this fungus and *H. glycines*. In pots infested with both the fungus and the nematode, a significant increase in nematode root penetration was observed compared with pots infested with *H. glycines* alone. Consequently, this led to an increase in nematode population in soil and roots at the end of the trials. The conclusion was that the significant interaction between *C. ilicicola* and *H. glycines* negatively impacted plant roots.

HETERODERA GLYCINES AND ABOVEGROUND PESTS IN THREE-WAY INTERACTIONS. Most of the interactions reviewed so far have involved soil-borne pathogens and the interactions were in the majority two-way interactions. There are, however, some studies that evaluated the effect of insects on the aboveground part of plants in the presence of cyst nematodes infecting root systems (Heeren *et al.*, 2012; McCarville *et al.*, 2012, 2014). McCarville *et al.* (2012) conducted microplot experiments evaluating the effect of 'single pest' (a cyst nematode, a fungus or an insect) to 'multiple pests', all organisms combined and applied to the host (Fig. 12.6). A significant increase in cyst nematode reproduction was observed in the presence of both the fungus and the insect (i.e. 'multiple pests' treatment). However, the individual effect each organism had on the increase of the nematode reproduction was not possible to separate due to the confounded effect under 'multiple pests' (McCarville *et al.*, 2012). Ultimately, understanding the intricate network of interaction that takes place in the field will help improve crop pest management; however, these studies bring huge challenges in terms of interpretation of results.

HETERODERA GLYCINES AND OOMYCETES. Adeniji *et al.* (1975) performed simultaneous and sequential infestation with the oomycete *Phytophthora sojae* (= *Phytophthora megasperma* var. *sojae*) and *H. glycines*. Abnormal development of SCN was observed when nematode-infected plants were later infected by *P. sojae*. They suggested that physiological changes in soybean caused by *P. sojae* infection could be the mechanism for the reduction in nematode reproduction and that the effect was additive. Kaitany *et al.* (2000) arrived at a similar conclusion: in fields heavily infested with *H. glycines*, some plots were treated with nematicides while others were not; they noted that *P. sojae* incidence was higher in plots with high levels of SCN. The soybean nutrient stress caused by the nematode, along with other abiotic factors, were the suggested conditions for higher *P. sojae* damage.

12.6.1.2 Interactions between Heterodera schachtii *and fungal pathogens*

The sugar beet cyst nematode, *H. schachtii*, is one of the most important sugar beet (*Beta vulgaris*)

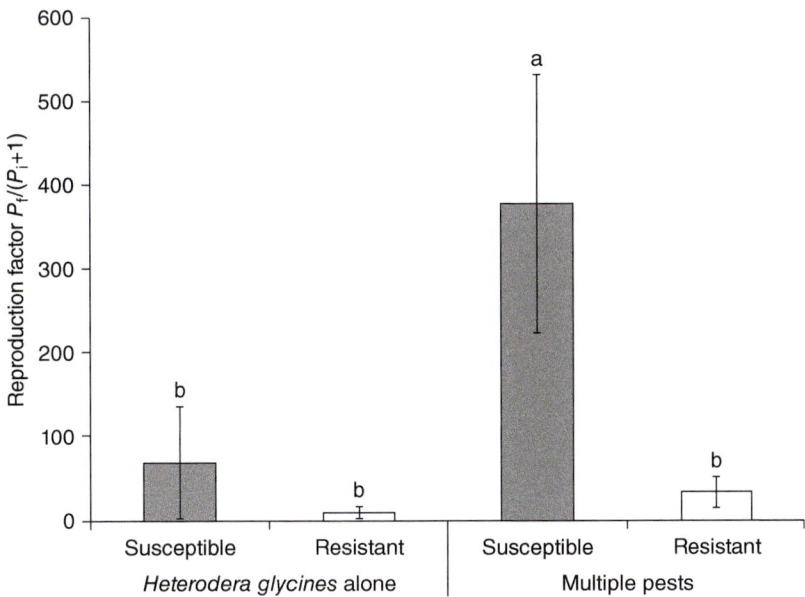

Fig. 12.6. Mean (± SEM) reproduction factors averaged over the *Heterodera glycines*-susceptible and resistant soybean cultivars for the 3 years of the study. Nematode resistance ($P = 0.026$) and pest treatment ($P = 0.010$) were found to be significant factors affecting *H. glycines* reproduction. The interaction between nematode resistance and pest treatment was non-significant ($P = 0.23$). Means capped with different letters are significantly different (*t*-test: $P < 0.05$). (From McCarville et al., 2012.)

pathogens, affecting production worldwide (Cooke, 1993; Evans and Rowe, 1998; Riggs and Schuster, 1998). In addition to sugar beet, *H. schachtii* can reproduce on certain weeds and has several economically important hosts, mainly in the Chenopodiaceae and Cruciferae (Steele, 1965). Some of these hosts are cabbage, canola, kale, broccoli and cauliflower, and yield reduction of up to 50% was reported in these hosts (Riggs and Schuster, 1998). A wide host range and the ability to survive in the soil for several years make this pathogen difficult to manage.

Similar to most soil-borne pathogens, the spatial distribution of *H. schachtii* in the field is aggregated (Hbirkou *et al.*, 2011; Hillnhütter *et al.*, 2011b). Under high inoculum levels, aboveground symptoms appear as chlorotic patches in the field. These groups of plants will eventually wilt under favourable environmental conditions (i.e. hot and dry). Infection of the nematodes stimulates production of secondary roots resulting in belowground symptoms referred to as 'bearded' roots (Cooke, 1993). Affected plants are also smaller in size and their dry weight is lower compared with healthy plants. Deformation and reduction in size of infected roots are also symptoms of *H. schachtii* damage. Whilst *H. schachtii* is responsible for significant yield reduction in sugar beet, interactions with other soil-borne pathogens have been reported and will be summarized in this section.

HETERODERA SCHACHTII AND RHIZOCTONIA SOLANI. Several studies were conducted to understand the mechanism of interaction between *H. schachtii* and *R. solani* (Polychronopoulos *et al.*, 1969; Hillnhütter *et al.*, 2011a, b, 2012). *Rhizoctonia solani* is considered a major fungal pathogen of sugar beets (Duffus and Ruppel, 1993). Isolates of *R. solani* are classified in anastomosis groups (AG) based on their ability to anastomose (fusion between branches of the same or different hyphae) or not when paired in culture media (Parmeter *et al.*, 1969; Sherwood, 1969; Herr and Roberts, 1980). The most damaging disease *R. solani* causes in sugar beet is Rhizoctonia crown and root rot (RCRR). This disease typically affects 6–8-week-old plants and fungal isolates from these plants are in the AG-2 (Herr and Roberts, 1980). Similar to *H. schachtii*, aboveground symptoms in the field occur in patches; RCRR causes chlorosis and sudden wilting with dead foliage attached to the crown (Duffus and Ruppel, 1993). These symptoms contribute to reduction in plant performance and ultimately yield loss. *Rhizoctonia solani* occasionally causes pre- and post-emergence damping-off of sugar beet seedlings and isolates from affected seedlings are classified in AG-4 (Herr and Roberts, 1980).

Polychronopoulos *et al.* (1969) examined the infection and progression of disease caused by *R. solani* in sugar beet seedlings infected with *H. schachtii*. Seedlings infected with *H. schachtii* were necrotic and killed 3 days after exposure to the fungus. Nematode-free seedlings, were still alive 6 days after exposure to *R. solani*. Isolates that belong to AG-4 cause damping-off, whereas those in AG-2 cause RCRR; this information was available only later in that same year (Parmeter *et al.*, 1969; Sherwood, 1969), but one could retrospectively suggest that Polychronopoulos *et al.* (1969) worked with *R. solani* AG-4. Polychronopoulos *et al.* (1969) additionally observed under the microscope that fungal mycelia grew towards syncytia initiated by nematodes, damaging them and further colonizing adjacent cells. This result is supporting evidence for subsequent research where a significant reduction in nematode population was observed in plants infected with *R. solani* (Dave, 1975; Hillnhütter *et al.*, 2011a, b, 2012; Frohning, 2013).

Hillnhütter *et al.* (2011a, b, 2012) conducted a series of studies to evaluate the interaction between *H. schachtii* and *R. solani* on sugar beet. They specifically used *R. solani* AG-2, causal agent of RCRR. In their first study, Hillnhütter *et al.* (2011a) conducted glasshouse experiments on sugar beet cultivars with different levels of resistance, and symptoms on plants were rated several times after infestation. Plant weights (roots, beets and shoots) were measured and MANOVA was used to analyse the data. In contrast to what Polychronopoulos *et al.* (1969) reported, plants infected with *H. schachtii* were not killed by *R. solani*, perhaps because older plants rather than seedlings were used, or more likely because of differences in *R. solani* pathogenicity (Hillnhütter *et al.*, 2011a). In pots co-infested with both pathogens, a synergistic interaction was observed in all sugar beet cultivars, except in the *H. schachtii*-resistant cultivar (Fig. 12.7).

Similar to the results of other studies, in the presence of *R. solani*, *H. schachtii* reproduction decreased when compared with the pots infested with the nematode alone (Hillnhütter *et al.*, 2011a).

Hillnhütter *et al.* (2011b) examined a field planted to sugar beet and naturally infested with both *H. schachtii* and *R. solani* AG-2. The synergistic interaction that these researchers previously observed in the glasshouse (Hillnhütter *et al.*, 2011a) was not detected in the field. This result is one of the many that demonstrates the complexity of studying interactions between pathogens in disease development: the environment is so complex that many factors that are not measured (omitted variables) could influence the relationships among variables measured in the study.

Hillnhütter *et al.* (2012) used nuclear magnetic resonance imaging (NMRI) to assess belowground damage to sugar beets plants infested with *H. schachtii*, *R. solani* AG-2 and both, to improve understanding of the mechanism involved in the synergistic interaction described previously (Hillnhütter *et al.*, 2011a). In co-infested pots, the fungus infected above and below the region where *H. schachtii* was delivered compared with infection only at or above the infestation site in pots with the fungus alone. The nematode, therefore, may be responsible for allowing the fungus to reach other areas of the rhizosphere. This result agrees with the suggestion by Polychronopoulos *et al.* (1969) that the fungus infection is stimulated by nematode damage. Finally, at the end of the experiment, *H. schachtii* population densities were always reduced in presence of *R. solani* (Hillnhütter *et al.*, 2012). This is a recurrent observation from different studies that provides a useful hint of focus for future interaction studies.

HETERODERA SCHACHTII AND VERTICILLIUM DAHLIAE. Oilseed rape (*Brassica napus*), also known as canola, is a host for *H. schachtii* but it is more severely affected by *V. dahliae*, the causal agent of Verticillium wilt. This fungus is an important disease in some oilseed rape production regions and screening for resistance to Verticillium wilt is part of many breeding programmes. The fungus is a soil-borne pathogen that overwinters as microsclerotia in soil or plant debris. It causes

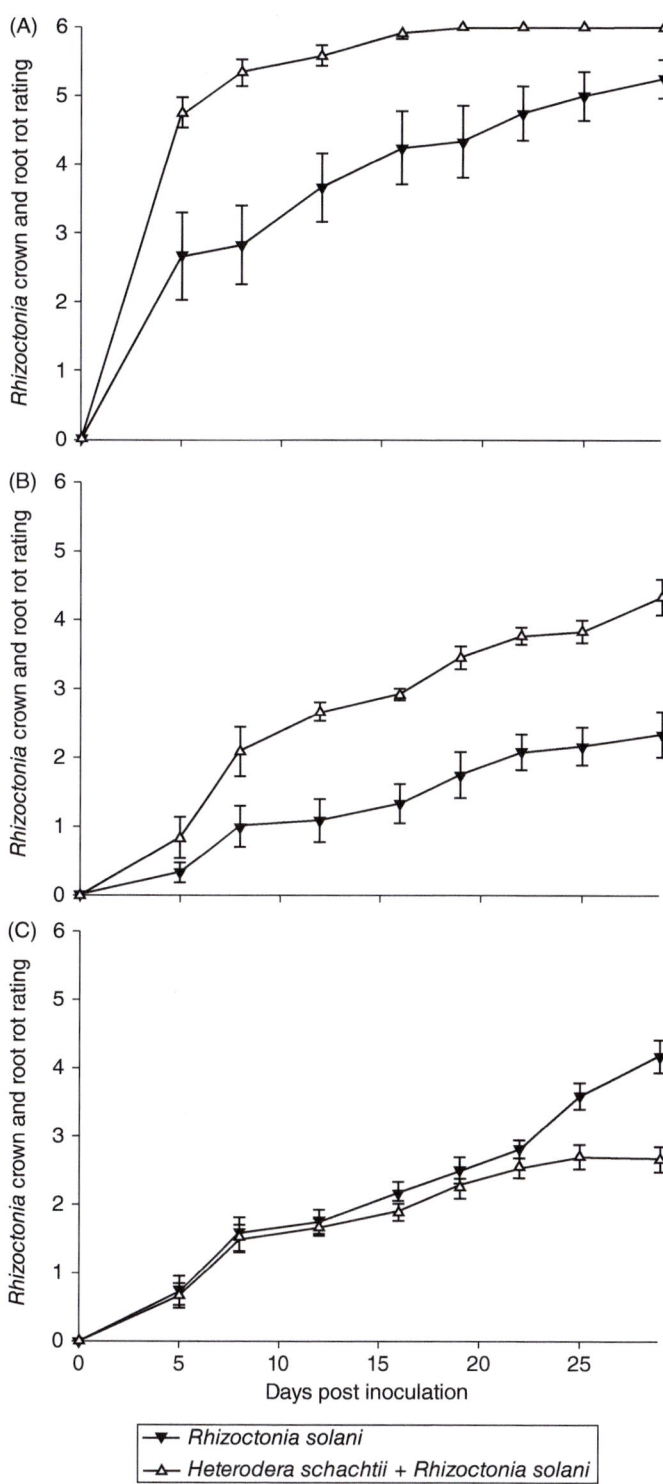

Fig. 12.7. Rhizoctonia crown rot and root rot rating of foliar leaf tissue of sugar beet susceptible cv. Alyssa (A), Rhizoctonia crown and root rot-tolerant cv. Calida (B) and *Heterodera schachtii*-resistant cv. Sanetta (C) in a time series of 29 days post-inoculation. Plants were inoculated simultaneously with *H. schachtii* and *Rhizoctonia solani*. Bars indicate standard error of the mean ($n = 12$). (From Hillnhütter et al., 2011a.)

premature foliar chlorosis, discoloration of the vascular system and wilting of plants; consequently, yield reduction takes place (Berlanger and Powelson, 2000). Screening for resistance to *V. dahliae* is not easy in the glasshouse due to late onset of clear symptoms (Gröntoft and Jonasson, 1992). Gröntoft and Jonasson (1992) evaluated the effect of co-infesting oilseed rape with both *H. schachtii* and *V. dahliae* and hypothesized that if a synergistic interaction were significant, Verticillium wilt symptoms would be more evident, develop faster and become easier to rate, shortening the resistant screening process. However, results from their glasshouse studies showed no synergistic interaction between these two organisms on wilt symptoms with the inoculum levels and cultivar they used (Gröntoft and Jonasson, 1992).

HETERODERA SCHACHTII AND OOMYCETES. Another major disease in sugar beet is black root or blackleg caused by the oomycete *Aphanomyces cochlioides* (Duffus and Ruppel, 1993; Windels, 2000). Whitney and Doney (1973) observed a synergistic interaction between *A. cochlioides* and *H. schachtii* at high infestation levels in killing sugar beet plants. A significant interaction was also measured in yield and percentage sugar reduction. These interactions were not observed or were weaker under low infestation levels. Researchers speculated that the mechanism of the interaction was that the nematode predisposes plants to infection by *A. cochlioides* because increasing nematode densities enhanced *A. cochlioides* damage (Whitney and Doney, 1973). In addition, *H. schachtii* population densities decreased in the presence of high levels of *A. cochlioides*.

Other damping-off pathogens that have been evaluated in sugar beet disease complexes with *H. schachtii* are *Pythium ultimum* and *P. aphanidermatum* (Whitney, 1974). In glasshouse studies, Whitney (1974) observed a synergistic interaction effect between *P. ultimum* and the nematode on damping-off of sugar beet. Rather than a synergistic interaction, an additive effect was detected between *P. aphanidermatum* and *H. schachtii* on damping-off (Whitney, 1974). The synergistic mechanism was suggested to be associated with the release of plant leachates or physiological modifications by the nematode infection stimulating the growth of *P. ultimum*.

Furthermore, *P. aphanidermatum* spreads in the soil by zoospores, whereas *P. ultimum* spreads primarily through mycelial growth; Whitney (1974) suggested this main difference may be important to consider in order to understand the interaction mechanism between *H. schachtii* and these oomycete pathogens.

12.6.1.3 Interactions between Globodera spp. and fungal pathogens

Three of the 12 species comprising the genus *Globodera* are economically important: *Globodera pallida*, *G. rostochiensis* and *G. tabacum* (Turner and Subbotin, 2013). Tobacco cyst nematode, *G. tabacum*, is an important pathogen of shade and broadleaf tobacco, causing yield losses of 15% on average (Turner and Subbotin, 2013). *Globodera pallida* and *G. rostochiensis* are considered the most important pathogens of potato and their distributions and damage to the crop is worldwide (Subbotin et al., 2010). *Globodera pallida* is becoming more prevalent and damaging in many regions where potatoes are grown with resistance solely to *G. rostochiensis* (Bhattarai et al., 2009). Interaction studies between these cyst nematodes and other organisms are discussed below.

GLOBODERA ROSTOCHIENSIS AND RHIZOCTONIA SOLANI. Grainger and Clark (1963) reported the interaction between *G. rostochiensis* and *R. solani* on potato. In a glasshouse study, yield was measured in a repeated experiment. After harvest, new potatoes were planted in the same pots; therefore, a natural increase in pathogen population was achieved. After three repetitions, the treatment with both pathogens and the one with the nematode alone showed reduction in yield, demonstrating that the interaction causes reduction in yield at a level below the *G. rostochiensis* damage threshold.

Several studies were conducted to evaluate the effect of potato cyst nematodes (*G. rostochiensis* and *G. pallida*) and *R. solani* (AG-3) on potato. *Rhizoctonia solani* (AG-3) is an important pathogen of potato (Tsror, 2010). The fungus causes stem and stolon cankers, which take place prior to emergence and can result in yield reduction, malformation of tubers and uneven emergence (Tsror, 2010). Later in the season, the fungus produces black scurf, characterized by the production of dark and irregular sclerotia on the tubers. Black scurf renders the tubers unmarketable (Back et al., 2006; Tsror, 2010).

Back et al. (2006) reported a synergistic interaction effect between G. rostochiensis and R. solani under field conditions. The studies took place in fields infested with the nematode and at a wide range of population densities. Linear regression analysis was used to determine the relationship between the nematodes inside the roots and the disease severity caused by the fungus; a positive relationship was reported (Back et al., 2006). The mechanism of the interaction is thought to be indirect because the two pathogens infect different parts of the plants.

Back et al. (2010) later evaluated the effect of root diffusates from G. rostochiensis-infected roots on R. solani growth in vitro where fungal radial growth was subsequently measured. Rhizoctonia solani grew significantly more in media amended with diffusates from G. rostochiensis-infected roots as compared with diffusates from non-infected roots (Back et al., 2010). This study supported the hypothesis previously suggested (Back et al., 2006) that infection by G. rostochiensis changed root diffusates that stimulated or enhanced R. solani growth in the potato rhizosphere.

Bhattarai et al. (2009) conducted glasshouse and field studies to evaluate the independent effects of G. pallida, G. rostochiensis and R. solani on plant growth, and the interaction effects of the nematodes and the fungus on disease severity. In the glasshouse, co-infestation with G. pallida and R. solani or G. rostochiensis and R. solani stimulated more fungal disease compared with pots infested with R. solani alone. In field studies, the effect of G. pallida and R. solani was evaluated in a 2-year study and Bhattarai et al. (2009) found that G. pallida in soil and roots increased R. solani disease severity. This was the first study that demonstrated an association between G. pallida and R. solani in increasing disease severity.

To understand further the mechanism of interaction between G. pallida and R. solani, Bhattarai et al. (2010) evaluated the interaction effect of R. solani (AG-3) with two G. pallida populations differing for hatching rates under controlled environmental conditions, with concomitant infestation of plants at different times after planting. Greater disease severity was observed when the nematode and fungus were applied at higher inoculum levels and at early stages of plant development (Bhattarai et al., 2010). In pots co-infested with the nematode and the fungus, greater R. solani disease severity occurred with the fast-hatching G. pallida population compared with the slow-hatching one (Bhattarai et al., 2010). The researchers suggested that faster hatching could lead to higher initial infection and more opportunity to affect infection by R. solani.

GLOBODERA SPP. AND VERTICILLIUM SPP. Verticillium wilt of potato, caused by either V. dahliae or V. albo-atrum, is an important disease that affects potato production around the world and can reduce yield by 30–50% (Berlanger and Powelson, 2000). The fungus is a soil-borne pathogen that infects roots and colonizes the cortical tissue, eventually invading the vascular system resulting in foliar chlorosis and plant wilt (Berlanger and Powelson, 2000). Several studies have been conducted to evaluate the interaction effect between potato cyst nematodes and V. dahliae.

Harrison (1971) found disease in potato was more severe where G. rostochiensis and V. dahliae concomitantly infected plants, but not when either pathogen was present alone. When pots were co-infested with both pathogens, leaf area and yield reduction exceeded the sum of the damage that each pathogen caused alone, demonstrating a synergistic interaction (Harrison, 1971).

Corbett and Hide (1971) conducted several experiments to evaluate the interaction between G. rostochiensis and V. dahliae. As initial inoculum of G. rostochiensis increased and reached a threshold, more severe disease symptoms were observed when the fungus was present. These symptoms appeared earlier than those in plants infested with V. dahliae alone (Corbett and Hide, 1971). Split-root systems were used to understand the mechanism of the interaction. The synergistic effect on the severity of symptoms was observed only when both G. rostochiensis and V. dahliae infected the same root, ruling out the suggestion of a systemic effect involved in the interaction effect (Corbett and Hide, 1971).

Evans (1987) evaluated the reactions of potato cultivars with different resistance levels to the potato cyst nematodes and V. dahliae. In the absence of the fungus, two cultivars showed tolerance to G. rostochiensis; however, when the fungus was present, one of the cultivars became susceptible to the nematode (Evans, 1987). In a different study, potato cultivars were evaluated in two fields, one infested with G. rostochiensis and V. dahliae and the other where infestation was not detected (Hide et al., 1984). Yields from non-infested fields were, on average, four-fold higher than those obtained from the infested field (Hide et al., 1984). Application of pesticides to plots infested with both pathogens decreased disease severity and yield reduction compared with control plots (Hide and Corbett, 1974; Hide et al., 1984). Results from these studies suggested that control of potato cyst nematode was the main reason severe disease symptoms were reduced.

Histological studies were conducted by Storey and Evans (1987) to understand the mechanism of the synergistic interaction between G. pallida and V. dahliae. Infection of roots by the nematode enhanced colonization of the root by the fungus. Penetration by the nematode allows the fungus to overcome the natural defence mechanism of potato cultivars with thick cortical cell walls (Storey and Evans, 1987). When V. dahliae reached the syncytia, an increase in mycelia growth was observed; however, affected syncytia did not deteriorate during the period of the study (Storey and Evans, 1987).

GLOBODERA TABACUM AND FUSARIUM OXYSPORUM. In glasshouse and microplot studies, LaMondia and Taylor (1987) evaluated the interaction effect of G. tabacum and F. oxysporum on tobacco. Fusarium oxysporum occasionally causes Fusarium wilt and has been reported to be more severe in fields infested with G. tabacum (LaMondia and Taylor, 1987). In a factorial design, Fusarium wilt disease severity increased and plant dry weight decreased with increasing G. tabacum population densities. Additionally, in a split-root system, disease severity increase took place only when both pathogens were present on the same root halves (LaMondia and Taylor, 1987). To understand the mechanism of the interaction better, LaMondia (1992) conducted a sequential infestation study in which tobacco plants were infected with G. tabacum before introducing F. oxysporum; a simultaneous infection with both pathogens was also done. Plants that were infected with the nematode weeks before the introduction of F. oxysporum (i.e. predisposed) revealed more severe disease symptoms and reduction in plant weight than occurred with simultaneously infected plants. LaMondia (1992) suggested that the mechanism for the synergistic interaction was a physiological rather than simply a physical one. In a different study, LaMondia (1995) evaluated the effect of Fusarium-resistant and -susceptible cultivars with simultaneous infection by both pathogens. Fungal root colonization and disease severity increased in the presence of G. tabacum only for susceptible plants. Resistance to F. oxysporum was not affected by the presence of G. tabacum, a result that preserved the use of resistant cultivars as a tactic to manage the disease complex in fields infested with both pathogens (LaMondia, 1995).

12.6.1.4 Interactions between cereal cyst nematodes and fungal pathogens

The cereal cyst nematode (CCN) group comprises several species of which H. avenae, H. filipjevi and H. latipons are considered to be the most economically important (Rivoal and Nicol, 2009; Nicol et al., 2011; Dababat et al., 2015). Interactions between CCN and fungal pathogens have been reported. For example, in glasshouse studies, Meagher and Chambers (1971) evaluated the interaction of H. avenae and R. solani, the causal agent of Rhizoctonia root rot on wheat. Each pathogen alone was able to cause significant plant and root growth reduction compared with the control, but the effect on plant and root growth reduction was more than additive when both pathogens were present. Moreover, plant height was reduced only when both pathogens infected the plants (Meagher and Chambers, 1971). These results and additional field studies (Meagher et al., 1978) suggested that the patches of stunted plants observed in the field may actually be manifestation of an interaction effect between H. avenae and R. solani.

Hassan et al. (2012) evaluated yield, nematode reproduction and disease severity when wheat plants were infected with H. avenae and F. culmorum. Their study demonstrated that each pathogen alone can cause significant damage to wheat. A synergistic effect on yield reduction and disease severity was observed when the nematodes were applied to pots 1 month before the fungus, suggesting that the plants are predisposed to damage by the second pathogen following infection by the nematode (Khan, 1993a; Back et al., 2002). An additive effect, however, was observed when the pathogens were simultaneously added to plants and an antagonistic effect when the fungus was present 1 month before the nematode (Hassan et al., 2012). The presence of the fungus reduced *H. avenae* population density, more so on treatments that received the fungus before the nematode. Reduction in population density could explain the antagonistic effect observed on yield, which resulted in less yield reduction than the sum of the damage caused by each pathogen alone (Hassan et al., 2012).

Hajihassani et al. (2013) investigated the interactions between *H. filipjevi* and two fungal pathogens (*F. culmorum* and *Bipolaris sorokiniana*) on wheat. The study was conducted with the same experimental design used by Hassan et al. (2012) and produced results regarding interaction between *H. filipjevi* and *F. culmorum* similar to those found between *H. avenae* and *F. culmorum*. In contrast to the findings of Hassan et al. (2012), the greatest yield reduction took place in the simultaneous infestation treatment with *H. filipjevi* and *F. culmorum* (Hajihassani et al., 2013). However, yield was not affected by any combination between *H. filipjevi* and *B. sorokiniana*. Similar to the results of Hassan et al. (2012), the combinations of each fungus with the nematode always reduced *H. filipjevi* population densities (Hajihassani et al., 2013).

12.6.1.5 Interactions between other cyst nematodes and fungal pathogens

Two important cyst nematodes that affect clover, *H. daverti* and *H. trifolii*, appear to interact with fungi (Sikora, 1977; Nordmeyer and Sikora, 1983a). In one study, severe clover decline was observed and reported from fields in Tunisia. From these symptomatic plants and rhizospheres, *F. oxysporum*, *F. avenaceum* and *H. trifolii* were consistently isolated. In other areas where the symptoms were not severe, the nematode but not the fungus was recovered. A positive correlation was observed between isolation of fungus from symptomatic plants and the presence and population density of the nematode. Based on these observations, Sikora (1977) suggested that a synergistic interaction between the nematode and fungi could exist. In a different study, under controlled conditions, the interactions of *H. daverti* with *F. oxysporum* and *H. daverti* with *F. avenaceum* were examined. Soil in pots was simultaneously infested with the nematode and each fungus; also, sequential infestation took place, in which either the nematode or the fungus was introduced 1 and 2 weeks before the other pathogen. A synergistic reduction in yield was observed only when the fungus was introduced before the nematode (Nordmeyer and Sikora, 1983a). Additionally, *H. daverti* population density increased when introduced after each fungus.

Two important soil-borne pathogens of pigeon pea (*Cajanus cajan*) are *F. udum* and *H. cajani*. Hasan (1984) observed wilt symptoms in fields from which *F. udum* was isolated; however, *H. cajani* was recovered only in fields with more severe wilt symptoms. In an interaction study, pots were infested simultaneously and sequentially with each pathogen. All plants infected with the nematode and the fungus had significantly increased wilt incidence compared with those treated with the fungus alone. The population density of *H. cajani* increased in all pots compared with the initial inoculum density; however, final population densities in the presence of the fungus were significantly lower than those on plants to which the nematode was applied alone (Hasan, 1984).

Sharma and Nene (1989) conducted glasshouse experiments to evaluate the relationship between *H. cajani* and *F. udum* on three pigeon pea cultivars with different levels of resistance to Fusarium wilt. *Heterodera cajani* enhanced wilt of susceptible plants but not of the resistant or tolerant cultivars in the presence of the fungus (Sharma and Nene, 1989). In the presence of *F. udum*, the number of nematodes was reduced on all cultivars; however, number of eggs per female was not affected (Sharma and Nene, 1989).

12.6.2 Interactions between cyst nematodes and other nematodes

Nematode communities are composed of multiple species (Niblack, 1989; Eisenback, 1993; Neher, 2010). Interactions among nematodes, therefore, take place at an ecological and at the plant (i.e. aetiological) level (Eisenback, 1985, 1993). Interactions between cyst nematodes and other nematodes were reported to be antagonistic (i.e. one inhibiting the development of the other) or neutral (neither had an effect on the other), and the host damage was reported to be additive or less than additive (Eisenback, 1993; Barker and McGawley, 1998; Bond and Wrather, 2004). Two of the first nematode–nematode interaction studies were those conducted by Ross (1959, 1964) involving *H. glycines* and the root-knot nematode, *M. incognita*, in soybean. The population density of *H. glycines* increased at medium levels of *M. incognita*; however, at high levels of each nematode the other one was inhibited (Ross, 1964). Similarly, Malakeberhan and Dey (2003) found that simultaneous infestation of soil by *H. glycines* and *M. incognita* reduced their infection rate in soybeans. In the same study, high population densities of SCN reduced invasion of soybean by a lesion nematode (*Pratylenchus penetrans*), whereas medium levels of SCN stimulated lesion nematode infection (Malakeberhan and Dey, 2003).

In two glasshouse studies (Brinkman *et al.*, 2004, 2005a) and one field trial (Brinkman *et al.*, 2005b) interactions between *Tylenchorhynchus ventralis*, *P. penetrans*, *H. arenaria* and *M. maritima* were evaluated along with their effects on *Ammophilia arenaria* (marram grass) biomass. Endoparasitic nematodes inhibited the development of the ectoparasitic (Brinkman *et al.*, 2004), agreeing with previous reports (Eisenback, 1985, 1993; Eisenback and Griffin, 1987). Interspecific competition was observed between the endoparasitic nematodes; whereas *H. arenaria* and *M. maritima* were equally strong competitors, *P. penetrans* was the strongest competitor (Brinkman *et al.*, 2005a). In this glasshouse study, *P. penetrans* and *H. arenaria* reduced biomass, whereas *M. maritima* did not do so, due to poor establishment (Brinkman *et al.*, 2005a). By contrast, only *M. maritima* reduced biomass in the field study, and when the three nematodes were present, host plant biomass reduction was alleviated (Brinkman *et al.*, 2005b). These studies demonstrated the importance of validating glasshouse results in the field, otherwise extrapolation of glasshouse results to field conditions would be misleading.

Interactions between cyst and other nematodes have been presented elsewhere (Eisenback 1993; Abawi and Chen, 1998; Barker and McGawley, 1998). Some examples presented in Barker and McGawley (1998) include *G. tabacum* (Miller, 1970), *H. avenae* (Lasserre *et al.*, 1994) and *H. cajani* (Sharma *et al.*, 1996). From the earliest to the most recent studies, results have not been consistent and usually were dependent on environmental conditions, nematode population densities, cropping history, nematode pathogenicity, presence of other organisms, etc. Moreover, due to the qualitative nature of the results, we lack models that can predict how nematode–nematode interactions will define final population densities and communities and consequently plant growth. In addition, as with other interaction studies, the mechanisms by which nematode–nematode interactions take place are not known.

12.6.3 Interactions between cyst nematodes and symbionts

12.6.3.1 Interactions between cyst nematodes and rhizobia

A group of soil bacteria commonly referred to as rhizobia are able to fix atmospheric nitrogen and make it available to their symbiotic hosts (e.g. legumes). In this well-documented mutualistic relationship (Long, 1989; Hirsch, 1992; Mathesius *et al.*, 1998; Grunewald *et al.*, 2009), the rhizobacteria infect legume plants and produce root nodules where the bacteria fix nitrogen, convert it to ammonia and provide it to the plant as a source of nitrogen. The plant provides the energy supply for the bacteria in the form of carbohydrates. This mutualistic symbiosis is negatively impacted by cyst nematodes. Most often, the cyst nematodes interact antagonistically by preventing nodule formation, resulting in plants with nitrogen-deficiency symptoms (i.e. chlorosis and stunted growth).

Early studies conducted by Taha and Raski (1969) evaluated the interaction between *H. trifolii* and *Rhizobium trifolii* in white clover (*Trifolium repens*) and found that the nematode could reproduce in nodules, reducing nodule numbers only because they reduced the size of the root system; the nitrogen fixation efficiency of nodules from nematode-infected plants was not affected. Later on, Yeates *et al.* (1977) concluded that *H. trifolii* reduced the nitrogen-fixing capability of nodules and the root dry weight of white clover, but the mechanism of this antagonistic interaction was not studied. Dalal and Bhatti (1996) evaluated the interaction between *H. cajani* and *Rhizobium* sp. in mung and cluster bean and noted that nitrogen fixation by nodules was reduced when soil in pots was simultaneously infested with both organisms.

Other studies examined the effect of *H. glycines* on the soybean rhizobium, *Bradyrhizobium japonicum*. *Heterodera glycines* interacts antagonistically with *B. japonicum* and can reduce or delay nodule formation, nitrogen fixation, and nodular efficiency in susceptible soybean cultivars (Lehman *et al.*, 1971; Barker *et al.*, 1972; Huang and Barker, 1983; Ko *et al.*, 1984, 1985, 1991). For example, in glasshouse studies, the effect on nodulation and nitrogen fixation was evaluated under simultaneous and sequential infestation of a susceptible soybean cultivar with different *H. glycines* isolates and *B. japonicum* (Lehman *et al.*, 1971; Barker *et al.*, 1972). The effect of the cyst nematode on the quality of the nodules was also studied based on the components of the nodules (i.e. starch granules and phytoferritin) and the content of leghemoglobin produced in nodules on nematode-infected and non-infected plants (Huang and Barker, 1983; Ko *et al.*, 1985). Differences in *H. glycines* isolates affecting nodule formation were observed in these early studies (Lehman *et al.*, 1971; Barker *et al.*, 1972). No effect of *B. japonicum* on cyst development was observed (Ko *et al.*, 1991) and it was unlikely – although not impossible – that *H. glycines* developed on nodules (Barker *et al.*, 1972).

Greater reduction in nodule formation and nitrogen fixation was observed when susceptible soybean cultivars were infected with what was known as *H. glycines* race 1 (in current terminology, HG Type 2−) compared with other nematode isolates (Lehman *et al.*, 1971; Barker *et al.*, 1972; Ko *et al.*, 1991). Even though there was no difference in penetration rate and development into cysts between race 1 (HG Type 2−) and race 3 (HG Type 0), a significant difference in the effect on nodulation by these isolates was repeatedly observed (Ko *et al.*, 1991). Almost complete suppression of nodule formation was observed in susceptible soybean cultivars; by contrast, resistant cultivars infected with *H. glycines* HG Type 2− did not differ in nodule formation from the non-infected control (Ko *et al.*, 1991). In studies in which only one isolate of *H. glycines* HG Type 2− (race 1) was used, then the difference reported could have been due to the isolate and not specifically to the 'race' identity.

Nodulation was not affected when a halt in nematode development took place in the resistant cultivars (Ko *et al.*, 1991). This result complements those found by Ko *et al.* (1984) in which a split-root technique was used to evaluate effects of cyst nematodes on nodulation in soybean roots. The suppression of nodule formation and nitrogen fixation by *H. glycines* were both localized and systemic. Plants that received the nematodes on one half of the root and the rhizobium on the other half had suppressed nodule formation and consequently nitrogen fixation. When the root half infected with *H. glycines* was excised, the nodulation took place on the other root half (Ko *et al.*, 1984).

The cyst nematode–host and the rhizobium–legume interactions have a lot in common, especially in their abilities to manipulate auxin transport (Grunewald *et al.*, 2009). Initial activation of syncytia (Goverse *et al.*, 2000) and rhizobium nodules (Mathesius *et al.*, 1998; Pacios-Bras *et al.*, 2003) requires a mobilization and concentration of auxin. Therefore, both rhizobia and cyst nematodes must take control of their host, manipulating auxin transport to generate new root structures (i.e. syncytia and nodules) that will successfully allow them to complete their life cycles (Mathesius *et al.*, 1998; Goverse *et al.*, 2000; Pacios-Bras *et al.*, 2003; Grunewald *et al.*, 2009). Cyst nematodes compete with rhizobia to control the same host machinery (i.e. auxin transport) and consequently an antagonistic interaction takes place. This may explain the systemic effect of nematodes on nodulation (Ko *et al.*, 1984) or the lack of effect on nodulation in nematode-resistant cultivars (Ko *et al.*, 1991).

12.6.3.2 *Interactions between cyst nematodes and mycorrhizae*

Vesicular arbuscular mycorrhizal (VAM), or simply arbuscular mycorrhizal fungi (AMF), are obligate symbionts (Hepper, 1987; Francl, 1993).

Similar to rhizobia, AMF infect compatible hosts, grow inter- and intracellularly in the root cortex and establish a mutualistic symbiosis with their plant hosts. The fungus obtains carbohydrates from the host through specialized fungal structures called vesicles, resembling those structures (i.e. haustoria) produced by obligate parasitic oomycetes (Francl, 1993). Mycelial growth facilitates the uptake of soil nutrients (especially phosphorus) and water. In addition, AMF could render their hosts tolerant to plant-parasitic nematodes attacks (Hussey and McGuire, 1987; Smith, 1987; Francl, 1993). Host tolerance to nematode damage (from a production stand point) or an antagonistic effect to nematodes by AMF are advantageous qualities. An antagonistic effect, however, towards AMF from cyst nematode parasitism could significantly jeopardize agricultural production. In this section we report on studies of interaction between AMF and cyst nematodes without covering studies on potential biological control of nematodes with AMF (see Chapter 11, this volume).

The effect of *H. cajani* on cowpea (*Vigna unguiculata*) in the presence of two AMF (*Glomus fasciculatum* and *G. epigaeus*) was evaluated by Jain and Sethi (1987). An increase in *H. cajani* population density resulted in reduction of root colonization by the fungi. The effect on *H. cajani* reproduction was detrimental (i.e. decreased nematode reproduction) in the presence of *G. fasciculatum* but not *G. epigaeus* (Jain and Sethi, 1987). The effects of *G. fasciculatum* on *H. cajani* were similar to those previously observed in *H. glycines* (Francl and Dropkin, 1985). The AMF was described as a weak SCN pathogen and the soybean plants infected with *G. fasciculatum* produced more biomass in the presence of *H. glycines* than did non-mycorrhizal plants (Francl and Dropkin, 1985).

The effects of AMF on *G. pallida* and *G. rostochiensis* in potato were evaluated in greenhouse studies (Ryan *et al.*, 2003). Increased nematode numbers were observed in mycorrhizal compared with non-mycorrhizal plants. These plants, however, produced more root mass, compensating for the burden of nematode infection (Ryan *et al.*, 2003). Also, AMF stimulated the relatively delayed hatching rate of *G. pallida*, synchronizing it with that of *G. rostochiensis* (Ryan *et al.*, 2000; Ryan and Jones, 2004; Deliopoulos *et al.*, 2008), which could potentially improve control of both nematodes in potato fields. These results reinforced the suggestion of plants becoming tolerant to nematodes in the presence of their mycorrhizal symbionts.

Other studies, however, documented an antagonistic effect of cyst nematodes on AMF and a lack of tolerance to nematode infection in mycorrhizal plants (Tylka *et al.*, 1991; Winkler *et al.*, 1994; Todd *et al.*, 2001). For example, Todd *et al.* (2001) found no evidence of tolerance to *H. glycines* in plants infected by *G. mosseae* in greenhouse studies. Colonization of soybean roots by the mycorrhizal fungus was inhibited by *H. glycines* (Todd *et al.*, 2001). Moreover, reproduction of nematodes was not affected by the presence of AMF. Winkler *et al.* (1994) showed that *H. glycines* reduced both yield in susceptible cultivars and colonization by the mycorrhizal fungus. These antagonistic interactions were shown to be localized rather than systemic by Tylka *et al.* (1991) in a split-root experiment.

12.7 Conclusions and Future Prospects

Over a century of data convinced the nematology community that plant-parasitic nematodes interact with other organisms producing more or less damage to the host than if they were acting alone. What these studies have still not answered is: *how* do nematodes and other organisms interact? Very few studies have gone a step beyond the typical four treatments (combined pathogens, pathogens alone and control) in greenhouse studies to the field or vice versa. Very few studies have complemented their results with histological evidence to explain the mechanism of interaction. None of the studies presented in this chapter had employed molecular technologies to unravel the molecular mechanisms of interaction between nematodes and other organisms.

Results from genomic, transcriptomic and proteomic studies are contributing to our understanding of the mechanisms by which plant-parasitic nematodes successfully infect and complete their life cycle in their plant hosts (Haegeman *et al.*, 2012; Goverse and Smant, 2014; Gardner *et al.*, 2015; Mantelin *et al.*, 2015). Some of these mechanisms involve degradation of host tissue, suppression of plant defences, and modification of plant signalling and hormone pathways to initiate and develop nematode feeding sites (syncytia, in the case of cyst nematodes). Similar approaches being used to unravel mechanisms of nematode–plant interactions could be used to understand how a nematode and another pathogen may interact in disease complex. Proteomic and metabolomic profiles could be used to identify potential factors that trigger the observed synergistic or antagonistic interaction effect observed in plants. Transcriptomic analyses can shed light on the regulation of genes when nematodes and other pathogens simultaneously infect the plant. More research is required that zooms in on the direct interaction between nematodes and other organisms to understand their effect on the host. Understanding the mechanisms by which organisms interact will allow us to understand the contradictory results in the literature when these interactions were measured indirectly (i.e. plant growth, disease severity, pathogen development, etc.). In addition, adequate experimental designs and statistical analyses are required.

Slinker (1998) commented that the intuition of researchers to set up interaction studies is reasonable; however, intuition sometimes fails in matching data analysis to experimental design. Spatio-temporal analyses are yet to be used in interaction studies; these models account for autocorrelation in space and time that will otherwise produce unbiased and inefficient parameters. For example, with one exception (Brinkman *et al.*, 2005b), repeated observations were made but not addressed in the analyses. As a result, the temporal component of disease development was not fully explored in most studies. Likewise, aggregation of soil-borne pathogens was not addressed via a spatial regression analysis when randomization of treatments was not feasible. Extrapolating statistical analyses from other fields in science, such as ecology (Piggott *et al.*, 2015), interaction studies in biocontrol (Xu *et al.*, 2011), interaction studies in pharmacology (Foucquier and Guedj, 2015), and public health studies with longitudinal (Fitzmaurice *et al.*, 2011) and spatial (Waller and Gotway, 2004) statistical analyses will help us to better understand the epidemiology of disease complexes and will allow us to better predict yield and economic losses, ultimately increasing food production sustainably to mitigate the problems of increasing global human populations.

12.8 References

Abawi, G.S. and Chen, J. (1998) Concomitant pathogen and pest interactions. In: Barker, K.R., Pederson, G.A. and Windham, G.L. (eds) *Plant Nematode Interactions*. Agronomy Monograph 36, 135–158. DOI: 10.2134/agronmonogr36.c7

Adeniji, M.O., Edwards, D.I., Sinclair, J.B. and Malek, R.B. (1975) Interrelationship of *Heterodera glycines* and *Phytophthora megasperma* var. *sojae* in soybeans. *Phytopathology* 65, 722–725. DOI: 10.1094/phyto-65-722

Ali, J.G. and Davidson-Lowe, E. (2015) Plant cues and factors influencing the behaviour of beneficial nematodes as a belowground indirect defense. *Advances in Botanical Research* 75, 191–214. DOI: 10.1016/bs.abr.2015.08.003

Allen, T.W., Bradley, C.A., Sisson, A.J. *et al.* (2017) Soybean yield loss estimates due to diseases in the United States and Ontario, Canada, from 2010 to 2014. *Plant Health Progress* 18, 19–27. DOI: 10.1094/PHP-RS-16-0066

Allington, W.B. and Chamberlain, D.W. (1948) Brown stem rot of soybean. *Phytopathology* 38, 793–802.

Aoki, T., O'Donnell, K., Homma, Y. and Lattanzi, A.R. (2003) Sudden-death syndrome of soybean is caused by two morphologically and phylogenetically distinct species within the *Fusarium solani* species complex: *F. virguliforme* in North America and *F. tucumaniae* in South America. *Mycologia* 95, 660–684. DOI: 10.2307/3761942

Aoki, T., O'Donnell, K. and Scandiani, M.M. (2005) Sudden death syndrome of soybean in South America is caused by four species of *Fusarium: Fusarium brasiliense* sp. nov., *F. cuneirostrum* sp. nov., *F. tucumaniae*, and *F. virguliforme*. *Mycoscience* 46, 162–183. DOI: 10.1007/s10267-005-0235-y

Aoki, T., Scandiani, M.M. and O'Donnell, K. (2012) Phenotypic, molecular phylogenetic, and pathogenetic characterization of *Fusarium crassistipitatum* sp. nov., a novel soybean sudden death syndrome pathogen from Argentina and Brazil. *Mycoscience* 53, 167–186. DOI: 10.1007/s10267-011-0150-3

Atkinson, G.F. (1892) Some diseases of cotton. *Alabama Polytechnical Institute and Agricultural Experiment Station Bulletin* No. 41, 61–65.
Back, M.A., Haydock, P.P.J. and Jenkinson, P. (2002) Disease complexes involving plant parasitic nematodes and soilborne pathogens. *Plant Pathology* 51, 683–697. DOI: 10.1046/j.1365-3059.2002.00785.x
Back, M., Haydock, P. and Jenkinson, P. (2006) Interactions between the potato cyst nematode *Globodera rostochiensis* and diseases caused by *Rhizoctonia solani* AG3 in potatoes under field conditions. *European Journal of Plant Pathology* 114, 215–223. DOI: 10.1007/s10658-005-5281-y
Back, M., Jenkinson, P., Deliopoulos, T. and Haydock, P. (2010) Modifications in the potato rhizosphere during infestations of *Globodera rostochiensis* and subsequent effects on the growth of *Rhizoctonia solani*. *European Journal of Plant Pathology* 128, 459–471. DOI: 10.1007/s10658-010-9673-2
Bais, H.P., Weir, T.L., Perry, L.G., Gilroy, S. and Vivanco, J.M. (2006) The role of root exudates in rhizosphere interactions with plants and other organisms. *Annual Review of Plant Biology* 57, 233–266. DOI: 10.1146/annurev.arplant.57.032905.105159
Barker, K.R. and McGawley, E.C. (1998) Interrelations with other microorganisms and pests. In: Sharma, S.B. (ed.) *The Cyst Nematodes.* Chapman & Hall, London, pp. 266–292. DOI: 10.1007/978-94-015-9018-1_11
Barker, K.R., Huisingh, D. and Johnson, S.A. (1972) Antagonistic Interaction between *Heterodera glycines* and *Rhizobium japonicum* on soybean. *Phytopathology* 62, 1201–1205. DOI: 10.1094/phyto-62-1201
Berenbaum, M.C. (1989) What is synergy? *Pharmacological Reviews* 41, 93–141.
Bergeson, B. (1972) Concepts of nematode-fungus disease complexes: a review. *Experimental Parasitology* 314, 301–314. DOI: 10.1016/0014-4894(72)90037-9
Berlanger, I. and Powelson, M.L. (2000) Verticillium wilt. *The Plant Health Instructor*. DOI: 10.1094/PHI-I 2000-0801-01
Bhattarai, S., Haydock, P.P.J., Back, M.A., Hare, M.C. and Lankford, W.T. (2009) Interactions between the potato cyst nematodes, *Globodera pallida, G. rostochiensis*, and soil-borne fungus, *Rhizoctonia solani* (AG3), diseases of potatoes in the glasshouse and the field. *Nematology* 11, 631–640. DOI: 10.1163/156854108X399173
Bhattarai, S., Haydock, P.P.J., Back, M.A., Hare, M.C. and Lankford, W.T. (2010) Interactions between field populations of the potato cyst nematode *Globodera pallida* and *Rhizoctonia solani* diseases of potatoes under controlled environment and glasshouse. *Nematology* 12, 783–790. DOI: 10.1163/138855410X12631974516235
Bond, J. and Wrather, J.A. (2004) Interactions with other plant pathogens and pests. In: Schmitt, D.P., Wrather, J.A. and Riggs R.D. (eds) *Biology and Management of the Soybean Cyst Nematode*, 2nd edn. Schmitt & Associates of Marceline, Missouri, pp. 111–129.
Brinkman, E.P., van Veen, J.A. and van der Putten, W.H. (2004) Endoparasitic nematodes reduce multiplication of ectoparasitic nematodes, but do not prevent growth reduction of *Ammophila arenaria* (L.) Link (marram grass). *Applied Soil Ecology* 27, 65–75. DOI: 10.1016/j.apsoil.2004.02.004
Brinkman, E.P., Duyts, H. and van der Putten, W.H. (2005a) Competition between endoparasitic nematodes and effect on biomass of *Ammophila arenaria* (marram grass) as affected by timing of inoculation and plant age. *Nematology* 7, 169–178. DOI: 10.1163/1568541054879647
Brinkman, E.P., Duyts, H. and van der Putten, W.H. (2005b) Consequences of variation in species diversity in a community of root-feeding herbivores for nematode dynamics and host plant biomass. *OIKOS Journal* 110, 417–427. DOI: 10.1111/j.0030-1299.2005.13659.x
Brzostowski, L.F., Schapaugh, W.T., Rzodkiewicz, P.A., Todd, T.C. and Little, C.R. (2014) Effect of host resistance to *Fusarium virguiliforme* and *Heterodera glycines* on sudden death syndrome disease severity and soybean yield. *Plant Health Progress*. DOI: 10.1094/PHP-RS-13-0100
Burrows, P.M. (1987) Interaction concepts for analysis of responses to mixtures of nematode populations. In: Veech, J.A. and Dickenson, D.W. (eds) *Vistas on Nematology*. Society of Nematologists, Hyattsville, Maryland, pp. 82–93.
Cabrera, J., Díaz-Manzano, F.E., Fenoll, C. and Escobar, C. (2015) Developmental pathways mediated by hormones in nematode feeding sites. *Advances in Botanical Research*, 73, 167–188. DOI: 10.1016/bs.abr.2014.12.005
Carris, L.M., Glawe, D.A., Smyth, C.A. and Edwards, D.I. (1989) Fungi associated with populations of *Heterodera glycines* in two Illinois soybean fields. *Mycologia* 81, 66–75. DOI: 10.2307/3759452
Cooke, D.A. (1993) Pests. In: Cooke, D.A. and Scott, R.K. (eds) *The Sugar Beet Crop*. Chapman & Hall, London, pp. 429–483. DOI:10.1007/978-94-009-0373-9_11

Corbett, D.C.M. and Hide, G.A. (1971) Interactions between *Heterodera rostochiensis* Woll. and *Verticillium dahliae* Kleb. on potatoes and the effect of CCC on both. *Annals of Applied Biology* 68, 71–80. DOI: 10.1111/j.1744-7348.1971.tb04639.x

Dababat, A.A., Imren, M., Erginbas-Orakci, G., Ashrafi, S., Yavuzaslanoglu, E., Toktay, H., Pariyar, S.R., Elekcioglu, H.I., Morgounov, A. and Mekete, T. (2015) The importance and management strategies of cereal cyst nematodes, *Heterodera* spp., in Turkey. *Euphytica* 202, 173–188. DOI: 10.1007/s10681-014-1269-z

Dalal, M.R. and Bhatti, D.S. (1996) Interaction between *Heterodera cajani* and *Rhizobium* on mung bean and cluster bean. *Nematologia Mediterranea* 24, 101–103.

Dave, G.S. (1975) Interrelationships of *Rhizoctonia solani* with *Heterodera glycines*, *Pratylenchus scribneri* and *Tylenchorhynchus martini* on 'Clark 63' soybeans. PhD thesis. University of Illinois, Urbana, Illinois.

Deliopoulos, T., Haydock, P.P.J. and Jones, P.W. (2008) Interaction between arbuscular mycorrhizal fungi and the nematicide aldicarb on hatch and development of the potato cyst nematode, *Globodera pallida*, and yield of potatoes. *Nematology* 10, 783–799. DOI: 10.1163/156854108786161427

Dinh, P.T.Y., Knoblauch, M. and Elling, A.A. (2014) Nondestructive imaging of plant-parasitic nematode development and host response to nematode pathogenesis. *Phytopathology* 104, 497–506. DOI: 10.1094/PHYTO-08-13-0240-R

Duffus, J.E. and Ruppel, E.G. (1993) Diseases. In: Cooke, D.A. and Scott, R.K. (eds) *The Sugar Beet Crop*. Chapman & Hall, London, pp. 347–427. DOI: 10.1007/978-94-009-0373-9_10

Eisenback, J.D. (1985) Interactions among concomitant populations of nematodes. In: Sasser, J.N. and Carter, C.C. (eds) *An Advanced Treatise on Meloidogyne, Vol. I.: Biology and Control*. North Carolina State University Press, North Carolina State University, Raleigh, North Carolina, pp. 193–213.

Eisenback, J.D. (1993) Interactions between nematodes in cohabitance. In: Khan, M.W. (ed.) *Nematode Interactions*. Chapman & Hall, London, pp. 134–174. DOI: 10.1007/978-94-011-1488-2_7

Eisenback, J.D. and Griffin, G.D. (1987) Interactions with other nematodes. In: Veech, J.A. and Dickenson, D.W. (eds) *Vistas on Nematology*. Society of Nematologists, Hyattsville, Maryland, pp. 313–320.

Endo, B.Y. (1965) Histological response of resistant and susceptible soybean varieties, and backcross progeny to entry and development of *Heterodera glycines*. *Phytopathology* 55, 375–381.

Escobar, C., Horowitz, S.B., and Mitchum, M.G. (2011) Transcriptomic and proteomic analysis of the plant response to nematode infection. In: Jones, J., Gheysen, G. and Fenoll, C. (eds) *Genomics and Molecular Genetics of Plant–nematode Interactions*. Springer, Dordrecht, The Netherlands, pp. 157–173. DOI: 10.1007/978-94-007-0434-3_9

Evans, K. (1987) The interactions of potato cyst nematodes and *Verticillium dahliae* on early and maincrop potato cultivars. *Annals of Applied Biology* 110, 329–339. DOI: 10.1111/j.1744-7348.1987.tb03263.x

Evans, K. and Haydock, P.P.J. (1993) Interactions of nematodes with root-rot fungi. In: Khan, M.W. (ed.) *Nematode Interactions*. Chapman & Hall, London, pp. 104–133. DOI: 10.1007/978-94-011-1488-2_6

Evans, K. and Rowe, J.A. (1998) Distribution and economic importance. In: Sharma, S.B. (ed.) *The Cyst Nematodes*. Chapman & Hall, London, pp. 1–30. DOI: 10.1007/978-94-015-9018-1_1

Evans, K. and Stone, A.R. (1977) A review of the distribution and biology of the potato cyst-nematodes *Globodera rostochiensis* and *G. pallida*. *International Journal of Pest Management* 23, 178–189. DOI: 10.1080/09670877709412426

Fawcett, H.S. (1931) The importance of investigations on the effects of known mixtures of organisms. *Phytopathology* 21, 545–550.

Fitzmaurice, G.M., Laird, N.M. and Ware, J.H. (2011) *Applied longitudinal analysis*, 2nd edn. John Wiley & Sons, Inc., Hoboken, New Jersey. DOI: 10.1080/10543406.2013.789817

Foucquier, J. and Guedj, M. (2015) Analysis of drug combinations: current methodological landscape. *Pharmacology Research and Perspective* 3, 1–11. DOI: 10.1002/prp2.149

Francl, L.J. (1993) Interaction of nematodes with mycorrhizae and mycorrhizal fungi. In: Khan, M.W. (ed.) *Nematode Interactions*. Chapman & Hall, London, pp. 203–216. DOI: 10.1007/978-94-011-1488-2_9

Francl, L.J. and Dropkin, V.H. (1985) *Glomus fasciculatum*, a weak pathogen of *Heterodera glycines*. *Journal of Nematology* 17, 470–475.

Francl, L.J. and Wheeler, T.A. (1993) Interactions of plant-parasitic nematodes with wilt-inducing fungi. In: Khan, M.W. (ed.) *Nematode Interactions*. Chapman & Hall, London, pp. 80–103. DOI: 10.1007/978-94-011-1488-2_5

Francl, L.J., Wyllie, T.D. and Rosenbrock, S.M. (1988) Influence of crop rotation on population density of *Macrophomina phaseolina* in soil infested with *Heterodora glycines*. *Plant Disease* 72, 760–764. DOI: 10.1094/pd-72-0760

Frohning, J.R. (2013) Evaluation of *Rhizoctonia solani–Heterodera glycines* interactions on soybean. MSc thesis. University of Illinois, Urbana, Illinois.

Gao, X., Jackson, T.A, Hartman, G.L. and Niblack, T.L. (2006) Interactions between the soybean cyst nematode and *Fusarium solani* f. sp. *glycines* based on greenhouse factorial experiments. *Phytopathology* 96, 1409–1415. DOI: 10.1094/PHYTO-96-1409

Gardner, M., Verma, A. and Mitchum, M.G. (2015) Emerging roles of cyst nematode effectors in exploiting plant cellular processes. *Advances in Botanical Research* 73, 259–291. DOI: 10.1016/bs.abr.2014.12.009

Gheysen, G. and Mitchum, M.G. (2009) Molecular insights in the susceptible plant response to nematode infection. In: Berg, R.H. and Taylor, C.G. (eds) *Cell Biology of Plant Nematode Parasitism*. Plant Cell Monographs 15. Springer, Heidelberg, Germany, pp. 45–81. DOI: 10.1007/7089_2008_35

Goverse, A. and Bird, D. (2011) The role of plant hormones in nematode feeding cell formation. In: Jones, J., Gheysen, G. and Fenoll, C. (eds) *Genomics and Molecular Genetics of Plant–Nematode Interactions*. Springer, Dordrecht, The Netherlands, pp. 325-347. DOI:10.1007/978-94-007-0434-3_16

Goverse, A. and Smant, G. (2014) The activation and suppression of plant innate immunity by parasitic nematodes. *Annual Review of Phytopathology* 52, 243–265. DOI:10.1146/annurev-phyto-102313-050118

Goverse, A., Overmars, H., Engelbertink, J., Schots, A., Bakker, J. and Helder, J. (2000) Both induction and morphogenesis of cyst nematode feeding cells are mediated by auxin. *Molecular Plant-Microbe Interactions* 13, 1121–1129. DOI: 10.1094/MPMI.2000.13.10.1121

Grainger, J. and Clark, M.R.M. (1963) Interactions of *Rhizoctonia* and potato root eelworm. *European Potato Journal* 6, 131–132. DOI: 10.1007/BF02365220

Gröntoft, M. and Jonasson, T. (1992) Influence of *Heterodera schachtii* on Verticllium wilt symptoms in oilseed rape. *Journal of Phytopathology* 134, 170–174. DOI: 10.1111/j.1439-0434.1992.tb01225.x

Grunewald, W., van Noorden, G., van Isterdael, G., Beeckman, T., Gheysen, G. and Mathesius, U. (2009) Manipulation of auxin transport in plant roots during rhizobium symbiosis and nematode parasitism. *The Plant Cell* 21, 2553–2562. DOI: 10.1105/tpc.109.069617

Haegeman, A., Mantelin, S., Jones, J.T. and Gheysen, G. (2012) Functional roles of effectors of plant-parasitic nematodes. *Gene* 492, 19–31. DOI: 10.1016/j.gene.2011.10.040

Hajihassani, A., Maafi, Z.T. and Hosseininejad, A. (2013) Interactions between *Heterodera filipjevi* and *Fusarium culmorum*, and between *Heterodera filipjevi* and *Bipolaris sorokiniana* in winter wheat. *Journal of Plant Diseases and Protection* 120, 77–84. DOI: 10.1007/BF03356457

Harrison, J.A.C. (1971) Association between the potato cyst-nematode, *Heterodera rostochiensis* Woll., and *Verticillium dahliae* Kleb. in the early-dying disease of potatoes. *Annals of Applied Biology* 67, 185–193. DOI: 10.1111/j.1744-7348.1971.tb02919.x

Hartman, G.L., Noel, G.R. and Gray, L.E. (1995) Occurrence of soybean sudden death syndrome in east-central Illinois and associated yield losses. *Plant Disease* 79, 314–418. DOI:10.1094/pd-79-0314

Hasan, A. (1984) Synergism between *Heterodera cajani* and *Fusarium udum* attacking *Cajanus cajan*. *Nematologia Mediterranea* 12, 159–162.

Hassan, G.A., Al-Assas, K. and Abou Al-Fadil, T. (2012) Interactions between *Heterodera avenae* and *Fusarium culmorum* on yield components of wheat, nematode reproduction and crown rot severity. *Nematropica* 42, 260–266.

Hbirkou, C., Welp, G., Rehbein, K., Hillnhütter, C., Daub, M., Oliver, M.A. and Pätzold, S. (2011) The effect of soil heterogeneity on the spatial distribution of *Heterodera schachtii* within sugar beet fields. *Applied Soil Ecology* 51, 25–34. DOI: 10.1016/j.apsoil.2011.08.008

Heeren, J.R., Steffey, K.L., Tinsley, N.A., Estes, R.E., Niblack, T.L. and Gray, M.E. (2012) The interaction of soybean aphids and soybean cyst nematodes on selected resistant and susceptible soybean lines. *Journal of Applied Entomology* 136, 646–655. DOI: 10.1111/j.1439-0418.2011.01701.x

Hepper, C.M. (1987) VAM spore germination and hyphal growth in vitro: Prospects for axenic culture, in Mycorrhizae in the next decade, practical applications and research priorities. In: Sylvia, D.M., Hung, L.L. and Graham, J.H. (eds) *Proceedings of the 7th North American Conference on Mycorrhizae*. Institute of Food and Agricultural Sciences, Gainesville, Florida, pp. 172–174.

Herr, L.J. and Roberts, D.L. (1980) Characterization of *Rhizoctonia* populations obtained from sugarbeet fields with differing soil textures. *Phytopathology* 70, 476–480. DOI: 10.1094/phyto-70-476

Hide, G.A. and Corbett, D.C.M. (1974) Field experiments in the control of *Verticillium dahliae* and *Heterodera rostochsiensis* on potatoes. *Annals of Applied Biology* 78, 295–307. DOI: 10.1111/j.1744-7348.1974.tb01509.x

Hide, G.A., Corbett, D.C.M. and Evans, K. (1984) Effects of soil treatments and cultivars on 'early dying' disease of potatoes caused by *Globodera rostochiensis* and *Verticillium dahliae*. *Annals of Applied Biology* 104, 277–289. DOI: 10.1111/j.1744-7348.1984.tb05612.x

Hillnhütter, C., Sikora, R.A. and Oerke, E.-C. (2011a) Influence of different levels of resistance or tolerance in sugar beet cultivars on complex interactions between *Heterodera schachtii* and *Rhizoctonia solani*. *Nematology* 13, 319–332. DOI: 10.1163/138855410X519398

Hillnhütter, C., Mahlein, A.-K., Sikora, R.A. and Oerke, E.-C. (2011b) Remote sensing to detect plant stress induced by *Heterodera schachtii* and *Rhizoctonia solani* in sugar beet fields. *Field Crops Research* 122, 70–77. DOI: 10.1016/j.fcr.2011.02.007

Hillnhütter, C., Sikora, R.A., Oerke, E.-C. and van Dusschoten, D. (2012) Nuclear magnetic resonance: a tool for imaging belowground damage caused by *Heterodera schachtii* and *Rhizoctonia solani* on sugar beet. *Journal of Experimental Botany* 63, 319–327. DOI: 10.1093/jxb/err273

Hiltpold, I., Jaffuel, G. and Turlings, T.C.J. (2015) The dual effects of root-cap exudates on nematodes: from quiescence in plant-parasitic nematodes to frenzy in entomopathogenic nematodes. *Journal of Experimental Botany* 66, 603–611. DOI: 10.1093/jxb/eru345

Hirrel, M.C. (1983) Sudden death syndrome of soybean – a disease of unknown eiology. *Phytopathology* 73, 501 (Abstr.).

Hirrel, M.C. (1987) Sudden death syndrome of soybean: new insights into its development, In: *Proceedings of Soybean Seed Conference*. American Seed Trade Association, 16th Soybean Research Conference, pp. 95–104.

Hirsch, A.M. (1992) Developmental biology of legume nodulation. *New Phytologist* 122, 211–237. DOI: 10.1111/j.1469-8137.1992.tb04227.x

Huang, J.-S. and Barker, K.R. (1983) Influence of *Heterodera glycines* on leghemoglobins of soybean nodules. *Phytopathology* 73, 1002–1004. DOI: 10.1094/phyto-73-1002

Hussey, R.S. and McGuire, J.M. (1987) Interaction with other organisms. In: Brown, R.H. and Kerry, B.R. (eds) *Principles and Practice of Nematode Control in Crops*. Academic Press, New York, pp. 313–320.

Impullitti, A.E. and Malvick, D.K. (2014) Anatomical response and infection of soybean during latent and pathogenic infection by type A and B of *Phialophora gregata*. *PLoS ONE* 9(5), e98311. DOI: 10.1371/journal.pone.0098311

Jain, R.K. and Sethi, C.L. (1987) Pathogenicity of *Heterodera cajani* on cowpea as influenced by the presence of VAM fungi, *Glomus fasciculatum* or *G. epigaeus*. *Indian Journal of Nematology* 17, 165–170.

Jolliffe P.A. (2000) The replacement series. *Journal of Ecology* 88, 371–385. DOI: 10.1046/j.1365-2745.2000.00470.x

Jorgenson, E.C. (1970) Antagonistic interaction of *Heterodera schachtii* Schmidt and *Fusarium* oxysporum (Woll.) on sugarbeets. *Journal of Nematology* 2, 393–398.

Kaitany, R., Melakeberhan, H., Bird, G.W. and Safir, G. (2000) Association of *Phytophthora sojae* with *Heterodera glycines* and nutrient stressed soybeans. *Nematropica* 30, 193–199.

Kerry, B.R. (2000) Rhizosphere interactions and the exploitation of microbial agents for the biological control of plant-parasitic nematodes. *Annual Review of Phytopathology* 38, 423–441. DOI: 10.1146/annurev.phyto.38.1.423

Khan, M.W. (ed.) (1993a) *Nematode Interactions*. Chapman & Hall, London. DOI: 10.1007/978-94-011-1488-2

Khan, M.W. (1993b) Mechanisms of interactions between nematodes and other plant pathogens. In: Khan, M.W. (ed.) *Nematode Interactions*. Chapman & Hall, London, pp. 55–78. DOI: 10.1007/978-94-011-1488-2_4

Khan, M.W. and Dasgupta, M.K. (1993) The concept of interaction. In: Khan, M.W. (ed.) *Nematode Interactions*. Chapman & Hall, London, pp. 42–54. DOI: 10.1007/978-94-011-1488-2_3

Ko, M.P., Barker, K.R. and Huang, J.-S. (1984) Nodulation of soybeans as affected by half-root infection with *Heterodera glycines*. *Journal of Nematology* 16, 97–105.

Ko, M.P., Huang, P.-Y., Huang, J.-S. and Barker, K.R. (1985) Accumulation of phytoferritin and starch granules in developing nodules of soybean roots infected with *Heterodera glycines*. *Phytopathology* 75, 159–164. DOI: 10.1094/phyto-75-159

Ko, M.P., Huang, P.-Y., Huang, J.-S. and Barker, K.R. (1991) Responses of nodulation to various combinations of *Bradyrhizobium japonicum* strains, soybean cultivars, and races of *Heterodera glycines*. *Phytopathology* 81, 591–595. DOI: 10.1094/phyto-81-591

Koenning, S.R. (2004) Population biology. In: Schmitt, D.P., Wrather, J.A. and Riggs, R.D. (eds) *Biology and Management of the Soybean Cyst Nematode*, 2nd edn. Schmitt & Associates of Marceline, Missouri, pp. 73–88.

Koenning, S.R., and Wrather, J.A. (2010) Suppression of soybean yield potential in the continual United States by plant diseases from 2006 to 2009. *Plant Health Progress.* DOI:10.1094/PHP-2010-1122-01-RS

Kutner, M.H., Nchtsheim, C.J. and Neter, J. (2004) *Applied linear regression*, 4th edn. McGraw-Hill Irwin, New York.

LaMondia, J.A. (1992) Predisposition of broadleaf tobacco to Fusarium wilt by early infection with *Globodera tabacum tabacum* or *Meloidogyne hapla. Journal of Nematology* 24, 425–431.

LaMondia, J.A. (1995) Influence of resistant tobacco and tobacco cyst nematodes on root infection and secondary inoculum of *Fusarium oxysporum* f. sp. *nicotianae. Plant Disease* 79, 337–340. DOI: 10.1094/pd-79-0337

LaMondia, J.A. and Taylor, G.S. (1987) Influence of the tobacco cyst nematode (*Globodera tabacum*) on Fusarium wilt of Connecticut broadleaf tobacco. *Plant Disease* 71, 1129–1132. DOI: 10.1094/pd-71-1129

Lasserre, F., Rivoal, R. and Cook, R. (1994) Interactions between *Heterodera avenae* and *Pratylenchus neglectus* on wheat. *Journal of Nematology* 26, 336–344.

Lauritis, J.A., Rebois, R.V., and Graney, I.S. (1983) Development of *Heterodera glycines* Ichinohe on soybean, *Glycine max* (L.) Merr., under gnotobiotic conditions. *Journal of Nematology* 15, 272–281.

Lehman, P.S., Huisingh, D. and Barker, K.R. (1971) The influence of races of *Heterodera glycines* on nodulation and nitrogen-fixing capacity of soybean. *Phytopathology* 61, 1239–1244. DOI: 10.1094/phyto-61-1239

Li, Y., Fester, T. and Taylor, C.G. (2009) Transcriptomic analysis of nematode infestation. In: Berg, R.H., Taylor, C.G. (eds) *Cell Biology of Plant Nematode Parasitism*. Plant Cell Monographs 15. Springer, Heidelberg, Germany, pp. 189–220. DOI: 10.1007/7089_2008_36

Long, S.R. (1989) Rhizobium–legume nodulation: life together in the underground. *Cell* 56, 203–214. DOI: 10.1016/0092-8674(89)90893-3

Lopez-Nicora, H., Dorrance, A. and Niblack, T. (2013) Evaluation of soybean fields infested with *Heterodera glycines* and *Macrophomina phaseolina* in southern Ohio. *Journal of Nematology* 45, 302 (Abstr.).

Lopez-Nicora, H.D., Diers, B.W., Dorrance, A.E. and Niblack, T.L. (2014) *Macrophomina phaseolina* and *Heterodera glycines* reducing soybean performance through early infection. *Phytopathology* 104, S3.72 (Abstr.).

Lorang, J.M., Tuori, R.P., Martinez, J.P. *et al.* (2001) Green fluorescent protein is lighting up fungal biology. *Applied and Environmental Microbiology* 67, 1987–1994. DOI: 10.1128/AEM.67.5.1987-1994.2001

Malakeberhan, H. and Dey, J. (2003) Competition between *Heterodera glycines* and *Meloidogyne incognita* or *Pratylenchus penetrans*: independent infection rate measurements. *Journal of Nematology* 35, 1–6.

Mantelin, S., Thorpe, P. and Jones, J.T. (2015) Suppression of plant defenses by plant-parasitic nematodes. *Advances in Botanical Research* 73, 325–337. DOI: 10.1016/bs.abr.2014.12.011

Marburger, D., Conley, S., Esker, P., MacGuidwin, A. and Smith, D. (2014) Relationship between *Fusarium virguliforme* and *Heterodera glycines* in commercial soybean fields in Wisconsin. *Plant Health Progress.* DOI: 10.1094/PHP-RS-13-0107.

Mathesius, U., Schlaman, H.R., Spaink, H.P., Of Sautter, C., Rolfe, B.G. and Djordjevic, M.A. (1998) Auxin transport inhibition precedes root nodule formation in white clover roots and is regulated by flavonoids and derivatives of chitin oligosaccharides. *The Plant Journal* 14, 23–34. DOI: 10.1046/j.1365-313X.1998.00090.x

McCarville, M.T., O'Neal, M., Tylka, G.L., Kanobe, C. and Macintosh, G.C. (2012) A nematode, fungus, and aphid interact via a shared host plant: implications for soybean management. *Entomologia Experimentalis et Applicata* 143, 55–66. DOI: 10.1111/j.1570-7458.2012.01227.x

McCarville, M.T., Soh, D.H., Tylka, G.L. and O'Neal, M.E. (2014) Aboveground feeding by soybean aphid, *Aphis glycines*, affects soybean cyst nematode, *Heterodera glycines*, reproduction belowground. *PLoS ONE* 9(1), e86415. DOI: 10.1371/journal.pone.0086415

McLean, K.S. and Lawrence, G.W. (1993a) Interrelationship of *Heterodera glycines* and *Fusarium solani* in sudden death syndrome of soybean. *Journal of Nematology* 25, 434–439.

McLean, K.S. and Lawrence, G.W. (1993b) Localized Influence of *Heterodera glycines* on sudden death syndrome of soybean. *Journal of Nematology* 25, 674–678.

McLean, K.S. and Lawrence, G.W. (1995) Development of *Heterodera glycines* as affected by *Fusarium solani*, the causal agent of sudden death syndrome of soybean. *Journal of Nematology* 27, 70–77.

Meagher, J.W. and Chambers, S.C. (1971) Pathogenic effects of *Heterodera avenae* and *Rhizoctonia solani* and their interaction on wheat. *Australian Journal of Agricultural Research* 22, 189–194. DOI: 10.1071/ar9710189

Meagher, J.W., Brown, R.H. and Rovira, A.D. (1978) The effects of cereal cyst nematode (*Heterodera avenae*) and *Rhizoctonia solani* on the growth and yield of wheat. *Australian Journal of Agricultural Research* 29, 1127–1137. DOI: 10.1071/ar9781127

Melgar, J., Roy, K.W. and Abney, T.S. (1994) Sudden death syndrome of soybean: etiology, symptomatology, and effects of irrigation and *Heterodera glycines* on incidence and severity under field conditions. *Canadian Journal of Botany* 72, 1647–1653. DOI: 10.1139/b94-202

Miller, P.M. (1970) Rate of increase of a low population of *Heterodera tabacum* reduced by *Pratylenchus penetrans* in the soil. *Plant Disease Reporter* 54, 25–26.

Minton, N.A., Parker, M.B. and Sumner, D.R. (1985) Nematode control related to Fusarium wilt in soybean and root rot and zinc deficiency in corn. *Journal of Nematology* 17, 314–321.

Neher, D.A. (2010) Ecology of plant and free-living nematodes in natural and agricultural soil. *Annual Review of Phytopathology* 48, 371–394. DOI: 10.1146/annurev-phyto-073009-114439

Niblack, T.L. (1989) Application of nematode community structure research to agricultural production and habitat disturbance. *Journal of Nematology* 21, 437–443.

Niblack, T.L. (2005) Soybean cyst nematode management reconsidered. *Plant Disease* 89, 1020–1026. DOI: 10.1094/PD-89-1020

Niblack, T.L., Baker, N.K. and Norton, D.C. (1992) Soybean yield losses due to *Heterodera glycines* in Iowa. *Plant Disease* 76, 943–948. DOI: 10.1094/pd-76-0943

Niblack, T.L., Lambert, K.N. and Tylka, G.L. (2006) A model plant pathogen from the kingdom Animalia: *Heterodera glycines*, the soybean cyst nematode. *Annual Review of Phytopathology* 44, 283–303. DOI: 10.1146/annurev.phyto.43.040204.140218

Nicol, J.M., Turner, S.J., Coyne, D.L., den Nijs, L., Hockland, S. and Maafi, Z.T. (2011) Current nematode threats to world agriculture. In: Jones, J., Gheysen, G. and Fenoll, C. (eds) *Genomics and Molecular Genetics of Plant–Nematode Interactions*. Springer, London, pp. 21–43. DOI: 10.1007/978-94-007-0434-3_2

Nordmeyer, D. and Sikora, R.A. (1983a) Studies on the interaction between *Heterodera daverti*, *Fusarium avenaceum* and *F. oxysporum* on *Trifolium subterraneum*. *Revue de Nématologie* 6, 193–198.

Nordmeyer, D. and Sikora, R.A. (1983b) Effect of a culture filtrate from *Fusarium avenaceum* on the penetration of *Heterodera daverti* into roots of *Trifolium subterraneum*. *Nematologica* 29, 88–94. DOI: 10.1163/187529283X00212

Overstreet, C. and McGawley, E.C. (1988) Influence of *Calonectria crotalariae* on reproduction of *Heterodera glycines* on soybean. *Journal of Nematology* 20, 457–467.

Overstreet, C., McGawley, E.C. and Russin, J.S. (1990) Interactions between *Calonectria* crotolariae and *Heterodera* gycines on soybean. *Journal of Nematology* 22, 496–505.

Pacios-Bras, C., Schlaman, H.R., Boot, K., Admiraal, P., Langerak, J.M., Stougaard, J. and Spaink, H.P. (2003) Auxin distribution in *Lotus japonicus* during root nodule development. *Plant Molecular Biology* 52, 1169–1180. DOI: 10.1023/B:PLAN.0000004308.78057.f5

Pacumbaba, R.P. (1992) Effects of induced epidemics of *Heterodera glycines*, *Diaporthe phaseolorum* var. *caulivora*, and *Pseudomonas syringae* pv. *glycinea* in single inoculations or in combinations on soybean yield and other agronomic characters. *Journal of Agronomy and Crop Science* 169, 176–183. DOI: 10.1111/j.1439-037X.1992.tb01024.x

Parmeter, J.R., Jr, Sherwood, R.T. and Platt, W.D. (1969) Anastomosis grouping among isolates of *Thanatephorus cucumeris*. *Phytopathology* 59, 1270–1278.

Pieterse, C.M., Van der Does, D., Zamioudis, C., Leon-Reyes, A. and Van Wees, S.C. (2012) Hormonal modulation of plant immunity. *Annual Review of Cell and Developmental Biology* 28, 489–521. DOI: 10.1146/annurev-cellbio-092910-154055

Piggott, J.J., Townsend, C.R. and Matthaei, C.D. (2015) Reconceptualizing synergism and antagonism among multiple stressors. *Ecology and Evolution* 5, 1538–1547. DOI: 10.1002/ece3.1465

Polychronopoulos, A.G., Houston, B.R. and Lownsbery, B.F. (1969) Penetration and development of *Rhizoctonia solani* in sugar beet seedlings infected with *Heterodera schachtii*. *Phytopathology* 59, 482–485.

Poromarto, S.H. and Nelson, B.D. (2010) Evaluation of northern-grown crops as hosts of soybean cyst nematode. *Plant Health Progress*. DOI: 10.1094/PHP-2010-0315-02-RS

Powell, N.T. (1971) Interactions between nematodes and fungi in disease complexes. *Annual Review of Phytopathology* 9, 253–274. DOI: 10.1146/annurev.py.09.090171.001345

Prior, A., Jones, J.T., Beauchamp, J., McDermott, L., Cooper, A. and Kennedy, M.W. (2001) A surface-associated retinol-and fatty acid-binding protein (Gp-FAR-1) from the potato cyst nematode *Globodera pallida*: lipid binding activities, structural analysis and expression pattern. *Biochemical Journal* 356, 387–394. DOI: 10.1042/bj3560387

Rasmann, S., Ali, J.G., Helder, J. and van der Putten, W.H. (2012) Ecology and evolution of soil nematode chemotaxis. *Journal of Chemical Ecology* 38, 615–628. DOI: 10.1007/s10886-012-0118-6

Riggs, R.D. and Schuster, R.P. (1998) Management. In: Sharma, S.B. (ed.) *The Cyst Nematodes*. Chapman & Hall, London, pp. 388–416. DOI: 10.1007/978-94-015-9018-1_16

Rivoal, R. and Nicol, J.M. (2009) Past research on the cereal cyst nematode complex and future needs. In: Riley, I.T., Nicol, J.M. and Dababat, A. (eds) *Proceedings of the First Workshop of the International Cereal Cyst Nematode Initiative*. CIMMYT, Antalya, Turkey, pp. 3–10.

Robert-Seilaniantz, A., Grant, M. and Jones, J.D. (2011) Hormone crosstalk in plant disease and defense: more than just jasmonate-salicylate antagonism. *Annual Review of Phytopathology* 49, 317–343. DOI: 10.1146/annurev-phyto-073009-114447

Ross, J.P. (1959) Interaction of *Meloidogyne incognita incognita* and *Heterodera glycines* on soybeans. *Phytopathology* 49, 549 (Abstr.).

Ross, J.P. (1964) Interaction of *Heteroda glycines* and *Meloidogyne incognita* on soybeans. *Phytopathology* 54, 304–307.

Ross, J.P. (1965) Predisposition of soybeans to Fusarium wilt by *Heteroda glycines* and *Meloidogyne incognita*. *Phytopathology* 55, 361–364.

Roy, K.W. (1997) *Fusarium solani* on soybean roots: nomenclature of the causal agent of sudden death syndrome and identity and relevance of *F. solani* form B. *Plant Disease* 81, 259–266. DOI: 10.1094/PDIS.1997.81.3.259

Roy, K.W., Lawrence, G.W., Hodges, H.H., McLean, K.S. and Killebrew, J.F. (1989) Sudden death syndrome of soybean: *Fusarium solani* as incitant and relation of *Heterodera glycines* to disease severity. *Phytopathology* 79, 191–197. DOI: 10.1094/phyto-79-191

Rupe, J.C., Robbins, R.T. and Gbur Jr, E.E. (1997) Effect of crop rotation on soil population densities of *Fusarium solani* and *Heterodera glycines* and on the development of sudden death syndrome of soybean. *Crop Protection* 16, 575–580. DOI: 10.1016/S0261-2194(97)00031-8

Rupe, J.C., Robbins, R.T., Becton, C.M., Sabbe, W.A. and Gbur Jr, E.E. (1999) Vertical and temporal distribution of *Fusarium solani* and *Heterodera glycines* in fields with sudden death syndrome of soybean. *Soil Biology and Biochemistry* 31, 245–251. DOI: 10.1016/S0038-0717(98)00108-4

Russin, J.S., Layton, M.B., Boethel, D.J., McGawley, E.C., Snow, J.P. and Berggren, G.T. (1989) Development of *Heterodera glycines* on soybean damaged by soybean looper and stem canker. *Journal of Nematology* 21, 108–114.

Ryan, A. and Jones, P. (2004) The effect of mycorrhization of potato roots on the hatching chemicals active towards the potato cyst nematodes, *Globodera pallida* and *G. rostochiensis*. *Nematology* 6, 335–342. DOI: 10.1163/1568541042360456

Ryan, N.A., Duffy, E.M., Cassells, A.C. and Jones, P.W. (2000) The effect of mycorrhizal fungi on the hatch of potato cyst nematodes. *Applied Soil Ecology* 15, 233–240. DOI: 10.1016/S0929-1393(00)00099-8

Ryan, N.A., Deliopoulos, T., Jones, P. and Haydock, P.P.J. (2003) Effects of a mixed-isolate mycorrhizal inoculum on the potato-potato cyst nematode interaction. *Annals of Applied Biology* 143, 111–119. DOI: 10.1111/j.1744-7348.2003.tb00275.x

Sharma, S.B. and Nene, Y.L. (1989) Interrelationship between *Heterodera cajani* and *Fusarium udum* in pigeonpea. *Nematropica* 19, 21–28.

Sharma, S.B., Rego, T.J., Mohiuddin, M. and Nageswara Rao, V. (1996) Regulation of densities of *Heterodera cajani* and other plant parasitic nematodes in semi-arid tropical production systems. *Journal of Nematology* 28, 244–251.

Sherwood, R.T. (1969) Morphology and physiology in four anastomosis groups of *Thanatephorus cucumeris*. *Phytopathology* 59, 1924–1929.

Sikora, R.A. (1977) *Heterodera trifolii* associated with Fusarium root rot of *Trifolium subterraneum* in northern Tunisia. *Nematologia Mediterranea* 5, 319–321.

Sikora, R.A. and Carter, W.W. (1987) Nematode interactions with fungal and bacterial plant pathogens – fact or fantasy. In: Veech, J.A. and Dickenson, D.W. (eds) *Vistas on Nematology*. Society of Nematologists, Hyattsville, Maryland, pp. 307–312.

Sikora, R.A., Schafer, K. and Dababat, A.A. (2007) Modes of action associated with microbially induced *in planta* suppression of plant-parasitic nematodes. *Australasian Plant Pathology* 36, 124–134. DOI: 10.1071/AP07008

Slinker, B.K. (1998) The statistics of synergism. *Journal of Molecular and Cellular Cardiology* 30, 723–731. DOI: 10.1006/jmcc.1998.0655

Smant, G. and Jones, J. (2011) Suppression of plant defences by nematodes. In: Jones, J., Gheysen, G. and Fenoll, C. (eds) *Genomics and Molecular Genetics of Plant-Nematode Interactions*. Springer, Dordrecht, The Netherlands, pp. 273–286. DOI: 10.1007/978-94-007-0434-3_13

Smith, G.S. (1987) Interactions of nematodes with mycorrhizal fungi. In: Veech, A. and Dickson, D.W. (eds) *Vistas on Nematology*. Society of Nematologists, Hyattsville, Maryland, pp. 292–300.

Smith, J.L., De Moraes, C.M. and Mescher, M.C. (2009) Jasmonate-and salicylate-mediated plant defense responses to insect herbivores, pathogens and parasitic plants. *Pest Management Science*, 65, 497–503. DOI: 10.1002/ps.1714

Steele, A.E. (1965) The host range of the Sugar Beet Nematode, *Heterodera schachtii* Schmidt. *Journal of the American Society of Sugar Beet Technologists* 13, 573–603. DOI: 10.5274/jsbr.13.7.573

Stetina, S.R., Russin, J.S. and McGawley, E.C. (1997) Replacement series: a tool for characterizing competition between phytoparasitic nematodes. *Journal of Nematology* 29, 35–42.

Stiles, C.M., Glawe, D.A., Noel, G.R. and Pataky, J.K. (1993) Reproduction of *Heterodera glycines* on soybean in nonsterile soil infested with cyst-colonizing fungi. *Nematropica* 23, 81–89.

Storey, G.W. and Evans, K. (1987) Interactions between *Globodera pallida* juveniles, *Verticillium dahliae* and three potato cultivars, with descriptions of associated histopathologies. *Plant Pathology* 36, 192–200. DOI: 10.1111/j.1365-3059.1987.tb02221.x

Subbotin, S.A., Mundo-Ocampo, M. and Baldwin, G.B. (2010) Systematics of cyst nematodes (Nematoda: Heteroderinae) *Nematology Monographs and Perspectives 8A*. (Series editors: Hunt, D.J. and Perry, R.N.). Brill, Leiden, The Netherlands, pp. 107–177. DOI: 10.1163/ej.9789004162259.i-352

Sugawara, K., Kobayashi, K. and Ogoshi, A. (1997) Influence of the soybean cyst nematode (*Heterodera glycines*) on the incidence of brown stem rot in soybean and adzuki bean. *Soil Biology and Biochemistry* 29, 1491–1498. DOI: 10.1016/S0038-0717(97)00033-3

Tabor, G.M., Tylka, G.L., Behm, J.E. and Bronson, C.R. (2003) *Heterodera glycines* infection increases incidence and severity of brown stem rot in both resistant and susceptible soybean. *Plant Disease* 87, 655–661. DOI: 10.1094/PDIS.2003.87.6.655

Tabor, G.M., Tylka, G.L. and Bronson, C.R. (2006) Soybean stem colonization by genotypes A and B of *Cadophora gregata* increases with increasing population densities of *Heterodera glycines*. *Plant Disease* 90, 1297–1301. DOI: 10.1094/PD-90-1297

Taha, A.H.Y. and Raski, D.J. (1969) Interrelationships between root-nodule bacteria, plant-parasitic nematodes and their leguminous host. *Journal of Nematology* 1, 201–211.

Taylor, C.E. (1990) Nematode interactions with other pathogens. *Annals of Applied Biology* 116, 405–416. DOI: 10.1111/j.1744-7348.1990.tb06622.x

Todd, T.C. (1993) Soybean planting date and maturity effects on *Heterodera glycines* and *Macrophomina phaseolina* in southeastern Kansas. *Supplement to the Journal of Nematology* 25, 731–737.

Todd, T.C., Pearson, C.A.S. and Schwenk, F.W. (1987) Effect of Heterodera glycines on charcoal rot severity in soybean cultivars resistant and susceptible to soybean cyst nematode. *Annals of Applied Nematology* 1, 35–40.

Todd, T.C., Winkler, H.E. and Wilson, G.W.T. (2001) Interaction of *Heterodera glycines* and *Glomus mosseae* on soybean. *Journal of Nematology* 33, 306–310.

Tsror, L. (2010) Biology, epidemiology and management of *Rhizoctonia solani* on potato. *Journal of Phytopathology* 158, 649–658. DOI: 10.1111/j.1439-0434.2010.01671.x

Tubajika, K.M., Tylka, G.L., Tachibana, H. and Yang, X.B. (1994) Incidence of brown stem rot as influenced by soybean cyst nematode. *Phytopathology* 84, 1101 (Abstr.). DOI: 10.1094/Phyto-84-1061

Turner, S.J. and Subbotin, S.A. (2013) Cyst nematodes. In: Perry, R. and Moens, M. (eds) *Plant Nematology*, 2nd edn. CAB International, Wallingford, UK, pp. 109–143. DOI: 10.1079/9781780641515.0109

Tylka, G.L., Hussey, R.S. and Roncadori, R.W. (1991) Interactions of vesicular-arbuscular mycorrhizal fungi, phosphorus, and *Heterodera glycines* on soybean. *Journal of Nematology* 23, 122–133.

Uehara, T., Sugiyama, S., Matsuura, H., Arie, T. and Masuta, C. (2010) Resistant and susceptible responses in tomato to cyst nematode are differentially regulated by salicylic acid. *Plant and Cell Physiology* 51, 1524–1536. DOI: 10.1093/pcp/pcq109

Wallace, H.R. (1978) The diagnosis of plant diseases of complex etiology. *Annual Review of Phytopathology* 16, 379–402. DOI: 10.1146/annurev.py.16.090178.002115

Wallace, H.R. (1983) Interactions between nematodes and other factors on plants. *Journal of Nematology* 15, 221–227.

Wallace, H.R. (1989) Environment and plant health: a nematological perception. *Annual Review of Phytopathology* 27, 59–75. DOI: 10.1146/annurev.phyto.27.1.59

Waller, L.A. and Gotway, C.A. (2004) *Applied Spatial Statistics for Public Health Data*. John Wiley & Sons, Inc., Hoboken, New Jersey. DOI: 10.1002/0471662682

Wang, J., Niblack, T.L., Tremain, J.A., Wiebold, W.J., Tylka, G.L., Marrett, C.C., Noel, G.R., Myers, O. and Schmidt, M.E. (2003) Soybean cyst nematode reduces soybean yield without causing obvious aboveground symptoms. *Plant Disease* 87, 623–628. DOI: 10.1094/PDIS.2003.87.6.623

Westphal, A., Li, C., Xing, L., McKay, A. and Malvick, D. (2014) Contributions of *Fusarium virguliforme* and *Heterodera glycines* to the disease complex of sudden death syndrome of soybean. *PLoS ONE* 9(6), e99529. DOI: 10.1371/journal.pone.0099529

Whitney, E.D. (1974) Synergistic effect of *Pythium ultimum* and the additive effect of *P. aphanidermatum* with *Heterodera schachtii* on sugarbeet. *Phytopathology* 64, 380–383. DOI: 10.1094/phyto-64-380

Whitney, E.D. and Doney, D.L. (1973) The effects of *Heterodera schachtii* and *Aphanomyces cochlioides* on root-rot of sugarbeet. *Journal of the American Society of Sugar Beet Technologists* 17, 240–245. DOI: 10.5274/jsbr.17.3.240

Wiesel, L., Newton, A.C., Elliott, I., Booty, D., Gilroy, E.M., Birch, P.R. and Hein, I. (2014) Molecular effects of resistance elicitors from biological origin and their potential for crop protection. *Frontiers in Plant Science* 5, 655. DOI: 10.3389/fpls.2014.00655

Windels, C.E. (2000) Aphanomyces root rot of sugar beet. *Plant Health Progress*. DOI: 10.1094/PHP-2000-0720-01-DG

Winkler, H.E., Hetrick, B.A.D. and Todd, T.C. (1994) Interactions of *Heterodera glycines*, *Macrophomina phaseolina*, and mycorrhizal fungi on soybean in Kansas. *Journal of Nematology* 26, 675–682.

Wrather, J.A., Anderson, T.R., Arsyad, D.M., Tan, Y., Ploper, L.D., Porta-Puglia, A., Ram, H.H. and Yorinori, J.T. (2001) Soybean disease loss estimates for the top ten soybean-producing countries in 1998. *Canadian Journal of Plant Pathology* 23, 115–121. DOI: 10.1080/07060660109506918

Xing, L. and Westphal, A. (2006) Interaction of *Fusarium solani* f. sp. glycines and *Heterodera glycines* in sudden death syndrome of soybean. *Phytopathology* 96, 763–770. DOI: 10.1094/PHYTO-96-0763

Xu, X.-M., Jeffries, P., Pautasso, M. and Jeger, M.J. (2011) Combined use of biocontrol agents to manage plant diseases in theory and practice. *Phytopathology* 101, 1024–1031. DOI: 10.1094/PHYTO-08-10-0216

Yeates, G.W., Ross, D.J., Bridger, B.A. and Visser, T.A. (1977) Influence of nematodes *Heterodera trifolii* and *Meloidogyne hapla* on nitrogen fixation by white clover under glasshouse conditions. *New Zealand Journal of Agricultural Research* 20, 401–413. DOI: 10.1080/00288233.1977.10427352

13 Field Management and Control Strategies

Matthew A. Back[1], Laura Cortada[2], Ivan G. Grove[1] and Victoria Taylor[3]

[1]Harper Adams University, Newport, UK; [2]International Institute of Tropical Agriculture, Nairobi, Kenya; [3]Arcis Biotechnology, Daresbury Innovation Centre, Sci-Tech Daresbury, Darebury, UK

13.1 Introduction	305
13.2 Cultural Control Methods for Cyst Nematodes	306
13.3 Biofumigation	311
13.4 Trap Cropping and Natural Resistance	313
13.5 Use of Plant Biomass, Oils and Extracts	314
13.6 Application of Biological Control Agents	315
13.7 Agrochemical Control of Cyst Nematodes	320
13.8 Conclusions and Future Prospects	327
13.9 References	328

13.1 Introduction

Globally, cyst nematode species are recognized for causing serious damage in a broad diversity of crops including, but not exclusively, soybean, potatoes, cereals, carrots, beet, peas, rice, brassicas, grapes and tobacco. In a review by Jones *et al.* (2013), cyst nematodes (*Globodera* and *Heterodera* spp.) were rated as the second most important group of plant-parasitic nematodes based on scientific and economic importance. Field management of cyst nematodes needs to be multifaceted and knowledge-based. Applications of granular nematicides are helpful in reducing the impact from invading juveniles soon after planting, but have limited effect on reducing populations in the longer term. Soil fumigation can be effective if undertaken in optimal conditions, but is often avoided due to being economically prohibitive and damaging to the soil environment. Long-term management requires crop managers to understand the host ranges and decline rates of cyst nematodes to plan rotations effectively. Similarly, the impact of selecting appropriate soil cultivations and weed management must be recognized.

The use of chemical compounds is important to maintain the productivity of agriculture around the world. The use of chemicals to control nematodes was first reported in 1881 when carbon disulfide was used against *H. schachtii* in sugar beet (Rich *et al.*, 2004). During the decades that followed, methyl bromide (MeBr) has been the most effective and widely used soil fumigant worldwide, thanks to its broad biocidal activity and effective action to control fungi, bacteria, insects, nematodes and weeds. However, due to its harmful effects on the environment (Watson *et al.*, 1992), a worldwide ban was imposed on MeBr for use in agriculture by the Montreal Protocol, with developed countries committing themselves to reduce consumption of MeBr until its complete phase out in 2005, except for critical use exemptions, while developing countries agreed to reduce consumption by 20% in 2005 and 100% in 2015. This, together with the increasing awareness of producers and consumers about the risks that chemical pesticides represent for human health, has stimulated a search for new sustainable and harmless methods to control pests, weeds and pathogens.

Resistant cultivars are available for a variety of cyst nematodes (see Chapters 8 and 9, this volume) and biological control may be a management option (see section 13.6 and Chapter 11, this volume). The use of integrated pest management (IPM) strategies has become crucial to achieve sustainable solutions to control plant-parasitic nematodes in general. This chapter considers field application of various control and management strategies that may be components of a broad IPM system. A summary of the available management strategies for species of cyst nematode is given in Table 13.1.

13.2 Cultural Control Methods for Cyst Nematodes

There are several methods used to control or manage the population densities of cyst nematodes that do not rely on the addition of chemicals or biological amendments. These cultural, or physical control methods are often used in an IPM system alongside chemical and/or biological nematicides. The methods outlined in this section are able to control more than one disease or pest, especially when used in combination with other control methods. Cultural practices such as rotation, the use of cover crops and incorporation of green and animal manures and other organic amendments provide benefits beyond pest control. These practices can enhance the soil structure and biodiversity, which are vital to plant health and optimal yield.

13.2.1 Crop rotation

Crop rotation is the most widely practised method of disease control in agriculture. By growing non-host, resistant, tolerant or no crops in between seasons of susceptible crops, the population of the pest falls below the economic damage threshold and low populations are prevented from increasing to damaging levels. The length of rotation and crops used are dependent on many factors, including nematode species and population density, host range, soil type, soil moisture, the economic damage threshold of susceptible crops, climate, available equipment, markets for specialized crops, other pathogens and pests present, and stipulations in contracts when renting land.

Potato cyst nematodes (PCN) (*Globodera pallida* and *G. rostochiensis*) have a very narrow host range, which makes rotation an ideal form of management. However, they are able to persist in the soil for up to 30 years and are dependent on hatching factors released from their host plants (see Chapter 3, this volume), with a spontaneous hatch rate of up to 26% per annum for *G. pallida* (Trudgill, 2014). When using non-resistant cultivars, Trudgill (2014) recommends rotations with non-hosts or resistant potato cultivars of up to 8 years to maintain a population of below 10 egg g^{-1} soil, whilst typical rotations for potatoes in Europe are 5–7 years.

Soybean cyst nematodes (SCN) (*Heterodera glycines*) also hatch more readily when exposed to hatching factors released from host plants (see Chapter 3, this volume, for information about hatching of cyst nematodes). They are able to persist in the soil for up to 11 years (Inagaki and Tsutsumi, 1971). Cultivars are available with resistance from three different sources (see Chapter 9, this volume), which can be combined in the rotation with non-host crops. Maize (*Zea mays*) is widely used in soybean rotations. Sasser and Uzzell (1991) found that susceptible soybean crops planted after 3 years in maize produced 1258 kg ha^{-1}, whereas the yield from continuous susceptible soybean dropped to 130 kg ha^{-1} in year 4. Two-year rotations evaluated by Kelley *et al.* (2003) produced a 13–24% increase in yield over a monoculture system, and over a 20-year study rotations with high residue-producing crops, such as wheat and grain sorghum, produced significant increases in total soil C + N (Kelley *et al.*, 2003). Although the population density of SCN is reduced after a 1-year break in the rotation, Miller *et al.* (2006) found that this limited rotation was not sufficient for increasing yield, and the population quickly rose back to an economically damaging level.

Table 13.1. A summary of options for the field management of major cyst nematode species.

Management strategy	Beet cyst nematode *Heterodera schachtii*	Cereal cyst nematode *Heterodera avenae* and *H. filipjevi*	Carrot cyst nematode *Heterodera carotae*	Pea cyst nematode *Heterodera goettingiana*	Potato cyst nematode *Globodera pallida* and *G. rostochiensis*	Soybean cyst nematode *Heterodera glycines*	Rice cyst nematode *Heterodera oryzicola* and *H. elachista*
Rotation – susceptible hosts	Chenopodiaceae (e.g. sugar beet) Brassicaceae (e.g. oilseed rape)	Poaceae (wheat, barley, oats, rye, maize, grasses)	Carrot	Fabaceae, causing most damage on peas	Solanaceae (potato, tomato, aubergine)	Fabaceae (soybean, cowpea, peas, beans)	Rice, plantain
Resistant cultivars	Resistance available	Resistance being developed in some countries (e.g. Australia)	None available	Soybean resistance reported in China	Partial resistance for *G. pallida* and complete resistance for *G. rostochiensis*	Resistance widely available	Resistance available
Biological control	*Pasteuria nishizawae* (seed treatment) Egg parasites such as *Paecilomyces lilacinus*	Egg parasites such as *Paecilomyces lilacinus*	Little or no evidence	*Pasteuria* spp.	*Pasteuria penetrans* Egg parasites such as *Pochonia chlamydosporia*	*Pasteuria nishizawae Myrothecium verrucaria* (egg parasite)	Little or no evidence
Biofumigation	Trap cropping has received greater attention	Little or no evidence	Little or no evidence	Little or no evidence	Some conflicting studies but generally well supported by the literature	Some evidence	Little or no evidence
Trap crops	*Raphanus sativus* (oilseed radish) and *Sinapis alba* (white mustard)	None reported	None reported	None reported	*Solanum sisymbriifolium* (sticky nightshade) *Solanum tuberosum* (resistant potatoes)	None reported	None reported
Chemical control options (non-volatile)	Carbamates	Seldom used	None available	None available	Organophosphates Oxime carbamates	Seed treatments such as clothianidin and abamectin	None available
Chemical control options (volatile)	Metam sodium 1,3-dichloropropene	Seldom used	None available	None available	Metam sodium Dazomet Methyl bromide 1,3-dichloropropene	Seldom used	None available

Beet cyst nematodes (BCN) (*H. schachtii*) have a broad host range including many brassica crops such as oilseed rape, turnip and cabbage. This reduces the number of available crops for rotation, which is typically 3–7 years (Flint and Roberts, 1988). The annual natural decline of *H. schachtii* is typically 50%, as no hatching factors are required from host plants. Resistant varieties of oil radish and mustards are frequently included in rotations to aid decline (see section 13.4).

Population densities of the cereal cyst nematodes (CCN), *H. avenae* and *H. filipjevi*, are also reduced by diverse rotations, which include a range of plant families, since the host range is limited to a range of cereal crops (wheat, barley, oats, rye and triticale) and grass species. Two consecutive seasons of a non-host crop is recommended for the effective management of CCN in the southern wheatbelt of Australia, as approximately 50–90% of second-stage juveniles (J2) hatch in each season (Vanstone *et al.*, 2008). Legume pastures have been used in rotations for control of CCN since the 1950s, while also providing fodder for livestock and contributing N to the soil (Meagher and Rooney, 1966). Suitable crops for rotation include oilseed rape (*Brassica napus*), chickpea (*Cicer arietinum*), field pea (*Pisum sativum*), lentil (*Lens culinaris*) and faba bean (*Vicia faba*) (Vanstone *et al.*, 2008). In Qinghai, China, rotations commonly include 1 year in a susceptible crop followed by 1 year in a non-host crop such as potatoes (*Solanum tuberosum*), faba beans or oilseed rape (Riley *et al.*, 2010). Conversely, rotations are more limited in Iraq where wheat and rice are alternated in the southern region, or wheat is followed by a short-term legume crop (2 months) and then maize in the middle of the country (G. Mahmood, 2017, pers. comm.). In the main wheat-growing regions of Saudi Arabia (Al-Kharj, Hail and Al-Qassim), *H. avaenae* can be problematic on the limited 'wheat–potato' rotations causing yield loss between 40 and 92% in susceptible wheat crops (Al-Hazmi and Dawabah, 2009). The same researchers report that CCN are being spread to uninfested fields by soil adhering to seed potatoes.

Pea cyst nematodes (*H. goettingiana*) have a broad host range that includes a number of leguminous crops, including faba bean and field pea (Tedford and Inglis, 1999). Non-host legumes include the soybean *Glycine max*, chick pea, alfalfa (*Medicago sativa*), lentil and sweet vetch (*Hedysarum coronarium*) (Tedford and Inglis, 1999). Natural annual decline has been reported to be 25–50% (Winslow, 1955; Brown, 1958; Moriarty, 1963). However, in Italy, where soil temperatures are high and soil moisture low in May to October, populations of *H. goettingiana* declined by 68% in the first year, and 85% after 4 years in fallow, wheat or oats (Di Vito and Greco, 1986). Di Vito and Greco (1986) recommended a rotation of 3–6 years to prevent significant yield loss.

Heterodera cajani is a nematode pest of pigeon pea (*Cajanus cajan*), as well as several other leguminous crops such as cowpea (*Vigna unguiculata*) and mungbean (*Vigna radiata*) (Sharma *et al.*, 1996). Summer fallow (February–May) is an important component of cropping systems in semi-arid India (Sharma *et al.*, 1996), where pigeon peas are grown; however, Sharma and Nene (1992) determined that summer fallow was not effective in reducing the population density of *H. cajani*. A fallow period of 305 days reduced the number of *H. cajani* cysts in vertisol (black cotton soil) by 91% (Sharma and Nene, 1992). Sorghum (*Sorghum bicolor*) and safflower (*Carthamus tinctorius*) are also grown in the same regions as pigeon pea, mung bean and cowpea, and their inclusion in rotation with hosts of *H. cajani* have been shown to reduce the population density by a factor of 8 (Sharma *et al.*, 1996).

The rice cyst nematodes *H. oryzae*, *H. oryzicola*, *H. elachista* and *H. sacchari* have a limited host range, the main hosts being rice (*Oryza sativa*) and banana (*Musa*) (Charles and Venkitesan, 1990). Nishizawa *et al.* (1972) found that cropping of soybeans or sweet potatoes decreased the number of *H. oryzae* cysts in the field; after 3 years in rice, soybean and sweet potato the number of cysts per 100 g of soil was 453.3, 2.3 and 1.7, respectively.

13.2.2 Soil cultivations (tillage)

The choice of tillage system can also affect the population density of cyst nematodes, as well as having an effect on other plant diseases. Roget *et al.* (1996) compared root damage, caused by CCN, in field plots with different tillage practices. Root damage was most severe in treatments that included one cultivation prior to sowing, whilst it was lowest in direct-drilled plots that did not disturb the soil below seed depth, although the incidence of the disease take-all (*Gaeumannomyces graminis* var. *tritici*) was highest in these plots. The population density of *H. elachista* was reduced during continuous cropping of rice (*Oryza sativa*) under a zero tillage system in comparison to conventional tillage system that utilized a mouldboard plough with rotary cultivation to a depth of 25–30 cm combined with a rotary cultivator with a working soil depth of 15 cm. However, soil populations increased when soybean was introduced into the rotation (Ito *et al.*, 2015). It is generally accepted that reduced tillage has other benefits for the soil environment. For example, Carter (1992) demonstrated that minimal tillage could increase soil moisture, reduce soil temperature and improve soil structure when compared with a range of tillage systems, including mouldboard ploughing.

13.2.3 Anaerobic soil disinfection, soil amendments and inundation

Anaerobic soil disinfection (ASD) is a term used to describe the induction of anaerobic conditions in the soil by incorporation of fresh organic matter to wet soil and covering with airtight plastic for several weeks (Blok *et al.*, 2000). The application of ASD causes a reduction in hatching and juvenile viability of plant-parasitic nematodes, including cyst nematodes, linked to the depletion of O_2 and an increase of the CO_2 and secondary toxic byproducts (e.g. short-chain fatty acids) in the soil, caused by the anaerobic decomposition of the organic carbon (Runia *et al.*, 2014; Ebrahimi *et al.*, 2016). Amendments with a higher C/N content have also proven to be effective but may be slower at controlling populations of plant-parasitic nematodes than those based on N, due to their limited capacity to produce secondary toxic compounds/metabolites. Typically, the production of secondary toxic compounds is more suppressive than the antagonistic/parasitic mechanisms achieved by the build-up of microbial populations. However, N-amended soils lose their suppressiveness in a very short period of time. Liquid swine manure, applied on fields where soil suppressiveness had been previously observed, was found to reduce the number of mobile J2 of *H. glycines* 45 days after planting but did not have an influence on egg production by the nematode, whilst rye residues applied as mulching have also been effective in increasing the parasitic ability of the egg parasite *Purpureocillium lilacinum* (= *Paecilomyces lilacinus*) (Timper, 2014).

Inundation involves the induction of anaerobic conditions in the soil by flooding for a prolonged period of time. This leads to various physical and chemical changes in the soil including the depletion of O_2, an increase in CO_2, reduced soil pH and production of organic acids, methane, ammonia and hydrogen sulfide (Inglett *et al.*, 2005; Runia *et al.*, 2012). The efficacy of indundation as a control method depends on a number of factors, including duration of flooding and temperature. Spaull *et al.* (1992) found that 14 weeks inundation at 20°C produced a 98% reduction in viability of PCN, but at 10°C 50 weeks were required for comparable efficacy. Ebrahimi *et al.* (2016) found that the addition of agro-industrial waste products reduced the required inundation time from 8 weeks to 4 weeks, and increased the control of PCN from 72% to 99.9%. Inundation has also been utilized in the control of other pests and pathogens present in the soil, such as *Pratylenchus penetrans* (root lesion nematode), *Meloidogyne hapla* (northern root-knot nematode), *Verticillium dahliae* (Verticillium wilt) (Runia *et al.*, 2012) and *Ralstonia solanacearum* (brown rot of potatoes) (van Overbeek *et al.*, 2014). Unlike *P. penetrans*, which can be controlled by anaerobic conditions (Runia *et al.*, 2012), Spaull *et al.* (1992) found that a reduction of PCN viability was more effective when the soil was flooded. The increased efficacy of inundation over ASD has been attributed to the formation of fatty acids (e.g. Hollis and Rodríguez-Kábana, 1966; López-Robles *et al.*, 2013) and H_2S (Spaull *et al.*, 1992). Runia *et al.* (2014) also found that inactivation occurs faster in soils that are relatively high in organic matter content, indicating that biological activity is essential (Oka *et al.*, 2010). Anaerobic incubation of organic amendments can be combined with induction of sublethal temperatures by solarization, as described by Melero-Vara *et al.* (2012).

13.2.4 Solarization

Solarization is a hydrothermal process that increases soil temperature by placing plastic sheeting over moist soil, and causes a number of physical and chemical changes within the soil system (Gaur and Perry, 1991). The decomposition of organic components during solarization increases the concentration of soluble mineral nutrients, such as ammonium and nitrate-nitrogen. This increase in available nutrients in the soil leads to increased plant health and reduces the dependence on fertilizers (Stapleton *et al.*, 1985; Stapleton, 2000). The broad spectrum lethal effects of solarization affect pests, pathogens and beneficial microorganisms. Soil-borne organisms are more likely to rapidly recolonize a sterilized soil system than plant-parasitic organisms, and this shift in the biological equilibrium provides a healthier environment for plant growth (Stapleton, 2000).

Cyst nematodes vary in their response to temperature. *Heterodera schachtii* and *G. rostochiensis* are both more tolerant to increased temperatures than *H. glycines*. Hatching of *H. schachtii* is inhibited after 4 h at 47.5°C (Steele, 1973), *G. rostochiensis* after 4 h at 50°C (Evans, 1991) and *H. glycines* after just 8 min at 52°C (Endo, 1962). According to D'Addabbo *et al.* (2005), *H. carotae* is less susceptible than *H. schachtii* to heat stress. When cysts of *H. schachtii* were exposed to 24 h at 45°C, a 98.5% loss of viability was observed, but 60 h at the same temperature was required to produce the same result with *H. carotae* (Greco *et al.*, 1998). There are, however, limitations to solarization. The cost and disposal of the plastic is an economical and environmental consideration. Soil temperatures of >40°C are required deep into the soil for a prolonged period, which limits use in countries in the northern hemisphere. A border effect has also been demonstrated by Grinstein *et al.* (1995), whereby the control of *H. avenae* was found to be effective up to 90 cm away from the edge of each plot. This border effect was less pronounced when combined with other control methods such as application of MeBr. Coyne and Plowright (1998) found that at 84 days after solarization, the population of *H. sacchari* had reduced by 80%, although at harvest there was no significant difference to the control. Although the soil temperature was increased from 41°C to 47°C under plastic sheeting, it was most likely that the reduction of *H. sacchari* was due to enhanced hatch, as the hatch is not stimulated by host diffusates. LaMondia and Brodie (1984) conducted two trials in consecutive years, and found that the effects of solarization on *G. rostochiensis* varied. In year 1, complete reduction of viable J2 was achieved at a depth of 5 cm, and cysts taken from 10 and 15 cm were 50% less viable than the control. In the second year, however, there was no significant difference between treatments at any depth. Only at 5 cm in the first year did a lethal time/temperature combination occur (47°C for 1 h). Sharma and Nene (1992) found that solarization reduced the *H. cajani* population density by 93% in pigeon pea, although this effect did not last for more than one season. They also found that during the second season of trials, the effects of solarization were not as great, as agroclimatic factors such as rain during solarization, solar radiation intensity and sunshine hours influence the effect of solarization.

13.2.5 Weed management

A number of weed species have been documented to serve as hosts to cyst nematodes, as well as reducing the efficacy of management techniques such as nematicide usage (Thomas *et al.*, 2005). For example, several winter annual weeds such as purple deadnettle (*Lamium purpureum*) are able to support SCN reproduction in both glasshouse and field conditions (Venkatesh *et al.*, 2004; Creech *et al.*, 2008). Creech *et al.* (2008) investigated the effect of weed control on the population of SCN and found that when the weed density is relatively low (<75 plants m^{-2}), control of weeds does not confer a reduction in SCN. Other studies have found no significant reduction in SCN population, even when winter annual SCN host weeds in non-treated plots reached 245 plants m^{-2} (Mock *et al.*, 2012). The application of herbicides reduced populations of SCN by 37% after spring application and suppressed multiplication after autumn application (while the control population increased by 34%) (Nelson *et al.*, 2006). Nelson *et al.* (2006) also found that in a corn system, weed management by herbicide did not significantly affect the SCN population density, although overseeding with winter rye did reduce the SCN population by 44%. The use of herbicides in weed management can interfere with management techniques used against cyst nematodes, such as nematicides. Certain herbicides have been shown to have nematicidal properties, such as suppression of SCN and PCN by thiocarbamates (Perry and Beane, 1989). As a general rule, it is a good practice to manage volunteer plants associated with specific cyst nematode species as these will support further multiplication in the absence of susceptible crop species.

13.3 Biofumigation

With a world-wide decline in chemical control options for managing cyst nematodes, there has been a rising interest in the use of biofumigants. Biofumigation typically refers to the use of biocidal compounds produced from the residues of freshly macerated *Brassica* species for pest, weed and disease management. *Brassica* spp. are selected on their glucosinolate content and their potential to produce biomass after a short cropping period of ~8–14 weeks. Glucosinolates are secondary metabolites that comprise glucose, sulphur, nitrogen, carbon, sulphate and a variable side chain (R). Each *Brassica* species has a unique glucosinolate profile, although the array of glucosinolates can vary in different plant tissues, particularly between shoots and roots. Glucosinolates are categorized structurally as being aliphatic (open chained), aromatic (contain an aromatic ring) or indole (aromatic and bicyclic) and are stored within the vacuoles of plant cells. Whilst glucosinolates are not directly biocidal, they are converted to toxic compounds when the tissues of *Brassica* spp. are disrupted through mechanical damage or herbivory. Cell damage results in glucosinolates being liberated from the vacuoles and in turn being hydrolysed by the thioglucosidase, myrosinase, which is found in separate cells (myrosin cells) that occur deeper within the tissue. Hydrolysis initially results in the separation of the side chain (R) to produce an unstable aglycone but this is then converted to a variety of volatile compounds including isothiocyanates, thiocyanates, nitriles, epithionitriles and oxazolidines. A summary of the biofumigation process is given in Fig. 13.1. Isothiocyanates are widely regarded as being the most important hydrolysis product due to their higher toxicity (Brown and Morra, 1997). Isothiocyanates cause cell death by interacting with proteins containing cysteines and also by increasing oxidative stress by depleting glutathione (Kawakishi and Kaneko, 1985). In general, the most popular *Brassica* species used for biofumigation are Indian mustard (*Brassica juncea*), oilseed radish (*Raphanus sativus*), Ethiopian mustard (*B. carinata*) and rocket or arugula (*Eruca sativa*).

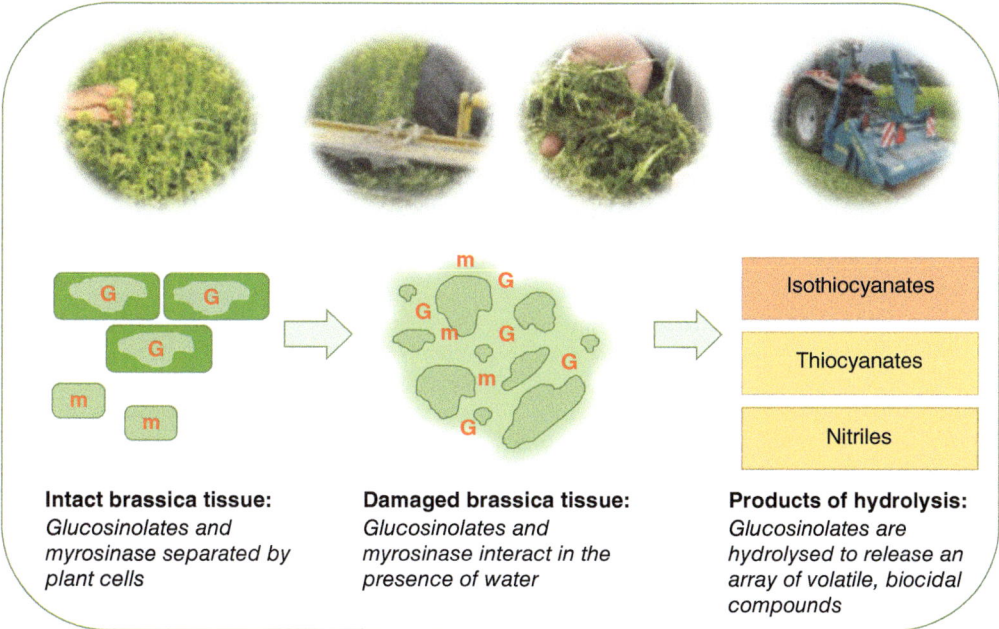

Fig. 13.1. A summary of the biofumigation process. (Courtesy of M.A. Back.)

Each *Brassica* species has a unique profile of glucosinolates in the foliage and the roots. For example, Indian mustard mainly produces 2-propenyl glucosinolate (Sinigrin), which hydrolyses to allyl isothiocyanate.

The process of biofumigation can be achieved in a number of different ways. Traditionally the technique involves growing short-term brassica cover crops during periods of the year where there is longer day length, higher radiation and warmer temperatures. Such conditions are conducive to glucosinolate accumulation and the development of higher quantities of biomass (Björkman *et al.*, 2011). The cover crops are macerated and incorporated soon after flowering to capitalize on peak glucosinolate levels and succulent biomass. Due to the rapid production of isothiocyanates, maceration and incorporation needs to be undertaken in quick succession to ensure that the volatiles are retained for as long as possible. Watts *et al.* (2015) conducted field studies to compare maceration and incorporation implements used for biofumigation and the subsequent effect on PCN population densities. This work highlighted the importance of using a rotary flail. More recently, alternative methods of biofumigation have been used such as the application of preserved and formulated brassica pellets (Lazzeri *et al.*, 2004) and defatted seed meals (DSMs), which can be used in a liquid formulation (De Nicola *et al.*, 2013).

A range of studies have investigated the use of biofumigation for the management of PCN. Research dates back to the first half of the 20th century where a string of investigations highlighted the effectiveness of mustard plants and isothiocyanates at inhibiting PCN hatching and increasing potato yield (Morgan, 1925; Triffit, 1929, 1930; Smedley, 1939; Ellenby, 1945a, b, 1951). Between 1951 and the early 1990s, the number of studies on biofumigation waned, perhaps due to the boom of synthetic pesticides that contributed to the 'Green Revolution'. However, the phase out of MeBr engendered a renewed interest from the mid-1990s onwards. In this second wave, several research groups considered the toxicity of hydrolysis products of glucosinolates to PCN under *in vitro* conditions. For example, Pinto *et al.* (1998) recorded 100% mortality of *G. rostochiensis* J2 when exposed to 1 mg ml^{-1} of 2-propenyl glucosinolate (sinigrin) in the presence of myrosinase. By contrast, the J2 were not killed when exposed to 2-propenyl glucosinolate without myrosinase. Similar findings were also recorded by Buskov *et al.* (2002), who expanded on this work by comparing eight different glucosinolates. In this study 2-phenylethyl (gluconasturtiin) caused 100% mortality of *G. rostochiensis* J2 within 16 h, whereas benzyl (glucotropaeolin) and 2-propenyl caused the same degree of mortality within 24 h and 48 h periods, respectively. Under glasshouse conditions Lord *et al.* (2011) recorded >95% reductions in the viability of unhatched, encysted J2 of *G. pallida* in pots where chopped *B. juncea* 'ISCI' ('Caliente') 99 had been incorporated and sealed with plastic. Similarly, Aires *et al.* (2009) in Portugal observed a significant reduction in the number of newly formed cysts of *G. rostochiensis* on potatoes 'Desiree' when the infested soil used for potting was pre-treated with brassica extracts (kale, broccoli, cauliflower, turnip, cabbage and watercress).

Ngala *et al.* (2014) undertook 2 years of field studies to evaluate the effect of biofumigant cover crops (*B. juncea* 'Caliente 99', *E. sativa* 'Nemat' and *R. sativus* 'Bento') against *G. pallida* to determine: (i) the performance (i.e. biomass production, glucosinolate accumulation and PCN suppression) of these biofumigant species/cultivars in different rotational positions; and (ii) to monitor *G. pallida* population densities pre-sowing, pre-incorporation and post incorporation of biofumigants, and additionally, post planting of potatoes. Generally, most suppression of PCN populations was observed when biofumigants were grown during the months of July to October. Using overwintering crops (i.e. September to April) was much less effective because, as the authors observed, biomass production and glucosinolate production was lower, particularly in *B. juncea*. Other authors have highlighted the importance of UV, day length and temperature in glucosinolate accumulation. Interestingly, the reductions in viable population densities were frequently recorded before biofumigant incorporation. This process is known as partial biofumigation and is due to the leaching of biofumigants during plant growth and their subsequent hydrolysis by myrosinase-producing microorganisms, for example, *Aspergillus* spp. (Sakorn *et al.*, 1999). Ngala *et al.* (2015) conducted further glasshouse experiments that help support this hypothesis. In particular, sinigrin degradation was greater when the soil microbial activity (measured as total microbial enzyme activity) increased.

Isothiocyanates have also been shown to be toxic to other cyst nematode species. For example, Schroeder and MacGuidwin (2010) investigated the *in vitro* sensitivity of quiescent and active *H. glycines* J2 to allyl, phenyl and benzyl isothiocyanate. Whilst all of the compounds increased mortality and decreased motility of *H. glycines* J2, benzyl was shown to be the most toxic.

Biofumigation is a crop protection technique with inherent variability. Care needs to be taken with the selection of species and cultivar. Additionally, crops require nitrogen and sulphur in an optimal ratio, typically 5:1. Climate or seasonal conditions will affect the production of biomass and glucosinolate accumulation. Crop maceration and incorporation needs to be undertaken as rapidly as possible, using a 'one pass' arrangement of implements, if possible. Isothiocyanate production has been shown to peak at 30 min post *Brassica* maceration/incorporation and the majority of ITCs dissipate within a period of 5 days (Gimsing and Kirkegaard, 2006). Soils need to be moist to enable hydrolysis of glucosinolates into secondary biocidal compounds, including ITCs.

13.4 Trap Cropping and Natural Resistance

In nematode management, trap crops are plant species that stimulate hatching and allow root invasion but prevent the completion of the life cycle. For example, development of the syncytium (feeding site) may be restricted, therefore preventing the production of females. Such plants are described as being poor hosts for the nematodes concerned. Trap cropping can also be achieved by using resistant cultivars of the main food crop.

Population densities of *H. schachtii* can be reduced with *Brassica* intercrops, such as oilseed radish (*R. sativus* var. *oleifera*) or white mustard (*S. alba*), which are poor hosts. A field experiment by Hafez (1994) showed 87–92% reductions in *H. schachtii* population densities following intercropping with cultivars of oilseed radish, whereas intercropping with white mustard resulted in reductions between 62 and 84%. However, this strategy can be variable as demonstrated by Kenter *et al.* (2014), who reported significant differences in population densities of *H. schachtii* following oilseed radish in only six out of 13 field experiments conducted in the sugar beet-growing regions of Germany. These differences were attributed to the sowing date of the radish with the best reductions observed with oilseed radish being sown in July as compared to August.

Trap cropping of PCN can be achieved using several approaches including the use of resistant potatoes or the solanaceous species *Solanum sisymbriifolium* (sticky nightshade). The former method can be highly effective as demonstrated by Scholte (2000), who conducted two long-term field-based studies to monitor populations of *G. pallida* in a cropping sequence whereby plots were sown with potato cultivars that were susceptible, moderately resistant or resistant to *G. pallida* before either cropping with resistant potatoes 'Kartel' or leaving fallow in the following year. In both experiments, trap cropping resulted in marked reductions of *G. pallida* populations particularly where the plots were previously cropped with either moderately resistant (92–93% reduction) or susceptible (95–97% reduction) potato cultivars. Turner *et al.* (2006) also found trap cropping to be effective at reducing PCN population densities in their experiments using tolerant and resistant solanaceous clones including wild species, breeder's hybrid lines or commercial potato cultivars with high resistance. In comparison to cropping with a susceptible 'Cara', resistant clones reduced the multiplication rate (P_f/P_i) of *G. pallida* by 62–88% where a granular nematicide was not used. Additionally, the authors investigated the effect of growing potatoes 'Cara' for either 8 or 11 weeks by defoliating the canopy with glyphosate at these times. In this case, the P_f/P_i was reduced by 78% and 89%, respectively, when compared to *G. pallida*-infested land where 'Cara' was grown for a period of 16 weeks. Whilst these studies appear to be encouraging, Ebrahimi *et al.* (2014) assert caution when using trap cropping with potatoes, as both *G. pallida* and *G. rostochiensis* have been seen to complete their life cycles earlier in seasons where the temperature is higher; *G. pallida* has a base temperature of 4°C, requiring 450 degree-days, and *G. rostochiensis* has a base temperature of 6°C, requiring 398 degree-days.

On the other hand, Kaczmarek et al. (2014) observed some variation in the base temperature of *G. pallida*; one population was estimated to have a base temperature of 7°C. Trap cropping is likely to be highly expensive when the cost of seed, fuel, labour, etc., and loss of food/cash crop is taken into account. However, as Turner et al. (2006) acknowledge, the technique does not require nematicides if tolerant cultivars are used, making it appealing for organic growers or where high PCN infestations occur.

Trap cropping for PCN using *S. sisymbriifolium* has been investigated in Europe (Malinowska et al., 2005; Timmermans et al., 2006, 2007a, b; Szymczak-Nowak et al., 2007) and more recently in the US (Dandurand and Knudsen, 2016). *Solanum sisymbriifolium* is a shrubby, annual weed species originating from South Africa. Its species nomenclature relates to the divided leaves produced that are very similar to the genus *Sisymbrium* of the Brassicaceae. This trap crop works in a similar way to resistant potato cultivars in that the plants stimulate hatching of PCN, but the invading J2 are unable to develop a satisfactory feeding site (syncytium). Scholte and Vos (2000) reported differences in the effect of using *S. sisymbriifolium* on *G. pallida* population densities of different ages; a 77% reduction was seen in 2-year-old populations, whilst 1-year-old populations were reduced by 60%. The authors suggest that the effect may have been due to a proportion of the juveniles in the 1-year-old population being in diapause.

Potato growers wishing to use *S. sisymbriifolium* need to follow specific agronomic guidelines to get the best effect from the control strategy. Research conducted in The Netherlands indicates that *S. sisymbriifolium* is best suited to summer cultivation under temperate conditions and should not be sown earlier than the end of April or when the soil temperature is lower than 9°C (Timmermans et al., 2007a). Additionally, seed rates of between 50 and 100 seeds m^{-2} appear to provide the most optimal plant density. In a further publication Timmermans et al. (2007b) calculated that a *S. sisymbriifolium* trap crop would need to produce 660 g m^{-2} of biomass (dry weight) to cause a 75% reduction in PCN population densities based on a 0.5-cm zone of influence around the roots. Under temperate conditions, *S. sisymbriifolium* should be grown for at least 10–14 weeks (Sasaki-Crawley, 2012) to maximize PCN suppression, but care should be taken to avoid crops with berries, which can produce unwanted volunteers. In the UK, it is commercial practice to top *S. sisymbriifolium* crops at the onset of fruiting.

Sasaki-Crawley et al. (2010) found no difference in the attraction of *G. pallida* J2 to root diffusates of *S. sisymbriifolium* and potato in *in vitro* assays but qualitative analysis (ESI-MS) showed that the hatching factor solanoeclepin-A (see Chapter 3, this volume) was found in potato but not the *S. sisymbriifolium* diffusate. Interestingly, the same author recorded that extracts from the foliage of *S. sisymbriifolium* could also induce hatching of *G. pallida* (Sasaki-Crawley, 2012), which suggests that other crop protection applications may be possible.

13.5 Use of Plant Biomass, Oils and Extracts

A wide variety of plant-derived products have been investigated for their potential suppressive effects on cyst nematode populations. Plants receiving specific interest include acacia (*Acacia nilotica*), eucalyptus (*Eucalyptus* spp.), garlic (*Allium sativum*), neem (*Azadirachta indica*), tobacco (*Nicotiana tabacum*) and sweet wormwood (*Artemisia annua*). Unfortunately, the vast majority of studies investigating plant-derived products are based on *in vitro* or controlled environment experiments that, although useful, do not provide a reliable indicator of field performance. A further criticism is the lack of detail on their exact mode of action/activity owing to the particular plant metabolites responsible for the observed inhibition. Finally, if the active substances have been identified, then there is a need to ensure a relatively consistent concentration in formulations sold as crop protection products.

Extracts, oil and biomass (cake) from the neem tree (*A. indica*) have received interest from both medical and agricultural researchers (for crop protection). The pesticidal properties of this plant are generally considered to be related to the tetranortriterpenoid, azadirachtin A (Silva et al., 2008). Whilst limited, there is some evidence that neem-based products have the potential to reduce populations of both PCN and SCN (Silva et al., 2008; Trifonova and Atanasov, 2011).

The nematicidal activity of garlic is most likely to be associated with the compound allicin (Gupta and Sharma, 1993), which is produced when garlic cloves are crushed or bruised. In this instance, the precursor alliin comes into contact with the enzyme allinase, which is stored in cell vacuoles, to produce pyruvate, ammonia and allylsulfonic acid; allicin is formed from the latter two compounds (Block, 1992). Garlic has been shown to be either effective or ineffective in the suppression of plant-parasitic nematodes, perhaps due to the variability of garlic products used (Danquah, 2012). It is possible that the biocidal metabolites are affected by the conditions in which the garlic was grown, the cultivar used or the method by which the extract was prepared and stored. Very few studies have investigated garlic extracts for managing cyst nematodes. However, Danquah et al. (2011) conducted in vitro assays on a garlic product, G8014S (Omex Agriculture Ltd, UK), to determine the effect of its constituents (garlic extract, 80%; salicylaldehyde, 14%; and a nonylphenol ethoxylate surfactant, Tergitol, 6%) on the hatching and viability of PCN eggs and the mortality of hatched J2. Interestingly, salicylaldehyde was the most toxic compound in G8014S to PCN J2, with an EC^{50} of 6.5 µl l^{-1} following 24 h exposure; salicylaldehyde is a precursor to salicylic acid and is often sourced from the bark of willow (*Salix alba*). The garlic extract was found to have a lower EC^{50} of 9836.5 µl l^{-1} following 24 h exposure and at concentrations 34.4–137.66.5 µl l^{-1} was found to stimulate hatch. In the UK, NEMguard PCN granules (MAPP no. 17377) is a registered 'garlic extract based' product for the reduction of PCN.

13.6 Application of Biological Control Agents

Biological control has been defined as the ability of any living organism (or biological control agent (BCA)) to reduce the population's density of a pest and/or a disease (target organism) (Eilenberg et al., 2001). From an ecological point of view, biological control should be considered as an approach for 'maintaining, restoring or enhancing the natural suppressive mechanisms that exist in the soil' (Stirling, 2011). This control method relies on the natural interactions existing within a healthy and auto-regulated soil's food web where plant-parasitic nematodes are not considered as a group of soil organisms in isolation, but as a key component of the soil as a whole. By boosting the microbial interactions of the food web of soils, biological control aims to reduce the initial population density of plant-parasitic nematodes in the soil either by: (i) promoting the presence of natural controllers; or (ii) by introducing new organisms in the root's ecosystem that can limit the capacity of the nematodes to parasitize plant roots. However, the level of efficacy of these two tactics seems to be quite broad and variable (Koenning and Barker, 1998). Biological control aims at reducing the presence of plant-parasitic nematodes below an economic damage threshold, rather than to eradicate them, in order to allow the BCAs to persist longer in the soil and become a durable and cost effective control solution (Bale et al., 2008).

Control using biological control can be achieved through two different modalities. The first one is through a direct interaction of the BCA and nematode through a parasitoid, predatory and/or pathogenic effect; in this case the biological cycles of the parasitic nematode and the BCA are synchronized. The second modality of biological control occurs when the interaction is based on an antagonist or competing mechanism, where the microorganism controlling nematodes at high nematode densities may have a saprophytic behaviour in the absence of their hosts. Beneficial microorganisms with an antagonistic effect usually produce metabolic byproducts, enzymes or toxins (antibiosis by antibiotics) that limit the nematode's infective capacity (e.g. by reducing hatching). Antagonism can also be based on a physical limitation and/or reduction of the nematode's penetration and attraction to the roots of the host plant. This can occur whenever there is: (i) a disruption on the usual patterns of emission of the root diffusates; (ii) a competition with nematodes for nutrients and/or space at the rhizosphere level; (iii) an alteration of plant growth promoters; and/or (iv) an enhancement of the systemic resistance of the plants, through an increased production of pathogenesis-related proteins and/or production of protective polymers (e.g. lignin). All these mechanisms have been described (see review on bacterial control by Siddiqui and Mahmood,

1999), although the control of nematodes by suppressive microbial communities has been mostly reported in controlled environments and under sterile conditions.

Specific examples of the BCAs controlling cyst nematodes under these two modalities are detailed in Chapter 11, this volume, and summarized here. Research on biological control of nematodes has been conducted mainly on microbial parasites (obligate and facultative) and rhizosphere bacteria, where successful control results have been observed especially among those agents parasitizing eggs and/or adult females (McDonald and Nicol, 2005). BCAs acting as obligate parasites have been described to control plant-parasitic nematodes, and in particular cyst nematodes. Some of these have proven to be effective in reducing the infective population of some nematode genera (mainly *Globodera* and *Heterodera* spp.) both under controlled and field conditions, although the registration and commercialization may be anecdotal or non-existent (Wilson and Jackson, 2013), which is the current scenario in most developing countries.

13.6.1 Direct antagonism through the application of biological control agents

Pasteuria spp. is an example of obligate microparasites that effectively reduce the infestations levels of *Globodera* and *Heterodera* spp. The genus *Pasteuria* consists of mycelial and endospore-forming, non-motile Gram-positive endoparastic bacteria. Of the four *Pasteuria* species described so far, *P. nishizawae* infects mainly cyst nematodes of the genera *Heterodera* and *Globodera* (Tian *et al.*, 2007). Spores of *P. nishizawae* have been observed parasitizing J2 of *H. glycines* but have not been observed infecting other stages (Noel *et al.*, 2005). It is only after J2 penetrate into the roots of soybean that the bacterium develops a germination tube that penetrates into the cuticle and through the body wall of the nematode, and allows reproduction of the bacterium. In other cyst nematode species, such as *G. rostochiensis*, *H. elachista*, *H. lespedezae*, *H. schachtii* and *H. trifolii*, endospores have been observed to adhere to the cuticle of the J2, but the capacity of the bacterium to parasitize and complete the life cycle in these species has not been confirmed even if they may act as its hosts (Noel *et al.*, 2005). The life cycle of *Pasteuria* has recently been re-evaluated (Mohan *et al.*, 2012). Despite parasitism by *Pasteuria* spp., female nematodes remain viable but they suffer a drastic reduction on their reproduction rate, which helps to reduce infective inoculum in the soil of the next nematode generation. Parasitism by *P. penetrans* has also been observed in *H. avenae*, *H. cajani* and *H. elachista* (Siddiqui and Mahmood, 1999).

In 2004 Pasteuria Bioscience developed a patent to produce *Pasteuria* spp. *in vitro* leading to the production of the first commercial products (Wilson and Jackson, 2013). In 2012 the company was acquired by Syngenta, and since 2014 *P. nizhiawae* has been commercialized in the USA under the name of CLARIVA as a seed treatment product for control of *H. glycines* and *H. schachtii* in soybean and sugar beet, respectively. Nematode control by *P. penetrans* has also been achieved by transferring inoculum from one suppressive soil to a non-suppressive one (Kariuki and Dickson, 2007) and by increasing the population of the bacterium to enable nematode control in the new site. The effectiveness of biological control by site transfer may depend on the specific agro-ecological conditions of the receiving soils plus the agronomic practices in each farm (e.g. use of pesticides); nevertheless, it is worth investigating for the control of cyst nematodes, especially in developing countries with limited access to registered biological control solutions.

Purpureocillium lilacinum is a ubiquitous nematophagous fungus naturally occurring in soils that has shown capacity to parasitize eggs, juveniles and females of plant-parasitic nematodes. This fungus produces a serine protease and chitinases that are responsible for the digestion of the eggshell (Khan *et al.*, 2004), allowing penetration of the fungal hyphae inside the eggs. Immature eggs seem to be a more vulnerable development stage to the protolithic attack of the fungus than the eggs containing juveniles (Bonants *et al.*, 1995).

Purpureocillium lilacinum strain 251 (PL251), originally isolated from nematode-infected eggs in the Phillipines, has been studied extensively. Several commercial formulations are available in the market based on water-dispersible granulate (WG) and/or water-dispersible powder (WP) (Moosavi and Askary, 2015) of conidiospores and it is currently the only strain registered by the European Union (EU) to control plant-parasitic nematodes (Pertot et al., 2015). As opposed to other *P. lilacinum* strains, PL251 does not produce mycotoxins or paecilotoxins, making it safer for end users. This is key considering that, beyond its pathogenic activity against insects and nematodes, this species has been found causing infections in immunocompromised humans (Luangsa-ard et al., 2011). *Purpureocillium lilacinum* strain 251 is able to penetrate through the surface of immature cysts to parasitize eggs of *H. avenae*, although infection does not occur on mature cysts (Khan et al., 2006) indicating that the efficacy could be limited to recent infestations and would not help to control cysts nematodes efficiently in fields having a previous history of infestation. Challenges related to the efficacy of the commercial formulated conidia of PL251 have been observed linked to the low persistence on the soil of the spores (reduction of 81.6% after 14 weeks) even in the presence of both the plant host and the nematode target (Rumbos et al., 2008), corroborating that facultative parasitic fungi are influenced by rhizosphere environmental factors and less so by the presence of the host and nematode density – as opposed to obligate parasitic BCAs. These findings indicate that *P. lilacinum* can be used as a reliable control agent, even if for a limited period of time. This same study shows the preference of the isolate for silty loam and/or clay soils as opposed to sandy substrates, where persistence of the fungi is lower, which could additionally explain the different levels of efficacy observed in different agro-ecological conditions and sites. The commercial product Bioact WG/Melocon-WG (Bayer Crop Sciences for Bioact WG; Certis in USA and Prophyta in Germany for Melocon-WG) of *P. lilacinum* PL251 has been registered to control *Heterodera* spp. and *G. rostochiensis* plus root-knot, burrowing, root lesion, citrus and reniform nematodes in vegetables, strawberries, citrus, nuts, peaches, grapevines and ornamental plants; up to three repeated applications are required (14 days before planting, at planting stage and at 6–8 weeks after planting) at a rate of 0.2 g plant^{-1} or 4 kg ha^{-1}, per season. These commercial formulations can be used through mechanical incorporation, spraying and drip irrigation systems, for conventional and organic farming; the formulation contains between 1×10^{10} (WG) and 1×10^{11} (WP) live spores per gram of formulation; effective nematode control requires the presence of water and the intimate contact of the fungi spores with the nematode. A secondary protective effect of this endophyte over plants has been observed in cotton, where the application of this fungus was also able to reduce the infestation levels of the cotton aphid *Aphis gossypii* (Castillo-López et al., 2014).

Pochonia chlamydosporia is a saprophytic fungus with facultative parasitic capacity on nematodes of economic importance, including *Globodera* and *Heterodera* spp. It can persist in the soil in the absence of its hosts, a desirable quality ensuring sustained nematode control. The growth rate and the parasitic/saprophytic activity of the fungus have been observed to depend also on the C:N ratio and pH of the soil through a regulatory mechanism of the VCP1 enzyme. Whereas low C:N ratios correlate positively with the pathogenic mode of the fungus, higher C:N contents and/or acidic pH in the soil trigger the production of conidia but compromises parasitism of nematode eggs (Ward et al., 2012; Manzanilla-López et al., 2013).

Pochonia chlamydosporia can also act as a plant growth promoter irrespective of whether nematodes are present in an agricultural soil or not (Ciancio et al., 2016) and can act as an endophyte both in monocotyledon and dicotyledon plants, promoting plant growth in several crops (Bordallo et al., 2002). The multifaceted spectrum of the feeding habits of *P. chlamydosporia*, ease of *in vitro* culturing, and the viability of multiple isolates/strains from fungal collections enhance its proven effectiveness as a biocontrol agent for cyst nematodes (Manzanilla-López et al., 2011a). A key aspect to maximize the efficacy of the fungus under field conditions is the presence of a host plant that supports colonization of the roots by *P. chlamydosporia* above 200 cfus cm^{-1} of root. Not all the isolates tested so far have the same capacity to colonize the plants' rhizosphere and different rates of root colonization between different crops and even cultivars have been found (Bourne et al., 1996). The fact that it has no phytotoxic effects makes it ideal for treating seedlings pre- and post-transplanting, applied as aqueous suspension of the spores directly to the soil and/or liquid sprays (Ciancio et al., 2016). In spite of extensive research on *P. chlamydosporia* (Manzanilla et al.,

2011b, 2013) the fungus has only very limited commercial uptake (Wilson and Jackson, 2013). Only recently a commercial product of *P. chlamydosporia* including *Arthrobotrys oligospora* and other plant growth-promoting microorganisms (*Glomus* spp. and *Bacillus* spp.) has been successfully tested to control root-knot nematodes in potato in open-field conditions (Pochar, Microspore, Larino, Italy; Sellitto et al., 2016). Other commercial products exist, such as Klamic (isolate *P. chlamydosporia* var. *catenulata*), a granulate preparation developed at the Centro Nacional de Sanidad Agropecuaria – CENASA in Cuba (Hernández and Hidalgo Díaz, 2008) and tested successfully with root-knot nematodes, and the PcMR-1 strain by Clamitec Myco-solutions (Portugal) (Moosavi and Askary, 2015), but all with limited commercial distribution. Although all these have been so far tested only with root-knot nematodes, these could represent a promising opportunity for the biological control of cyst nematodes.

Myrothecium verrucaria was first found associated with cysts of *H. glycines*. The strain AARC-0255 is currently commercialized as DiTera WP (Valent Biosciences Corp., USA), and has been registered for the biological control of several groups of plant-parasitic nematodes, including cyst nematodes, on a wide range of annual and perennial crops. In Chile it is registered for the control of *H. trifolii* in grapes, whilst in Kenya the product has been registered so far only for flower farms. This bionematicide contains 90% w/w (900 g kg^{-1}) of the fermentation byproduct of the fungus but no living spores. The product reduced infection of potato plants by J2 of *G. rostochiensis* that could be attributed to: (i) a hampered movement of J2; (ii) lack of perception of the potato root diffusates by J2; and (iii) reduced stylet protraction of the J2 (Twomey et al., 2002; see Chapter 3, this volume). This commercial product can be applied to the soil as a pre-plant, planting and/or post-plant soil treatment alone, or mixed with water; the mixed suspension can be applied through drip or border irrigation systems, although mist sprayer or aerial spray equipment should be avoided because DiTera should not be applied to areas where surface water is present as it is toxic to fish and aquatic invertebrates. There is an increased interest in this product from end users (Wilson and Jackson, 2013).

Bacillus spp. bacteria have also been extensively studied as a potential BCA for plant-parasitic nematodes. *Bacillus thuringiensis*, *B. cereus* and *B. firmus* have shown promising nematicidal properties against cyst nematodes, mainly from the *Heterodera* genus (Geng et al., 2016; Zhang, et al., 2016; Zheng et al., 2016). The plant growth-promoting bacterium, *B. cereus* (09B18), isolated from the cereal cyst nematode *H. filipjevi*, has shown nematicidal activity against the latter in field trials where increased yields were observed in plots where *Bacillus* seed-coated seeds were planted (Zhang et al., 2016). This study showed a correlation between higher cereal yields and a decrease in the viability of J2, together with a reduction of hatching; however, the positive effect of *B. cereus* as a plant growth promoter may influence increased yield. *Bacillus firmus* (isolate DS-1) also displays nematicidal capacity against *H. glycines*, linked to the secretion of a serine protease (Sep1) that degrades multiple intestinal cuticle-associated nematode proteins, causing J2 death (Geng et al., 2016). This protease has been found in more than 100 species of bacteria and fungi, so it is not unique to the *Bacillus* genus and could help explain the capacity of many saprophytic microorganisms to contribute to control of plant-parasitic nematodes whenever the specific agro-ecological conditions of a particular soil allow it. Only Bayer is currently commercializing a *Bacillus*-based product, VOTIVO, for the control of *Heteodera* spp. in maize, with a rate of 2×10^6 *B. firmus* seed^{-1} or 1.5×10^{11} *B. firmus* ha^{-1} (Wilson and Jackson, 2013). The company is commercializing the product as a cocktail of plant growth-promoting rhizobacteria and it is one of the few registered products targeting a range of crops and cereals.

Trichoderma spp. is a genus of free-living fungi with saprophytic and mycoparasitic activity. This fungus can act as an endophyte, promoting plant growth and development, and can activate both localized and systemic induced resistance responses of the colonized plants (Harman et al., 2004). Species such as *T. harzianum*, *T. kovingii*, *T. virens*, *T. asperellum*, *T. lignorum*, *T. viride* and *T. atroviride* exert direct and/or indirect effects on eggs, females and/or juveniles of plant-parasitic nematodes (Sharon et al., 2011), but mainly parasitize nematode eggs. *Trichoderma harzianum* parasitizes *G. rostochiensis* (Saifullah and Thomas, 1996), whilst *T. longibrachiatum* suppressed *H. avenae* in wheat under *in vitro* conditions and in glasshouse experiments (Zhang et al., 2014). As observed with *V. lecanii* (Meyer and Wergin, 1998), parasitism of *H. avenae* cysts by *T. longibrachiatum* is achieved mainly through the secretion of exogenous proteolytic enzymes (mainly chitinolytic proteins), rather than by mechanical penetration by the fungal hyphae.

Root colonization by *T. harzianum* is known to trigger the defence system in cucumber (*Cucumis sativus*) (Yedidia et al., 1999) and *T. atroviridae* is able simultaneously to activate, both locally and systemically, the salicylic acid and jasmonate/ethylene pathways, the oxidative burst and the camalexin defence-related genes in *Arabidopsis thaliana* against microorganisms of different pathogenic nature (Salas-Marina et al., 2011). *Trichoderma* spp. can also enhance plant protection conferred by other microorganisms. For example, a synergistic relationship between *T. harzianum* and *Pseudomonas fluorescens* was found to enhance control of root-knot nematodes in tomato (Siddiqui and Shaukat, 2004).

Despite the potential benefits of *Trichoderma* spp. for management of plant-parasitic nematodes, the vast majority of *Trichoderma*-based products are being registered for biological control of economically important soil-borne fungal diseases, such as *Botrytis cinerea* and other damping-off diseases. A search on the Directory of Biopesticides of the Agriculture and Agri-Food Canada Agency (AAFC) and on the EC Health and Food Safety Directorate-general website reveals that, as of 2016, within the group of the Organisation for Economic Co-operation and Development (OECD) and the EU countries, there are no *Trichoderma*-based products specifically registered for the control of plant-parasitic nematodes, whilst in other non-OECD/EU countries there are several *Trichoderma* spp. products that are being marketed for nematode control, although not all of them are registered under the national plant protection organizations (NPPOs) in spite of their commercialization.

Several *Trichoderma* spp. strains have been isolated in West and East Africa and have showed biological control potential against nematodes (Affokpon et al., 2011; Bogner et al., 2016). In Kenya and Tanzania, a few commercial products based on *Trichoderma* spp. have been registered, although only one *T. asperellum* strain, TR900 (commercialized as Real Trichoderma, from Real IPM), is registered for the control of root-knot nematodes in French beans (PCPB, 2016). Real Trichoderma is being sold as a suspension of spores in vegetable oil (200 ml ha^{-1}, with cfus 1×10^9 ml^{-1}) or as a granular product (1 kg granular product ha^{-1}, with cfus 1.7×10^9 g^{-1}). This product was experimentally tested during 2016 for the control of *G. rostochiensis* in Kenya in two potato fields with high infestation levels of the nematode; preliminary results indicate that the oil solution at a dose of 5 l ha^{-1} applied in the potato drills, is able to control *G. rostochiensis* and reduce the final population densities compared to the untreated controls (D. Coyne and L. Cortada, 2016, pers. comm.). *In vitro* assays show compatibility between this fungal strain and the abamectin commercial formulate Tervigo (Syngenta), indicating that these two could be used in combination to develop effective IPM packages to control PCN.

13.6.2 Antagonists and endophytes: suppressing cyst nematodes through the enhancement of the soil food web

Most recent advances on biological control for plant-parasitic nematodes are focused mainly on the enhancement of the natural suppressive capacities of agricultural soils (Mendes et al., 2011). The interaction among all the organisms that compose the micro, meso and macro biota of the soil food web is fragile and strongly influenced by agro-ecological conditions and farming practices. It may rapidly change in time and space, and the actual mechanisms that are responsible for making soils suppressive are not well understood. Suppressiveness of a given soil may be conditioned by the long-term presence of nematodes. Control methods that may eradicate nematodes from the soil could prevent further development and conservation of BCAs, hampering the suppressive effects of a particular soil: the key is how to keep an intricate and complex food web in the soil that could prevent plant-parasitic nematodes from becoming predominant. Fungi are perhaps the most important parasites and predators of nematodes but the impact of organic matter on their predatory activity is poorly understood (Stirling, 2011).

Even if biological control of cyst nematodes could be achieved through the application of both endogenous and exogenous agents into a particular agro-ecosystem, control conducted via a native/indigenous BCA in a specific soil may have a better chance of achieving an ecological equilibrium within the soil food web, as opposed to those applied de novo. However, a massive broadcast application of BCAs is often not economically profitable either for commercial farmers or for smallholders, and it is therefore crucial to ensure a targeted application in time and space. Thus, promoting endogenous control agents in the soil (i.e. natural supressiveness) is considered as the most feasible alternative to achieve a sustained and profitable biological control of plant-parasitic nematodes (Moosavi and Askary, 2015). Two types of suppression exist: (i) organic matter-mediated general suppression; and (ii) specific suppression. The presence of sufficient and good quality organic matter is crucial for the soil food web and as a carbon source for beneficial microorganisms, in order to maintain high levels of suppressiveness. Activity against nematodes derives from the production of secondary metabolites in the soil (e.g. short-chain fatty acids or ammonia) and less from the proliferation of microbial communities with antagonistic capacity. Unfortunately, the outcomes are difficult to predict and are subject to many external factors that can alter the sensitive balance of such microbial communities (Costa et al., 2011). Application of external sources of C may not be cost effective in subsistence farming; nevertheless, the use in combination of the three basic pillars of conservation agriculture (permanent soil coverage, crop rotation with leguminous species and minimum tillage), together with a reduced period of bare fallow, may be the most effective method for sustaining and increasing C soil content in the medium to long term. In such agro-ecosystems, the presence of mycorrhiza, rhizobacteria and/or endophytes, is thereby promoted, helping to control nematodes and other soil-borne diseases (Govaerts et al., 2008). The capacity of BCAs to suppress cyst nematodes is strongly dependent on the indigenous microorganisms present in each soil. For example, in an experiment conducted to promote egg parasitism of H. schachtii by nematophagous fungi using green manure, fields in monoculture had a greater capacity to sustain nematode parasitism, compared to those receiving the same amendments under a crop rotation scheme, due to a higher microbial diversity in the latter that made their microbiological communities more resilient (Pyrowolakis et al., 1999).

13.7 Agrochemical Control of Cyst Nematodes

In the past 100 years our understanding of factors leading to crop failure has identified plant-parasitic nematodes as serious contributory organisms. Some of the key cyst nematodes identified were H. glycines, G. tabacum, G. rostochiensis, G. pallida, H. schachtii and H. avenae. At the same time as these connections were made between crop failure and nematodes, so did our understanding and ability to produce synthetic agrochemicals for nematode control increase. In the early part of the process manufactured versions of MeBr, registered in 1932, provided soil sterilization as a broad spectrum means of control. As mentioned in section 13.1, this chemical and many others have now been banned in many countries due to their adverse environmental impact; however, some are still available in specific countries or in specific cases. These substances include 1,3-dichloropropene, MeBr, chloropicrin, carbofuran (www.pesticideinfo.org and http://apvma.gov.au) and Temik (aldicarb), where they are used for control of a range of soil-borne nematodes or used in research as reliable control/non-infested treatments to compare with potential replacements. The EU does not support these substances in crop protection (EC, n.d.). Chemicals used to control nematodes are generally referred to as nematicides, a term commonly meaning a substance lethal to nematodes. In reality, however, the term nematostat is probably more suited to most substances as they often paralyse or prevent some necessary behaviour, such as feeding (Kearn, 2015), rather than killing directly. The ultimate fate may be nematode death but the chemical is not considered lethal per se at the prescribed rate of application (Evans and Wright, 1982). For a nematicide to work it needs to be applied in the correct concentration and period, so as to persist long enough for the nematode to deplete its lipid food reserves before it can find a host. The main aims of these products are to protect the yield of the crop, prevent or reduce nematode reproduction and prevent or reduce the transmission of the nematode-borne virus to the plant (Whitehead, 1968).

For chemical control of cyst nematodes there are several target options, the cyst-containing eggs in soil, the hatched J2 in transit through the soil and feeding on or within the plant itself. The control of cyst nematodes before hatching is complicated by the protective wall of the cyst, which provides additional protection to the eggs and the unhatched J2. This defensive barrier, formed by the tanned swollen cuticle of the dead female, is the first barrier that substances must penetrate before encountering the eggshell as a second barrier and then the cuticle of juvenile nematodes. It is not certain how substances enter the cysts as there are few natural openings except small holes left from natural opening such as vulval/anal fenestrae from which larvae exit the cyst (Shepherd, 1962). The cyst wall of *G. rostochiensis* is a complex biochemical structure containing a high percentage (72%) of proteins, 2% lipids and various other compounds (Clarke, 1968). The structure of the eggshell and cuticle is covered in detail by Davies and Curtis (2011) and in Chapter 14 of this volume. As a barrier to chemical agents it is sufficient to say that the selectively permeable layers of the cyst wall, the eggshell and the juvenile cuticle will influence how, or if, nematicidal compounds pass through this barrier. However, water and PRD are suggested to diffuse through it and the vulval/anal openings, which allow for juveniles to exit (Shepherd, 1962), would provide an entry point for both liquids and gases. As the cyst itself ages, degradation of the cyst wall would be expected to occur and may positively influence chemical ingress, although no evidence of cyst wall degradation appears to be published. Once the J2 exit the cyst and migrate through the soil, the protection is solely from the cuticle, and the J2 is an easier target due to active contact of the nematode cuticle with soil solution. At this point the active substances penetrate the body wall of the nematode and, in the case of organophosphates and carbamates, act as acetyl cholinesterase inhibitors, which interfere with normal nerve impulse transmission causing paralysis, disorientation and ultimately death (Noling, 1997).

13.7.1 Active substances currently in use

The chemicals currently in use (Table 13.2) can be split broadly into three categories: fumigant, non-fumigant and seed treatment. Non-fumigant nematicides are generally dust-free, 1 mm diameter micro-granules formed from materials such as brick dust or clay minerals such as sepiolite, onto which is added the active substance. Currently these granular nematicides include oxamyl as Vydate 10G (10% oxamyl; Du Pont) and fosthiazate as Nemathorin (10% fosthiazate; Syngenta AG), ethoprophos as Mocap 15G (15% ethoprophos, Amvac; Certis) and fluensulfone as Nimitz 15G (Adama Agriculture), a fluoroalkenyl systemic product. Fenamiphos (as Nemacur 100G; Bayer) is listed for plant-parasitic nematode

Table 13.2. Active substances for nematode control in widespread use in 2016.

Active substances	Type	Group	Example	Manufacturer
1,3-dicloropropene	Liquid fumigant	Halogenated hydrocarbon	Telone C-35	Dow Agrosciences
Metam sodium	Liquid fumigant	Methyl isothiacyanate liberator	Metam 510	Taminco (Certis Europe)
Fosthiazate	Microgranule	Organophosphorus	Nemathorin	Syngenta
Oxamyl	Microgranule or liquid	Oxime carbamate	Vydate 10g l^{-1}	Du Pont
Ethoprophos	Microgranule	Organophosphorus	Mocap 15G	Amvac
Fluensulfone	Microgranule	Fluoralkenyl	Nimitz	Adama
Carbofuran	Microgranule	Carbamate	Furadan	FMC corp.

control but not specifically for cyst nematodes; similarly, carbofuran (as Furadan 10G; FMC) is listed only for root lesion nematode, *Pratylenchus zeae*, in Australia. Avermectins, also classed as nematicides, are macrocyclic lactone byproducts that are derived from soil bacterium *Streptomyces* to produce abamectin, but these will not be covered further because they are a natural fermentation product of the bacterium *Streptomyces avermitilis*. Although not currently registered in the EU for use on cyst nematodes, Nimitz 15G is a non-fumigant granular and Nimitz EC is a liquid formulation containing fluensulfone (Adama Agriculture). The active ingredient has been shown to provide control of *G. rostochiensis* and *G. pallida* (Norshie *et al.*, 2016). The EC formulation was sent for registration in the USA in 2014 and is considered as posing a reduced risk to human health compared with the six restricted-use nematicides currently registered, MeBr, metam sodium, metam potassium, 1,3-dichloropropene, dimethyl disulfide and chloropicrin, and the one non-fumigant, oxamyl (EPA, 2014). It is also registered in Australia as Nimitz 480 EC, but it is used for the control of root-knot nematode and not cyst nematodes.

Fluensulfone has a novel mode of action. Kearn (2015) demonstrated various effects including metabolic impairment (also see Kearn *et al.*, 2017), feeding behaviour disruption and inhibition of hatch of *G. pallida*. Fluensulfone has a relatively short DT_{50} (time to 50% dissipation) of ~24 days, which indicates that it has a lower risk to the environment but may be less effective against *G. pallida* populations with a protracted period of hatch (Norshie *et al.*, 2017). Oximecarbamates and organophosphates are believed to act as cholinesterase inhibitors, whereby they disrupt the enzyme acetylcholinesterase in nerve tissues, which then impairs neuromuscular activity disrupting nematode movement (Evans and Wright, 1982). Oxamyl, an oxime carbamate, works by direct contact with the nematode and through its systemic action within the plant, which provides an ingestion route while nematodes feed. This is an added benefit as it can lead to reduced feeding and moult development. Fosthiazate and ethoprophos are organophosphates that work by contact only and, therefore, target the cyst and the migrating J2.

The fumigant nematicides are based either on halogenated hydrocarbons or methyl isothiocyanate (MITC) release. Metam sodium decomposes rapidly in water and forms the secondary product MITC, which, once inside the nematode body interferes with enzymatic, nervous and respiratory systems; death is normally rapid (Noling, 1997). Granular nematicides such as oxamyl, fosthiazate and ethoprophos need incorporation into the soil and their role includes hatch inhibition and/or suppression of J2 activity in the soil after hatching for up to 28 days, athough the reality is between 3 and 28 days. Ideally they should persist for long enough to offer good control of the nematode but their degradation must be rapid enough to not leave residues in the crop or environment. Good incorporation and adequate soil moisture is needed in order for the chemical to be in the soil solution around the cyst or J2. Unfortunately, there is limited work that considers the direct effect of soil moisture on nematicide activity across a range of soil moistures. Regrettably, soil moisture data to quantify the soil moisture status at any point in nematology research are seldom collected, with the exception of nematicide leaching work, which does not relate specifically to nematicide efficacy. Also the interaction between volumetric soil moisture, nematicide application and crop yield generally focuses on moisture effect on crop yield rather than nematicide activity. The results of Griffin (1977) illustrate this issue. Where soil contains limited moisture at application the nematicide will not be adequately released into a soil solution and poor control will occur, though no reports were found that directly state a minimum soil moisture for granular nematicides. The threshold for their use is often simply the detection of the nematode in soil samples and there are generally no guidelines for doses linked to ranges of nematode infestation. Oxamyl, as Vydate 10G, does currently give an option of reduced dosage at PCN populations of <10 eggs g^{-1} soil on very light, light, medium and organic or peat soils (Du Pont, n.d.). If excess rainfall occurs soon after application the chemicals can be leached from the soil leaving a sublethal concentration that will not provide sufficient control of the nematodes. Similarly, other soil conditions can also greatly affect leaching potential. Karpouzas *et al.* (2007) reported that fosthiazate is more prone to leaching under acidic low organic matter content soils than with alkaline high organic matter soil.

Incorporation of granular nematicides into the soil needs to be carried out using suitable equipment (Figs 13.2 and 13.3) and at the manufacturer's specified depth. This becomes evident when the application is considered relative to the soil volumes being used. A full rate 55 kg ha^{-1} application of oxamyl (as Vydate 10G) incorporated to a depth of 15 cm would be equivalent to 5.5 kg oxamyl in 1500 m^3 soil, approximately 1800–2250 tonnes (t) of soil. Woods (1997) demonstrated how various types of nematicide incorporation equipment used in potato production could result with very different granule distribution profiles.

Liquid non-fumigants are also available for nematode control, for example, oxamyl (as Vydate L; Du Pont) and carbofuran (as Furadan 360 flowable; FMC). However, the targets are mostly burrowing, lesion, dagger or sting nematodes. The products can often be applied as a foliar spray or used in chemigation, a process whereby the chemical is injected into the irrigation water of suitable equipment such as drip irrigation.

Fumigant nematicides are available as liquid or granular but are fewer in number than the non-fumigant. Only Metam sodium 510 (Certis Europe) is licenced in the EU and is used as a soil sterilant, specifically for PCN control in the UK, but would also be effective for all stages of soil-borne endoparasitic as well as ectoparasitic nematodes. Telone II, C-35, C-17, EC and Inline (1,3-dichloropropene; Dow AgroSciences) are licenced in some states in the USA for control of *Heterodera* and *Globodera* spp. in a wide range of crops. Telone is also licenced for use in South Africa for unspecified nematodes. There has also been research to determine its use as a MeBr replacement in China (Mao *et al.*, 2012). All formulations of 1,3-D have been removed from the approved list in the EU. Australia currently has MeBr, chloropicrin, 1,3-dichloropropene and metam sodium products listed that have been used for cyst nematode control around the world but they are reviewed annually by the Australian Pesticides and Veterinary Medicines Authority (APVMA), which could see their removal at any time. Liquid fumigants are applied using several methods. Metam sodium is applied using a specialized subsoiler type configuration with broad soil lifting wings (Fig. 13.3) under which the fumigant is injected and from which it quickly volatilizes and spreads laterally, vertically upwards

Fig. 13.2. Rear-mounted nematicide applicator with specialized rotovator incorporation. (Courtesy of I. Grove.)

Fig. 13.3. Nematicide hopper fitted to a web de-stoner. Not an ideal position especially if the soil is very dry as the nematicide will be applied on top of the soil surface. (Courtesy of I. Grove.)

and to a small extent downwards. Telone II, C-35 and C-17 can also be applied in this manner. The Telone EC formulation is normally applied as chemigation. For the soil-injected applications to work effectively the soil moisture needs to be around 60% of field capacity and at a temperature above 10°C; guidelines are available from manufacturers. This will allow the chemical to volatilize effectively and spread throughout the soil zone. A soil seal (Fig. 13.4) formed with a contra-rotating roller or a sealing tarpaulin is required to prevent early loss of the fumigant. As these chemicals are phytotoxic there is normally a 4–8-week interval between fumigation and planting and a 'cress test' is suggested if earlier planting is required or to ensure that it is safe to plant. To perform a cress test a sample of the fumigated soil is taken, cress (*Lepidium sativum*) is sown and should grow quickly under normal room temperatures (~20°C); if the cress fails to grow, the soil is not suitable for planting crops. There is a more scientific method, which was proposed by Ohba *et al.* (1986). A granular fumigant, dazomet (as Basamid, 97% dazomet; Certis Europe) is available in the EU for control of *G. rostochiensis* and *G. pallida*. This product is normally applied well in advance of planting to soils at temperatures between 10 and 25°C. At temperatures above 25°C the fumigant is rapidly lost to the atmosphere unless a soil seal is used, whereas at <5°C inversion may occur that takes the chemical further down the soil profile than is required. The latter situation delays planting whilst the fumigant dissipates. Soils also need to be moist – between 60 and 70% field capacity is optimum – to ensure activation of the fumigant. The granules should be spread with appropriate applicators and then rotovated into soils quickly (BASF, n.d.).

Soil-applied nematicides can be significantly affected by the soil environment in which they are operating. Factors such as soil texture, structure, organic matter and moisture content can interact to allow or impede the passage of the fumigant through the soil (Bromilow, 1973). Coarse-textured soils offer a more open structure and can normally be fumigated more easily than fine-textured (clay) soils. However, the nature of these soils also means that they are more open and thus may be more difficult to seal. Clay soils can form clods, which are difficult for gases to penetrate, whereas high organic matter content can both adsorb active substances and also make soils difficult to consolidate and seal.

Seed treatments are currently being used for *H. glycines* control (Vitti *et al.*, 2014) and these include chemicals normally classed as insecticides. Poncho/Votiva contains clothianidin (40.3%; Bayer CropScience), Avicta complete beans 500 contains abamectin 22.2% and the

Fig. 13.4. Application of soil fumigants, such as metam sodium, need surface sealing to retain the fumigant within the soil profile. (Courtesy of I. Grove.)

insecticide thiamethoxam (11.1%; Syngenta Crop Protection), and AerisTM (Bayer CropScience) contains thiodicarb 24% and imadocloprid 24%, both insecticides.

13.7.2 Nematicide persistence

For nematicide substances to be effective against cyst nematodes it is important that they can persist for periods long enough to protect the plant during key growth phases after the time of application. For annual crops this timing is often at the time of planting from which the substance will then begin the degradation period that limits the useful life of the substance.

Smelt and Leistra (1992) suggested that transformation rates of pesticides can be used to estimate the useful period of nematode control but noted that actual values are often not directly comparable as transformation rate is affected by soil factors including soil temperature and soil moisture content. Consequently in their review, they only considered studies that either used soil temperatures of, or corrected to, 15°C and −10 to −33 kPa soil moisture, the latter being close to field capacity in all soils. These transformation periods were reported as, carbofuran 20–167 days, ethoprophos 16–120 days, fenamiphos 70–190 days (total toxic residue) and oxamyl 7–39 days, which show a substantial range that reflects the variability between soils. If the substances were to persist indefinitely, then problems with residues in the crop and environmental effects would prevent their usefulness and registration. In relation to nematicide persistence or plant protection a useful but seldom used value from experimental work that demonstrates the duration of suppression is the quantity of nematodes invading the plant over time. For example Norshie et al. (2016) demonstrated substantial suppression of invasion of potato roots by G. pallida up to 44 days after planting from the use of fluensulfone, oxamyl and fosthiazate.

In contrast to oxamyl, which is hydrophilic (Whitehead, 1988) and not adsorbed onto organic matter, the organophosphates are lipophilic and are adsorbed onto organic matter, so soils with high organic matter will reduce their effectiveness (Bromilow, 1980). Poor performance of nematicides has been reported on several occasions and, assuming optimal application had been achieved, it was postulated that this loss of performance could be attributed to biotic and abiotic factors. This has included nematicide-degrading bacteria utilizing the substance as a carbon source, adsorption onto organic matter, interactions with soil pH or using the nematicides repeatedly on the same soil thus predisposing the soil to enhanced degradation. Osborn (2005) demonstrated enhanced degradation in some soils for the oxime carbamates aldicarb and oxamyl and the organophosphate fosthiazate, and Ou *et al.* (1991) found similar results with fenamiphos in turfgrass in Florida, USA, and Chin-Pampillo *et al.* (2015) reported it for carbofuran and ethoprophos in tropical soils. The phenomenon is therefore widespread and is a complex relationship not simply related to the number of applications, as some soils reduced the nematicides half-life with no previous history of application of this or similar chemicals. The enhanced degradation is not restricted to the granular nematicides but has also been reported for the soil fumigants 1,3-D and MeBr (Ou, 1998).

13.7.3 Nematicide registration and stewardship

In the EU currently there are few nematicides registered for use on either cyst or free-living stages of plant-parasitic nematodes; the few remaining, such as oxamyl, fosthiazate, ethoprophos and metam sodium, are under constant threat of licence revocation. The details of how the registration process works and a complete list of pesticide approvals can be found on the EU pesticide website (see EC, n.d.).

The product authorizations are dealt with on a coordinated basis in accordance with EU Regulation (EC) No. 1107/2009. Similarly, there are few nematicides now registered for general use in the USA and of those few there are differences between the states within the USA. The Environmental Protection Agency (EPA) pesticides website provides the most information (see EPA, n.d.). As part of the EPA pesticide guidelines, certified applicators are required to complete site-specific fumigation management plans before the application is carried out and then post application summaries. The Australian government website (http://apvma.gov.au) provides details and product labels of all registered and approved products. In the majority of countries where pesticides are routinely used, similar legislative controls are in place and individual countries have specified registration procedures. For example, China has the Chemical Inspection and Registration Service (CIRS) and Egypt has the Agricultural Pesticides Committee (APC). Africa is slightly different because many of the countries within Africa have their own requirements and registration, but several are coming together to harmonize regulations between countries, for example, Benin, Côte d'Ivoire, Guinea and Ghana.

In the UK applying granular nematicides requires that staff are suitably trained and hold National Proficiency Training certificates, PA4 or PA4G. By 2017 they will also have to have completed an industry stewardship training module and all applicators must be fitted with a device in the tractor cab that allows the operator to shut off nematicide granule flow at least 3 m from the end of each row. The full programme, which will probably evolve, can be viewed on the Nematicide Stewardship Programme (NSP) website (see NSP, n.d.). Similar stewardship guidelines are available for products from Dow Agroscience for Telone in the USA and from Du Pont for best practice use of Vydate 10G (Du Pont, 2015).

In response to the operator concerns from the use of granular nematicides, manufacturers such as Syngenta and Du Pont have introduced closed transfer systems that greatly reduce the potential for operator contamination. The Surefil (www.syngenta-crop.co.uk) and Ecolite (www2.dupont.com) systems are sealed product containers that need to be locked into place on applicators before they are opened. Similarly, the containers can be relocked before removal to prevent spillage of any remaining product.

13.7.4 Precision aspects of nematode management

Nematode control has traditionally been based around a complete field or growing area approach, especially in respect of agrochemical use, for three reasons: (i) the physical time to intensively sample and separate areas was prohibitive; (ii) the expense of analysing multiple samples was similarly prohibitive; and (iii) no other method of detection was readily available at economic cost, for example, aerial views of growing crops. Ultimately this led to only one combined sample being used as a representative of the whole area and the management practices applied accordingly. However, with the advent of affordable geo-referencing equipment utilizing global navigation satellite systems (GNSS) geo-referenced sampling can be done on quad bikes with automated sampling equipment. Although the expense of the sample analysis remains similar, the resulting maps generated from the samples do allow for 'zoning' of infested areas and thus targeted agrochemical application can be achieved as reported for PCN by Haydock and Evans (1995). Evans and Barker (2004) suggested that there was a sound economic argument for site-specific application of nematicides but warned that there was a danger of substantial population increases in areas where PCN were undetected and thus had received no control treatment. However, their conclusion was that it would be more pertinent to fumigate any 'hotspots' detected and then apply a blanket granular nematicide!

GNSS satellites and the low-cost aerial platforms, such as multirotor and fixed wing drones, also have the potential to provide field images in standard photographic format, which highlight patches of poor crop growth, or multispectral images that can indicate potential nematode infestations. Heath (2003) reported a significant reduction in the amount of 550 nm wavelength reflected from PCN-infested potato plants; Hillnhütter *et al.* (2011) found spectral vegetation indices and spectroradiometer findings were significantly correlated to BCN damage and Nutter *et al.* (2002) found similar correlations for the SCN. For all of these nematodes the images could be geo-referenced and used to target sampling or problem identification during the growth of the crop or for post-crop sampling. A good synopsis of information in this area can be found in Hillnhütter *et al.* (2010).

13.8 Conclusions and Future Prospects

Cyst nematodes pose a significant threat to a wide variety of important crop species around the world. Taking this impact into account, the large volume of research that has been undertaken on field management of these pests is fully justified. However, there is much more to be done, particularly with the continued loss of agrochemical options, which are gradually being phased out due to concerns over environmental pollution and food safety. The longevity of chemical control options could be extended if stewardship programmes are implemented and IPM schemes are used more routinely. On the other hand, many of the alternative strategies require further development to make them viable options for crop managers. For example, there has been a significant volume of research undertaken on *P. chlamydosporia* but the availability of commercial products is exceptionally limited. Perhaps this highlights the need for greater interaction between the academic community and industry in the development of crop protection options. Nevertheless, the biopesticides sector is growing as evidenced by the inclusion of 'biologicals' in the portfolios of the six major agrochemical suppliers (Syngenta, Bayer CropScience, BASF, Dow Agrosciences, Monsanto and Du Pont).

A number of the emerging crop protection strategies require further research to allow their full potential to be realized. For instance, biofumigation is a process that is affected by many agronomic factors such as nutrient availability and crop destruction. Such factors require full investigation to ensure that crop managers receive the necessary guidelines for optimizing the approach. The application of plant extracts, oils and biomass is another area that could receive greater attention in the future, as highlighted by the commercialization of neem- and garlic-based products. To avoid unnecessary scepticism, the application of these products needs to be supported by empiricism and good knowledge transfer to ensure appropriate industry uptake.

Future management of cyst nematodes could include precision agriculture practices to a greater degree, such as the use of tractor-mounted remote sensing technology to identify nematode population 'hotspots' within a given field. Such approaches may result in greater reductions of plant-parasitic nematodes whilst limiting damage to the environment through reduced nematicide usage. It is clear from this chapter that each crop protection strategy has limitations and the only sustainable way forward is to follow an integrated management approach. Only with a complete understanding of the biology of cyst nematode species can control strategies be deployed effectively. Additionally, prior to the selection of management options, cyst nematode populations need to be identified using an appropriate sampling plan and diagnostic method (see Chapter 7, this volume). Knowledge of field populations will help growers to plan rotations, select appropriate cultivars, and make decisions on the application of nematicides, biofumigants, trap crops and biopesticides. Whilst this chapter did not include cultivar resistance (see Chapters 8 and 9, this volume), it is obvious that this aspect is a crucial component of IPM and an important area for further development.

13.9 References

Affokpon, A., Coyne, D.L., Htay, C.C., Agbèdè, R.D., Lawouin, L. and Coosemans, J. (2011) Biocontrol potential of native *Trichoderma* isolates against root-knot nematodes in West African vegetable production systems. *Soil Biology and Biochemistry* 43, 600–608. http://dx.doi.org/10.1016/j.soilbio.2010.11.029

Aires, A., Carvalho, R., Da Conceição Barbosa, M. and Rosa, E. (2009) Suppressing potato cyst nematode, *Globodera rostochiensis*, with extracts of Brassicaceae plants. *American Journal of Potato Research* 86, 327–333. DOI: 10.1007/s12230-009-9086-y

Al-Hazmi, A.S. and Dawabah, A.A.M. (2009) Present status of the cereal cyst nematode (*Heterodera avenae*) in Saudi Arabia. In: Riley, I.T., Nicol, J.M. and Dababat, A.A. (eds) *Cereal Cyst Nematodes: Status, Research and Outlook*. CIMMYT, Ankara, Turkey, pp. 56–60.

Bale, J.S., van Lenteren, J.C. and Bigler, F. (2008) Biological control and sustainable food production. *Philosophical Transactions of the Royal Society* 363, 761–766. DOI: 10.1098/rstb.2007.2182

BASF (n.d.) Directions for use. Available at: www.green-tech.co.uk/filedepository/productdocuments/insecticides/basamidlabel.pdf (accessed 19 September 2017).

Björkman, M., Klingen, I., Birch, A.N.E. et al. (2011) Phytochemicals of Brassicaceae in plant protection and human health – influences of climate, environment and agronomic practice. *Phytochemistry* 72, 538–556. DOI: 10.1016/j.phytochem.2011.01.014

Block, E. (1992) The organosulfur chemistry of the genus *Allium* – implications for the organic chemistry of sulfur. *Angewandte Chemie* 31, 1135–1178.

Blok, W.J., Jamers, J.G., Termorshuizen, A.J. and Bollen, G.J. (2000) Control of soilborne plant pathogens by incorporating fresh organic amendments followed by tarping. *Phytopathology* 90, 253–259. http://dx.doi.org/10.1094/PHYTO.2000.90.3.253

Bogner, C.W., Kariuki, G.M., Elashry, A., Sichtermann, G., Buch, A.-K., Mishra, B., Thines, M., Grundler, F.M.W. and Schouten, A. (2016) Fungal root endophytes of tomato from Kenya and their nematode biocontrol potential. *Mycological Progress* 15, 1–30. DOI: 10.1007/s11557-016-1169-9

Bonants, P., Fitters, P., Thijs, H., Belder, E., Waalwijk, C. and Henfling, J. (1995) A basic serine protease from *Paecilomyces lilacinus* with biological activity against *Meloidogyne hapla* eggs. *Microbiology* 41, 775–784. DOI: 10.1099/13500872-141-4-775

Bordallo, J.J., López-Llorca, L.V., Jansson, H.-B., Salinas, J., Persmark, L. and Asensio, L. (2002) Colonization of plant roots by egg-parasitic and nematode-trapping fungi. *New Phytologist* 154, 491–499. DOI: 10.1046/j.1469-8137.2002.00399.x

Bourne, J.M., Kerry, B.R. and De Leij, F.A.A.M. (1996) The importance of the host plant on the interaction between root-knot nematodes (*Meloidogyne* spp.) and the nematophagus fungus, *Verticillium chlamydosporium*. *Biocontrol Science and Technology* 6, 539–548. http://dx.doi.org/10.1080/09583159631172

Bromilow, R.H. (1973) Breakdown and fate of oximecarbamates in crops and soil. *Annals of Applied Biology* 75, 473–479.

Bromilow, R.H. (1980) Behaviour of nematicides in soil and plants. In: *Factors Affecting the Application and Use of Nematicides in Western Europe – AAB Nematology Group Workshop*. Association of Applied Biologists, Wellesbourne, UK, pp. 87–107.

Brown, E.B. (1958) Pea root eelworm in the eastern counties of England. *Nematologica* 3, 257–268. DOI: 10.1163/187529258X00012

Brown, P.D. and Morra, M.J. (1997) Control of soil-borne plant pests using glucosinolate containing plants. *Advances in Agronomy* 61, 167–231. DOI: 10.1016/s0065-2113(08)60664-1

Buskov, S., Serra, B., Rosa, E., Sorensen, H. and Sorensen, J.C. (2002) Effects of intact glucosinolates and products produced from glucosinolates in myrosinase catalyzed hydrolysis on the potato cyst nematode (*Globodera rostochiensis* cv. Woll). *Journal of Agricultural and Food Chemistry* 50, 690–695. DOI: 10.1021/jf010470s

Carter, M.R. (1992) Influence of reduced tillage systems on organic matter, microbial biomass, macro-aggregate distribution and structural stability of the surface soil in a humid climate. *Soil Tillage Research* 23, 361–372. DOI: 10.1016/0167-1987(92)90081-L

Castillo-López, D., Zhu-Salzman, K., Ek-Ramos, M.J. and Sword, G.A. (2014) The entomopathogenic fungal endophytes *Purpureocillium lilacinum* (formerly *Paecilomyces lilacinus*) and *Beauveria bassiana* negatively affect cotton aphid reproduction under both greenhouse and field conditions. *PLoS ONE* 9, e103891. DOI.org/10.1371/journal.pone.0103891

Charles, J.S.K. and Venkitesan T.S. (1990) Host records of the rice cyst nematode, *Heteroderea oryzicola*. *Indian Journal of Nematology* 20, 222–224.

Chin-Pampillo, J.S., Carazo-Rojas, E., Pérez-Rojas, G., Castro-Gutiérrez, V. and Rodríguez-Rodríguez, C.E. (2015) Accelerated biodegradation of selected nematicides in tropical crop soils from Costa Rica. *Environmental Science and Pollution Research* 22, 1240–1249. DOI: 10.1007/s11356-014-3414-6

Ciancio, A., Colagiero, M., Pentimone, I. and Rosso, L.C. (2016) Formulation of *Pochonia chlamydosporia* for plant and nematode management. In: Arora, N.K., Mehnaz, S. and Balestrini, R. (eds) *Bioformulations for Sustainable Agriculture*. Springer, India, pp. 177–197. DOI: 10.1007/978-81-322-2779-3_10

Clarke, A.J. (1968) The chemical composition of the cyst wall of the potato cyst nematode *Heterodera rostochiensis*. *Journal of Biochemistry* 108, 221–225.

Costa, S.R., van der Putten, W.H. and Kerry, B.R. (2011) Microbial ecology and nematode control in natural ecosystems. In: Davies, K. and Spiegel, Y. (eds) *Biological Control of Plant-Parasitic Nematodes: Building Coherence Between Microbial Ecology and Molecular Mechanisms*. Springer, Dordrecht, The Netherlands, pp. 39–64. DOI: 10.1007/978-1-4020-9648-8_2

Coyne, D. and Plowright, R.A. (1998) Use of solarisation to control *Heterodera sacchari* and other plant parasitic nematodes. *International Journal of Nematology* 8, 81–84.

Creech, J.E., Westphal, A., Ferris, V.R., Faghihi, J., Vyn, T.J., Santini, J.B. and Johnson, W.G. (2008) Influence of winter annual weed management and crop rotation on soybean cyst nematode (*Heterodera glycines*) and winter annual weeds. *Weed Science* 56, 103–111. DOI: http://dx.doi.org/10.1614/WS-07-084.1

D'Addabbo, T., Sasanelli, N., Greco, N., Stea, V. and Brandonisio, A. (2005) Effect of water, soil temperatures, and exposure times on the survival of the sugar beet cyst nematode, *Heterodera schachtii*. *Phytopathology* 95, 339–344. http://dx.doi.org/10.1094/PHYTO-95-0339

Dandurand, L.M. and Knudsen, G.R. (2016) Effect of the trap crop *Solanum sisymbriifolium* and two bio-control fungi on reproduction of the potato cyst nematode, *Globodera pallida*. *Annals of Applied Biology* 169, 180–189. DOI: 10.1111/aab.12295

Danquah, W.B. (2012) The use of plant derived compounds in the management of the potato cyst nematode, *Globodera pallida*. PhD thesis. Harper Adams University, Newport, Shropshire, UK.

Danquah, W.B., Back M.A., Grove, I.G. and Haydock, P.P.J. (2011) *In vitro* nematicidal activity of a garlic extract and salicylaldehyde to the potato cyst nematode. *Nematology* 13, 869–885. DOI: 10.1163/138855411X560959

Davies, K.G. and Curtis R.H.C. (2011) Cuticle surface coat of plant-parasitic nematodes. *Annual Review of Phytopathology* 49, 135–156. DOI: 10.1146/annurev-phyto-121310-111406

De Nicola, G.R., D'Avino, L., Curto, G., Malaguti, L., Ugolini, L., Cinti, S., Patalano, G. and Lazzeri, L. (2013) A new biobased liquid formulation with biofumigant and fertilising properties for drip irrigation distribution. *Industrial Crops and Products* 42, 113–118. http://dx.doi.org/10.1016/j.indcrop.2012.05.018

Di Vito, M. and Greco, N. (1986) The pea cyst nematode. In: Lamberti, F. and Taylor, C.E. (eds) *Cyst Nematodes*. Plenum Press, London, pp. 321–332.

Du Pont (n.d.) Crop protection products that help farmers succeed. Available at: www.dupont.co.uk/products-and-services/crop-protection.html (accessed 19 September 2017).

Du Pont (2015) Nematicide stewardship: a guide to best practice 2015. Available at: www.dupont.co.uk/content/dam/assets/industries/agriculture/assets/Nematicide%20Stewardship%20-%20A%20Guide%20to%20Best%20Practice%202015_LR.pdf (accessed 7 March 2017).

Ebrahimi, N., Viaene, N., Demeulemeester, K. and Moens, M. (2014) Observations on the life cycle of potato cyst nematodes, *Globodera rostochiensis* and *G. pallida*, on early potato cultivars. *Nematology* 16, 937–952. DOI: 10.1163/15685411-00002821

Ebrahimi, N., Viaene, N., Aerts, J., Debode, J. and Moens, M. (2016) Agricultural waste amendments improve inundation treatment of soil contaminated with potato cyst nematodes, *Globodera rostochiensis* and *G. pallida*. *European Journal of Plant Pathology* 145, 755–775. DOI: 10.1007/s10658-016-0864-3

EC (n.d.) EU pesticides database. Available at: http://ec.europa.eu/food/plant/pesticides/eu-pesticides-database/public (accessed 19 September 2017).

Eilenberg, J., Hajek, E.A. and Lomer, C. (2001) Suggestions for unifying the terminology in biological control. *BioControl* 46, 387–400. DOI: 10.1023/A:1014193329979

Ellenby, C. (1945a) Control of the potato-root eelworm, *Heterodera rostochiensis* Wollenweber, by allyl isothiocyanate, the mustard oil of *Brassica nigra* L. *Annals of Applied Biology* 32, 237–239. DOI: 10.1111/j.1744-7348.1945.tb06242.x

Ellenby, C. (1945b) The influence of crucifers and mustard oil on the emergence of larvae of the potato-root eelworm, *Heterodera rostochiensis* Wollenweber. *Annals of Applied Biology* 32, 67–70. DOI: 10.1111/j.1744-7348.1945.tb06761.x

Ellenby, C. (1951) Mustard oils and control of the potato-root eelworm, *Heterodera rostochiensis* Wollenweber: further field and laboratory experiments. *Annals of Applied Biology* 38, 859–875. DOI:10.1111/j.1744-7348.1951.tb07856.x

Endo, B.J. (1962) Lethal time–temperature relations for *Heterodera glycines*. *Phytopathology* 52, 992–997.

EPA (n.d.) Pesticides. Available at: https://www.epa.gov/pesticides (accessed 19 September 2017).

EPA (2014) Registration decision for fluensulfone. USA Environmental Protection Agency. Available at: www.regulations.gov/document?D=EPA-HQ-OPP-2012-0629-0014 (accessed 7 March 2017).

Evans, K. (1991) Lethal temperatures for eggs of *Globodera rostochiensis*, determined by staining with New Blue R. *Nematologica* 37, 225–229. DOI: 10.1163/187529291X00204

Evans, K. and Barker, A.D.P. (2004) Economies in nematode management from precision agriculture – limitations and possibilities. In: Cook, R. and Hunt, D.J. (eds) *Nematology Monographs and Perspectives Vol 2. Proceedings of the Fourth International Congress of Nematology, 8–13 June 2002, Tenerife, Spain*. Brill, Leiden, The Netherlands, pp. 23–32.

Evans, S.G. and Wright, D.J. (1982) Effects of the nematicide oxamyl on life cycle stages of *Globodera rostochiensis*. *Annals of Applied Biology* 100, 511–519. DOI: 10.1111/j.1744-7348.1982.tb01417.x

Flint, M.L. and Roberts, P.A. (1988) Using crop diversity to manage pest problems: some California examples. *American Journal of Alternative Agriculture* 3, 163–167. DOI: 10.1017/S0889189300002447

Gaur, H.S. and Perry, R.N. (1991) The use of soil solarization for control of plant parasitic nematodes. *Nematological Abstracts* 60, 153–167.

Geng, C., Nie, X., Tang, Z., Zhang, Y., Lin, J., Sun, M. and Peng, D. (2016) A novel serine protease, Sep1, from *Bacillus firmus* DS-1 has nematicidal activity and degrades multiple intestinal-associated nematode proteins. *Scientific Reports* 6, 25012. DOI: 10.1038/srep25012

Gimsing, A.L. and Kirkegaard, J.A. (2006) Glucosinolate and isothiocyanate concentration in soil following incorporation of Brassica biofumigants. *Soil Biology and Biochemistry* 38, 2255–2264. http://dx.doi.org/10.1016/j.soilbio.2006.01.024

Govaerts, B., Mezzalama, M., Sayre, K., Crossa, J., Lichter, K., Troch, V., Vanherck, K., Corte, P. and Deckers, J. (2008) Long-term consequences of tillage, residue management, and crop rotation on selected soil micro-flora groups in the subtropical highlands. *Applied Soil Ecology* 38, 197–210. http://dx.doi.org/10.1016/j.apsoil.2007.10.009

Greco, N., D'Addabbo, T., Sasanelli, N., Seinhorst, J.W., Stea, V. and Brandonisio, A. (1998) Effects of temperature and length of exposure on the mortality of the carrot cyst nematode, *Heterodera carotae*. *International Journal of Pest Management* 44, 99–107. http://dx.doi.org/10.1080/096708798228392

Griffin, G.D. (1977) Effects of soil moisture on control of *Heterodera schachtii* with aldicarb. *Journal of Nematology* 9, 211–215.

Grinstein, A., Kritzman, G., Hetzroni, A., Gamliel, A., Mor, M. and Katan, J. (1995) The border effect of soil solarization. *Crop Protection* 14, 315–320. DOI: 10.1016/0261-2194(94)00005-S

Gupta, R. and Sharma, K. (1993) A study of the nematicidal activity of allicin – an active principal in garlic, *Allium sativum* L., against root-knot nematode, *Meloidogyne incognita* (Kofoid and White, 1919) Chitwood, 1949. *International Journal of Pest Management* 39, 390–392. http://dx.doi.org/10.1080/09670879309371828

Hafez, S.L. (1994) The use of green manure crops in a sugarbeet rotation for sugarbeet cyst nematode management. *Journal of Nematology* 26, 548 (Abstr.).

Harman, G.E., Howell, C.R., Viterbo, A., Chet, I. and Lorito, M. (2004) *Trichoderma* species – opportunistic, avirulent plant symbionts. *Nature Reviews/Microbiology* 2, 43–56. DOI: 10.1038/nrmicro797

Haydock, P.P.J. and Evans, K. (1995) The potential use of global positioning satellite (GPS) technology in the mapping and management of potato cyst nematode populations. In: Bartley, M.R., Basford, W.D., Brain, P., Christensen, C., Lancashire, P.D., Sparks, T.H. and Welham, S.J. (eds) *Aspects of Applied Biology 43, Field Experiment Techniques*. Association of Applied Biologists, Warwick, UK, pp. 125–128.

Heath, W.L. (2003) The detection of potato cyst nematode (PCN) infestation using remotely sensed imagery. PhD thesis. Open University, Milton Keynes, UK.

Hernández, M.A. and Hidalgo Díaz, L. (2008) KlamiC®: bionematicida agrícola producido a partir del hongo *Pochonia chlamydosporia* var. *catenulata*. Nota técnica. *Revista Protección Vegetal* 23, 131–134.

Hillnhütter, C., Schweizer, A., Kühnhold, V. and Sikora, A. (2010) Remote sensing for detection of soil-borne plant parasitic nematodes and fungal pathogens. In: Oerke, E.C., Gerhards, R., Menz, G. and Sikora, R.A. (eds) *Precision Crop Protection – the Challenge and use of Heterogeneity*. Springer, Heidelberg, Germany, pp. 151–165.

Hillnhütter, C., Mahlein, A.K., Sikora, R.A. and Oerke, E.C. (2011) Remote sensing to detect plant stress induced by *Heterodera schachtii* and *Rhizoctonia solani* in sugar beet fields. *Field Crops Research* 122, 70–77. http://dx.doi.org/10.1016/j.fcr.2011.02.007

Hollis, J. and Rodríguez-Kábana, R. (1966) Rapid kill of nematodes in flooded soil. *Phytopathology* 56, 1015–1019.

Inagaki, T. and Tsutsumi, M. (1971) Survival of the soybean cyst nematode, *Heterodera glycines* Ichinohe (Tylenchida: Heteroderidae) under certain storing conditions. *Applied Entomological Zoology* 6, 156–162. DOI: ORG/10.1303/AEZ.6.156

Inglett, P.W., Reddy, K.R. and Corstanje, R. (2005) Anaerobic soils. In: Hillel, D. (ed.) *Encyclopedia of Soils in the Environment*. Elsevier, London, pp. 72–78.

Ito, T., Araki, M. and Komatsuzaki, M. (2015) No-tillage cultivation reduces rice cyst nematode (*Heterodera elachista*) in continuous upland rice (*Oryza sativa*) culture and after conversion to soybean (*Glycine max*) in Kanto, Japan. *Field Crops Research* 179, 44–51. http://doi.org/10.1016/j.fcr.2015.04.008

Jones, J.T., Haegeman, A., Danchin, E. *et al.* (2013) Top 10 plant parasitic nematodes in molecular plant pathology. *Molecular Plant Pathology* 14, 946–961. DOI: 10.1111/mpp.12057

Kaczmarek, A., MacKenzie, K., Kettle, H. and Blok, V. (2014) Influence of soil temperature on *Globodera rostochiensis* and *G. pallida*. *Phytopathologia Mediterranea* 53, 396–405. http://dx.doi.org/10.1016/j.fcr.2015.04.008

Kariuki, G.M. and Dickson, D.W. (2007) Transfer and development of *Pasteuria penetrans*. *Journal of Nematology* 39, 55–61.

Karpouzas, D.G., Pantelelis, I., Menkissoglu-Soiroudi, U., Golia, E. and Tsiropoulos, N.G. (2007) Leaching of the organophosphorus nematicide fosthiazate. *Chemosphere* 68, 1359–1364. http://dx.doi.org/10.1016/j.chemosphere.2007.01.023

Kawakishi, S. and Kaneko, T. (1985) Interaction of oxidized glutathione with allyl isothiocyanate. *Phytochemistry* 24, 715–718. https://doi.org/10.1016/S0031-9422(00)84882-7

Kearn, J. (2015) Mode of action studies on the nematicide fluensulfone. PhD thesis. University of Southampton, Southampton, UK.

Kearn, J., Lilley, C., Urwin, P., O'Connor, V. and Holden-Dye, L. (2017) Progressive metabolic impairment underlies the novel nematicidal action of fluensulfone on the potato cyst nematode *Globodera pallida*. *Pesticide Biochemistry and Physiology* 142, 83–90. DOI: 10.1016/j.pestbp.2017.01.009

Kelley, K.W., Long, J.H. Jr and Todd, T.C. (2003) Long-term crop rotations affect soybean yield, seed weight, and soil chemical properties. *Field Crops Research* 83, 41–50. DOI: 10.1016/S0378-4290(03)00055-8

Kenter, C., Lukashyk, P., Daub, M. and Ladewig, E. (2014) Population dynamics of *Heterodera schachtii* Schm. and yield response of susceptible and resistant sugar beet (*Beta vulgaris* L.) after cultivation of susceptible and resistant oilseed radish (*Raphanus sativus* L.). *Journal für Kulturpflanzen* 66, 289–299. DOI: 10.5073/JFK.2014.09.01

Khan, A., Williams, K.L. and Nevalainen, H.K.M. (2004) Effects of *Paecilomyces lilacinus* protease and chitinase on the eggshell structures and hatching of *Meloidogyne javanica* juveniles. *Biological Control* 31, 346–352. http://dx.doi.org/10.1016/j.biocontrol.2004.07.011

Khan, A., Williams, K.L. and Navaleinen, H.K.M. (2006) Infection of plant-parasitic nematodes by *Paecilomyces lilacinus* and *Monacrosporium lysipagum*. *BioControl* 51, 659–678. DOI: 10.1007/s10526-005-4242-x

Koenning, S.R. and Barker, K.R. (1998) Survey of *Heterodera glycines* races and other plant-parasitic nematodes on soybean in North Carolina. *Journal of Nematology* (supplement) 30, 569–576.

LaMondia, J.A. and Brodie, B.B. (1984) Control of *Globodera rostochiensis* by solar heat. *Plant Disease* 68, 474–476. DOI: 10.1094/PD-69-474

Lazzeri, L., Leoni, O. and Manici, L.M. (2004) Biocidal plant dried pellets for biofumigation. *Industrial Crops and Products* 20, 59–65. http://dx.doi.org/10.1016/j.indcrop.2003.12.018

López-Robles, J., Olalla, C., Rad, C., Díez-Rojo, M.A., López-Pérez, J.A., Bello, A. and Rodríguez-Kábana, R. (2013) The use of liquid swine manure for the control of potato cyst nematode through soil disinfestation in laboratory conditions. *Crop Protection* 49, 1–7. DOI: 10.1016/j.cropro.2013.03.004

Lord, J.S., Lazzeri, L., Atkinson, H.J. and Urwin, P.E. (2011) Biofumigation for control of pale potato cyst nematodes: activity of *Brassica* leaf extracts and green manures on *Globodera pallida in vitro* and in soil. *Journal of Agricultural and Food Chemistry* 59, 7882–7890. DOI: 10.1021/jf200925k

Luangsa-ard, J., Houbraken, J., Doorn, T., Hong, S., Borman, A., Hywel-Jones, N. and Samson, R. (2011) *Purpureocillium*, a new genus for the medically important *Paecilomyces lilacinus*. *FEMS Microbiological Letters* 321, 141–149. DOI: 10.1111/j.1574-6968.2011.02322.x

Malinowska, E., Tyburski, J., Rychcik, B. and Szymczak-Nowak, J. (2005) The influence of *Solanum sisymbriifolium* on potato cyst nematode population reduction. In: Haverkort, A.J. and Struik, P.C (eds) *Potato in Progress: Science Meets Practice*. Wageningen Academic Press, Wageningen, The Netherlands, pp. 239–241. DOI: 10.3920/978-90-8686-562-8

Manzanilla-López, R.H., Esteves, I., Powers, S.J. and Kerry, B.R. (2011a) Effects of crop plants on abundance of *Pochonia chlamydosporia* and other fungal parasites of root-knot and potato cyst nematodes. *Annals of Applied Biology* 159, 118–129. DOI: 10.1111/j.1744-7348.2011.00479.x

Manzanilla-López, R.H., Esteves, I. and Finetti-Sialer, M.M. (2011b) *Pochonia chlamydosporia*: biological, ecological and physiological aspects in the host-parasite relationship of a biological control agent of nematodes. In: Boeri, F. and Chung, J.A. (eds) *Nematodes: Morphology, Functions and Management Strategies*. Nova Science Publishers Inc., New York, pp. 267–300.

Manzanilla-López, R.H., Esteves, I., Finetti-Sialer, M.M., Hirsch, P.R., Ward, E., Devonshire, J. and Hidalgo Díaz, L. (2013) *Pochonia chlamydosporia*: advances and challenges to improve its performance as a biological control agent of sedentary endo-parasitic nematodes. *Journal of Nematology* 45, 1–7.

Mao, L.-G., Wang, Q.-X., Yan, D.-D., Xie, H.-W., Li, Y., Guo, M.-X. and Cao, A.-C. (2012) Evaluation of the combination of 1,3-dichloropropene and dazomet as an efficient alternative to methyl bromide for cucumber production in China. *Pest Management Science* 68, 602–609. DOI: 10.1002/ps.2303

McDonald, A.H. and Nicol, J.M. (2005) Nematode parasites of cereals. In: Luc, M., Sikora, R.A. and Bridge, J. (eds) *Plant Parasitic Nematodes in Tropical and Subtropical Agriculture*, 2nd edn. CAB International, Wallingford, UK, pp. 131–191.

Meagher, J.W. and Rooney, D.R. (1966) The effect of crop rotations in the Victorian Wimmera on the cereal cyst nematode (*Heterodera avenae*) nitrogen fertility and wheat yield. *Australian Journal of Experimental Agriculture* 6, 425–431.

Melero-Vara, J.M., López-Herrera, C.J., Basallote-Ureba, M.J., Prados, A.M., Vela, M.D., Macias, F.J., Flor-Peregrín, E. and Talavera, M. (2012) Use of poultry manure combined with soil solarization as a control method for *Meloidogyne incognita* in carnation. *Plant Disease* 96, 990–996. DOI: org/10.1094/PDIS-01-12-0080-RE

Mendes, R., Kruijt, M., Brujin, I., Dekkers, E., van der Voort, M., Schneider, J.H.M., Piceno, Y.M., DeSantis, T.Z., Andersen, G. L., Bakker, P.A.H.M. and Raaijmakers, J.M. (2011) Deciphering the rhizosphere microbiome for disease-suppressive bacteria. *Science* 332, 1097–1099. DOI: 10.1126/science.1203980

Meyer, S.L.F. and Wergin, W.P. (1998) Colonization of soybean cyst nematode females, cysts, and gelatinous matrices by the fungus *Verticillium lecanii*. *Journal of Nematology* 30, 436–450.

Miller, D.R., Chen, S.Y., Porter, P.M., Johnson, G.A., Wyse, D.L., Stetina, S.R., Klossner, L.D. and Nelson, G.A. (2006) Rotation crop evaluation for management of the soybean cyst Nematode in Minnesota. *Agronomy Journal* 98, 569–578. DOI: 10.2134/agronj2005.0185

Mock, V.A., Creech, J.E., Ferris, V.R., Faghihi, J., Westphal, A., Santini, J.B. and Johnson, W.G. (2012) Influence of winter annual weed management and crop rotation on soybean cyst nematode (*Heterodera glycines*) and winter annual weeds: years four and five. *Weed Science* 60, 634–640. DOI: 10.1614/WS-D-11-00192.1

Mohan, S., Mauchline, T.H., Rowe, J., Hirsch, P.R. and Davies, K.G. (2012) *Pasteuria* endospores from *Heterodera cajani* (Nematoda: Heteroderidae) exhibit inverted attachment and altered germination in cross-infection studies with *Globodera pallida* (Nematoda: Heteroderidae). *FEMS Microbiology Ecology* 79, 675–684. DOI: 10.1111/j.1574-6941.2011.01249.x

Moosavi, M.R. and Askary, T.H. (2015) Nematophagous fungi: commercialization. In: Hassan, A.T. and Martinelli, P.R.P. (eds) *Biocontrol Agents of Phytonematodes*. CAB International, Wallingford, UK, pp. 187–202.

Morgan, D.G. (1925) Investigation on eelworm in potatoes in South Lincolnshire. *Journal of Helminthology* 3, 185–192. DOI: 10.1017/S0022149X00002017

Moriarty, F. (1963) The decline of a pea root eelworm (*Heterodera gottingiana* Liebscher) population in the absence of host plants. *The Journal of Agricultural Science* 61, 6–9. DOI: 10.1017/S0021859600013794

Nelson, K.A., Johnson, W.G., Wait, J.I.M.D. and Smoot, R.L. (2006) Winter-annual weed management in corn (*Zea mays*) and Soybean (*Glycine max*) and the impact on soybean cyst nematode (*Heterodera glycines*) egg population densities. *Weed Technology* 20, 965–970. DOI: 10.1614/WT-05-119.1

Ngala, B.M., Haydock, P.P.J., Woods, S. and Back, M.A. (2014) Biofumigation with *Brassica juncea*, *Raphanus sativus* and *Eruca sativa* for the management of field populations of the potato cyst nematode *Globodera pallida*. *Pest Management Science* 71, 759–769. DOI: 10.1002/ps.3849

Ngala, B.M., Woods, S. and Back, M.A. (2015) Sinigrin degradation and *G. pallida* suppression in soil cultivated with brassicas under controlled environmental conditions. *Applied Soil Ecology* 95, 9–14. DOI: 10.1016/j.apsoil.2015.05.009

Nishizawa, T., Shimizu, K. and Nagashima, T. (1972) Chemical and cultural control of the rice cyst nematode, *Heterodera oryzae* Luc et Berdon Brizuela, and hatching responses of the larvae to some root extracts. *Japanese Journal of Nematology* 2, 27–32.

Noel, G.R., Atibalentja, N. and Domier, L.L. (2005) Emended description of *Pasteuria nishizawae*. *International Journal of Systematic and Evolutionary Microbiology* 55, 1681–1685. DOI: 10.1099/ijs.0.63174-0

Noling, J.W. (1997) Movement and toxicity of nematicides in the plant root zone – Fact Sheet ENY-041. Available at: http://edis.ifas.ufl.edu/pdffiles/NG/NG00200.pdf (accessed 9 March 2017).

Norshie, P.M., Grove, I.G. and Back, M.A. (2016) Field evaluation of the nematicide fluensulfone for control of the potato cyst nematode *Globodera pallida*. *Pest Management Science* 72, 2001–2007. DOI: 10.1002/ps.4329.

Norshie, P.M., Grove, I.G. and Back, M.A. (2017) Persistence of the nematicide fluensulfone in potato (*Solanum tuberosum* ssp. *tuberosum*) beds under field conditions. *Nematology* 19, 739–747. DOI: 10.1163/15685411-00003085

NSP (n.d.) Best practice. Available at: http://nspstewardship.co.uk/best-practice/ (accessed 19 September 2017).

Nutter, F.W., Tylka, G.L., Guan, J., Moretra, A.J.D., Marett, C.C., Rosburg, T.R., Basart, J.P. and Chong, C.S. (2002) Use of remote sensing to detect soybean cyst nematode-induced plant stress. *Journal of Nematology* 34, 222–231.

Ohba, K., Hirao, F., Ishiuro, T. and Hayashi, Y. (1986) A simple and reliable device for monitoring fumigant residues in soil. *Journal of Nematology* 18, 421–422.

Oka, Y. (2010) Mechanisms of nematode suppression by organic soil amendments – a review. *Applied Soil Ecology* 44, 101–115. DOI: 10.1016/j.apsoil.2009.11.003

Osborn, R.K. (2005) Identification of microbes degrading nematicides and the development of a diagnostic assay for nematicide persistence in soil. PhD thesis. Harper Adams University, Newport, UK.

Ou, L.T. (1998) Enhanced degradation of the volatile fumigant nematicides 1,3-D and methyl bromide in soil. *Journal of Nematology* 30, 56–64.

Ou, L.T., Thomas, J.E. and Dickson, D.W. (1991) Enhanced biodegradation of the nematicide fenamiphos in soil. In: Linn, D.L. (ed.) *Sorption and Degradation of Pesticides and Organic Chemicals in Soil*. Soil Science Society of America and American Society of Agronomy, Denver, Colorado, pp. 253–260. DOI: 10.2136/sssaspecpub32.c14

Perry, R.N. and Beane, J. (1989) Effects of certain herbicides on the *in-vitro* hatch of *Globodera rostochiensis* and *Heterodera schachtii*. *Revue de Nématologie* 12, 191–196.

Pertot, I., Alabouvette, C., Hinarejos Esteve, E. and Franca, S. (2015) Mini-paper – the use of microbial biocontrol agents against soil-borne diseases. Available at: http://ec.europa.eu/eip/agriculture/sites/agrieip/files/8_eip_sbd_mp_biocontrol_final.pdf (accessed 1 November 2016).

Pest Control Products Board (PCPB) (2016) Products Registered for Use on Crops. Available at: http://www.pcpb.or.ke/cropproductsviewform.php (accessed 14 February 2018).

Pinto, S., Rosa, E. and Santos, S. (1998) Effect of 2-propenyl glucosinolate and derived isothiocyanate on the activity of the nematodes *Globodera rostochiensis* (Woll.). *Acta Horticulturae* 459, 323–327.

Pyrowolakis, A., Schuster, R.-P. and Sikora, R. (1999) Effect of cropping pattern and green manure on the antagonistic potential and the diversity of egg pathogenic fungi in fields with *Heterodera schachtii* infection. *Nematology* 1, 165–171. DOI: 10.1163/156854199508135

Rich, J.R., Dunn, R.A. and Noling, J.W. (2004) Nematicides: past and present uses. In: Chen, Z.X., Chen, S.Y. and Dickson, D.W. (eds) *Nematology, Advances and Perspectives. Volume 2: Nematode Management and Utilization*. CAB International, Wallingford, UK, pp. 1181–1200.

Riley, I.T., Hou, S. and Chen, S. (2010) Crop rotational and spatial determinants of variation in *Heterodera avenae* (cereal cyst nematode) population density at village scale in spring cereals grown at high altitude on the Tibetan Plateau, Qinghai, China. *Australasian Plant Pathology* 39, 424–430. DOI: 10.1071/AP10084

Roget, D.K., Neat, S.M. and Rovira, A.D. (1996) Effect of sowing point design and tillage practice on the incidence of rhizoctonia root rot, take-all and cereal cyst nematode in wheat and barley. *Animal Production Science* 36, 683–693. DOI: 10.1071/EA9960683

Rumbos, C., Mendoza, A., Sikora, R. and Kiewnick, S. (2008) Persistence of the nematophagous fungus *Paecilomyces lilacinus* strain 251 in soil under controlled conditions. *Biocontrol Science Technology* 18, 1041–1050. http://dx.doi.org/10.1080/09583150802526979

Runia, W., Molendijk, L., Ludeking, D. and Schomaker, C. (2012) Improvement of anaerobic soil disinfestation. *Communications in Agricultural and Applied Biological Sciences* 77, 753–762.

Runia, W.T., Thoden, T.C., Molendijk, L.P.G., van den Berg, W., Termorshuizen, A.J., Streminska, M.A., van der Wurff, A.W.G., Feil, H. and Meints, H. (2014) Unravelling the mechanism of pathogen inactivation during anaerobic soil disinfestation. *Acta Horticulturae* 1044, 177–193. DOI: 10.17660/ActaHortic.2014.1044.21

Saifullah, S.M. and Thomas, B.J. (1996) Studies on the parasitism of *Globodera rostochiensis* by *Trichoderma harzianum* using low temperature scanning electron microscopy. *Afro-Asian Journal of Nematology* 6, 117–122.

Sakorn, P., Rakariyathamb, N., Niamsupb, H. and Kovitayac, P. (1999) Sinigrin degradation by *Aspergillus* sp. NR-4201 in liquid culture. *ScienceAsia* 25, 189–194.

Salas-Marina, M., Silva-Flores, M., Uresti-Rivera, E., Castro-Longoria, E., Herrera-Estrella, A. and Casas-Flores, S. (2011) Colonization of *Arabidopsis* roots by *Trichoderma atroviride* promotes growth and enhances systemic disease resistance through jasmonic acid/ethylene and salicylic acid pathways. *European Journal of Plant Pathology* 131, 15–26. DOI: 10.1007/s10658-011-9782-6

Sasaki-Crawley, A. (2012) Signalling and behaviour of *Globodera pallida* in the rhizosphere of the trap crop *Solanum sisymbriifolium*. PhD thesis. University of Plymouth, Plymouth, UK.

Sasaki-Crawley, A., Curtis, A., Birkett, M., Powers, S., Papadopoulis, A., Blackshaw, R. and Kerry, B.R. (2010) Signalling and behaviour of potato cyst nematode in the rhizosphere of the trap crop, *Solanum sisymbriifolium*. *Aspects of Applied Biology* 103, 45–51.

Sasser, N. and Uzzell, G. (1991) Control of the soybean cyst nematode by crop rotation in combination with a nematicide. *Journal of Nematology* 23, 344–347.

Scholte, K. (2000) Effect of potato used as a trap crop on potato cyst nematodes and other soil pathogens and on the growth of a subsequent main potato crop. *Annals of Applied Biology* 136, 229–238. DOI: 10.1111/j.1744-7348.2000.tb00029.x

Scholte, K. and Vos, J. (2000) Effects of potential trap crops and planting date on soil infestation with potato cyst nematodes and root-knot nematodes. *Annals of Applied Biology* 137, 153–164. DOI: 10.1111/j.1744-7348.2000.tb00047.x

Schroeder, N.E. and MacGuidwin, A.E. (2010) Mortality and behavior in *Heterodera glycines* juveniles following exposure to isothiocyanate compounds. *Journal of Nematology* 42, 194–200.

Sellitto, V., Curto, G., Dallavalle, E., Ciancio, A., Colagiero, M., Pietrantonio, L., Bireescu, G., Stoleru, V. and Storari, M. (2016) Effect of *Pochonia chlamydosporia*-based formulates on the regulation of root-knot nematodes and plant growth response. *Frontiers in Life Science* 9, 1–5. http://dx.doi.org/10.1080/21553769.2016.1193827

Sharma, S.B. and Nene, Y.L. (1992) Spatial and temporal dynamics of plant-parasitic nematodes on pigeonpea in alfisols and vertisols. *Nematropica* 22, 13–20.

Sharma, S.B., Rego, T.J., Mohiuddin, M. and Rao, V.N. (1996) Regulation of population densities of *Heterodera cajani* and other plant-parasitic nematodes by crop rotations on vertisols in semi-arid tropical production systems in India. *Journal of Nematology* 28, 244–251.

Sharon, E., Illan, C. and Spiegel, Y. (2011) *Trichoderma* as biological control agent. In: Davies, K. and Spiegel, Y. (eds) *Biological Control of Plant Parasitic Nematodes: Building Coherence between Microbial Ecology and Molecular Mechanisms.* Springer, Dordrecht, The Netherlands, pp. 183–201. DOI: 10.1007/978-1-4020-9648-8_8

Shepherd, A.M. (1962) *The Emergence of Larvae from Cysts in the Genus* Heterodera. Commonwealth Agricultural Bureaux, St. Albans, UK.

Siddiqui, Z.A. and Mahmood, I. (1999) Role of bacteria in the management of plant parasitic nematodes – a review. *Bioresource Technology* 69, 167–179. https://doi.org/10.1016/S0960-8524(98)00122-9

Siddiqui, I.A. and Shaukat, S.S. (2004) *Trichoderma harzianum* enhances the production of nematicidal compounds in vitro and improves biocontrol of *Meloidogyne javanica* by *Pseudomonas fluorescens*. *Letters in Applied Microbiology* 38, 169–175. DOI: 10.1111/j.1472-765X.2003.01481.x

Silva, J.C.T., Oliveira, R.D.L., Jham, G.N. and Naylor, D.C.A. (2008) Effect of neem seed extracts on the development of the soybean cyst nematode. *Tropical Plant Pathology* 33, 171–179. DOI: 10.1590/S1982-56762008000300001

Smedley, E.M. (1939) Experiments on the use of isothiocyanates in the control of the potato strain of *Heterodera schachtii* (Schmidt). *Journal of Helminthology* 17, 31–38. DOI: 10.1017/S0022149X00031011

Smelt, J.H. and Leistra, M. (1992) Availability, movement and (accelerated) transformation of soil-applied nematicides. In: Gommers, F.J. and Maas, P.W.T. (eds) *Nematology from Molecule to Ecosystem – Proceedings 2nd International Nematology Congress*, Veldhoven, The Netherlands.

Spaull, A.M., Trudgill, D.L. and Batey, T. (1992) Effects of anaerobiosis on the survival of *Globodera pallida* and possibilities for control. *Nematologica* 38, 88–97. DOI: 10.1163/187529292X00072

Stapleton, J.J. (2000) Soil solarization in various agricultural production systems. *Crop Protection* 19, 837–841. http://dx.doi.org/10.1016/S0261-2194(00)00111-3

Stapleton, J.J., Quick, J. and Devay, J.E. (1985) Soil solarization: effects on soil properties, crop fertilization and plant growth. *Soil Biology and Biochemistry* 17, 369–373. DOI: 10.1016/0038-0717(85)90075-6

Steele, A.E. (1973) The effects of hot water treatments on survival of *Heterodera schachtii*. *Journal of Nematology* 5, 81–84.

Stirling, G.R. (2011) Biological control of plant parasitic nematodes: an ecological perspective, a review of progress and opportunities for further research. In: Davies, K. and Spiegel, Y. (eds) *Biological Control of Plant-Parasitic Nematodes: Building Coherence Between Microbial Ecology and Molecular Mechanisms*. Springer, Dordrecht, The Netherlands, pp. 1–38. DOI: 10.1007/978-1-4020-9648-8_1

Szymczak-Nowak, J., Malinowska, E., Tyburski, J. and Rychcik, B. (2007) Influence of *Solanum sisymbriifolium* on potato cyst nematode population reduction. *Progress in Plant Protection* 47, 224–226.

Tedford, E.C. and Inglis, D.A. (1999) Evaluation of legumes common to the Pacific Northwest as hosts for the pea cyst nematode, *Heterodera goettingiana*. *Journal of Nematology* 31, 155–163.

Thomas, S.H., Schroeder, J. and Murray, L.W. (2005) The role of weeds in nematode management. *Weed Science* 53, 923–928. DOI: http://dx.doi.org/10.1614/WS-04-053R.1

Tian, B., Yang, J. and Zhang, K.Q. (2007) Bacteria used in the biological control of plant parasitic nematodes: populations, mechanisms of action, and future prospects. *FEMS Microbiology Ecology* 61, 197–213. DOI: 10.1111/j.1574-6941.2007.00349.x

Timmermans, B.G.H., Vos, J., Stomph, T.J., Van Nieuwburg, J. and Van der Putten, P.E.L. (2006) Growth duration and root length density of *Solanum sisymbriifolium* (Lam.) as determinants of hatching of *Globodera pallida* (Stone). *Annals of Applied Biology* 148, 213–222. DOI: 10.1111/j.1744-7348.2006.00056.x

Timmermans, B.G.H., Vos, J., Van Nieuwburg, J., Stomph, T.J., Van der Putten, P.E.L. and Molendijk, P.G. (2007a) Field performance of *Solanum sisymbriifolium*, a trap crop for potato cyst nematodes. I. Dry matter accumulation in relation to sowing time, location, season and plant density. *Annals of Applied Biology* 150, 89–97. DOI: 10.1111/j.1744-7348.2006.00112.x

Timmermans, B.G.H., Vos, J., Stomph, T.J., Van Nieuwburg, J. and Van der Putten, P.E.L. (2007b) Field performance of *Solanum sisymbriifolium*, a trap crop for potato cyst nematodes. II. Root characteristics. *Annals of Applied Biology* 150, 99–106. DOI: 10.1111/j.1744-7348.2006.00113.x

Timper, P. (2014) Conserving and enhancing biological control of nematodes. *Journal of Nematology* 46, 75–89.

Triffit, M.J. (1929) Preliminary researches on mustard as a factor inhibiting cyst formation in *Heterodera schachtii*. *Journal of Helminthology* 7, 81–92. DOI: 10.1017/S0022149X00002431

Triffit, M.J. (1930) On the bionomics of *Heterodera schachtii* on potatoes, with special reference to the influence of mustard on the escape of larvae from the cysts. *Journal of Helminthology* 8, 19–48. DOI: 10.1017/S0022149X00002509

Trifonova, Z. and Atanasov, A. (2011) Control of potato cyst nematode *Globodera rostochiensis* with some plant extracts and neem products. *Bulgarian Journal of Agricultural Science* 17, 623–627.

Trudgill, D.L., Phillips, M.S. and Elliott, M.J. (2014) Dynamics and management of the white potato cyst nematode *Globodera pallida* in commercial potato crops. *Annals of Applied Biology* 164, 18–34. DOI: 10.1111/aab.12085

Turner, S.J., Martin, T.J.G., McAleavey, P.B.W. and Fleming, C.C. (2006) The management of potato cyst nematodes using resistant Solanaceae potato clones as trap crops. *Annals of Applied Biology* 149, 271–280. DOI: 10.1111/j.1744-7348.2006.00089.x

Twomey, U., Rolfe, R., Warrior, P. and Perry, R.N. (2002) Effects of the biological nematicide, DiTera®, on movement and sensory responses of second-stage juveniles of *Globodera rostochiensis* and stylet activity of *G. rostochiensis* and fourth-stage juveniles of *Ditylenchus dipsaci*. *Nematology* 4, 909–915. DOI: 10.1163/156854102321122520

van Overbeek, L.S., Runia, W., Kastelein, P. and Molendijk, L. (2014) Anaerobic disinfestation of tare soils contaminated with *Ralstonia solanacearum* biovar 2 and *Globodera pallida*. *European Journal of Plant Pathology* 138, 323–330. DOI: 10.1007/s10658-013-0331-3

Vanstone, V.A., Hollaway, G.J. and Stirling, G.R. (2008) Managing nematode pests in the southern and western regions of the Australian cereal industry: continuing progress in a challenging environment. *Australasian Plant Pathology* 37, 220–234. DOI: 10.1071/AP08020

Venkatesh, R., Harrison, S.K., Regnier, E.E. and Riedel, R. (2004) Purple deadnettle effects on soybean cyst nematode populations in no-till soybean. In: Hartzler, R.G. and Hartzler, A.N. (eds) *Proceedings of the North Central Weed Science Society Vol 59*. North Central Weed Science Society, Champaign, Illinois, p. 56.

Vitti, A.J., Rezende Neto, U. da R., de Araújo, F.G., Santos, L.C., Barbosa, K.A.G. and da Rocha, M.R. (2014) Effect of soybean seed treatment with abamectin and thiabendazole on *Heterodera glycines*. *Nematropica* 44, 74–80.

Ward, E., Kerry, B.R., Manzanilla-López, R.H., Mutua, G., Devonshire, J., Kimenju, J. and Hirsch, P.R. (2012) The *Pochonia chlamydosporia* serine protease gene *vcp1* is subject to regulation by carbon, nitrogen and pH: implications for nematode biocontrol. *PLoS ONE* 7, e35657.

Watson, R.T., Albritton, D.L., Anderson, S.O. and Lee-Bapty, S. (1992) Methyl bromide: Its atmospheric science, technology and economics. *Montreal Protocol Assessment Supplement, United Nations Environmental Programme on Behalf of the Contracting Parties to the Montreal Protocol*, Nairobi, Kenya.

Watts, W.D.J., Grove, I.G., Hand, P. and Back, M.A. (2015) *Brassica* residue incorporation technique for optimized biofumigation of potato cyst nematodes. *Aspects of Applied Biology* 130, 91–99.

Whitehead, A.G. (1968) Chemical control: a soil treatment. In: Southey J.F. (ed.) *Plant Nematology*. HMSO, London, pp. 283–296.

Whitehead, A.G. (1988) *Plant Nematode Control*. CAB International, Wallingford, UK.

Wilson, M.J. and Jackson, T.A. (2013) Progress in commercialisation of bionematicides. *BioControl* 58, 715–722. DOI: 10.1007/s10526-013-9511-5

Winslow, R.D. (1955) The hatching responses of some root eelworms of the genus *Heterodera*. *Annals of Applied Biology* 43, 19–36. DOI: 10.1111/j.1744-7348.1955.tb02450.x

Woods, S.R. (1997) The placement, fate and effectiveness of granular nematicides in potato beds infested with the potato cyst nematode *Globodera pallida* (Stone). PhD thesis, Open University, Milton Keynes, UK.

Yedidia, I., Benhamou, N. and Chet, I. (1999) Induction of defense responses in cucumber plants (*Cucumis sativus* L.) by the biocontrol agent *Trichoderma harzianum*. *Applied and Environmental Microbiology* 65, 1061–1070.

Zhang, S., Gan, Y. and Xu, B. (2014) The parasitic and lethal effects of *Trichoderma longibrachiatum* against *Heterodera avenae*. *Biological Control* 72, 1–8. http://dx.doi.org/10.1016/j.biocontrol.2014.01.009

Zhang, J., Li, Y., Yuan, H., Sun, B. and Li, H. (2016) Biological control of the cereal cyst nematode (*Heterodera filipjevi*) by *Achromobacter xylosoxidans* isolate 09X01 and *Bacillus cereus* isolate 09B18. *Biological Control* 92, 1–6. http://dx.doi.org/10.1016/j.biocontrol.2015.08.004

Zheng, Z., Zheng, J., Zhang, Z., Peng, D. and Sun, M. (2016) Nematicidal spore-forming bacilli share similar virulence factors and mechanisms. *Scientific Reports* 6, 31341. DOI: 10.1038/srep31341

14 General Morphology of Cyst Nematodes

James G. Baldwin[1] and Zafar A. Handoo[2]

[1]University of California, Riverside, California, USA; [2]USDA, ARS, BARC-West, Beltsville, Maryland, USA

14.1 Introduction	337
14.2 Egg and Embryo	338
14.3 Second- to Fourth-stage Juveniles	340
14.4 Males	346
14.5 Females	347
14.6 Cysts	350
14.7 Techniques	352
14.8 Minimal Standards for Species Descriptions	357
14.9 Conclusions and Future Prospects	359
14.10 References	360

14.1 Introduction

Within Tylenchomorpha, including most plant-parasitic nematodes, cyst nematodes (Heteroderinae) are morphologically distinctive among Hoplolaimidae[1] consistent with adaptations for sedentary parasitism and the capacity for dormancy or suspended development. Most important among these adaptations is the cyst, a structure that evolved within heteroderids[2] and that has been defined as 'a persistent tanned sac which retains eggs and is derived from some or all components of the mature female body wall' (Luc *et al.*, 1986). Although this type of cyst is unique to heteroderids, convergent morphological adaptations for sedentary parasitism throughout diverse groups of Tylenchomorpha, including *Meloidogyne*, *Rotylenchulus*, *Tylenchulus* and *Nacobbus*, have been the basis for many misunderstandings of relationships. However, these controversies have been largely addressed by molecular phylogenetics and by detailed morphology that underscores the uniqueness of features particular to cyst nematodes of Heteroderinae. These unique features throughout the group point to specialized functions and taxon-specific expressions that are crucial for classical identification and for understanding phenotypic evolution of the group (Baldwin and Mundo-Ocampo, 1991).

The life history of heteroderids, described in detail by Raski (1950), is consistent with other

---

[1] We follow De Ley and Blaxter (2002) with placement of Heteroderinae within Hoplolaimidae.
[2] Although some heteroderids (Heteroderinae) lack cysts, this feature is plesiomorphic within the group (Baldwin and Schouest, 1990; Baldwin, 1992).

Tylenchomorpha with respect to having five (juvenile and adult) stages separated by four moults (Fig. 14.1). The embryo, first stage and first moult occur exclusively within the egg and, potentially following a period of dormancy, hatching is as an infective second-stage juvenile (J2) (see Chapter 3, this volume). While J2 destined to be males or females all include a body wall, digestive, reproductive, nervous and secretory-excretory system, these systems differ in J2 and subsequent stages based on adaptations of particular phases of the life history with respect to the stage being active or dormant, migratory or sedentary, sexually immature or mature and parasitic or free-living. Once the J2 establishes a feeding site in the host the J2 begins to swell and then undergoes successive moults. Adults are sedentary obese egg-producing females and, in most cases, also present are migratory non-feeding males. Earlier juvenile stages developing as males are swollen and sedentary but development proceeds with metamorphosis to a vermiform fourth-stage juvenile (J4) coiled within the sac-like ovoid casing of earlier stages from which it emerges as an elongate migratory adult (Figs 14.1 and 14.7). Adult males do not feed and die soon after mating. Adult females mature, die and transform into cysts.

14.2 Egg and Embryo

Eggs vary little among Heteroderinae and generally they are not distinctive from those of other Tylenchomorpha. Although suggestions of taxon-specific variability in size are often included in species descriptions, studies suggest size is probably influenced by external factors and it has not proven to be useful to identification. Notably some species of *Cactodera* (Heteroderinae) have diagnostic surface punctations, as seen with light microscopy, and they are further

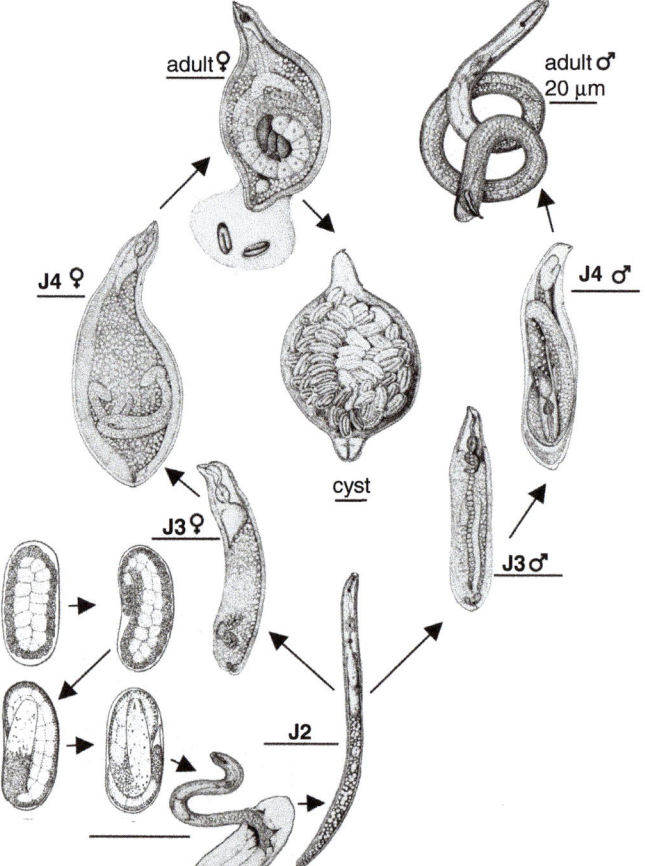

Fig. 14.1. Life cycle of cyst nematode (*Heterodera schachtii*) on sugar beet. (Adapted from C. Papp in Raski, 1950.)

shown by scanning electron microscopy (SEM) to be tubercle-like protuberances. Eggs are transparent such that within them embryonic, first-stage juveniles (J1) and pre-hatch J2 can be examined (Figs 14.1 and 14.2B). Within the egg, the J1 has three to five folds and not until after the first moult does a stylet form and cuticle striation is expressed.

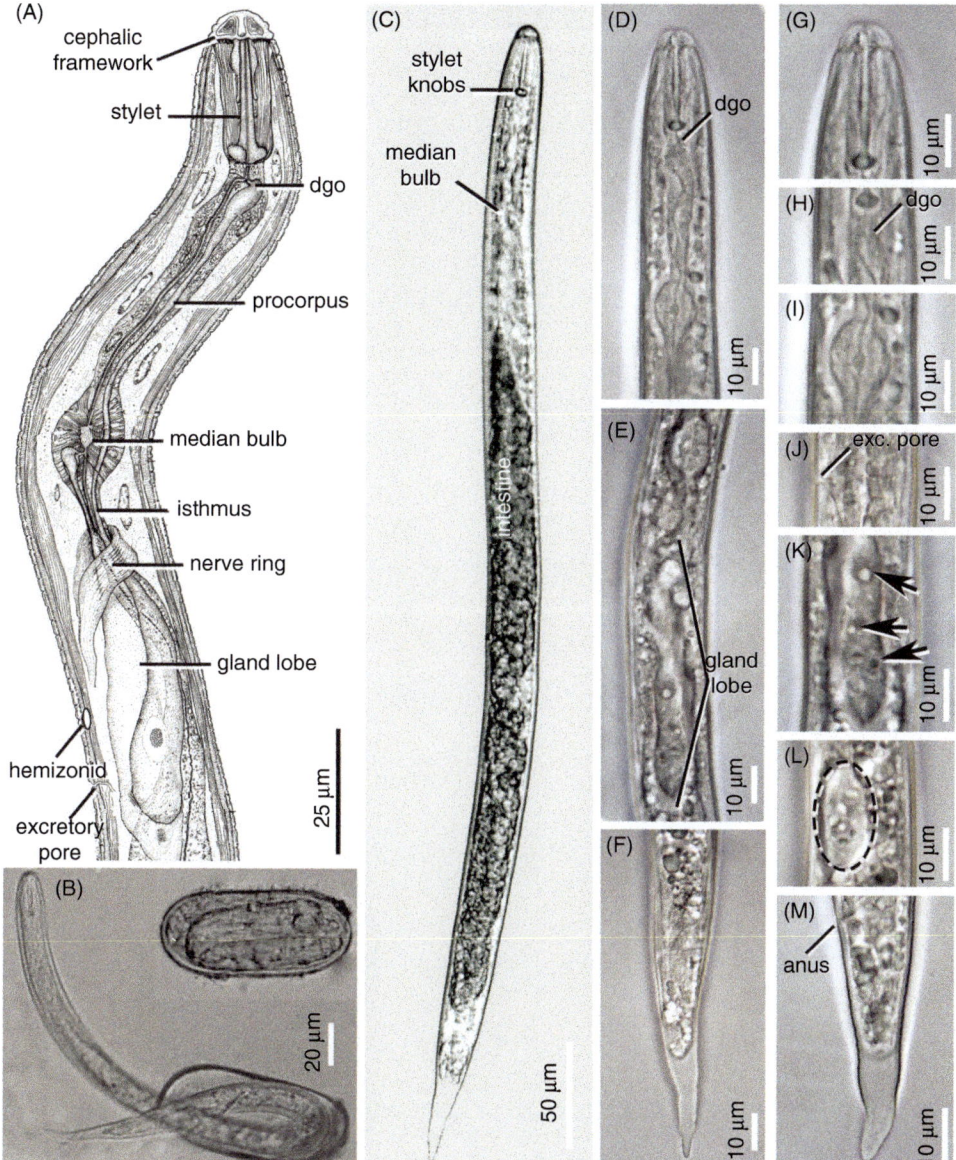

Fig. 14.2. Second-stage juvenile of cyst nematodes. (A) Anterior end including pharyngeal region of *Heterodera glycines* (modified from Endo, 1979, courtesy of USDA Nematode Laboratory). (B) Egg and hatching juvenile of *H. sojae* (courtesy of Kang *et al.*, 2016 and *Journal of Nematology*). (C) Entire juvenile, lateral view. (D) Anterior end including median bulb. (E) Pharyngeal region including gland lobe. (F) Tail region showing hyaline tip. (G) Stylet enlarged from 2D. (H) Enlargement of dorsal gland orifice (dgo) from 2D. (I) Median bulb enlarged from 2D. (J) Excretory pore. (K) Gland lobe showing position of nuclei enlarged from 2E. (L) Primordium. (M) Tail region showing anus.

14.3 Second- to Fourth-stage Juveniles

Eggs with J2 ready to hatch may occur within cysts, or in some species they may also be deposited within eggsacs (Figs 14.1 and 14.6F); after hatching J2 may be found within host roots or in the soil. Recognition of distinctive heteroderid J2 in the soil is often the 'first alert' to then examine adjacent roots for the association of females and cysts and the more specific morphological identification they afford. Where dormancy occurs, it is as the pre-hatch J2 within the cyst (see Chapter 3, this volume). Upon hatching the J2 morphology, including somatic musculature and cuticular features conferring strength/flexibility, reflects the temporary role of the J2 as a free-living migratory stage, whereas its fully formed protrusible stylet and robust digestive system are indicative of an infective role in penetrating the host and establishing a feeding site (Fig. 14.3). Once a feeding site is established the J2 becomes sedentary and undergoes significant changes as it swells and commences development of the reproductive system, and the rapid sequence of moults through J3 and J4 to the adult. Generation time is highly variable among species and dependent on environmental conditions, but often it is completed in less than 30 days.

The heteroderid J2 is distinguished by its size of 300–700 µm, overall shape including a rounded head continuous with the body contour and tail that tapers to a point; the tail also includes a notable hyaline region, the presence of which is

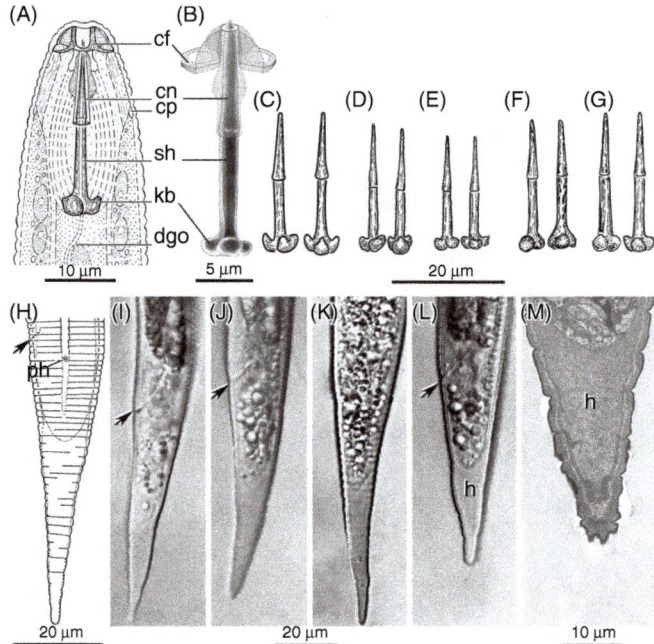

Fig. 14.3. The stylet and tail region of second-stage juveniles. (A) Diagrammatic interpretation of stylet components (after Hirschmann, 1956). (B) 3D dorsal view of stylet in relation to labial framework and vestibule extension (after Baldwin from Subbotin et al., 2010). (C–G) Diagrammatic interpretation of stylets (left is lateral view, right is dorsal view). (C) Heterodera mani. (D) H. trifolii. (E) H. zeae stylet. (F) Globodera rostochiensis. (G) G. pallida. (H) Diagrammatic interpretation of tail (after Wouts and Weischer, 1977). (I) H. zeae tail. (J) H. schachtii tail. (K) H. trifolii tail. (L) H. glycines tail. (M) Transmission electron micrograph of H. glycines tail (after Endo, courtesy USDA Nematode Laboratory). Arrow indicates position of anus. Abbreviations: cf = cephalic framework, cn = cone of stylet, cp = cephalid, dgo = dorsal gland orifice, sh = shaft of stylet, h = hyaline region of tail, kb = knobs of stylet, ph = phasmid. C–G after Wouts and Baldwin (1998), I–L after Mulvey and Golden, 1983, courtesy of USDA Nematology Laboratory and Journal of Nematology.

distinctive in heteroderids relative to outgroups. The size and shape, tail and hyaline region length may be useful in species diagnostics (Fig. 14.3). The body wall of the J2 comprises an outer layer of cuticle underlain by epidermis (= hypodermis) and internally by a layer of platymyarian somatic muscles (Fig. 14.4). The cuticle is superficially striated and striae are interrupted on each lateral side with a field (lateral field) marked by what appears in the light microscope as the diagnostic feature of three vs four longitudinal lines, the 'lines' being an expression of incisures between alae (in this case small longitudinal elevations) (Figs 14.4 and 14.5I, J). Interspecific variations in the patterns include those of the anterior and posterior termini and extent of areolation of lateral lines. Post infection, cuticular striations diminish through developmental stages and lateral

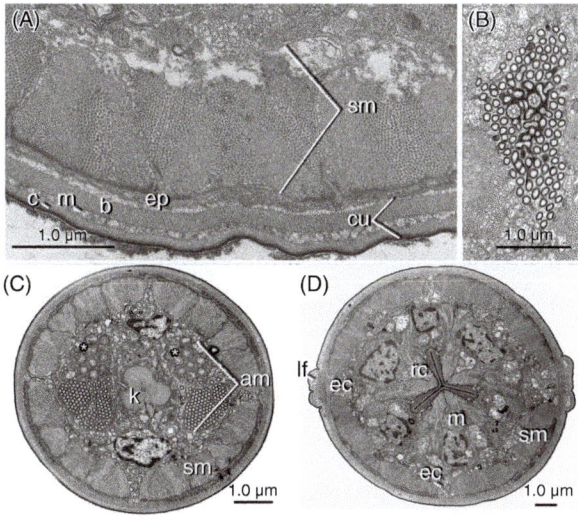

Fig. 14.4. Transmission electron micrographs of transverse sections of second-stage juveniles of *Heterodera glycines* (after Endo, courtesy of USDA Nematology Laboratory). (A) Body wall showing platymyarian somatic muscles (sm), epidermis (ep) and cuticle (cu) comprising cortex (c), medial (m) and basal (b) layers. (B) amphid including microvilli of the finger cell. (C) Level of the stylet knobs (k) including amphids (am). Asterisks (*) denote position of ciliary region of amphid sensilla. (D) Level of the median bulb including radial muscle (rc) and marginal (m) cells. Note epidermal chords (ec) and lateral field (lf).

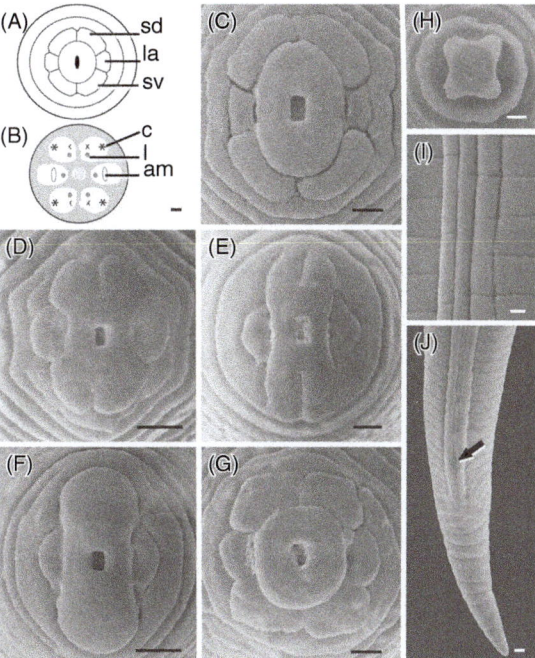

Fig. 14.5. Surface patterns of second-stage juveniles, unless otherwise indicated. (A) Diagrammatic representation of the basic pattern of the lip region including labial (la), subdorsal (sd) and subventral (sv) lips. (B) Transverse view of sensilla of the labial region including cephalic (c) and labial (l) papillae and amphid (am) opening. (C) *Cactodera cacti* lip pattern. (D) *Heterodera glycines* lip pattern. (E) *H. trifolii* lip pattern. (F) *H. avenae* lip pattern. (G) *H: cruciferae* lip pattern. (H) female lip pattern of *H. glycines*. (I) Lateral field of *Punctodera chalcoensis*. (J) Lateral view of tail of *H. fici*. Arrow indicates phasmid opening. (A–D, F–J after Baldwin in Subbotin et al., 2010). All scale bars are 1.0 µm.

lines are lost early in the sedentary transformations. By J3 there is evidence of transformation of surface patterns from striated towards the more irregular patterns characteristic of adults (Hesling, 1978).

Transmission electron microscopy (TEM) reveals that the body wall cuticle comprises three basic layers of a cortex (with a thin osmophilic external region), a 'spongy' medial and an internal striated basal layer; this basic three-layered pattern, widespread throughout the Chromodorea, is interpreted as plesiomorphic (Fig. 14.4A) (Baldwin, 1992). The three cuticle layers persist through the sedentary J2 and J3, but it is unknown to what extent they are expressed through the J4 in transition to the highly modified cuticle that characterizes adult females.

A key expression of surface cuticular patterns in J2 is that of the lip region, particularly as viewed with SEM, and specific variations of these patterns often have diagnostic value (Fig. 14.5) (Stone, 1975; Othman et al., 1988; Baldwin and Schouest, 1990). Components of these patterns are demarcated by lines of shallow indentations. The basic lip pattern, widespread throughout Heteroderinae, is a labial disc surrounded by six lip sectors, two lateral as well as subdorsal and subventral pairs (Fig. 14.5A). Variations occur through partial or complete fusion of the labial disc with certain lip sectors and/or with fusion of adjacent lip sectors (Fig. 14.5C–H). The head region also varies by the number of additional annules as well as the possibility of partial or complete additional longitudinal incisures posterior to the lips.

The cellular/syncytial epidermal layer underlies the cuticle and during moults it plays a key role in dissolution and secretion of the cuticle (Fig. 14.4C). It occurs as a thin 'interchordal' region and especially laterally, as well as (to a lesser extent) dorsally and ventrally, it expands to thickened chords that extend most of the length of the J2 (Fig. 14.4D). Cuticle adjacent to lateral chords and expressing the lateral field may be slightly thickened and layering may be modified. Anteriorly the epidermis underlies and secretes a cuticular cephalic framework (see description below) surrounding the stoma creating a conduit for the protrusible stylet and anterior point of attachment of stylet protractors (Figs 14.2, 14.3A, B and 14.5B). Posteriorly, the epidermis extends to the tail tip but in the J2 it terminates just anterior to the cuticular hyaline region (Fig. 14.3H–M). The epidermis, particularly of J2, is closely associated with specialized cells that form cuticularized ducts that penetrate the cuticle including the secretory-excretory duct and the rectum/anus, as well as with the socket cells that form cuticular connections of sensory structures. During the J4 and final moult the epidermis plays a key role in formation of secondary sexual cuticular structures associated with the developing vulva/vagina and male cloaca.

The innermost component of the body wall is the single layer of longitudinally spindle-shaped overlapping platymyarian somatic muscle cells organized in four fields separated by the epidermal chords (Fig. 14.4A–C). As is the case for platymyarian muscles, the contractile region underlies the epidermis, whereas a non-contractile region including the cell nucleus expands into the body cavity and one or more innervation processes extend from the non-contractile region to synapses with neurons. The contractile region is composed of bands of thin and thick myofilaments organized consistent with obliquely striated musculature (Fig. 14.4A). Whilst somatic muscles are clearly essential during the migratory stages, once a feeding site is established somatic musculature deteriorates apparently then being primarily limited to the head region of feeding stages and subsequently being re-established in migratory males.

The digestive system, including the stomatostylet, pharynx, intestine, rectum and anus, is well-formed beginning in the J2 infective stage (Fig. 14.2). Detailed information on digestive and other systems are given in an excellent book chapter 'Nematode morphology, sensory structure and function', by Baldwin and Perry (2004). The stylet, appearing in the first moult primarily as a product of arcade syncytia, is an expression of the stoma as a protrusible needle-like cuticular structure that includes a lumen through which food passes posteriorly into the pharynx. However, prior to a role in feeding, the J2 stylet has an important role in perforating the egg shell and hatching (Doncaster and Shepherd, 1967; see Chapter 3, this volume). The J2 stylet is somewhat flexible (Zunke and Eisenback, 1998), but how this flexibility functions in relation to its role in hatching or feeding needs to be further explored. Anteriorly the stylet is a cone that tapers to a point and thus is adapted to penetrating the egg shell and then host cells (Fig. 14.2A, C, D, G and 14.3A–G). Posterior to the cone the stylet also includes a columnar shaft that further posteriorly merges into three (dorsal and two subventral) enlarged knobs (Fig. 14.4C). The stylet knobs are embedded in and produced by the anterior end of the pharynx from which three corresponding stylet protractor muscles extend anteriorly to the cuticularized cephalic framework and body wall. Contraction of the protractors moves the stylet anteriorly through the guiding cuticular conduit including a tube-like vestibule that posteriorly is continuous with a vestibule extension; the vestibule, lying just posterior to the lip region and stoma opening, is anchored by six cuticular radii extending from the vestibule/hub to the body wall and it is the combination of the vestibule and radii that comprise the cephalic framework. Contraction of stylet protractors counteracts elasticity of the adjacent alimentary tract; thus, the stylet retracts when protractors relax. Details of the J2 stylet may be useful in species diagnosis. These include stylet length, from 16–28 µm, overall robustness, and details of the size and shape of stylet knobs, including rounded or anchor-shaped (Fig. 14.3A–G). These features, however, often vary intra-specifically so that assessing differences typically requires examining a significant number of individuals and consideration in the context of additional diagnostic features.

Posterior to the stylet is the pharynx, including corpus, isthmus and glandular lobe; the corpus is further divided into a procorpus and metacorpus (Fig. 14.2). The lumen of the stylet is continuous with the cuticularized lumen of the pharynx; the stylet protractor muscles posteriorly are expressed as non-contractile regions that primarily comprise the procorpus. Within the procorpus, which is non-muscular, the lumen is round in cross-section and anteriorly it is penetrated by a cuticularized dorsal gland duct (dgo) (Figs 14.2A, D and 14.3A). This duct is the anterior terminus from the gland lobe and its position relative to the base of the stylet knobs (4–12 µm) varies among species and thus has diagnostic value.

Posterior to the procorpus the corpus enlarges to an oval muscular median bulb (metacorpus) and its anterior end is demarcated by 'constraining muscles' (Endo, 1984) (Fig. 14.2A, I). Within the median bulb the cuticularized lumen is triradiate and surrounded by non-muscular marginal cells peripheral to the apicies. Between the apicies and marginal cells are muscular radial cells oriented such that upon contraction they open the triradiate lumen (Fig. 14.4C). This opening creates pressure that results in ingesting food through the stylet and pharynx lumen. In subventral positions near the posterior end of the median bulb are a pair of cuticularized ducts associated with anterior processes of subventral glands; these open into the lumen of the median bulb. Relative to J2 the median bulb increases in size in J3, J4 and it reaches maximum size in adult females (Fig. 14.6A, B).

The median bulb, with its triradiate lumen, posteriorly merges with the narrower elongate isthmus that is primarily comprised of dorsal and subventral gland processes embedded within a framework of constraining and radial muscles (Fig. 14.2A). Associated with radial muscles, the cuticularized lumen of the isthmus is triradiate suggesting that, as in the median bulb, the lumen can be opened by contracting muscles. The pharyngeal lumen extends posteriorly into the gland lobe where it merges with the intestinal lumen. The transition with the lumen is a pair of cells that comprise the pharyngeal-intestinal valve. Within the gland lobe the dorsal gland predominates anteriorly and its large nucleus typically occurs at about the level of the pharyngeal-intestinal valve. Posteriorly, the lobe is primarily composed of a pair of subventral glands and typically the nucleus of one subventral gland is positioned anteriorly to the other (Fig. 14.2K). All the pharyngeal glands appear to be robust and functional in feeding juvenile stages although they may change in size and shape; Endo (1984) describes changes in apparent activity of subventral glands in the transition from J2 to J3.

The cuticle lining of the digestive system ends posteriorly at the pharyngeal-intestinal valve where it transitions to the intestinal lumen. The intestine is composed of large epithelial cells and in cyst nematodes the lumen is poorly defined; unlike most other nematodes it apparently lacks a border of microvilli (Endo, 1988; Borgonie et al., 1995). The intestine terminates posteriorly in a cuticle-lined rectum and anus that appears to be present and functional throughout all stages (Figs 14.2A and 14.3H–J, L); an exception is the last steps of the J4 moult to the male in which the anus is not observed and the region is

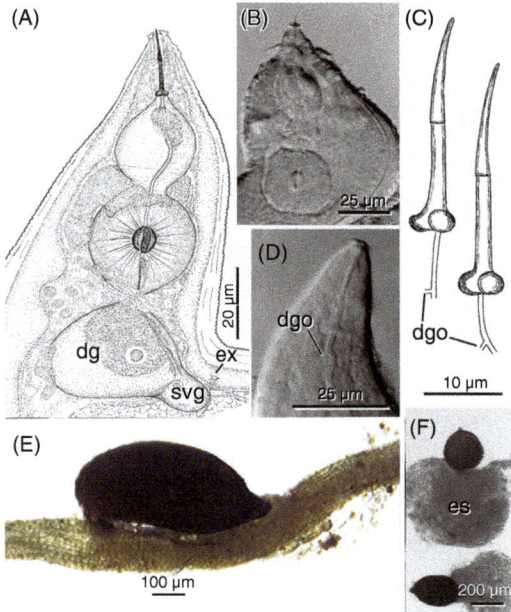

Fig. 14.6. Females of cyst nematodes. (A) Diagrammatic representation of anterior end of *Vittatidera zeaphila* showing pharynx including dorsal (dg) and subvental (svg) glands in relation to excretory pore (ex). Modified from Bernard *et al.* (2010), courtesy of *Journal of Nematology*. (B) Anterior end of female (*Heterodera zeae*) including pharyngeal bulb. (C) Diagrammatic representation of lateral view of stylet showing variation in shape (left, *H. trifolii*; right, *H. galeopsidis*; after Hirschmann and Triantaphyllou, 1979). (D) Anterior end of female (*H. zeae*) including dorsal gland orifice (dgo) (photographs for B and D modified from Mulvey and Golden (1983), courtesy of USDA Nematology Laboratory and *Journal of Nematology*). (E) Feeding on root surface (*H. goldeni*). (F) Female of *V. zeaphila* with large eggsac (es).

transformed by the development of a cloaca and male secondary accessory structures.

The reproductive system of the infective J2 is an oval primordium positioned about in midbody (Fig. 14.2L). It is composed of a sheath of two epithelial (cap) cells that enclose a pair of germ cells. By J3 of developing females the primordium develops into two gonad branches that become anteriorly directed as they migrate posteriorly in the J4 and attach to the developing vagina and vulva by the final moult. However, in males only a single gonad branch develops and by the final moult it has migrated posteriorly, joining with the cloaca lining and associated accessory structures (Raski, 1950) (Fig. 14.7).

The nervous system of Heteroderinae, as understood by light microscopy, is apparently conserved among species and is not known to vary among juvenile stages with respect to a nerve ring encircling the anterior portion of the isthmus as well as labial and caudal sensory organs (Figs 14.2A, 14.3H, 14.4B, C and 14.5B, J). Additional commissures are embedded within the interchordal epidermis and these may be characterized, including in J2, as hemizonid (anterior to the excretory pore), hemizonion (posterior to the excretory pore), cephalids (near the lip region) and caudalids (in the tail region) (Hirschmann, 1956) (Figs 14.2A and 14.3A).

Although the position of these structures relative to numbers of annules is often reported in descriptions, they are not known to be of reliable diagnostic value. Additional commissures occur surrounding the rectum and within the pharynx, but these are not readily observed by light microscopy.

Surrounding the stoma opening are six inner labial sensilla expressed with external openings and thus presumed to be chemoreceptive. In addition four slightly elevated cephalic papillae, probably mechanoreceptors, occur peripherally and submedially in the lip region together with a pair of lateral oval amphid openings (Fig. 14.5B). Each amphid opening leads posteriorly to a cuticle-lined canal with seven sensilla receptors and enclosed by a socket cell. Posteriorly the sensilla of each amphid are embedded in a sheath cell that also encloses two finger cells, characterized by microvilli termini and presumed to be thermal receptors (Fig. 14.4B, C). In the tail region of J2 a pair of phasmids, each with a single opening occurring on each lateral side, are the predominant sensory organs (Figs 14.3H and 14.5J). These include a single sensillum that penetrates a duct, opening to the outside (Baldwin, 1985). In some cases near the opening, the duct has a small ampulla that is visible with light microscopy as a conspicuous 'lens-like' structure,

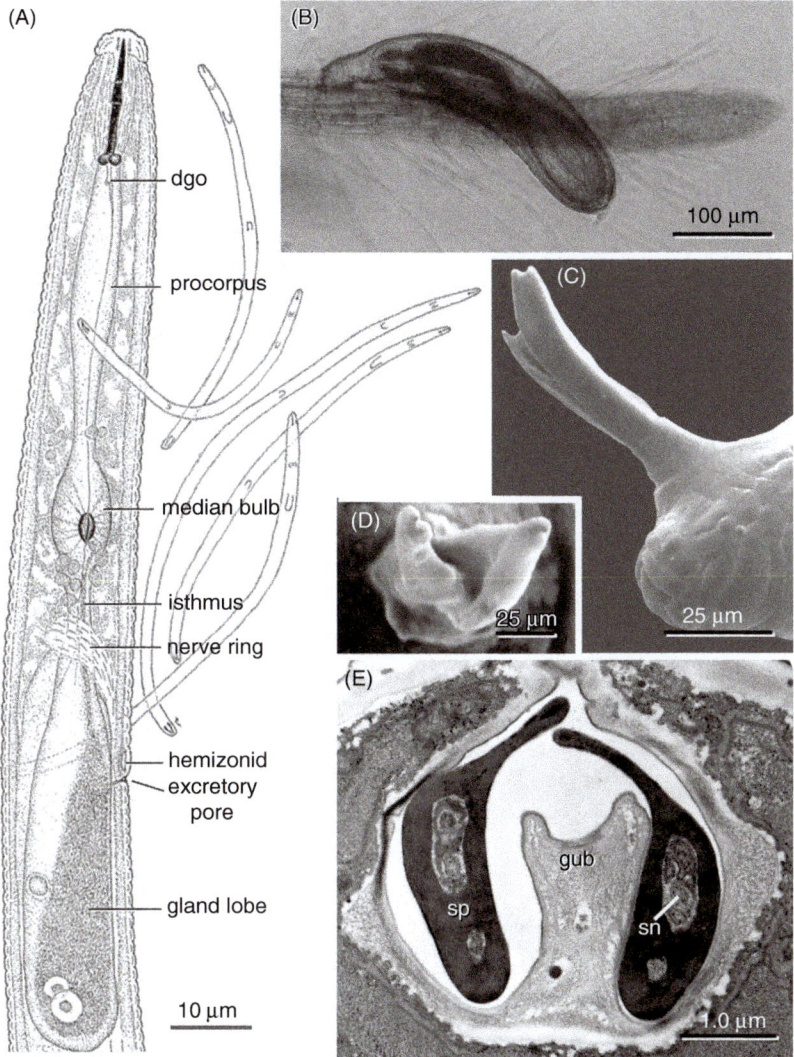

Fig. 14.7. Males of cyst nematodes. (A) Anterior end of male of *Vittatidera zeaphila* (modified after Bernard *et al*. (2010), courtesy of *Journal of Nematology*). (B) Fourth-stage male (*Heterodera schachtii*) (by permission, Ole Becker). (C) Scanning electron micrograph of lateral view of spicules showing bifid tip (*H. schachtii*). (D) Scanning electron micrograph of ventral view of spicule tips (*H. schachtii*). (E) Transmission electron micrograph of transverse view of spicules (sp) with embedded sensilla (sn) and in relation to gubernaculum (gub).

but more often the opening is small; some species are characterized by having obscure (absent?) phasmid openings. Although phasmids are expressed in J2, there is evidence that they deteriorate once a feeding site is established and they are not expressed in J3 and J4 (Carta and Baldwin, 1990). In J2 the position of the phasmids relative to the anus and tail terminus may vary among species and this position may have some diagnostic value.

The secretory-excretory system in J2 is expressed as a ventral cuticle-lined duct opening at about the level of the pharyngeal isthmus (Fig. 14.2A), although the position varies more specifically among species and in some cases it may have diagnostic value in both J2 and adults.

The duct leads to a cell body and single lateral canal, the latter extending in association throughout much of the length of the lateral chord. The status of the excretory system, including the canal and pore, is not known in J3 and J4.

14.4 Males

Most species of Heteroderinae, reproducing by amphimixis, have abundant males; more rarely species are facultative parthenogenetic such that males are rare (Triantaphyllou and Hirschmann, 1980). As is the case for other taxa of Tylenchomorpha with enlarged sedentary adult females, Heteroderinae is characterized by a striking level of sexual dimorphism (Figs 14.1, 14.7 and 14.8). As noted above, in cyst nematodes once J2 establish a feeding site, subsequent juvenile stages are sedentary until the final moult in which the J4 undergoes a metamorphosis resulting in the migratory adult male emerging from the cuticle of the sedentary juvenile stage (Figs 14.1 and 14.7). These males are vermiform with little tapering

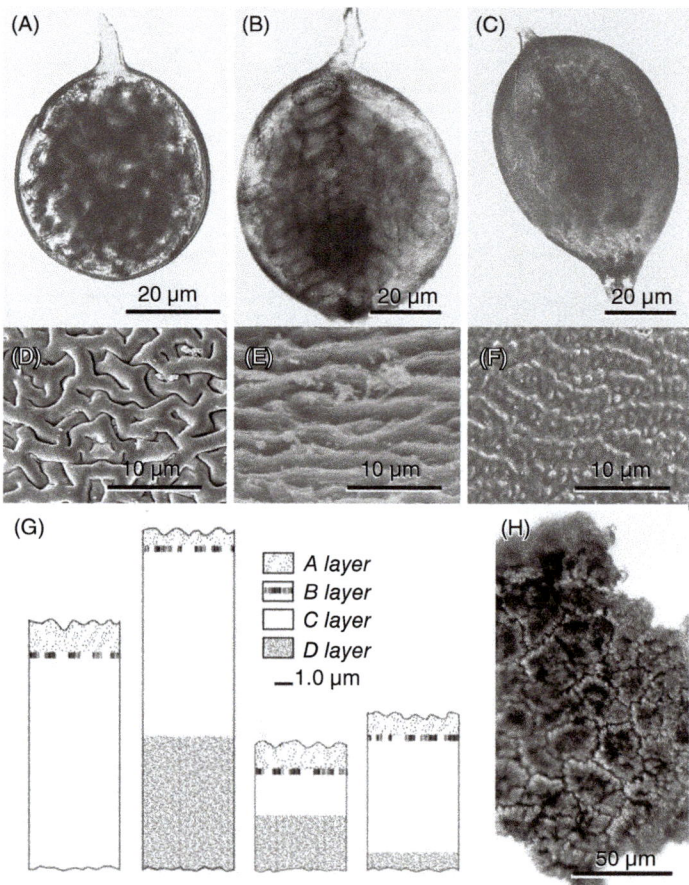

Fig. 14.8. Features of cysts. (A–C) Profiles. (A) Spherical lacking cone. (B) Nearly spherical but with small cone. (C) Lemon-shaped with pronounced cone. (D–F) Scanning electron micrographs of cyst surface patterns. (D) Rugose from mid-body (*Globodera tabacum*). (E) Striated from mid-body (*Punctodera punctata*). (F) Punctate posterior to mid-body (*G. tabacum*) of zigzag cuticular pattern. (G) Diagrammatic representation of cuticular layering without (far left) and with D layer. (H) subcrystalline layer (*Heterodera avenae*). (A–C and H by Mulvey and Golden (1983), courtesy of USDA Nematode Laboratory and *Journal of Nematology*; D–G after Subbotin *et al.* (2010).)

along the body length and rounded at both the anterior and posterior end. They are highly variable in size, even within a given species or isolate, but generally they range from 900–1600 μm. Males are present in most species of cyst nematodes and, in contrast to stationary females, they must be mobile for successful mating. They live about 10 days and are commonly found in soil or within the gelatinous matrix of an eggsac at the posterior terminus of the female.

The mobility of the males is reflected in the well-developed body wall where the cuticle, epidermis and platymyarian muscles are almost identical to those of J2. The lip patterns and lateral field with three to four lines and with or without areolation, are external expressions of the body wall. The tail region of males is characterized by a nearly terminal cuticle-lined cloacal opening through which protractable spicules may protrude (Fig. 14.7C–E), but otherwise the tail lacks caudal alae. This is in contrast to many species with vermiform females where in corresponding males caudal alae are present and may be functionally important in wrapping around the female for transferring sperm during copulation; in the case of the nearly globose sedentary females of cyst nematodes caudal alae would not likely to be functionally important. The nervous system of males is likely to play an important role in locating females and in mating. Morphologically, the system resembles that of J2 with respect to the nerve ring, hemizonids, hemizonian and other interchordal commissures as well as sensory organs of the lip region (Fig. 14.7A). In contrast to J2, phasmids apparently are not expressed in males of many species (Carta and Baldwin, 1990; Sturhan, 2016); however, males are distinctive by the presence of sensilla that extend within each spicule and that are exposed to the exterior through a minute pore (Clark *et al.*, 1973) (Fig. 14.7C). The excretory system, including the ventral position of the cuticle-lined pore, is not known to differ from that of the J2 (Fig. 14.7A).

Although males of cyst nematodes are not known to feed they have a well-developed digestive system (Fig. 14.7A). The cephalic framework and stylet are larger than those of J2 and, while the stylet may be used less often for diagnostics compared to J2, stylet length and shape of knobs may vary among species, as may the distance of the dorsal gland orifice from the base of the stylet knobs. Remarkably, considering the putative absence of feeding, the pharynx and pharyngeal-intestinal transition of males differs little from that of infective J2.

The male reproductive system includes a single testis that extends anteriorly from a vas deferens that connects posteriorly in a cloaca shared with the digestive system. Posteriorly the cloaca is enlarged to enclose a pair of spicules that protract and retract by attached muscles and are guided by a thickening of the dorsal wall, the gubernaculum (Fig. 14.7C–E). The gubernaculum includes a partition that separates the spicules from one another (Clark *et al.*, 1973). Spicules typically extend through a tubular sheath (tubus) and this sheath may vary with respect to its terminal position as well as the presence or absence of surface annulation. Each spicule has a concave surface that faces the other such that the two spicules together provide a conduit through which sperm pass. Spicules, while having a high degree of intraspecific variation, may vary among species with respect to overall length and details of the tip (Clark *et al.*, 1973) that may be rounded, flat, bifid or tridentate. Cyst sperm are amoeboid with filopodia and although diversity of form has not been broadly studied there is some evidence of morphological variation among taxa (Shepherd *et al.*, 1974; Walsh and Shepherd, 1983; Cares and Baldwin, 1995).

14.5 Females

Being sedentary and gradually enlarging from the infective feeding stage (J2) onwards, mature females are generally swollen/rounded with a narrow neck (Fig. 14.1). These white or cream-coloured enlarged females are typically found on the surface of host roots (Fig. 14.6E). Details of shape include various taxon-specific modifications; for example, a posterior cone may be present and prominent, reduced in prominence or absent (Fig. 14.8A–C). To some extent shape may be a function of body wall, including differential cuticle thickness, elasticity or composition in an individual body region, and/or the persistence, arrangement and points of attachment of vaginal muscles in the terminal region. Heteroderids with a prominent cone, such as *Heterodera schachtii* and *H. trifolii* are said to be lemon-shaped, whereas those that lack a cone, such as

Globodera and *Punctodera* are said to be globose; intermediates with small cones include, for example, *H. carotae*; *Cactodera* also typically have a small cone (Fig. 14.8A–C). Details of shape and size may be influenced by environmental factors including the host, nutrition, age and crowding; these variables may confound diagnostics, leading some to suggest that interspecific comparisons are best made across the largest individuals of a population (Hirschmann, 1956; Thorne, 1961).

In contrast to the body wall of vermiform juveniles and males, that of females is especially modified by increased thickness and complexity of the cuticle (Fig. 14.8G). Furthermore, body wall somatic musculature in these sedentary stages is mostly lost, being primarily limited to the somewhat-mobile head region (Shepherd and Clark, 1978). External to the cuticle, in some cases, as in certain species of *Heterodera*, presence of a 'white, flakey' subcrystalline layer is considered to be somewhat diagnostic (Liebscher, 1892; Fuchs, 1911; Kirjanova, 1969) (Fig. 14.8H). The origin of the layer has been controversial even including the question of whether or not it is of nematode or fungal origin (Brown *et al.*, 1971; Zunke, 1986; Endo and Wyss, 1992; Endo, 1993; Zunke and Eisenback, 1998).

The cuticle of mature white females retains the cortex, including typical outer and inner components. Underlying the cortex is what appears to be a homologue of the basal striated region that is recognized in J2 and males, but in females the layer is interrupted by regions that lack striation and these gaps in striations may facilitate expanding girth (Fig. 14.8G). Posterior to the basal region, the cuticle of mature females is primarily thickened by a deep homogeneous layer designated C and this pattern of cortex, basal and C layers is taxon-specific, characterizing, for example, females of all species examined of *Heterodera*. By contrast, species of *Globodera*, *Cactodera* and *Punctodera* all have an additional layer internal to the C layer. This layer, designated D, is composed of helical layers of fibres such that in transverse section they appear to form a herringbone pattern (Shepherd *et al.*, 1972; Cliff and Baldwin, 1985). These taxon-specific variations in cuticular layering of mature females are informative for interpreting phylogenetic relationships among heteroderids, including relationships between cyst- and non-cyst-forming taxa (Baldwin and Schouest, 1990). Notably these patterns first described by TEM can also be interpreted from light microscope sections and particularly so by staining with toluidine blue (Shepherd and Clark, 1978; Baldwin, 1983).

Cuticular surface patterns of mature females range from striations to variations of irregular patterns often designated by ambiguous and inconsistently applied descriptors, including rugose, zigzag, reticulate, striated, punctate or lace-like (Fig. 14.8D–F). These patterns show some species-specific distinctions of diagnostic value, but they also may be subject to intraspecific variability and even within an individual they may differ by position on the body and age of the female. For example, posterior to the head region they may be transverse ridges that at midbody are zigzag and further posteriorly, whorled ridges (Zunke and Eisenback, 1998). Cuticular surface patterns of the lip region of females include a somewhat irregular oblong to rectangular elevated labial disc (Fig. 14.5H). Surrounding and posterior to the disc the lips are fused to form a circular plate. Immediately posterior to the lips transverse rows of protuberances have been reported in females of *Globodera* but these were not observed in other genera of Heteroderinae (Othman *et al.*, 1988).

Surface patterns are modified relative to the rest of the body in the terminal region of females. In a region surrounding the vulva the cuticle is thin, composed of a loose mesh of fibres that eventually ruptures in the cyst-forming fenestrae. Cuticular patterns external to the thin region are distinctive among taxa (Fig. 14.9A–C) (Green, 1975; Mulvey and Golden, 1983). In globose females, such as *Globodera*, the area surrounding the vulva may become sunken and is aptly named the vulval basin demarcated by a thickened rim. A discrete region peripheral to the rim may be characterized by a distinctive pattern, in some cases including tubercles (Green, 1971) (Fig. 14.9C). By contrast, in the tip of the cone of mature lemon-shaped females the basin is distinguished by a region on each side of the vulva slit that in younger females typically includes a fine surface pattern of rivulets that are finer in texture relative to the overall cuticular pattern (Fig. 14.9A). As the lemon-shaped female matures the region immediately surrounding the vulval slit remains elevated as a vulval bridge, thus bisecting the sunken basin (Fig. 14.9B). The width and breadth of this

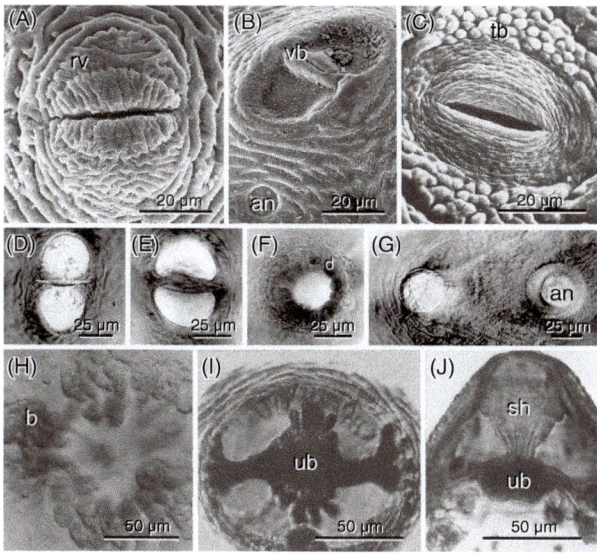

Fig. 14.9. Structures of the terminal region of cyst nematodes. (A–C) Scanning electron micrograph end views of terminal region of females. (A) Mature *Heterodera glycines* showing developing semifenestrae including rivulets (rv). (B) Young *H. glycines* showing developing vulval bridge (vb) and anus (an). (C) *Globodera tabacum* with vulval region surrounded by tubercles (tb). (D–G) Light micrographs of fenestrae of terminal region of cysts. (D) Bifenestrate *H. humuli*. (E) Ambifenestrate *H. trifolii*. (F) Circumfenestrate *Cactodera weissi* showing vulval denticles (d). (G) Circumfenestrate with separate anal fenestrate, *Punctodera punctata*. (H–J) Light micrographs of structures within terminal region. (H) Bullae (b) of *H. iri*. (I) Underbridge (ub) of *H. leucilyma*. (J) Dorsoventral view of *H. trifolii* showing vulval sheath (sh) and underbridge (ub). (A, B after Subbotin *et al.* (2010); C–J after Mulvey and Golden (1983); C–I courtesy of USDA Nematode Laboratory and *Journal of Nematology*.)

bridge, together with the length of the vulval slit are of diagnostic value. In both round and lemon-shaped females the distance of the vulva from the anus opening and the cuticular pattern between the vulva and the anus are diagnostic for some taxa.

The digestive system of females is not known to vary among species, with the exception of the length of the often-curved stylet, the size and shape of knobs and posteriorly the position of the dorsal gland orifice (Fig. 14.6A–D). The pharynx, compared to J2 and males, has a large median bulb and gland lobe, but fine structural detail of this region is not available, in part because of technical difficulties in processing globose females for TEM. The intestine is also not well understood although there is evidence that it functions primarily as a storage organ; consistent with this possibility, Raski (1950) observed that the rectum of the adult female is not well developed. The anus occurs on the cuticle surface dorsal to the vulva (Figs 14.9A, G and 14.10).

The didelphic coiled reproductive tracts of heterodid females extend from the terminal vulva where anteriorly each gonad has a short germinal zone and proximally an extended growth zone followed posteriorly by a short narrow oviduct; in amphimictic species there is a pronounced sperm-filled spermatheca and this joins, at a sharp bend, with the uterus (Triantaphyllou and Hirschmann, 1962). The reproductive tract terminates posteriorly as a vulval slit and internal to the slit is a complex morphology including the cuticle-lined vagina. The vagina is lined by dilator muscles that, in taxa with cones, attach to the body wall. In *Heterodera* species that have been examined, these include 48 muscles as four levels of six on each side (Fig. 14.10), but in other taxa, such as *Cactodera*, the musculature is greatly reduced (Cordero and Baldwin, 1991; Cordero *et al.*, 1991). Proximally the lining of the vagina is enclosed by a thick sphincter muscle and in some species immediately distal to this muscle in the mature female the cuticle

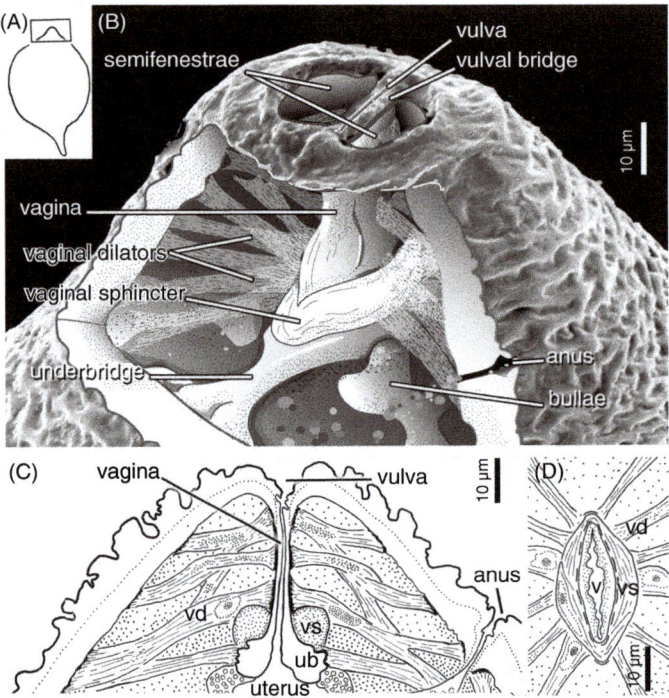

Fig. 14.10. Diagrammatic representation of basic structures of terminal region of cyst nematodes. (A) Overall profile showing cone. Box indicates terminal region as enlarged in B and C. (B) Ventrolateral 3D with cut away showing internal structures of female and/or cyst. (C) Right lateral view through cone; vd = vaginal dilator, vs = vaginal sphincter, ub = underbridge. (D) Cross-section through vagina (v) at level of vulval sphincter. (A, B after Subbotin et al. (2010); C, D after Cordero and Baldwin (1991) courtesy of Journal of Nematology.)

lining thickens (Fig. 14.10). There is some evidence that development of vaginal musculature is associated with the capacity of females to lay eggs in a matrix vs retaining all eggs into the cyst stage; that is, species vary with respect to retention of eggs within the body of the female and cyst. Some species retain all the eggs through maturation of the female into the cyst, whereas others produce an eggsac depositing at least some eggs externally in a gelatinous matrix (Fig. 14.6F). The source of the gelatinous matrix, while uncertain, is most likely the uterus and even taxa that retain eggs may nevertheless produce a small amount of gelatinous exudate to which males may be attracted (Hesling, 1978; Turner and Subbotin, 2013).

The nervous system and excretory system of females are not well understood. Yet, with SEM, depending on preparation techniques, amphid and some inner labial sensory openings are generally observed in the head region of females and the excretory pore is typically present ventrally at about the level of the gland lobe (Fig. 14.6A). In particular, insight into female sensory systems and differences relative to corresponding organs in migratory stages could provide important specific insight into function.

14.6 Cysts

The distinguishing feature of cyst nematodes is that mature egg-filled females, upon death, are transformed into persistent tanned 'vessels' that retain and provide protection for viable eggs. Tanning is the result of polyphenoloxidase (Ellenby, 1946; Awan and Hominick, 1982) and following death of the female tanning transitions to final levels of pigmentation ranging in intensity from light tan to nearly black (Fig. 14.11). Some cysts, prior to reaching their

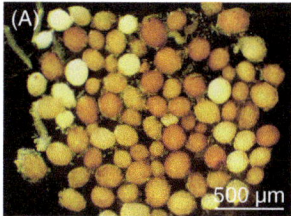 

Fig. 14.11. Cysts showing variation in colour. (A) *Heterodera zeae*. (B) *H. glycines*.

darkest colour, pass through a phase described as golden and the presence or absence of this yellowish stage is considered to be diagnostic. Franklin (1951) suggested that the yellow colour of some *Globodera* cysts might be attributed to changes in the state of internal contents of the cyst, prior to later full pigmentation of the persistent cuticle. Regardless of pigment, the whitish subcrystalline surface, described in females of certain taxa, may persist in the cyst phase.

Cuticular layering of the young cyst is at first recognizable as consistent with that of white females, although Cordero and Baldwin (1990) note that in some young cysts additional layers of the cuticle may appear internally towards the epidermis. Whether cuticular layers described in white females remain distinguishable in mature cysts is not known. However, cuticular transformations, including with respect to layers, during cyst development have been described for the terminal region of some cysts (Cordero and Baldwin, 1990, 1991; Cordero *et al.*, 1991). Cuticular surface patterns are typically not modified from those of mature females except, as noted, fenestrae open in the terminal region corresponding in females, to a discrete area defined by a loose mesh of fibres (Figs 14.9A–G and 14.10B). This may be expressed as semifenestrae, where two openings are separated by a retained cuticular 'vulval bridge' that includes a persistent vulval slit (Figs 14.9B, D, E and 14.10B). Semifenestrae may be expressed as ambifenestrae, where each opening is semicircular separated by a narrow vulval bridge, or as bifenestrae, where the pair of semifenestrae are round and separated by a relatively stout vulval bridge (Wouts and Baldwin, 1998). In circumfenestrae individuals a vulval slit and cuticular vulval bridge is not retained and there is a singular circular fenestral opening. In certain circumfenestrae taxa, specifically in *Punctodera*, there is an additional fenestrae area corresponding to the anal region (Fig. 14.9D–G). Types of fenestrae are taxon-specific. In addition to *Punctodera*, circumfenestrate cysts characterize *Globodera* and *Cactodera*; species within these genera may be further characterized by variations in size and shape of this opening, as well as by the distance of the fenestrae to the anus (e.g. see the ratio of Granek, 1955, and as further defined by Hesling, 1973). By contrast, semifenestrae, either ambifenestrae or bifenestrae, characterize most lemon-shaped cysts of *Heterodera* and the specific size and shape of these openings may further characterize species. In general ambifenestrae are associated with a persistent relatively long vulval slit and an associated bridge, in contrast to bifenestrae in which the pair of openings is separated by a persistent shorter vulval slit. Notably some cyst taxa do not form fenestrae as is reported to be the case for *Heterodera (Afenestrata) africana*.

As the cyst matures, cuticular transformations of the body wall are particularly apparent in the terminal region and many of these are important in diagnostics. Bullae, apparently noted by Franklin (1939) but described and named by Cooper (1955), are irregularly shaped, sometimes branched, large darkened cuticular deposits that extend inward from the body wall cuticle. Variation in shape, size, position and pattern of bullae may have diagnostic value (Figs 14.9H and 14.10B). For example, *H. schachtii* is characterized by the presence of bullae, whereas bullae are said to be absent in the closely related *H. cruciferae*. Where present within *Heterodera*, bullae of the *Schachtii* group (see Chapter 16, this volume) tend to be scattered and positioned near the underbridge (see below) (Hesling, 1978), whereas those of the *H. avenae* complex tend to be particularly prominent, more specific in structure, and positioned closer to the cone terminus. Bullae are common among species of *Punctodera* in contrast to *Globodera* and *Cactodera*. However, many taxa lacking bullae may have comparable tooth-like pigmented structures such as those termed 'vulval denticles' (Fig. 14.9F). Unlike bullae of cuticular origin, vulval denticles and perhaps also vulval bodies (Wilson, 1968; Mulvey, 1973, 1974) and 'Mulvey's bridge', are understood to be remnants of vulval musculature (Mulvey, 1959; Golden and Raski, 1977; Cordero, 1989; Cordero and Baldwin, 1991; Cordero et al., 1991).

In cyst development, internally there is a transformation of the vaginal region. In some *Heterodera* species, the cuticle thickens in the proximal region of the vagina and this area becomes anchored to the lateral cone wall by a pair of branched or forked projections that comprise the cyst underbridge (Figs 14.9I and 14.10B, C). The underbridge is so-called because it is oriented internally and aligned parallel to the superficial vulval bridge. The presence, shape and orientation of the underbridge are diagnostically significant. In some cases persistent remnants of the vagina transverse between the bridge and underbridge and these may be referred to as a 'sheaf' (Golden, 1986) (Fig. 14.9J).

14.7 Techniques

14.7.1 Light microscopy (LM)

There is a rapidly expanding array of tools applicable for understanding heteroderid structure, yet traditional LM remains essential to the discovery and understanding of morphology. Even at the relatively low magnifications of a dissecting microscope, in an aqueous solution with transmitted light an initial assessment can be made of key features of heteroderid J2 and males. These include an estimate of overall length, relative length of the tail, hyaline region and stylet. Similarly, the dissecting microscope, particularly equipped with reflected light, is a primary tool for revealing cyst shape and colour. These initial low power assessments, especially combined with knowledge of the host, provide a simple but powerful starting point to morphological understanding and diagnostics. Clearly, the dissecting microscope is also the starting point for many other morphological techniques, including preparing LM slides as well as processing material for SEM and TEM.

The dissecting microscope is essential but limited by its relatively low magnification. Assessment of key aspects of heteroderid morphology also requires a compound light microscope equipped with high resolution oil immersion lens and condenser and field diaphragm suitable for focusing/optimizing transmitted bright field or differential interference contrast (DIC) optics. Techniques for preparing permanent microscope slides of heteroderid J2 and males follow standard approaches for tylenchids and other small nematodes (Thorne, 1961; Goodey, 1963; Golden, 1986; Eisenback and Zunke, 1998). Specimens are typically killed and 'relaxed' (straightened) by gentle heat, followed by aldehyde fixation and then dehydrated through increasing concentrations of ethanol, which is finally replaced by anhydrous glycerin. Specimens mounted in glycerin are sealed under a cover glass supported so as to not crush specimens. Notably, for some types of evaluation, temporary slides are suitable and these can be prepared by mounting living, heat-killed or anaesthetized specimens directly in water or on a thin layer of water agar (Sulston and Hodgkin, 1988).

Some light microscope methods have been developed to interpret unique aspects of heteroderid females and cysts. Although whole individuals can be mounted on slides, provided adequate support of cover glasses, the thickness of the specimens may not allow for adequate transmitted light, and particularly so for viewing surface patterns. Such cases typically require dissecting areas of cuticle that can be separately mounted on slides for viewing. For certain studies there is value in dissecting female heads for separate mounting and viewing of features such as the stylet, DGO position and excretory pore. Previously, some investigators dissected and mounted head regions for *en face* viewing, but this approach is now used less often in lieu of broader availability of SEM of lip patterns. One of the most important cyst nematode-specific procedures for LM is that of dissecting the terminal region for detailed interpretation of morphology (Golden, 1986; Mulvey, 1973; Hesling, 1978). For some structures, including bullae, the underbridge and position of the anus, lateral side mounts are particularly informative. More difficult, but essential for viewing posterior cuticular patterns, vulval slits, the bridge and fenestrae, are terminal mounts (Riggs, 1990; Subbotin et al., 2010). Such patterns are prepared under a dissecting microscope by cutting off the terminal region with a fine knife or scalpel (Fig. 14.12).

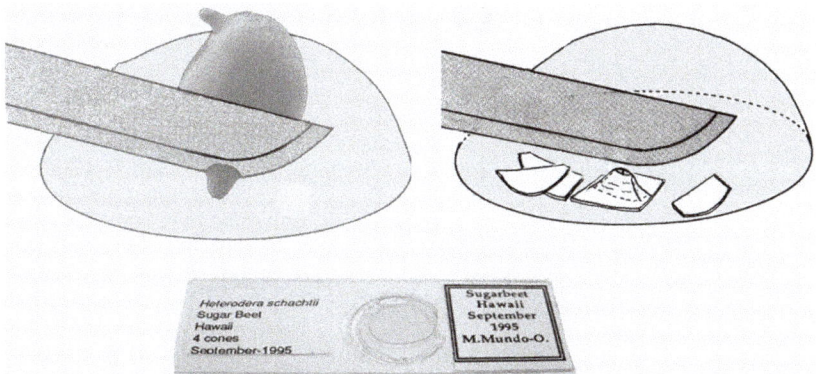

Fig. 14.12. Diagrammatic representation of the procedure for excising, trimming and mounting the terminal region of a female or cyst.

Using fine needles or an eyelash mounted on a handle the end can then be cleaned of debris, oriented end up and, if needed, the periphery can be further trimmed. To provide support, mounting can be in Canada balsam (Goodey, 1963), a glycerin agar block (Correia and Abrantes, 1997) or in glycerin jelly (Thorne, 1961; Subbotin et al., 2010).

Particularly crucial to heteroderid morphology, especially with regard to cone patterns, are applications of LM that allow recording of enhanced depth of focus including multiple-focus photomicroscopy (Eisenback, 1988). Similarly, through-focus videos enable high definition and high magnification digital recording through fresh or preserved specimens of all stages, and digitally archiving and enhancing access to records of morphological details (De Ley and Bert, 2002). Beyond depth of focus, video microscopy is also employed as a tool for understanding cyst nematode morphology and function including time lapse or 4D recording of development and pathogenesis (Wyss and Zunke, 1986a, b; Cordero and Baldwin, 1990, 1991). Additional enhancements of LM applicable to cyst nematodes are achieved with specialized staining, florescence and confocal techniques. For example, fluorescent phalloidin has been shown to be an efficient complement to TEM in interpreting dilator muscles of the cyst cone (Cordero and Baldwin, 1991) and fluorescent antibodies and confocal microscopy have been used to evaluate specific cyst nematode gland activity (Willats et al., 1995). Fluorescent antibodies and confocal microscopy, used to visualize cell boundaries, also show promise for applications to understanding heteroderid morphology (Burr and Baldwin, 2016).

14.7.2 Scanning electron microscopy (SEM)

SEM is crucial to interpreting surface details of cyst nematodes including many of diagnostic importance (Eisenback, 1991). Often morphological understanding gained with SEM, can subsequently 'open our eyes' to understand, anticipate and then interpret those same features using routine LM. Key among details best resolved by SEM are lip patterns, especially of J2. Also SEM is an important supplement to LM in resolving aspects of J2 with respect to numbers, areolation and anterior/posterior terminus of lateral lines. Sensory openings often best resolved by SEM with respect to shape and position include those of the inner labial papilla, amphids and phasmids of J2 and males; in some cases (as when phasmid openings are positioned within the fold of a lateral line) these may be difficult to observe using LM. In males, presence/absence of phasmids vary among taxa and this diversity is best examined by SEM. Details of the morphology of the male tail and particularly diagnostic aspects of the spicule tip also can best be resolved with SEM. In females and cysts, beyond the lip region, SEM is of particular importance in resolving more general details of surface patterns that may have diagnostic importance. In the terminal region patterns of the basin, vulval bridge, vulva, anus and surrounding cuticle patterns are ideally resolved by SEM, not obscured, relative to LM, by underlying focal planes. Beyond surface structures, SEM of cutaways of cyst cones may be useful for interpreting underlying structures such as the underbridge and bullae. SEM also has special importance in observing, on eggs, the presence of punctations/tubercles that characterize some species of *Cactodera*.

Preparation of specimens for SEM typically requires cleaning, fixation, dehydration, mounting, treating with a conductive coating and imaging. The first step of cleaning is often best accomplished prior to fixation and by repeated rinsing in clean water; it may be helpful to use an eyelash mounted on a supporting handle to brush away any specs of debris. In the Baldwin laboratory at UCR (University of California, Riverside, USA) on occasion we have used gentle sonication in water to promote clean specimens but usually this is not necessary. Mature air-dried cysts may be directly suitable for mounting, coating and imaging but living material requires fixation and particular care during dehydration to minimize distortion. Fixation for SEM usually is by aldehydes. The approach at UCR is to place the living specimens in 12 ml of filtered tap water in a 25-ml vial. The vial with specimens is then partly immersed in a hot water bath (60°C), relaxing vermiform stages, and to this is added 10 ml of 10% buffered (pH 7.0) formalin solution, or alternatively 10% buffered glutaraldehyde solution. Although fixation overnight is adequate, it has been observed that longer periods enhance results. For some particularly fragile specimens prone to collapsing, post-fixation with osmium tetroxide is advantageous, resulting in improved rigidity of delicate features and slightly increased conductivity. Specimens are placed in 2% buffered osmium textroxide solution overnight and subsequently repeatedly rinsed in buffer.

The first step in dehydration is processing specimens through a series of aqueous ethanol solutions ranging from 10–100%. To avoid collapsed specimens it is crucial that the series culminates in absolute ethanol, and a consideration is that opened bottles of 100% ethanol are hydroscopic and may quickly become slightly diluted. In this regard a freshly opened bottle of ethanol is recommended for the final steps of dehydration. Dehydrated specimens are then dried in a critical point dryer. This instrumentation allows replacing ethanol (or in some procedures, acetone) with liquid CO_2 (or freon) and then regulating temperature and pressure to dry material while minimizing the impact of liquid surface tension that would otherwise damage tissue.

Using a dissecting microscope, critical point dried specimens are mounted on SEM stubs. At UCR the preference is aluminium stubs using double sticking copper (conductive) tape (Fig. 14.13A). Males or J2 are typically mounted on wrinkles in the tape so that heads, for example, can project beyond the tape for optimal viewing. Excised cones can be oriented directly on the tape, and often whole females or cysts can be oriented for optimal viewing of the terminal region. Critical point dried specimens generally have sufficient charge to allow picking and orienting them on tape using an eyelash mounted on a holder. Prior to viewing, specimens are coated with an ultrathin (25 nm thick) coating of conductive surface such as gold-palladium. This is accomplished using sputter coater instrumentation.

Alternatives to critical point drying of cyst nematodes include direct SEM viewing of specimens that are infiltrated with dehydrated glycerin by a methodology typically used for preparing permanent slides (Sher and Bell, 1975). Advantages of this method are its simplicity, its accommodation to using archived specimens from permanent slides and that it minimizes risk of specimens' surface artefacts associated with other forms of drying. Disadvantages include vulnerability of the specimens to overheating and collapse in elevated temperatures under the beam as well as some potential for contaminating the SEM column with glycerin. To minimize these effects, specimens infiltrated with glycerin must typically be viewed at very low accelerating voltages of 5–15 KV, but such low voltage also may reduce resolution. Additional potential alternatives to critical point drying may include applications of low vacuum SEM instrumentation (Sammons and Marquis, 1997); this allows viewing specimens with reduced preparation including the option of omitting a conductive coating. At UCR a range of procedures using SEM instrumentation designed for such low vacuum have been tested, including on cyst nematodes, but results (particularly resolution) remain unsatisfactory.

One of the challenges of processing specimens for both SEM and TEM is that of transferring nematodes in solutions without loss of

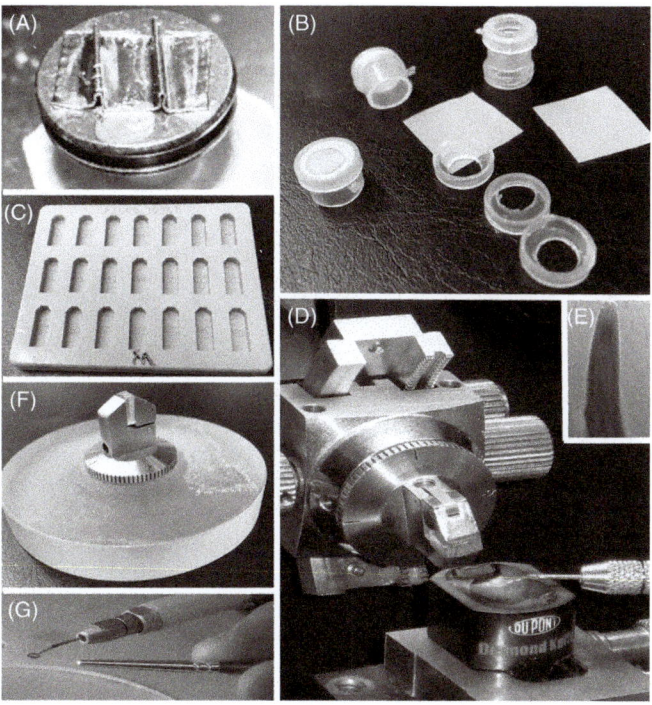

Fig. 14.13. Procedures for preparing specimens for scanning (SEM) and transmission (TEM) microscopy. (A) SEM stub with folded foil tape upon which specimens are mounted. (B) Containers modified from Beem® capsules for processing specimens. The base is excised from the capsule and each end is covered with a lid from which the centre is removed using a hole punch. Each lid holds in place a fine nylon mesh fabric. (C) Moulds designed to polymerize specimens in epoxy for subsequent TEM sectioning. (D) An epoxy block with specimens mounted on an ultramicrotome with a diamond knife for producing ribbons of sections. Sections are being removed using a loop. (E) Osmium tetroxide-fixed specimen embedded in a layer of epoxy. (F) A lucent block developed to secure the specimen holder for trimming an epoxy block. (G) Using a mounted eyelash to manipulate a ribbon of sections on a loop. (A, D and E after Subbotin et al. (2010).)

specimens and without exposure of the researcher to toxic materials including fixatives. With this in mind a number of chambers have been proposed (McClure and Stowell, 1978; Bumbarger et al., 2006). In our experience, the most helpful container for this purpose is an adaptation of the BEEM® or Eppendorf capsule container modified with very fine (<10 µm mesh) nylon mesh cloth to facilitate exchange of liquids (Fig. 14.13B) (Eisenback, 1991).

14.7.3 Transmission electron microscopy (TEM)

TEM is invaluable for elucidating details of nematode morphology. As noted for SEM, often once TEM is used to clarify detailed structures those details can sometimes subsequently be interpreted from LM. For example, cuticular layers of females first clearly resolved by TEM could also be visualized using LM sections (Shepherd et al., 1972). In the 1970s, heteroderid J2 and males were among the first tylenchids in which details of body wall, sensory and digestive system were described together with host parasite relationships (Baldwin and Hirschmann, 1975a, b, 1976; Baldwin et al., 1977; Endo, 1980, 1983, 1984, 1985, 1991). Later TEM was applied to females and cysts with respect to the body wall and terminal region as well as to the male tail and sperm diversity. TEM has been central to addressing questions of origin, function, diversity and homologies of classical features of heteroderids; often these features, otherwise, based on the limited resolution of LM, are subject to misinterpretation. Examples include the taxon-specific diversity in cuticle-layering of females and of nurse cells as well as, for example, understanding the nature of bullae (Baldwin, 1983; Cliff and Baldwin, 1985; Cordero and Baldwin, 1991; Cordero et al., 1991; Mundo-Ocampo and Baldwin, 1984, 1992). Although approaches may vary with different applications and goals, TEM procedures typically include fixation, dehydration, infiltration, embedding/polymerization, trimming, sectioning, staining, coating and imaging (Carta, 1991).

Excellent fixation is essential for successful TEM, yet typically success of this first step cannot be fully evaluated until the final step of imaging. A challenge with preparing heteroderids is that fixatives may penetrate slowly and this often results in inadequate preservation, in part probably associated with catabolism. Early attempts to address poor preservation often included prolonged fixation times, but this may have resulted in artefacts including those associated with leaching of cytoplasm. Leaching, particularly of cytoplasm, may result in images with impressive contrast but nevertheless loss of information. Where fixation is too slow to prevent catabolism, results can include artefacts of broken discontinuous cell membranes and this confounds understanding cell topology. In some cases cutting specimens in buffer or fixative can improve fixation but this can be technically difficult (especially safely handling specimens in fixatives) and it can result in distorted topologies and leaching. The approach that provides the best preservation to date is that of rapidly freezing specimens in liquid nitrogen under high pressure followed by chemical fixation by substitution within an autofreeze substitution apparatus (Bumbarger et al., 2006). Classic chemical preservation typically first includes a buffered solution of glutaraldehyde, paraformaldehyde or a mixture (e.g. Karnovsky, 1965) followed by post-fixation in osmium tetroxide. In some applications, where the emphasis is on preservation of membranes, aldehydes may be omitted in favour of only using osmium tetroxide. At UCR a standard procedure, including for heteroderids, is fixation in 3.5% glutaraldehyde followed by post-fixation in 2% osmium tetroxide; pH is regulated at 7.2 with 0.2 M phosphate buffer.

Thin sectioning requires fixed specimens to be infiltrated with epoxy that is then polymerized to provide a suitable substrate. Since most epoxies are not water soluble, infiltration typically requires complete dehydration in an intermediate solvent miscible with the epoxy of choice; usually this requires dehydration by a series of aqueous buffered solutions to 100% ethanol or acetone. As for SEM, it is crucial that hydroscopic properties of solutions be considered to ensure complete dehydration. With dehydration, complete infiltration is accomplished by slowly introducing epoxy into solvent and subsequently allowing all solvent to evaporate resulting in 100% liquid (but highly viscose) infiltrated epoxy. Often 72 h are needed to meet the principle of complete infiltration, although timing is variable in relation to conditions including the epoxy of choice as well as properties of the tissue. Epoxys commonly used for TEM of heteroderids include Epon 812 and mixtures of Epon with Araldite 6005, Durcupan and Spurr.

Following infiltration and prior to polymerization of epoxy, specimens must be oriented to optimize their position for subsequent trimming. Depending on the tissue and goals the typical approach at UCR is orientate and embed the specimens using block moulds optimized in shape for trimming (Fig. 14.13C). The dark colour of osmium-treated specimen aids visualization for orientation (Fig. 14.13E). Specimens can be placed and oriented directly in the moulds filled with fresh epoxy, but a disadvantage is that specimens sink to a suboptimal position at the bottom of the mould. A solution is to first embed the specimens in a small block of 2% water agar prior to infiltration. This block containing the specimen can then be oriented so that the specimen is at an intermediate level from the bottom of the mould. An alternative to an agar block is to first embed and polymerize specimens in a thin plate-like layer of epoxy using a flat slide-shaped mould (Kolotuev et al., 2010). Once removed from the mould, using a compound microscope, specimens can be observed within the thin layer to make an initial assessment of quality of preservation, and to prepare a LM photo record of the specimen (Fig. 14.13E). Individual specimens can then be cut from this layer, oriented and re-embedded within a block mould (Fig. 14.13C, F). Polymerization is typically accomplished with heat; optimal temperatures for this process vary with the epoxy formulation. After polymerization the epoxy block is trimmed to a trapezoid shape, ideally only slightly larger than the specimen. Trimming can be accomplished with a single edge razor blade, shaving thin layers under a dissecting microscope (Fig. 14.13F); the trimming can be further refined using an old diamond knife with the epoxy block mounted on a microtome. Although some have had success with glass knives, successful microtome sectioning generally requires a flawless-edged diamond knife where the goal is to achieve a ribbon of sections at about 70 nm thickness. These float from the knife edge into a boat filled with filtered distilled water and from this ribbons of sections are transferred onto a grid (a slot grid for serial sections) coated with a commercially available electron-lucent plastic film. Transfer is done by using a fine loop in which the ribbon is supported by a film spanning the loop and created by water surface tension from the knife's boat (Fig. 14.13D, E).

Sections of fixed specimens inherently are of very low contrast in TEM but this is addressed by staining with heavy metals that differentially impede penetration of electrons in subcellular components. Typically these stains are heavy metals through treatments with uranyl acetate and lead citrate. Procedures for uranyl acetate include staining sections directly on grids and/or (additionally or alternatively) the stain may be introduced at the time of fixation (Bumbarger et al., 2007, 2009). At UCR uranyl acetate staining on grids is typically done by floating or submerging sections in a saturated solution in ethanol, followed by rinsing in ethanol. Staining with lead is particularly challenging because of the risk of introducing residues including products of interaction of the stain with atmospheric carbon dioxide. Such residues can be minimized by staining below room temperature, by using one of various commercial tools designed for submerging grids to minimize exposure to air and by careful rinsing following staining. Following staining and prior to TEM viewing and recording images, sections and the supporting plastic film are rendered less fragile and more stable under the electron beam by first treating them with a thin electron-transparent coat of carbon. This is done by transferring a current through carbon rods under high vacuum using a vacuum evaporator.

Although not yet applied to heteroderids, digital images from serial sections have been demonstrated to be useful for creating 3D reconstructions of nematodes (Bumbarger et al., 2006, 2007, 2009; Ragsdale et al. 2008, 2009; Subbotin et al., 2010). These approaches may be particularly promising for interpreting morphological transformations, including of phylogenetically informative structures such as the posterior region of heteroderid females/cysts. After many years of technical stagnation TEM is undergoing a transformation towards automated sectioning, simultaneous sectioning of multiple specimens and high throughput imaging (D.J. Bumbarger, 2016, pers. comm.). These transformations result in improved cost and labour efficiency to support the application of 3D reconstruction and phylogenetic comparisons of heteroderids.

14.8 Minimal Standards for Species Descriptions

For heteroderids, as for other taxa, the question of standards for species descriptions is integral to the question of *what is a species*. Distilled from ongoing discussion of species concepts, is that a species, as an *independent evolutionary trajectory*, cannot be delineated separately from a phylogenetic context (Adams, 1998; Nadler, 2002; De Queiroz, 2007). The 'ideal' of *what is a species* is further confounded because what is perceived as a present 'independent lineage' might be challenged by future reticulation with other putative lineages. Consequently, species descriptions are *testable hypotheses* and the strength of such hypotheses depends on the degree to which those tests continue to be supportive of lineage independence. An important question, relevant to cyst nematodes, is *what is the minimum strength of support for a species hypothesis recommended as a basis for description, naming and publication*. Minimally, description as a new cyst species should briefly introduce the author's statement of species concept, what operational parameters are being used to test species status, and what are the limitations of the methods used for assessing that status. Relevant to establishing independence and uniqueness of the candidate species, there must be clear evidence that there has been a thorough review of extant species (literature and in some cases direct examination of type specimens). Precise collecting site(s), preferably presented as GPS coordinates, and host(s) should be specified.

The classical ideas of 'typological species' have been largely rejected (Adams, 1998), and there is a growing role of reverse taxonomy[3] and molecular markers in species discovery (see section 14.9). Nevertheless, morphological distinctiveness is often the first test used to predict species status of cyst nematodes. Since a goal is comparison with extant species, descriptions must consider features and morphometrics for J2, males, females and cysts as they are used in comparable extant descriptions, and to do so with adequate representation (e.g. numbers of individuals) to allow for statistical assessment of intra- and interspecific variation. It may be possible to demonstrate that certain morphological features are particularly convincing as indicators of species independence. For example, distinguishing morphological characters that track evolution versus those that are not informative (e.g. stochastic or mosaic) might be based on mapping such features on existing molecular-based phylogenetic trees of Heteroderinae. Minimally, morphology should be represented by drawings and/or informative photographs that target diagnostic features. SEM and through-focus videos (see section 14.9) are optional but potentially powerfully informative additions (De Ley and Bert, 2002) with videos being referenced as supplementary material accessible through websites. Morphometrics are typically best presented in tables.

Delineating species solely by morphology may be problematic as suggested by lineages that are phenotypically plastic (including at the intraspecific level) as well as by cryptic species, where independence is reflected by parameters (i.e. molecular) other than morphology. The best tests of independence and species status include examination of the new candidate in the context of extant heteroderid species through constructing phylogenies based on multiple genes. However, minimally it is practical to consider ribosomal DNA molecular markers (e.g. ITS and 28S) that are widely reported and comparable among existing cyst species. When relying on such markers to support species status it is also important to acknowledge the limitations relative to assessing phylogenetic monophyly and the potential for misinterpretation, for example, as a result of paralogy due to gene duplications of ribosomal DNA as a multicopy gene (Nyaku *et al.* 2013; Pereira and Baldwin, 2016).

In addition to morphological and molecular characters, reproductive isolation is often proposed as a test of species status and indeed in this regard it is advantageous that many cyst species are cross-fertilizing. However, there are widely discussed arguments for rejecting reproductive isolation as a criterion for defining species (Mallet, 2008, 2010). Lineages essentially independent under natural conditions may retain the capacity to cross under rare or artificial conditions and, notably, fertile crosses have been reported across cyst genera (Miller, 1983).

Consistent with the International Code on Zoological Nomenclature (Anonymous, 1999) minimal standards of species description must include reporting deposit of type specimens in one or more curated and assessable collections; for cyst nematodes this should include J2, males, females and cysts (including mounts of the terminal region) appropriately prepared on microscope slides. While not minimal requirements, there is certainly strong justification for supplementing conservation of types with through-focus videos and with sources of molecular data for types (see section 14.9).

Whereas meeting minimal standards for description of cyst nematodes should be possible under nearly every circumstance, we note arguments for exceptions. For example, it may be especially valuable to phylogenetic studies to report limited information on unique cyst specimens collected through rare opportunities in exotic localities. Reports of such new linages may have broad significance, and arguments for species status may be considered even under circumstances where criteria that are normally expected cannot be met (numbers of individuals are limited, certain stages are missing, sequencing efforts are not successful). The quest for worldwide discovery of cyst nematode diversity and broad phylogenetic representation of cyst nematodes must engage global efforts.

---

[3] The approach of *reverse taxonomy* is prospecting for new species with molecular markers that can then be linked to other parameters including morphology (De Ley *et al.*, 2005; Markmann and Tautz, 2005).

To this end participation must not be unnecessarily restrictive and standards must be applied with a degree of flexibility in order to drive the process with broad-based access, communication and collaboration (Eyualem-Abebe et al., 2006).

14.9 Conclusions and Future Prospects

Morphology of cyst nematodes has long held a central place for understanding biology, diversity, evolution, physiology, pathogenicity and management/regulation. Technological advances are transforming ways in which new growing morphological understanding interfaces with complementary avenues of investigation. These include applications of morphology in the context of genomics towards understanding the regulation of phenotypic expression, to pinpoint gene expression at specific morphological sites as in understanding regulation of diapause, the role of specific neurons in host/mate-seeking or the role of particular glands in hatching or establishing feeding sites.

Molecular phylogenetics and diagnostics have been at the forefront of advances that intersect with classical and new approaches in morphology. Morphological phylogenetics, while informative when applied under rigorous models (Baldwin and Schouest, 1990; Baldwin, 1992), is often confounded by misunderstood homology such as similar phenotypes based, not on shared evolution, but on convergence. Misinterpretations of relationships, including of cyst nematodes, may also stem from phenotypic plasticity, cryptic species and characters poorly understood because they are at the limits of LM resolution. Molecular phylogenetics (see Chapter 16, this volume) is becoming increasingly sophisticated in providing an independent character set against which to test classical morphology-based hypotheses of phylogeny and, in so doing, sometimes challenging classical classification systems. In the search for congruence between morphology and molecular character sets, there is a challenge for an increasingly sophisticated understanding of morphology. For example, the classical proposal of a unique shared ancestry between cyst and root-knot nematodes based on features linked to sedentary parasitism have been challenged and revised (Baldwin, 1992; Baldwin et al., 2004). Such revisions provide new insight into morphological evolution. For example, absence of fenestra was previously considered definitive to a cyst genus, *Afenestrata*, until molecular phylogeny helped to demonstrate that this distinctive feature and genus was embedded within the clade that defines *Heterodera* (Baldwin and Bell, 1985; Mundo-Ocampo et al., 2008). Revisions enlightened by molecular phylogeny also provide a basis for more predictable and meaningful applications of model systems. For example, investigating specific pathogenesis of cyst nematodes will more likely benefit from comparisons to the phylogenetic context of hoplolaimids than to convergent and phylogenetically distant root-knot nematodes.

We have noted that in diagnostics increasingly morphology is complementary to the use of molecular markers, and this is underscored by the very broad and growing representation of cyst species on GenBank. Linkage with molecular markers has been well verified for widespread agricultural pests, but for some taxa the process of verification continues including addressing, as noted in section 14.8, issues of intra- and interspecific variability. Are the samples to which the marker is linked co-specific with the types (e.g. as often best verified by topotypes) designated for the species? Do problems with the initial species designation (e.g. lack of monophyly) confound use of markers? There is growing recognition of an interdependence of morphological and molecular technology in addressing these challenges.

Whereas diagnostics of cyst nematodes is often considered in relation to agricultural regulation and management, it is also relevant to biodiversity, species discovery and bioprospecting, including of cyst nematodes from non-agricultural sites. Such discoveries are essential to broad taxon representation in developing a sound phylogenetic framework and classification for the Heteroderinae. Previously, species discovery was driven primarily by morphological novelties. For cyst nematodes this typically meant recognizing heterodid J2 in a soil sample and then resampling the site to search for females, cysts and possible host associations. The added steps to the process of discovery probably have resulted in under-representation of Heteroderinae in species discovery from non-agricultural sites. However, we have noted that increasingly sequences (*versus* morphology) are becoming the most efficient driver for discovering species diversity. Through reverse taxonomy, once sequence-delineated species are discovered, these 'species hypotheses' are then tested with additional data including morphological methods.

With broader access to increasingly powerful computers, software and internet, digital through-focus videos (DeLey and Bert, 2002) will play a powerful role in advancing understanding of cyst nematodes. These video records can be applied not only to vermiform specimens but also to other vouchers (e.g. cones) specific to cyst nematodes. Here morphological and molecular technologies are again linked where imaging provides a morphological voucher of a live or freshly mounted individual that is then linked to DNA product/sequences of that same individual (De Ley *et al.*, 2005). Digital through-focus imaging can exceed quality of slide-preserved vouchers and particularly so with software that allows knitting into a single movie 100× components from head to tail of a specimen. This can be accomplished with an automated software-controlled microscope stage. With growing internet capabilities, a desktop computer-controlled stage in one location can also engage global expertise for worldwide remote morphological examination of specimens.

The role of classical morphology and morphology-based species has driven priorities of taxonomic repositories (museums) and voucher specimens and for heteroderids this has meant the preservation and deposit of types, vouchers and other slides including J2, males, cysts, females and cone/terminal mounts. However, advances in the ways we discover species must also drive new ways in which we conserve information. Modern taxonomic museums, in addition to curating slides and wet collections, must also conserve and make accessible LM digital virtual specimens and SEM. Beyond morphology there is a growing need for collections to conserve sources of DNA (DNA extraction, frozen and/or dried specimens) as vouchers linked to morphology and with global access through relational databases linked across collections.

14.10 References

Adams, B.J. (1998) Species concepts and the evolutionary paradigm in modern nematology. *Journal of Nematology* 30, 1–21.
Anonymous (1999) *International Code of Zoological Nomenclature*, 4th edn. International Trust for Zoological Nomenclature, Natural History Museum, London.
Awan, F.A. and Hominick, W.H. (1982) Observation on tanning of the potato cyst nematode, *Globodera rostochiensis*. *Parasitology* 85, 61–71. DOI: 10.1017/S0031182000054159
Baldwin, J.G. (1983) Fine structure of body wall cuticle of females of *Meloidodera charis, Atalodera lonicerae* and *Sarisodera hydrophila* (Heteroderidae). *Journal of Nematology* 15, 370–381.
Baldwin, J.G. (1985) Fine structure of the phasmid of second-stage juveniles of *Heterodera schachtii* (Tylenchida: Nematoda). *Canadian Journal of Zoology* 63, 534–542.
Baldwin, J.G. (1992) Evolution of cyst and non-cyst forming Heteroderinae. *Annual Review of Phytopathology* 30, 271–290. DOI: 10.1146/annurev.py.30.090192.001415
Baldwin, J.G. and Bell, A.H. (1985) *Cactodera eremica* n. sp., and *Afenestrata africana* (Luc *et al.*, 1973) n. gen., n. comb. and an emended diagnosis of *Sarisodera* Wouts & Sher, 1971 (Heteroderidae). *Journal of Nematology* 17, 187–201.
Baldwin, J.G. and Hirschmann, H. (1975a) Fine structure of cephalic sense organs in *Heterodera glycines* males. *Journal of Nematology* 7, 40–53.
Baldwin, J.G. and Hirschmann, H. (1975b) Body wall fine structure of the anterior region of *Meloidogyne incognita* and *Heterodera glycines* males. *Journal of Nematology* 7, 175–193.
Baldwin, J.G. and Hirschmann, H. (1976) Comparative fine structure of the stomatal region of males of *Meloidogyne incognita* and *Heterodera glycines*. *Journal of Nematology* 8, 1–17.
Baldwin, J.G. and Mundo-Ocampo, M. (1991) Heteroderinae, cyst- and non-cyst forming nematodes. In: Nickel, W.R. (ed.) *Manual of Agricultural Nematology*. Marcel Dekker Inc., New York, pp. 275–363.
Baldwin, J.G. and Perry, R.N. (2004) Nematode morphology, sensory structure and function. In: Chen, Z.X., Chen, S.Y. and Dickson, D.W. (eds) *Nematology Advances and Perspectives Volume 1 (Nematode Morphology, Physiology and Ecology)*. CAB International, Wallingford, UK, pp. 175–257.

Baldwin, J.G. and Schouest, L.P. Jr (1990) Comparative detailed morphology of Heteroderinae Filip'ev & Schuurmans Stekhoven, 1941, *sensu* Luc *et al.* (1988) phylogenetic systematics and revised classification. *Systematic Parasitology* 15, 81–106.

Baldwin, J.G., Hirschmann, H. and Triantaphyllou, A.C. (1977) Comparative fine structure of the esophagus of males of *Heterodera glycines* and *Meloidogyne incognita*. *Nematologica* 23, 239–252. DOI: 10.1163/187529277X00598

Baldwin, J.G., Nadler, S.A. and Adams, B.J. (2004) Evolution of plant parasitism among nematodes. *Annual Review of Phytopathology* 42, 83–105. DOI: 10.1146/annurev.phyto.42.012204.130804

Bernard, E.C., Handoo, Z.A., Powers, T.O., Donald, P.A. and Heinz, R.D. (2010) *Vittatidera zeaphila* (Nematoda: Heteroderidae). A new genus and species of cyst nematode parasitic on corn (*Zea mays*). *Journal of Nematology* 42, 139–150.

Borgonie, G., Claeys, M., De Waele, D. and Coomans, A. (1995) Ultrastructure of the intestine of the bacteriophagous nematodes *Caenorhabditis elegans*, *Panagrolaimus superbus* and *Acrobeloides maximus* (Nematoda: Rhabditida). *Fundamental and Applied Nematology* 16, 47–56.

Brown, G., Callow, R.K., Green, C.D., Jones, F.G.W., Rayner, J.J., Shepherd, A.M. and Williams, T.D. (1971) The structure, composition and origin of the sub-crystalline layer in some species of the genus *Heterodera*. *Nematologica* 17, 591–599. DOI: 10.1163/187529271X00305

Bumbarger, D.J., Crum, J., Ellisman, M.H. and Baldwin, J.G. (2006) Three-dimensional reconstruction of the nose epidermal cells in the microbial feeding nematode, *Acrobeles complexus* (Nematoda: Rhabditida). *Journal of Morphology* 267, 1257–1272.

Bumbarger, D.J., Crum, J., Ellisman, M.H. and Baldwin, J.G. (2007) Three-dimensional fine structural reconstruction of the nose sensory structures of *Acrobeles complexus* compared to *Caenorhabdtis elegans* (Nematoda: Rhabditida). *Journal of Morphology* 268, 649–663.

Bumbarger, D., Wijeratne, S., Carter, C., Crum, J., Ellisman, M. and Baldwin, J.G. (2009) Three-dimensional reconstruction of the amphid sensilla in the microbial feeding nematode, *Acrobeles complexus* (Nematoda: Rhabditida). *Journal of Comparative Neurology* 512, 271–181. DOI: 10.1002/cne.21882

Burr, A.J. and Baldwin, J.G. (2016) The nematode stoma: homology of cell architecture with improved understanding by confocal microscopy of labeled cell boundaries. *Journal of Morphology* 277, 1168–1186.

Cares, J.E. and Baldwin, J.G. (1995) Comparative fine structure of sperm in *Heterodera schachtii* and *Punctodera chalcoensis*, with phylogenetic implications for Heteroderinae (Nematoda). *Canadian Journal of Zoology* 73, 309–320.

Carta, L.K. (1991) Preparation of nematodes for transmission electron microscopy. In: Nickel, W.R. (ed.) *Manual of Agricultural Nematology*. Marcel Dekker Inc., New York, pp. 97–106.

Carta, L.K. and Baldwin, J.G. (1990) Phylogenetic implications of phasmid absence in males of three genera in Heteroderinae. *Journal of Nematology* 22, 386–394.

Clark, S.A., Shepherd, A.M. and Kempton, A. (1973) Spicule structure in some *Heterodera* spp. *Nematologica* 19, 242–247. DOI: 10.1163/187529273X00367

Cliff, G. and Baldwin, J.G. (1985) Fine structure of body wall cuticle of females of eight genera of Heteroderidae. *Journal of Nematology* 17, 286–296.

Cooper, B.A. (1955) A preliminary key to British species of *Heterodera* for use in soil examination. In: Kevan, D.K.McE. (ed.) *Soil Zoology*. Butterworths, London, pp. 269–280.

Cordero, D.A. (1989) Comparative morphology and development of the cone of *Heterodera schachtii* and *Cactodera cacti*. PhD thesis, University of California, Riverside, California.

Cordero, D.A. and Baldwin, J.G. (1990) Effect of age on body wall cuticle morphology of *Heterodera schachtii* Schmidt females. *Journal of Nematology* 22, 356–361.

Cordero, D.A. and Baldwin, J.G. (1991) Fine structure of the posterior cone of *Heterodera schachtii* Schmidt (Heteroderinae) with emphasis on musculature and fenestration. *Journal of Nematology* 23, 110–121.

Cordero, D.A., Baldwin, J.G. and Mundo-Ocampo, M. (1991) Fine structure of the posterior cone of females of *Cactodera cacti* Filip'ev and Schuurmans Stekhoven (Nemata: Heteroderinae). *Revue de Nématologie* 14, 455–465.

Correia, J.S. and Abrantes, I.M. (1997) An improved technique for mounting *Heterodera* cysts for light microscopy. *Nematologica* 43, 507–509. DOI: 10.1163/005125997X00101

De Ley, P. and Bert, W. (2002) Video capture and editing as a tool for the storage, distribution and illustration of morphological characters of nematodes. *Journal of Nematology* 34, 296–302.

De Ley, P. and Blaxter, M.L. (2002) Systematic position and phylogeny. In: Lee, D.L. (ed.) *The Biology of Nematodes*. Taylor and Francis, New York, pp. 1–30.

De Ley, P., De Ley, I.T., Morris, K. et al. (2005) An integrated approach to fast and informative morphological vouchering of nematodes for applications in molecular barcoding. *Philosophical Transactions of the Royal Society* B 360, 1945–1958.

De Queiroz, K. (2007) Species concepts and species delimitation. *Systematic Biology* 56, 879–886.

Doncaster, C.C. and Shepherd, A.M. (1967) The behaviour of second-stage *Heterodera rostochiensis* larvae leading to their emergence from the egg. *Nematologica* 13, 476–478. DOI: 10.1163/187529267X00797

Eisenback, J.D. (1988) Multiple focus and exposure photomicroscopy of nematodes for increased depth of field. *Journal of Nematology* 20, 333–334.

Eisenback, J.D. (1991) Preparation of nematodes for scanning electron microscopy. In: Nickel, W.R. (ed.) *Manual of Agricultural Nematology*. Marcel Dekker Inc., New York, pp. 87–96.

Eisenback, J.D. and Zunke, U. (1998) Extraction, culturing and microscopy. In: Sharma, S.B. (ed.) *The Cyst Nematodes*. Kluwer Academic Publishers, Dordrecht, The Netherlands, pp. 141–155.

Ellenby, C. (1946) Nature of the cyst wall of the potato-root eelworm, *Heterodera rostochiensis* Wollenweber, and its permeability to water. *Nature* 157, 302–303.

Endo, B.Y. (1979) The ultrastructure and distribution of an intracellular bacterium-like microorganism in tissues of larvae of the soybean cyst nematode, *Heterodera glycines*. *Journal of Ultrastructure Research* 67, 1–14.

Endo, B.Y. (1980) Ultrastructure of the anterior neurosensory organs of the larvae of the soybean cyst nematode, *Heterodera glycines*. *Journal of Ultrastructural Research* 72, 349–366.

Endo, B.Y. (1983) Ultrastructure of the stomatal region of the juvenile stage of the soybean cyst nematode, *Heterodera glycines*. *Proceedings of the Helminthological Society of Washington* 50, 43–61.

Endo, B.Y. (1984) Ultrastructure of the esophagus of larvae of the soybean cyst nematode, *Heterodera glycines*. *Proceedings of the Helminthological Society of Washington* 51, 1–24.

Endo, B.Y. (1985) Ultrastructure of the head region of molting second-stage juveniles of *Heterodera glycines* with emphasis on stylet formation. *Journal of Nematology* 17, 112–123.

Endo, B.Y. (1988) Ultrastructure of the intestine of second and third juvenile stages of the soybean cyst nematode, *Heterodera glycines*. *Proceedings of the Helminthological Society of Washington* 55, 117–131.

Endo, B.Y. (1991) Ultrastructure of initial responses of susceptible and resistant soybean roots to infection by *Heterodera glycines*. *Revue de Nématologie* 14, 73–91.

Endo, B.Y. (1993) Ultrastructure of cuticular exudates and related cuticular changes on juveniles in *Heterodera glycines*. *Journal of the Helminthological Society of Washington* 60, 76–88.

Endo, B.Y. and Wyss, U. (1992) Ultrastructure of cuticular exudations in parasitic juvenile *Heterodera schachtii*, as related to cuticle structure. *Protoplasma* 166, 67–70.

Eyualem-Abebe, E., Baldwin, J.G., Adams, B. et al. (2006) A position paper on the electronic publication of nematode taxonomic manuscripts. *Journal of Nematology* 38, 305–311.

Franklin, M.T. (1939) On the structure of the cyst wall of *Heterodera schachtii* (Schmidt). *Journal of Helminthology* 17, 127–134.

Franklin, M.T. (1951) *The cyst-forming species of Heterodera*. Commonwealth Agricultural Bureaux, Farnham Royal, Farnham, UK.

Fuchs, O. (1911) Beitrage zur Biologie der Rubennematoden. *Zeitschrift für das landwirschaftliche Versuchswesen in Deutsch Österreich* 14, 923–949.

Golden, A.M. (1986) Morphology and identification of cyst nematodes. In: Lamberti, F. and Taylor, C.E. (eds) *Cyst Nematodes*. Plenum Press, New York, pp. 23–79.

Golden, A.M. and Raski, D.J. (1977) *Heterodera thornei* n. sp. (Nematoda: Heteroderidae) and a review of related species. *Journal of Nematology* 9, 93–112.

Goodey, J.B. (1963) *Laboratory Methods for Work with Plant and Soil Nematodes*. Technical Bulletin No. 2. Her Majesty's Stationery Office, Ministry of Agriculture, Fisheries and Food, London.

Granek, I. (1955) Additional morphological differences between the cysts of *Heterodera rostochiensis* and *Heterodera tabacum*. *Plant Disease Reporter* 39, 716–718.

Green, C.D. (1971) The morphology of the terminal area of the round-cyst nematodes S. G. *Heterodera rostochiensis* and allied species. *Nematologica* 17, 34–46. DOI: 10.1163/187529271X00396

Green, C.D. (1975) The vulval cone and associated structures of some cyst nematodes (genus *Heterodera*). *Nematologica* 21, 134–144. DOI: 10.1163/187529275X00491

Hesling, J.J. (1973) The estimation of Granek's ratio in round-cyst *Heteroderas*. *Nematologica* 19, 119–120. DOI: 10.1163/187529273X00213

Hesling, J.J. (1978) Cyst nematodes: morphology and identification of *Heterodera*, *Globodera* and *Punctodera*. In: Southey, J.F. (ed.) *Plant Nematology*. Her Majesty's Stationery Office, London, pp. 125–155.

Hirschmann, H. (1956) Comparative morphological studies on the soybean cyst nematode, *Heterodera glycines* and the clover cyst nematode, *H. trifolii* (Nematoda: Heteroderidae). *Proceedings of the Helminthological Society of Washington* 23, 140–151.

Hirschmann, H. and Triantaphyllou, A.C. (1979) Morphological comparison of members of the *Heterodera trifolii* species complex. *Nematologica* 25, 458–481. DOI: 10.1163/187529279X00613

Karnovsky, M.J. (1965) A formaldehyde-glutardehyde fixative of high osmolarity for use in electron microscopy. *Journal of Cell Biology* 27,137–138A.

Kang, H., Eun, G., Ha, J., Kim, Y., Park, N. and Kim, D. (2016) New cyst nematode, *Heterodera sojae* n. sp. (Nematoda: Heteroderidae) from soybean in Korea. *Journal of Nematology* 48, 280–289.

Kirjanova, E.S. (1969) [On the structure of the subcrystalline layer of the nematode genus *Heterodera* (Nematoda: Heteroderidae) with a description of two new species.] *Parazitologiya* 3, 81–91.

Kolotuev, I., Schwab, Y. and Labouesse, M. (2010) A precise and rapid mapping protocol for correlative light and electron microscopy of small invertebrate organisms. *Biology of the Cell* 102, 121–132.

Liebscher, G. (1892) Beobachtungen über das Auftreten eines Nematoden an Erbseb, *Journal für Landwirtschaft* 40, 357–368.

Luc, M., Weischer, B., Stone, A.R. and Baldwin, J.G. (1986) On the definition of heterodid cyst. *Revue de Nématologie* 9, 418–421.

Mallet, J. (2008) A century of evolution: Ernst Mayr (1904–2005) Mayr's view of Darwin: was Darwin wrong about speciation. *Biological Journal of the Linnean Society* 95, 3–16.

Mallet, J. (2010) Why was Darwin's view of species rejected by twentieth century biologists? *Biology and Philosophy* 25, 497–527.

Markmann, M. and Tautz, D. (2005) Reverse taxonomy: an approach towards determining the diversity of meiobenthic organisms based on ribosomal RNA signature sequences. *Philosophical Transactions of the Royal Society B* 360, 1917–1924.

McClure, M.A. and Stowell, L.J. (1978) A simple method of processing nematodes for electron microscopy. *Journal of Nematology* 10, 376–377.

Miller, L.I. (1983) Diversity of selected taxa of *Globodera* and *Heterodera* and their interspecific and intergeneric hybrids. In: Stone, A.R., Pratt, H.M. and Khali, L.F. (eds) *Concepts in Nematode Systematics*. The Systematics Association Special Volume No. 22. Academic Press, New York, pp. 207–220.

Mulvey, R.H. (1959) Investigations on the clover cyst nematode, *Heterodera trifolii* (Nematoda: Heteroderidae). *Nematologica* 4, 147–156. DOI: 10.1163/187529259X00138

Mulvey, R.H. (1973) Morphology of the terminal areas of white females and cysts of the genus *Heterodera* (s.g. *Globodera*). *Journal of Nematology* 5, 303–311.

Mulvey, R.H. (1974) Cone top morphology of white females and cysts of the genus *Heterodera* (subgenus *Heterodera*), a cyst-forming nematode. *Canadian Journal of Zoology* 52, 77–81.

Mulvey, R.H. and Golden, A.M. (1983) An illustrated key to the cyst-forming genera and species of Heteroderidae in the Western Hemisphere with species morphometerics and distribution. *Journal of Nematology* 15, 1–29.

Mundo-Ocampo, M. and Baldwin, J.G. (1984) Comparison of host response of *Cryphodera utahensis* with other Heteroderidae and a discussion of phylogeny. *Proceedings of the Helminthological Society of Washington* 51, 25–31.

Mundo-Ocampo, M. and Baldwin, J.G. (1992) Comparison of host response of *Ekphymatodera thomasoni* with other Heteroderinae. *Fundamental and Applied Nematology* 15, 63–70.

Mundo-Ocampo, M., Troccoli, A., Subbotin, S.A., Del Cid, J., Baldwin, J.G. and Inserra, R.N. (2008) Synonymy of *Afenestrata* with *Heterodera* supported by phylogenetics with molecular and morphological characterization of *H. koreana* comb. n. and *H. orientalis* comb. n. (Tylenchida: Heteroderidae). *Nematology* 10, 611–632. DOI: 10.1163/156854108785787190

Nadler, S.A. (2002) Species delimitation and nematode biodiversity: phylogenies rule. *Nematology* 4, 615–625.

Nyaku, S.T., Sripathi, V.R., Kantety, R.V., Gu. U.Q., Lawrence, K. and Sharma, G.C. (2013) Characterization of the two intra-individual sequence variants in the 18S rRNA gene in the plant-parasitic nematode, *Rotylenchulus reniformis*. *PLOS ONE* 8, e60891. DOI: 10.1371/journal.pone.0060891

Othman, A.A., Baldwin, J.G. and Mundo-Ocampo, M. (1988) Comparative morphology of *Globodera*, *Cactodera*, and *Punctodera* spp. (Heteroderidae) with scanning electron microscopy. *Revue de Nématologie* 11, 53–63.

Pereira, T.J. and Baldwin, J.G. (2016) Contrasting evolutionary patterns of 28S and ITS rRNA genes reveal high genomic variation in *Cephalenchus* (Nematoda): implications for species delimitation. *Molecular Phylogenetics and Evolution* 98, 244–260. DOI: 10.1016/j.ympev.2016.02.016

Ragsdale, E.J., Crum, J., Ellisman, M.H. and Baldwin, J.G. (2008) Three-dimensional reconstruction of the stomatostylet and anterior epidermis in the nematode *Aphelenchus avenae*: Nematoda: Aphelenchidae) with implications for the evolution of plant parasitism. *Journal of Morphology* 269, 1181–1196. DOI: 10.1002/jmor.10651.

Ragsdale, E.J., Ngo, P., Crum, J., Ellisman, M.H. and Baldwin, J.G. (2009) Comparative, three-dimensional anterior sensory reconstruction of *Aphelenchus avenae* (Nematoda: Tylenchomorpha). *Journal of Comparative Neurology* 517, 616–632. DOI: 10.1002/cne.22170

Raski, D.J. (1950) The life history and morphology of the sugar-beet nematode, *Heterodera schachtii* Schmidt. *Phytopathology* 40, 135–152.

Riggs, R.D. (1990) Making perineal patterns of root-knot nematodes and vulval cones of cyst nematodes. In: Zuckerman, B.M., Mai, W.F. and Krusberg, L.R. (eds) *Plant Nematology Laboratory Manual*, revised edn. The University of Massachusetts Agricultural Experiment Station, Amherst, Massachusetts, pp. 103–106.

Sammons, R. and Marquis, P. (1997) Application of the low vacuum scanning electron microscope to the study of biomaterials and mammalian cells. *Biomaterials* 18, 81–86.

Shepherd, A.M. and Clark, S.A. (1978) Cuticle structure and 'cement' formation at the anterior end of female cyst-nematodes of the genera *Heterodera* and *Globodera* (Heteroderidae: Tylenchida). *Nematologica* 24, 201–208. DOI: 10.1163/187529278X00407

Shepherd, A.M., Clark, S.A. and Dart, P.J. (1972) Cuticle structure in the genus *Heterodera*. *Nematologica* 18, 1–17. DOI: 10.1163/187529272X00197

Shepherd, A.M., Clark, S.A. and Kempton, A. (1974) Spermatogenesis and sperm ultrastructure in some cyst nematodes, *Heterodera* spp. *Nematologica* 19, 551–560. DOI: 10.1163/187529273X00574

Sher, S.A. and Bell, A.H. (1975) Scanning electron micrographs of the anterior region of some species of Tylenchoidea (Tylenchida: Nematoda). *Journal of Nematology* 7, 69–83.

Stone, A.R. (1975) Head morphology of second-stage juveniles of some Heteroderidae (Nematoda: Tylenchoidea). *Nematologica* 21, 81–88. DOI: 10.1163/187529275X00374

Sturhan, D. (2016) On the presence or absence of phasmids in males of Heteroderidae (Tylenchida). *Nematology* 18, 23–27. DOI: 10.1163/15685411-00002939

Subbotin, S.A., Mundo-Ocampo, M. and Baldwin, J.G. (2010) *Systematics of cyst nematodes (Nematoda: Heteroderinae)*. Nematology Monographs and Perspectives, 8A. (Series eds: Hunt, D.J. and Perry, R.N.). Leiden, The Netherlands, Brill.

Sulston, J. and Hodgkin, J. (1988) *Methods*. In: Wood, W.B. (ed.) *The Nematode* Caenorhabditis elegans. Cold Springs Harbor Press, New York, pp. 587–606.

Thorne, G. (1961) *Principals of Nematology*. McGraw Hill, New York.

Triantaphyllou, A.C. and Hirschmann, H. (1962) Oogenesis and mode of reproduction in the soybean cyst nematode, *Heterodera glycines*. *Nematologica* 7, 235–241. DOI: 10.1163/187529262X00224

Triantaphyllou, A.C. and Hirschmann, H. (1980) Cytogenetics and morphology in relation to evolution and speciation of plant-parasitic nematodes. *Annual Review of Phytopathology* 18, 333–359.

Turner, S.J. and Subbotin, S.A. (2013) Cyst nematodes. In: Perry, R.N. and Moens, M. (eds) *Plant Nematology*. CAB International, Wallingford, UK, pp. 91–122.

Walsh, J.A. and Shepherd, A.M. (1983) A further observation on sperm structure in a *Heterodera* sp. *Revue de Nématologie* 6, 148–150.

Willats, W.G., Atkinson, H.J. and Perry, R.N. (1995) The immunofluorescent localization of subventral pharyngeal gland epitopes of preparasitic juveniles of *Heterodera glycines* using laser scanning confocal microscopy. *Journal of Nematology* 27, 135–142.

Wilson, E.M. (1968) Vulval bodies in certain species of *Heterodera*. *Nature* 217, 879.

Wouts, W.M. and Baldwin, J.G. (1998) Taxonomy and identification. In: Sharma, S.B. (ed.) *The Cyst Nematodes*. Kluwer Academic Publishers, Dordrecht, The Netherlands, pp. 83–122.

Wouts, W.M. and Weischer, B. (1977) Eine Klassifizierung von fünfzehn in Westeuropa häufigen Arten der Heteroderinae auf Grund von Larvenmerkmalen. *Nematologica* 23, 289–310. DOI: 10.1163/187529277X00039

Wyss, U. and Zunke, U. (1986a) The potential of high resolution video-enhanced contrast microscopy in nematological research. *Revue de Nématologie* 9, 91–94.

Wyss, U. and Zunke, U. (1986b) Observations on the behaviour of second-stage juveniles of *Heterodera schachtii* inside host roots. *Revue de Nématologie* 9, 153–165.

Zunke, U. (1986) Zur Bildung der subkristallinen Schicht bei *Heterodera schachtii* unter aseptischen Bedingungen. *Nematologica* 31, 117–120. DOI: 10.1163/187529285X00157

Zunke, U. and Eisenback, J.D. (1998) Morphology and ultrastructure. In: Sharma, S.B. (ed.) *The Cyst Nematodes*. Kluwer Academic Publishers, Dordrecht, The Netherlands, pp. 31–56.

15 Taxonomy, Identification and Principal Species

Zafar A. Handoo[1] and Sergei A. Subbotin[2,3]

[1]USDA, ARS, Beltsville, Maryland, USA; [2]California Department of Food and Agriculture, Sacramento, California, USA; [3]Center of Parasitology of A.N. Severtsov Institute of Ecology and Evolution, Moscow, Russia

15.1	Introduction	365
15.2	Identification	366
15.3	Systematic Position	368
15.4	Subfamily Heteroderinae Diagnosis	369
15.5	Genus *Heterodera* Schmidt, 1871	369
15.6	Subfamily Punctoderinae Diagnosis	384
15.7	Genus *Globodera* Skarbilovich, 1959	385
15.8	Genus *Punctodera* Mulvey & Stone, 1976	390
15.9	Genus *Cactodera* Krall & Krall, 1978	390
15.10	Genus *Dolichodera* Mulvey & Ebsary, 1980	391
15.11	Genus *Betulodera* Sturhan, 2002	392
15.12	Genus *Paradolichodera* Sturhan, Wouts & Subbotin, 2007	393
15.13	Genus *Vittatidera* Bernard, Handoo, Powers, Donald & Heinz, 2010	393
15.14	Conclusions and Future Prospects	393
15.15	Acknowledgements	393
15.16	References	394

15.1 Introduction

15.1.1 History

The first cyst-forming nematode was discovered on sugar beets by Schacht (1859) and was later described by Schmidt (1871), who established the first new genus of cyst nematode, *Heterodera*, and named this nematode as *Heterodera schachtii* in honour of Hermann Schacht. Presently, the genus *Heterodera* includes species with cyst bodies more or less lemon-shaped and ambifenestrate, bifenestrate vulval fenestration or without fenestration. To accommodate the potato cyst nematodes and related species having round cysts, Skarbilovich (1959) erected the subgenus *Globodera*, which later was elevated to generic status by Behrens (1975). Around the same time, Mulvey and Stone (1976) proposed recognizing *Globodera* at generic level. These authors also described the genus *Punctodera* for species having extensive anal fenestration. Krall and Krall (1978) erected the genus *Cactodera* for species with cysts that have a posterior protuberance and circumfenestrate vulva region. Sturhan (2002) established a new genus, *Betulodera*, to accommodate *Cactodera betulae* based on peculiarities of its host range and cyst morphology. Later three genera, *Dolichodera*, *Paradolichodera* and *Vittatidera*, were described with morphological peculiarities; each of these genera presently contains only single species. The introduction of cyst nematodes and detailed historical background of various cyst-forming nematode genera is given in several books and chapters (Franklin, 1951; Baldwin and Mundo-Ocampo, 1991; Ferris, 1998; Wouts and Baldwin, 1998; Siddiqi, 2000; Baldwin and Perry, 2004; Subbotin *et al.*, 2010; Subbotin and Franco, 2012; Turner and Subbotin, 2013).

The cyst-forming nematode species have been instrumental in stimulating growth of nematology worldwide. Presently this nematode group contains eight genera: *Heterodera* (85 species), *Globodera* (14 species), *Cactodera* (14 species), *Dolichodera* (1 species), *Paradolichodera* (1 species), *Betulodera* (1 species), *Punctodera* (4 species) and *Vittatidera* (1 species), with a total number of 121 valid species.

15.1.2 Major reference sources

Cyst nematode taxonomy and specific diagnostic variability were studied by a number of authors, with the earlier monographs and major reference sources discussed in great detail in several book chapters and articles by Mulvey (1959, 1972, 1973), Franklin (1972), Stone (1973a, b, 1975, 1979), Stone and Rowe (1976), Mulvey and Ebsary (1980), Stone and Hill (1982), Mulvey and Golden (1983), Vovlas (1985), Vovlas *et al.* (1985), Golden (1986), Baldwin and Mundo-Ocampo (1991), Wouts and Baldwin (1998), Zunke and Eisenback (1998), Siddiqi (2000), Sturhan (2002, 2010) and Sturhan *et al.* (2007). The importance of the cone mounts in morphology and identification of cyst nematode was first stressed by Oostenbrink and den Ouden (1954), followed by Cooper (1955), Mulvey (1957, 1960, 1972) and Green (1971). In the book chapter on morphology and identification of cyst nematodes, Golden (1986) gives detailed information on cyst wall pattern, cyst posterior, vulval cone and terminal region, together with other characters useful for identification of males and second-stage juveniles. Other major sources include the two-volume treatise written by Subbotin *et al.* (2010). The latest chapters by Subbotin and Franco (2012) and Turner and Subbotin (2013) contain a brief introduction to the cyst-forming nematode groups, together with information on life cycle and behaviour and general morphology of the subfamily Heteroderidae, including diagnosis and descriptions of the principal species.

15.2 Identification

Morphology is the essential basis for identification of cyst nematodes (Fig. 15.1). Molecular approaches in taxonomy are more widely used now and will be used in future to supplement and extend our morphological knowledge, and will no doubt provide new understanding of the identity and systematics of various populations and species. Accurate and rapid identification of cyst and other nematodes is crucial for the implementation of effective control measures. In addition, sound decisions regarding quarantine of imported and exported plant material and commodities also demand timely and accurate diagnostics. However, the identification of cyst nematodes to species level is fraught with difficulty. In cyst nematodes, conserved morphology, variable morphometrics, host effects, intraspecific variation and existence of cryptic species complicate the situation, and the ever-increasing number of described species, of which the diagnosis and relationships of many vary from less than ideal to doubtful (as is also the case with root-knot nematodes) causes problems.

Verification of mixed populations and/or detection of rare species requires identification techniques, including morphological (cone mounts of posterior end of cysts; male, female and second-stage juvenile (J2) labial region shape, and stylet morphology; length and shape of J2 tail) and biochemical or molecular methodologies. Detailed diagnostic characters differentiating various genera of cyst-forming nematode species have been given by authors such as Mulvey and Golden (1983), Golden (1986), Baldwin and Mundo-Ocampo (1991), Ferris (1998), Wouts and Baldwin (1998), Zunke and Eisenback (1998), Siddiqi (2000), Subbotin *et al.*

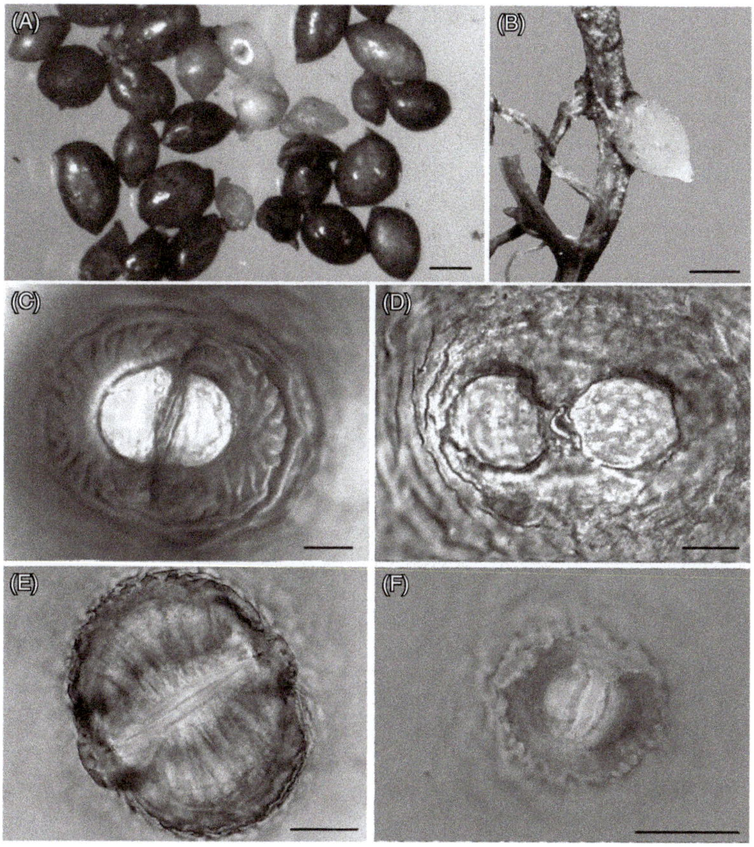

Fig. 15.1. (A) Cysts of *Heterodera cruciferae*. (B) White female of *H. cruciferae* on a root. Vulval plates. (C) *H. medicaginis* (ambifenestrate). (D) *Heterodera* sp. (bifenestrate). (E) *H. orientalis* (no fenestration). (F) *Cactodera rosae* (circumfenestrate). A–D, Courtesy of V.N. Chizhov; E, after Mundo-Ocampo *et al.*, 2008; F, courtesy of I. Cid del Prado Vera. Scale bars: A, B = 200 μm; C = 15 μm; D, E, F = 25 μm.

(2000, 2010), Handoo (2002), Skantar *et al.* (2007, 2011), Nakhla *et al.* (2010), Subbotin and Franco (2012) and Turner and Subbotin (2013). They are also detailed in in Chapter 14, this volume, under 'Conclusions and future prospects'.

15.2.1 General techniques

For morphological observation, J2 and males can be recovered from fresh infected roots or cysts/eggs incubated in Petri dishes with a small amount of water. They may also be recovered from soil by sieving and Baermann funnel techniques. Cysts and white females are dissected from infected roots after fixation overnight in 3% formaldehyde. Procedures for measuring and preparing specimens are given in Golden and Birchfield (1972). Females have the anterior and posterior ends cut with a sharp knife, cleaned with a dental root canal file and mounted permanently on a glass slide in a drop of lactophenol solution (see Chapter 14, this volume). Photomicrographs of cone mounts, J2 and males can be done with a digital camera attached to a dissecting microscope.

For more details on killing, fixing, processing nematodes to glycerin, preparing temporary and permanent slide mounts and preserving nematode structures in a life-like manner, the reader is referred to Whitehead (1968), Hooper (1970, 1986, 1990), Golden (1990) and Carta (1991).

15.2.2 Cone mounts

The character most frequently used for cyst species identification is the morphology of the cone, which is located in the posterior body region of the cyst. To investigate fully the posterior part of the cyst (perinea), it must first be mounted on a slide. Oostenbrink and den Ouden (1954) were the first to use cone top structures, including fenestra, bullae and underbridge, in the separation of cyst-forming nematodes. Cooper (1955) provided detailed information and later Mulvey (1957, 1960), Fenwick (1959) and Hesling (1965) added more information. The identification of *Heterodera* cysts by terminal and cone top structures was discussed by Mulvey (1972) in a detailed examination of 39 of the 53 species of *Heterodera* described at that time to provide adequate basis for identification of these species; brief information on making vulval cones of cyst nematodes is given by Riggs (1990). A protocol for preparation of the vulval cones of cyst nematodes plus figures showing techniques for preparing cyst cones for light microscope observation, are given in Appendix II of Subbotin *et al.* (2010). A more detailed account on cone mounts is covered in Chapter 14, this volume.

15.2.3 Root staining

Many methods have been developed for staining and clearing nematode-infected root tissues. Staining with acid fuchsin-lactophenol or lacto-glycerol are the most widely used methods. In addition, a method that utilizes chlorine bleach as a prestaining treatment has proved very reliable and is relatively simple to use (Byrd *et al.*, 1983). For more detail, see McBeth *et al.* (1941), Hooper (1986, 1990) and Hussey (1990).

15.2.4 Scanning electron microscopy

For scanning electron microscopy (SEM), living specimens are fixed in 3% glutaraldehyde solution buffered with 0.05 M phosphate (pH 6.8), dehydrated in a graded series of ethanol, critical point dried from liquid CO_2 and sputter-coated with a 20–30-nm layer of gold-palladium. For more detail, the papers by Eisenback (1991) and Charchar and Eisenback (2000), and Chapter 14, this volume, are recommended.

15.2.5 Diagnostic characters

The most important diagnostic features used for identification of cyst nematodes include: **Cyst**: Size, shape and colour, nature of cone (ambifenestrate, bifenestrate, circumfenestrate or no fenestration; Fig. 15.1C–F). Fenestra length and width, presence or absence of bullae, underbridge, vulval slit length, nature of cyst wall pattern, Granek's ratio (distance from the anus to the nearest edge of the fenestra divided by the length of the fenestra). **Female**: Shape of body, labial region, stylet length, shape of stylet cone, shaft and basal knobs, and excretory pore/stylet length ratios (EP/ST). **Male**: Size, height and shape of labial cap, the number of annulations, diameter of the labial region as compared to the first body annule, stylet length, form of stylet cone, shaft and basal knobs, distance of the dorsal gland orifice (DGO) from the stylet base and length and form of spicule and gubernaculum. **J2**: Body and stylet length, form of labial region and shape of stylet knobs, location of the hemizonid in relation to the excretory pore, distance of DGO from stylet base, number of incisures in the lateral field and shape and length of the tail and hyaline terminus. For more details about these and other differentiating characters, see Baldwin and Mundo-Ocampo (1991), Golden (1986), Mulvey and Golden (1983), Golden (1986), Subbotin *et al.* (2010), Turner and Subbotin (2013) and Chapters 14 and 16, this volume.

15.3 Systematic Position

Because of certain similarities in morphology and biology, root-knot nematodes and cyst-forming nematodes have often been thought to be closely related. As a consequence, in many old systematic schemes both groups were often placed in a single family or subfamily, the Heteroderidae or Heroderinae, respectively, closely related to the hoplolaimids. However, a growing suspicion indicated that the two groups had evolved separately and had achieved their similarities via the process of convergent evolution. According to this view, the cyst and root-knot nematodes each justify their own family or subfamilies, but the root-knot nematodes are closer to the pratylenchids than to the hoplolaimids (Hunt and Handoo, 2009). Systematics of the family Heteroderidae and its genera are given in great detail by Subbotin *et al.* (2010) as well as the classification scheme followed after Siddiqi (2000) presented here in Table 15.1. The taxonomic history and problems of Heteroderinae are discussed by Hesling (1965), Krall and Krall (1978), Golden, (1986), Baldwin and Mundo-Ocampo (1991) and Siddiqi (2000).

The advances in molecular methodologies have facilitated a better understanding of the phylogeny of nematodes. We have proposed here a new improved classification for the family Heteroderidae on the basis of up-to-date information collected on the classification of cyst nematodes (Table 15.1) (Subbotin *et al.*, 2017). Regrouping of some genera under these subfamilies has been undertaken, wherever it was felt necessary and these genera have been shifted to their appropriate subfamiles. These additions seemed desirable and necessary to provide more useful groupings. The changes are incorporated to make the classification easier to handle. Classification of the family Heteroderidae by other authors, including our current proposed one, is given in Table 15.1. Cyst-forming nematodes are placed in two subfamilies: Heteroderinae, Filipjev & Schuurmans Stekhoven, 1941, and Punctoderinae, Krall & Krall, 1978.

15.4 Subfamily Heteroderinae Filipjev & Schuurman Stekhoven, 1941

Diagnosis: Heteroderidae: Filipjev & Schuurmans Stekhoven, 1941. **Mature female and cyst**: body more or less lemon-shaped, with a posterior cone. Female cuticle not annulated, changing colour after death. Vulval slit terminal or sunken into cone, anus on dorsal vulval lip. Vulval slit 6–68 µm in length. Cuticle surface with zigzag or lace-like pattern of ridges. Vulval fenestration: ambifenestrate, bifenestrate or absent. D-layer in cuticle rudimentary or absent. Subcrystalline layer present or absent. Anus without fenestration. Bullae and underbridge present or absent. Eggs retained in body, in some cases eggsac also present. **Male**: developed through metamorphosis, labial region annulated, lateral field with four or three incisures, tail short, hemispherical, without bursa. **J2**: lateral field marked by three or four incisures, phasmids punctiform or lens-like. **Type and only genus**: *Heterodera* Schmidt, 1871

15.5 Genus *Heterodera* Schmidt, 1871

Diagnosis: (Heteroderinae) is the same as for the subfamily.
Genus: *Heterodera* Schmidt, 1871
= *Tylenchus* (*Heterodera* Schmidt, 1871)
= *Heterodera* (*Heterodera* Schmidt, 1871)
= *Heterobolbus* Railliet, 1896
= *Bidera* Krall & Krall, 1978
= *Ephippiodera* Shagalina & Krall, 1981
= *Afenestrata* Baldwin & Bell, 1985
= *Afrodera* Wouts, 1985
= *Brevicephalodera* Kaushal & Swarup, 1989

Molecular and morphological data support the division of most *Heterodera* species into several groups: *Afenestrata*, *Avenae*, *Bifenestra*, *Cardiolata*, *Cyperi*, *Goettingiana*, *Humuli*, *Sacchari* and *Schachtii* (Table 15.2). Key features for study include the formation of the fenestra. These are classified as without fenestration (*Afenestrata* group), ambifenestrate (two openings divided by a narrow vulval bridge) or bifenestrate (two openings separated by a much wider vulval bridge) (Fig. 15.1C–D). The length of the vulval slit varies. In the *Avenae* group it is very short at 8–10 µm, whereas members of the *Schachtii* group have a much longer slit averaging 45 µm in length.

15.5.1 List of species and synonyms

Type species:

1. Heterodera schachtii A. Schmidt, 1871

= *Tylenchus schachtii* (A. Schmidt) Örley, 1880
= *Heterodera schachtii minor* O. Schmidt, 1930

Table 15.1. Classification of the family Heteroderidae Filipjev & Schuurman Stekhoven, 1941.

Krall and Krall (1978)	Wouts (1985)	Siddiqi (2000)	Handoo and Subbotin (this chapter)
Heteroderinae Filipjev & Schuurman Stekhoven, 1941	Heteroderinae Filipjev & Schuurman Stekhoven, 1941	Heteroderinae Filipjev & Schuurman Stekhoven, 1941	Heteroderinae Filipjev & Schuurman Stekhoven, 1941
Bidera Krall & Krall, 1978 *Cactodera* Krall & Krall, 1978 *Globodera* Skarbolovich, 1959 *Heterodera* Schmidt, 1871	*Afrodera* Wouts, 1985 *Bidera* Krall & Krall, 1978 *Heterodera* Schmidt, 1871 *Hylonema* Luc, Taylor & Cadet, 1978	*Afenestrata* Baldwin & Bell, 1985 *Cactodera* Krall & Krall, 1978 *Dolichodera* Mulvey & Ebsary, 1980 *Globodera* Skarbolovich, 1959 *Heterodera* Schmidt, 1871 *Punctodera* Mulvey & Stone, 1976	*Heterodera* Schmidt, 1871
Punctoderinae Krall & Krall, 1978	Punctoderinae Krall & Krall, 1978		Punctoderinae Krall & Krall, 1978
Punctodera Mulvey & Stone, 1976	*Cactodera* Krall & Krall, 1978 *Dolichodera* Mulvey & Ebsary, 1980 *Globodera* Skarbolovich, 1959 *Punctodera* Mulvey & Stone, 1976		*Betulodera* Sturhan, 2002 *Cactodera* Krall & Krall, 1978 *Dolichodera* Mulvey & Ebsary, 1980 *Globodera* Skarbolovich, 1959 *Paradolichodera* Sturhan, Wouts & Subbotin, 2007 *Punctodera* Mulvey & Stone, 1976 *Vittatidera* Bernard, Handoo, Powers, Donald & Heinz, 2010
	Meloidoderinae Golden, 1971 Cryphoderinae Cooman, 1978 Ataloderinae Wouts, 1973 Verutinae Esser, 1981	Meloidoderinae Golden, 1971	Meloidoderinae Golden, 1971
		Ataloderinae Wouts, 1973	Ataloderinae Wouts, 1973 Verutinae Esser, 1981
Sarisoderinae Krall & Krall, 1978			
Sarisodera Wouts & Sher, 1971			

Table 15.2. Morphological characterizations of Heterodera groups.

Stage and Characters Group	Shape	Cyst				J2		Host
		Fenestration	Bullae	Underbridge	Vulval slit	Lateral field		
Afenestrata	Lemon or rounded	Absent	Absent	Absent or weak	Long	3 or 4		Monocotyledons
Avenae	Lemon	Bifenestrate	Well developed	Absent or present	Short	4		Monocotyledons
Bifenestra	Lemon	Bifenestrate	Absent	Absent	Short	3		Monocotyledons
Cardiolata	Lemon	Ambifenestrate	Absent	Present	Long	3		Monocotyledons
Cyperi	Lemon or rounded	Ambifenestrate	Absent or present	Absent or present	Long	3 or 4		Monocotyledons
Goettingiana	Lemon	Ambifenestrate	Absent or present	Weak	Long	4		Dicotyledons
Humuli	Lemon	Bifenestrate (ambifenestrate for H. fici)	Absent or present	Weak	Long	4		Dicotyledons
Sacchari	Lemon or rounded	Ambifenestrate	Finger-like	Strong	Long	3		Monocotyledons
Schachtii	Lemon	Ambifenestrate	Well developed	Strong	Long	4		Dicotyledons

Other species:

2. *Heterodera africana* (Luc, Germani & Netscher, 1973) Mundo-Ocampo, Troccoli, Subbotin, Del Cid, Baldwin & Inserra, 2008

= *Sarisodera africana* Luc, Germani & Netscher, 1973
= *Afenestrata africana* (Luc, Germani & Netscher, 1973) Baldwin & Bell, 1985
= *Afrodera africana* (Luc, Germani & Netscher, 1973) Wouts, 1985

3. *H. agrostis* Kazachenko, 1993
4. *H. amygdali* Kirjanova & Ivanova, 1975
5. *H. arenaria* Cooper, 1955

= *Bidera arenaria* (Cooper, 1955) Krall & Krall, 1978

6. *H. aucklandica* Wouts & Sturhan, 1995
7. *H. australis* Subbotin, Rumpenhorst, Sturhan & Moens, 2002
8. *H. avenae* Wollenweber, 1924

= *Heterodera schachtii* var. *avenae* Wollenweber, 1924
= *Bidera avenae* (Wollenweber, 1924) Krall & Krall, 1978
= *Heterodera schachtii major* O. Schmidt, 1930
= *Heterodera major* O. Schmidt, 1930

9. *H. axonopi* (Souza, 1996) Mundo-Ocampo, Troccoli, Subbotin, Del Cid, Baldwin & Inserra, 2008

= *Afenestrata axonopi* Souza, 1996

10. *Heterodera bamboosi* (Kaushal & Swarup, 1988) Wouts & Baldwin, 1998

= *Brevicephalodera bamboosi* Kaushal & Swarup, 1988
= *Afenestrata bamboosi* (Kaushal & Swarup, 1989) Siddiqi, 2000

11. *H. bergeniae* Maqbool & Shahina, 1988
12. *H. betae* Wouts, Rumpenhorst & Sturhan, 2001
13. *H. bifenestra* Cooper, 1955

= *Bidera bifenestra* (Cooper, 1955) Krall & Krall, 1978
= *Heterodera longicaudata* Seidel, 1972
= *Bidera longicaudata* (Seidel, 1972) Krall & Krall, 1978

14. *H. cajani* Koshy, 1967

= *Heterodera vigni* Edward & Misra, 1968

15. *H. canadensis* Mulvey, 1979
16. *H. cardiolata* Kirjanova & Ivanova, 1969

= *Heterodera cynodontis* Shahina & Maqbool, 1989

17. *H. carotae* Jones, 1950a
18. *H. ciceri* Vovlas, Greco & Di Vito, 1985
19. *H. circeae* Subbotin & Sturhan, 2004
20. *H. cruciferae* Franklin, 1945
21. *H. cyperi* Golden, Rau & Cobb, 1962
22. *H. daverti* Wouts & Sturhan, 1978
23. *H. delvii* Jairajpuri, Khan, Setty & Govindu, 1979
24. *H. elachista* Ohshima, 1974
25. *H. fengi* Wang, Zhuo, Ye, Zhang, Peng & Liao, 2013
26. *H. fici* Kirjanova, 1954
27. *H. filipjevi* (Madzhidov, 1981) Stelter, 1984

= *Bidera filipjevi* Madzhidov, 1981

28. *H. galeopsidis* Goffart, 1936

= *Heterodera schachtii galeopsidis* Goffart, 1936

29. *H. gambiensis* Merny & Netscher, 1976
30. *H. glycines* Ichinohe, 1952
31. *H. glycyrrhizae* Narbaev, 1987
32. *H. goettingiana* Liebscher, 1892
33. *H. goldeni* Handoo & Ibrahim, 2002
34. *H. graminis* Stynes, 1971
35. *H. graminophila* Golden & Birchfield, 1972
36. *H. guangdongensis* Zhuo, Wang, Zhang & Liao, 2014
37. *H. hainanensis* Zhuo, Wang, Ye, Peng & Liao, 2013
38. *H. hordecalis* Andersson, 1975

= *Bidera hordecalis* (Andersson, 1975) Krall & Krall, 1978

39. *H. humuli* Filipjev, 1934
40. *H. johanseni* (Sharma, Kaushal, Singh, Pande, Pokharel & Upreti, 2001) Sturhan, 2002

= *Cactodera johanseni* Sharma, Kaushal, Singh, Pande, Pokharel & Upreti, 2001

41. *H. kirjanovae* Narbaev, 1988
42. *H. koreana* (Vovlas, Lamberti & Choo, 1992) Mundo-Ocampo, Troccoli, Subbotin, Del Cid, Baldwin & Inserra, 2008

= *Afenestrata koreana* Vovlas, Lamberti & Choo, 1992

43. *H. latipons* Franklin, 1969

= *Bidera latipons* (Franklin, 1969) Krall & Krall, 1978
= *Ephippiodera latipons* (Franklin, 1969) Shagalina & Krall, 1981

44. *H. lespedezae* Golden & Cobb, 1963
45. *H. leuceilyma* Di Edwardo & Perry, 1964
46. *H. litoralis* Wouts & Sturhan, 1996
47. *H. longicolla* Golden & Dickerson, 1973
48. *H. mani* Mathews, 1971

= *Bidera mani* (Mathews, 1971) Krall & Krall, 1978

49. *H. medicaginis* Kirjanova in Kirjanova & Krall, 1971
50. *H. mediterranea* Vovlas, Inserra & Stone, 1981
51. *H. menthae* Kirjanova & Narbaev, 1977
52. *H. mothi* Khan & Husain, 1965
53. *H. orientalis* (Kazachenko, 1989) Mundo-Ocampo, Troccoli, Subbotin, Del Cid, Baldwin & Inserra, 2008

= *Afenestrata orientalis* Kazachenko, 1989

54. *H. oryzae* Luc & Brizuela, 1961
55. *H. oryzicola* Rao & Jayaprakas, 1978
56. *H. pakistanensis* Maqbool & Shahina, 1986
57. *H. persica* Tanha Maafi, Sturhan, Subbotin & Moens, 2006
58. *H. phragmitidis* Kazachenko, 1986
59. *H. plantaginis* Narbaev & Sidikov, 1987
60. *H. pratensis* Gäbler, Sturhan, Subbotin & Rumpenhorst, 2000
61. *H. raskii* Basnet & Jayaprakash, 1984
62. *H. ripae* Subbotin, Sturhan, Rumpenhorst & Moens, 2003

= *Heterodera riparia* Subbotin, Sturhan, Waeyenberge & Moens, 1997 [= junior homonym] nec Kazachenko, 1993

63. *H. riparia* (Kazachenko, 1993) Subbotin, Sturhan, Rumpenhorst & Moens, 2003

= *Bidera riparia* Kazachenko, 1993

64. *H. rosii* Duggan & Brennan, 1966
65. *H. sacchari* Luc & Merny, 1963
66. *H. saccharophila* Mundo-Ocampo, Troccoli, Subbotin, Del Cid, Baldwin & Inserra, 2008

= *Afenestrata sacchari* Kaushal & Swarup, 1988

67. *H. salixophila* Kirjanova, 1969
68. *H. scutellariae* Subbotin & Sturhan, 2004
69. *H. sinensis* Chen & Zheng, 1994
70. *H. skohensis* Kaushal, Sharma & Singh, 2000
71. *H. sojae* Kang, Eun, Ha, Kim, Park & Choi, 2016
72. *H. sonchophila* Kirjanova, Krall & Krall, 1976
73. *H. sorghi* Jain, Sethi, Swarup & Srivastava, 1982
74. *H. spinicauda* Wouts, Shoemaker, Sturhan & Burrows, 1995
75. *H. spiraeae* Kazachenko, 1993
76. *H. sturhani* Subbotin, 2015
77. *H. swarupi* Sharma, Siddiqi, Rahaman, Ali & Ansari, 1998
78. *H. trifolii* Goffart, 1932

= *Heterodera schachtii* var. *trifolii* Goffart, 1932
= *Heterodera scleranthii* Kaktina, 1957
= *Heterodera rumicis* Poghossian, 1961
= *Heterodera paratrifolii* Kirjanova, 1963

79. *H. turangae* Narbaev, 1988
80. *H. turcomanica* Kirjanova & Shagalina, 1965

= *Bidera turcomanica* (Kirjanova & Shagalina, 1965) Krall & Krall, 1978
= *Ephippiodera turcomanica* (Kirjanova & Shagalina, 1965) Shagalina & Krall, 1981

81. *H. urticae* Cooper, 1955
82. *H. ustinovi* Kirjanova, 1969

= *Bidera ustinovi* (Kirjanova, 1969) Krall & Krall, 1978
= *Heterodera iri* Mathews, 1971

83. *H. uzbekistanica* Narbaev, 1980
84. *H. vallicola* Eroshenko, Subbotin & Kazachenko, 2001
85. *H. zeae* Koshy, Swarup & Sethi, 1971

Species inquirendae:
H. aquatica Kirjanova, 1971
H. chaubattia Gupta & Edward, 1973
H. graduni Kirjanova in Kirjanova & Krall, 1971
H. limonii Cooper, 1955
H. methwoldensis Cooper, 1955
H. oxiana Kirjanova, 1962
H. polygoni Cooper, 1955
H. tajikistanica Kirjanova & Ivanova, 1966

Nomen nudum:
H. indocyperi Husain & Khan, 1964

15.5.2 Principal species

15.5.2.1 European cereal cyst nematode, H. avenae

Presently, several *Heterodera* species are included under the common name 'cereal cyst nematodes' (CCN) comprising European cereal cyst nematode (ECCN) *H. avenae*; Filipjev cereal cyst nematode *H. filipjevi*; Australian cereal cyst nematode *H. australis*; Sturhan or Chinese cereal cyst nematode *H. sturhani*; and Mediterranean cereal cyst nematode *H. latipons*.

ECCN was first recorded by Kühn (1874) as a parasite of cereals in Germany and was later found in other countries. This nematode is now reported in most wheat-growing regions of Europe, Asia, North Africa and North America. ECCN is an important pest of cereals and is the principal nematode species on temperate cereals. In Europe, more than 50% of the fields in major cereal-growing areas are infected by this nematode (Rivoal and Cook, 1993), with annual yield losses reaching £3 million (Nicol and Rivoal, 2008). At least US$3.4 million is estimated to be lost annually in wheat production in the states of Idaho, Oregon and Washington, USA, because of the cereal cyst nematodes *H. avenae* and *H. filipjevi*. The yield losses it causes on wheat range from 15–20% in Pakistan, and 40–92% on wheat and 17–77% on barley in Saudi Arabia. Hosts of *H. avenae* include species of cereals and grasses from the following genera: *Agropyron*, *Agrostis*, *Alopecurus*, *Anisantha*, *Arrhenatherum*, *Avena*, *Brachypodium*, *Bromus*, *Dactylis*, *Echinochloa*, *Festuca*, *Hordeum*, *Koeleria*, *Lolium*, *Phalaris*, *Phleum*, *Poa*, *Polypogon*, *Secale*, *Setaria*, *Sorghum*, *Trisetum*, *Triticum*, *Vulpia*, *Zerna* and *Zea* (Williams and Siddiqi, 1972). In several regions with cereals *H. avenae* occurs in a mixture with *H. filipevi*. *Heterodera avenae* develops only one generation per year, with J2 hatch from the eggs determined largely by temperature (Rivoal and Cook, 1993).

Description: Cyst: L = 518–801 μm; W = 432–744 μm; L/W ratio = 0.8–1.8; fenestral length = 32–55 μm; vulval slit = 7–12 μm. **Male**: L = 1020–1590 μm; stylet = 27–33 μm; spicules = 33–38 μm; gubernaculum = 10–13 μm. **J2**: L = 505–598 μm; stylet = 24–27.5 μm; tail hyaline region = 34–50 μm; tail = 52–79 μm. **Cyst**: lemon-shaped, with prominent neck and vulval cone (Fig. 15.2). Subcrystalline layer conspicuous, sloughing off with formation of dark brown cyst. Bifenestrate, bullae prominent, crowded beneath vulval cone (Figs 15.3A and 15.6A). **J2**: stylet well developed, with large, anteriorly flattened to concave basal knobs (Fig. 15.4A). Lateral field with four incisures. Tail with a sharply pointed terminus (Fig. 15.5A).

ECCN belongs to the *Avenae* group and is a member of the *H. avenae* species complex. It differs from closely related species (*H. australis*, *H. aucklandica*, *H. riparia*, *H. sturhani*, *H. pratensis* and *H. arenaria*) by morphometrical characters of the J2 and cysts and sequences of the internal transcribed spacer (ITS)-rRNA and mitochondrial cytochrome oxidase I (COI mtDNA) genes. Molecular analysis revealed that the world populations of the ECCN can be divided into two types: (i) European and North American populations of *H. avenae* – type A; (ii) Asian and African populations of *H. avenae* – type B.

15.5.2.2 Yellow beet cyst nematode, H. betae

The yellow beet cyst nematode (YBCN) was first observed by Maas *et al.* (1976) during examination of samples collected from the southern sugar beet regions of The Netherlands and it has been considered as a biotype of the clover cyst nematode, *H. trifolii*. After comparative morphological and molecular analysis of several populations of the YBCN with closely related species, Wouts *et al.* (2001) described it as a new species, *H. betae*. Further studies showed that this species was preferentially distributed in the warm environments of southern Europe. YBCN has been reported in several European countries and Morocco. Recently, *H. betae* was found in natural conditions on *B. vulgaris* ssp. *maritima* along the Atlantic and North Sea coastlines from northern France to Spain and Portugal (Gracianne *et al.*, 2014). *Heterodera betae* is considered another important cyst nematode after *H. schachtii*, causing considerable damage to sugar beet. YBCN induces yield losses mainly on sandy soils. In several locations this species co-occurs with *H. schachtii*. *Heterodera betae* reproduces on Brassicaceae (*Brassica* spp., *Sinapis alba*), Amaranthaceae (*Betae vulgare*, *Spinacia oleracea*), Popygonaceae (*Rumex* spp.), Caryophyllaceae (*Stellaria media*), Fabaceae (*Trifolium* spp.) and Solanaceae (*Solanum esculentum*) (Ambrogioni *et al.*, 2004). Cabbage and broad beans are also good hosts for *H. betae*. YBCN may complete three or four generations during the vegetation period. The species can multiply on sugar beet varieties with resistance genes to *H. schachtii*. YBCN has 35 or 36 chromosomes and reproduces by mitotic parthenogenesis, indicating that this species is

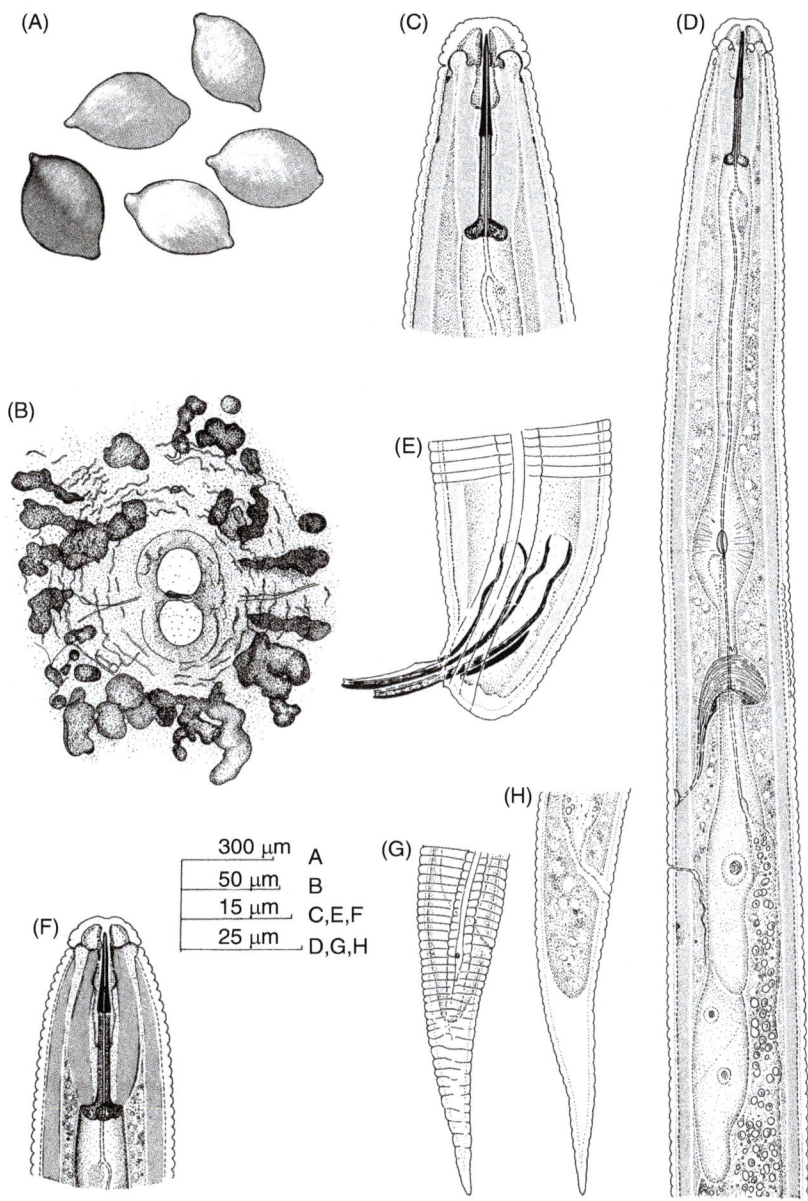

Fig. 15.2. *Heterodera avenae*. (A) Cysts. (B) Vulval cone. (C) Anterior region of male. (D) Pharyngeal region of male. (E) Tail of male. (F) Anterior region of J2. (G, H) Tail of second-stage juvenile. (After Williams and Siddiqi, 1972.)

potentially tetraploid (Steele and Whitehand, 1984).

Description: **Cyst**: L = 475–1160 µm; W = 168–702 µm; L/W ratio = 1.4–2.5; fenestral length = 30–73 µm; vulval slit = 48–67 µm. **Male**: not found. **J2**: L = 525–672 µm; stylet = 25–33 µm; tail hyaline region = 32–50 µm; tail = 64–84 µm. **Female**: passes through a distinct yellow phase. **Cyst**: lemon-shaped, often asymmetrical. Vulval cone ambifenestrate (Fig. 15.3B), in young cysts surrounded by eggsac. Vulval bridge narrow. Underbridge distinct,

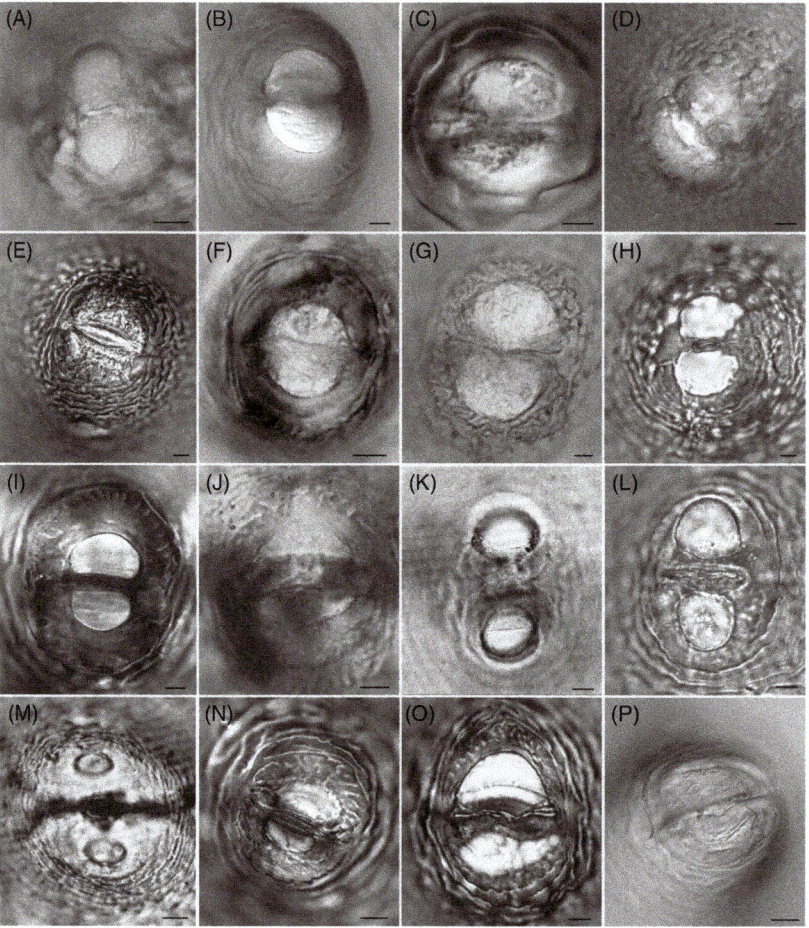

Fig. 15.3. Vulval plates. (A) *Heterodera avenae*. (B) *H. betae*. (C) *H. cajani*. (D) *H. carotae*. (E) *H. cruciferae*. (F) *H. elachista*. (G) *H. fici*. (H) *H. filipjevi*. (I) *H. glycines*. (J) *H. goettingiana*. (K) *H. hordecalis*. (L) *H. humuli*. (M) *H. latipons*. (N) *H. schachtii*. (O) *H. trifolii*. (P) *H. zeae*. B, C, D, P, after Subbotin *et al.* (2010) with modifications; K, after Andersson (1975); M, after Franklin (1969); E, H, I, L, N, O, courtesy of V.N. Chizhov. Scale bars = 10 μm.

long, heavily pigmented, with bifurcate ends. Heavily pigmented bullae present, mainly anterior to underbridge, globular. **J2**: slightly curved ventrally. Stylet robust, knobs heavy, deeply concave anteriorly, convex posteriorly. Lateral field with four incisures. Tail gradually tapering to slender rounded terminus.

Heterodera betae belongs to the *Schachtii* group. It differs from other related species owing to its longer body length, stylet and tail of J2. This species can be differentiated from other species by polymerase chain reaction-internal transcribed spacers-restriction fragment length polymorphism (PCR–ITS–RFLP).

15.5.2.3 Pigeonpea cyst nematode, H. cajani

The pigeonpea cyst nematode (PPCN) was described from roots of pigeon pea, *Cajanus cajan*, and later found parasitizing cowpea, *Vigna sinensis*, in India. The list of host plants includes several dozen plant species of the families Fabaceae (*Vigna* spp. and others) and Pedaliaceae (*Sesamum indicum*). *Heterodera cajani* is now present in all major pigeon pea-growing regions of India and Pakistan and is considered as the most important nematode pathogen of this crop. More than nine generations may be developed during

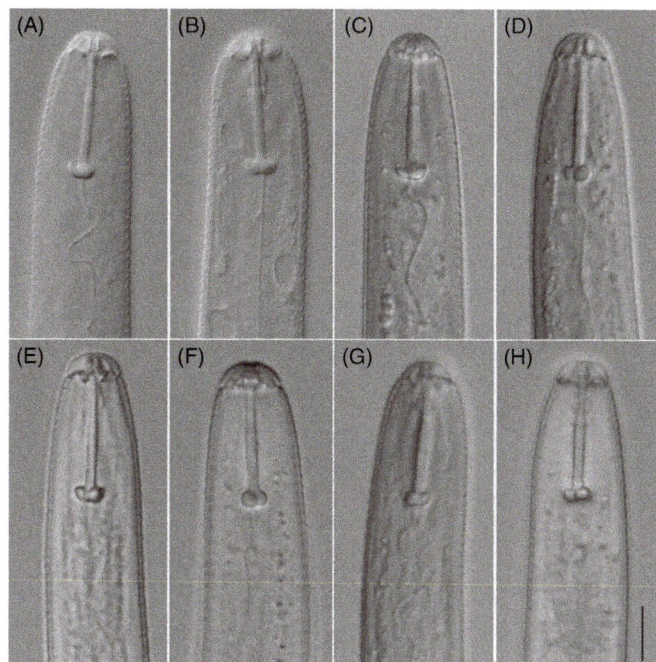

Fig. 15.4. Heads of second-stage juveniles. (A) *Heterodera avenae*. (B) *H. filipjevi*. (C) *H. schachtii*. (D) *H. cruciferae*. (E) *H. humuli*. (F) *Globodera rostochiensis*. (G) *Betulodera* sp. (H) *Punctodera punctata*. A, B, after Subbotin et al. (2003). Scale bar, A–H = 10 μm.

a vegetation season. Three races of PPCN are distinguished by their ability to multiply on cluster bean (*Cyamopsis tetragonolobus*) and sunn hemp (*Crotalaria juncea*) (Walia and Bajaj, 1986, 1988; Siddiqi and Mahmood, 1993). Walia and Bajaj (2000) found these races could be distinguished by vulval cone structure and male morphology.

Description: **Cyst**: L = 390–690 μm; W = 175–510 μm; L/W ratio = 1.0–2.5; fenestral length = 31–69 μm; vulval slit = 31–60 μm. **Male**: L = 780–1280 μm; stylet = 27–30 μm; spicules = 31–41 μm; gubernaculum = 8–12 μm. **J2**: L = 324–515 μm; stylet = 20–31 μm; tail hyaline region = 17–40 μm; tail = 32–64 μm. **Cyst**: light to dark brown, typically lemon-shaped with protruding neck and vulva. Subcrystalline layer present on young cysts. Ambifenestrate with strong underbridge (Fig. 15.3C). Bullae prominent dark brown, located beneath underbridge. **J2**: body tapering anteriorly and posteriorly but more so posteriorly. Stylet strong, well developed with anteriorly directed or rounded knobs. Lateral field with four incisures. Tail with bluntly rounded and narrow terminus.

Heterodera cajani belongs to the *Schachtii* group. It differs from several species of this group by a smaller cyst size. The ITS-rRNA gene sequence clearly distinguishes this species from all other species of the *Schachtii* group.

15.5.2.4 Carrot cyst nematode, H. carotae

Heterodera carotae occurs throughout the carrot-growing areas of Europe and is also found in the USA, Canada and South Africa. The species has a rather narrow list of host plants. It was found to parasitize only several subspecies of *Daucus carota* and several wild Apiaceae, such as *Torilis* spp. Up to four generations may occur in favourable growing conditions. In infected fields, irregular plant growth was observed, the patches enlarging over time. Foliage is stunted, reddish and may dry out when infestations are heavy. Carrots are usually small, abnormally developed and have numerous radicles, which gives the roots a characteristic bearded appearance, referred to as 'hairy root'. Carrot cyst nematodes have caused significant crop losses.

Description: **Cyst**: L = 218–625 μm; W = 165–500 μm; L/W ratio = 1.9–2.8; fenestral length = 27–36 μm; vulval slit = 43–51 μm. **Male**: L = 1090–1220 μm; stylet = 31–38 μm;

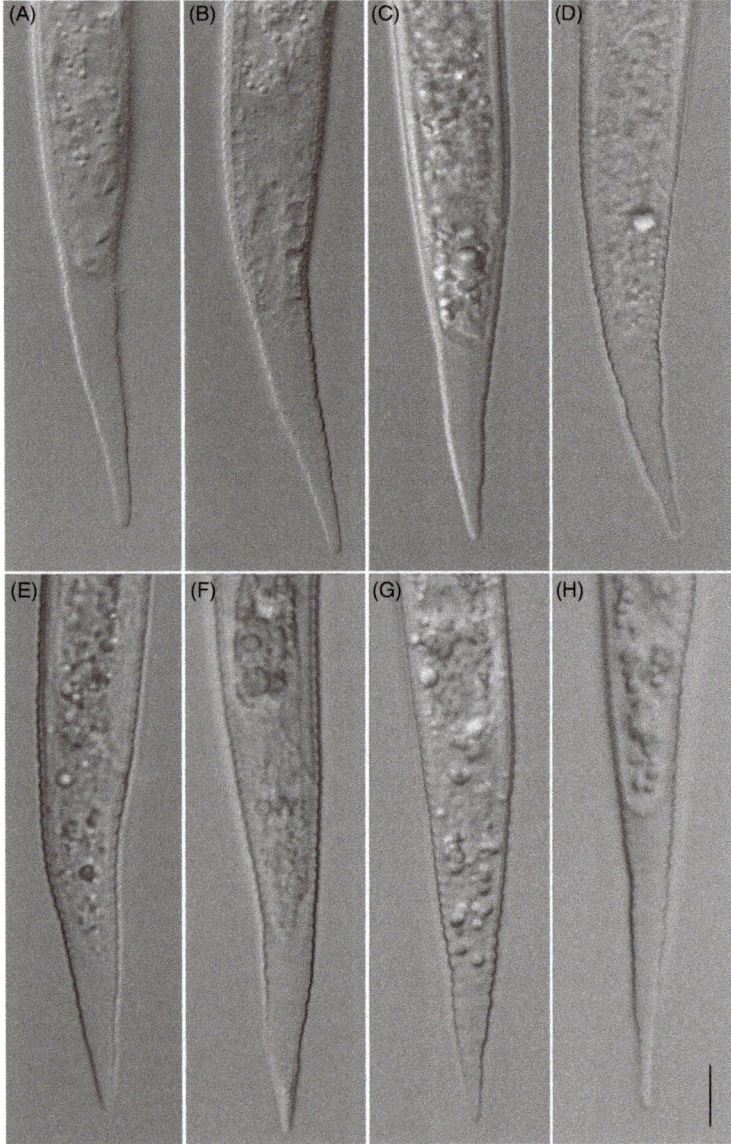

Fig. 15.5. Tails of second-stage juveniles. (A) *Heterodera avenae*. (B) *H. filipjevi*. (C) *H. schachtii*. (D) *H. cruciferae*. (E) *H. humuli*. (F) *Globodera rostochiensis*. (G) *Betulodera* sp. (H) *Punctodera punctata*. A, B, after Subbotin *et al.* (2003). Scale bar A–H = 10 µm.

spicules = 31–36 µm; gubernaculum = 10–13 µm. **J2:** L = 375–452 µm; stylet = 22–25 µm; tail hyaline region = 20–31.8 µm; tail = 43.5–59 µm. **Cysts:** small, lemon-shaped without bullae and slender underbridge. Fenestration ambifenestrate (Fig. 15.3D). **J2:** labial region slightly offset with four indistinct annuli. Stylet knobs with concave anterior faces. Lateral field with four incisures. Tail with a pointed terminus.

Heterodera carotae belongs to the Goettingiana group. It most closely resembles *H. cruciferae*, but molecular and morphological differentiation of these species is problematic.

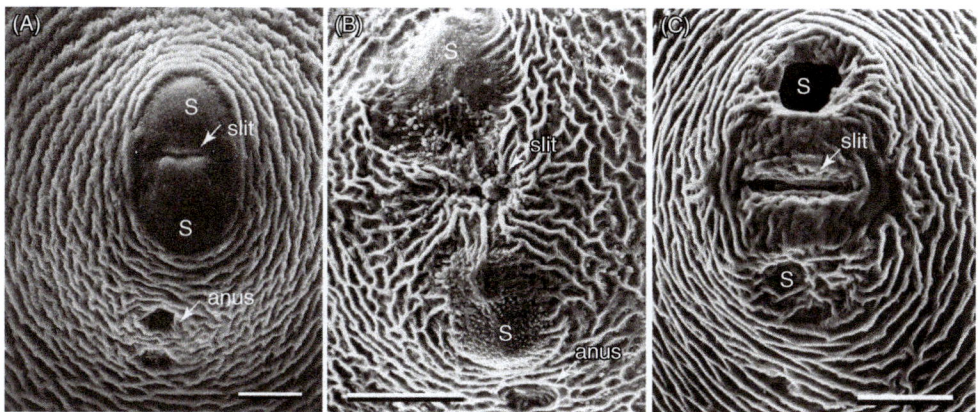

Fig. 15.6. SEM micrographs of the vulval area. (A) *Heterodera avenae*. (B) *H. latipons*. (C) *H. hordecalis*. S – semifenestrae. Scale bars = 20 μm. After Greco *et al.* (2002).

15.5.2.5 Cabbage cyst nematode, H. cruciferae

The cabbage cyst nematode is known in several cabbage growing areas in Europe, Asia, Australia and North America, where it is considered to be a major pest of cabbage crops. Plants infected by this nematode have short and bushy roots, due to secondary root production, resulting in patches of plants showing severe decline and stunting. This species is often associated with *H. schachtii*. *Heterodera cruciferae* parasitizes cool-weather or winter-grown crops, so that the number of generations completed in a season depends on the growing period; up to three generations may occur in Europe. The cabbage cyst nematode has a narrower host range than *H. schachtii*, but apparently infects all species of *Brassica*.

Description: Cyst: L = 355–690 μm; W = 288–571 μm; L/W ratio = 0.6–1.8; fenestral length = 22–59 μm; vulval slit = 29–55 μm. **Male**: L = 718–1343 μm; stylet = 20–28 μm; spicules = 16–38 μm; gubernaculum = 7–11 μm. **J2**: L = 333–504 μm; stylet = 20–25 μm; tail hyaline region = 16–30 μm; tail = 26–58 μm. **Cyst**: broad, almost spherical to lemon-shaped, ambifenestrate (Fig. 15.3E), without bullae, with very low semifenestral arches separated by a narrow vulval bridge, semifenestrae unobstructed in mature cysts. **J2**: labial region offset with three or four annuli and a dorsoventrally elongated oral disc flanked by lateral lips bearing amphidial apertures. Stylet knobs anterior face flat, rounded or slightly concave (Fig. 15.4D). Lateral field with four incisures. Tail tapering uniformly to a fine rounded terminus (Fig. 15.5D).

Heterodera cruciferae belongs to the Goettingiana group, and is closely related to and morphologically similar to *H. carotae*.

15.5.2.6 Japanese cyst nematode, H. elachista

This nematode species was described from Japan on upland rice fields and later in Iran, China and Italy. In addition to *Oryza sativa*, host plants include *Zea mays* and *Carex* spp. This nematode causes 7–19% rice yield loss. Greater yield losses were observed when the roots were invaded by the nematodes before tillering. Nematodes had the most severe impact during the later stages of plant growth. Corn plants infected by the nematode were stunted with a marked proliferation of short lateral roots, resulting in plant decline and infested patches in the field.

Description: Cyst: L = 278–586 μm; W = 207–540 μm; L/W ratio = 1.1–1.9; fenestral length = 23–50 μm; vulval slit = 26–55 μm. **Male**: L = 820–940 μm; stylet = 20–21 μm; spicules = 26–29 μm; gubernaculum = 9–13 μm. **J2**: L = 330–535 μm; stylet = 16–25 μm; tail hyaline region = 25–50 μm; tail = 44–87 μm. **Cyst**: light to dark brown, spherical to lemon-shaped. Subcrystalline layer present. Ambifenestrate (Fig. 15.3F), vulval cone with semifenestrae almost as wide as long. Narrow vulval bridge and weak underbridge, and prominent dark brown bullae. **Male**: very rare. **J2**: body slightly curved ventrally. Lip region hemispherical with three annuli. Stylet well developed with rounded or anteriorly concave knobs. Lateral field with three incisures. Tail long, narrowly tapering to a very fine, rounded terminus.

Heterodera elachista belongs to the *Cyperi* group and is morphologically close to *H. oryzae*, *H. sacchari* and *H. leuceilyma*. It differs from *H. sacchari* and *H. leuceilyma* as it lacks finger-shaped projections on the slender underbridge of the cysts and from *H. oryzae* as it has smaller cysts and a shorter body of the J2.

15.5.2.7 Fig cyst nematode, H. fici

The fig cyst nematode, *H. fici*, was first described from roots of the plant *Ficus elastica*, which was imported from Harbin, China, to Russia. Subsequent surveys revealed that this nematode is widely distributed in natural conditions in the Mediterranean and other countries on roots of *F. carica* and ornamental species of the genus *Ficus*. The nematode can be considered a potential threat in fig nurseries. Plants infected by this nematode showed symptoms of retarded growth and yellowing of leaves.

Description: **Cyst**: L = 340–697 µm; W = 272–560 µm; L/W = 1.0–1.7; fenestral length = 42–74 µm; vulval slit = 35–56 µm. **Male**: L = 760–1002 µm; stylet = 26–32 µm; spicules = 27–32 µm; gubernaculum = 6–9 µm. **J2**: L = 320–470 µm; stylet = 20–25 µm; tail hyaline region = 18–33 µm; tail = 40–61 µm. **Cyst**: body light to dark brown, basically lemon-shaped, neck and vulval cone distinct. Neck protruding, curved posteriorly. Fenestra ambifenestrate, sometimes top of cone appearing bifenestrate (Fig. 15.3G). Bullae dome shaped, small, scattered around underbridge plane. Underbridge weakly developed, with furcate ends. **J2**: stylet well developed, basal knobs rounded, directed slightly anteriorly. Lateral field with four incisures. Tail terminus rounded.

Heterodera fici belongs to the *Humuli* group. It differs from other members of the *Humuli* group (*H. humuli*, *H. ripae*, *H. vallicola* and *H. litoralis*) as it has ambifenestrate rather than bifenestrate cysts and a longer vulval slit.

15.5.2.8 Filipjev cereal cyst nematode, H. filipjevi

This species was described by Madzhidov (1991) from Tajikistan and later identified from many European and Asian countries and the USA. Presently, *H. filipjevi* is considered to be an important worldwide pest of cereals. The list of plant hosts of this species includes more than 20 species of cereals and grasses. One generation develops during the growing season. At least two pathotypes of this species are recognized.

Description: **Cyst**: L = 455–936 µm; W = 306–792 µm; L/W = 1.0–1.9; fenestral length = 38–66 µm; vulval slit = 6–12 µm. **Male**: L = 1160–1400 µm; stylet = 27–29 µm; spicules = 28–34 µm. **J2**: L = 431–614 µm; stylet = 22–30 µm; hyaline region = 28–41 µm; tail = 49–64 µm. **Cyst**: lemon-shaped with prominent vulval cone. Subcrystalline layer present. Colour varying from light to dark brown. Bifenestrate with massive underbridge (Fig. 15.3H). Bullae large and numerous. **J2**: labial region offset with two faint annuli. Stylet with anteriorly projecting knobs (Fig. 15.4B). Lateral field with four incisures. Tail conical (Fig. 15.5B).

Heterodera filipjevi belongs to the *Avenae* group and the *H. avenae* complex. It differs from most species owing to the presence of an underbridge. The ITS-rRNA and COI gene sequences clearly distinguishes this species from other species of the *Avenae* group.

15.5.2.9 Soybean cyst nematode, H. glycines

Soybean cyst nematode is a major pest of soybean in Asia, North and South America and is found in most countries of the world where soybean is produced. In Japan, the yield loss was estimated to be 10–75% (Ichinohe, 1988), in China 10–20% and in the USA annual losses in north central states attributable to *H. glycines* parasitism were estimated to be worth over $200 million (Doupnik, 1993). In six north central states (Illinois, Indiana, Iowa, Minnesota, Missouri and Ohio) surveyed for the presence of *H. glycines*, 47–83% of soybean hectarage was found to be infested (Workneh *et al*., 1999). In central China *H. glycines* was also recognized as a pest of tobacco (Shi and Zheng, 2013).

A cyst nematode parasitizing soybean plants and causing 'yellow dwarf' symptoms was recorded from Japan, in 1915. Ichinohe (1952) was the first to make careful morphological comparisons with other *Heterodera* species and to give a specific name to and brief description of this nematode. *Heterodera glycines* has a broad host range, especially Fabaceae, but also on other families. More than 66 weed species of nine families are suitable hosts. Riggs (1992) provided a list of non-fabaceous hosts comprising species from 22 families (e.g. Boraginaceae, Capparaceae, Caryophyllaceae, Chenopodiaceae, Brassicaceae, Lamiaceae, Fabaceae, Scrophulariaceae and Solanaceae). In field conditions, *H. glycines* was also found in several other plants, including henbit (*Lamium amplexicaule*), purple deadnettle (*Lamium purpureum*), mouse-ear chickweed (*Cerastium holosteoides*) and common chickweed (*Stellaria media*) (Riggs, 1992). Three to five generations develop during the cropping season.

Heterodera glycines disrupts root growth, interferes with nodulation and causes early yellowing of soybean plants. The above-ground symptoms of damage on individual plants and appearance of infested fields are usually not sufficiently specific to allow direct identification. Infected plants are predisposed to *Fusarium* wilt (see Chapter 12, this volume). Sudden death syndrome is a soil-borne disease of soybean caused by the fungus *Fusarium solani* in association with *H. glycines*. However, soybean cyst nematode can reduce soybean yields without causing above-ground symptoms (Wang *et al.*, 2003).

Description: Cyst: L = 340–920 µm; W = 200–688 µm; L/W ratio = 1.0–2.4; fenestral length = 35–72 µm; vulval slit = 36–60 µm. **Male**: L = 911–1400 µm; stylet = 24–27 µm; spicules = 28–45 µm; gubernaculum = 8–13 µm. **J2**: L = 345–504 µm; stylet = 21–25 µm; tail hyaline region = 18–36 µm; tail = 35–59 µm. **Cyst**: mainly lemon-shaped, sometimes round with a protruding neck and cone. Ambifenestrate, bullae prominent, located at or anterior to underbridge, extending into vulval cone from interior of body wall cuticle (Fig. 15.3I). Shape varying from round to finger-like, round bullae differently sized, finger-like bullae of variable length and thickness. Underbridge well developed. **J2**: body vermiform with regularly annulated cuticle. Stylet with anteriorly protruding knobs. Tail tapering uniformly to a finely rounded terminus.

Heterodera glycines belongs to the *Schachtii* group and is distinguished from similar species by a combination of morphological and morphometric characteristics. It differs from *H. schachtii* in the shape of the stylet knobs of J2 (slightly convex *vs* moderately or strongly concave), shorter average J2 stylet length and longer average fenestral length. The ITS-rRNA and *COI* gene sequences distinguish this species from other species of the *Schachtii* group.

15.5.2.10 Pea cyst nematode, H. goettingiana

Liebscher described the pea cyst nematode in 1892 and also reported symptoms of infection and yield loss of pea (*Pisum sativum*) and vetch (*Vicia sativa*) caused by this nematode in fields of the Agricultural Institute at Göttingen, Germany. Presently, *H. goettingiana* has been found in many countries of Europe, North Africa, Asia and the USA. Infested pea fields show sharply delineated patches with dwarfed, poorly branched and yellowing plants that die prematurely. Infected plants either fail to flower or flower too early. The root system is poorly developed. Hosts of *H. goettingiana* also include several species of the genera belonging to the Fabaceae family: *Cicer*, *Glycine*, *Lathyrus*, *Lens*, *Lupinus*, *Medicago*, *Pisum*, *Trifolium* and *Vicia*. The number of generations produced depends on environmental conditions. One or two generations occur during the growing season in the UK, and three generations may develop in southern Italy.

Description: Cyst: L = 400–780 µm; W = 310–540 µm; L/W ratio = 1.3–2.2; fenestral length = 43–71 µm; vulval slit = 43–61 µm. **Male**: L = 1270 µm; stylet = 27 µm; spicules = 27 µm; gubernaculum = 12 µm. **J2**: L = 408–519 µm; stylet = 23–26 µm; tail hyaline region = 27–38 µm; tail = 54–74 µm. **Cyst**: lemon-shaped with light to dark brown cyst wall. Vulval cone ambifenestrate (Fig. 15.3J). In some old cysts, vulval bridge ruptured, fenestrae joining to form a large oval fenestrum. Bullae absent, although bullae-like structures and vulval denticles present. Underbridge very weak. **J2**: body curved ventrally after fixation. Labial region hemispherical, with two to five annuli, slightly offset from body. Lateral field with four incisures, not areolated. Stylet knobs rounded, slightly projecting anteriorly. Tail tapering uniformly to a finely rounded terminus.

Heterodera goettingiana belongs to the *Goettingiana* group. It differs from several other representatives of the *Goettingiana* group (*H. cruciferae, H. carotae, H. circeae, H. scutellariae* and others) owing to longer average J2 body, longer tail and longer hyaline region. The ITS-rRNA gene sequence distinguishes this species from other species of the *Goettingiana* group.

15.5.2.11 Barley cyst nematode, H. hordecalis

The barley cyst nematode was found and described from a barley field in Halland province, Sweden in 1975. Presently, *H. hordecalis* is reported from several central and north European and Asian countries. The nematode also parasitizes rye, winter cereals and many grasses from the genera *Ammophila, Leymus, Bromus, Calamagrostis, Dactylis, Festuca, Lolium* and *Phleum*.

Description: **Cyst**: L = 330–950 µm; W = 255–680 µm; L/W ratio = 1.0–1.5; fenestral length = 47–80 µm. **Male**: L = 805–1390 µm; stylet = 24–29 µm; spicules = 34–42 µm; gubernaculum = 9–13 µm. **J2**: L = 410–550 µm; stylet = 21–26 µm; tail hyaline region = 29–46 µm; tail = 44–60 µm. **Cyst**: body ovoid, with distinct neck and vulval cone. Bifenestrate, semifenestrae widely separated (Figs 15.3K and 15.6C). In centre of vulval bridge, a rigid, often dumb-bell like structure, surrounding anterior end of vagina. Underbridge extremely strong, sometimes with pronounced thickening in middle, and with ends basically bifurcate, each branch consisting of a number of cords, although often irregularly splayed. Bullae absent. **J2**: labial region slightly offset with three or four annuli. Lateral field with four incisures. Tail tip finely rounded.

Heterodera hordecalis belongs to the *Avenae* group and is most similar to *H. latipons*. The most striking distinguishing features are those of the vulval slit, which is much longer in *H. hordecalis* than in *H. latipons*. The ITS-rRNA and *COI* gene sequences clearly differentiate this species from all others.

15.5.2.12 Hop cyst nematode, H. humuli

Heterodera humuli was described by Filipjev (1934) from hop plants based on morphological data provided by several researchers. Presently, the hop cyst nematode is reported from the hop-growing areas in European and Asian countries, South Africa, Australia, New Zealand, the USA and Canada. In an experimental study (Hafez *et al.*, 1999), the nematode significantly reduced plant height and fresh and dry weight of shoots. Infected plants showed more severe nutrient deficiency symptoms. A negative correlation has been demonstrated between the numbers of J2 in soil and cone yield in hop plantations. The growth of nematode-infested hop cultivars was significantly reduced when the plant pathogenic fungus *Verticillium alboatrum* was also present. Hosts include *Humulus lupulus, Urtica dioica* and *U. urens*. Only one generation is produced per year.

Description: **Cyst**: L = 290–610 µm; W = 245–450 µm; L/W ratio = 1.2–1.7; fenestral length = 49–76 µm; vulval slit = 33–43 µm. **Male**: L = 670–1000 µm; stylet = 20–28 µm; spicules = 29–33 µm; gubernaculum = 7–8 µm. **J2**: L = 336–468 µm; stylet = 21–25 µm; tail hyaline region = 22–30 µm; tail = 42–53 µm. **Cyst**: lemon-shaped, occasionally nearly spherical. Abullate, thin walled, light coloured, darkening with age. Bifenestrate, vulval bridge broad, semifenestrae circular or subcircular, often obscured by thin membrane with a fine fingerprint-like pattern (Fig. 15.3K). Underbridge slender, weak, with furcate ends. **J2**: labial region rounded, offset with two to four annuli (Fig. 15.4E). Lateral field with four incisures, not areolated. Tail tapering, terminus often constricted and irregularly shaped (Fig. 15.5E).

Heterodera humuli belongs to the *Humuli* group and is similar to *H. ripae* and *H. vallicola*. It differs from *H. ripae* as it has a longer J2 tail and hyaline region and from *H. vallicola* as it has a more slender cyst and longer fenestral length. The ITS-rRNA gene sequence distinguishes this species from other species of the *Humuli* group.

15.5.2.13 Mediterranean cereal cyst nematode, H. latipons

This species was collected from roots of stunted wheat plants in Israel and described by Franklin (1969). Presently, *H. latipons* has been recorded in many countries of the Mediterranean basin, North Africa, the Near East and Japan. Hosts include *Hordeum vulgare, Avena sativa, Secale cereale*,

Phalaris minor, *P. paradoxa* and *Elytrigia repens*. Yield losses as high as 50% were reported on barley in Cyprus. In Syria, the nematode causes average yield losses of 20 and 30% in barley and durum wheat, respectively, and the nematode was more damaging under water stress conditions (Scholz, 2001). Moreover, damage is more severe in fields infested concomitantly by *H. latipons* and the fungus *Cochliobolus sativus*. In all areas studied, *H. latipons* completed only one life cycle during the growing season.

Description: **Cyst**: L = 300–700 μm; W = 310–560 μm; L/W ratio = 0.6–1.7; fenestral length = 45–76 μm; vulval slit = 6–11 μm. **Male**: L = 960–1406 μm; stylet = 22–29 μm; spicules = 32–36 μm; gubernaculum = 8 μm. **J2**: L = 401–598 μm; stylet = 22–26 μm; tail hyaline region = 20–36 μm; tail = 42–72 μm. **Cyst**: dark to mid brown covered with white subcrystalline layer. Bifenestrate, semifenestrae separated by a distance greater than fenestral width, vulval slit short (Figs 15.3M and Fig. 15.6B). Strong underbridge with pronounced thickening in middle and with ends splayed. Bullae usually absent, sometimes present at underbridge level. **J2**: labial region with three annuli. Stylet with well developed, anteriorly concave knobs. Tail with finely rounded terminus.

Heterodera latipons belongs to the *Avenae* group and closely resembles *H. hordecalis* and *H. turcomanica*. These nematodes share similar circular semifenestrae separated by a distance longer than the semifenestra diameter and a rather typical underbridge but with a pronounced enlargement underlying the vulval slit. The most important differentiating character between *H. latipons* and *H. hordecalis* is the vulval slit, which in *H. latipons* is much shorter. The ITS-rRNA and *COI* gene sequences clearly differentiate this species from all others.

15.5.2.14 Sugar beet cyst nematode, H. schachtii

The sugar beet cyst nematode has been recognized as a plant pathogen since 1859, when it was associated with stunted and declining sugar beet in Germany. *Heterodera schachtii* is found in all major sugar beet production areas of the world and is widespread in Europe, the USA and Canada, and considered as one of the most serious pests of sugar beet. Annual yield loss in EU countries based upon world market sugar prices was estimated in 1999 to be up to US$90 million (Müller, 1999). *Heterodera schachtii* parasitizes more than 200 different plants, mainly from the families Amaranthaceae (including many species of *Beta* and *Chenopodium*) and Brassicaceae (*Brassica oleracea*, *B. napus*, *B. rapa*, *Raphanus sativus* and many others including a diversity of common weeds). Some plants from Polygonaceae, Scrophulariaceae, Caryophyllaceae and Solanaceae are susceptible to nematode infection. In some climates, three to five generations may complete development on sugar beet in one season.

Description: **Cyst**: L = 480–960 μm; W = 396–696 μm; L/W ratio = 0.9–2.0; fenestral length = 28–51 μm; vulval slit = 33–54 μm. **Male**: L = 1038–1638 μm; stylet = 27–30 μm; spicules = 27–39 μm; gubernaculum = 10–11 μm. **J2**: L = 400–512 μm; stylet = 23–28 μm; tail hyaline region = 17–33 μm; tail = 40–56 μm. **Cyst**: colour light to dark brown. Ambifenestrate (Fig. 15.3N), within cone, remnants of vagina attached to side walls by underbridge and a number of irregularly arranged, dark brown molar-shaped bullae situated a short distance beneath the vulval bridge. **J2**: labial region offset, with four indistinct annuli (Fig. 15.4C). Stylet moderately heavy with prominent, forwardly directed knobs. Tail acutely conical with rounded tip (Fig. 15.5C).

Heterodera schachtii belongs to the *Schachtii* group and is distinguished from closely related species (*H. trifolii*, *H. glycines*, *H. betae* and others) by a combination of morphological and morphometric characteristics. The PCR–ITS–RFLP, ITS-rRNA and *COI* sequences distinguish this species from other species in the *Schachtii* group.

15.5.2.15 Clover cyst nematode, H. trifolii

The clover cyst nematode is considered to be a pest of diverse agricultural crops and several pasture plants, especially parasitizing white clover, *Trifolium repens*. *Heterodera trifolii* is a cosmopolitan species and reported from all continents. Some authors propose considering species *H. trifolii* as a conglomerate of independently evolved mitotic parthenogenetic populations, comprising polyploidy and aneuploid forms with 3n = 24–28 and 4n = 34–35 chromosomes, and host races with more or less extended host ranges. Host plants include many species of Fabaceae and other families. Several generations may occur during a vegetative period. This nematode may cause a reduction in yield, nitrogen fixation and persistence of clover plants in pastures. Several studies have shown yield losses and growth suppression of white clover due to *H. trifolii* infection.

Description: **Cyst**: L = 360–1020 µm; W = 195–680 µm; L/W = 1.2–2.7; fenestral length = 40–80 µm; vulval slit = 39–66 µm. **Male**: not found. **J2**: L = 461–678 µm; stylet = 23–31 µm; tail hyaline region = 27–45 µm; tail = 49–78 µm. **Cyst**: brown to dark brown, ambifenestrate with strong underbridge and elongated bullae (Fig. 15.3O). Long vulval slit and strongly pigmented underbridge with bifurcate ends. **J2**: labial region offset, with three or four annuli. Lateral field with four incisures. Stylet robust, anterior surfaces of knobs concave. Tail conoid, tapering uniformly to a finely rounded terminus.

Heterodera trifolii is a member of the *H. trifolii* species complex including *H. betae*, *H. daverti*, *H. lespedezae*, *H. medicaginis* and *H. galeopsidis*, and belongs to the *Schachtii* group. The clover cyst nematode can be found to parasitize clover in a mixture with *H. daverti*; the latter species has males. Molecular characterization of this species complex is given by Subbotin *et al*. (2010) and more recently by Vovlas *et al*. (2015).

15.5.2.16 *Maize cyst nematode,* H. zeae

This species was described in India, where it is considered to be the most important nematode problem in maize, causing yield loss to the maize crop of 21–29% (Srivastava and Chawla, 2005). It is presently also found in Egypt, Pakistan, Afganistan, Thailand, Portugal and Greece. *Heterodera zeae* was first detected in the western hemisphere in North America, in Kent County, Maryland in 1981 (Sardanelli *et al*., 1981). The maize cyst nematode *H. zeae* can cause serious losses in maize yields, especially under conditions meeting its high optimum temperature requirements (Krusberg *et al*., 1997). This species was found mainly on *Zea mays*; however, other cultivated or wild Poaceae species are also considered as suitable host plants including sorghum, rice, wheat, barley, foxtail millet, barnyard millet, finger millet, little millet, oat, rye, sugarcane and khus-khus grass. *Heterodera zeae* may complete six to seven generations during a maize-growing season. Laboratory experiments showed that the host races of *H. zeae* can be differentiated on the basis of their ability to reproduce on maize and vetiver (*Vetiveria zizanioides*) (Bajaj and Gupta, 1994).

Description: **Cyst**: L = 342–805 µm; W = 245–551 µm; L/W ratio = 1.0–2.2; fenestral length = 35–58 µm; vulval slit = 29–58 µm. **Male**: L = 641–994 µm; stylet = 24–25 µm; spicules = 25–32 µm; gubernaculum = 8–11 µm. **J2**: L = 350–484 µm; stylet = 19–25 µm; tail hyaline region = 17–30 µm; tail = 32–50 µm. **Cyst**: light brown, basically lemon-shaped, cuticle thin walled, ambifenestrate. Semifenestra separated by fairly wide vulval bridge (Fig. 15.3P), fenestral length and width variable, basin wide but generally poorly defined. Bullae prominent, immediately beneath to underbridge and characteristically arranged as four finger-like bullae in a distinct formation. Underbridge simple, short, and thin, found in all but a few of cysts examined, lacking forking at ends. **J2**: labial region slightly offset, rounded, with low profile, with three to five annuli, labial framework moderately developed. Lateral field with four distinct lines. Stylet strongly developed with round or slightly anteriorly directed knobs. Tail short, tapering conically, with acutely rounded terminus.

Heterodera zeae differs from other species by the location of the bullae at two levels, that is, four, finger-like bullae located immediately beneath a short, thin, underbridge and with many randomly located bullae further below. The ITS-rRNA sequence distinguishes this species from other species.

15.6 Subfamily Punctoderinae Krall & Krall, 1978

Diagnosis: Heteroderidae Filipjev & Schuurmans Stekhoven, 1941.

Mature female and cyst: body spherical or with vulva cone. Perineal area subterminal or terminal. Female cuticle not annulated, changing colour after death. Vulval lips absent, vulval slit short. Vulval fenestration circumfenestrate. Cuticle thick, D- layer present or absent. Bullae absent or present. Underbridge absent. Subcrystalline layer

present or absent. **Male**: developed through metamorphosis, labial region annulated, lateral field with four incisures, tail short, hemispherical, without bursa. **Second-stage juvenile**: with lateral field marked by four incisures, phasmids punctiform without lens-like structure.
Type genus: Genus *Globodera* Skarbilovich, 1959
Other genera:
Punctodera Mulvey & Stone, 1976
Cactodera Krall & Krall, 1978
Dolichodera Mulvey & Ebsary, 1980
Betulodera Sturhan, 2002
Paradolichodera Sturhan, Wouts & Subbotin, 2007
Vittatidera Bernard, Handoo, Powers, Donald & Heinz, 2010

15.7 Genus *Globodera* Skarbilovich, 1959

Diagnosis: (Punctoderinae) (after Baldwin and Mundo-Ocampo, 1991 and Siddiqi, 2000).
Mature female and cyst: spheroid, lacking terminal cone. Vulval area circumfenestrate. Vulva located in a cavity beneath outline of body, vulval slit <15 µm. No anal fenestra. Vaginal remnants, underbridge and bullae rarely present. Cuticle with distinct D-layer. All eggs retained in body, eggsac absent. **Male**: lateral field with four lines, spicules >30 µm, distally pointed. **J2**: with four incisures in lateral field. Tail conical, pointed, phasmids punctiform. **Egg**: surface smooth.

15.7.1 List of species and synonyms

Type species:

1. *Globodera rostochiensis* (Wollenweber, 1923) Skarbilovich, 1959

= *Heterodera schachtii rostochiensis* Wollenweber, 1923
= *Heterodera rostochiensis* Wollenweber, 1923
= *Heterodera (Globodera) rostochiensis* (Wollenweber, 1923) Skarbilovich, 1959
= *Heterodera schachtii solani* Zimmermann, 1927
= *Heterodera solani* Zimmermann, 1927
= *Heterodera pseudorostochiensis* Kirjanova, 1963
= *Globodera pseudorostochiensis* (Kirjanova, 1963) Mulvey & Stone, 1976
= *Globodera arenaria* Chizhov, Udalova & Nasonova, 2008

Other species:

2. *G. agulhasensis* Knoetze, Swart, Wentzel & Tiedt, 2017
3. *G. artemisiae* (Eroshenko & Kazachenko, 1972) Behrens, 1975

= *Heterodera artemisiae* Eroshenko & Kazachenko, 1972
= *Globodera hypolysi* Ogawa, Ohshima & Ichinohe, 1983

4. *G. bravoae* Franco, Cid del Prado & Lamothe-Argumedo, 2000
5. *G. capensis* Knoetze, Swart & Tiedt, 2013
6. *G. ellingtonae* Handoo, Carta, Skantar, & Chitwood, 2012
7. *G. leptonepia* (Cobb & Taylor, 1953) Skarbilovich, 1959

= *Heterodera leptonepia* Cobb & Taylor, 1953
= *Heterodera (Globodera) leptonepia* (Cobb & Taylor, 1953) Skarbilovich, 1959

8. *G. mali* (Kirjanova & Borisenko, 1975) Behrens, 1975

= *Heterodera mali* Kirjanova & Borisenko, 1975
= *Globodera mali* (Kirjanova & Borisenko, 1975) Mulvey & Stone, 1976

9. *G. mexicana* Subbotin, Mundo-Ocampo & Baldwin, 2010

= *Heterodera mexicana* Campos-Vela, 1967 (= *nomen nudum*)

10. *G. millefolii* (Kirjanova & Krall, 1965) Behrens, 1975

= *Heterodera millefolii* Kirjanova & Krall, 1965
= *Heterodera (Globodera) millefolii* (Kirjanova & Krall, 1965) Mulvey, 1973
= *Globodera achilleae* (Golden & Klindie, 1973) Behrens, 1975
= *Heterodera achilleae* Golden & Klindie, 1973

11. *G. pallida* Stone, 1973

= *Heterodera pallida* Stone, 1973
= *Heterodera (Globodera) pallida* Stone, 1973

12. *G. sandveldensis* Knoetze, Swart, Wentzel & Tiedt, 2017

13. *G. tabacum tabacum* (Lownsbery & Lownsbery, 1954) Skarbilovich, 1959

= *Heterodera tabacum* Lownsbery & Lownsbery, 1954

= *Globodera tabacum* (Lownsbery & Lownsbery, 1954) Behrens, 1975
= *G. tabacum solanacearum* (Miller & Gray, 1972) Behrens, 1975
= *Heterodera solanacearum* Miller & Gray, 1972
= *Heterodera tabacum solanacearum* (Miller & Gray, 1972) Stone, 1983
= *Globodera solanacearum* (Miller & Gray, 1972) Behrens, 1975
= *G. tabacum virginiae* (Miller & Gray, 1968) Stone, 1973
= *Heterodera virginiae* Miller & Gray, 1968
= *Heterodera tabacum virginiae* (Miller & Gray, 1968) Stone, 1983
= *Globodera virginiae* (Miller & Gray, 1968) Behrens, 1975

14. *G. zelandica* Wouts, 1984

15.7.2 Principal species

15.7.2.1 Golden potato cyst nematode, G. rostochiensis

Presently, the potato cyst nematodes (PCN) include three species, two species of them, *G. rostochiensis* and *G. pallida*, are reported from many countries and are considered to be economically important pests of potato. The recently described *G. ellingtonae* was found parasitizing potatoes in the USA and some regions of South America (Handoo et al., 2012; Zasada et al., 2013, 2015; Lax et al., 2014).

The golden potato cyst nematode (GPCN), *G. rostochiensis*, is a serious pest of potatoes around the world and is a target of strict regulatory actions in many countries. The GPCN was first found associated with potato plant, *Solanum tuberosum*, from Rostock, Germany in 1881 and was considered to be *H. schachtii*, this being the only known species of cyst nematode at that time. During the early 1900s, GPCN became more widely known throughout Europe and was formally described in 1923. Presently, GPCN are reported in most potato-producing regions of Europe, Africa, Asia, North, Central and South America, and Oceania. Hosts of GPCN include potato (*S. tuberosum*), tomato (*S. lycopersicum*) and eggplant (*S. melongena*). Other hosts include many *Solanum* spp., *Datura* spp., *Hyoscyamus niger*, *Nicotiana acuminata*, *Physalis* spp., *Physochlaina orientalis*, *Salpiglossis* spp., *Capsicum annuum* and *Saracha jaltomata*.

In temperate regions, *G. rostochiensis* usually completes only one generation, although a second generation may be initiated but not completed. In subtropical regions two generations might occur. Development of one generation requires 6–10 weeks. The J2 can go into diapause and remain viable for many years, hatching continuing for 25 or more years (see Chapter 3, this volume). The main routes of spread are infested seed potatoes and movement of soil through farm machinery. Heavily infected plants become yellow and stunted. Infected plants have reduced root systems, which are abnormally branched and brownish in colour. Symptoms in the field first appear in small patches. At low nematode densities tuber sizes are reduced, whereas at higher densities both number and size of tubers can be reduced. At 8 and 64 eggs (g soil)$^{-1}$, yield losses of about 20 and 70%, respectively, can be expected. The damaging effect of PCN is not only determined by nematode density, but also by such factors as cultivar, crop husbandry and environmental conditions.

Description: **Cyst**: L = 450–990 μm; W = 250–810 μm; L/W ratio = 0.9–1.8; fenestral diameter = 14–21 μm; number of ridges between anus and fenestra = 16–24; Granek's ratio = 2.0–7.0. **Male**: L = 860–1406 μm; stylet = 26–27 μm; spicules = 32–39 μm; gubernaculum = 10–14 μm. **J2**: L = 366–502 μm; stylet = 19–23 μm; tail hyaline region = 18–30 μm; tail = 37–57 μm. **Female**: colour changing from white to yellow to light golden as female matures to cyst stage. **Cyst**: brown, ovate to spherical in shape with protruding neck, circumfenestrate, abullate (Fig. 15.8). Fenestra circular (Fig. 15.7A), anus conspicuous at apex of a V-shaped subsurface cuticular mark. **J2**: body tapering at both extremities but more at posterior end. Stylet well developed, with prominent rounded knobs as viewed laterally (Fig. 15.4F). Lateral fields with four lines extending for most of body length. Tail tapering to small, rounded terminus (Fig. 15.5F).

Globodera rostochiensis is morphologically similar to *G. pallida*, *G. ellingtonae*, *G. mexicana* and *G. tabacum*. It differs from *G. pallida* as it has yellow or gold *vs* cream coloured maturing females, a higher number of ridges between the vulva and anus, a larger mean for Granek's

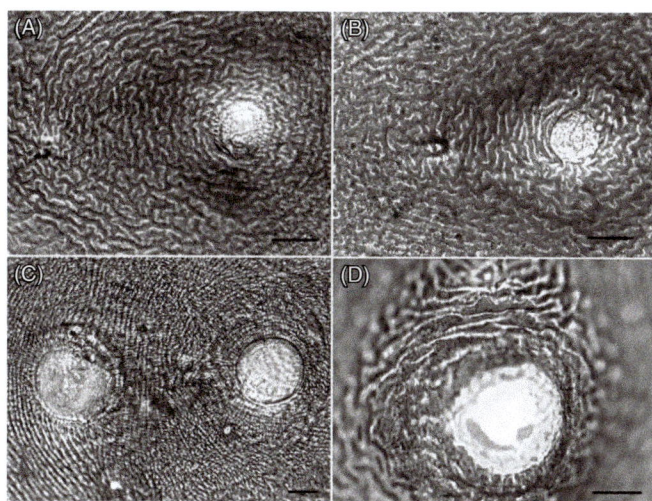

Fig. 15.7. Vulval plates.
(A) *Globodera rostochiensis*.
(B) *G. pallida*. (C) *Punctodera punctata*. (D) *Cactodera cacti*.
Courtesy of V.N. Chizhov. Scale bars = 15 μm.

ratio, a different stylet knob shape, shorter average stylet length and bluntly rounded *vs* more pointed J2 tail terminus. The ITS-rRNA and *COI* gene sequences clearly distinguish GPCN from all other PCN and *Globodera* species.

15.7.2.2 Ellington potato cyst nematode, G. ellingtonae

Ellington potato cyst nematode (EPCN) was recently recognized as a new species from Oregon and Idaho, USA. Glasshouse experiments have demonstrated that this nematode can reproduce on potatoes (Handoo *et al.*, 2012) and biological information, including hatch in potato root diffuste, was provided by Zasada *et al.* (2013, 2015), who confirmed that this nematode reproduces well on various cultivars of potato and tomato. In 2014, EPCN was also identified on roots of Andean potatoes collected from the Salta region of Northern Argentina (Lax *et al.*, 2014). Isolates of *Globodera* sp. from Chile also showed a high degree of molecular similarity with populations of *G. ellingtonae* from Argentina and the USA, suggesting that this nematode also occurs in Chile (see also Chapter 7, this volume).
Description: **Cyst**: L = 370–860 μm; W = 320–890 μm; L/W ratio = 0.9–1.4; fenestral diameter = 12–42.5 μm; number of ridges between anus and fenestra = 8–25; Granek's ratio = 0.9–5.9. **Male**: L = 717–1368 μm; stylet = 21–27 μm; spicules = 30–44 μm; gubernaculum = 8–15 μm. **J2**: L = 365–526 μm; stylet = 19–24 μm; hyaline region = 19–33 μm; tail = 39–56 μm. **Female**: body white becoming yellow to pale brown as eggs mature, ovate to rounded or subspherical in shape with elongate, protruding neck. **Cysts**: light brown to brown in colour, generally spherical to occasionally oval with a protruding neck. Vulval region fenestrated with a single circumfenestrate opening occupying all or part of vulval basin. Cyst wall pattern ridge-like to irregular, wavy to whorled with heavy punctations. **J2**: lip region with three or four complete or incomplete cephalic annules. Stylet short, robust; basal knobs rounded posteriorly; the anterior surface of three shapes: rounded, forward projection anteriorly or flattened, the latter being the most frequent. Tail tapering uniformly but abruptly narrowing with a pronounced to slight constriction near the posterior third of the hyaline portion, ending with a peg-like, finely rounded to pointed terminus.

EPCN differs morphologically from the related species, *G. pallida, G. rostochiensis, G. tabacum* and *G. mexicana* in the distinctive J2 tail that uniformly tapers but abruptly narrows with a pronounced to slight constriction near the posterior third of the hyaline portion, ending with a peg-like, finely rounded to pointed terminus. Detailed differentiating characters between other PCN and *Globodera* species are given in Handoo *et al.* (2012) and Lax *et al.* (2014). The ITS-rRNA and *COI* gene sequences clearly distinguish EPCN from all other PCN and *Globodera* species.

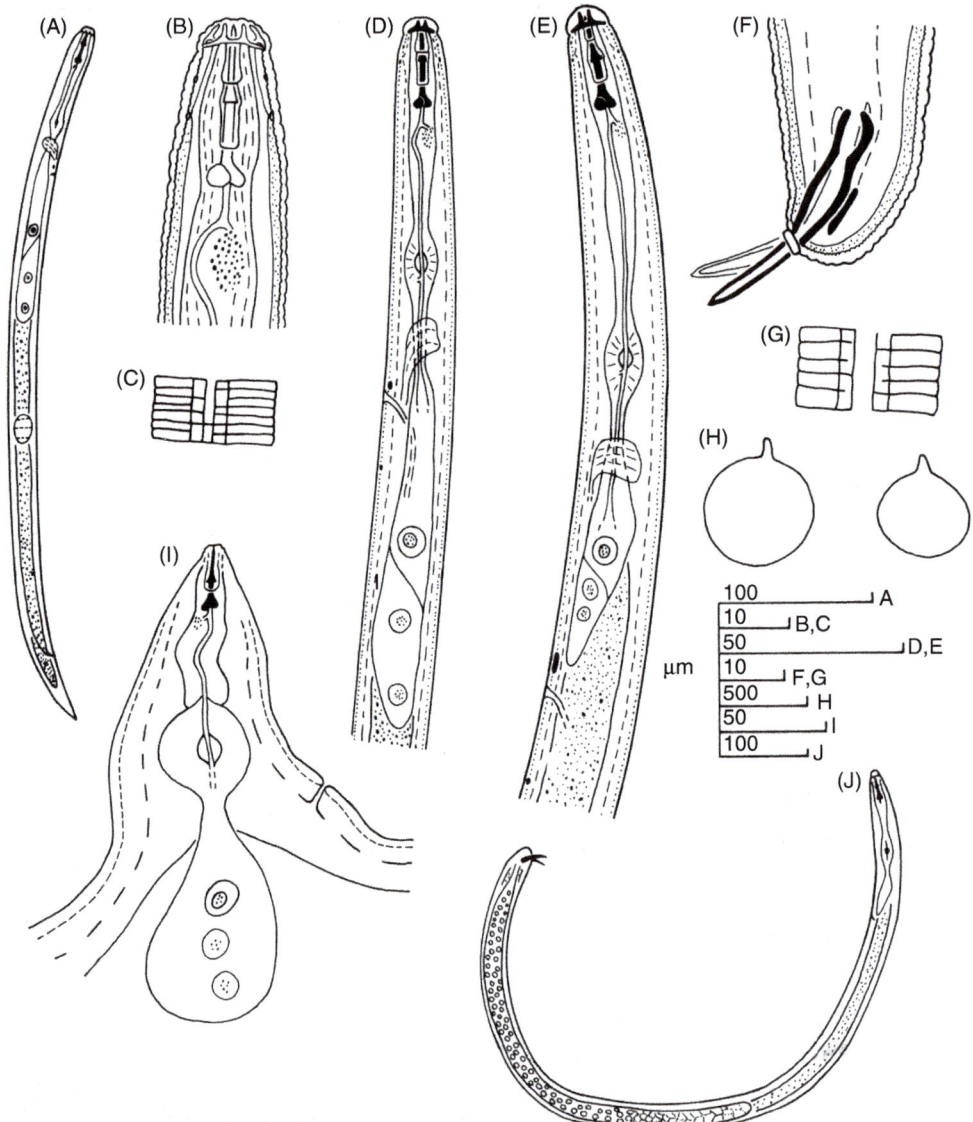

Fig. 15.8. *Globodera rostochiensis*. (A–D) J2. (A) Entire body. (B) Lip region. (C) Lateral field. (D) Pharyngeal region. (E–G) Male. (E) Pharyngeal region. (F) Tail. (G) Lateral field. (H) Entire cysts. (I) Anterior region of female. (J) Entire male. After Stone (1973a).

15.7.2.3 Pale potato cyst nematode, G. pallida

The pale potato cyst nematode (PPCN) is considered to be a major pest of potato crops in cool temperate climates. PPCN is reported from several counties in Europe, Asia, Africa and South America. *Globodera pallida* has not been recorded in Australia, but is found in New Zealand. In Central and North America *G. pallida* has been reported in Panama, the USA and Canada, but in the last two countries *Globodera* species on potato have a rather restricted distribution, with only small infested areas, because of rigorous phytosanitary regulations and seed potato certification programmes. Recently, mtDNA analysis has been used to study genetic relationships among Peruvian populations of *G. pallida*, thus identifying the origin of western European populations of this species (Picard *et al.*, 2007; Plantard *et al.*, 2008). Using the mtDNA gene, cytochrome *b* (*cytb*) sequences and microsatellite loci, Plantard *et al.* (2008) showed that the *G. pallida* presently distributed in Europe derived from a single restricted area in the extreme south of Peru, located between the north shore of Lake Titicaca and Cusco. *Globodera pallida* develops one generation for a vegetation season. This species is adapted to cool temperatures and is able to hatch earlier in the year and develop at temperatures 2°C cooler than those required by *G. rostochiensis* (Langeslag *et al.*, 1982). The symptoms of attack by *G. pallida* are similar to those for *G. rostochiensis* and the damage threshold is 1–2 eggs (g soil)$^{-1}$. Hosts include potato (*S. tuberosum*), eggplant (*S. melongena*), tomato (*S. lycopersicum*), many other species of *Solanum* and black henbane (*Hyoscyamus niger*).

Description: **Cyst**: L = 420–748 μm; W = 400–685 μm; fenestral diameter = 17.5–25 μm; number of ridges between anus and fenestra = 7–17; Granek's ratio = 1.2–3.6. **Male**: L = 1198 μm; stylet = 27 μm; spicules = 36 μm; gubernaculum = 11 μm. **J2**: L = 380–533 μm; stylet = 22.5–25 μm; tail hyaline region = 20–31 μm; tail = 40–57 μm. **Female**: white in colour, some populations passing, after 4–6 weeks, through a cream stage, turning glossy brown when dead. **Cyst**: vulval region intact or fenestrated with single circumfenestrate opening occupying all or part of vulval basin, abullate (Fig. 15.7B). **J2**: lateral field with four incisures but with three anteriorly and posteriorly, occasionally completely areolated. Stylet well developed, basal knobs with distinct anterior projection as viewed laterally. Tail tapering uniformly with a finely rounded point, hyaline region is about half of tail region.

Globodera pallida is most closely related to *G. mexicana*, *G. rostochiensis*, *G. tabacum* and *G. ellingtonae*. It differs from *G. rostochiensis* as it has cream coloured females *vs* yellow or gold, smaller number of ridges between the vulva and anus, a smaller mean for Granek's ratio, a different stylet knob shape, longer stylet length, tail terminus and presence of refractive bodies on the hyaline part of tail (usually four to seven refractive bodies *vs* absence) in J2. The ITS-rRNA and *COI* gene sequences clearly distinguish PPCN from all other PCN and *Globodera* species.

15.7.2.4 Tobacco cyst nematode, G. tabacum

Tobacco cyst nematode (TCN) is considered as a serious and important pest of shade and broadleaf tobacco. It is recorded from several countries in Europe, Asia, Africa, South and North America. *Globodera tabacum* was considered to be a polytypic species containing the following subspecies: *G. tabacum tabacum* (Lownsbery & Lownsbery, 1954), *G. tabacum virginiae* (Miller & Gray, 1968) and *G. tabacum solanacearum* (Miller & Gray, 1972). All three subspecies develop on tobacco and horse nettle (*Solanum carolinense*), but otherwise differ in host preference. *Globodera tabacum* parasitizes *Nicotiana tabacum*, *S. carolinense*, tomato and other species of the genera *Nicotiana* and *Solanum*, as well as *Atropa belladona*, *Hyoscyamus niger*, *Nicandra physalodes* and *Capsicum annuum*. TCN may have four to five generations on tobacco under field conditions in the USA and Italy. Infected tobacco plants have small root systems and stunting of the aerial parts. TCN infection is often associated with increased damage from bacterial wilt and black shank. Farmers in Virginia, USA, have recorded complete crop failures, but losses generally average 15%. A high density of nematode populations early in the growing season can reduce flue-cured tobacco yield by 25–50%, although tobacco may escape significant losses from moderate populations, especially under favourable growing conditions.

Description: **Cyst**: L = 337–937 μm; W = 232–812 μm; L/W ratio = 0.9–1.5; fenestral diameter = 13–36 μm; number of ridges between anus and fenestra = 5–15; Granek's ratio = 1.0–4.2. **Male**: L = 710–1450 μm; stylet = 24–29 μm; spicules = 26–35 μm; gubernaculum = 9–12 μm. **J2**: L = 410–621 μm; stylet = 20–27 μm; tail hyaline region = 17–35 μm; tail = 33–64 μm. **Female**: body ovate to spherical with elongate neck, white, becoming yellow. **Cyst**: light shiny brown, circumfenestrate, abullate. **J2**: with well developed rounded basal knobs. Labial region offset by a slight constriction, marked by four striae. Terminus of tail finely rounded.

Globodera tabacum differs from *G. rostochiensis* as it has J2 with longer mean values of body length, mean stylet and cysts with a smaller mean number of cuticular ridges. It differs from *G. mexicana* as it has J2 with longer mean body length and from *G. pallida* owing to cysts with a smaller mean number of cuticular ridges and J2 with longer mean body length. The ITS-rRNA gene sequence clearly distinguishes TCN from all other *Globodera* species.

15.8 Genus *Punctodera* Mulvey & Stone, 1976

Diagnosis: (Punctoderinae) (after Siddiqi, 2000). **Mature females and cysts**: spherical, pear-shaped or ovoid, with short projecting neck and heavy subcrystalline layer. Cuticle reticulate, subcuticle with punctations. D-layer present. Terminal region not cone-shaped; cyst light to dark brown. Vulval slit extremely short (<5 μm), anus at a short distance from vulval fenestra. Circumfenestrate, fenestra surrounding vulva 16–40 μm in diameter, anus offset toward ventral margin of fenestra, an anal fenestra of similar shape and size to vulval fenestra present. Underbridge and perineal papilla-like tubercles absent. Bullae present or absent. Egg retained in body, no eggsac. Parasites of monocotyledonous plants.

15.8.1 List of species and synonyms

Type species:

1. *Punctodera punctata* (Thorne, 1928) Mulvey & Stone, 1976

= *Heterodera punctata* Thorne, 1928
= *Heterodera (Globodera) punctata* (Thorne, 1928) Skarbilovich, 1959
Other species

2. *P. chalcoensis* Stone, Sosa Moss & Mulvey, 1976
3. *P. matadorensis* Mulvey & Stone, 1976
4. *P. stonei* Brzeski, 1998

15.8.2 Principal species

15.8.2.1 Grass cyst nematode, P. punctata

This species was described by Thorne in 1928 based on specimens from heavily infected wheat roots from a field in the Humboldt area, Saskatchewan, Canada. Subsequently, *P. punctata* was also reported as a common species infecting grasses from Europe, the USA and Canada. However, all attempts to infect wheat or other cereals with these nematodes failed to give any positive results. It has been suggested *P. punctata* might represent a complex of several closely related species. Several grasses are good hosts of this nematode. Only a single generation occurs each year.

Description: **Female and cyst**: L = 330–901 μm; W =170–720 μm; L/W ratio = 1.2–2.0; vulval fenestral diameter = 16–33 μm; anal fenestral diameter = 11–42 μm. **Male**: L = 910–1270 μm; stylet = 23–33 μm; spicules = 28–36 μm; gubernaculum = 8–10 μm. **J2**: L = 438–680 μm; stylet = 23–32 μm; DGO = 3.5–6.5 μm; tail hyaline region = 37–64 μm; tail = 63–93 μm. **Female and cyst**: ovoid, pear- or flask-shaped without vulval cone, white. Anal fenestra and vulval fenestra present (Fig. 15.7C). Vulva slit bordered by thickened ridges, set in a subcircular translucent area of cuticle. Newly formed cysts with conspicuous subcrystalline layer. **J2**: with well developed projecting anteriorly basal knobs (Fig. 15.4H). Conspicuous hyaline region at least twice as long as stylet, distal third of tail tapering, ending in a rounded point (Fig. 15.5H).

Punctodera punctata differs from other *Punctodera* species owing to the pear-shaped cysts and the absence of bullae.

15.9 Genus *Cactodera* Krall & Krall, 1978

Diagnosis: (Punctoderinae) (after Sturhan, 2002 with modification).
Mature females and cysts: lemon-shaped to spherical, with posterior protuberance. Vulva terminal, vulval slit <30 μm, fenestra circumfenestrate. Anus without fenestration. Bullae and underbridge absent, vulval denticles usually present. Cuticle with D-layer. The eggs are usually retained within the cyst body. Eggsac present or absent. Eggshell surface smooth or punctate. **J2**: have a lateral field with four incisures, phasmid openings punctiform.

15.9.1 List of species and synonyms

Type species:

1. *Cactodera cacti* (Filipjev & Schuurmans Stekhoven, 1941) Krall & Krall, 1978

= *Heterodera cacti* Filipjev & Schuurmans Stekhoven, 1941

Other species:

2. *C. acnidae* (Schuster & Brezina, 1979) Wouts, 1985

= *Heterodera acnidae* Schuster & Brezina, 1979

3. *C. amaranthi* (Stoyanov, 1972) Krall & Krall, 1978

= *Heterodera amaranthi* Stoyanov, 1972

4. *C. eremica* Baldwin & Bell, 1985
5. *C. estonica* (Kirjanova & Krall, 1963) Krall & Krall, 1978

= *Heterodera estonica* Kirjanova & Krall, 1963

6. *C. evansi* Cid del Prado & Rowe, 2000
7. *C. galinsogae* Tovar Soto, Cid del Prado, Nicol, Evans, Sandoval Islas & Martinez Garza, 2003
8. *C. milleri* Graney & Bird, 1990
9. *C. radicale* Chizhov, Udalova & Nasonova, 2008
10. *C. rosae* Cid del Prado & Miranda, 2008
11. *C. salina* Baldwin, Mundo-Ocampo & McClure, 1997
12. *C. thornei* (Golden & Raski, 1977) Mulvey & Golden, 1983

= *Heterodera thornei* Golden & Raski, 1977

13. *C. torreyanae* Cid del Prado Vera & Subbotin, 2014
14. *C. weissi* (Steiner, 1949) Krall & Krall, 1978

= *Heterodera weissi* Steiner, 1949

15.9.2 Principal species

15.9.2.1 Cactus cyst nematode, C. cacti

Cyst nematode-infecting cacti, *Discocactus akkermannii* and *Cereus speciosus*, both of which were expressing growth-declining symptoms, were first recorded and described from Maartensdijk, near Utrecht, The Netherlands. The cactus cyst nematode is distributed worldwide, mainly on plants of the family Cactaceae grown in glasshouses as ornamentals. The dispersal of *C. cacti* from native regions in Mexico is associated with the international trade of infested ornamental cactus plants around the world. The cactus cyst nematode damages certain cacti grown as foodcrops in Mexico and various ornamental cacti. It has been associated with succulent plants belonging to three families: Cactaceae: *Cereus*, *Cleistocactus*, *Coryphantha*, *Discocactus*, *Echinocactus*, *Echinopsis*, *Echinocereus*, *Epiphyllum*, *Gymnocalycium*, *Hatiora*, *Heliocereus*, *Hylocereus*, *Leuchtenbergia*, *Mammillaria*, *Melocactus*, *Notocactus*, *Nopalea*, *Notocactus*, *Opuntia*, *Oreocereus*, *Rebutia*, *Rhipsalis*, *Schlumbergera*, *Selenicereus* and *Thelocactus*; Umbelliferae: *Apium*; and Euphorbiaceae: *Euphorbia*. Infected plants may exhibit various symptoms including branched roots and increased numbers of rootlets. Plants become reddish-brown to yellow in colour, wilted and stunted, with reduced flower production and shortening of the flowering period. The life cycle takes around 30 days at 22°C.

Description: Cyst: L = 328–780 µm; W = 240–598 µm; L/W ratio = 1.1–2.0; fenestral diameter = 16–48 µm. **Male**: L = 910–1113 µm; stylet = 22–29 µm; spicules = 30–37 µm; gubernaculum = 10–15 µm. **J2**: L = 344–584 µm; stylet = 21–26 µm; tail hyaline region = 12–23 µm; tail = 34–60 µm. **Female**: body lemon-shaped to almost spherical, pearly white, yellow or golden. **Cyst**: usually lemon-shaped, but may be rounded with protruding neck and vulva, light or medium brown, sometimes reddish-brown. Vulval denticles generally present, visible beneath fenestral surface. Cone tops abullate, circumfenestrate (Fig. 15.7D). **Males**: have been reported as rare. **J2**: vermiform with hyaline region often shorter than stylet. Eggshells heavily punctuate.

Cactodera cacti resembles: *C. weissi*, *C. acnidae*, *C. milleri* and *C. galinsogae*. It differs from *C. weissi* and *C. acnidae* in having eggshells heavily punctate *vs* shells without visible markings, and J2 with larger stylet. *Cactodera cacti* can be clearly differentiated from other species by sequences of the ITS-rRNA gene.

15.10 Genus *Dolichodera* Mulvey & Ebsary, 1980

Diagnosis: (Punctoderinae) (after Siddiqi, 2000, with modifications). **Females and cysts**: body

elongate to oval, without terminal protuberance (Fig. 15.9A), white, swollen part 400–500 μm long, 140–270 μm wide, 2.0–2.8 times as long as wide, neck moderately long. Cuticle not annulated but with fine irregular striae. Vulval area terminal or just subterminal, circumfenestrate, fenestra approx. 20 μm in diameter, bullae present, perineal tubercles absent (Fig. 15.9B). Anus pore-like, lacking a fenestra, located 10–13 μm dorsal to vulval fenestral margin. Cyst with several large bullae. Perineal tubercles absent. Vulva circumfenestrate, underbridge absent. **Male**: not found. **J2**: with long tail (95–120 μm). Lateral field with three incisures, inner one faint. Labial region hemispherical, offset, with two annuli. Tail tip narrowly rounded. Phasmid openings lacking a lens-like ampulla, located about one anal body diameter posterior to anus.
Type and only species: *Dolichodera fluvialis* Mulvey & Ebsary, 1980

15.11 Genus *Betulodera* Sturhan, 2002

Diagnosis: (Punctoderinae) (after Sturhan, 2002). **Cysts**: lemon-shaped, pear-shaped or spheroid with insignificant, obtuse vulval cone. Cyst wall thick, with irregular network-like pattern, D-layer absent (no punctations in inner, deeper layers of cyst wall), subcrystalline layer heavily developed. Vulva terminal, surrounded by

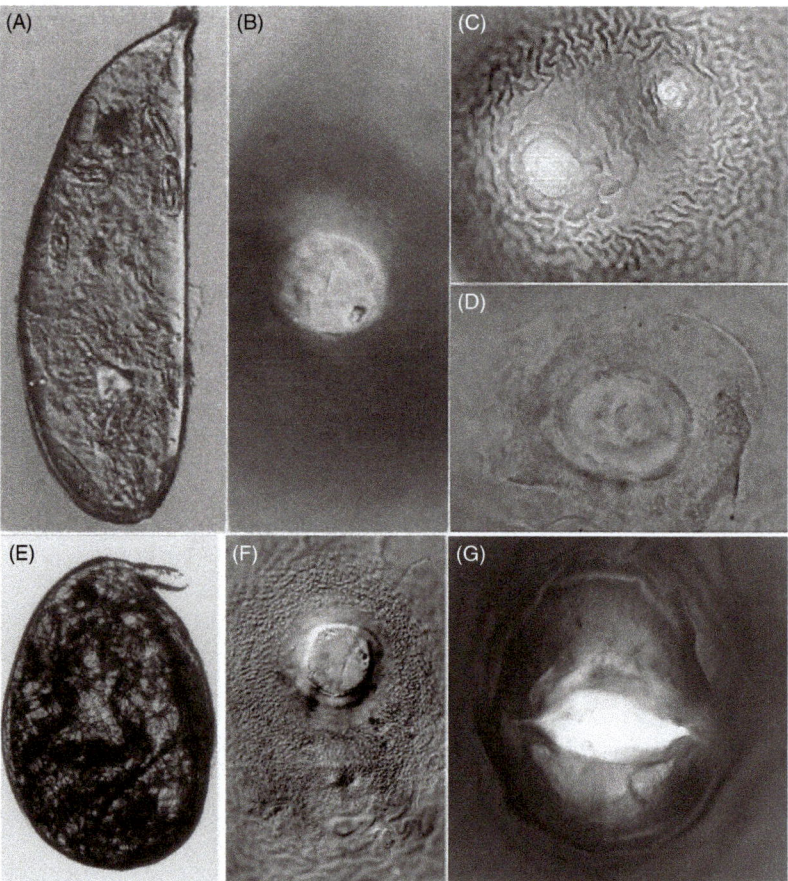

Fig. 15.9. *Dolichodera fluvialis.* (A) Cyst. (B) Fenestral area. *Betulodera betulae.* (C) Surface view of cyst cone top with single fenestra (large opening) and anus (small opening). (D) Fenestral area of white female during early stage of fenestration. *Paradolichodera tenuissima.* (E) Cyst. (F) Cyst posterior end with vulval fenestra and anus (below). *Vittatidera zeaphila.* (G) Fenestral area. A, B, After Mulvey and Ebsary (1980); C, D, Hirschmann and Riggs (1969); E, F, after Sturhan *et al.* (2007); G, Bernard *et al.* (2010).

circumfenestration, vulval slit short (<10 µm), underbridge absent, denticles occasionally present, anus without fenestration (Fig. 15.9C, D). **Male**: body twisted, no cloacal tube, spicules with bifid distal tips, phasmid openings punctiform. **J2**: has lateral field with three incisures, phasmid openings punctiform, without lens-like structure, labial region with three or four labial annuli and labial disc fused with submedial lips (Figs 15.4G and 15.5G).
Type and only species:
Betulodera betulae (Hirschmann & Riggs, 1969) Sturhan, 2002
= *Heterodera betulae* Hirschmann & Riggs, 1969
= *Cactodera betulae* (Hirschmann & Riggs, 1969) Krall & Krall, 1978

15.12 Genus *Paradolichodera* Sturhan, Wouts & Subbotin, 2007

Diagnosis: (Punctoderinae) (after Sturhan *et al.*, 2007).
Female and cyst: elongate to ovoid, with rounded posterior end (Fig. 15.9E). Cuticle transparent, with faint transverse striations on anterior part of body and posteriorly mostly with faint irregular ridges superimposed on distinct punctations. Cuticle turning yellowish to light brown on death, covered by a subcrystalline-like film. Eggs retained in body, eggsac not observed. Labial disc squarish. Vulva terminal or subterminal, vulval slit short, circumfenestrate (Fig. 15.9F). Anus lacking fenestration. **Male**: body not twisted, lateral field with four incisures. Phasmids lacking. **J2**: long, extremely slender for family, lateral fields indistinct. Stylet short (<20 µm). Dorsal gland orifice located more than half stylet length posterior to stylet base, pharyngeal glands long, filling body cavity. Tail long, slender, phasmid openings punctiform.
Type and only species:
Paradolichodera tenuissima Sturhan, Wouts & Subbotin, 2007

15.13 Genus *Vittatidera* Bernard, Handoo, Powers, Donald & Heinz, 2010

Diagnosis: (Punctoderinae) (after Bernard *et al.*, 2010). **Cysts**: orange-brown to brown, lemon-shaped with short necks and vulval cone. Vulval cone slightly protuberant, membranous vulval lips persistent; vulval aperture circular to rhomboid, circumfenestrate, with irregular denticle-like protuberances around the periphery of orifice (Fig. 15.9G). Bullae, vulval bridge, vulval underbridge and internal denticles absent. Anus subterminal. **Male**: variable length, stylet knobs rounded. **J2**: with stylet less than 18 µm, conoid tail with narrowly rounded tip, phasmid apertures pore-like. Lateral field with four incisures. **Eggshell**: smooth.
Type and only species:
Vittatidera zeaphila Bernard, Handoo, Powers, Donald & Heinz, 2010

15.14 Conclusions and Future Prospects

For diagnostics of cyst nematodes, morphology is complementry to the use of molecular markers Accordingly, for any future proposals of new species in cyst nematode genera a blend of both morphological (including SEM), morphometric and molecular data is essential and desirable, with some of the genera rapidly approaching close to 121 valid species. The future prospects in cyst nematode taxonomy and diagnostics are dependent on molecular-based methodologies that will discriminate not only at the species level but also at the level of host races and pathotypes, thereby opening up opportunities for more focused management strategies. These techniques will offer rapid diagnostics and help resolve the present problems associated with relatively morphologically conserved organisms. Once these techniques are widely employed no doubt a number of the current nominal species will be shown to be junior synonyms, whilst others will be shown to be species complexes, possibly of sibling species. Molecular characterization will also enhance our understanding of the phylogeny of the genus and its relationship with other plant-parasitic nematodes.

15.15 Acknowledgements

ZAH is grateful to Ellen Lee for technical help.

15.16 References

Andersson, S. (1975) *Heterodera hordecalis* n. sp. (Nematoda: Heteroderidae), a cyst nematode of cereals and grasses in southern Sweden. *Nematologica* 20 (1974), 445–454. DOI: 10.1163/187529274X00078

Ambrogioni, L.C., Carletti, B., Cotroneo, A. and Caroppo, S. (2004) Host range of an Italian population of *Heterodera betae* Wouts, Rumpenhorst et Sturhan (Nematoda Heteroderidae). *Redia* 87, 23–25.

Bajaj, K. and Gupta, D.C. (1994) Existence of host races in *Heterodea zeae* Koshy *et al. Fundamental and Applied Nematology* 17, 389–390.

Baldwin, J.G. and Mundo-Ocampo, M. (1991) Heteroderinae, cyst and non cyst-forming nematodes. In: Nickle, W.R. (ed.) *Manual of Agricultural Nematology*. Marcel Dekker Inc., New York, pp. 275–362.

Baldwin, J.G. and Perry, R.N. (2004) Nematode morphology, sensory structure and function. In: Chen, Z.X, Chen, S.Y and Dickson, D.W (eds) *Nematology Advances and Perspectives Volume 1 (Nematode Morphology, Physiology and Ecology)*. CAB International, Wallingford, UK, pp. 175–257.

Behrens, E. (1975) *Globodera* Skarbilovich, 1959, eine selbständige Gattung in der Unterfamilie Heteroderinae Skarbilovich, 1947 (Nematoda: Heteroderidae). *Vortragstagung zu Aktuellen Problemen der Phytonematologie am 29.5.1975 in Rostock. Manuskriptdruck der Vorträge. Biologische Gesellschaft der DDR, Sektion Phytopathologie und Universität, Rostock*, pp. 12–26.

Bernard, E.C., Handoo, Z.A., Powers, T.O., Donald, P.A. and Heinz, R.D. (2010) *Vittatidera zeaphila* (Nematoda: Heteroderidae), a new genus and species of cyst nematode parasitic on corn (*Zea mays*). *Journal of Nematology* 42, 139–150.

Byrd Jr, D.W., Kirkpatrick, T. and Barker, K.R. (1983) An improved technique for clearing and staining plant tissues for detection of nematodes. *Journal of Nematology* 15, 142.

Carta, L.K. (1991) Part 3 Preparation of nematodes for transmission electron microscopy. In: Nickle, W.R. (ed.) *Manual of Agricultural Nematology*. Marcel Dekker Inc., New York, pp. 97–106.

Charchar, J.M. and Eisenback, J.D. (2000). An improved technique to prepare perineal patterns of root-knot nematodes for SEM. *Nematologia Brasileira* 24, 245–247.

Cooper, B.A. (1955) A preliminary key to British species of *Heterodera* for use in soil examination. In: Kevan, D.K.M. (ed.) *Soil Zoology*. Butterworths, London, pp. 269–280.

Doupnik, B., Jr. (1993) Soybean production and disease loss estimates for north central United States from 1989 to 1991. *Plant Disease* 77, 1170–1171. DOI: 10.1094/PD-77-1170

Eisenback, J.D. (1991) Preparation of nematodes for scanning electron microscopy. In: Nickle, W.R. (ed.) *Manual of Agricultural Nematology*. Marcel Dekker Inc., New York, pp. 87–96.

Fenwick, D.W. (1959) Red ring of coconut, a problem for the nematologist. *Indian Coconut Journal* 12, 82–86.

Ferris, V.R. (1998) Evolution, phylogeny and systematics. In: Sharma, S.B. (ed.) *Cyst Nematodes*. Kluwer Academic Publishers, Dordrecht, The Netherlands, pp. 57–82.

Filipjev, I.N. (1934) [Nematodes harmful and beneficial to agriculture.] OGIZ-Selkhozgiz, Moscow and Leningrad, Russia.

Franklin, M.T. (1951) *The Cyst-forming Species of Heterodera*. Commonwealth Agricultural Bureaux, Farnham Royal, UK, 147 pp.

Franklin, M.T. (1969) *Heterodera latipons* n. sp., a cereal cyst nematode from the Mediterranean region. *Nematologica* 15, 535–542.

Franklin, M.T. (1972) *Heterodera schachtii*. In: *CIH Descriptions of plant-parasitic nematodes*, Set 1, No. 1. Commonwealth Agricultural Bureaux, Farnham Royal, UK, 4 pp. DOI: 10.1163/187529269X00867

Golden, A.M. (1986) Morphology and identification of cyst nematodes. In: Lamberti, F. and Taylor, C.E. (eds) *Cyst Nematodes*. Plenum Press, New York and London, pp. 23–45. DOI: 10.1007/978-1-4613-2251-1_2

Golden, A.M. (1990) Preparation and mounting nematodes for microscopic observation. In: Zuckerman, B.M., Mai, W.F. and Krusberg, L.R. (eds) *Plant Nematology Laboratory Manual*. University of Massachusetts Agricultural Experiment Station, Amherst, MA, pp. 197–205.

Golden, A.M. and Birchfield, W. (1972) *Heterodera graminophila* n. sp. (Nematoda: Heteroderidae) from grass with a key to closely related species. *Journal of Nematology* 4, 147–154.

Gracianne, C., Petit, E.J., Arnaud J.-F., Porte, C., Renault, R., Fouville, D., Rouaux, C. and Fournet, S. (2014) Spatial distribution and basic ecology of *Heterodera schachtii* and *H. betae* wild populations developing on sea beet, *Beta vulgaris* ssp. *maritima*. *Nematology* 16, 797–805. DOI: 10.1163/15685411-00002809

Greco, N., Vovlas, N., Troccoli, A. and Inserra, R.N. (2002) The Mediterranean cereal cyst nematode, *Heterodera latipons*: a menace to cool season cereals of the United States. *Nematology Circular*

No. 221, Florida Department of Agriculture and Consumer Services, Division of Plant Industry, Gainesville, FL, USA, 4 pp.

Green, C.D. (1971) The morphology of the terminal area of the round-cyst nematodes, S.G. *Heterodera rostochiensis* and allied species. *Nematologica* 17, 34–46. DOI: 10.1163/187529271X00396

Hafez, S.L., Sundararaj, P. and Barbour, J. (1999) Impact of *Heterodera humuli* on growth and mineral nutrition composition of hops, *Humulus lupulus* cv. Cascade. *International Journal of Nematology* 9, 23–26.

Handoo, Z.A., Carta, L.K., Skantar, A.M. and Chitwood, D.J. (2012) Description of *Globodera ellingtonae* n. sp. (Nematoda: Heteroderidae) from Oregon. *Journal of Nematology* 44, 40–57.

Hesling, J.J. (1965) Heterodera: morphology and identification. In: Southey, J.F. (ed.) *Plant Nematology.* Ministry of Agriculture, Fisheries and Food, London, pp. 103–130.

Hirschmann, H. and Riggs, R.D. (1969) *Heterodera betulae* n. sp. (Heteroderidae), a cyst-forming nematode from river birch. *Journal of Nematology* 1, 169–179.

Hooper, D.J. (1970) Extraction of nematodes from plant material. In: Southey, J.F. (ed.) *Laboratory Methods for Work with Plant and Soil Nematodes.* Her Majesty's Stationary Office, London, pp. 34–38. DOI: 10.1111/epp.12077

Hooper, D.J. (1986) Extraction of free-living stages from soil. In: Southey, J.F. (ed.) *Laboratory Methods for Work with Plant and Soil Nematodes.* Ministry of Agriculture, Fisheries and Food, London, pp. 5–30.

Hooper, D.J. (1990) Extraction and processing of plant and soil nematodes. In: Luc, M., Sikora, R.A. and Bridge, J. (eds) *Plant Parasitic Nematodes in Subtropical and Tropical Agriculture.* CAB International, Wallingford, UK, pp. 45–68.

Hunt, D.J. and Handoo, Z.A. (2009) Taxonomy, identification and principal species. In: Perry, R.N., Moens, M. and Starr, J.L. (eds) *Root-knot Nematodes.* CAB International, Wallingford, UK, pp. 55–88.

Hussey, R.S. (1990) Staining nematodes in plant tissue. In: Zuckerman, B.M., Mai, W.F. and Krusberg, L.R. (eds) *Plant Nematology Laboratory Manual.* University of Massachusetts Agricultural Experimental Station, Amherst, Massachusetts, pp. 190–193.

Ichinohe, M. (1952) On the soybean nematode, *Heterodera glycines* n. sp., from Japan. *Magazine of Applied Zoology* 17, 1–4.

Ichinohe, M. (1988) Current research on the major nematode problems in Japan. *Journal of Nematology* 20, 184–190.

Krall, E.L. and Krall, K.A. (1978) Revision of the plant nematodes of the family Heteroderidae on the basis of the trophic specialization of these parasites and their co-evolution with their host plants. *Fito-gel' mintologicheskie issledovaniya*, Nauka, Moscow, Russia, pp. 39–56.

Kühn, J. (1874) Über das Vorkommen von Rübennematoden an den Wurzeln der Halmfrüchte. *Landwirtschaftliche Jahrbücher* 3, 47–50.

Krusberg, L.R., Sardanelli, S. and Grybauskas, P. (1997) Damage potential of *Heterodera zeae* to *Zea mays* as affected by edaphic factors. *Fundamental and Applied Nematology* 20, 593–599.

Langeslag, M., Mugniéry, D. and Fayet, G. (1982) Développement embryonaire de *Globodera rostochiensis et G. pallida* en fonction de la temperature, en conditions contrôlées et naturelles. *Revue de Nématologie* 5, 103–109.

Lax, P., Dueñas, J.C.R., Franco-Ponce, J., Gardenal, C.N. and Doucet, M.E. (2014) Morphology and DNA sequence data reveal the presence of *Globodera ellingtonae* in the Andean region. *Contributions to Zoology* 83, 227–243.

Liebscher, G. (1892) Beobachtungen über das Aufreten eines Nematoden an Erbsen. *Journal für Landwirtschaft* 40, 357–368.

Maas, P.W.T., Schoemaker, A. and Stemerding, S. (1976) Een cysteaaltje bij suikerbieten in Brabant. *Gewasbescherming* 7, 10–11.

Madzhidov, A.R. (1991) The cyst forming nematodes of the family Heteroderidae and their significance for the cereal crops of Tadzhikistan. PhD thesis. Skrjabin's Institute of Helminthology, Moscow, Russia.

McBeth, C.W., Taylor, A.L. and Smith, A.L. (1941) Note on staining nematodes in root tissues. *Proceedings of the Helminthological Society of Washington* 8, 1–26.

Miller, L.I. and Gray, B.J. (1968) Horsenette cyst nematode, *Heterodera virginiae* n. sp., a parasite of solanaceous plants. *Nematologica* 14, 535–543.

Miller, L.I. and Gray, B.J. (1972) *Heterodera solanacearum* n. sp. a parasite of solanaceaous plants. *Nematologica* 18, 404–413.

Mulvey, R.H. (1957) Taxonomic value of the cone top and the underbridge in the cyst-forming nematodes *Heterodera schactii, H. schactii* var. *trifolii* and *H. avenae* (Nematoda: Heteroderidae). *Canadian Journal of Zoology* 35, 421–423. DOI: 10.1139/z57-032

Mulvey, R.H. (1959) Investigation on the clover cyst nematode, *Heterodera trifolii* (Nematoda: Heteroderidae). *Nematologica* 4, 147–159. DOI: 10.1163/187529259X00138

Mulvey, R.H. (1960) The value of cone top and underbridge structures in the separation of some cyst-forming nematodes. In: Sasser, J.N. and Jenkins, W.R. (eds) *Nematology*. Chapel Hill University North Carolina Press, Chapel Hill, NC, pp. 212–215.

Mulvey, R.H. (1972) Identification of *Heterodera* cysts by terminal and cone top structures. *Canadian Journal of Zoology* 50, 1277–1292. DOI: 10.1139/z72-173

Mulvey, R.H. (1973) Morphology of the terminal areas of white females and cysts of the genus *Heterodera* (s.g., *Globodera*). *Journal of Nematology* 5, 303–311.

Mulvey, R.H. and Ebsary, B.A. (1980) *Dolichodera fluvialis* n. gen., n. sp. (Nematoda: Heteroderidae) from Québec, Canada. *Canadian Journal of Zoology* 58, 1697–1702. DOI: 10.1139/z80-232

Mulvey, R.H. and Golden, A.M. (1983) An illustrated key to the cyst-forming genera and species of Heteroderidae in the western hemisphere with morphometrics and distribution. *Journal of Nematology* 15, 1–59.

Mulvey, R.H. and Stone, A.R. (1976) Description of *Punctodera matadorensis* n. gen., n. sp. (Nematoda: Heteroderidae) from Saskatchewan, with lists of species and generic diagnosis of *Globodera* (n. rank), *Heterodera* and *Sarisodera*. *Canadian Journal of Zoology* 54, 772–785. DOI: 10.1139/z76-087

Müller, J. (1999) The economic importance of *Heterodera schachtii* in Europe. *Helminthologia* 36, 205–213.

Nakhla, M.K., Owens, K.J., Li, W. and Wei, G. (2010) Multiplex Real-time PCR assays for the identification of the potato cyst and tobacco cyst nematodes. *Plant Disease* 94, 959–965. DOI: 10.1094/PDIS-94-8-0959

Nicol, J. and Rivoal, R. (2008) Global knowledge and its application for the integrated control and management of nematodes on wheat. In: Ciancio, A. and Mukerji, K.G. (eds) *Integrated Management and Biocontrol of Vegetable and Grain Crops Nematodes*. Springer, The Netherlands, pp. 251–294. DOI: 10.1007/978-1-4020-6063-2_13

Oostenbrink, M. and den Ouden, H. (1954) De struuctuur van de kegeltop als taxonomisch kenmerk bij *Heterodera* - soorten met citroenvormige cysten. *Tijdschrift Over Plantenziekten* 60, 146–151. DOI: 10.1007/BF01988487

Picard, D., Sempere, T. and Plantard, O. (2007) A northward colonisation of the Andes by the potato cyst nematode during geological times suggests multiple host-shifts from wild to cultivated potatoes. *Molecular Phylogenetics and Evolution* 42, 308–316. DOI: 10.1016/j.ympev.2006.06.018

Plantard, O., Picard, D., Valette, S., Scurrah, M., Grenier, E. and Mugniery, D. (2008) Origin and genetic diversity of Western European populations of the potato cyst nematode *Globodera pallida* inferred from mitochondrial sequences and microsatellite loci. *Molecular Ecology* 17, 2208–2218.

Riggs, R.D. (1990) Making perineal patterns of root-knot nematodes and vulval cones of cyst nematodes. In: Zuckerman, B.M., Mai, W.F., Krusberg, L.R. (eds) *Plant Nematology Laboratory Manual*. University of Massachusetts Agricultural Experiment Station, Amherst, MA, pp. 103–106. DOI: 10.1111/j.1365-294X.2008.03718.x

Riggs, R.D. (1992) Host range. In: Riggs, R.D. and Wrather, J.A. (eds) *Biology and Management of the Soybean Cyst Nematode*. American Phytopathological Society, St. Paul, MN, pp. 107–114.

Rivoal, R. and Cook, R. (1993) Nematode pests of cereals. In: Evans, K., Trudgill, D.L. and Webster, J.M. (eds) *Plant Parasitic Nematodes in Temperate Agriculture*. CAB International, Wallingford, UK, pp. 259–303.

Sardanelli, S., Krusberg, L.R. and Golden, A.M. (1981) Corn cyst nematode, *Heterodera zeae*, in the United States. *Plant Disease* 65, 622.

Schacht, H. (1859) Über einige Feinde der Rübenfelder. *Zeitschrift des Vereines für die Rübenzucker-Industrie im Zollverein* 9, 175–179.

Schmidt, A. (1871) Über den Rübennematoden (*Heterodera schachtii* A. S.). *Zeitschrift des Vereines für die Rübenzucker-Industrie im Zollverein* 21, 1–19.

Scholz, U. (2001) Biology, pathogenicity and control of the cereal cyst nematode *Heterodera latipons* Franklin on wheat and barley under semiarid conditions, and interactions with common root *Bipolaris sorokiniana* (Sacc.) Shoemaker (teleomorph: *Cochliobolus sativus* (Ito et Kurib.) Drechs. ex Dastur.). PhD thesis, Bonn University, Germany.

Shi, H. and Zheng, J. (2013) First report of soybean cyst nematode (*Heterodera glycines*) on tobacco in Henan, central China. *Plant Disease* 97, 852. DOI: 10.1094/PDIS-10-12-0926-PDN

Siddiqi, M.R. (2000) *Tylenchida Parasites of Plants and Insects*, 2nd edn. CAB International, Wallingford, UK, 833 pp. http://dx.doi.org/10.1079/9780851992020.0000

Siddiqui, Z.A. and Mahmood, I. (1993) Occurrence of races of *Heterodera cajani* in Uttar Pradesh, India. *Nematologia Mediterranea* 21, 185–186.

Skantar, A.M., Handoo, Z.A., Carta, L.K. and Chitwood, D.J. (2007) Morphological and molecular identification of *Globodera pallida* associated with potatoes in Idaho. *Journal of Nematology* 39, 133–144.

Skantar, A.M., Handoo, Z.A., Carta, L.K., Chitwood, D.J. (2011) Molecular and morphological characterization of *Globodera* populations from Oregon and Idaho. *Journal of Nematology* 43, 282. http://dx.doi.org /10.1094/PHYTO-01-10-0010

Skarbilovich, T.S. (1959) On the structure of the systematic of nematode order Tylenchida Thorne, 1949. *Acta Parasitologica Polonica* 7, 117–132.

Srivastava, A.N. and Chawla, G. (2005) *Maize cyst nematode,* Heterodera zeae, *a key nematode pest of maize and its management*. IARI, New Delhi, India.

Steele, A.E. and Whitehand, L. (1984) Comparative morphometrics of eggs and second-stage juveniles of *Heterodera schachtii* and a race of *Heterodera trifolii* parasitic on sugarbeet in the Netherlands. *Journal of Nematology* 16, 171–177.

Stone, A.R. (1973a) *Heterodera rostochiensis. CIH Descriptions of Plant-parasitic Nematodes*, Set 2, No 16. Commonwealth Agricultural Bureaux, Farnham Royal, UK, 3 pp.

Stone, A.R. (1973b) *Heterodera pallida* n. sp. (Nematoda: Heteroderidae), a second species of potato cyst nematode. *Nematologica* 18, 591–606. http://dx.doi.org /10.1163/187529272X00179

Stone, A.R. (1975) Lip morphology of second-stage juveniles of some Heteroderidae (Nematoda: Tylenchoidea). *Nematologica* 21, 81–88. http://dx.doi.org /10.1111/j.1365-2338.1985.tb00212.x

Stone, A.R. (1979) Co-evolution of nematodes and plants. *Symbolae Botanicae Upsaliensis* 22, 46–61.

Stone, A.R. and Hill, A.J. (1982) Some problems posed by the *Heterodera avenae* complex. *EPPO Bulletin* 12, 317–320. http://dx.doi.org /10.1111/j.1365-2338.1982.tb01808.x

Stone, A.R. and Rowe, J.A. (1976) *Heterodera cruciferae*. In: *CIH Descriptions of Plant-parasitic Nematodes*, Set 6, No. 90. Commonwealth Agricultural Bureaux, Farnham Royal, UK, 4 pp.

Sturhan, D. (2002) Notes on the genus *Cactodera* Krall and Krall, 1978 and proposal of *Betulodera betulae* gen. nov., comb. nov. (Nematoda: Heteroderidae). *Nematology* 4, 875–882. DOI: 10.1163/156854102760402649

Sturhan, D. (2010) Notes on morphological characteristics of 25 cyst nematodes and related Heteroderidae. *Russian Journal of Nematology* 18, 1–8.

Sturhan, D., Wouts, W.M. and Subbotin, S.A. (2007) An unusual cyst nematode from New Zealand, *Paradolichodera tenuissima* gen. n., sp. n. (Tylenchida, Heteroderidae). *Nematology* 9, 561–571. DOI: 10.1163/156854107781487314

Subbotin, S.A. and Franco, J. (2012) Cyst nematodes. In: Manzanilla, R.H. and Marban-Mendoza, N. (eds) *Practical Plant Nematology*. Mundi Prensa-CP, Mexico City, Mexico, pp. 299–357.

Subbotin, S.A., Waeyenberge, L. and Moens, M. (2000) Identification of cyst forming nematodes of the genus *Heterodera* (Nematoda: Heteroderidae) based on the ribosomal DNA-RFLPs. *Nematology* 2, 153–164.

Subbotin, S.A., Sturhan, D., Rumpenhorst, H.J. and Moens, M. (2003) Molecular and morphological characterisation of the *Heterodera avenae* species complex (Tylenchida: Heteroderidae). *Nematology* 5, 515–538. DOI: 10.1163/156854103322683247

Subbotin, S.A., Mundo-Ocampo, M. and Baldwin, J.G. (2010) *Systematics of Cyst Nematodes (Nematoda: Heteroderinae)*. *Nematology Monographs and Perspectives 8A and 8B* (series editors: Hunt, D.J. and Perry, R.N.). Brill, Leiden, The Netherlands.

Subbotin, S.A., Akanwari, J., Nguyen, C.N., Cid del Prado Vera, I., Chitambar, J.J., Inserra, R.N. and Chizhov, V.N. (2017) Molecular characterization and phylogenetic relationships of cystoid nematodes of the family Heteroderidae (Nematoda: Tylenchida). *Nematology* 19, 1065–1081. DOI: 10.1163/15685411-00003107

Turner, S.J. and Subbotin, S.A. (2013) Cyst nematodes. In: Perry, R.N. and Moens, M. (eds) *Plant Nematology*. CAB International, Wallingford, UK, pp. 109–143.

Vovlas, N. (1985) Morphology and histopathology of the cereal cyst nematode (*Heterodera avenae* Woll.) attacking wheat, oats, and barley in Italy. *Nematologica Mediterranea* 13, 87–96.

Vovlas, N., Greco, N. and Di Vito, M. (1985) *Heterodera ciceri* sp. n. (Nematoda: Heteroderidae) on *Cicer arietinum* L from northern Syria. *Nematologia Mediterranea* 13, 239–252.

Vovlas, N., Vovlas, A., Leonetti, P., Liébanas, G., Castillo, P., Subbotin, S.A. and Palomares Rius, J.E. (2015) Parasitism effects on white clover by root-knot and cyst nematodes and molecular separation of *Heterodera daverti* from *H. trifolii*. *European Journal of Plant Pathology* 143, 833–845. DOI: 10.1007/s10658-015-0735-3

Walia, R.K. and Bajaj, H.K. (1986) Existence of host races of pigeon-pea cyst nematode, *Heterodera cajani* Koshy. *Nematologica* 32, 117–119.

Walia, R.K. and Bajaj, H.K. (1988) Further studies on the existence of races in pigeon-pea cyst nematode, *Heterodera cajani*. *Indian Journal of Nematology* 18, 269–272.

Walia, R.K. and Bajaj, H.K. (2000) Morphological and morphometric variations in two races of *Heterodera cajani* Koshy. *Indian Journal of Nematology*, 30 124–128.

Wang, J., Niblack, T.L., Tremaine, J.N., Wiebold, W.J., Tylka, G.L., Marett, C.C., Noel, G.R., Myers, O. and Schmidt, M.E. (2003) The soybean cyst nematode reduces soybean yield without causing obvious symptoms. *Plant Disease* 87, 623–628. DOI: 10.1094/PDIS.2003.87.6.623

Whitehead, A.G. (1968) Taxonomy of *Meloidogyne* (Nematoda: Heteroderidae) with description of four new species. *Transactions of the Zoological Society of London* 31, 263–401. DOI: 10.1111/j.1096-3642.1968.tb00368.x

Williams, T.D. and Siddiqi, M.R. (1972) *Heterodera avenae*. In: *CIH Descriptions of Plant-parasitic Nematodes*, Set 1, No. 2. Commonwealth Agricultural Bureaux, Farnham Royal, UK.

Workneh, F., Tylka, G.L., Yang, X.B., Faghihi, J. and Ferris, J.M. (1999) Regional assessment of soybean brown stem rot, *Phytophthora sojae*, and *Heterodera glycines* using area-frame sampling: prevalence and effects of tillage. *Phytopathology* 89, 204–211. DOI: 10.1094/PHYTO.1999.89.3.204

Wouts, W.M. (1985) Phylogenetic classification of the family Heteroderidae (Nematoda: Tylenchida). *Systematic Parasitology* 7, 295–328. DOI: 10.1007/BF00009997

Wouts, W.M. and Baldwin, J.G. (1998) Taxonomy and identification. In: Sharma, S.B. (ed.) *Cyst Nematodes*. Kluwer Academic Publishers, Dordrecht, The Netherlands, pp. 83–122.

Wouts, W.M., Rumpenhorst, H.J. and Sturhan, D. (2001) *Heterodera betae* sp. n., the yellow beet cyst nematode (Nematoda: Heteroderidae). *Russian Journal of Nematology* 9, 33–42. DOI: 10.1080/13921657.2004.10512573

Zasada, I.A., Peetz, A., Wade, N., Navarre, R.A. and Ingham, R.E. (2013) Host status of different potato (*Solanum tuberosum*) varieties and hatching in root diffusates of *Globodera ellingtonae*. *Journal of Nematology* 45, 195–201.

Zasada, I.A., Phillips, W.S. and Ingham, R.E. (2015) Biological insights into *Globodera ellingtonae*. *Aspects of Applied Biology* 130, 1–9.

Zunke, U. and Eisenback, J.D. (1998) Morphology and ultrastructure. In: Sharma, S.B. (ed.) *Cyst Nematodes*. Kluwer Academic Publishers, Dordrecht, The Netherlands, pp. 31–56.

16 Molecular Taxonomy and Phylogeny

Sergei A. Subbotin[1,2] and Andrea M. Skantar[3]

[1]*Plant Pest Diagnostic Center, California Department of Food and Agriculture, California, USA;* [2]*Center of Parasitology of A.N. Severtsov Institute of Ecology and Evolution, Moscow, Russia;* [3]*USDA, ARS, Beltsville, Maryland, USA*

16.1	Introduction	399
16.2	Nuclear Ribosomal RNA Genes	400
16.3	Nuclear Protein-coding Genes	400
16.4	Mitochondrial DNA Genome Organization	403
16.5	Origin and Phylogeny of Heteroderidae	407
16.6	Phylogeny and Phylogeography of Punctoderinae	409
16.7	Phylogeny and Phylogeography of *Globodera*	410
16.8	Co-evolution of Cyst Nematodes with their Host Plants	412
16.9	Conclusions and Future Prospects	413
16.10	References	414

16.1 Introduction

For many years evolutionary relationships among species and genera of cyst nematodes were estimated by classical comparison of morphological characters. Since the 1990s molecular information, such as nucleotide, amino acid sequences and restriction fragment length polymorphism (RFLP), has become increasingly available for inferring phylogenetic relationships and for reconstructing phylogenetic trees of cyst nematodes.

The availability of molecular data has had a significant impact on the systematics of cyst nematodes, reshaping concepts of their relationships at both the species and genus level. Phylogenetic analysis of sequences has allowed the validity of some taxa to be tested, has supported or rejected species synonymization or the erection of several new genera and, finally, has helped to arrange all taxa in a natural classification. In recent years, molecular phylogenetic trees have become increasingly valuable to taxonomists as the information is integrated with morphological characters and biological particularities in evolutionary analysis.

Although molecular studies of cyst nematodes have yielded a substantial amount of data on their genomes, transcriptomes and proteomes, at present only small fractions of these datasets have been used for phylogenetic reconstructions. Molecular data currently used for cyst nematode phylogeny have come primarily from nuclear ribosomal RNA genes, and nuclear and mitochondrial protein-coding genes. Reconstructing the phylogeny from genes is not straightforward and requires several methodical steps to be followed with certain assumptions and further careful verification of results. Phylogenetic trees that illustrate the relationship among the aligned gene sequences are considered as gene trees. Whether these gene trees can be interpreted as representing the relationship among species depends on whether the genes provided for the alignment are truly orthologous, having evolved from a common ancestral gene by speciation.

16.2 Nuclear Ribosomal RNA Genes

The rRNA genes are the main genes traditionally used for phylogenetic studies of cyst nematodes. These include 18S, 28S and, especially, the internal transcribed spacer 1 (ITS1) and internal transcribed spacer 2 (ITS2), which are interspersed between the 18S, 5.8S and 28S rRNA genes, respectively. The ITS regions are subjected to a higher mutation rate, thus containing sufficient signals to resolve phylogenies within subfamilies or genera of cyst nematodes better than, for example, the 18S rRNA gene. The partial and whole ITS rRNA gene phylogenies of cyst nematodes have been published in many articles (Ferris, 1998; Ferris et al., 1993, 1995, 1998, 1999a, b, 2004; Sabo et al., 2001, 2002; Subbotin et al., 2001, 2006, 2010, 2017; Tanha Maafi et al., 2003; Madani et al., 2004; Sturhan et al., 2007; Mundo-Ocampo et al., 2008; Bernard et al., 2010; De Luca et al., 2013; Knoetze et al., 2013; Wang et al., 2013; Zhuo et al., 2013, 2014 and others) (Fig. 16.1). Phylogenies reconstructed using 18S rRNA gene sequences (Ferris et al., 2004; Mundo-Ocampo et al., 2008; van Megen et al., 2009; Bernard et al., 2010; De Luca et al., 2013) and the D2-D3 expansion fragment of 28S rRNA gene sequences (Subbotin et al., 2006, 2010; Mundo-Ocampo et al., 2008; Bernard et al., 2010; De Luca et al., 2013; Wang et al., 2013; Zhuo et al., 2013, 2014) were more appropriate to study relationships among subfamilies and genera of Heteroderidae.

16.3 Nuclear Protein-coding Genes

Nuclear protein-coding genes can serve as a rich source of genetic data for phylogenetic analysis of many organisms, including nematodes. Genomic, cDNA or amino acid sequences from the same gene locus provide flexibility in the information content that is available for phylogenetic reconstruction of cyst nematode relationships at the genus and species levels. Several nuclear protein-coding genes, including heat shock protein 90 (Skantar and Carta, 2004; Mundo-Ocampo et al., 2008), actin (Kovaleva et al., 2005a; Mundo-Ocampo et al., 2008; Toumi et al., 2013a), fructose-bisphosphate aldolase (Kovaleva et al., 2005b) and beta-tubulin (Sabo and Ferris, 2004) have also been used for characterization and reconstruction of phylogenies of cyst nematodes. Phylogenetic analysis of combined 47 protein-coding gene sequences recovered from the EST data of *Heterodera glycines*, *Globodera pallida* and *G. rostochiensis* with several species of root-knot nematodes was conducted by Scholl and Bird (2005). This study illustrated, for the first time, that genomic analyses using EST data mining methods can lead to interesting and useful results.

16.3.1 Heat shock protein 90

Heat shock protein 90 (*hsp90*) encodes a molecular chaperone protein that assists other proteins to fold properly, and serves as a buffer against deleterious variation, particularly under heat stress. Skantar and Carta (2004) used degenerate primers to obtain partial *hsp90* gene sequences from several plant-parasitic nematodes including *H. glycines*, and demonstrated that this gene was present in a single copy in the genome of this species. This is also the case for the model nematode *Caenorhabditis elegans*, in which *hsp90* is encoded by the *daf-21* gene (Birnby et al., 2000), as well as the animal-parasitic nematode *Brugia pahangi* (Devaney et al., 2005) and the pine wood nematode *Bursaphelenchus xylophilus* (Wang et al., 2012). Moreover, *hsp90* EST sequences were included in the multigene phylogeny of Scholl and Bird (2005), which was based upon several single-copy genes. *Hsp90* protein sequence alignments were used to construct higher order taxonomic trees, whilst genomic DNA and coding region nucleotide alignments resolved several species with high bootstrap support (Skantar and Carta, 2004). Mundo-Ocampo et al. (2008) and Skantar et al. (2012) characterized partial *hsp90* from several *Heterodera* species,

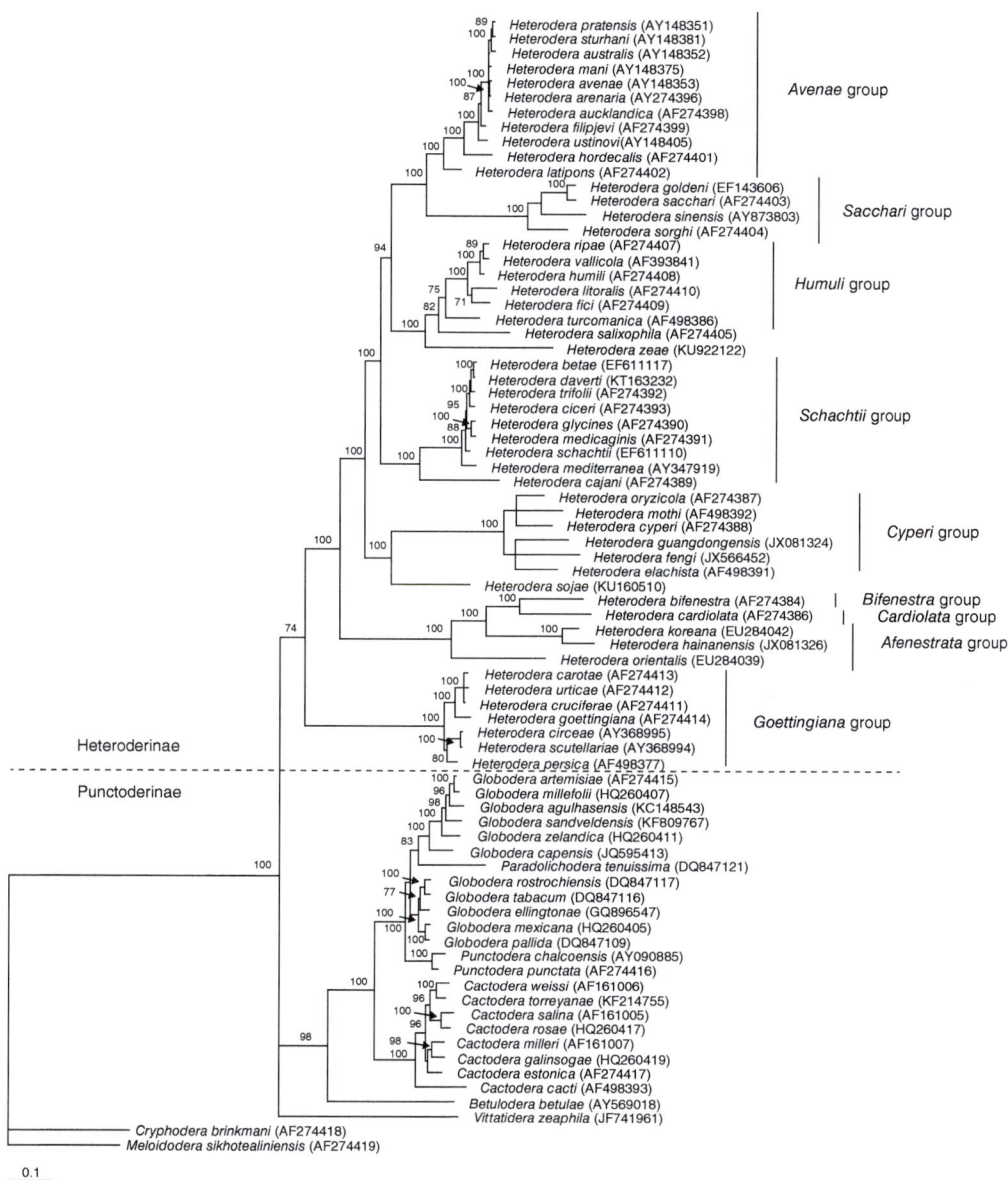

Fig. 16.1. Phylogenetic relationships within cyst nematodes as inferred from Bayesian analysis of the ITS1-5.8S-ITS2 rRNA gene sequence alignment under the GTR + I + G model. Posterior probabilities more than 70% are given for appropriate clades.

and Madani *et al.* (2011) provided some sequences for several *Globodera* species. Phylogenetic trees based on *hsp90* sequences showed equal resolution and in most cases were congruent with those inferred from ribosomal markers. The phylogenetic relationships of cyst nematodes are given in Figure 16.2.

16.3.2 Actin

Actin is a highly conserved structural, multifunctional protein that forms microfilaments and is ubiquitously expressed in eukaryotic cells. Multicellular eukaryotes contain multiple actin genes, the sequences of which are

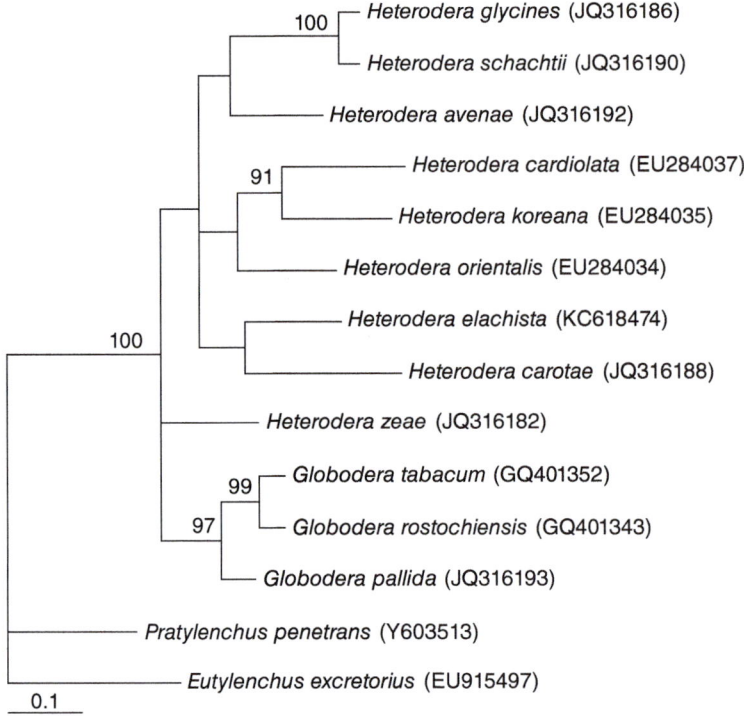

Fig. 16.2. Phylogenetic relationships within cyst nematodes as inferred from Bayesian analysis of the *hsp90* gene sequence alignment under the GTR + I + G model. Posterior probabilities more than 70% are given for appropriate clades.

highly conserved, with pairwise percentage identities of actins in the range of 88–98%. The existence of multiple actin genes in eukaryotes is thought to provide a higher copy number to cope with the demands for actin. The *C. elegans* genome encodes five actin genes (*act-1*, *act-2*, *act-3*, *act-4* and *act-5*). Kovaleva *et al.* (2005a) characterized three genes encoding actin from *H. glycines* and *G. rostochiensis*, and partial sequences of this gene from six other cyst nematodes. Actin sequence variation within the Heteroderidae ranged from 0.3 to 13.0%, allowing separation of species similar to that achieved by other ribosomal markers and *hsp90* (Mundo-Ocampo *et al.*, 2008). Partial sequences of actin genes from several species were also published by Toumi *et al.* (2013a). The actin phylogenetic tree for *Heterodera* is similar with those for ribosomal RNA genes (Fig. 16.3). Amino acid sequences were similar for the cyst nematodes, ranging from 99–100% identity, and apparently do not contain phylogenetic signals.

16.3.3 Fructose-bisphosphate aldolase

Fructose-bisphosphate aldolase, often just named as aldolase, is an enzyme catalysing a reversible cleavage of fructose 1,6-bisphosphate into the triose phosphates dihydroxyacetone phosphate and glyceraldehyde 3-phosphate. Two distinct types of cDNAs for aldolase, *ce-1* and *ce-2*, have been isolated from *C. elegans* (Inoue *et al.*, 1997). Kovaleva *et al.* (2005b) were the first to characterize full cDNA and genomic aldolase sequences for *H. glycines* and *G. rostochienis* and fragments of this gene from several other cyst nematodes. Based upon intron characteristics, it appeared that *ce-2* was more similar to the cyst nematodes aldolase genes than *ce-1*. In addition, *C. elegans ce-2* aldolase had a higher amino acid similarity to the cyst nematodes (72%) than does *ce-1* (60%). Whilst it has not been used extensively to date, the aldolase gene can be considered as a prospective candidate marker for reconstruction of phylogeny within cyst nematode subfamilies.

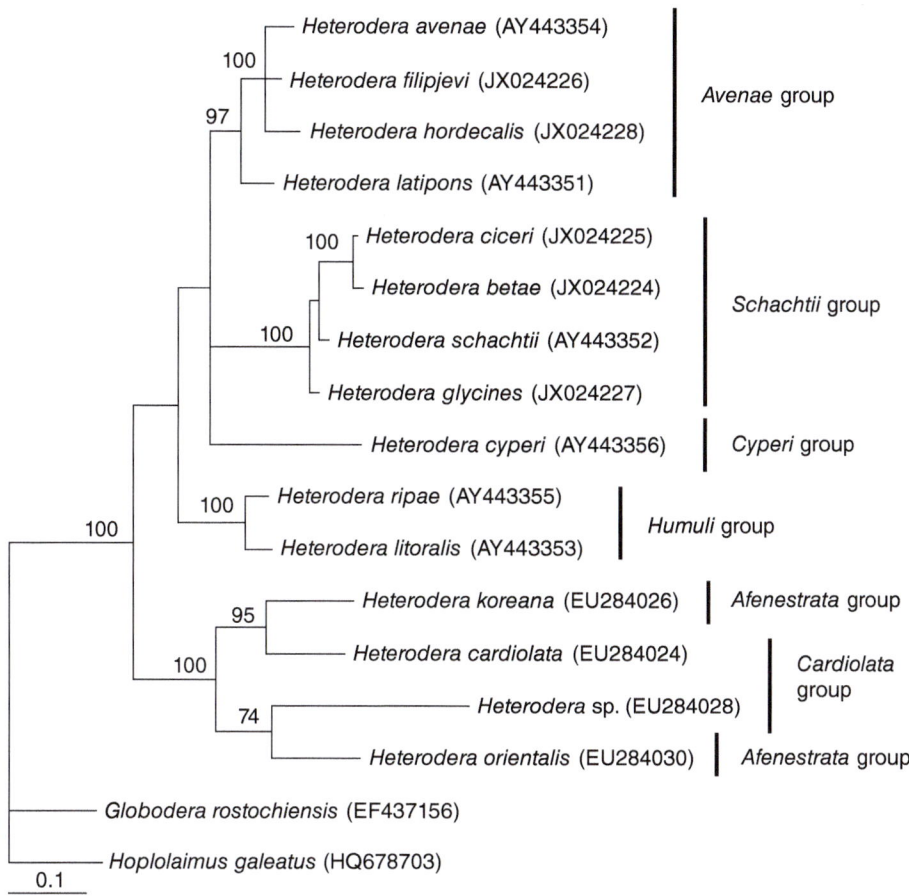

Fig. 16.3. Phylogenetic relationships within the genus *Heterodera* as inferred from Bayesian analysis of the actin gene sequence alignment under the GTR + I + G model. Posterior probabilities more than 70% are given for appropriate clades.

16.3.4 β-tubulin gene

Sabo and Ferris (2004) attempted to reconstruct the cyst nematode phylogeny using nuclear and amino acid sequences of β-tubulin gene. These studies resulted in an unresolved tree reconstructed from six putative paralogues of this gene, thus leading to the conclusion that this gene is not suitable for phylogenetic study of this nematode group.

16.4 Mitochondrial DNA Genome Organization

At present, mitochondrial genomes have been characterized in detail for only four cyst nematode species: *G. pallida* (Armstrong et al., 2000; Gibson et al., 2007a), *G. rostochiensis* (Gibson et al., 2007b), *G. ellingtonae* (Phillips et al., 2016) and *H. glycines* (Gibson et al., 2011). The studies revealed two types of mitochondrial DNA genome organizations for cyst nematodes: (i) one circular, double-stranded DNA molecule presently found in *H. glycines*; and (ii) several circular, double-stranded DNA molecules with different gene contents found in representatives of the genus *Globodera*. The mechanism by which the multipartite genome in *Globodera* arose is still uncertain.

The mitochondrial genome of *H. glycines* was characterized by Gibson et al. (2011), who estimated that the entire single circular mtDNA should be around 21–22 kb containing 12 protein-coding genes, two rRNA genes and 22 tRNA genes (Fig. 16.4A) The genome contained a major

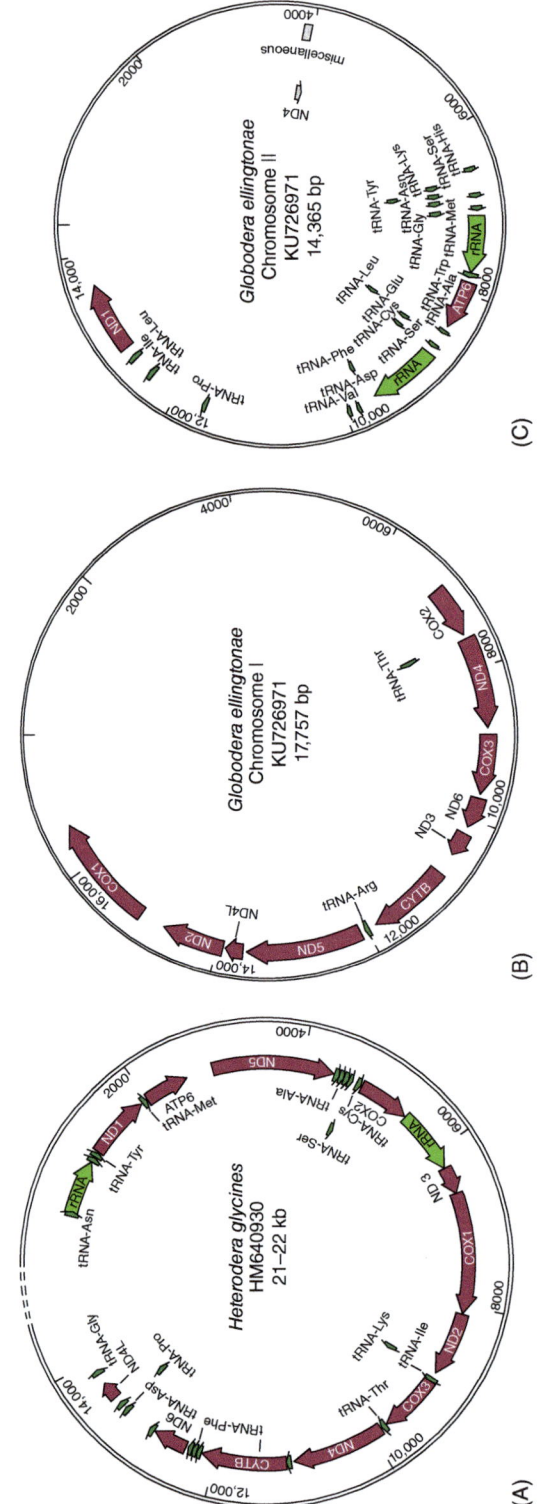

Fig. 16.4. Mitochondrial genome organization: *Heterodera glycines* (A) and chromosome I and II of *Globodera ellingtonae* (B and C, respectively).

non-coding and unsequenced region (between *trnG* and *trnN*), which was estimated to be 7–8 kb long. There were also smaller non-coding tracts between the *atp6* and *nad5* genes (330 nucleotides), the *trnR* and *trnD* genes (335 nucleotides), and between the *nad4L* and *trnG* genes (130 nucleotides). The organization of the mitochondrial genome of *Radopholus similis*, a migratory endoparasitic nematode, showed the highest level of similarity to that of *H. glycines*. Only minor differences were found when the *H. glycines* mtDNA genome was compared with partly sequenced mtDNA genomes of *H. cardiolata* and *Punctodera chalcoensis*.

Armstrong *et al*. (2000) were the first to analyse mitochondrial DNA genome organization for *G. pallida* mtDNA. The researchers provided evidence that this species was unusual among the metazoa and had a multipartite mtDNA structure or subgenomes. Small, circular mitochondrial DNAs (*Gpa*-scmtDNAs I-VI) ranging from 6.3 to 9.5 kb, have been discovered by polymerase chain reaction (PCR) in a British population of *G. pallida*, although additional components of the *G. pallida* mtDNA still remain uncharacterized. All of these *G. pallida* mtDNAs contained sequences similar to known mitochondrial genes, with most containing sequences that showed highest sequence similarity to previously described nematode mitochondrial genes. Armstrong *et al*. (2000) also discovered that these *Gpa*-scmtDNAs were present together in populations of *G. pallida*, although their relative frequencies may vary considerably between populations. The analysis revealed a degree of redundancy in gene content, suggesting that not all *Gpa*-scmtDNAs may be required to compose a functional mtDNA. For example, *Gpa*-scmtDNA I (*coxII*, *nad4*, *coxIII*, *nad6*, *nad1*, *nad3*) was found to include genes duplicated on *Gpa*-scmtDNAs II (*nad1*, *coxII*) and III (*nad3*, *cytb*). This was consistent with the observation that the Luffness population of *G. pallida* was found to contain only *Gpa*-scmtDNA I, and not *Gpa*-scmtDNA II and III, whereas the Gourdie population was found to contain *Gpa*-scmtDNAs II and III, with *Gpa*-scmtDNA I being detectable only by PCR (Armstrong *et al*., 2000). Later studies also revealed evidence for recombination of sequence variants of *Gpa*-scmtDNA IV for South American P4A population (Armstrong *et al*., 2000). Gibson *et al*. (2007a) also found that three of the *Gpa*-scmtDNAs (I, II and III) of *G. pallida* were mosaics, composed predominantly of multigenic fragments found on other scmtDNA molecules, and suggested that this mosaic pattern was an indication of the operation of inter-mt recombination.

Subgenomic organization was also found to occur in *G. rostochiensis*, which is a close relative of *G. pallida* (Gibson *et al*., 2007b). A comparison of subgenomic organization between these two *Globodera* species revealed a considerable degree of overlap between them, showing that the two subgenomes were identical to that reported for *G. pallida*. However, other subgenomes were unique to *G. rostochiensis* (*Gro*-scmtDNA VI and VII), although some of these have blocks of genes comparable to those in *G. pallida*. Comparisons of pairs of subgenomes from *G. rostochiensis* indicated that the different subgenomes shared fragments with high sequence identity. Gibson *et al*. (2007b) interpreted this as evidence that recombination is operating in the mitochondria of *G. rostochiensis*.

Several genes, *nad2*, *nad4L*, *nad5*, 12S rRNA and two tRNA, normally found in nematode mitochondrial genomes, were not identified in any of the five subgenomes completely sequenced in *G. pallida* (Armstrong *et al*., 2000; Gibson *et al*., 2007a). Similarly, *nad2*, *nad6*, 12S rRNA and two tRNA were also not found in any of the subgenomes sequenced from *G. rostochiensis* (Gibson *et al*., 2007b). These missing genes may therefore reside on other as yet uncharacterized subgenomes.

A recent study of *G. ellingtonae* revealed that its mitochondrial genome is unique and distinct from other multipartite mitochondrial genomes. The genetic content of the genome was disproportionately divided between the two circles, although they shared a ~6.5 kb non-coding region. The 17.8 kb circle (*Gel*-mtDNA-I) contained ten protein-coding genes and two tRNA genes, whereas the 14.4 kb circle (*Gel*-mtDNA-II) contained two protein-coding genes, 20 tRNA genes and both rRNA genes (Fig. 16.4B, C). The copy number of mtDNA-II was more than fourfold that of mtDNA-I in individual nematodes. The difference in copy number increased between second-stage and fourth-stage juveniles (Phillips *et al*., 2016).

The representatives of the genus *Punctodera* are closely related to those from the genus

Globodera. Punctodera chalcoensis was studied by Gibson *et al.* (2011), who found that its genome was larger than the multipartite mitochondrial genomes of *G. pallida* and *G. rostochiensis* and might also have minicircular mitochondrial genome organization.

16.4.1 Mitochondrial DNA genes

MtDNA genes have advantages over rRNA genes for studies of phylogenetic relationships at the genus and species level, due in part to the higher possibility of discovering intraspecific polymorphism resulting from faster accumulation of substitutions within mtDNA genes. The cytochrome c oxidase subunit I, or *coxI* gene, was successfully used by several researchers (Toumi *et al.*, 2013a, b; De Luca *et al.*, 2013; Vovlas *et al.*, 2015; Subbotin, 2015; Sekimoto *et al.*, 2017) for diagnostics and reconstruction of phylogenetic relationships within several groups of cyst nematodes. This gene is known as the standard molecular barcode for many animals. The *coxI* gene and other mtDNA genes are also presently used as markers for phylogeographical analysis of cyst nematode populations. For phylogeographical studies of *G. pallida* another mtDNA gene, cytochrome b, or *cytb* was also successfully applied (Picard *et al.*, 2007; Plantard *et al.*, 2008; Madani *et al.*, 2010; Eves-van den Akker *et al.*, 2015) (Fig. 16.5). It is interesting to note that intraspecific divergence in mtDNA sequences for cyst nematodes can reach 10%, which is higher than for other organisms. Such a high degree of polymorphism may also limit the ability of researchers to develop sets of universal primers for amplification of mtDNA genes in cyst nematodes.

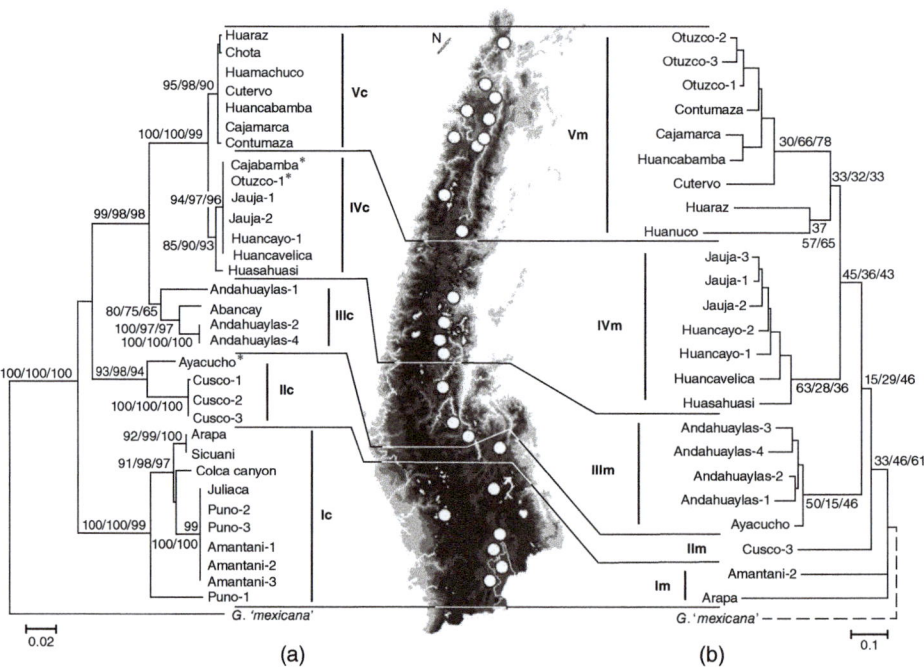

Fig. 16.5. Relationships between populations of *Globodera pallida* along the Andean Cordillera as inferred from the analysis of cytochrome *b* partial gene (a) and eight microsatellite loci (b). Asterisks (*) symbolize haplotypes located outside the clade corresponding to their geographical origin; bootstrap values are indicated for the nodes common to the tree analyses (neighbour-joining/maximum parsimony/maximum likelihood). (b) Neighbour-joining analysis, with an Infinite alleles model (IAM) genetic distance D_A, or Nei *et al.* (1983) distance, based on eight microsatellite loci; numbers by nodes represent the node between clades common to the three IAM genetic distance methods. (After Picard *et al.*, 2007.)

16.5 Origin and Phylogeny of Heteroderidae

The relationships of cyst nematodes with other Tylenchida have been the subjects of several studies. Paramonov (1967) was the first to propose that cyst nematodes might represent a particular phylogenetic lineage of hoplolaimids based upon feeding adaptations on host plant roots that appears to suggest a transition of this group to sedentary endoparasitism. This idea was later supported by Wouts and Sher (1971) and Krall and Krall (1978). Several evolutionary schemes that represent historical developments for cyst nematodes were proposed by Stone (1975, 1979), Krall and Krall (1978), Ferris (1979, 1985), Coomans (1979, 1983) and Baldwin and Schouest (1990) using morphological characters. Wouts (1985) was the first to develop a detailed phylogenetic classification of the family Heteroderidae, in which a putative cladogram consisted of six main lineages corresponding to six subfamilies: Verutinae, Meloidoderinae, Cryphoderinae, Heteroderinae, Ataloderinae and Punctoderinae. According to Wouts (1985), the hypothetical heterodrid ancestor developed from a nematode that resembled modern members of the family Hoplolaimidae and was possibly closely related to *Rotylenchulus*. The genus *Verutus* was considered the most primitive genus within heterodrids and was characterized by a very large equatorial vulval slit. The genus *Meloidodera* was considered to be further derived by a reduction in the length of the vulva. Genera, interpreted as developing later, exhibited a subterminal vulval slit and progressive loss of annulation of the female cuticle. Wouts (1985) also proposed that genera that developed later exhibit a subterminally located vulval slit and progressively lost the annulation of the female cuticle. In this process four evolutionary lines emerged: (i) a posterior shift of the vulva and the formation of more or less distinct vulval lips gave rise to the genera *Zelandodera* and *Cryphodera*; (ii) changes in the lip configuration of the second-stage juvenile gave rise to the genera *Hylonema* and *Heterodera*; (iii) changes in the composition of the female cuticle resulted in the genera *Thecavermiculatus*, *Atalodera*, *Sherodera*, *Sarisodera* and *Bellodera*; and (iv) a reduction in vulval slit size led to the development of the genera *Dolichodera*, *Globodera*, *Cactodera* and *Punctodera*. Wouts's phylogeny proposes independent development of cysts in two different lineages.

New insight into the phylogeny of Heteroderidae has been provided by the analyses of ribosomal RNA gene sequences as represented by Ferris (1998), Ferris *et al.* (1999a, b), Sabo *et al.* (2001), Subbotin *et al.* (2001, 2006, 2010), Tanha Maafi *et al.* (2003), Sturhan *et al.* (2007), Mundo-Ocampo *et al.* (2008), Bernard *et al.* (2010), De Luca *et al.* (2013), Zhuo *et al.* (2014) and others. Molecular phylogeny strongly supports close relationships between Heteroderidae, Hoplolaimidae and Rotylenchulidae and the hypothesis of Wouts (1985) that *Verutus* and *Rotylenchulus* are placed in the basal positions to all sedentary nematodes of Heteroderidae (Subbotin *et al.*, 2010). Molecular phylogenies also support: (i) the division of cyst nematodes into two subfamilies, the Heteroderinae and Punctoderinae; (ii) the monophylies of *Cactodera*, *Punctodera*, *Globodera* and *Heterodera*; and (iii) the validities of *Paradolichodera*, *Betulodera* and *Vittatidera*. The molecular analysis indicates that cyst formation appeared only once during the evolution of Heteroderidae.

16.5.1 Phylogeny of Heteroderinae

The subfamily Heteroderinae includes only a single genus *Heterodera*. Historically, species of the genus *Heterodera* were separated into three groups based on vulval cone structures, that is, the *Schachtii*, *Goettingiana* and *Avenae* groups (Mulvey and Golden, 1983; Baldwin and Mundo-Ocampo, 1991). Stone (1975) also recognized several groups based on juvenile lip morphology. Several additional, partly overlapping, groups have also been proposed: *Humuli* (Mathews, 1971; Subbotin *et al.*, 1997), *Fici-humuli* (Ferris, 1979), *Oryzae*, *Cruciferae* and *Graminis* (Stone, 1979), *Latipons* (Wouts and Sturhan, 1995) and *Bifenestra* (Sturhan, 2006). Subbotin *et al.* (2001), considering a combination of molecular data with the morphology of vulval structures and the number of incisures in the lateral field of juveniles, supported the recognition of the *Schachtii*, *Goettingiana*, *Avenae* and *Humuli* groups, albeit with a modified species composition and the erection of two new groups, *Cyperi* and *Sacchari*.

After synonymization of the genus *Afenestrata* with the genus *Heterodera*, the *Afenestrata* group was also proposed (Mundo-Ocampo et al., 2008). Morphological and molecular analysis further supports the division of *Heterodera* into more or less distinct species groups recognized as follows: *Afenestrata, Avenae, Bifenestra, Cardiolata, Cyperi, Goettingiana, Humuli, Sacchari* and *Schachtii*. Although most valid species can be accommodated in one of the above-mentioned groups, the position of some *Heterodera* species (e.g. *H. salixophila* and *H. zeae*) remains undefined or uncertain (Subbotin et al., 2010).

There are several hypotheses proposed about the earliest group to arise within the genus *Heterodera*: amphimictic species *H. schachtii, H. glycines* and, perhaps, *H. salixophila* (Krall and Krall, 1978); or a *H. cruciferae*-like form parasitizing dicots (Stone, 1979); or the cyst-forming species previously placed in *Afenestrata* (Ferris, 1979; Wouts, 1985). The phylogenetic analyses presented by Subbotin et al. (2001, 2010), Tanha Maafi et al. (2003) and De Luca et al. (2013) resolved the basal relationships within *Heterodera*, and usually placed the *Goettingiana* group (*H. goettingiana, H. carotae, H. urticae, H. cruciferae, H. scutellariae, H. circeae* and *H. persica*) as an earliest diverged group within the genus. Obviously, the molecular data lend significant support to Stone's (1979) hypothesis. Although relationships between heteroderid groups are not well resolved, nevertheless four groups, *Avenae, Sacchari, Schachtii* and *Humuli*, are often placed in derived positions in most phylogenetic trees. The *Avenae* and *Sacchari* groups have a strong sister relationship in all trees. The relationships between *Avenae + Sacchari, Schachtii* and *Humuli* groups remain unresolved, indicating the possible rapid evolutionary radiation of these nematode lineages.

Some inconsistencies between molecular phylogeny and previously proposed phylogenetic hypotheses or groupings may be attributed to homoplastic evolution. For example, according to molecular analysis a bifenestral vulval cone developed independently in three groups, *Avenae, Humuli* and *Bifenestra*, during the evolution of cyst nematodes. Likewise, the presence of three incisures in the lateral field of second-stage juveniles of the cyst nematodes seems to have also arisen three times independently (*Afenestrata, Cyperi* and *Sacchari* groups). Finally, a short vulval slit developed independently in the *Avenae* and *Bifenestra* groups within Heteroderinae.

The molecular data suggest an early divergence between tropical and temperate heteroderid species. Krall and Krall (1978) stated that the centre of origin for any organism group is likely to be the extant area with the highest species diversity. On this basis they suggested that the Mediterranean and Central Asia regions could be the centre of origin for the genus *Heterodera*. However, the analysis of presently known patterns of geographical distributions of the species occupying basal positions (*H. cajani, H. mothi, H. zeae, H. latipons* and *H. persica*) in phylogenetic trees from different groups suggest that most probably a centre of origin of *Heterodera* is in south and western Asia, although the *Afenestrata* group may have originated from the east Asia region.

Phylogenetic relationships of species from the genus *Heterodera* have also been intensively studied within the groups, especially within *Avenae* and *Schachtii* groups, to which the most agricultural important species belong. Phylogenetic relationships with the *Avenae* group were analysed using the ITS-rRNA gene sequences described in detail by Subbotin et al. (2003). When *H. latipons* and *H. hordecalis* were used as outgroup taxa, the species having cysts with an underbridge (*H. ustinovi* and *H. filipjevi*) occupied basal positions in the tree. Other species were distributed into two main clades: (i) *H. avenae, H. arenaria, H. aucklandica* and *H. mani*; and (ii) *H. australis, H. pratensis* and *H. sturhani*. However, the ITS-rRNA gene sequences did enable *H. avenae* to be distinguished from *H. arenaria*, and *H. pratensis* from *H. sturhani*. Partial *coxI* mtDNA gene sequences provided better discrimination and resolution for species relationships. Phylogenetic relationships between cyst nematode species of the *Avenae* group as inferred from the analysis of *coxI* mtDNA gene sequences were recently analysed by Subbotin (2015). This gene revealed distinct differences between *H. australis* and *H. avenae*, between *H. pratensis* and *H. sturhani*, and between European and Asian populations of *H. avenae* (type A and type B). The partial *coxI* mtDNA gene sequences also discriminate *H. avenae* type A from *H. arenaria*.

Vovlas et al. (2015) recently presented phylogenetic relationships between several species of the *Schachtii* group as inferred from the analysis of the ITS-rRNA gene sequences. The tropical species *H. cajani* is distributed in India, Pakistan and Egypt, and occupies a basal position in the *Schachtii* group. *Heterodera glycines* had sister relationships with *H. medicaginis*, whereas relationships between *H. trifolii*, *H. daverti*, *H. betae* and *H. ciceri* were not well resolved. It is interesting that some ITS-rRNA clones of *H. schachtii* were also clustered with these species, possibly indicating incomplete lineage sorting within species of this group. Madani *et al.* (2007) showed that intraspecific divergence for the ITS-rRNA gene sequence can reach 2.5% for *H. schachtii*, whereas for other cyst nematode species it does not exceed 1.6–1.8% (Subbotin *et al.*, 2000; Tanha Maafi *et al.*, 2003; Madani *et al.*, 2004). Recent analysis showed that the *coxI* mtDNA gene sequences provided better resolution for the study of relationships within this group (Vovlas *et al.*, 2015; Sekimoto *et al.*, 2017).

Phylogenetic relationships within species of other groups were analysed by Mundo-Ocampo *et al.* (2008) for *Afenestrata* group, Tanha Maafi *et al.* (2007) for *Sacchari* group, Eroshenko *et al.* (2001) for *Humuli* group and Zhuo *et al.* (2014) for *Cyperi* group.

16.6 Phylogeny and Phylogeography of Punctoderinae

The subfamily Punctoderinae constitutes six genera, *Betulodera*, *Cactodera*, *Dolichodera*, *Globodera*, *Paradolichodera* and *Punctodera*, which share the character of possessing a vulval circumfenestra. Although Siddiqi (1986, 2000) and Luc *et al.* (1988) did not accept the subdivision of cyst nematodes into two subfamilies, Heteroderinae and Punctoderinae, the phylogenetic analyses of rRNA gene sequences gave evidence that circumfenestrate nematodes represent one of two separate major lineages within cyst nematodes (Subbotin *et al.*, 2001, 2010, 2011; Tanha Maafi *et al.*, 2003).

Phylogenetic relationships among circumfenestrate cyst nematodes were inferred from analyses of ITS-rRNA gene sequences in studies conducted by several researchers (Ferris *et al.*, 1999a, b, 2004; Subbotin *et al.*, 2000, 2001; Sabo *et al.*, 2002; Skantar *et al.*, 2007; Bernard *et al.*, 2010; Madani *et al.*, 2010; Handoo *et al.*, 2012; Lax *et al.*, 2014). The molecular phylogenetic tree of Punctoderinae consists of six major clades corresponding to the division into genera (Fig. 16.1). Monophylies of *Punctodera*, *Cactodera* and *Betulodera* were always highly supported, whereas monophyly of *Globodera* was not always evident. It has been previously remarked that the observed paraphyly of *Globodera* may simply reflect the general tendency of phylogenetic algorithms to produce unbalanced trees rather than to define any true evolutionary history of a group with high rates of evolution. Genetic divergence within this genus is reflected by the species groupings based on geographical origin and host plant speciation (Subbotin *et al.*, 2001). *Betulodera* with *Vittatidera* clades occupy basal positions on all trees. Relationships of *Paradolichodera* and *Punctodera* with other genera are not well resolved and may vary among trees (Subbotin *et al.*, 2011).

The phylogeographic analysis using rRNA gene sequences was used to test possible origin and patterns of dispersal of circumfenestrate cyst nematodes. Subbotin *et al.* (2011, 2017) suggested that North America appeared to be a centre of early evolution and the area from which subsequent dispersal of Punctoderinae occurred. Indeed, in North America the greatest number of related circumfenestrate cyst nematode species and genera occur: 12 of the 13 known *Cactodera* species, three of four known *Punctodera*, known *Betulodera*, *Dolichodera* and *Vittatidera* species and several *Globodera* species. Of six valid Punctoderinae genera, only *Paradolichodera* was described from New Zealand and has never been recorded in North America or in any other region of the world.

Another proposed approach to determine a centre of origin is to identify the area in which the most evolutionarily primitive representatives, closest in form to the supposed ancestral group, occur, with the assumption that they are not likely to have dispersed far from the centre of origin. Phylogenetic analyses using morphological (Krall and Krall, 1978; Wouts, 1985; Baldwin and Schouest, 1990) and molecular (Subbotin *et al.*, 2006, 2010) datasets revealed that extant Ataloderinae and Meloidoderinae share common ancestors with cyst nematodes. The species of these subfamilies parasitize a rather wide spectrum of plants, including gymnosperms and angiosperms. The ability of some species of *Meloidodera*, *Cryphodera* (Meloidoderinae) and

Rhizonemella sequoiae (Ataloderinae) to parasitize plants from *Pinus* and *Sequoia*, respectively, is consistent with the argument of Krall and Krall (1978) that suggests an ancient origin for these genera in the New World. The distribution of extant Meloidoderinae includes North America, Europe, Asia and Oceania, whilst most representatives of Ataloderinae are found in North America (*Ekphymatodera, Sarisodera, Rhizonemella, Bellodera* and *Atalodera*). However, some species of *Atalodera* are not restricted to North America and have been described from South America and Asia. In addition, only two Ataloderinae genera, *Hylonema* and *Camelodera*, are reported from Africa and Asia, respectively. Thus, the general pattern of modern distribution of non-cyst nematodes sharing a common ancestor with Punctoderinae favours the hypothesis of a North American origin of circumfenestrate cyst nematodes. Although *Cactodera* and *Punctodera* putatively originated and began to diversify in North America, results of the phylogeographical analysis, in which several new findings of undescribed species of *Globodera* from New Zealand and South Africa were included, revealed that South America or Africa appears to be a centre of origin of *Globodera*. Thus, Stone's (1979) hypothesis of an 'out-of-the-west' Gondwana origin of *Globodera* with subsequent dispersal of the species of this genus to Europe, North America, Asia and Oceania found some support from the rRNA datasets.

The absence of a fossil record together with unequal rates of rRNA gene evolution do not allow reliable application of molecular clock approaches to estimate divergence time for different cyst nematode lineages. The ages of cyst nematode divergences have been estimated using different approaches. Picard *et al.* (2008) considered the age of the *Cactodera* – (*Punctodera* + *Globodera*) divergence to be close to the Heteroderinae–Punctoderinae divergence, which might have occurred after separation of Laurasia and Gondwana, that is, 173–130 million years ago. The divergence of the South American and Laurasian *Globodera* lineages was considered to have occurred between 80 and 60 million years ago after the break of a temporary connection between North and South America in the Palaeocene. However, if we consider the estimated ages of host plants presently associated with non-cyst and cyst nematodes, slightly younger dates of divergence between nematode lineages could be suggested. For example, the age of Meloidoderinae associated with *Pinus* could be proposed as the time when *Pinus* diverged from the other genera, that is, 140 million years ago. For Ataloderinae parasitizing *Sequoia*, it might be the late Cretaceous (99–65 million years ago) and Tertiary, periods from which *Sequoia* fossil records are primarily known. The late Cretaceous was also considered as a possible time period for origin, diversification and spread of various Poaceae (Prasad *et al.*, 2005), which are reported to be hosts for some Ataloderinae. From this line of reasoning one could speculate on the Heteroderinae–Punctoderinae divergence occurring between 90 and 70 million years ago. The divergence of the two main *Globodera* lineages is associated with the time of origin for the Solanaceae, that is, 65–51 million years ago (Wikström *et al.*, 2001; Paape *et al.*, 2008). Thus, the split of the two *Globodera* lineages might have occurred subsequent to the breakup of Gondwana and the Africa and South America split in the Mid-Cretaceous, and it seems that the evolution of Punctoderinae cannot be explained solely by the separation of the continents and diffusion expansion. Recent advances in phylogenetics and, in particular, molecular dating, indicate that transoceanic dispersal has played an important role in shaping plant and animal distributions, thereby obscuring any effect of tectonic history (Vanderpoorten *et al.*, 2010). Subbotin *et al.* (2011) also suggested that 'jump dispersal' might play a significant role in the biogeographic history of cyst nematodes and hypothesized that long-distance dispersal might even have occurred repeatedly during cyst nematode evolution. This idea suggested a scenario whereby a modern *Globodera* lineage arose from introduction of the ancestral Punctoderinae to Gondwana via long-distance dispersal from North America.

16.7 Phylogeny and Phylogeography of *Globodera*

The genus *Globodera* displays two main clades in phylogenetic trees: (i) *Globodera* from South and North America parasitizing plants from Solanaceae; and (ii) *Globodera* from Africa, Europe,

Asia and New Zealand parasitizing plants from Asteraceae and other families. The first main clade includes the first subclade with *G. pallida* and *G. mexicana* and the second one with *G. ellingtonae*, *G. tabacum* and *G. rostochiensis*. The second main clade consists of three subclades: (i) *G. capensis* from South Africa; (ii) *G. zelandica* and two undescribed species from New Zealand; and (iii) *G. artemisiae*, *G. millefolii* from Europe and Asia, two recently described species from South Africa and one undescribed species from Portugal. It has been hypothesized that centres of diversification and speciation in each main clade are associated with mountain regions (Subbotin *et al.*, 2016). The centre of diversification for *Globodera* parasitizing Solanaceae occurs in the Andes (Grenier *et al.*, 2010), known as one of 35 world biodiversity hotspots. The *Globodera* belonging to the second main clade have several centres of diversification: mountains of the Western Cape in South Africa, South Island mountains in New Zealand and, perhaps, mountain ranges of Portugal (Subbotin *et al.*, 2016).

The pale potato cyst nematode, *G. pallida*, is considered a major worldwide pest of potatoes. Several studies revealed that this species originated from South America, from where it has been introduced to many parts of the world, particularly to Europe, Asia, and also to the USA, Canada and New Zealand. In South America, this cyst nematode is mainly found between 2000 and 4000 m above sea level, with the heaviest infestations between 2900 and 3800 m elevation (Franco, 1977). Distinct genetic differences among *G. pallida* populations spanning Europe and those found in South America were revealed by sequence and phylogenetic analyses of the ITS-rRNA (Blok *et al.*, 1998; Subbotin *et al.*, 2000, 2011; Madani *et al.*, 2010), *cytb* (Picard *et al.*, 2007; Plantard *et al.*, 2008; Pylypenko *et al.*, 2008; Madani *et al.*, 2010), PCR-ITS-RFLP (Ayub and Rumpenhorst, 2000; Grenier *et al.*, 2001), simple sequence repeat primer analysis (Blok and Phillips, 1995), PCR-RAPD (Blok *et al.*, 1997; Bendezu *et al.*, 1998; Grenier *et al.*, 2001; Rumpenhorst and Ayub, 2001), isoelectric focusing (IEF) of proteins (Rumpenhorst and Ayub, 2001), 2-DGE of proteins (Grenier *et al.*, 2001) and satellite DNA analysis (Grenier *et al.*, 2001; Plantard *et al.*, 2008). The analysis of partial *cytb* gene sequences of *G. pallida* collected in different regions allowed Plantard *et al.* (2008) to identify the origin of western European populations with a high degree of certainty (Fig. 16.5). This analysis showed that all of these populations originated from a single restricted area in the extreme south of Peru, located between the north shore of the Lake Titicaca and Cusco. Plantard *et al.* (2008) found that only four *cytb* haplotypes are reported in western Europe, one of them also being found in some populations of this area of southern Peru. After studying the USA and Canadian *G. pallida*, Madani *et al.* (2010) concluded that they belong to the western European haplotype and, thus, North American populations resulted from the continued spread of *G. pallida* from western Europe to other countries and continents, and were unlikely to be the result of a separate introduction to North America directly from South America.

The highest ITS-rRNA and *cytb* gene sequence diversity for *G. pallida* was reported in Peru and the Andes. Picard *et al.* (2007) used 42 populations sampled along a 1500-km north–south transect in Peru genotyped with a partial *cytb* and seven nuclear microsatellite loci, and described a clear phylogeographical pattern among Peruvian populations revealing five distinct clades (Fig. 16.5). The clade containing the southern populations was genetically more diverse and forms the most basal branch. The large divergence among *cytb* haplotypes suggested that they diverged before human domestication of potato. Investigations by Picard *et al.* (2007) also clearly illustrated a northward expansion of populations from the south of Peru (around Lake Titicaca) to the north. It has been hypothesized that this south-to-north pattern took place during the uplift of the Andes beginning 20 million years ago and followed the same direction, and reflected the colonization of progressively emerging favourable areas of the Andes by wild potatoes and the co-evolution of *G. pallida* with its host plant (Picard *et al.*, 2007; Grenier *et al.*, 2010). Grenier *et al.* (2010) also suggested that northward from Peru (i.e. in Ecuador and Columbia) this pattern of decreasing genetic variability and speciation might continue. It is also noted that *G. mexicana*, exhibiting a lower genetic variability, was only reported in Mexico. Under the assumption of a northward expansion from south Peru of *G. pallida* populations on wild solanaceous hosts, Grenier *et al.* (2010) hypothesized that *G. mexicana* would represent a distinct speciation event that was initiated through G. pallida. Close relationships of G. mexicana with representatives of one of the G. pallida clade have been strongly supported by the phylogenetic analysis of the ITS-rRNA gene sequences (Subbotin et al., 2011).

Phylogenetic analysis with limited samplings for representatives of the second subclade of Globodera parasitizing Solanaceae also shows a pattern of their northward expansion in the Andes. The potato cyst nematode G. ellingtonae, recently described from the USA and evidently introduced from a central Andes location (Grenier et al., 2010; Handoo et al., 2012; Lax et al., 2014), occupies a basal position in the subclade, whereas G. tabacum and G. rostochiensis, having a sister relationship, are distributed more in the north.

The golden cyst nematode G. rostochiensis, native to South America, has been introduced in many parts of the world, including Europe and North America. Boucher et al. (2013) analysed 12 new microsatellite markers in order to characterize the genetic links between 15 globally distributed G. rostochiensis populations, including the populations found in Quebec. The results revealed that the five populations from South America, especially those from Bolivia, had a higher genetic diversity than those originating from Europe and North America. The results also indicated that the Bolivian populations were distinct from other populations and suggested that a minimum of two introductions with different origins would have to have occurred in North America: (i) one in Quebec and/or Newfoundland directly or indirectly from European populations; and (ii) one in British Columbia and/or New York from a currently unidentified location, as inferred by the fact that these populations make up an independent genetic cluster.

The tobacco cyst nematode G. tabacum is a polytypic species containing the following subspecies: G. tabacum tabacum, G. tabacum virginiae and G. tabacum solanacearum, which are poorly differentiated by morphology although separable by host preference. Relationships within this complex were not resolved using the ITS-rRNA gene sequence analysis; however, PCR-RAPD (Thiéry et al., 1997) and AFLP (Marché et al., 2001) techniques, and sequences of genes coding CLE peptides and cell wall degrading enzymes (Alenda et al., 2013) were able to differentiate the subspecies. Moreover, Marché et al. (2001) recognized a fourth subspecies within G. tabacum from Mexico, named G. tabacum 'azteca'; however, Alenda et al. (2013) later discarded this subspecies. The analysis of six genes did not support the fourth subspecies as all the Mexican populations clustered either with G. tabacum virginiae or G. tabacum solanacearum.

Picard et al. (2008) estimated divergence dates using nuclear ribosomal (ITS1-5.8S-ITS2) and cytb sequences, and proposed that G. tabacum most probably diverged from G. rostochiensis 24.7–18.5 million years ago (Late Oligocene–Early Miocene), and that G. pallida diverged from G. mexicana 20.6–15.4 million years ago (Early Miocene). Within G. pallida, basal divergence of clades of the cytb dendrogram occurred at some time in the 17.8–13.4 million years ago (Middle Miocene) and 11.8–8.8 million years ago (Late Miocene) intervals. Alenda et al. (2013) assumed that subspecies of G. tabacum appeared less than 15 million years ago.

One of the important practical implications of these phylogeographical studies is that the reference sequence databases of cytb, ITS-rRNA and other genes for Globodera populations parasitizing Solanaceae have been created. These databases comprise a valuable resource for accurate identification of suspected new Globodera from quarantine samples, providing early detection that could prevent introduction and spread of new nematode isolates or species. As was the case for G. ellingtonae, it is possible that new nematodes would be genetically very divergent from known species, requiring additional plant-resistance sources to control them, and would enhance the already high adaptive potential of established local Globodera populations.

16.8 Co-evolution of Cyst Nematodes with their Host Plants

Krall and Krall (1978) and Krall (1990) argued the importance of a comparative ecological approach to study the evolution of cyst nematodes. They proposed a hypothesis of co-evolution of heteroderids with plants using the concept of flowering plant evolution developed by

Grossheim (1945). Additional aspects of hypotheses of co-evolution of cyst nematodes and their plant hosts were developed and discussed by Stone (1979, 1985) and Sturhan (2000). Molecular phylogeny of cyst nematodes allowed Subbotin et al. (2001, 2010) to evaluate the hypothesis of co-evolution of heteroderids with their hosts as proposed by Krall and Krall (1978) and Stone (1979). The results tended to support the idea that different cyst nematode groups co-evolved with hosts belonging to single or closely related families of plants. Nevertheless, some species were able secondarily to colonize ecologically convergent plant species from unrelated families (Subbotin et al., 2010).

In relation to the taxonomy according to current plant classification (Angiosperm Phylogeny Group, 2016), cyst nematode species have been described from the type plant hosts belonging to the following orders: the genus *Heterodera* – Poales (46 species), Fabales (10), Caryophyllales (5), Rosales (7), Lamiales (4), Mapighiales (3), Apiales (2), Brassicales (2), Saxifragales (1), Myrtales (1), Fagales (1), Sapindales (1), Asterales (1); the genus *Globodera* – Solanales (7), Asterales (2), Rosales (1), Myrtales (1), Caryophyllales (1); the genus *Cactodera* (14 species) – Caryophyllales (11), Asterales (1), Poales (1); the genus *Punctodera* – Poales (4); the genus *Dolichodera* – Poales (1); the genus *Paradolichodera* – Poales (1); the genus *Betulodera* – Fagales (1) and *Vittatidera* – Poales (1). Thus, half of known cyst nematode species have been described from plants in the order Poales, one of the largest orders of flowering plants. Six main lineages are evident within *Heterodera* – three of them with monocotyledons and three with dicotyledons. The *Goettingiana* group includes species parasitizing plants belonging to various taxonomic groups of dicotyledons. The *Sacchari*, *Avenae*, *Afenestrata*, *Bifenestra*, *Cardiolata* and *Cyperi* groups all co-evolved with Poales. The *Schachtii* group primarily co-evolved with plants from the order Fabales and then colonized Lamiales, Caryophyllales and Asterales. Another lineage includes species of the *Humuli* group associated with hosts from Caryophyllales and Rosales. Two species related to this group but different in morphology, *H. salixophila* and *H. zeae*, parasitize plants from Malpighiales and Poales, respectively. It is plausible to hypothesize that some of these species are associated with changes in host specialization relatively recently, jumping to phylogenetically unrelated but ecologically similar host plants

Molecular phylogenetics further suggest that within Punctoderinae there are several distinct lineages: (i) basally derived *Betulodera* having hosts from Fagales; (ii) *Vittatidera* parasitizes Poales; (iii) *Cactodera* associating with Caryophyllales (Amaranthaceae, Caryophyllaceae, Portulacaceae) and, perhaps, relatively recently with Asterales and Poales; (iv) *Punctodera* and *Paradolichodera* parasitizing Poales; (v) *Globodera* having hosts from Asterales and Myrtales; and (vi) *Globodera* parasitizing Solanales (Subbotin et al., 2010).

16.9 Conclusions and Future Prospects

In recent years, genome-scale data or phylogenomics has become an increasingly powerful tool for phylogenetic inference of many organisms and for resolving difficult phylogenetic and phylogeographic questions; however, these approaches have not yet found widespread practical application for the study of cyst nematode evolution. The main strength of phylogenomics is the drastic reduction of errors in tree inference that may arise from a single gene dataset when replaced by the use of large multigene datasets. Genome data afford a much broader selection of genes that are appropriate for phylogenetic analysis, and reduction of non-phylogenetic signals from the large multigene datasets is a challenging area of future research, which enables difficult issues of relationships in these nematodes to be resolved.

Another topic of molecular phylogenetics that has been recently influenced by population genetics is species delimitation by tree-based and non-tree-based methods that rely upon molecular sequence data. A robust hypothesis testing for species limits that incorporates information on the rRNA and mtDNA genes together with application of coalescent theory, which deals with complex models incorporating phenomena such as migration, selection and recombination (Rosenberg and Nordborg, 2002), has enormous promise.

16.10 References

Armstrong, M.R., Blok, V.C. and Phillips, M.S. (2000) A multipartite mitochondrial genome in the potato cyst nematode *Globodera pallida*. *Genetics* 154, 181–192.

Alenda, C., Gallot-Legrand, A., Fouville, D. and Grenier, E. (2013) Sequence polymorphism of nematode effectors highlights molecular differences among the subspecies of the tobacco cyst nematode complex. *Physiological and Molecular Plant Pathology* 84, 107–114. DOI: 10.1016/J.PMPP.2013.08.004

Angiosperm Phylogeny Group (2016) An update of the Angiosperm Phylogeny Group classification for the orders and families of flowering plants: APG IV. *Botanical Journal of the Linnean Society* 181, 1–20. DOI: 10.1111/boj.12385

Ayub, M. and Rumpenhorst, H.J. (2000) Genetic variation in potato cyst nematode (*Globodera pallida*) from Europe and South America as revealed by PCR-RFLP of ITS regions. In: *Abstracts der 28 Tagung des Arbeitskreises Nematologie, 1–2 März 2000, Veitshöchheim*, pp. 36–37.

Baldwin, J.G. and Mundo-Ocampo, M. (1991) Heteroderinae, cyst and non-cyst-forming nematodes. In: Nickle, W.R. (ed.) *Manual of Agricultural Nematology*. Marcel Dekker Inc., New York, pp. 275–362.

Baldwin, J.G. and Schouest Jr, L.P. (1990) Comparative detailed morphology of the Heteroderinae Filipjev and Schuurmans Stekhoven, 1941, *sensu* Luc *et al.* (1988): phylogenetic systematics and revised classification. *Systematic Parasitology* 15, 81–106.

Bendezu, I.F., Evans, K., Burrows, P.R., De Pomerai, D. and Canto-Saenz, M. (1998) Inter and intraspecific variability of the potato cyst nematodes *Globodera pallida* and *G. rostochiensis* from Europe and South America using RAPD-PCR. *Nematologica* 44, 49–61. DOI: 10.1163/005225998X00064

Bernard, E.C., Handoo, Z.A., Powers, T.O., Donald, P.A. and Heinz, R.D. (2010) *Vittatidera zeaphila* (Nematoda: Heteroderidae), a new genus and species of cyst nematode parasitic on corn (*Zea mays*). *Journal of Nematology* 42, 139–150.

Birnby, D.A., Link, E.M., Vowels, J.J., Tian, H., Colacurcio, P.L. and Thomas, J.H. (2000) A transmembrane guanylyl cyclase (DAF-11) and Hsp90 (DAF-21) regulate a common set of chemosensory behaviors in *Caenorhabditis elegans*. *Genetics* 155, 85–104.

Blok, V.C. and Phillips, M.S. (1995) The use of repeat sequence primers for investigating genetic diversity between populations of potato cyst nematodes with differing virulence. *Fundamental and Applied Nematology* 18, 575–582.

Blok, V.C., Phillips, M.S. and Harrower, B.E. (1997) Comparison of British populations of potato cyst nematodes with populations from continental Europe and South America using RAPDs. *Genome* 40, 286–293. DOI: 10.1139/g97-040

Blok, V.C., Malloch, G., Harrower, B., Phillips, M.S. and Vrain, T.C. (1998) Intraspecific variation in ribosomal DNA in populations of the potato cyst nematode *Globodera pallida*. *Journal of Nematology* 30, 262–274.

Boucher, A.C., Mimee, B., Montarry, J., Bardou-Valette, S., Bélair, G., Moffett, P. and Grenier, E. (2013) Genetic diversity of the golden potato cyst nematode *Globodera rostochiensis* and determination of the origin of populations in Quebec, Canada. *Molecular Phylogenetics and Evolution* 69, 75–82. DOI: 10.1016/j.ympev.2013.05.020

Coomans, A. (1979) General principles of systematics with particular reference to speciation; 1–19. In: Lamberti, F. and Taylor, C.E. (eds) *Root-knot Nematodes* (Meloidogyne *Species*). *Systematics, Biology and Control*. Academic Press, London, pp. 1–19.

Coomans, A. (1983) General principles for the phylogenetic systematics of nematodes. In: Stone, A.R., Platt, H.M. and Khalil, L.F. (eds) *Concepts in Nematode Systematics*. Academic Press, New York, pp. 1–10.

Devaney, E., O'Neill, K., Harnett, W., Whitesell, L. and Kinnaird, J.H. (2005) Hsp90 is essential in the filarial nematode *Brugia pahangi*. *International Journal of Parasitology* 35, 627–636.

De Luca, F., Vovlas, N., Lucarelli, G., Troccoli, A., Radicci, V., Fanelli, E., Cantalapiedra-Navarrete, C., Palomares-Rius, J. and Castillo, P. (2013) *Heterodera elachista* the Japanese cyst nematode parasitizing corn in Northern Italy: integrative diagnosis and bionomics. *European Journal of Plant Pathology* 136, 857–872. DOI: 10.1007/s10658-013-0212-9

Eroshenko, A.S., Subbotin, S.A. and Kazachenko, I.P. (2001) *Heterodera vallicola* sp. n. (Tylenchida: Heteroderidae) from elm trees, *Ulmus japonica* in the Primorsky territory, the Russian Far East, with rDNA identification of closely related species. *Russian Journal of Nematology* 9, 9–17.

Eves-van den Akker, S., Lilley, C.J., Reid, A., Pickup, J., Anderson, E., Cock, P.J.A., Blaxter, M., Urwin, P.E., Jones, J.T. and Blok, V.C. (2015) A metagenetic approach to determine the diversity and distribution of cyst nematodes at the level of the country, the field and the individual. *Molecular Ecology* 24, 5842–5851. DOI: 10.1111/mec.13434

Ferris, V.R. (1979) Cladistic approaches in the study of soil and plant parasitic nematodes. *American Zoologist* 19, 1195–1215. DOI: 10.1093/icb/19.4.1195

Ferris, V.R. (1985) Evolution and biogeography of cyst-forming nematodes. *EPPO Bulletin* 15, 123–129. DOI: 10.1111/j.1365-2338.1985.tb00211.x

Ferris, V.R. (1998) Evolution, phylogeny and systematics. In: Sharma, S. (ed.) *The Cyst Nematodes*. Dordrecht, The Netherlands, Kluwer Academic Publishers, pp. 57–82.

Ferris, V.R., Ferris, J.M. and Faghihi, J. (1993) Variation in spacer ribosomal DNA in some cyst-forming species of plant-parasitic nematodes. *Fundamental and Applied Nematology* 16, 177–184.

Ferris, V.R., Miller, L.I., Faghihi, J. and Ferris, J.M. (1995) Ribosomal DNA comparisons of *Globodera* from two continents. *Journal of Nematology* 27, 273–283.

Ferris, V.R., Riggs, R.D., Sabo, A., Faghihi, J. and Ferris, J.M. (1998) Relationships of *Cactodera betulae* to other cyst nematodes based on ribosomal DNA. *Phytopathology* 88, S28.

Ferris, V.R., Krall, E., Faghihi, J. and Ferris, J.M. (1999a) Relationships of *Globodera millefolii, G. artemisiae* and *Cactodera salina* based on ITS region of ribosomal DNA. *Journal of Nematology* 31, 498–507.

Ferris, V.R., Subbotin, S.A., Ireholm, A., Spiegel, Y., Faghihi, J. and Ferris, J.M. (1999b) Ribosomal DNA sequence analysis of *Heterodera filipjevi* and *H. latipons* isolates from Russia and comparisons with other nematode isolates. *Russian Journal of Nematology* 7, 121–125.

Ferris, V.R., Sabo, A., Baldwin, J.G., Mundo-Ocampo, M., Inserra, R.N. and Sharma, S. (2004) Phylogenetic relationships among selected Heteroderoideae based on 18S and ITS ribosomal DNA. *Journal of Nematology* 36, 202–206.

Franco, J. (1977) *Studies on the Taxonomy and Biology of Potato Cyst Nematodes* Globodera *spp*. Ph.D. Dissertation, University of London, London.

Gibson, T., Blok, V.C., Phillips, M.S., Hong, G., Kumarasinghe, D., Riley, I.T. and Dowton, M. (2007a) The mitochondrial subgenomes of the nematode *Globodera pallida* are mosaics: evidence of recombination in an animal mitochondrial genome. *Journal of Molecular Evolution* 64, 463–471. DOI: 10.1007/s00239-006-0187-7

Gibson, T., Blok, V.C. and Dowton, M. (2007b) Sequence and characterization of six mitochondrial subgenomes from *Globodera rostochiensis*: multipartite structure is conserved among close nematode relatives. *Journal of Molecular Evolution* 65, 308–315. DOI: 10.1007/s00239-007-9007-y

Gibson, T., Farrugia, D., Barrett, J., Chitwood, D., Rowe, J., Subbotin, S.A. and Dowton, M. (2011) The mitochondrial genome of the soybean cyst nematode, *Heterodera glycines*. *Genome* 54, 565–574. DOI: 10.1139/g11-024

Grenier, E., Bossis, M., Fouville, D., Renault, L. and Mugniéry, D. (2001) Molecular approaches to the taxonomic position of Peruvian potato cyst nematodes and gene pool similarities in indigenous and imported populations of *Globodera*. *Heredity* 86, 277–290. DOI: 10.1046/j.1365-2540.2001.00826.x

Grenier, E., Fournet, S., Pettit, E. and Anthoine, G. (2010) A cyst nematode 'species factory' called the Andes. *Nematology* 12, 163–169. DOI: 10.1163/138855409X12573393054942

Grossheim, A.A. (1945) [On question of graphical image of the system of flowering plants.] *Sovetskaya Botanika* 13, 3–27.

Handoo, Z.A., Carta, L.K., Skantar, A.M. and Chitwood, D.J. (2012) Description of *Globodera ellingtonae* n. sp. (Nematoda: Heteroderidae) from Oregon. *Journal of Nematology* 44, 40–57.

Inoue, T., Yatsuki, H., Kusakabe, T., Joh, K., Takasaki. Y., Nikoh, N., Miyata, T. and Hori, K. (1997) *Caenorhabditis elegans* has two isozymic forms, CE-1 and CE-2, of fructose-1,6-bisphosphate aldolase which are encoded by different genes. *Archives of Biochemistry and Biophysics* 339, 226–234.

Knoetze, R., Swart, A. and Tiedt, L.R. (2013) Description of *Globodera capensis* n. sp. (Nematoda: Heteroderidae) from South Africa. *Nematology* 15, 233–250. DOI: 10.1163/15685411-00002673

Kovaleva, E.S., Subbotin, S.A., Masler, E.P. and Chitwood, D.J. (2005a) Molecular characterization of the actin gene from cyst nematodes in comparison to those from other nematodes. *Comparative Parasitology* 72, 39–49. DOI: 10.1654/4138

Kovaleva, E.S., Masler, E.P., Subbotin, S.A. and Chitwood, D.J. (2005b) Molecular characterization of aldolase from *Heterodera glycines* and *Globodera rostochiensis*. *Journal of Nematology* 37, 292–296.

Krall, E. (1990) Different approaches to and recent developments in the systematics and co-evolution of the family Heteroderidae (Nematoda: Tylenchida) with host plants. *Proceedings of the Estonian Academy of Sciences, Biology* 39, 259–270.

Krall, E.L. and Krall, H.A. (1978) [Revision of the plant nematodes of the family Heteroderidae on the basis of trophic specialization of these parasites and their co-evolution with their host plants.] In: *Fitogel'mintologicheskie Issledovaniya.* Nauka, Moscow, USSR, pp. 39–56.

Lax, P., Rondan Dueñas, J.C., Franco-Ponce, J., Gardenal, C.N. and Doucet, M.E. (2014) Morphology and DNA sequence data reveal the presence of *Globodera ellingtonae* in the Andean region. *Contributions to Zoology* 83, 227–243.

Luc, M., Maggenti, A.R. and Fortuner, R. (1988) A reappraisal of Tylenchina (Nemata). 9. The family Heteroderidae Filipjev and Schuurmans Stekhoven, 1941. *Revue de Nématologie* 11, 159–176.

Madani, M., Vovlas, N., Castillo, P., Subbotin, S.A. and Moens, M. (2004) Molecular characterization of cyst nematode species (*Heterodera* spp.) from the Mediterranean basin using RFLPs and sequences of ITS-rDNA. *Journal of Phytopathology* 152, 229–234. DOI: 10.1111/j.1439-0434.2004.00835.x

Madani, M., Kyndt, T., Colpaert, N., Subbotin, S.A., Gheysen, G. and Moens, M. (2007) Polymorphism among sugar beet cyst nematode *Heterodera schachtii* populations as inferred from AFLP and ITS-rRNA gene analyses. *Russian Journal of Nematology* 15, 117–128.

Madani, M., Subbotin, S.A., Ward, L.J. and De Boer, S.H. (2010) Molecular characterization of Canadian populations of potato cyst nematodes, *Globodera rostochiensis* and *G. pallida* using ribosomal nuclear RNA and cytochrome b genes. *Canadian Journal of Plant Pathology* 32, 252–263.

Madani, M., Ward, L.J. and De Boer, S.H. (2011) Hsp90 gene, an additional target for discrimination between the potato cyst nematodes, *Globodera rostochiensis* and *G. pallida*, and the related species, *G. tabacum tabacum*. *European Journal of Plant Pathology* 130, 271–285. DOI: 10.1007/s10658-011-9752-z

Marché, L., Valette, S., Grenier, E. and Mugniéry, D. (2001) Intraspecies DNA polymorphism in the tobacco cyst nematode complex (*Globodera tabacum*) using AFLP. *Genome* 44, 941–946. DOI: 10.1139/g01-091

Mathews, H.J.P. (1971) Two new species of cyst nematode, *Heterodera mani* n. sp. and *H. iri* n. sp., from Northern Ireland. *Nematologica* 17, 553–565. DOI: 10.1163/187529271X00279

Mulvey, R.H. and Golden, A.M. (1983) An illustrated key to the cyst forming genera and species of Heteroderidae in the western hemisphere with morphometrics and distribution. *Journal of Nematology* 15, 1–59.

Mundo-Ocampo, M., Troccoli, A., Subbotin, S.A., Del Cid, J., Baldwin, J.G. and Inserra, R.N. (2008) Synonymy of *Afenestrata* with *Heterodera* supported by phylogenetics with molecular and morphological characterisation of *H. koreana* comb. n. and *H. orientalis* comb. n. (Tylenchida: Heteroderidae). *Nematology* 10, 611–632. DOI: 10.1163/156854108785787190

Nei, M., Tajima, F. and Tateno, Y. (1983) Accuracy of estimated phylogenetic trees from molecular data. *Journal of Molecular Evolution* 19, 153–170.

Paape, T., Igic, B., Smith, S.D., Olmstead, R., Bohs, L. and Kohn, J.R. (2008) A 15-myr old genetic bottleneck. *Molecular Biology and Evolution* 25, 655–663. DOI: 10.1093/molbev/msn016

Paramonov, A.A. (1967) [A critical review of the suborder Tylenchina (Filip'jev, 1934) (Nematoda: Secernentea).] *Trudy Gel'mintologicheskoi Laboratorii Akademii Nauk SSSR* 18, 78–101.

Phillips, W.S., Brown, A.M.V., Howe, D.K., Peetz, A.B., Blok, V.C., Denver, D.R. and Zasada, I.A. (2016) The mitochondrial genome of *Globodera ellingtonae* is composed of two circles with segregated gene content and differential copy numbers. *BMC Genomics* 17, 706. DOI: 10.1186/s12864-016-3047-x

Picard, D., Sempere, T. and Plantard, O. (2007) A northward colonisation of the Andes by the potato cyst nematode during geological times suggests multiple host-shifts from wild to cultivated potatoes. *Molecular Phylogenetics and Evolution* 42, 308–316. DOI: 10.1016/j.ympev.2006.06.018

Picard, D., Sempere, T. and Plantard, O. (2008) Direction and timing of uplift propagation in the Peruvian Andes deduced from molecular phylogenetics of highland biotaxa. *Earth and Planetary Science Letters* 271, 326–336. DOI: 10.1016/j.epsl.2008.04.024

Plantard, O., Picard, D., Valette, S., Scurrah, M., Grenier, E. and Mugniéry, D. (2008) Origin and genetic diversity of Western European populations of the potato cyst nematode *Globodera pallida* inferred from mitochondrial sequences and microsatellite loci. *Molecular Ecology* 17, 2208–2218. DOI: 10.1111/j.1365-294X.2008.03718.x

Prasad, V., Stromberg, C.A.E., Alimohammadian, H. and Sahni, A. (2005) Dinosaur coprolites and the early evolution of grasses and grazers. *Science* 310, 1177–1180. DOI: 10.1126/science.1118806

Pylypenko, L.A., Phillips, M.S. and Blok, V.C. (2008) Characterisation of two Ukrainian populations of *Globodera pallida* in terms of their virulence and mtDNA, and the biological assessment of a new resistant cultivar Vales Everest. *Nematology* 10, 585–590. DOI: 10.1163/156854108784513798

Rosenberg, N.A. and Nordborg, M. (2002) Genealogical trees, coalescent theory and the analysis of genetic polymorphisms. *Nature Reviews Genetics* 3, 380–390. DOI: 10.1038/nrg795

Rumpenhorst, H.J. and Ayub, M. (2001) Gibt es eine dritte Art beim Kartoffelnematoden? Die 29 Tagung des Arbeitskreises Nematologie, Monheim, Germany, 14–15 March 2001.

Sabo, A. and Ferris, V.R. (2004) β-tubulin paralogs provide a qualitative test for a phylogeny of cyst nematodes. *Journal of Nematology* 36, 440–448.

Sabo, A., Vovlas, N. and Ferris, V.R. (2001) Phylogenetic relationships based on ribosomal DNA data for four species of cyst nematodes from Italy and one from Syria. *Journal of Nematology* 33, 183–190.

Sabo, A., Reis, L.G.L., Krall, E., Mundo-Ocampo, M. and Ferris, V.R. (2002) Phylogenetic relationships of a distinct species of *Globodera* from Portugal and two *Punctodera* species. *Journal of Nematology* 34, 263–266.

Scholl, E.H. and Bird, D.M.K. (2005) Resolving tylenchid evolutionary relationships through multiple gene analysis derived from EST data. *Molecular Phylogenetics and Evolution* 36, 536–545. DOI: 10.1016/j.ympev.2005.03.016

Sekimoto, S., Uehara, T. and Mizukubo, T. (2017) Characterisation of populations of *Heterodera trifolii* Goffart, 1932 (Nematoda: Heteroderidae) in Japan and their phylogenetic relationships with closely related species. *Nematology* 19 543–558. DOI: 10.1163/15685411-00003067

Siddiqi, M.R. (1986) *Tylenchida: Parasites of Plants and Insects*. Commonwealth Agricultural Bureaux, Farnham Royal, Farnham, UK.

Siddiqi, M.R. (2000) *Tylenchida: Parasites of Plants and Insects*, 2nd edn. CAB International, Wallingford, UK. DOI: 10.1079/9780851992020.0000

Skantar, A.M. and Carta, L.K. (2004) Phylogenetic evaluation of nucleotide and protein sequences from the heat shock protein 90 gene of selected nematodes. *Journal of Nematology* 36, 136–145.

Skantar, A.M., Handoo, Z.A., Carta, L.K. and Chitwood, D.J. (2007) Morphological and molecular identification of *Globodera pallida* associated with potato in Idaho. *Journal of Nematology* 39, 133–144.

Skantar, A.M., Handoo, Z.A., Zanakis, G.N. and Tzortzakakis, E.A. (2012) Molecular and morphological characterization of the corn cyst nematode, *Heterodera zeae*, from Greece. *Journal of Nematology* 44, 58–66.

Stone, A.R. (1975) Head morphology of second-stage juveniles of some Heteroderidae (Nematoda: Tylenchoidea). *Nematologica* 21, 81–88. DOI: 10.1163/187529275X00374

Stone, A.R. (1979) Co-evolution of nematodes and plants. *Symbolae Botanicae Upsaliensis* 22, 46–61. DOI: 10.1163/187529275X00374

Stone, A.R. (1985) Co-evolution of potato cyst nematodes and their hosts: implications for pathotypes and resistance. *EPPO Bulletin* 15, 131–137.

Sturhan, D. (2000) Wirts-Spezifität bei Zystennematoden und anderen Heteroderiden. *Mitteilungen aus der Biologischen Bundesanstalt für Land und Forstwirtschaft* 376, 299.

Sturhan, D. (2006) Zystenbildende Nematoden und andere Heteroderiden in Deutschland. *Mitteilungen aus der Biologischen Bundesanstalt für Land- und Forstwirtschaft Berlin-Dahlem* 404, 18–30.

Sturhan, D., Wouts, W.M. and Subbotin, S.A. (2007) An unusual cyst nematode from New Zealand, *Paradolichodera tenuissima* gen. n., sp n. (Tylenchida: Heteroderidae). *Nematology* 9, 561–571. DOI: 10.1163/156854107781487314

Subbotin, S.A. (2015) *Heterodera sturhani* sp. n. from China, a new species of the *Heterodera avenae* species complex (Tylenchida: Heteroderidae). *Russian Journal of Nematology* 23, 145–152.

Subbotin, S.A., Sturhan, D., Waeyenberge, L. and Moens, M. (1997) *Heterodera riparia* sp. n. (Tylenchida: Heteroderidae) from common nettle, *Urtica dioica* L., and rDNA-RFLP separation of species from the *H. humuli* group. *Russian Journal of Nematology* 5, 143–157.

Subbotin, S.A., Halford, P.D., Warry, A. and Perry, R.N. (2000) Variations in ribosomal DNA sequences and phylogeny of *Globodera* parasitising Solanaceae. *Nematology* 2, 591–604.

Subbotin, S.A., Vierstraete, A., De Ley, P., Rowe, J., Waeyenberge, L., Moens, M. and Vanfleteren, J.R. (2001) Phylogenetic relationships within the cyst-forming nematodes (Nematoda, Heteroderidae) based on analysis of sequences from the ITS regions of ribosomal DNA. *Molecular Phylogenetics and Evolution* 21, 1–16. DOI: 10.1006/mpev.2001.0998

Subbotin, S.A., Sturhan, D., Rumpenhorst, H.J. and Moens, M. (2003) Molecular and morphological characterisation of the *Heterodera avenae* complex species (Tylenchida: Heteroderidae). *Nematology* 5, 515–538. DOI: 10.1163/156854103322683247

Subbotin, S.A., Sturhan, D., Chizhov, V.N., Vovlas, N. and Baldwin, J.G. (2006) Phylogenetic analysis of Tylenchida Thorne, 1949 as inferred from D2 and D3 expansion fragments of the 28S rRNA gene sequences. *Nematology* 8, 455–474. DOI: 10.1163/156854106778493420

Subbotin, S.A., Mundo-Ocampo, M. and Baldwin, J.G. (2010) *Systematics of Cyst Nematodes (Nematoda: Heteroderinae). Nematology Monographs and Perspectives 8A* (series editors: Hunt, D.J. and Perry, R.N.). Brill, Leiden, The Netherlands.

Subbotin, S.A., Cid Del Prado Vera, I., Mundo-Ocampo, M. and Baldwin, J.G. (2011) Identification, phylogeny and phylogeography of circumfenestrate cyst nematodes (Nematoda: Heteroderidae) as inferred from analysis of ITS-rDNA. *Nematology* 13, 805–824. DOI: 10.1163/138855410X552661

Subbotin, S.A., Knoetze, R. and Cid Del Prado Vera, I. (2016) *Globodera* species: current systematics and phylogeography. *Journal of Nematology (Abstract)* 48, 373–374.

Subbotin, S.A., Akanwari, J., Nguyen, C.N., Cid del Prado Vera, I., Chitambar, J.J., Inserra, R.N. and Chizhov, V.N. (2017) Molecular characterization and phylogenetic relationships of cystoid nematodes of the family Heteroderidae (Nematoda: Tylenchida). *Nematology* 19, 1065–1081. DOI: 10.1163/15685411-00003107

Tanha Maafi, Z., Subbotin, S.A. and Moens, M. (2003) Molecular identification of cyst-forming nematodes (Heteroderidae) from Iran and a phylogeny based on the ITS sequences of rDNA. *Nematology* 5, 99–111. DOI: 10.1163/156854102765216731

Tanha Maafi, Z., Sturhan, D., Handoo, Z., Mordehai, M., Moens, M. and Subbotin, S.A. (2007) Morphological and molecular studies of *Heterodera sacchari*, *H. goldeni* and *H. leuceilyma* (Nematoda: Heteroderidae). *Nematology* 9, 483–497. DOI: 10.1163/156854107781487242

Thiéry, M., Fouville, D. and Mugniéry, D. (1997) Intra- and interspecific variability in *Globodera*, parasites of solanaceous plants, revealed by Random Amplified Polymorphic DNA (RAPD) and correlation with biological features. *Fundamental and Applied Nematology* 20, 495–504.

Toumi, F., Waeyenberge, L., Viaene, N., Dababat, A., Nicol, J.M., Ogbonnaya, F. and Moens, M. (2013a) Development of two species-specific primer sets to detect the cereal cyst nematodes *Heterodera avenae* and *Heterodera filipjevi*. *European Journal of Plant Pathology* 136, 613–624. DOI: 10.1007/s10658-013-0192-9

Toumi, F., Waeyenberge, L., Viaene, N., Dababat, A., Nicol, J.M., Ogbonnaya, F. and Moens, M. (2013b) Development of a species-specific PCR to detect the cereal cyst nematode, *Heterodera latipons*. *Nematology* 15, 709–717. DOI: 10.1163/15685411-00002713

van Megen, H., van den Elsen, S., Holterman, M., Karssen, G., Mooyman, P., Bongers, T., Holovachov, O., Bakker, J. and Helder, J. (2009) A phylogenetic tree of nematodes based on about 1200 full-length small subunit ribosomal DNA sequences. *Nematology* 11, 927–950. DOI: 10.1163/156854109X456862

Vanderpoorten, A., Gradstein, S.R., Carine, M.A. and Devos, N. (2010) The ghosts of Gondwana and Laurasia in modern liverwort distributions. *Biological Reviews of the Cambridge Philosophical Society* 85, 471–487. DOI: 10.1111/j.1469-185X.2009.00111.x

Vovlas, N., Vovlas, A., Leonetti, P., Liébanas, G., Castillo, P., Subbotin, S.A. and Palomares-Rius, J.E. (2015) Parasitism effects on white clover by root-knot and cyst nematodes and molecular separation of *Heterodera daverti* from *H. trifolii*. *European Journal of Plant Pathology* 143, 833–845. DOI: 10.1007/s10658-015-0735-3

Wang, F., Wang, Z., Li, D. and Chen, Q. (2012) Identification and characterization of a *Bursaphelenchus xylophilus* (Aphelenchida: Aphelenchoididae) thermotolerance-related gene: Bx-HSP90. *International Journal of Molecular Sciences* 13, 8819–8833.

Wang, H.H., Zhuo, K., Ye, W., Zhang, H.L., Peng, D. and Liao, J.L. (2013) *Heterodera fengi* n. sp. (Nematoda: Heteroderinae) from bamboo in Guangdong Province, China – a new cyst nematode in the *Cyperi* group. *Zootaxa* 3652, 179–192. DOI: 10.11646/zootaxa.3652.1.7

Wikström, N., Savolainen, V. and Chase, M.W. (2001) Evolution of the angiosperms: calibrating the family tree. *Proceedings of the Royal Society of London, B, Biological Sciences* 268, 2211–2220. DOI: 10.1098/rspb.2001.1782

Wouts, W.M. (1985) Phylogenetic classification of the family Heteroderidae (Nematoda: Tylenchida). *Systematic Parasitology* 7, 295–328. DOI: 10.1007/BF00009997

Wouts, W.M. and Sher, S.A. (1971) The genera of the subfamily Heteroderinae (Nematoda: Tylenchoidea) with a description of two new genera. *Journal of Nematology* 3, 129–144.

Wouts, W.M. and Sturhan, D. (1995) *Heterodera aucklandica* sp. n. (Nematoda: Heteroderidae) from a New Zealand native grass, with notes on the species of the *H. avenae* group. *New Zealand Journal of Zoology* 22, 199–207.

Zhuo, K., Wang, H.H., Ye, W., Peng, D.L. and Liao, J.L. (2013) *Heterodera hainanensis* n. sp. (Nematoda: Heteroderinae) from bamboo in Hainan Province, China – a new cyst nematode in the *Afenestrata* group. *Nematology* 15, 303–314. DOI: 10.1163/15685411-00002678

Zhuo, K., Wang, H.H., Zhang, H.L. and Liao, J.L. (2014) *Heterodera guangdongensis* n. sp. (Nematoda: Heteroderinae) from bamboo in Guangdong Province, China – a new cyst nematode in the *Cyperi* group. *Zootaxa* 3881, 488–500. DOI: 10.11646/zootaxa.3881.5.4

17 Biochemical and Molecular Identification

Lieven Waeyenberge
ILVO, Flanders Research Institute for Agriculture, Fisheries and Food, Merelbeke, Belgium

17.1 Statutory and Non-statutory Issues	419
17.2 Biochemical and Molecular Identification	421
17.3 Conclusions and Future Prospects	434
17.4 References	436

17.1 Statutory and Non-statutory Issues

The International Plant Protection Convention (IPPC) was founded on 6 December 1951 at the 6th Conference of the Food and Agriculture Organization of the United Nations (FAO-UN). Its mission is 'to secure cooperation among nations in protecting global plant resources from the spread and introduction of pests of plants, in order to preserve food security, biodiversity and to facilitate trade' (www.ippc.int). To consolidate the objectives of the IPPC, National Plant Protection Organizations (NPPOs) were initiated or further fortified. These organizations developed their own plant health regulations in agreement with the principles of the World Trade Organizations Agreement on the Application of Sanitary and Phytosanitary Measures (WTO-SPS) of the IPPC. These principles include: (i) 'minimal impact' on movement of people and commodities; (ii) 'non-discrimination' between countries of the same phytosanitary status; and (iii) 'transparency', 'necessity' and 'technical justification', meaning that regulations should be published, thoroughly explained, only implemented when really needed and scientifically supported by a pest risk analysis (PRA). Many countries throughout the world became aware of the necessity to work more closely together, especially to be able to prevent pests spreading. As a result, NPPOs grouped into so-called Regional Plant Protection Organizations (RPPOs). RPPOs function as coordinating institutions between their members and the FAO. Supported by the NPPOs, the RPPOs identify pests, which may become a risk for its entire region or for parts within its region. These pests are added to an alert list and after an evaluation period removed again from this list when evaluated as without or minimal risk, or shifted to the A1 or A2 quarantine list for pests when, respectively, absent or locally present in the region but under official control. The RPPOs further promote the development of relevant international standards for phytosanitary measures (ISPMs), negotiate to harmonize regulations among their members and try to organize interregional cooperation. In 2015, 182 countries had already joined the IPPC and nine RPPOs were installed (Table 17.1). EPPO, the European and Mediterranean Plant Protection Organization, is the oldest RPPO (www.eppo.int). Founded in 1951, it has grown from originally 15 to 50 member governments (as of 31 March 2016), including almost

Table 17.1. List of Regional Plant Protection Organizations (RPPOs) currently installed or otherwise mentioned (as of 31 March 2016).

APPPC	Asia and Pacific Plant Protection Commission	www.apppc.org
CAN	Comunidad Andina	www.comunidadandina.org
COSAVE	Comité de Sanidad Vegetal	www.cosave.org
CPPC	Caribbean Plant Protection Commission	Abolished in 2014
IAPSC/CPI	Inter-African Phytosanitary Council/Conseil Phytosanitaire Interafricain	www.ippc.int/en/external-cooperation/regional-plant-protection-organizations/interafricanphytosanitarycouncil
NAPPO	North American Plant Protection Organization	www.nappo.org
NEPPO	Near East Plant Protection Organization	www.neppo.org
OEPP/EPPO	European and Mediterranean Plant Protection Organization/Organisation Européenne et Méditerranéenne pour la Protection des Plantes	www.eppo.int
OIRSA	Organismo Internacional Regional de Sanidad Agropecuaria	portal.oirsa.org/
PPPO	Pacific Plant Protection Organization	www.spc.int/lrd/plantprotection organisation-/pppo

all countries of Europe and the Mediterranean region. An important fact is that RPPOs cannot impose recommendations with legal status, the national governments should provide for this taking into consideration the regulations accepted internationally in the IPPC-WTO-SPS and RPPOs.

Cyst nematodes contain species that are considered as being the most destructive pests of different, economically important crops worldwide. Because of the potential impact on economical welfare, several species of cyst nematodes are subjected to international regulations. *Heterodera glycines* causes reduction of yield greater than that caused by any other disease on soybean (Wrather *et al.*, 2001), a crop important as a source of protein foods and edible oil. *Heterodera glycines* is therefore considered of quarantine significance for IAPSC and NAPPO. Although soybean production is limited in the EPPO region, *H. glycines* is also present on the EPPO A2 quarantine list of pests because it is believed that production of soybeans in Europe and surrounding areas will increase. The potato cyst nematodes (PCN) *Globodera rostochiensis* and *G. pallida* parasitize potato, reducing yield and quality of tubers (Greco *et al.*, 1982). Potatoes are considered as being the fourth-largest food source following maize, wheat and rice. As a consequence, *G. pallida* and *G. rostochiensis* are subjected to legislation around the world. Both species are listed as quarantine pests for APPPC, COSAVE, EPPO and PPPO. In addition, *G. rostochiensis* is included as a quarantine pest for IAPSC. Wheat yield is reduced considerably because of infection by *H. avenae* (Nicol, 2002). However, control of this species is believed to be easy: crop rotation is effective, maize is not a host plant and can be grown safely. Therefore, this species is only recognized as a quarantine organism in Brazil and Jordan, countries growing wheat in monoculture systems. Some cyst nematode species parasitize crops that are grown and traded more locally. They are more commonly regulated by a few RPPOs and countries (Lehman, 2004). For example, *H. schachtii* is mentioned on the A1 list in Argentina, Brazil and the IAPSC. Additionally, the species is also included on the A2 list in China and APPPC. More information about regional restricted regulations for other cyst nematode species can be found on the EPPO Global Database (https://gd.eppo.int).

In Europe, the Council of the European Union (EU) in consultation with the national governments, issued plant health regulations that were legalized under the Council Directive 2000/29/EC (eur-lex.europa.eu). This Directive aims at: (i) regulating the introduction of listed plants and plant products into the EU from non-EU countries by means of a phytosanitary certificate, guaranteeing that the plants or the plant products were properly inspected, are free from quarantine pests, are practically free from other harmful organisms and are in line with the plant health regulations of the importing country; (ii) regulating transport of listed plants and plant products within the EU by means of a plant passport, which may replace the phytosanitary certificate; (iii) imposing co-financed eradication or containment measures in case of an outbreak; and (iv) placing obligations on non-EU countries wanting to export plants or plant products into the EU (ec.europa.eu/food/plant/index_en.htm). On the lists with phytopathogens only two cyst nematode species are mentioned: the PCN species *G. rostochiensis* and *G. pallida*. As a consequence, the introduction and spread of PCN within the EU is prohibited. The addition of PCN to the list is due to the fact that the potato industry is one of the most important, still expanding, economic sectors of Europe (and beyond), and PCN pose a threat because they can easily spread in infested soil, whether or not adhering to potatoes, infested (seed) potatoes and other plants. Moreover, research proved that these cyst nematode species also can be easily spread by soil adhering to agricultural machines, and by washing water and waste from processing activities (EPPO-conference, Lille, France, 2014). Therefore, both species are additionally subjected to legislative measures described in the EU Council Directive 2007/33/EC (eur-lex.europa.eu). This Directive takes into account new insights about the biology of PCN, the distribution across the EU and practices within the potato industry. Both Directives together (Council Directive 2000/29/EC and Council Directive 2007/33/EC) require an official investigation to ensure that seed potatoes and potato plants intended for transplanting, moved into or transported within the EU, are produced only on non-infested fields. When fields are found to be infested they should be subjected to an official control programme ensuring that only fully resistant varieties or early varieties harvested before the cysts mature and not intended for the production of seed potatoes, or other crops after disinfection and without posing any risk of further spreading PCN (during at least 4 years), can be grown. When seed potatoes are found to be infested, they should be decontaminated before being distributed. Additionally, an annual screening of potato-producing fields, and by extension other crops that are hosts for PCN, is obligated. Finally, the Directives aim at harmonizing the investigations and sampling for PCN. To achieve this, a protocol for sampling and resistance testing is proposed.

When economically important cyst nematode species are the subject of a research project, regulations are not always that strict. Some RPPOs or NPPOs issue permits that enable researchers to send or receive samples so that these species can be used in experiments in the laboratory, glasshouse or even in the open field as long as measures are taken to control and prevent their spread. Although many cyst-forming nematode species are subjected to (non-) statutory regulations, as mentioned before, many more species exist for which regulations do not apply. These species possess no threat to any commercial activity because they simply do not parasitize an important agricultural crop. Instead, they can be found in natural habitats like river banks, grassland biotopes and woodlands (Subbotin *et al.*, 1997; Subbotin and Sturhan, 2004).

For research purposes, all cyst nematode isolates preferably need to be identified to species level. Moreover, statutory and non-statutory regulations ask for formalized identification protocols. For many years, identification of plant-parasitic nematodes was based on morphological and morphometrical data. However, the number of experts skilled in this classical identification and diagnosis is declining rapidly. To address this problem, molecular tools are being developed to assist the classical identification. These methods are especially needed in cases where morphological identification is difficult due to the lack of clear diagnostic characters or impossible due to the interception of only immature, diagnostically non-valuable life stages. Nowadays, diagnostic molecular protocols have been developed for a few regulated species only. This means that identification and diagnosis by a morphological specialist remain a necessity.

17.2 Biochemical and Molecular Identification

17.2.1 Introduction

In many reviews, books and other publications, identification of nematodes is divided into biochemical and molecular methods. The first focuses on proteins and, to lesser extent, on carbohydrates and lipids, the latter on nucleic acids, but because nucleic acids are in fact also biochemical components, molecular identification should be thought of as being part of biochemical identification. However, before the existence of polymerase chain reaction (PCR), nucleic acids were very difficult to investigate and proteins were the most important molecules to be used for identification. A number of years after the discovery of PCR (Mullis *et al.*, 1986), when its broad-range applicability became clear and well-known, a shift occurred from protein-based towards DNA-based identification techniques for almost all organisms, including nematodes. The separation of molecular from biochemical identification methods can in fact be considered as related to this shift, molecular identification methods being PCR-based whilst biochemical identification uses other techniques, mainly with proteins.

17.2.2 Protein-based identification

17.2.2.1 Isoelectric focusing

Electrophoresis is the migration of electrically charged particles or ions due to an applied electric field. Since 1950, the use of electrophoresis increased because of the introduction of migration of molecules in solid media like agarose or polyacrylamide gel. Moreover, after 1960, isoelectric focusing (IEF) improved separation resolution (Vesterberg, 1989). IEF is an electrophoretic method capable of separating amphoteric molecules. These molecules have the distinct characteristic that they can chemically react as an acid and a base. An example of such molecules is proteins. When proteins are loaded on a matrix, usually a polyacrylamide gel, containing a stable pH gradient increasing progressively from the anode to the cathode, they will migrate according to their surface charge, irrespective of their starting position in the electric field, until they reach an equilibrium position where their net charge is zero. At that position, called the isoelectric point (pI), the proteins will stop migrating. All proteins with the same pI will join and concentrate into a sharp band on the gel. So IEF is an equilibrium technique in which the effects of diffusion are overcome. It is therefore the electrophoretic technique with the highest resolution in which components that differ by 0.001 of a pH unit, or less, can be resolved (Prats, 1992).

In 1985, the possibilities for using protein variability visualized by polyacrylamide gel electrophoresis to determine species and races of cyst nematodes were reviewed (Bergé and Dalmasso, 1985). Differences in protein banding patterns and the appearance of species-specific bands among species, showed this approach to be diagnostically informative. However, variation in protein patterns due to differences in the experimental procedure could not be avoided. Marks and Fleming (1985) confirmed the conclusions of Bergé and Dalmasso in a review on the identification of the economically important PCN, *G. rostochiensis* and *G. pallida*, using the electrophoretic technique of IEF. The general protein banding patterns for both species on a thin polyacrylamide gel with a pH ranging from 3.5 to 9.5 showed clear differences at a number of positions. Two bands in particular were found to be useful for distinguishing both species (Fleming and Marks, 1983; Fox and Atkinson, 1984). A clear protein band at pI 5.9 was unique for *G. rostochiensis*; another at pI 5.7 was unique for *G. pallida*. This result was further demonstrated on a wider range of PCN populations indicating the usefulness of the technique on a global scale (Zaheer *et al.*, 1992, 1996; Subbotin *et al.*, 1999a; Da Cunha *et al.*, 2004; Molinari *et al.*, 2010). Also for *Heterodera*, general protein banding patterns by IEF proved to be useful in separating species within a morphologically similar species complex. Two new *Heterodera* species, *H. pratensis* and *H. australis*, previously both identified as *H. avenae*, were described (Gäbler *et al.*, 2000; Subbotin *et al.*, 2002). The IEF analysis produced consistent banding patterns that were distinctly different from each other and from those of the other species of the *H. avenae* complex. In 2004, Holgado *et al.* showed that some Scandinavian cyst nematode populations had different protein profiles compared with profiles from the *H. avenae* and *H. filipjevi* populations. They concluded the populations might be a separate species, or a so far unknown *H. avenae* pathotype.

IEF was further improved by the introduction of a miniaturized electrophoresis system, the so-called PhastSystem (GE Healthcare Life Sciences, Uppsala, Sweden). The PhastSystem is a high-speed, automated electrophoresis system consisting of a separation-control unit for multi-step programming and electrophoresis, and a development unit for gel staining. Two pre-cast polyacrylamide gels can be loaded and run simultaneously and fully automatically on a temperature controlled plate. The pI of sample proteins is estimated with calibration marker proteins (Pharmacia broad pI calibration kit 3.5–9.3). By combining the electrophoresis with a sensitive staining procedure using silver nitrate, the apparatus requires only small amounts of soluble proteins derived from a single *Globodera* cyst for diagnosis. Even an old cyst from soil samples or an air-dried cyst from collection material is sufficient to obtain a clear result. *Globodera pallida* and *G. rostochiensis* protein patterns show the typical major band at pI 5.7 and pI 5.9, respectively, and one additional weaker species-specific band at pI 8.7 and pI 6.9, respectively (Karssen et al., 1995). IEF has rarely been investigated on other *Globodera* species. One research group in Japan applied IEF in combination with the PhastSystem on single mature females and cysts of *G. rostochiensis*, *G. tabacum* and *G. hypolysi* (Sumiya et al., 2002). Although this method is applicable for potato cyst species identification in a simple, standardized and reproducible manner, and is still being used routinely for diagnostics of PCN in a few nematology laboratories, the PhastSystem has been taken out of production and thus can only be replaced by an alternative electrophoretic apparatus (Ibrahim et al., 2000; Da Cunha et al., 2004).

General protein patterns obtained after IEF are sensitive enough to distinguish between *Globodera* and *Heterodera* species; however, they are less useful for the detection of genetic variation among populations of the same species. Specific enzyme staining has more potential, for example, for pathotype designation. Fleming and Marks (1983) and Marks and Fleming (1985) demonstrated variation in phosphoglucomutase and phosphoglucose isomerase stained patterns between different pathotypes of *G. pallida*. This preliminary work was extended by investigations on a wider range of soluble enzymes (Fox and Atkinson, 1984, 1985a; Zaheer et al., 1992). The majority of the enzymes were common to all pathotypes examined, but the intra-species variation visible in phosphoglucomutase and phosphoglucose isomerase patterns within *G. pallida* correlated with different pathotypes. Additionally, the *G. rostochiensis* pathotypes Ro2/3 and Ro4/5 could be separated from pathotype Ro1 on the basis of the appearance of extra bands on the superoxide dismutase isozyme patterns (Fox and Atkinson, 1984; Molinari et al., 2010).

Despite all these and other investigations (Zaheer et al., 1992, 1996), it was not possible to discriminate all pathotypes of PCN.

Specific enzyme staining was also applied for *Globodera* species identification but not as extensively compared to *Heterodera*. Peptidase and esterase isozyme banding patterns could clearly differentiate both PCN species (Fox and Atkinson, 1984, 1988; Molinari et al., 2010). Comparative non-specific esterase banding patterns using IEF was conducted on 19 species of *Heterodera* from diverse geographical origins by Ibrahim and Rowe (1995). All species could be separated based on 26 bands of esterase isozymes of which some were common to some species, while others were species-specific. The results for most species correlated well with previously obtained isozyme phenotypes (and molecular identification methods) indicating that these species are well differentiated (Pozdol and Noel, 1984; Ferris et al., 1993). However, for other species, the positions of the bands were not in agreement with previously conducted studies, suggesting the presence of intraspecific variation (Ganguly et al., 1990). An investigation on species within the cereal cyst nematode complex by means of five different isozymes using IEF, confirmed previous morphological and biochemical characterizations separating *H. avenae*, *H. filipjevi*, *H. latipons* and *H. mani*. The genetic diversity between *H. avenae* and the Gotland strain, for many years also known as *H. avenae* 'race 3' or 'pathotype 3', was determined and confirmed the Gotland strain as being *H. filipjevi* (Andrés et al., 2001). However, the same study also revealed further intraspecific dissimilarities. The highest degree of polymorphisms was found in the *H. avenae* complex. Also, within *H. glycines*, intraspecific diversity was detected using phosphoglucose isomerase staining after IEF (Radice et al., 1988). The authors concluded that the genetic diversity is complex. This fact can potentially cause problems for diagnostics as not all isozyme phenotypic variants have yet been characterized.

For many years, IEF was used for studying protein polymorphisms between and within species of cyst-forming nematodes. Although it was considered as a simple and efficient method for nematode diagnostics, it has disadvantages. The most important are the influence on the banding patterns of sample preparation, sample storage and the developmental stages of the nematode samples. Possible improvements like addition of protein protectants or substrates for visualization directly into the gel to improve the sensitivity (Radice et al., 1988) were never established. Eventually, the technique was replaced by more sensitive DNA-based methods.

17.2.2.2 Antibody mediated detection and identification

The protein composition can also be studied by serology. For each protein an antibody can be produced. This antibody recognizes that protein, called the antigen, and reacts with it. Because of the relative specificity of the antigen–antibody reaction, serological techniques were also recognized as being very reliable in determining differences between proteins of different organisms (Landsteiner, 1936). Initially, polyclonal antibodies, recognizing multiple antigenic determinants or epitopes, were used to differentiate nematode species, especially *Globodera*, *Heterodera* and *Meloidogyne* species. Webster and Hooper (1968) were the first to separate serologically *H. schachtii*, *H. trifolii* and *G. rostochiensis* (at that time still named as *Heterodera rostochiensis*) from *H. cruciferae*, *H. goettingiana* and *H. carotae*, but within these two groups no differentiation could be demonstrated. Later, Scott and Riggs (1971) and Riggs et al. (1982) could separate *H. betulae* from a number of other *Heterodera* and *Globodera* species. Wharton et al. (1983) and Fox and Atkinson (1985b) separated *G. rostochiensis* from *G. pallida* using the more refined and sensitive crossed immuno-electrophoresis technique. This technique combines separation of proteins by electrophoresis with characterization of proteins with antibodies. However, cross-reactivity of the antisera with some *Heterodera* species was also observed. Cross-reactivity and the fact that polyclonal antibodies were often only family- or genus-specific demonstrated that the use of such antibodies is of limited value for species diagnosis. This led to the development of monoclonal antibodies exhibiting greater specificity.

Since the development of the hybridoma technique (Köhler and Milstein, 1975), it has been possible to produce large quantities of monoclonal antibodies (MAbs) routinely and with unprecedented uniformity. In brief, the technique starts by injecting a mouse with an antigen, for example, a protein from the nematode to be investigated. The antigen provokes an immune response: antibodies that can bind the antigen are produced by white blood cells (B cells). These cells can be harvested and fused with B cancer cells to produce a hybrid cell line called a hybridoma. During this process, lots of hybridoma cells are raised that need to be screened. To aid the screening process, enzyme linked immunosorbent assay (ELISA) was developed (Engvall and Perlmann, 1971). In this technique, the antibody is linked to an enzyme. When the enzyme's substrate is added, the subsequent reaction produces a detectable signal, most commonly a colour. The intensity of the colour correlates with the number of MAbs bound to the antigens. This means that the technique is quantitative. Atkinson et al. (1988) used the hybridoma technique to produce MAbs to react with antigens from second-stage juveniles (J2) and adult females from *H. glycines*. Although never tested, it is possible that some of them were species-specific and thus might be useful for diagnostic purposes. MAbs were also produced against proteins purified from eggs of the PCN species (Schots et al., 1989). Three and two hybridomas turned out to be producing antibodies that reacted preferentially with proteins from *G. pallida* and *G. rostochiensis*, respectively; unfortunately, the *G. rostochiensis* hybridomas also cross-reacted with other cyst-forming nematodes, especially *Heterodera* species. However, by combining several antibodies, the authors could differentiate the two species as each antibody had a different affinity for the antigens present in different nematode species. Robinson et al. (1993) characterized a 34-kD protein from PCN using MAbs. The protein from *G. rostochiensis* showed a different isoelectric point compared to the protein from *G. pallida*. Thus, two different antibodies could be produced to react with the different proteins and in this way discriminate both species from each other. Further research revealed that these antibodies recognized the same diagnostic marker at pI 5.7 for *G. pallida* and pI 5.9 for *G. rostochiensis* as those identified by Fleming and Marks (1983) by IEF. As both proteins are likely to have a similar structure and physiological function, they are probably present in similar concentrations in both species. As a consequence, the two MAbs, which showed no cross-reactivity with a wide range of soil nematode species, could be used for a quantitative immunoassay. The procedure of extraction of the nematode's antigen and the ELISA technique itself were further improved. Different chemical extractants and several extraction methods were tested. Cetyl trimethylammonium bromide (CTAB) and especially 1% urea were shown to be more effective than the standard phosphate buffered saline (PBS) in releasing antigens from cysts. Sonication was more effective compared to chemical and other physical procedures. Sonication of crude Fenwick can extracts containing cysts released enough antigen to detect approximately 3×10^{-3} cysts (g soil)$^{-1}$ (Evans et al., 1995). A double antibody sandwich enzyme linked immunosorbent assay (DAS-ELISA) showed improved sensitivity compared to the standard ELISA method. An equivalent of 0.5–3.6 eggs (g soil)$^{-1}$ could be detected in soil samples containing up to 20% of organic matter after a DAS-ELISA analysis (Curtis et al., 1997).

Immunological diagnostic protocols using monoclonal antibodies were developed to combine specificity, speed and simplicity at a relative low cost. Therefore, immunoassays were considered as very suitable for routine applications like nematode identification. However, time and finances are required to select and develop useful MAbs. Moreover, the cell lines from which they are produced are not stable (Hadas and Theilen, 1987; Neil and Urnovitz, 1988). Additionally, nematode structures (e.g. the reproductive system) and composition (e.g. the cuticle) change with development (Bird and Bird, 1991), which has consequences for the specificity of the MAbs. From several specific MAbs against antigens of J2 and young females of *H. glycines*, only one remained specific to the J2 just before invasion and one to the cuticle of the adult female (Atkinson and Harris, 1989). This led eventually to the replacement of antibodies by species-specific DNA-based methods. However, these results can be advantageous. Stage-specific antibodies can be used for research on parasitic development on host plants. Backett et al. (1993) tested three stage-specific MAbs for detecting J2 and young females of *G. pallida* without cross-reactivity with other nematodes and the host plant. It allowed the authors to detect plant invasion and facilitate the discrimination of resistant and susceptible plant lines in potato plant breeding programmes.

17.2.3 DNA-based identification and detection

17.2.3.1 rDNA markers

The rDNA (ribosomal DNA) cistron is the genomic DNA region that has been targeted the most for diagnostics of cyst nematodes. It consists of three rRNA genes (18S, 28S and 5.8S) separated by internal transcribed spacers (ITS 1 and 2). This region is tandemly repeated and each repetition is connected by the non-transcribed spacer (NTS) also called the intergenic spacer (IGS). The ITS regions have a higher mutation rate compared to the rRNA genes and possess more sequence variability to distinguish cyst nematodes at species level.

In 1993, Ferris et al. published the first ITS-rDNA sequences from several isolates of cyst nematodes belonging to the genus *Heterodera*. Universal primers used for PCR enabled the amplification of the complete ITS1, 5.8S gene and ITS2 of the rDNA region, including parts of the 18S and 28S genes adjacent to the spacer regions. The sequences for five geographic isolates of *H. glycines* were very similar to one another but dissimilar compared to sequences of other *Heterodera* species. These findings confirmed the usefulness of the rDNA region for identification. A few years later, Thiéry and Mugniéry (1996) and Szalanski et al. (1997) applied sequencing and PCR-restriction fragment length polymorphism (RFLP) of the ITS-rDNA to differentiate *Globodera* and *Heterodera* species, respectively. The PCR-RFLP method uses restriction enzymes, after the amplification process, to cut the amplicon (the PCR-amplified DNA fragment) into smaller fragments in order to create RFLP profiles. The number of fragments depends on the number of recognition sites for the enzyme within the amplicon's sequence. Although ITS1 possesses more variation than the ITS2 region (Ferris et al., 1993, 1994), it became clear that both regions together made it possible to create more species-specific RFLP patterns simply because of the larger size of the amplicon and thus the potential presence of more recognition sites for the restriction enzymes. Between 1997 and 2010, a number of articles contributed to the expansion of ITS-RFLP profiles and provided sequences of more than 40 different *Heterodera* species, five different *Globodera* species and a few species from related genera

(Subbotin et al., 2010). Three major conclusions could be drawn from these studies: (i) in most cases, a combination of RFLP profiles derived from different restriction enzymes is needed in order to separate certain cyst nematode species (Fig. 17.1); (ii) in some cases, it was not possible to discriminate species despite the large number of restriction enzymes tested; and (iii) heterogeneity is present in several cyst nematode species resulting in composite RFLP profiles, thus jeopardizing the identification. However, Amiri et al. (2002) demonstrated that the composite MvaI RFLP profile was constant over different populations of H. schachtii, making the composite RFLP profile eventually reliable for identification of the sugar beet cyst nematode.

The available cyst nematode ITS-rDNA sequences in public DNA databases (e.g. NCBI, National Center for Biotechnology Information, www.ncbi.nlm.nih.gov) made it possible for research groups to design species-specific primers to be used in a PCR. It is the intention to design primers that specifically target a sequence present in only one cyst nematode species. By using these primers in a PCR, it became possible to detect one species in a mixture of species without the need for an enzyme digestion step. When these species-specific primers are well designed (e.g. no inter-primer complementarity, equal annealing temperature) and their DNA targets are strategically well chosen (the resulting amplicons should have different sizes making it possible

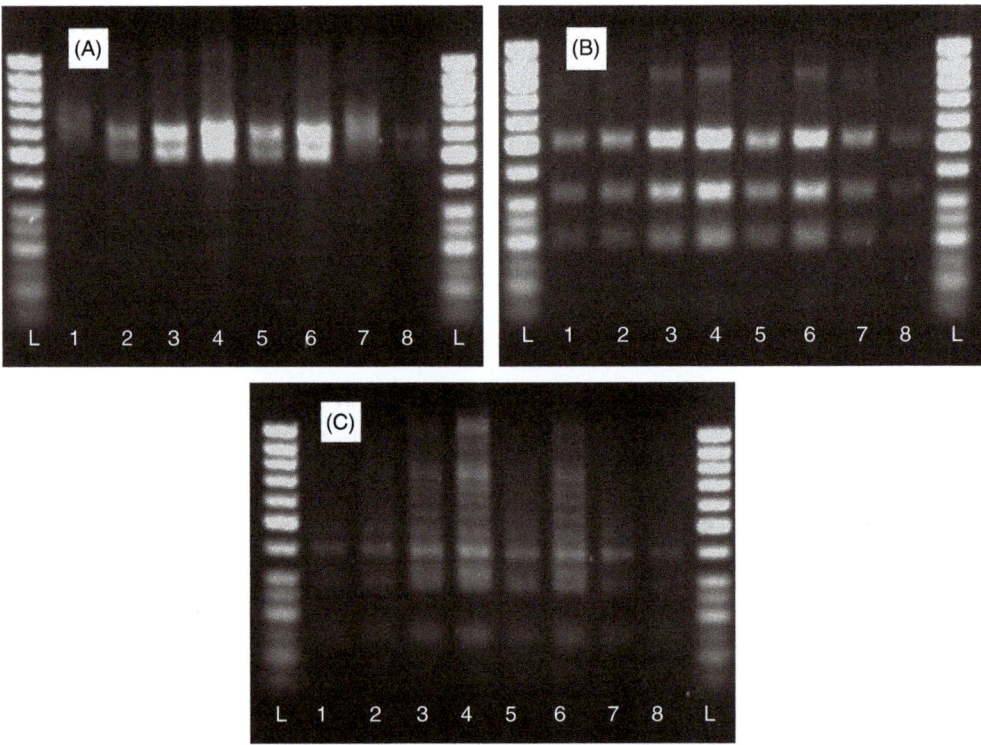

Fig. 17.1. Restriction fragment length polymorphism (RFLP) profiles from internal transcribed spacer (ITS)-rDNA amplicons derived from eight (1–8) cysts isolated from a field sample. The cysts were confirmed as being *Heterodera avenae* type B according to the RFLP profiles in Gäbler et al. (2000) and Subbotin et al. (1999b). (A) *Alu*I RFLP profile showing two bands of approximately 500 and 600 bp equals the pattern of *H. pratensis*, *H. aucklandica* and *H. avenae* type B. (B) *Hin*fI RFLP profile showing three bands of approximately 550, 350 and 225 bp equals the pattern of *H. avenae* type A and B and *H. aucklandica*. (C) *Taq*I RFLP profile showing multiple bands of which the brightest are of approximately 400, 275, 150 bp equals the pattern of *H. pratensis* and *H. avenae* type A and B. Ladder (L): 50 bp DNA ladder (Fermentas).

to visualize them separately after electrophoresis), they can even be combined in a multiplex PCR enabling detection of two or three cyst nematodes species by a single PCR test (Table 17.2). A multiplex PCR can also be used to avoid false negative results. If no amplicon is visible after electrophoresis on the gel, this can be due to failure of the preceding DNA extraction or not having added the DNA template. In this case, the operator can have the false impression that the target species is absent in that sample. Mixing a universal primer set with the species-specific primer set will always show at least one amplicon after electrophoresis when DNA extraction was successful and the DNA template was not omitted. Two amplicons will be visible when additionally the target species is present (Fig. 17.2). Multiplexing can significantly decrease time and costs.

Real-time PCR (RT-PCR) or quantitative PCR (qPCR) is more sensitive than end-point PCR due to the alternative method of signal detection: qPCR applies fluorescent DNA molecules (e.g. MGB TaqMan probe) or DNA-binding dyes (e.g. SYBR Green I). It is also faster since it eliminates the time-consuming post-PCR agarose gel electrophoresis because the detection happens during the run (in 'real time', hence the name). Additionally, the technique is quantitative: the number of PCR cycles (the threshold cycle (Ct) or quantitative cycle (Cq)) needed to be able to detect a fluorescent signal exceeding the background noise, is inversely proportional to the number of DNA molecules in the sample at the start. Thus, when a sample contains many nematodes, more DNA will be extracted and the number of cycles needed to detect the signal will be fewer. When the sample contains few nematodes, the number of cycles to detect fluorescence will be greater. By including DNA samples representing different known number of nematodes, a standard curve can be plotted on which samples with unknown number of nematodes can be quantified (Fig. 17.3). Bates et al. (2002), Goto et al. (2009), Madani et al. (2005), Quader et al. (2008) and Toyota et al. (2008) all applied SYBR Green to detect and quantify G. rostochiensis, G. pallida, H. glycines or H. schachtii. Bates et al. (2002) were able to semi-quantitatively measure the relative proportions of G. pallida and G. rostochiensis mixed in a sample by determining the peak heights of the respective amplicons in the melting curve after amplification and comparing them with standard samples containing varying proportions of cysts of each species. The sensitivity was such that 2% of cysts from one species could be detected in a mixture. Madani et al. (2005) measured the fluorescent signal of SYBR Green I itself to detect and quantify G. pallida or H. schachtii. The assay was able to quantify accurately J2 of G. pallida ranging in numbers from 1 to 500. For H. schachtii the sensitivity was lower than 100% for samples containing fewer than five J2. Toyota et al. (2008) could quantify between 1 and 250 J2 of G. rostochiensis mixed with a total of 500 nematode individuals from another species. The proposed assay also could detect one J2 of G. rostochiensis within 1000 free-living nematode individuals. The other authors focused on improving the DNA extraction method to give more consistent results (Quader et al., 2008) or to be able to extract DNA directly from the soil (Goto et al., 2009).

Quantification with probes is more accurate compared to fluorescent DNA-binding dyes because probes are designed to bind solely to DNA of a particular species. By contrast, DNA-binding dyes attach to all double-stranded DNA, whether it is the amplicon, a secondary unwanted amplicon, primer dimers, etc. Several species-specific qPCR probes have been developed for different cyst nematode species (Table 17.3). The assays were able to detect DNA from a 1/10 dilution of a single cyst (Nowaczyk et al., 2008) to one to five copies of the target ITS template (Nakhla et al., 2010). Madani et al. (2008) used DNA probes with locked nucleic acids (LNA). The increased stability, affinity and specificity of the hybridization process of these LNA probes with their respective targets made it possible to detect three cyst nematode species, G. rostochiensis, G. pallida and G. tabacum, simultaneously in a single assay. Finally, Papayiannis et al. (2013) compared five different extraction methods and demonstrated that a crude DNA preparation enables easy and cost-effective testing of cyst samples with similar accuracy to the DNA purification methods.

17.2.3.2 mtDNA markers

Mitochondrial DNAs (mtDNA) are circular DNA molecules located in the mitochondrion where they are present in high copy numbers. They are

Table 17.2. List of species-specific end-point polymerase chain reaction (PCR) assays to detect species of *Heterodera* and *Globodera*.

Species	Primer codes	Primer sequences (5'–3')	DNA region	Amplicon size (bp)	Reference
G. pallida [a]	PITSp4 ITS5	ACAACAGCAATCGTCGAG GGAAGTAAAAGTCGTAA-CAAGG	ITS-rDNA	265	1
G. pallida	- -	TGTCCATTCCTCTCCACCAG CCGCTTCCCCATTGCTTTCG	SCAR	798	2
G. pallida [a]	Fpa2 Rpa1	TCAACAATGTATGGACAGCG GGCACGTACGACATGGAA-TA	ITS-rDNA	239	3
G. pallida [a]	UNI GPA1	GCAGTTGGCTAGGGATCT-TC GGTGACTCGACGATTGCT-GT	ITS-rDNA	391	4
G. rostochiensis [a]	PITSr3 ITS5	AGCGCAGACATGCCGCAA GGAAGTAAAAGTCGTAA-CAAGG	ITS-rDNA	434	1
G. rostochiensis	- -	GCAAGCCCAGCGTCAG-CAAC GAACATCAACCTCCTATCGG	SCAR	315	2
G. rostochiensis [a]	Fro1 Rro1	ACACATGCCCGCTGTG-TATG AAAGATGGGAAAAAGCT-GGCC	ITS-rDNA	411	3
G. rostochiensis [a]	UNI GRO5	GCAGTTGGCTAGGGATCT-TC ATGTTGTACGTGCCG-TACCTT	ITS-rDNA	239	4
G. tabacum [a]	PITSt3 ITS5	AGCGCAGATATGCCGCGG GGAAGTAAAAGTCGTAA-CAAGG	ITS-rDNA	434	5
G. tabacum [a]	PITSt4 ITS5	ACAGCAGCAATCGTCGGC GGAAGTAAAAGTCGTAA-CAAGG	ITS-rDNA	265	5
H. avenae	AVEN-COIF AVEN-COIR	GGGTTTTCGGTTATTTGG CGCCTATCTAAATCTATAC-CA	COI-mtDNA	109	6
H. avenae	HaITS-F6 HaITS-R4	ATGCCCCGTCTGCTGA GAGCGTGCTCGTCCAAC	ITS-rDNA	242	7
H. filipjevi	HfF1 HfR1	CAGGACGAAACTCAT-TCAACCAA AGGGCGAACAGGAGAA-GATTAGA	SCAR	646	8
H. filipjevi	FILI-COIF FILI-COIR	GTAGGAATAGATTTAGATA-GTC TGAGCAACAACATAATAAG	COI-mtDNA	245	6
H. filipjevi	HfITS-F1 HfITS-R1	CCCGTCTGCTGTTGAGA ACCTCAGGCTTTTATTAT-CAC	ITS-rDNA	170	7
H. glycines [b]	SCNFI SCNRI	GGACCCTGACCAAAAAGT-TTCCGC GGACCCTGACGAGTTATG-GGCCCG	SCAR	477	9

Continued

Table 17.2. Continued.

Species	Primer codes	Primer sequences (5′–3′)	DNA region	Amplicon size (bp)	Reference
H. glycines (b)	GlyF1 rDNA2	TTACGGACCGTAACTCAA TTTCACTCGCCGTTACTAA-GG	ITS-rDNA	181	10
H. latipons	HlatactF HlatactR	ATGCCATCATTATTCCTT ACAGAGAGTCAAATTGTG	Actin gene	204	11
H. schachtii (b)	TW81 SHF6	GTTTCCGTAGGTGAACCTGC GTTCTTACGTTACTTCCA	ITS-rDNA	200	12

ᵃDuplex PCR, combination of two species-specific primer sets to detect two nematode species simultaneously. ᵇDuplex PCR, combination of a species-specific primer set and a universal primer set; the latter to control whether DNA extraction was successful. (Without letter) Singleplex PCR to detect solely the species for which the species-specific primer was designed.
References: 1: Bulman and Marshall (1997); 2: Fullaondo et al. (1999); 3: Vejl et al. (2002); 4: Zouhar et al. (2000); 5: Skantar et al. (2007); 6: Toumi et al. (2013a); 7: Yan et al. (2013); 8: Peng et al. (2013); 9: Ou et al. (2008); 10: Subbotin et al. (2001a); 11: Toumi et al. (2013b); 12: Amiri et al. (2002).

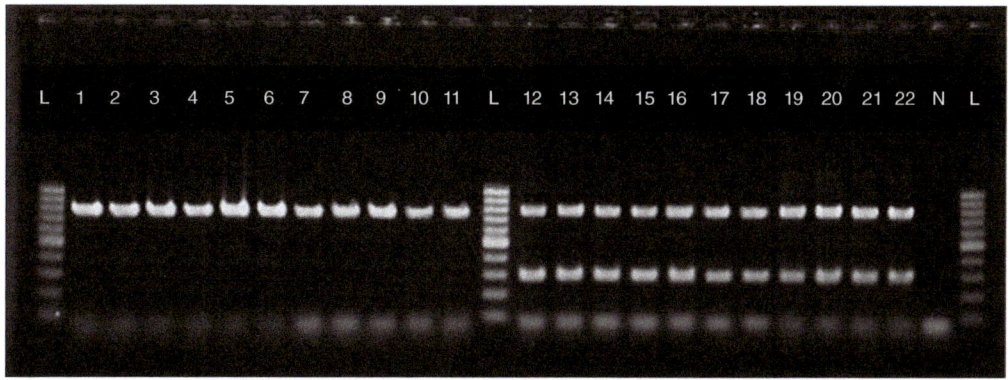

Fig. 17.2. Duplex species-specific polymerase chain reaction (PCR) for the detection of *Heterodera latipons* (F. Toumi, 2010, pers. comm.). (L) 100 bp DNA ladder. (1) *H. avenae*. (2, 3) *H. filipjevi*. (4) *H. cicero*. (5) *H. mani*. (6) *H. hordecalis*. (7) *H. betae*. (8, 9) *H. schachtii*. (10, 11) *H. glycines*. (12–22) *H. latipons*. (N) negative control. The upper band (approximately 800 bp) shows the amplicon derived from part of the 28S rRNA gene using universal primers, the smaller band (approximately 300 bp) shows the amplicon derived from part of the actin gene for *H. latipons* using a species-specific primer set.

inherited solely from the maternal parent and contain genes that are suitable for diagnosis of races, pathotypes and populations of cyst nematodes because of their relatively higher mutation rate compared to the rRNA genes (Hyman, 1988). For example, analysis of cytochrome *b* sequences enabled the identification of different *G. pallida* haplotypes, and detailed their distribution and spread in South America, Europe, the USA and Canada (Picard et al., 2007; Plantard et al., 2008; Madani et al., 2010). However, care is needed when using mtDNA genes for population studies on PCN. The mitochondrial genome of PCN exists as a population of small, circular DNAs (scmtDNAs). This unusual structural organization is believed to be unique among higher metazoans (Armstrong et al., 2000; Gibson et al., 2007). This complexity reduces its utility as a population genetic marker. Indeed, nothing is known about the variation in frequency of the different scmtDNAs between populations.

The higher mutation rate and, in the case of PCN, the multipartite structure, make it more difficult to find conserved regions for the design of universal primers. For this reason, mtDNA has not received extensive application in nematology

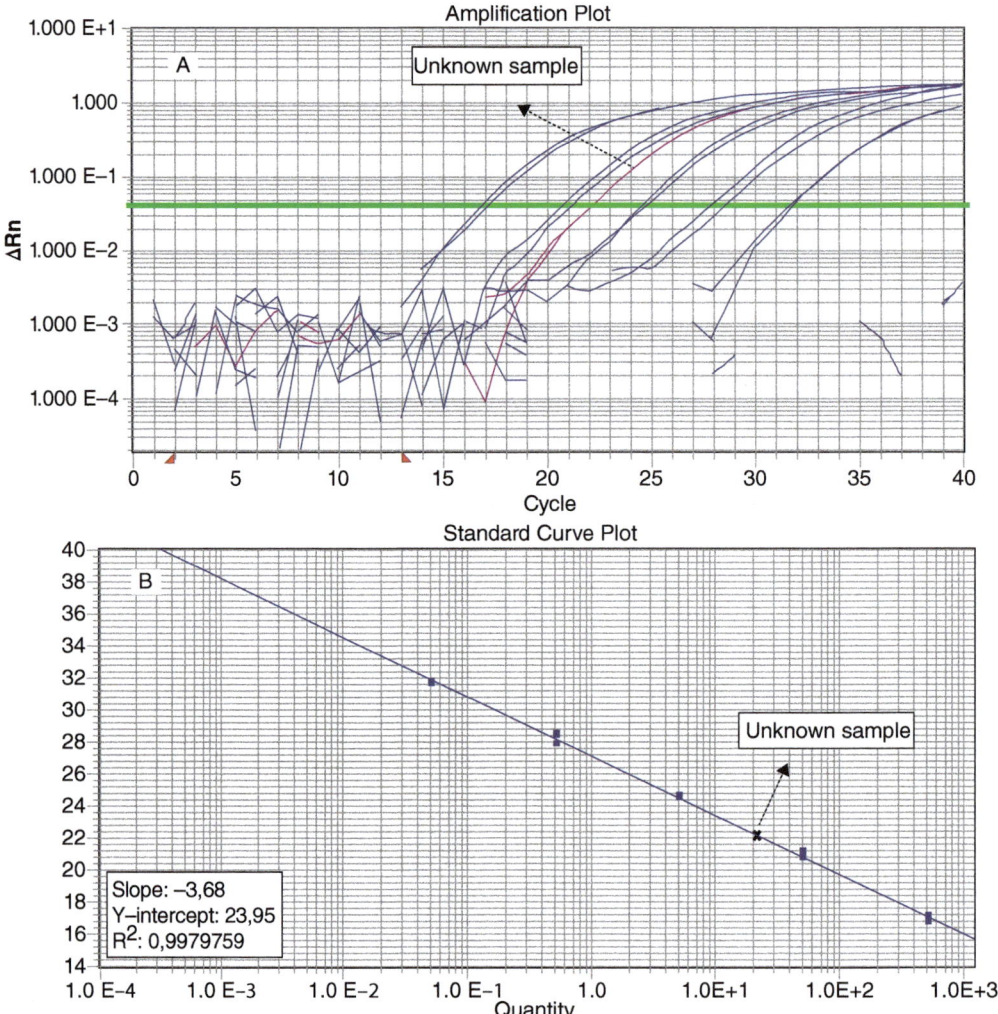

Fig. 17.3. (A) qPCR amplification plot of *Globodera pallida* samples containing DNA extracted from five different but known number of second-stage juveniles (J2) (in duplicate). (B) The Ct values of these samples were plotted against the known number of J2 in a standard curve. A DNA sample with unknown number of J2 was run simultaneously and quantified by means of the standard curve.

diagnosis. However, mtDNA has proven to be useful for designing species-specific diagnostic markers for *Heterodera* species (Tables 17.2 and 17.3). Toumi *et al.* (2013a) developed two species-specific primer sets targeting the cytochrome oxidase subunit 1 (*COI*) gene to detect the cereal cyst nematodes *H. avenae* and *H. filipjevi*. Two years later, the authors developed a qPCR assay to detect the cereal cyst nematode species *H. avenae* and *H. latipons* targeting the same mtDNA region (Toumi *et al.*, 2015). Both assays were able to detect the respective target species among 14 other *Heterodera* species and a *Punctodera punctata* population. The assays were also able to detect one J2 of the respective target species mixed with 100 J2 of a non-target *Heterodera* species.

17.2.3.3 sDNA markers

Satellite DNA (sDNA) are highly repetitive, non-coding sequences organized in very long

Table 17.3. List of species-specific quantitative polymerase chain reaction (qPCR) assays to detect *Heterodera* and *Globodera* species.

Species	Primer and probe codes	Primer and probe sequences (5′–3′)	DNA region	Reference
G. artemisiae	qGa1 qGa2 TibA	CACTGCGCCAACAGAGGTAG TAGCACACAAACGCCGACATG TGCTGACATGGAGTGTGTAGGCT-TC	ITS-rDNA	1
G. pallida[a]	Gfor PITS-pall Gpall	GTGTAACCGATGTTGGTGGCC AGCGCAGACATGCCGCTG ATCGTCGAGTCACCCATT	ITS-rDNA	2
G. pallida[a]	PITSpf PITSp4 GFAMp	ACGGACACATGCCCGCTA ACAACAGCAATCGTCGAG ACATGAGTGTTGGGGTGTAAC	ITS-rDNA	3
G. pallida[a]	GLOBOFOR PALLIREV PALLITAQ	CACATGCCTCCGTTTGTTGT GCGCTGTCCATACATTGTTGA CACATGCCCGCTATGTTTGGGCTG	ITS-rDNA	4
G. rostochiensis[a]	Gfor Grev Grost	GTGTAACCGATGTTGGTGGCC GGACGTAGCACACAAGCGCA GCTTCCTCCGTTGGCG	ITS-rDNA	2
G. rostochiensis[a]	PGrtf Prostor GTETp	TCTGTGCGTCGTTGAGC CGCAGACATGCCGCAA CGCAGATATGCTAACATGGAGTG-TAG	ITS-rDNA	3
G. rostochiensis	qGr1 qGr2 TibR	GTTGTTGCGCCTTGCGTAGA TAGCACACAAGCGCAGACATG CTAACATGGAGTGTAGCTGC-TACTC	ITS-rDNA	1
G. rostochiensis[a]	GLOBOFOR ROSTOREV ROSTOTAQ	CACATGCCTCCGTTTGTTGT GGCGCTGTCCGTACATTGTT CATATGCCCACTGTGTATGGGCT-GGC	ITS-rDNA	4
G. tabacum[a]	Gfor Grev Gtab	GTGTAACCGATGTTGGTGGCC GGACGTAGCACACAAGCGCA ATATGCCGCGGGGTACG	ITS-rDNA	2
G. tabacum[a]	Ptab-rt2f PITSt3mr GTETp	TCGTTGAGCGGTTGTTGC AGCGCAGATATGCCGCG CGCAGATATGCTAACATGGAGTG-TAG	ITS-rDNA	3
H. avenae	AvenF-COI AvenR-COI AvenProbe-COI	CTGGTTTGAGCACATCATA CCGGTAGGAATTGCAATA CCGCCTATCTAAATCTATACCAAC-CAC	COI-mtDNA	5
H. glycines[a]	SCNrtF SCNrtR SCNrtP	AAATTCCAGGCCGCTATCTC CGTGGACTGAACTGGACAAAG TGGGCTGGGGCTTCTAGAACTTTT	SCAR	6
H. latipons	LatF-COI LatR-COI LatProbe-COI	TTGGGCTCATCATATATTTG GTTGGAATTGCAATAATTATAGTA TAGGCTCGTCTATCCAAATCTAT-TCCA	COI-mtDNA	5

[a]Multiplex qPCR, combination of several species-specific probes to detect more than one nematode species simultaneously. Without letter: singleplex qPCR to detect solely the species for which the species-specific probe was designed.
References: 1: Nowaczyk *et al.* (2008); 2: Madani *et al.* (2008); 3: Nakhla *et al.* (2010); 4: Papayiannis *et al.* (2013); 5: Toumi *et al.* (2015); 6: Ye (2012).

arrays. The length of each repeat can vary from one or a few base pairs (microsatellites) to several thousands of base pairs. They have been characterized for a number of plant-parasitic nematodes and it was found that a few of them were potentially species-specific. This could make them useful diagnostic markers (Grenier et al., 1997). However, after screening the genomic library of a G. pallida population, primers were designed to amplify 14 microsatellite loci (Thiéry and Mugniéry, 2000). Eight were TC repeats, five TG repeats and one TGG repeats. The primers could also successfully amplify DNA from four other G. pallida populations, while the amplification was weaker or absent for G. rostochiensis, G. tabacum and G. mexicana, with the exception of three loci, which provided clear amplicons for all the four species. The authors concluded that although some of the primers seemed to be G. pallida-specific, microsatellite DNA would be extremely useful for research on population genetic structure and genetic drift. In that context, Plantard et al. (2008) applied microsatellite loci to study the origin and genetic diversity of western European populations of G. pallida.

Despite the promising diagnostic applicability, sDNA has only recently been applied for diagnostics of cyst nematodes at the species level. After screening hundreds of microsatellite primer combinations, presently possible due to the availability of microsatellite-enriched libraries obtained through the new technique of next generation sequencing (NGS, Montarry et al., 2015), Gamel et al. (2017) succeeded in developing a multiplex species-specific qPCR assay, based on microsatellite loci, simultaneously to detect and identify G. pallida, G. rostochiensis and H. schachtii. The latter can co-exist with PCN in agricultural fields on which the production of sugar beet is rotated with potato, and considering the growing trade of seed potato worldwide, H. schachtii was included in the assay because it is a regulated nematode in some parts of the world.

17.2.3.4 Other DNA markers

There are two main approaches to developing DNA markers for species identification: (i) explore known DNA regions to find variation separating species; and (ii) screen the whole genome to find DNA fragments that are unique for a particular taxon. The first approach has been demonstrated clearly in the sections of this chapter about rDNA, mtDNA and, to a lesser extent, sDNA markers. Next to these, other known DNA regions can be explored for the purpose of developing species-specific DNA markers. Toumi et al. (2013b) developed a species-specific primer set amplifying part of the actin gene to detect H. latipons. The assay was evaluated against 14 other Heterodera species and a P. punctata population, and was able to detect one J2 of H. latipons mixed with 100 J2 of a non-target Heterodera species.

Screening the whole genome in pursuit of species-specific markers can be achieved by PCR-based techniques such as random amplified polymorphic DNA (RAPD) or amplified fragment length polymorphism (AFLP). Both techniques create a multi-banding pattern of randomly amplified fragments from the entire genome visualized on agarose (sufficient for RAPD) or polyacrylamide (needed for AFLP) gels after electrophoresis. However, RAPD was the only technique used for diagnostics of cyst nematodes to species level. As it was demonstrated that the reproducibility of this technique was a limiting factor, the banding patterns themselves were only occasionally used for species characterization (Caswell-Chen et al., 1992; Bendezu et al., 1998; Amiri et al., 2003; Subbotin et al., 2003; Venkatesan et al., 2004). On the other hand, the technique made it possible to isolate, clone and sequence particular bands only appearing in patterns from certain nematode species. The resulting sequences were then used to design species-specific primers, the so-called sequence-characterized amplified region (SCAR) primers (Tables 17.2 and 17.3). RAPD fragments specific for G. rostochiensis and G. pallida were identified and used to develop two species-specific SCAR primer combinations, one for each species (Fullaondo et al., 1999). The primers were evaluated against 39 populations of PCN containing different pathotypes and including some South American isolates. The assays were capable of detecting separately both species in a mixed sample. Ou et al. (2008) combined SCAR primers with universal primers in a multiplex PCR to detect specifically H. glycines. The universal primer set was added to be used as a DNA amplification control to avoid false negative results. The same SCAR marker was also used to develop a DNA probe to be applied in a qPCR, evidently specific for *H. glycines* (Ye, 2012). A total of 44 populations including cyst and non-cyst-forming nematode species were tested. An accurate positive result was achieved when one cyst of *H. glycines* was mixed with ten cysts of another species. An additional primer set and a DNA probe was developed to amplify, in a separate run, part of the 18S rRNA gene. This amplicon functioned as an internal control in analogy to the universal primer set in the assay of Ou *et al.* (2008). Finally, Peng *et al.* (2013) created a SCAR primer to detect specifically *H. filipjevi* from wheat roots and soil. The PCR assay was capable of detecting directly a single nematode in 0.5 g of soil.

17.2.3.5 Detection in soil

The use of MAbs for the differentiation and quantification of cyst nematode species from plant roots has been demonstrated as explained in section 17.2.2.2 above. However, whether the quantification of nematode population densities directly from soil could be determined based on these immunological reactions was not investigated. Two factors played an important role: (i) the interaction between the antigen and soil fumigants; and (ii) the effect on the ELISA of organic material remaining in the cyst sample obtained from different types of soil, in particular regarding the determination of the upper and lower thresholds of the assay influenced by inhibition. Schots *et al.* (1992) applied a computer program (KINETICS) to predict the reactivity of a MAb with an antigen. From the resulting curves, it was possible to calculate the optimal quantity of the antibody required in an ELISA and to determine the detection limit of the assay under various conditions. This means that comprehensive empirical tests were no longer necessary. Nevertheless, attempts to quantify PCN directly in soil samples were not subsequently conducted. In fact, there was no need for this as cysts could easily be isolated first from soil samples using various extraction methods.

For a long time, the most efficient method to extract DNA from the whole microbial community from different soil types was bead beating, as described by Burgmann *et al.* (2001). This method used less than 10 g of soil, thus representing only a small fraction of the soil sample. In 2013, Peng *et al.* used a commercial DNA extraction method to purify DNA directly from soil to detect specifically *H. filipjevi*; however, an even smaller amount of soil (maximum 0.5 g) was required. For this reason, nematologists preferred to isolate cysts or J2 from a soil sample first before DNA extraction and subsequent detection and quantification. Exceptionally, Goto *et al.* (2009) introduced a novel detection method to quantify *H. glycines* directly from soil. A manually operated compacter could increase DNA yield and improve the sensitivity of qPCR by reducing the volume of a soil core to 1.4 g cm^{-3} prior to DNA extraction. As a consequence, 20 g of soil could be processed as a sample for DNA extraction and qPCR. A significant correlation between Ct values and number of eggs ranging from 10 to 3000 could be demonstrated, supporting the idea that most cysts and eggs were destroyed or at least damaged by the compactor. Shortly after, the method was further tested and compared to ball-milling (Goto *et al.*, 2010). Ball-milling uses a big metal bullet to pulverize as much as 20 g of air-dried soil. It was demonstrated that ball-milling was more or equally efficient compared to the compacter, depending on the soil type. It was possible to detect as few as five J2 of *G. rostochiensis* in 20 g of soil. No further attempts to quantify PCN directly in soil samples were conducted. As for antibody mediated quantification, there was no need for this as cysts or J2 could easily be isolated first from soil samples with various extraction methods.

17.2.3.6 Barcoding

The expanding number of sequences obtained by the collective efforts of contributors doing research on soil-borne nematodes will make public DNA databases (like GenBank, www.ncbi.nlm.nih.gov, and DDBJ (www.ddbj.nig.ac.jp/index-e.html) a valuable tool for DNA barcoding. Like all DNA-based identification methods, DNA barcoding was designed for situations where the morphology-based approach proved problematic. DNA barcoding is nothing more than sequencing a single piece of DNA, the DNA barcode, to identify all organisms (Floyd *et al.*, 2002). The ultimate goal is to set up reference libraries with sequences, also called molecular 'operational taxonomic units' (OTUs), of as many different organisms as possible. By comparing the sequences from unidentified organisms with these reference OTUs, their identities can be determined. Hebert *et al.* (2003) demonstrated that *COI* of the mitochondrial DNA can be used as a DNA barcode to identify accurately species across broad divisions of the animal kingdom. As a result, *COI* has been widely used as a standard barcode marker for metazoans, including marine nematodes (Derycke *et al.*, 2010). Actually, nematodes were among the first organisms used to test the barcode concept because of their more than 800 million years of evolution (Blaxter, 1998) and their dominance both in terms of densities and diversity in a wide range of different habitats (Bernard, 1992). Surprisingly, not the *COI* but the 18S and 28S rRNA genes have been applied mainly to develop DNA barcodes for soil-borne nematodes (Floyd *et al.*, 2002). This is probably due to: (i) the availability of more conserved regions within the rRNA genes for universal primer design; (ii) the abundance of 18S and 28S nematode sequences in public databases making it easier to match sequences for identification; and (iii) the economic importance of plant-parasitic nematodes.

For cyst nematodes, the situation is somewhat different. Sequencing efforts focused mainly on the ITS regions of the rDNA. Of all the cyst nematode rDNA sequences available in GenBank more than 80% are ITS-rRNA sequences. The extensive work on cyst nematodes by different research groups (Ferris *et al.*, 1993, 1994; Thiéry and Mugniéry, 1996; Szalanski *et al.*, 1997; Subbotin *et al.*, 1999b, 2000a, b, 2001b, 2003; Zheng *et al.*, 2000; Amiri *et al.*, 2002; Sabo *et al.*, 2002; Skantar *et al.*, 2007; Madani *et al.*, 2010) contributed to the creation of an extensive database of cyst nematode ITS sequences in GenBank. These sequences provided the basis for the development of diagnostic methods including RFLP-PCR, species-specific end-point PCR and qPCR (see section 17.2.3.1). However, intraspecific sequence variation in the ITS region of 0.5 to 1.5% is not uncommon (Powers, 2004). How this variation is distributed within individuals, among individuals, populations and species is not yet well documented. Sequence polymorphisms can have an impact on the performance of molecular diagnostic assays (Lefever *et al.*, 2013) or can even lead to false negative results, as has been demonstrated for microbial pathogens (Whiley *et al.*, 2008). So, there is still a need to increase the number of ITS sequences from more species and isolates. Alternatively, the mtDNA gene *COI* would be a good candidate for additional confidence in species delimitation and ultimately successful DNA barcoding for cyst nematode species. It would complement ITS as a locus unlinked to the nuclear ribosomal gene region (Powers, 2004). Also, the 18S gene is surprisingly informative at species level for plant-parasitic nematodes in general and for cyst nematodes in particular (Holterman *et al.*, 2009).

17.3 Conclusions and Future Prospects

Climate change and global increase of trade facilitate spreading of nematodes and their introduction into new territories. This is especially the case for cyst nematode species because cysts are difficult to detect and can survive for years. It is essential that the causative species of plant infection is detected and identified correctly in order to implement effective management strategies. The detection methods also need to be sensitive as the presence of a single individual can lead to the rejection of an entire shipment of plants and parts of them intended for export. Traditional methods are time-consuming and the reliability and accuracy depends on the experience and skill of the worker. Moreover, the number of specialists is declining rapidly. This makes the demand for alternative, fast, sensitive, accurate, specific and broadly applicable methods to identify nematode pests more urgent.

Proteins are products from genes that are being expressed. The expression of genes is altered by environmental conditions and developmental stages. This means that the composition of proteins is not constant and, as a consequence, protein-based diagnostics is tricky. This is not the case for DNA. DNA can be regarded as a permanent storage of all genes, and by extension the complete genome, containing a blueprint for protein production. Because of this function, the DNA is well protected in the cells of an organism and is far less influenced by external factors or developmental stages. This makes it more suitable for diagnostic purposes. However, for many years it was far less accessible compared with proteins. This disadvantage vanished because of the introduction of PCR (Mullis *et al.*, 1986).

PCR made DNA available to many researchers as it was now possible to amplify any DNA region of interest *in vitro*. Compared to protein-based assays, PCR-based identification methods have been demonstrated to be more sensitive (Ibrahim *et al.*, 2001). PCR requires very little material: DNA from single eggs or juveniles is sufficient, whilst protein-based methods need larger numbers of viable eggs to obtain enough protein for diagnosis. PCR is also very versatile. A wide range of samples, for example, soil, infected plant material like roots and tubers, dried material or individuals mounted on permanent slides, can be used for DNA amplification and subsequent identification. Hence, molecular diagnostics of nematodes progressed enormously, with particular improvements in DNA amplification and sequencing methods. This has made it possible to accumulate substantial amounts of genetic data with sufficient information on sequence divergence that can aid in reliable and easy identification of nematodes (Blok, 2005).

At first glance, PCR-based methods seem perfect. Unfortunately, they are not. One important factor is that the reliability can be affected by inhibition (Bessetti, 2007). This was discovered relatively soon and solutions to overcome inhibition have been proposed (De Boer *et al.*, 1995). However, inhibition is often ignored and a worrying aspect in this regard is that inhibition does not affect all PCR reactions to the same extent; in other words, some assays are more susceptible to inhibition than others (Huggett *et al.*, 2008). This means that inhibition is a source of error for PCR-based molecular diagnostic assays, especially quantitative assays like qPCR. It is believed that digital PCR (dPCR) will become a serious competitor to qPCR (Huggett *et al.*, 2015). Although the technique already existed before qPCR, it had to await the development of suitable instrumentation. dPCR allows precise quantification of DNA molecules by partitioning a sample of DNA into many individual, parallel PCR reactions. During amplification, dye-labelled probes are used to detect sequence-specific targets. Some of the reactions contain the target molecule, while others do not. When no target sequence is present, no signal accumulates. Following PCR analysis, the fraction of positive reactions is used to generate an absolute count of the number of target molecules in the sample, without the need for standards to create a calibration curve. This is a big advantage compared to qPCR. Another advantage is that although dPCR is also a PCR-based method, surprisingly, it seems to be less susceptible to inhibitors (Hoshino and Inagaki, 2012).

Whilst DNA barcoding and most of the other PCR-based identification methods diagnose individuals or single cysts, or try to detect one to two or three species in one nematode extract from soil or plant tissue, DNA metabarcoding can sequence a target DNA region from thousands of nematodes simultaneously (amplicon sequencing) and identify the nematodes by comparing the sequences with a DNA database (Porazinska *et al.*, 2010). Characterizing nematode communities in the soil is considered as the key component in the new vision of integrated pest management (Neher, 2001). Alternatively, DNA metabarcoding allows detection of the same nematode species in thousands of different samples simultaneously (Eves-van den Akker *et al.*, 2015). The latter could be of particular interest in relation to the regulated, annual survey of potato-producing fields in Europe to ensure the absence of PCN species. However, there are still a number of challenges associated with DNA metabarcoding: (i) a DNA marker covering the required taxonomic range of nematodes is yet to be found, albeit, the 18S rRNA gene seems to be on its way to becoming such a marker (Holterman *et al.*, 2009); (ii) DNA databases contain mistakes in linking sequences to a correct species; and (iii) amplicon sequencing relies on PCR, which can introduce errors in the sequences and form chimeras. The latter are constructed when an incomplete amplicon from one cycle anneals to DNA of an unrelated taxon. This construct is completed and further amplified in the subsequent cycles.

At present, other problems of DNA metabarcoding are speed and costs. Customers such as farmers and governmental bodies now expect faster and cheaper diagnostic results. One promising technique, in this regard, is loop-mediated isothermal amplification (LAMP) (Notomi *et al.*, 2000). This technique not only makes it possible to analyse samples on-the-spot using small, mobile equipment, but also it has the potential to develop into a 'do-it-yourself' method. LAMP can species-specifically amplify a particular DNA region at a single temperature, avoiding the need for a thermocycler.

After the amplification, a positive sample can appear coloured while negative samples remain transparent. This makes the interpretation straightforward and easier for the operator. Already LAMP assays are under development, but they still need to be further optimized, validated properly and their usefulness in the field tested.

17.4 References

Amiri, S., Subbotin, S.A. and Moens, M. (2002) Identification of the beet cyst nematode *Heterodera schachtii* by PCR. *European Journal of Plant Pathology* 108, 497–506. DOI: 10.1023/A:1019974101225

Amiri, S., Subbotin, S.A. and Moens, M. (2003) Comparative morphometrics and RAPD studies of *Heterodera schachtii* and *H. betae* populations. *Russian Journal of Nematology* 11, 91–99.

Andrés, M.F., Romero, M.D., Montes, M.J. and Delibes, A. (2001) Genetic relationships and isozyme variability in the *Heterodera avenae* complex determined by isoelectrofocusing. *Plant Pathology* 50, 270–279. DOI: 10.1046/j.1365-3059.2001.00543.x

Armstrong, M.R., Blok, V.C. and Philips, M.S. (2000) A multipartite mitochondrial genome in the potato cyst nematode *Globodera pallida*. *Genetics* 154, 181–192.

Atkinson, H.J. and Harris, P.D. (1989) Changes in nematode antigen recognized by monoclonal antibodies during early infections of soya beans with the cyst nematode *Heterodera glycines*. *Parasitology* 98, 479–487. DOI: 10.1017/S0031182000061576

Atkinson, H.J., Harris, P.D., Halk, E.J. *et al.* (1988) Monoclonal antibodies to the soya bean cyst nematode, *Heterodera glycines*. *Annals of Applied Biology* 112, 459–469. DOI: 10.1111/j.1744-7348.1988.tb02083.x

Backett, K.D., Atkinson, H.J. and Forrest, J.M.S. (1993) Stage-specific monoclonal antibodies to the potato cyst nematode *Globodera pallida* (Stone) Behrens. *Journal of Nematology* 25, 395–400.

Bates, J.A., Taylor, E.J.A., Gans, P.T. and Thomas, J.E. (2002) Determination of relative proportions of *Globodera* species in mixed populations of potato cyst nematodes using PCR product melting peak analysis. *Molecular Plant Pathology* 3, 153–161. DOI: 10.1046/j.1364-3703.2002.00107.x

Bendezu, I.F., Evans, K., Burrow, P.R., De Pomerai, D. and Canto-Saenz, M. (1998) Inter and intra-specific genomic variability of the potato cyst nematodes *Globodera pallida* and *G. rostochiensis* from Europe and South America using RAPD-PCR. *Nematologica* 44, 49–61. DOI: 10.1163/005225998X00064

Bergé, J.B. and Dalmasso, A. (1985) Possibilités et limites de l'utilisation de la variabilité protéique dans la determination des espèces et races de nématodés à kystes. *Bulletin OEPP/EPPO Bulletin* 15, 151–154. DOI: 10.1111/j.1365-2338.1985.tb00215.x

Bernard, E.C. (1992) Soil nematode biodiversity. *Biology and Fertility of Soils* 14, 99–103. DOI: 10.1007/BF00336257.

Bessetti, J. (2007) An introduction to PCR inhibitors. *Profiles in DNA* 10, 9–10.

Bird, A.F. and Bird, J. (1991) *The Structure of Nematodes*, 2nd edn. Academic Press, San Diego, California.

Blaxter, M. (1998) *Caenorhabditis elegans* is a nematode. *Science* 282, 2041–2046. DOI: 10.1126/science.282.5396.2041

Blok, V. (2005) Achievements in and future prospects for molecular diagnostics of plant-parasitic nematodes. *Canadian Journal of Plant Pathology* 27, 176–185. DOI: 10.1080/07060660509507214

Bulman, S.R. and Marshall, J.W. (1997) Differentiation of Australasian potato cyst nematode (PCN) populations using the polymerase chain reaction (PCR). *New Zealand Journal of Crop and Horticultural Science* 25, 123–129. DOI: 10.1080/01140671.1997.9513998

Burgmann, H., Pesaro, M., Widmer, F. and Zeyer, J. (2001) A strategy for optimizing quality and quantity of DNA extracted from soil. *Journal of Microbiological Methods* 45, 7–20. DOI: 10.1016/S0167-7012(01)00213-5

Caswell-Chen, E.P., Williamson, V.M. and Wu, F.F. (1992) Random amplified polymorphic DNA analysis of *Heterodera cruciferae* and *H. schachtii* populations. *Journal of Nematology* 24, 343–351.

Curtis, R.H.C., Al-Hinai, M.S., Diggines, A.E.R. and Evans, K. (1997) Serological identification and quantification of *Heterodera avenae* from processed soil samples. *Nematologica* 43, 199–213. DOI: 10.1163/004825997X00079

Da Cunha, M.J.M., Da Conceição, I.L.P.M., De O. Abrantes, I.M., Evans, K. and De A. Santos, M.S. (2004) Characterisation of potato cyst nematode populations from Portugal. *Nematology* 6, 55–58. DOI: 10.1163/156854104323072928

De Boer, S.H., Ward, L.J., Li, X. and Chittaranjan, S. (1995) Attenuation of PCR inhibition in the presence of plant compounds by addition of BLOTTO. *Nucleic Acid Research* 23, 2567–2568. DOI: 10.1093/nar/23.13.2567

Derycke, S., Vanaverbeke, J., Rigaux, A., Backeljau, T. and Moens, T. (2010) Exploring the use of cytochrome oxidase subunit 1 (COI) for DNA barcoding of free-living marine nematodes. *PLoS ONE* 5, e13716. DOI: 10.1371/journal.pone.0013716

Engvall, E. and Perlmann, P. (1971) Enzyme-linked immunosorbent assay (ELISA) quantitative assay of immunoglobulin G. *Immunochemistry* 8, 871–874. DOI: 10.1016/0019-2791(71)90454-X

Evans, K., Curtis, R.H., Robinson, M.P. and Yeung, M. (1995) The use of monoclonal antibodies for the identification and quantification of potato cyst nematodes. *Bulletin OEPP/EPPO Bulletin* 25, 357–365. DOI: 10.1111/j.1365-2338.1995.tb01478.x

Eves-van den Akker, S., Lilley, C.J., Reid, A. *et al.* (2015) A metagenetic approach to determine the diversity and distribution of cyst nematodes at the level of the country, the field and the individual. *Molecular Ecology* 24, 5842–5851. DOI: 10.1111/mec.13434

Ferris, V.R., Ferris, J.M. and Faghihi, J. (1993) Variation in spacer ribosomal DNA in some cyst-forming species of plant parasitic nematodes. *Fundamental and Applied Nematology* 16, 177–184.

Ferris, V.R., Ferris, J.M., Faghihi, J. and Ireholm, A. (1994) Comparisons of isolates of *Heterodera avenae* using 2-D PAGE protein patterns and ribosomal DNA. *Journal of Nematology* 26, 144–151.

Fleming, C.C. and Marks, R.J. (1983) The identification of the potato cyst nematodes *Globodera rostochiensis* and *G. pallida* by isoelectric focusing of proteins on polyacrylamide gels. *Annals of Applied Biology* 103, 277–281. DOI: 10.1111/j.1744-7348.1983.tb02765.x

Floyd, R., Abebe, E., Papert, A. and Blaxter, M. (2002) Molecular barcodes for soil nematode identification. *Molecular Ecology* 11, 839–850. DOI: 10.1046/j.1365-294X.2002.01485.x

Fox, P.C. and Atkinson, H.J. (1984) Isoelectric focusing of general protein and specific enzymes from pathotypes of *Globodera rostochiensis* and *G. pallida*. *Parasitology* 88, 131–139. DOI: 10.1017/S0031182000054408

Fox, P.C. and Atkinson, H.J. (1985a) Enzyme variation in pathotypes of the potato cyst nematodes *Globodera rostochiensis* and *G. pallida*. *Parasitology* 91, 499–506. DOI: 10.1017/S0031182000062740

Fox, P.C. and Atkinson, H.J. (1985b) Immunochemical studies on pathotypes of the potato cyst nematodes, *Globodera rostochiensis* and *G. pallida*. *Parasitology* 90, 471–483. DOI: 10.1017/S0031182000055475

Fox, P.C. and Atkinson, H.J. (1988) Non-specific esterase variation in field populations of the potato cyst nematodes *Globodera rostochiensis* and *G. pallida*. *Nematologica* 34, 156–163. DOI: 10.1163/002825988X00242

Fullaondo, A., Barrena, E., Viribay, M. *et al.* (1999) Identification of potato cyst nematode species *Globodera rostochiensis* and *G. pallida* by PCR using specific primer combinations. *Nematology* 1, 157–163. DOI: 10.1163/156854199508126

Gäbler, C., Sturhan, D., Subbotin, S.A. and Rumpenhorst, H.J. (2000) *Heterodera pratensis* sp. n., a new cyst nematode of the *H. avenae* complex (Nematoda: Heteroderidae). *Russian Journal of Nematology* 8, 115–126.

Gamel, S., Letort, A., Fouville, D., Folcher, L. and Grenier, E. (2017) Development and validation of real-time PCR assays based on novel molecular markers for the simultaneous detection and identification of *Globodera pallida*, *G. rostochiensis* and *Heterodera schachtii*. *Nematology* 19, 789–804. DOI: 10.1163/15685411-00003086

Ganguly, A.K., Dasgupta, D.R. and Rajasekhar, S.P. (1990) b-Esterase variation in three common species of *Heterodera*. *Indian Journal of Nematology* 20, 113–114.

Gibson, T., Blok, V.C. and Dowton, M. (2007) Sequence and characterization of six mitochondrial subgenomes from *Globodera rostochiensis*: multipartite structure is conserved among close nematode relatives. *Journal of Molecular Evolution* 65, 308–315. DOI: 10.1007/s00239-007-9007-y

Goto, K., Sato, E. and Toyota, K. (2009) A novel detection method for the soybean cyst *nematode Heterodera glycines* Ichinohe using soil compaction and real-time PCR. *Japanese Journal of Nematology* 39, 1–7. DOI: 10.3725/jjn.39.1

Goto, K., Sato, E., Gang, L.F., Toyota, K. and Sugito, T. (2010) Comparison of calibration curves prepared by soil compaction and ball milling methods for direct quantification of the potato cyst nematode *Globodera rostochiensis* in soil. *Japanese Journal of Nematology* 40, 41–45. DOI: 10.3725/jjn.40.41

Greco, N., Di Vito, M., Brandonisio, A., Giordano, I. and De Marinis, G. (1982) The effect of *Globodera pallida* and *G. rostochiensis* on potato yield. *Nematologica* 28, 379–386. DOI: 10.1163/187529282X00187

Grenier, E., Castagnone-Sereno, P. and Abad, P. (1997) Satellite DNA sequences as taxonomic markers in nematodes of agronomic interest. *Trends in Parasitology* 13, 398–401. DOI: 10.1016/S0169-4758(97)01113-7

Hadas, E. and Theilen, G. (1987) Production of monoclonal antibodies: the effect of hybridoma concentration on the yield of antibody producing clones. *Journal of Immunologal Methods* 96, 3–6. DOI: 10.1016/0022-1759(87)90359-0

Hebert, P.D.N., Cywinska, A., Ball, S.L. and de Waard, J.R. (2003) Biological identifications through DNA barcodes. *Proceedings of the Royal Society of London B* 270, 313–321. DOI: 10.1098/rspb.2002.2218

Holgado, R., Rowe, J., Andersson, S. and Magnusson, C. (2004) Electrophoresis and biotest studies on some populations of cereal cyst nematode, *Heterodera* spp. (Tylenchida: Heteroderidae). *Nematology* 6, 857–865. DOI: 10.1163/1568541044038551

Holterman, M., Karssen, G., van den Elsen, S. *et al.* (2009) Small subunit rDNA-based phylogeny of the Tylenchida sheds light on relationships among some high-impact plant-parasitic nematodes and the evolution of plant feeding. *Phytopathology* 99, 227–235. DOI: 10.1094/PHYTO-99-3-0227

Hoshino, T. and Inagaki, F. (2012) Molecular quantification of environmental DNA using microfluidics and digital PCR. *Systematic and Applied Microbiology* 35, 390–395. DOI: 10.1016/j.syapm.2012.06.006

Huggett, J.F., Novak, T., Garson, J.A. *et al.* (2008) Differential susceptibility of PCR reactions to inhibitors: an important and unrecognized phenomenon. *BMC Research Notes* 1, 70. DOI: 10.1186/1756-0500-1-70

Huggett, J.F., O'Grady, J. and Bustin, S. (2015) qPCR, dPCR, NGS – a journey. *Biomolecular Detection and Quantification* 3, A1–A5. DOI: 10.1016/j.bdq.2015.01.001

Hyman, B.C. (1988) Nematode mitochondrial DNA: anomalies and applications. *Journal of Nematology* 20, 523–531.

Ibrahim, S.K. and Rowe, J.A. (1995) Use of isoelectric focusing and polyacrylamide gel electrophoresis of nonspecific esterase phenotypes for the identification of cyst nematodes *Heterodera* species. *Fundamental and Applied Nematology* 18, 189–196.

Ibrahim, S.K., Saad, A.T., Haydock, P.P.J. and Al-Masri, Y. (2000) Occurrence of the potato cyst nematode *Globodera rostochiensis* in Lebanon. *Nematology* 2, 125–128. DOI: 10.1163/156854100508926

Ibrahim, S.K., Minnis, S.T., Barker, A.D.P. *et al.* (2001) Evaluation of PCR, IEF and ELISA techniques for the detection and identification of potato cyst nematodes from field soil samples in England and Wales. *Pest Management Science* 57, 1068–1074. DOI: 10.1002/ps.397

Karssen, G., van Hoenselaar, T., Verkerk-Bakker, B. and Janssen, R. (1995) Species identification of cyst and root-knot nematodes from potato by electrophoresis of individual females. *Electrophoresis* 16, 105–109. DOI: 10.1002/elps.1150160119

Köhler, G. and Milstein, C. (1975) Continuous cultures of fused cells secreting antibody of predefined specificity. *Nature* 256, 495–497. DOI: 10.1038/256495a0

Landsteiner, K. (1936) *The Specificity of Serological Reactions*, 1st edn. Charles C. Thomas, Springfield, Illinois.

Lefever, S., Pattyn, F., Hellemans, J. and Vandesompele, J. (2013) Single-nucleotide polymorphisms and other mismatches reduce performance of quantitative PCR assays. *Clinical Chemistry* 59, 1470–1480. DOI: 10.1373/clinchem.2013.203653

Lehman, P.S. (2004) Cost-benefits of nematode management through regulatory programs. In: Chen, Z.X., Chen, S.Y. and Dickson, D.W. (eds) *Nematology: Advances and Perspectives. Volume 2: Nematode Management and Utilization*. CAB International, Wallingford, UK, pp. 1133–1177. DOI: 10.1079/9780851996462.1133

Madani, M., Subbotin, S.A. and Moens, M. (2005) Quantitative detection of the potato cyst nematode, *Globodera pallida*, and the beet cyst nematode, *Heterodera schachtii*, using real-time PCR with SYBR green I dye. *Molecular and Cellular Probes* 19, 81–86. DOI: 10.1016/j.mcp.2004.09.006

Madani, M., Ward, L.J. and De Boer, S.H. (2008) Multiplex real-time polymerase chain reaction for identifying potato cyst nematodes, *Globodera pallida* and *Globodera rostochiensis*, and the tobacco cyst nematode, *Globodera tabacum*. *Canadian Journal of Plant Pathology* 30, 554–564. DOI: 10.1080/07060660809507555

Madani, M., Subbotin, S.A., Ward, L.J. and De Boer, S.H. (2010) Molecular characterization of Canadian populations of potato cyst nematodes, *Globodera rostochiensis* and *G. pallida* using ribosomal nuclear RNA and cytochrome b genes. *Canadian Journal of Plant Pathology* 32, 142–153. DOI: 10.1080/07060661003740033

Marks, R.J. and Fleming, C.C. (1985) The use of isoelectric focusing as a tool in the identification and management of potato cyst nematode populations. *Bulletin OEPP/EPPO Bulletin* 15, 289–297. DOI: 10.1111/j.1365-2338.1985.tb00231.x

Molinari, S., Greco, N. and Zouhar, M. (2010) Superoxidase dismutase isoelectric focusing patterns as a tool to differentiate of *Globodera* spp. *Nematology* 12, 751–758. DOI: 10.1163/138855410X12628646275961

Montarry, J., Jan, P.L., Gracianne, C., Overall, A.D.J., Bardou-Valette, S., Olivier, E., Fournet, S., Grenier, E. and Petit, E.J. (2015) Heterozygote deficits in cyst plant-parasitic nematodes: possible causes and consequences. *Molecular Ecology* 24, 1654–1677. DOI: 10.1111/mec.13142

Mullis, K., Faloona, F., Scharf, S. *et al.* (1986) Specific enzymatic amplification of DNA *in vitro*: the polymerase chain reaction. *Cold Spring Harbor Symposium in Quantitative Biology* 51, 263–273. DOI: 10.1101/SQB.1986.051.01.032

Nakhla, M.K., Owens, K.J., Li, W. *et al.* (2010) Multiplex real-time PCR assays for the identification of the potato cyst and tobacco cyst nematodes. *Plant Disease* 94, 959–965. DOI: 10.1094/ PDIS-94-8-0959

Neher, D.A. (2001) Nematode communities as ecological indicators of agroecosystem health. In: Gliessman, S.R. (ed.) *Agroecosystem Sustainability: Developing Practical Strategies*. CRC Press, Boca Raton, Florida, pp. 105–120. DOI: 10.1201/9781420041514.ch7

Neil, G.A. and Urnovitz, H.B. (1988) Recent improvements in the production of antibody-secreting hybridoma cells. *Trends in Biotechnology* 6, 209–213. DOI: 10.1016/0167-7799(88)90075-3

Nicol, J.M. (2002) Important nematode pests. In: Curtis, B.C., Rajaram, S. and Macpherson, H.G. (eds) *Bread Wheat Improvement and Production*. FAO, Rome, pp. 345–366.

Notomi, T., Okayama, H., Masubuchi, H. *et al.* (2000) Loop-mediated isothermal amplification of DNA. *Nucleic Acid Research* 28, e63. DOI: 10.1093/nar/28.12.e63

Nowaczyk, K., Dobosz, R., Kornobis, S. and Obrepalska-Steplowska, A. (2008) TaqMan REAL-Time PCR-based approach for differentiation between *Globodera rostochiensis* (golden nematode) and *Globodera artemisiae* species. *Parasitology Research* 103, 577–581. DOI: 10.1007/s00436-008-1012-6

Ou, S., Peng, D., Liu, X., Li, Y. and Moens, M. (2008) Identification of *Heterodera glycines* using PCR with sequence characterized amplified region (SCAR) primers. *Nematology* 10, 397–403. DOI: 10.1163/ 156854108783900212

Papayiannis, L.C., Christoforou, M., Markou, Y.M. and Tsaltas, D. (2013) Molecular typing of cyst-forming nematodes *Globodera pallida* and *G. rostochiensis*, using real-time PCR and evaluation of five methods for template preparation. *Journal of Phytopathology* 161, 459–469. DOI: 10.1111/jph.12091

Peng, H., Qi, X., Peng, D. *et al.* (2013) Sensitive and direct detection of *Heterodera filipjevi* in soil and wheat roots by species-specific SCAR-PCR assays. *Plant Disease* 97, 1288–1294. DOI: 10.1094/ PDIS-02-13-0132-RE

Picard, D., Sempere, T. and Plantard, O. (2007) A northwards colonization of the Andes by the potato cyst nematode during geological times suggest multiple host-shifts from wild to cultivated potatoes. *Molecular Phylogenetics and Evolution* 42, 308–316. DOI: 10.1016/j.ympev.2006.06.018

Plantard, O., Picard, D., Valette, S. *et al.* (2008) Origin and genetic diversity of western European populations of the potato cyst nematode (*Globodera pallida*) inferred from mitochondrial sequences and microsatellite loci. *Molecular Ecology* 17, 2208–2218. DOI: 10.1111/j.1365-294X.2008.03718.x

Porazinska, D.L., Giblin-Davis, R.M., Sung, W. and Thomas, W.K. (2010) Linking operational clustered taxonomic units (OCTUs) from parallel ultra sequencing (PUS) to nematode species. *Zootaxa* 2427, 55–63. DOI: 10.11646/zootaxa.2427.1.6

Powers, T. (2004) Nematode molecular diagnostics: from bands to barcodes. *Annual Review of Phytopathology* 42, 367–383. DOI: 10.1146/annurev.phyto.42.040803.140348

Pozdol, R.F. and Noel, G.R. (1984) Comparative electrophoretic analyses of soluble proteins from *Heterodera glycines* races 1–4 and three other *Heterodera* species. *Journal of Nematology* 16, 332–340.

Prats, M. (1992) Minireview: isoelectric focusing. *Biochemical Education* 20, 109–111. DOI: 10.1016/0307-4412(92)90118-6

Quader, M., Nambiar, L. and Cunnington, J. (2008) Conventional and real-time PCR-based species identification and diversity of potato cyst nematodes (*Globodera* spp.) from Victoria, Australia. *Nematology* 10, 471–478. DOI: 10.1163/156854108784513860

Radice, A.D., Riggs, R.D. and Huang, F.H. (1988) Detection of intraspecific diversity of *Heterodera glycines* using isozyme phenotypes. *Journal of Nematology* 20, 29–39.

Riggs, R.D., Rakes, L. and Hamblen, M.L. (1982) Morphometric and serologic comparison of a number of populations of cyst nematodes. *Journal of Nematology* 14, 190–199.

Robinson, M.P., Butcher, G., Curtis, R.H., Davies, K.G. and Evans, K. (1993) Characterisation of a 354 kD protein from potato cyst nematodes, using monoclonal antibodies with potential for species diagnosis. *Annual Applied Biology* 123, 337–347. DOI: 10.1111/j.1744-7348.1993.tb04096.x

Sabo, A., Reis, L.G.L., Krall, E., Mundo-Ocampo, M. and Ferris, V.R. (2002) Phylogenetic relationships of a distinct species of *Globodera* from Portugal and two *Punctodera* species. *Journal of Nematology* 34, 263–266.

Schots, A., Hermsen, T., Schouten, S., Gommers, F.J. and Egberts, E. (1989) Serological differentiation of the potato-cyst nematodes *Globodera pallida* and *G. rostochiensis*: II. Preparation and characterization of species specific monoclonal antibodies. *Hybridoma* 8, 401–413. DOI: 10.1089/hyb.1989.8.401

Schots, A., Gommers, F.J. and Egberts, E. (1992) Quantitative ELISA for the detection of potato cyst nematodes in soil samples. *Fundamental and Applied Nematology* 15, 55–61.

Scott, H.A. and Riggs, R.D. (1971) Immunoelectrophoretic comparisons of three plant-parasitic nematodes. *Phytopathology* 61, 751–752. DOI: 10.1094/Phyto-61-751

Skantar, A.M., Handoo, Z.A., Carta, L.K. and Chitwood, D.J. (2007) Morphological and molecular identification of *Globodera pallida* associated with potato in Idaho. *Journal of Nematology* 39, 133–144.

Subbotin, S.A. and Sturhan, D. (2004) *Heterodera circeae* sp. n. and *H. scutellariae* sp. n. (Tylenchida: Heteroderidae) from Germany, with notes on the *goettingiana* group. *Nematology* 6, 343–355. DOI 10.1163/1568541042360582

Subbotin, S.A., Sturhan, D., Waeyenberge, L. and Moens, M. (1997) *Heterodera riparia* sp. n. (Tylenchida: Heteroderidae) from common nettle, *Urtica dioica* L., and rDNA-RFLP separation of species from the *H. humuli* group. *Russian Journal of Nematology* 5, 143–157.

Subbotin, S.A., Halford, P.D. and Perry, R.N. (1999a) Identification of populations of potato cyst nematodes from Russia using protein electrophoresis, rDNA-RFLPs and RAPDs. *Russian Journal of Nematology* 7, 57–63.

Subbotin, S.A., Waeyenberge, L., Molakanova, I.A. and Moens, M. (1999b) Identification of *Heterodera avenae* group species by morphometrics and rDNA-RFLP. *Nematology* 1, 195–207. DOI: 10.1163/156854199508018

Subbotin, S.A., Waeyenberge, L. and Moens, M. (2000a) Identification of cyst forming nematodes of the genus *Heterodera* (Nematoda: Heteroderidae) based on the ribosomal DNA-RFLP. *Nematology* 2, 153–164. DOI: 10.1163/156854100509042

Subbotin, S.A., Halford, P., Warry, A. and Perry, R.N. (2000b) Variations in ribosomal DNA sequences and phylogeny of *Globodera* parasitizing solanaceous plants. *Nematology* 2, 591–604. DOI: 10.1163/156854100509484

Subbotin, S.A., Peng, D. and Moens, M. (2001a) A rapid method for the identification of the soybean cyst nematode *Heterodera glycines* using duplex PCR. *Nematology* 3, 365–371. DOI: 10.1163/156854101317020286

Subbotin, S.A., Vierstraete, A., DeLey, P. et al. (2001b) Phylogenetic relationships within the cyst-forming nematodes (Nematoda: Heteroderidae) based on analysis of sequences from the ITS region of ribosomal DNA. *Molecular Phylogenetics and Evolution* 21, 1–16. DOI: 10.1006/mpev.2001.0998

Subbotin, S.A., Sturhan, D., Rumpenhorst, H.J. and Moens, M. (2002) Description of the Australian cereal cyst nematode *Heterodera australis* sp. n. (Tylenchida: Heteroderidae). *Russian Journal of Nematology* 10, 139–148.

Subbotin, S.A., Sturhan, D., Rumpenhorst, H.J. and Moens, M. (2003) Molecular and morphological characterization of the *Heterodera avenae* species complex (Tylenchida: Heteroderidae). *Nematology* 5, 515–538. DOI: 10.1163/1568541033226832 47

Subbotin, S.A., Mundo-Ocampo, M. and Baldwin, J.G. (2010) *Systematics of Cyst Nematodes (Nematoda: Heteroderinae)*. Nematology Monographs and Perspectives volume 8A, 1st edn. Brill, Leiden, The Netherlands. DOI: 10.1163/ej.9789004162259.i-352

Sumiya, T., Hirata, K. and Yaegashi, T. (2002) Identification of three *Globodera* species in Japan by using isoelectric focusing method. *Research Bulletin of the Plant Protection Service Japan* 2, 49–51.

Szalanski, A.L., Sui, D.D., Harris, T.S. and Powers, T.O. (1997) Identification of cyst nematodes of agronomic and regulatory concern with PCR-RFLP of ITS1. *Journal of Nematology* 29, 255–267.

Thiéry, M. and Mugniéry, D. (1996) Interspecific rDNA restriction fragment length polymorphism in *Globodera* species, parasites of Solanaceaous plants. *Fundamental and Applied Nematology* 19, 472–479.

Thiéry, M. and Mugniéry, D. (2000) Microsatellite loci in the phytoparasitic nematode *Globodera*. *Genome* 43, 160–165. DOI: 10.1139/g99-106

Toumi, F., Waeyenberge, L., Viaene, N. et al. (2013a) Development of two species-specific primer sets to detect the cereal cyst nematodes *Heterodera avenae* and *Heterodera filipjevi*. *European Journal of Plant Pathology* 136, 613–624. DOI: 10.1007/s10658-013-0192-9

Toumi, F., Waeyenberge, L., Viaene, N. et al. (2013b) Development of a species-specific PCR to detect the cereal cyst nematode, *Heterodera latipons*. *Nematology* 15, 709–717. DOI: 10.1163/15685411-00002713

Toumi, F., Waeyenberge, L., Viaene, N. et al. (2015) Development of qPCR assays for quantitative detection of *Heterodera avenae* and *H. latipons*. *European Journal of Plant Pathology* 143, 305–316. DOI: 10.1007/s10658-015-0681-0

Toyota, K., Shirakashi, T., Sato, E., Wada, S. and Min, Y.Y. (2008) Development of a real-time PCR method for the potato-cyst nematode *Globodera rostochiensis* and the root-knot nematode *Meloidogyne incognita*. *Soil Science and Plant Nutrition* 54, 72–76. DOI: 10.1111/j.1747-0765.2007.00212.x

Vejl, P., Skupinova, S., Sedlák, P. and Domkářová, J. (2002) Identification of PCN species (*Globodera rostochiensis*, *G. pallida*) by using of ITS-1 region's polymorphism. *Rostlinná Vyroba* 48, 486–489.

Venkatesan, P., Meher, H.C., Srivastava, A.N. and Singh, G. (2004) Isozyme, RAPD and microsatellite markers as quick diagnostic of four *Heterodera* species. *Annals of Plant Protection Sciences* 12, 99–105.

Vesterberg, O. (1989) History of electrophoretic methods. *Journal of Chromatography A* 20, 3–19. DOI: 10.1016/S0021-9673(01)84276-X

Webster, J.M. and Hooper, D.J. (1968) Serological and morphological studies on the inter- and intraspecific differences of the plant-parasitic nematodes *Heterodera* and *Ditylenchus*. *Parasitology* 58, 879–891. DOI: 10.1017/S0031182000069651

Wharton, R.J., Storey, R.M.J. and Fox, P.C. (1983) The potential of some immunochemical and biochemical approaches to the taxonomy of potato cyst nematodes. In: Stone, A.R., Platt, H.M. and Khalil, L.K. (eds) *Concepts in Nematode Systematics*. Academic Press, London, pp. 235–248.

Whiley, D.M., Lambert, S.B., Bialasiewicz, S. et al. (2008) False-negative results in nucleic acid amplification test – do we need to routinely use two genetic targets in all assays to overcome problems caused by sequence variation? *Critical Reviews in Microbiology* 34, 71–76. DOI: 10.1080/10408410801960913

Wrather, J.A., Anderson, T.R., Arsyad, D.M. et al. (2001) Soybean disease loss estimates for the top ten soybean-producing countries in 1998. *Canadian Journal of Plant Pathology* 23, 115–121. DOI: 10.1080/07060660109506918

Yan, G.P., Smiley, R.W., Okubara, P.A. and Skantar, A.M. (2013) Species-specific PCR assays for differentiating *Heterodera filipjevi* and *H. avenae*. *Plant Disease* 97, 1611–1619. DOI: 10.1094/PDIS-01-13-0064-RE

Ye, W. (2012) Development of primetime-real-time PCR for species identification of soybean cyst nematode (*Heterodera glycines* Ichinohe, 1952) in North Carolina. *Journal of Nematology* 44, 284–290.

Zaheer, K., Fleming, C., Turner, S.J., Kerr, J.A. and McAdam, J. (1992) Genetic variation and pathotype response in *Globodera pallida* (Nematoda: Heteroderidae) from the Falkland Islands. *Nematologica* 38, 175–189. DOI: 10.1163/187529292X00153

Zaheer, K., Fleming, C., Turner, S.J. and Philis, J. (1996) Genetic variation and pathotype response in potato cyst-nematodes from Cyprus. *Nematologia Mediterranea* 24, 161–167.

Zheng, J., Subbotin, S.A., Waeyenberge, L. and Moens, M. (2000) Molecular characterization of Chinese *Heterodera glycines* and *H. avenae* populations based on RFLPs and sequences of rDNA-ITS regions. *Russian Journal of Nematology* 8, 109–113.

Zouhar, M., Ryšánek, P. and Kocová, M. (2000) Detection and differentiation of the potato cyst nematodes *Globodera rostochiensis* and *Globodera pallida* by PCR. *Plant Protection Science* 36, 81–84.

Genes Index

Note: bold page numbers indicate figures and tables.

18S 425, 433, 434, 435
28S 425, 434

ap2c1 156
ARSK1 225, 229
arx 95
atp6 405
AtS40-3 223
Avr 156–157, **176**, 178, 199, 200, **203**

β-tubulin 400, 403

CaMi 219
CaMV35S 220, 228–229
Cf-2 157, 158, **158**, 159, 161
COI mtDNA 374, 434
COI sequence 381, 383, 387, 434
coxI 406, 408
coxII 405
coxIII 405
Cre genes 161, 181, 185
cytb 13, 216, 389, **404**, 405, **406**, 411, 412, 429

daf-12 92
daf-21 400
DMR6 224
DOG (DOrsal Gland) box 34

far 94
flp 62, 95–96

Gel-mtDNA-I 405
Gel-mtDNA-II 405
Gm-DEA1 163
GmSAMT1 197, 226
Gp-flp-32/32R 95–96
Gpa-scmtDNAs 405
Gpa2 79, 80, 157, 158–159, **158**, 160, 161, 188, 218
Gpa5 188, 191
Gpa6 188
$GpaIV^s_{adg}$ 188, 191
gr-ams-1 58
Gro1 188, 218
Gro1-4 157, **158**, 188
Grp1 161, 188

H1 14, **15**, 18, 34, 131, 157, 161, 165–166, 187, 188, 189, 191
H2 188
H3 188
HEL 166
Hero 157–158, **158**, 161, 163, 166, 218–219
HS1 199
$Hs1^{pat-1}$ 199
$Hs1^{pro-1}$ 157, **158**, 159, 161, 199
$Hs1^{pro-7}$ 199
$Hs1^{pro-8}$ 199

$HS1^{RPH}$ 198
$Hs1^{web-1}$ 199
$Hs1^{web-7}$ 199
$Hs1^{web-8}$ 199
HS2 199
$Hsa-1^{Og}$ 201–202
hsp90 (heat shock protein 90) 400–401, **402**
Hs^{Bvm-1} 200

ins 62, 96
ITS-rRNA sequence 377, 380, 381, 382, 383, 384, 387, 389, 390, 400, **401**, 408–409, 411, 412, 434
ITS1 425
ITS2 425

MDK4-20 221, 225–227, **226**, 228
Mi 81
Mi1 219
MIOX4 228
MIOX5 228
mitochondrial DNA (mtDNA) 13, 35, 36, 374, 388–389, 403–406, **404**, 408, 413, 427–430, 432, 434
MLO/mlo 223, 224
mpk-3 156, 165
mpk-6 156, 165
mRNA 147, 221

nad1 405
nad2 405
nad3 405
nad4 405
nad4L 405
nad5 405
nad6 405
NHL 165
nlp 96
NodL 37

Os11N3 224

Pdf2.1 228
pfn 95
Prb1 251

RanGAP2 160
RAP2.6 156
Rha1 14, 181, 185
Rha2 14, 162, 181, 185
Rha3 14, 181
Rha4 181–182, 185
RhaE 182
Rhg 193
Rhg1 **158**, 162, 163, 164, 166, 167, 219
Rhg4 **158**, 162, 163, 167, 193, 216–217
Rpi-amr3i 219
RPL16A 229
rRNA (ribosomal RNA) 4, 374, 399, 400, **401**, 402, 403, 405, 407, 409, 410, 413, 429, 434
Rx 80
Rx1 159, 160, 161

SW5 80
Sw5B 159
Sw5F **158**

trnD 40
trnG 405
trnN 405
trnR 40
TUB-1 229

VAP-1 **158**, 159, 167

Nematodes Index

Note: bold page numbers indicate figures and tables.

Acrobeloides sp. 240
Afenestrata group 369, **371**, **401**, 408, 409, 413
Aphelenchoides spp. 78, 241
Aphelenchoides besseyi **32**
Aphelenchoides composticola 243
Ascaris 92
Ascaris lumbricoides 61, 241
Atalodera 1, 407, 410
Ataloderinae **370**, 407, 409–410
Avenae group 4, 9, 351, 369, **371**, 374, 380, 382, 383, **401**, 407, 408, 413, 422

barley cyst nematode *see Heterodera hordecalis*
Bellodera 1, 407, 410
Belonolaimus 241
Betulodera 1, 3, 103, 366, **370**, **377**, **378**, 392–393, 407, 409, 413
Betulodera betulae **392**, 393
Bidera 369, **370**
Bifenestra group 369, **371**, **401**, 407, 408, 413
Brugia pahangi 400
Bursaphelenchus spp. 78
Bursaphelenchus xylophilus **32**, 35, 37, 78, 248, 400

cabbage cyst nematode *see Heterodera cruciferae*
Cactodera 1, 3, 4, 103, 338–339, 348, 349, 351, 354, 365–366, **370**, 390–391, 407, 409, 410, 413
Cactodera acnidae 391
Cactodera betulae 366, **401**
Cactodera cacti **387**, 391, **401**
Cactodera eremica 391
Cactodera estonica 391, **401**
Cactodera evansi 391
Cactodera galinsogae 391, **401**
Cactodera milleri 117, 391, **401**
Cactodera radicale 391
Cactodera rosae **367**, 391, **401**
Cactodera salina 391, **401**
Cactodera thornei 391
Cactodera torreyanae 391, **401**
Cactodera weissi 117, **349**, 391, **401**
cactus cyst nematode *see Cactodera cacti*
Caenorhabditis elegans 28, 35, 56, 58, 61, 62, 92, 95, 96, 246, 247, 249–250, 252, 400, 402
Cardiolata group 369, **371**, **401**, 408, 413
CCN (cereal cyst nematodes) 2, 8–10, 13, 17, 162, 174, 175, 290, **307**, 308–309, 373–374, 380, 382–383, 423
  hatching/dormancy in 9–10
  infection process of 10
  pathotypes for 14, **15**
  resistance breeding for 181–187
  see also Heterodera avenae; *Heterodera filipjevi*; *Heterodera latipons*
clover cyst nematode *see Heterodera trifolii*
Cruciferae group 407
Cryphodera 407, 409–410
Cryphoderinae **370**, 407
cryptic species 358, 359, 366
Cyperi group 4, 369, **371**, 380, **401**, 407, 408, 409, 413

Ditylenchus destructor **32**
Dolichodera 1, 3, 4, 103, 366, **370**, 391–392, 407, 409, 413
Dolichodera fluvialis 392, **392**

Ellington potato cyst nematode (EPCN) *see Globodera ellingtonae*
European cereal cyst nematode (ECCN) *see Heterodera avenae*

Fici-humuli group 407
fig cyst nematode *see Heterodera fici*
Filipjev cereal cyst nematode *see Heterodera filipjevi*

Globodera spp. 1, 2, 3, 4, 6, 8, 29, 36, 44, 46, 47, 53, 59, 93, 94, 103, 107–112, **108**, 166, 219, 220, 229, 241, 242, 249, 252, 288–290, 316, 317, 323, 348, 351, 365, 366, **370**, 385–390, 401, 403, 407, 409, 410–412, 413, 423, 424, 425
Globodera agulhasensis 385, **401**
Globodera artemisiae 385, **401**, 411, **431**
Globodera bravoae 385
Globodera capensis 385, **401**, 411
Globodera ellingtonae 12, 13, 29, **31**, 36, 51, 108, 109, 110, 130, 187, 385, 386, 387, 389, **401**, 403, **404**, 405, 411, 412
Globodera hypolysi 385, 423
Globodera leptonepia 385
Globodera mali 385
Globodera mexicana 385, 386, 387, 389, 390, **401**, 411–412, 432
Globodera millefolii 385, **401**, 411
Globodera pallida 2, 3, 4, 5, 8, 12, 13, 14, **15**, 17–18, 29, 30, **31**, 33, 35, 36, 37, 38, 45, **45**, 46, 47, 49, 55, 57, 58, 59, 60, 62, 63, 75, 77, **77**, 78, 79, 80, 81, 90, 91, 94, 95–96, 107, 108, 109, 111, 130, **130**, 131, **133–134**, 143, 146, 157, 159, 160, 161, 166, 187, 188, 189, 200, 216, 217, 220, 221–222, 225, **226**, 228, 228, 240, 244–246, 247, 248–249, 288, 289, 290, 294, 306, **307**, 312, 313, 314, 320, 322, 324, 325, 385, 386–387, **387**, 388–389, 390, 400, **401**, **402**, 403, **404**, 405, **406**, 411, 412, 420, 421, 422, 423, 424–425, 427, **428**, **429**, **430**, **431**, 432
Globodera rostochiensis 2, 3, 4, 5, **7**, 8, 12, 13, 14, **15**, 17–18, 27–28, 29, 30, **31**, 34–35, 36, 38, 45, **45**, 46, 47, 48–49, 54, 56–57, 58, 59–61, 63, 76, 77, **77**, 78, 80, 81, 82, 90, 91, 93, 94, 96, 107, 108, 109–110, 111, 130, **130**, 131, **133–134**, 143, 157, **158**, 160, 161, 187, 188, 189, 190, 216, 218, 228, 240, 242, 244, 246, 248, 279, 288–289, 290, 294, 306, **307**, 310, 312, 313, 316, 317, 318, 319, 320, 321, 322, 324, **377**, **378**, 385, 386–387, **387**, **388**, 389–390, 400, **401**, 402, **402**, 403, 405, 411, 412, 420, 421, 422, 423, 424, 427, **428**, **431**, 432, 433
Globodera sandveldensis 385, **401**
Globodera tabacum 3, 288, 290, 292, 320, **346**, **349**, 386, 387, 389–390, **401**, **402**, 411, 412, 423, **428**, **431**, 432
Globodera tabacum sensu lato **130**
Globodera tabacum solanacearum 18, 90, 91, 92, 386, 389, 412
Globodera tabacum tabacum 18, 385, 389, 412
Globodera tabacum virginiae 389, 412
Globodera zelandica 386, **401**
Goettingiana group 4, 369, **371**, 378, 379, 382, **401**, 407, 408, 413
golden potato cyst nematode (GPCN) *see Globodera rostochiensis*
Graminis group 407
grass cyst nematode *see Punctodera punctata*

Helicotylenchus multicinctus 226
Heterodera spp. 1, 2, 3, 4, 8, 50, 54, 103, 216, 220, 240, 241, 316, 317, 318, 323, 337–338, 341, 346, 347–348, 352, 365, 366, 368, 369–385, **370**, **371**, 400, 407–408, 413, 422, 423, 424, 425, 430, 432
Heterodera 1, 2, 3, 4, 8, 50, 54, 103, 240, 241
Heterodera africana 351, 372
Heterodera agrostis 372
Heterodera amygdali 372
Heterodera arenaria 292, 372, 374, **401**, 408
Heterodera aucklandica 372, 374, **401**, 408, **426**
Heterodera australis 13, 14, **16**, 181, 183, 186, 372, 374, **401**, 408, 422
Heterodera avenae 2, 3, 4, 5, 6, 9, 10, 13, 14, **16**, 17, 18, **31**, 45, **45**, 54–55, 60, 62, 93, 94, 95, 96, 117, 118, **130**, 161, 164, 181, 182, 186, 238, 240, 242, 290–291, 292, 308, 310, 316, 319, 320, **341**, **346**, 351, 372, 373–374, **375**, **376**, **377**, **378**, 379, **401**, **402**, 408, 420, 422, 423, **426**, **428**, **429**, 430, **431**
Heterodera axonopi 372
Heterodera bamboosi 372
Heterodera bergeniae 372
Heterodera betae 8, 372, 374–376, **376**, 383, 384, **401**, 409, **429**
Heterodera betulae 393, 424
Heterodera bifenestra 372, **401**
Heterodera cajani 3, 4, 6, 10–11, 18, **45**, 50, **130**, 202, 244, 248, 249, 291, 292, 293, 294, 308, 310, 316, 372, 376–377, **376**, **401**, 408, 409

Heterodera canadensis 372
Heterodera cardiolata 372, **401**, **402**, 405
Heterodera carotae 5, 18, **45**, 46, 50, 60, 117, 202, 307, 310, 348, 372, **376**, 377–378, 379, 382, **401**, **402**, 408, 424
Heterodera ciceri 372, **401**, 409, **429**
Heterodera circeae 372, 382, **401**, 408
Heterodera cruciferae 3, 5, 6, 8, 18, 45, **45**, 60, **130**, 246, **341**, 351, **367**, 372, **376**, **377**, 378, **378**, 379, 382, **401**, 408, 424
Heterodera cyperi 372, **401**
Heterodera daverti 291, 372, 383, **401**, 409
Heterodera delvii 372
Heterodera elachista **307**, 308, 309, 316, 372, **376**, 379–380, **401**, **402**
Heterodera fengi 372, **401**
Heterodera fici **130**, **341**, 372, **376**, 380, **401**
Heterodera filipjevi 2, 3, 9–10, 13, 14, **16**, 94, 181, 186, 246, 290, 291, 308, 372, 373–374, **376**, **377**, **378**, 380, **401**, 408, 422, 423, **428**, **429**, 430, 433
Heterodera galeopsidis 5, **45**, **344**, 372, 383
Heterodera gambiensis 372
Heterodera glycines 2, 3, 4, 5, 6, 7, 13, 14, 17–18, 29, 33, 34, **45**, 46, 50–53, **52**, 58, 59, 60, 61–62, 79, 82, 90, 91, 92, 93, 96, 103, 112–116, 117, 128, 130, **130**, 131, 147, 156, **158**, 165, 167, 175, 191–197, 216–217, 219, 220, 221–222, 224, 226, 229, 238, 241, 243, 244, 246, 247, 248, 254–255, 256, 257, 278, 281–284, **285**, 292, 293, 294, 306, **307**, 309, 310–311, 312, 314, 316, 318, 320, 324–325, 327, **341**, **349**, 372, **376**, 380–381, 383, 400, **401**, 402, **402**, 403, **404**, 408, 409, 420, 425, 427, **428–429**, **431**, 432–433
Heterodera glycyrrhizae 372
Heterodera goettingiana 2, 5, **45**, 46, 48, 50, 60, 117, **130**, 202, **307**, 308, 372, **376**, 381–382, **401**, 408, 424
Heterodera goldeni 11, **344**, 372, **401**
Heterodera graminis 4, 372
Heterodera graminophila 372
Heterodera guangdongensis 372, **401**
Heterodera hainanensis 372, **401**
Heterodera hordecalis **130**, 372, **376**, **379**, 382, 383, **401**, 408, **429**
Heterodera humuli 5, **45**, 117, **349**, 372, **376**, **377**, **378**, 380, 382, **401**
Heterodera johanseni 372
Heterodera kirjanovae 372
Heterodera koreana 372, **401**, **402**
Heterodera latipons 3, 4, 9, 10, **130**, 181, 290, 372, 374, **376**, **379**, 382–383, **401**, 408, 423, **429**, 430, **431**
Heterodera lespedezae 248, 316, 373, 383

Heterodera leuceilyma 11, **349**, 373, 380
Heterodera litoralis 373, 380, **401**
Heterodera longicolla 373
Heterodera mani 373, **401**, 408, 423, **429**
Heterodera medicaginis **367**, 373, 383, **401**, 409
Heterodera mediterranea 373
Heterodera menthae 373
Heterodera mexicana 60
Heterodera mothi 373, **401**, 408
Heterodera orientalis 117, **367**, 373, **401**, **402**
Heterodera oryzae 3, 11, 12, 18, 90, **130**, 308, 373, 380
Heterodera oryzicola 3, 7, 11–12, 18, 46, 201, **307**, 308, 373, **401**
Heterodera pakistanensis 373
Heterodera persica 373, **401**, 408
Heterodera phragmitidis 373
Heterodera plantaginis 373
Heterodera pratensis 373, 374, **401**, 408, 422, **426**
Heterodera punctata 2
Heterodera raskii 373
Heterodera ripae 373, 380, 382, **401**
Heterodera riparia 373, 374
Heterodera rosii 373
Heterodera sacchari 3, 11, 12, 18, 46, 201–202, 308, 310, 373, 380, **401**
Heterodera saccharophila 373
Heterodera salixophila 4, 373, **401**, 408, 413
Heterodera schachtii 2, 3, 4, 5, 6–7, 8, 17–18, 29, **31**, 44–45, **45**, 48, 49–50, 51, 56, 58, 59, 60, 61, 78, 79, 81, 82, 83, 90, 92–93, 94, 103–107, 117, 130, **130**, 148, 156, 157, 159, 163, 165, 166, 175, 197–201, 218, 220, 221, 222, 223, 225, 228, 238, 240, 241, 242, 243, 246, 247, 248, 251, 256, 284–288, 305, **307**, 308, 310, 313, 316, 320, 338, **345**, 347, 351, 365, 369, **371**, 374, **376**, **377**, **378**, 381, 383, 386, **402**, 408, 409, 420, 424, 426, 427, **429**, 432
Heterodera scutellariae 373, 382, **401**, 408
Heterodera sinensis 373, **401**
Heterodera skohensis 373
Heterodera sojae 192, 373, **401**
Heterodera sonchophila 373
Heterodera sorghi 3, 4, 6, **45**, 50, 373, **401**
Heterodera spinicauda 373
Heterodera spiraeae 373
Heterodera sturhani 181, 373, 374, **401**, 408
Heterodera swarupi 373
Heterodera tabacum 60
Heterodera trifolii 3, 5, 8, 10, **45**, 60, 117, 202, 248, 291, 292–293, 316, 318, **341**, **344**, 347, **349**, 373, 374, **376**, 383, **401**, 409, 424
Heterodera trifolii complex 384
Heterodera turangae 373
Heterodera turcomanica 4, 373, 383, **401**
Heterodera urticae 373, **401**, 408

Heterodera ustinovi 117, 373, **401**, 408
Heterodera uzbekistanica 373
Heterodera vallicola 373, 380, 382, **401**
Heterodera zeae 3, 4, 8, 18, 45, 90, 92, **130**, 202, **344**, **351**, **376**, 384, **401**, **402**, 408, 413
Heteroderinae 1, 2, 3–4, 102, 116–117, 337–338, 340, 342, 344, 346, 348, 359–360, 366, 369, **401**, 402, 407–409
Heterorhabditis 241
Hirschmanniella oryzae **32**
hop cyst nematode *see Heterodera humuli*
Hoplolaimidae 337, 368–369, 407
Hoplolaimus 241
Humuli group 4, 369, **371**, 380, 382, **401**, 407, 408, 409, 413
Hylonema 407

Japanese cyst nematode *see Heterodera elachista*

Latipons group 407

maize cyst nematode *see Heterodera zeae*
Mediterranean cereal cyst nematode *see Heterodera latipons*
Meloidodera 1, 409–410
Meloidoderinae **370**, 407, 409–410
Meloidogyne spp. 2, 48, 240, 241, 243, 253, 337, 424
Meloidogyne arenaria 29, 240
Meloidogyne chitwoodi 239
Meloidogyne floridensis 29, **31**
Meloidogyne graminicola **31**
Meloidogyne hapla 29, **31**, 37, 61, 240, 309
Meloidogyne incognita 29, **31**, 35, 37, 58, 61, 219, 220, 223, 239, 240, 246, 247, 255, 281, 292
Meloidogyne javanica 29, **31**, 61, 239, 240, 252
Meloidogyne maritima 292
Meloidogynidae 369
Mesocriconema 241

Nacobbus 337
Nacobbus aberrans **31**, 33

Oryzae group 407

pale potato cyst nematode (PPCN) *see Globodera pallida*
Panagrellus redivivus 243, 246, 248
Panagrolaimus rigidus 60
Paradolichodera 1, 3, 103, 366, **370**, 393, 407, 409, 413
Paradolichodera tenuissima **392**, 393, **401**
Paratrichodorus 29

PCN (potato cyst nematodes) 2, 12–14, **15**, 17, 91, 106, 107–112, 129, 130–131, 132–150, **133–134**, **158**, 161, 175, 178, 180, 187–191, 201, 216, 221, 225–226, **226**, 230, 306, **307**, 309, 310–311, 312, 313–314, 319, 323, 327, 365, 386–389, 420, 421, 422–423
  see also Globodera ellingtonae; Globodera pallida; Globodera rostochiensis
pea cyst nematode *see Heterodera goettingiana*
pigeon pea cyst nematode *see Heterodera cajani*
PPCN (pale potato cyst nematode) *see Globodera pallida*
Pratylenchus 243
Pratylenchus brachyurus 61
Pratylenchus coffeae **32**
Pratylenchus neglectus **32**
Pratylenchus pacificus 245
Pratylenchus penetrans **32**, 292, 309, 316, **402**
Pratylenchus scribneri 245
Pratylenchus thornei **32**
Pratylenchus zeae **32**, 322
Pseudomonas syringae 224
Punctodera 1, 3, 4, 103, 348, 351, 365, **370**, 390, 405–406, 407, 409, 410, 413
Punctodera chalcoensis **341**, 390, **401**, 405, 406
Punctodera matadorensis 390
Punctodera punctata 117, **349**, **377**, **378**, **387**, 390, **401**, 430, 432
Punctodera stonei 390
Punctoderinae 369, **370**, 384–385, **401**, 407, 413

Radopholus 243
Radopholus similis 226, 405
Rhabditis rainai 245
Rhizonemella sequoiae 410
rice cyst nematodes *see Heterodera oryzae; Heterodera oryzicola*
root lesion nematodes 309, 317, 322
root-knot nematodes 2, 7–8, 29, 37, 75, 77, 78, 82, 219, 220, 222, 223, 228, 229, 240, 242, 251, 252, 253, 271, 292, 309, 318, 319, 359, 368–369
Rotylenchulus 241, 337, 407
Rotylenchulus reniformis **31**, 58

Sacchari group 4, 11, 369, **371**, **401**, 407, 408, 409, 413
Sarisodera **370**, 407, 410
Schachtii group 4, 8, 10, 351, 369, 377, 383, 384, **401**, 407, 408–409, 413
Solanum americanum 219
Solanum pimpinellifolium 218

soybean cyst nematode (SCN) *see Heterodera glycines*
Steinernema 241
sugar beet cyst nematode *see Heterodera schachtii*
sugarcane cyst nematode *see Heterodera sacchari*

tobacco cyst nematode (TCN)
    see Globodera tabacum
Tylenchomorpha 337, 338
Tylenchorhynchus ventralis 292
Tylenchulus 337, 369
Tylenchulus semipenetrans 240

Verutinae **370**, 407
Verutus 1, 407

Vittatidera 1, 3, 103, 366, 393, 407, 409, 413
Vittatidera zeaphila 117, **344**, **345**, **392**, 393, **401**

Xanthomonas gardneri 224
Xanthomonas perforans 224
Xiphinema 241
Xiphinema americanum 30, **32**
Xiphinema index 29–30, **32**

yellow beet cyst nematode (YBCN)
    see Heterodera betae

Zelandodera 407

General Index

Note: bold page numbers indicate figures and tables.

13β-hydroxysteroid dehydrogenase 92

αSNAP 162, 164, 219
abamectin **307**, 319, 322, 324
acacia (*Acacia nilotica*) 314
acetylcholinesterase 220, 221, 322
Achromobacter xylosoxidans 246
Acinetobacter sp. 248–249
Acremonium 242
actin 95, 401–402, **403**, 432
additive effect 274, 275, **276**, **277**, 284, 288, 290, 291, 292
Aegilops ventricosa 117
AFLP (amplified fragment length polymorphism) 412, 432
Africa 8, 11, 197, 319, 326, 386, 388, 389, 410
    see also North Africa; South Africa; West Africa; and see specific countries
Agrobacterium tumefaciens 217–218
agrochemical control 320–327
    application of 323–324, **323**, **324**
    in EU 320, 322, 323, 324, 326
    fumigant *see* soil fumigation
    persistence of 325–326
    precision aspects of 326–327
    seed treatments 324–325
Aizoaceae 104
aldicarb 183, 249, 320, 326
aldolase 400, 402
alfalfa (*Medicago sativa*) 308
Algeria 55, 202
Alium sativum 314, 315

Amaranthaceae 8, 104, 374, 383, 413
amino acids 55, 58, 59–60, 251, 351, 402
ammonia 247, 251, 256, 257, 292, 309, 315, 320
Ammophilia arenaria 292, 383
amphids 49, 57–58, 62–63, 353
amplified fragment length polymorphism (AFLP) 412, 432
Andes 12, 131, 411
antagonistic fungi **239**, 242–245, 316–320
    arbuscular mycorrhiza (AMF) 47, 243–245, 293–294
    endophytic 242–243
antagonistic interactions 242–245, 276, **277**, 280
antibodies 27, 424–425, 433
anus **339**, **340**, 342, 343, 345, 349, **349**, **350**, 352, 365
Aphanomyces cochloides 288
aphids 81, 115, 227, 317
Aphis glycines 115
Aphis gossypii 317
Apiaceae 202, 377
apoplast 75, 78, 155, **158**
Arabidopsis 156, 163, 164–165, 166, 197, 220, 221, 222, 225–226, **226**, 228
Arabidopsis thaliana 29, 61, 92–93, 94, 156, 223, 228, 240, 243, 319
arbuscular mycorrhizal fungi (AMF) 47, 243–245, 293–294
Argentina 3, 10, 13, 113, **130**, 192, 387, 420
Artemisia annua 314
Arthrobacter 248–249

451

Arthrobotrys spp. 256
Arthrobotrys cladodes 240
Arthrobotrys dactyloides 239
Arthrobotrys irregularis 254
Arthrobotrys oligospora 239, **239**, 240, 318
Arthrobotrys robusta 254
Arthrobotrys superba 239
ascarosides 61–62, 92–93, 257
Ascomycota 238, 240, 241, 242
ASD (anaerobic soil disinfection) 309
Asia 9, 10, 11, 202, 374, 379, 380, 381, 382, 386, 388, 389, 408, 410, 411
   see also specific countries
Aspergillus spp. 312–313
assays *see* bioassays
Atropa belladona 389
Australia 8, 9, 54, 117, 131, 134, 186, 197, **307**, 308, 322, 323, 326, 379, 382
auxin/auxin transport 7, 82–83, 218, 223, 293
auxin-inducible 83
Avena sativa 382
Azadirachta indica 314

β1.3 glucanase 166
β1.4 endoglucanase 27–28
Bacillus spp. 246, 252, 253, 318
Bacillus anthracis 253
Bacillus cereus 246, 252, 253, 318
Bacillus firmus 246, **255**, 318
Bacillus nematocida 246
Bacillus sphericus 249
Bacillus subtilis 246, 249
Bacillus thuringiensis 220, 246–247, 252, 253, 318 *see also* Bt crops
backcrossing 185, 186, 200, 202
bacteria 74, 77, 81, 154, 238, 247–249, 252, 305, 316, 322, 326
   chitinase-producing 247
   nematode-feeding 240, 241
   Pasteuria spp. 248, **248**, 252
   symbiotic 292–293
   see also rhizobacteria
Bahrain **130**
ball-milling 433
banana 3, 117, 201, 227
barley 14, **16**, 17, 45, 117, 162, 181–182, 186, 223, 242, 253, 256, **307**, 308, 374, 383, 384
Basidiomycota 238, 242–243
BCAs (biological control agents) *see* biological control
beans 46, 113–114, **307**
Beta 197, 198
Beta patellaris 199
Beta procumbens 198, 201, 218
Beta vulgaris 2, 8, 103–106, **106**, **107**, 157, **158**, 174, 197–201, 242, 248–249, 256, 279, 284–286, 288, 305, **307**, 374, 383

Beta vulgaris ssp. *maritima* 198, 200, 201, 374
Beta webbiana 199
Bioact WG **255**, 317
bioassays 46, 59, 96, 115, 147, 180, 186–187, 189–190, 220, 225–226
   glasshouse trials 194–195, 225–226, 228
   see also pot trials
biochemical/molecular identification 4, 216, 421–436
   and antibody-mediated detection 424–425
   and barcoding 433–434, 435
   and detection in soil 433
   DNA-based 425–434
   future prospects for 434–436
   and IEF 422–424
   and LAMP 435–436
   and mtDNA markers 427–430, 432
   and PCR *see* PCR
   protein-based 422–425
   and rDNA markers 425–427, **428–429**, **430**, 432
   and sDNA markers 430–432
biochemistry 4, 30, 35–36, 53, 78, 89–96
   ascarosides 61–62, 92–93
   carbohydrates 93, 94, 116
   fatty acids 91
   future research on 96
   hydrocarbons 92
   lipids 90–91, 94
   proteins *see* proteins
   research gap in 89, 96
   sterols 92
biofumigation *see* soil fumigation
bioinformatic analysis 74, 89, 215, 227–228
biological control 237–258, **307**, 315–320
   and C:N ratio 256–257, 306, 309, 317
   and direct antagonism 316
   future prospects for 257–258
   management of 254–257
   products 253–254, **255**
   and soil organic matter 256
   three approaches to 237–238, 257
   and tillage 254–256, 257, 308–309
   and transgenic engineering 94, 218–219, 220, 229–230, 258
   see also bacteria; fungi
BioPotatoes Ltd 230
biosafety/biosecurity 129, **133**, 189, 225, 227–229
   see also phytosanitary status/certification; quarantine
biotechnology 215–230
   future prospects for 229–230
   genetic engineering *see* genetic engineering
   PCR/next generation sequencing 216–217, 219
   and plant–nematode interactions 217
Bipolaris sorokiniana 291
black gram (*Vigna mungo*) 10, 202

black root/blackleg 288
Bolivia 12, 13, 109, 412
Botrytis cinerea 319
Bradyrhizobium japonicum 249, 293
Brassica spp. 311–312, 313, 379
Brassica carinata 311
Brassica juncea 311, 312
Brassica napus 5–6, 198, 199, 256, 285, 286–288, 308, 383
Brassicaceae 8, **307**, 308, 314, 374, 381, 383
Brazil 3, 10, 113, **130**, 192, 195, **255**, 420
breeding 218–219
Brevibacillus laterosporus 246, 248
Britain (UK) 13, 18, 108, 110, 138, 140, 200, 230, 241, 248, 323, 381
    nematicide regulations in 326
brown stem rot (*Cadophora gregata/Phialophora gregata*) 115, 281, 282–283
Bt crops 220, 229, 230
bullae 351–352, 356, **371**, 384
Burkina Faso 11

C:N ratio 256–257, 257, 306, 309, 317
cabbage 240–241, 285, 308, 374, 379
Cactaceae 391
Cadophora gregata see Phialophora gregata
Cajanus cajan 10, 202, 244, 291, 308
Cajanus platycarpus 202
Calonectria ilicicola/Calonectria crotalariae 281, 284
Canada 3, 8, 10, 109, 112, 113, 131, 134, **145**, 192, 193, 377, 382, 383, 388, 390, 411, 412, 429
canister tests 189–190
canola *see* oilseed rape/canola
Capsicum annuum 219, 386, 389
carbamates **307**, 321, **321**, 322, 326
carbofuran 320, **321**, 322, 323, 325, 326
carbohydrate binding module (CBM) **77**, 78–79
carbohydrates 93, 94, 116
carrot 117, 202, 305, **307**
Caryophyllaceae 8, 104, 374, 381, 383, 413
Catenaria auxiliaris 238, **239**, 240
cDNA 27–28, **31**, 33, 400, 402
cell wall degrading enzymes (CWDEs) 75, 77–79
cellulases 36–37, 77–78, 94, 156
Cerastium holosteoides 381
cereals 3, 305, **307**, 308
    see also specific crops
Cereus speciosus 391
chemoreception 59, 62–63, 344
Chenopodiaceae 104, 285, **307**, 381
chickpea (*Cicer arietinum*) 308
chickweed (*Cerastium holosteoides/Stellaria media*) 381
Chile 13, 109, **130**, 318, 387
China 3, 9, 10, 54, 117, **130**, 131, 181, 191, 192, 193, 201, 241, **307**, 326, 380, 420

chitin/chitinase 93, 94, 95, 166, 240, 242, 247, 251, 252, 257, 319
chloropicrin 320, 322, 323
cholesterol 92
cholinesterase inhibitors 249, 321, 322
chorismate mutase 37, 75, 81–82, 166
Chromobacterium sp. 247
chromosomes 161, 181–182, 185, 188, 199, 202, 374, **404**
Chrysodeixis includens 115
Chytridiomycota 238
Cicer arietinum 308
cisgenesis 218
Cladosporium fulvum 81, **158**, 159
CLARIVA 316
CLE peptides 76–77, 82, 412
climate change 175, 434
cloaca 342, 344, 347
clover, white (*Trifolium repens*) 202, 291, 292–293, 383
cluster bean (*Cyamopsis tetragonolobus*) 10, 377
co-evolution 7, 79, 131, 154, 160, 161, 166, 252, 258, 412–413
    see also resistance
Cochliobolus sativus 383
Codex Alimentarius Commission 228
Colombia 3, 10
confined field trials (CFTs) 227
Congo 3, 11
conservation agriculture 320
convergent evolution 34, 238
Coprinus comatus **239**, 242
copy number variation **158**, 162, 193, 219
corn *see Zea mays*
corn rootworm 221
Cornell University breeding programme 190, 191
Côte d'Ivoire 11, 326
cotton 227, 229, **255**, 271
cotton aphid (*Aphis gossypii*) 317
cover crops 105, 106, 118, 175, 256, 306, 312
cowpea (*Vigna unguiculata*) 10, 202, 244, 294, **307**, 308
cress (*Lepidium sativum*) 324
CRISPR/CRISPR-Cas9 224, 230
crop rotation 2, 18, 103, 104, 106, 112, 117, **133**, **134**, 175, 176, 238, 306–308, **307**, 420
Crotalaria juncea 377
Cruciferae 3, 104, 285
cryobiosis 53, 54
crystal proteins (Cry) 220, 247
cucumber (*Cucumis sativus*) 319
cuticle 341–342, **341**, 343, 348, 351
CWDEs (cell wall degrading enzymes) 75, 77–79
Cyamopsis tetragonolobus 10, 377
Cyperaceae 201
Cyprus 9, 202, 383
cystatins 220, 225–226, 227–228, 230
cysteines 94–95, 159, 217, 220, 225, 311

cysts 3–4, **5**, 6, 11, 54, 337, 350–352, 365, **367**, 368, **375**, **392**
    bullae of 351–352
    colour of 350–351, **351**
    cuticular layering of 351
    vaginal transformation in 352
cytokinin/cytokinesis 83, 95, 197

Dactylella oviparasitica **239**, 240
Dactylella pseudoclavata **239**
DAMPs (damaged host cells) 155, 165
DAPG (2,4-diacetylphloroglucinol) 245–246
data collection/monitoring 118
Datura spp. 386
Daucus spp. 202
Daucus carota 377
dauer formation/regulation 53, 61, 62, 96
daumone pheromones 61–62
Dazomet **307**, 324
de novo assembly 33, 47, 320
desiccation 11, 53, 54, 62
diapause 45, 51, 54–56, **55**
    facultative 5, 9, 54, 56
    obligatory 9, 54
Diaporthe phaseolorum 281, 284
Dickeya ssp. 36
digestive system 37, 342–343
Diplenteron sp. 240
Discocactus akkermannii 391
dissection/dissecting microscope 352–353, **353**, 356–357
distribution of cyst nematodes 3
DNA analysis 47, 215, 229–230
    see also biochemical/molecular identification
DNA barcoding 433–434, **435**
DNA cloning 216, 217
DNA databases//germplasm collections 112, 178, 182, 188, 192–193, 360, 426, 433
dormancy 5, 9, 50–56, **52**, 62, 340
    and diapause *see* diapause
    and quiescence 52–53, **55**
Drechslera dactyloides **239**
drought 5, 8, 9, 17, 45
Dryad (database) 38
dsRNA 221–222, **222**, 227

economic importance of cyst nematodes 1–2, 3, 174, 181, 187, 192, 198, 281, 284–285, 305
effector triggered immunity (ETI) 79–81, **80**, 162
effectors 30, 33–34, 74–83, 94, 217, 219, 222, 223, 250
    CLE peptide 76–77, 82
    defined/genomic analyses of 74–75
    and degradation of plant cell wall 77–79, **77**
    DOG (DOrsal Gland box) 34

effectors, HYP 33, 58, 75, 76
    and feeding site induction 82–83
    functions/actions of 75–77, **76**
    future research areas 83
    sequence homologues in 227
    sources of 75
    and suppression of host defences 79–82, **80**
    see also CWDEs
eggplant (*Solanum melongena*) 12, 386, 389
eggsac 6, 10, 18, 240, 337, 340
eggs 1, 4, **5**, 6, 11–12, 18, 47–50, 90, 116, 199, 316, 318, 321, 338–339, 350
    change in permeability of 48–49, 54, 93
    chitin in *see* chitin/chitinase
    cold tolerance of 56–57
    decline rates 109–110
    and dehydration/water loss 11, 54
    and dormancy *see* dormancy
    and fungi/bacteria 240, 241–242, 251, 254, 256, 280
    produced outside body 10, 128
    and soil sampling 140–144, **141**, **142**, **143**, **144**, 146, 148–150, 182
    and trehalose 93
Egypt 3, 10, 202, 326, 384, 409
electrophoresis 422–423, 424, 427, 432
electrophysiology 59, 60, 63
ELISA techniques 424, 425, 433
Elytrigia repens 383
endoparasitic fungi 240–241
endophytic fungi 242–243, 320
England *see* Britain (UK)
enrichment index 227
enzymes 75, 77–79, 156, 164, 197, 239, 248, 250–251, 315, 317, 319, 402, 412, 423, 425–426
    proteinase 251–252, 258
    see also protease
EPPO *see* European and Mediterranean Plant Protection Organization
Eruca sativa 311, 312
EST (expressed sequence tag) 30, 400
esterase isoenzyme patterns 4
ethoprophos 321, **321**, 322, 325, 326
ethylene 156, 164, 224, 319
ETI (effector triggered immunity) 79–81, **80**, 162
eucalyptus (*Eucalyptus* spp.) 314
Euphorbia 391
Europe 8, 9, 131, **145**, 185, 186, 197–198, 199, 200, 202, 314, 374, 377, 379, 381, 382, 383, 386, 388, 389, 390, 410, 411, 412, 429, 432
    see also specific countries
European and Mediterranean Plant Protection Organization (EPPO) **130**, 131, 132, 146, 147, 188, 189, 419–420, **420**

European Union (EU) 132, 135, 139, **139**,
    148, 317, 319, 320, 322, 323, 324, 326,
    420–421

faba bean (*Vicia faba*) 308
Fabaceae 10, **307**, 374, 376, 381
FAO (Food and Agriculture Organization) 129–130,
    129, 132, 230, 419
fatty acids 57, 61, 90, 91, 92, 94, 309, 320
female index 14, **17**, 193, 195
females, morphology of 338, **344**, 347–350, 368
    cone 3–4, 348–349, 353
        nervous/excretory system 350
        reproductive tract 349–350
        terminal region 342, 348–349, **349**, **350**
        'white, flaky' layer 348
Fenamiphos 321–322, 325, 326
fenestrae 4, 321, 348, **349**, 351, 352, 381
fibronectin 252, **253**
Ficus carica 380
Ficus elastic 380
field studies/trials 183, 225–228, **226**, 229,
    273–274
    confined 226
flooding/flooded soil 11, 116, 134, 309
FLPs (FMRFamide-like peptides) 62, 95–96
fluensulfone 321, **321**, 322, 325
Food and Agriculture Organization 230
formalin 238, 354
fosthiazate 321, **321**, 322, 325, 326
France 9, 197–198, 374
Frankliniella occidentalis 166
freezing/ice 53, 54, 56–57
fructose biphosphate aldolase 400, 402
fungi 18, 63, 74, 81, 115, 154, 183, 238–245, **239**,
    308–309, 316–319, 383
    and biocontrol products 253–254
    endoparasitic 240–241
    endophytic 242–243, 320
    in interactions 271, 278, 279, 280–291
    see also antagonistic fungi; nematophagous
        fungi
Fusarium spp. 242, 281
Fusarium avenaceum 291
Fusarium culmorum 290–291
Fusarium oxysporum 243, 281, 290, 291
Fusarium solani 381
Fusarium udum 249, 291
Fusarium virguliforme 115, 281–282, **282**
Fusarium wilt 290, 291, 381

Gaeumannomyces tritici var. *tritici* 308–309
garlic (*Alium sativum*) 314, 315
GenBank 359, 433, 434
gene cloning 27–28, 218–219

gene expression 7, 32–33, 49, 55, 58, 79, 92, 93,
    94, 95, 225, 228–229, 251, 434
    and resistance 162–164, 165–166, 167,
        225, 280
gene promoters 220, 221–222, **222**, 224, 225,
    226, 228
gene-for-gene concept (*Avr/avr*) 157, **176**, 178, 199,
    200, **203**
gene/genome editing 187, 224, 258
genebanks 112, 178, 182, 188, 192–193, 359,
    433, 434
genetic engineering 191, 215–216, 217–228
    and biosafety 227–228
    gene/genome editing 187, 224, 258
    GM field trials 225–227, **226**
    and nematode feeding/digestion 219–220
    and peptide repellents 220–221
    and resistance 218–219, 225
    and RNAi 221–222, **222**, 224, 225, 227
    and susceptibility genes 222–225
genetic transformation 217–218
genomics/transcriptomics 27–39, 63, 89, 162–164,
    165, 227, 253, 257, 280, 295
    cost of DNA sequencing 28, **28**
    current status of 29–30, **31–32**
    and data accessibility 38
    and effectors *see* effectors
    future prospects for 38–39
    and horizontal gene transfer (HGT) 36–38
    identifying genes of interest in 30
    identifying pathways/targets for control
        35–36
    limitations of 29
    mitochondrial 36
    and population genetics/metagenetics
        34–35
    and promoter motifs 34, 229
    sequence similarity searches 30–32, 33
    spatial transcriptomic profiling 33–34
    temporal transcriptional profiling
        32–33
Germany 3, 9, 131, 197–198, 199, 200,
    241, **255**, 313, 374, 381, 383, 386
germplasm collections *see* genebanks
giant cells 7–8, 37, 229
Gibson assembly 230
gland cells 27, 33–34, 74
    pharyngeal 33, 75, 83, 343
gland lobe **339**, 343, **345**, 349, 350
glasshouse trials 194–195, 225–226, 228
Glomeromycota 243
Glomus epigaeus 244, 294
Glomus fasciculatum **239**, 244, 249, 294
Glomus mosseae 294
glucosinolates 165, 311, **311**, 312
glutathione synthase 35, 94
glycine 59–60

Glycine max 2, 3, 7, 10, 14, 50–51, 59, 95, 112–116, **158**, 163, 167, 174, 191–197, 202, 216, 219, 220, 222, 223, 224, 225, 226, 229, 246, **255**, 283, 293, 305, 306, 380–381, 420
    'Lee' 14
    'PI 88788'('Bedford') 14, **17**, 113, 163, 163–164, 192, 196–197, 281
    'PI 90763' 14, **17**, 192, 197
    'PI 548402'('Peking') 14, 163, 192, 197
    'Pickett' 14, **17**
Glycine soja 197
Glycine tomatella 197
GMOs (genetically modified organisms) 217, 227–228
Golden Gate cloning 230
GPCRs (G-protein coupled receptors) 95–96
Graminae 201
Granek's ratio 351, 368, 386–387, 389
grapes/grapevine 256, 305, 317, 318
grasses 11, 12, 45, 292, **307**, 374, 380, 382, 390
green manure 254, 256, 306, 320
growth chambers 226, 272–273

hairpin construct 221–222, **222**
hairy root cultures 200, 220, 225, 377
hatch inhibitors (HI) 47
hatching 44–57, 59, 62, 91, 94–95, 279
    and chemical cues 108–109
    and cold tolerance 56–57
    and diapause 45, 54–56, **55** *see* diapause
    and dormancy/durability of hatching 50–53, **52**
    effect of microorganisms on 47
    and fungi/bacteria 241, 244, 245–246
    mechanism of 47–49, **48**
    multiple generations per year 6–7, 10, 49–50, 105
    and osmotic regulation 56
    and responses to water 5–6, 9–10, 45, 48–49, 51
    and root diffusates *see* root diffusates/leachates
    and soil types 45
    and survival 53–54
    and temperature *see* temperature factor
    see also population dynamics
hatching factor stimulants (HS) 47
Hedysarum coronarium 308
henbane, black (*Hyoscyamus niger*) 389
henbit (*Lamium amplexicaule*) 381
HG Type Test 14, 115, 193, 194, 195
HIGS (host-induced gene silencing) 221, **222**
Hirsutella spp. 254
Hirsutella minnesotensis **239**, 240, 241, 255, 256
Hirsutella rhossiliensis **239**, 240, 241, 255, 256
hops 382
Hordeum vulgare 382

horizontal gene transfer (HGT) 36–38
hormones 82, 83, 92, 164–166, 280
host range 2, 3, 6, 7, 10, 17, 44–45, 54, 107, 117, 161, 175, 202, 285, 305, 306, 308, 366, 379, 381
    field studies 104
host-induced gene silencing (HIGS) 221, **222**
Humulus lupulus 382
hydrocarbons 92
    halogenated **321**, 322
Hyoscyamus niger 386, 389
HYP (hyper-variable extracellular) effectors 33, 58, 75, 76

identification of cyst nematodes 3–4, 366–368, 419–436
    biochemical/molecular *see* biochemical/molecular identification
    diagnostic features in 368
    morphological 216
    and regulations/standards 419–421, **420**
    and root staining 368
    and scanning electron microscopy *see* SEM
    second-stage juveniles 4
    techniques 367
    see also morphology
IEF (isoelectric focusing) 4, 422–424
IgE tests 228
Illumina RNA sequencing (RNAseq) 33, 38
India 3, 9, 10, 11, 131, 186, 201, 202, 243, 248, **255**, 376, 384
Indonesia 3, 10
inoculum 50, 116, 177, 180, 182–183, 189, 199, 244, 272, 273, 281, 291, 316
    see also pathotypes
insecticides **107**, 220, 230, 324–325
insects 59, 63, 154, 230, 248, 284, 305
insulin-like peptides (ILPs) 62, 95, 96
integrated pest management (IPM) 306, 328
interactions 217, 271–295
    and aboveground pests 284
    antagonistic 242–245, 276, **277**, 280
    and CCN 290–291
    defined/use of term 272, 274
    evidence of 274–276, **275**, **276**
    fungi/oomycetes 271, 278, 279, 281–291
    future prospects for 294–295
    and *Globodera* spp. 288–290
    and *Heterodera glycines* 281–284, **285**
    and *Heterodera schachtii* 284–288, **287**
    histological studies on 278, 290
    mechanisms of 276–280, **277**
    methodologies for investigating 272–274
    and 'modified rhizosphere' effect 279
    and mycorrhizae 293–294
    nematode–nematode 291–292
    and 'physiological changes' effect 278–279

and 'resistance breaker' effect 279–280
and resistant cultivars 286, **287**, 290, 293
and rhizobia 292–293
sequential/simultaneous 278, 290
and split root technique 247, 273, 278, 283, 289, 290, 293
and symbionts 292–294
synergistic 247, 257, 276, 277–280, **277**
and 'wounding agent' effect 277–278
see also bacteria
internal transcribed spacer (ITS) 4, 374
International Plant Protection Convention (IPPC) 129, 132, 419–420
International Standards on Phytosanitary Measures (ISPM) 128, 129, 130, 132, 419
introgression 175, 178, 186, 187, 199, 218–219
introns 37, 221–222, **222**
IPM (integrated pest management) 306, 328
Iran 3, 9, 10
Ireland 110, 200
irrigation 11, 115, **255**, 317, 318, 323
isobologram analysis 275–276
isoelectric focusing (IEF) 4, 422–424
isothiocyanates 311, **311**, 312
isozymes 423
ISPM *see* International Standards on Phytosanitary Measures
Israel 9, **130**, 382
Italy 3, 109, 131, 201
Ivory Coast 3, 11

Japan 3, 9, 10, 116, 192, 201, 379, 381, 382, 423
jasmonic acid (JA) 164, 165, 166, 280, 319
Jordan 9, **130**, 420
JR Simplot 230
juveniles, second- to fourth-stage (J2-J4) 4–5, **5**, 6, 48, 49, 50, 58, 59, 82, 90, 182, 240, 243, **248**, 318, 321, 338, **339**, 340–346, 368, **375**
cuticle 341–342, **341**, 343
detection of, in laboratory diagnosis 145–146
heads **377**
nervous/sensory system 344–345
reproductive system 344
secretory-excretory system 345–346
stylet/tail region 340, **340**, 342–343
tails **378**

Kenya **255**, 318, 319
Klamic **255**, 318
Koch's postulates 258, 281
Korea 3, 10, 192

Lactuca sativa (lettuce) 219
Labiatae 104

laboratory diagnosis 144–147, 150
automated 147–148, **149**
and carousel rinsing machine 148, **149**
cysts from soil 146
DNA/RNA trehalose methods 147, 148
endoparasitic/sedentary stages from roots 145–146
extraction techniques 144–145
free-living stages from soil 145
future prospects for 150–151
and identification/viability assessment 146–147
PCR techniques 147, 148
SASA process 148
Lamium amplexicaule 381
Lamium purpureum 310, 381
LAMP (loop-mediated isothermal amplification) 435–436
late blight disease *see Phytophthora infestans*
leafhoppers 227
legumes 3, 308
lentil (*Lens culinaris*) 308
Lepidium sativum 324
life cycle 1–2, 4–8, **5**, 10–11, 44, 108, 137, 180, 277, **338**
dormancy in *see* dormancy
effect of abiotic factors on 8
eggs *see* eggsac; eggs
hatching *see* hatching
and host range 6
mating 6
number of generations per year 6–7, 10, 49–50
and sensory perception *see* sensory perception
syncytia *see* syncytia
light microscopy (LM) 352–353, **353**, 360
linear interaction effect 274
lipids 90–91, 94, 109, 321
lipoprotein membrane 48, 49
Loewe additivity 275–276
Lolium multiflorum 90
Longidorus elongatus 29–30, **32**
loop-mediated isothermal amplification (LAMP) 435–436
lumen 342, 343

MABs (monoclonal antibodies) 424–425, 433
Macrophomina phaseolina 281, 283, **283**
Magnaporthe oryzae (rice blast fungus) 224
maize *see Zea mays*
males, morphology of 338, **345**, 346–347, 368
MAMPs (microbe-associated molecular patterns) 257–258
management/control strategies 17–18, 175, 187, 198, 201, 305–328
ASD/soil amendments/inundation 309
biocidal chemicals *see* agrochemical control
and biosafety 227–229

management/control strategies (*continued*)
    and certified planting material 17
    and crop rotation *see* crop rotation
    in EU 317, 319, 320, 322, 323, 324, 326
    future prospects for 327–328
    and GM crops 229–230
    green manure 254, 256, 306, 320
    Heterodera schachtii 106, **106, 107**
    IPM 306, 328
    and nematicides *see* nematicides
    overview 305–306, **307**
    and PCR 216–217
    and peptide repellents 220–221, 225–226, **226**, 227–228, 230
    and plant biomass/oils/extracts 314–315
    and promoter choice 228–229
    and proteinase inhibitors 217, 219–220, 225, 227, 228
    and RNAi 221–222, **222**, 224, 225, 227
    solarization 309–310
    and susceptibility genes 222–225
    tillage 254–256, 257, 308–309
    trap cropping 2, 106, **133**, 134, 175, **307**, 313–314
    and weeds 310–311
    see also biological control; quarantine; resistance breeding
manure 256, 306, 309
    green 254, 256, 306, 320
MAPKs (mitogen-activated protein kinases) 156, 162, 164
MapMan 36
marram grass (*Ammophilia arenaria*) 292, 383
MAS (marker assisted selection) 180, 185, 186, 187, 188, 195, 202
median bulb/metacorpus **339, 341**, 343, **345**, 349
Medicago sativa 308
Mediterranean 3, 9, 54–55, 200, 202, 380, 382, 408
Melocon-WG **255**, 317
Mendel, Gregor 218
metabolomics 279, 280, 295
metagenetics 34–35
metam sodium **307, 321**, 322, 323–324, **325**, 326
methyl bromide (MeBr) 305–306, **307**, 310, 312, 320, 322, 326
Mexico 391, 411
microarray analysis 7, 55, 163, 224, 228, 229
Microbacterium nematophilum 250
microbe-associated molecular patterns (MAMPs) 257–258
microbes 237–258
    attachment/surface recognition by 252–253
    diversity of 250–251
    see also bacteria; biological control; fungi
microplots 51, 104–105, 180, 184, 244, 272, 273
Middle East 3, 8
millet 384
miRNA 95

mitogen-activated protein kinases (MAPKs) 156, 162, 164
Mocap 321, **321**
molecular identification *see* biochemical/molecular identification
molecular markers 188, 189, 191
molluscs 241, 248
Monacrosporium cionopagum 238, 240
Monacrosporium doedycoides 239
Monacrosporium ellipsosporum 240
Monacrosporium lysipagum 240
monoclonal antibodies (MABs) 424–425, 433
monoculture systems 9, 241, 306, 320, 420
Monsanto 220, 327
morphology 337–360, 366, 421
    cysts *see* cysts
    egg/embryo 338–339
    females *see* females, morphology of
    future prospects for 359–360
    Granek's ratio 351, 368, 386–387, 389
    and LM techniques 352–353, **353**, 360
    males **345**, 346–347
    pharynx 342, 343, **344**, 347, 349
    reproductive system 344, 347
    second- to fourth-stage juveniles **339**, 340–346, **341**
    secretory-excretory system 345–346, 347
    and SEM techniques 350, 352, 353–355, **355**, 360
    and species description standards 357–359
    and TEM techniques 349, 352, 354–357
    terminal region 342, 348–349, **349, 350**
    see also taxonomy
multipartism 36, 403, 405, 406, 429
mungbean (*Vigna radiata*) 10, 202, 308
Musa paradisiaca 12, 227
mustard 256
    Ethiopian (*Brassica carinata*) 311
    Indian (*Brassica juncea*) 311, 312
    white (*Sinapis alba*) 59, 198, 199, 374
Myanmar 3
Myb-related 163
mycorrhizae 293–294
myrosin/myrosinase 311, **311**, 312
Myrothecium verrucaria **255, 307**, 318

nAChRs 220–221, 226, **226**, 227
national plant protection organizations (NPPOs) 319, 419, 421
NB-LRR 78, 79, 157, 159–160, 161–162, 163
neem (*Azadirachta indica*) 314
NemaDecide 110, 111, **111**
Nemathorin 321, **321**
nematicides 18, 63, **133**, 176, 183, 221, 227, 242, 249, 305, 305–306, 306, **307**, 320–327
    application of 323–324, **323, 324**
nematophagous fungi 238–242, **239**, 250–251

egg/female parasitic 241–242, 243
endoparasitic 240–241
nematode-trapping 238–240, 243, 256, 257
toxin producing 242
Nematophthora gynophila 18, 238
Nematroctonus robustus **239**
nervous system 344, 347, 350
Netherlands 131, 137, 148, 198, 314, 374, 391
neuropeptides 95–96, 217, 221
New Zealand 3, 109, 202, 382, 388, 409, 410, 411
next generation sequencing 216–217, 219
Nicandra physalodes 389
Nicotiana acuminata 386
Nicotiana benthamiana 159, 160, 167
Nicotiana tabacum 305, 314, 380, 389
Nimitz 15G **321**, 322
NLPs (neuropeptide-like proteins) 95, 96
NLR (nucleotide-binding, leucine-rich repeat) proteins 159, 219
nodule formation 278, 292–293
North Africa 9, 200, 374, 381, 382
North America 2, 131, 132, 193, 197, 374, 379, 380, 386, 388, 389, 409, 410, 412
see also specific countries
Norway 3, 9, **130**
NPPOs (national plant protection organizations) 319, 419, 421

oats 2, **16**, 17, 45, 181, 182, 186, **307**, 308, 384
oil seed rape 5–6
oilseed radish (*Raphanus sativus*) 198, 199, **307**, 311, 313, 383
oilseed rape/canola (*Brassica napus*) 5–6, 198, 199, 256, 285, 286–288, **307**, 308
Onagraceae 104
1,3-dichloropropene **307**, 320, 322
oomycetes 74, 154, 281, 284, 288
organophosphates **307**, 321, **321**, 322, 326
ornamental plants 227, 317, 380, 391
Oryza breviligulata 201
Oryza elachista 201
Oryza glaberrima 201–202
Oryza sativa 3, 11–12, 94, 166, 174, 201–202, 305, **307**, 308, 309, 379, 384
oryzacystatin 94, 220
osmotic regulation 48–49, 53, 54, 56
oxamyl 18, 90, 321, **321**, 322, 323, 325–326
oxime carbamates 18, **307**, **321**, 322, 326

Paecilomyces lilacinus 238, **307**
Pakistan 3, 9, 201, 202, 374, 376, 384, 409
PAMPs (pathogen associated molecular patterns) 61, 79, **80**, 93, 155
Paraguay 3, 10, **130**
Pasteuria spp. 248, **248**, 252, 253, **307**, 316
Pasteuria nishizawae 248, 255, **255**, **307**, 316
Pasteuria penetrans 238, 253, 255, 258, 292, **307**
Pasteuria ramosa 253
pathogen associated molecular patterns (PAMPs) 61, 79, **80**, 93, 155
pathotypes 9, 13–14, **15–16**, 34–35, 111, 131, 178–179, 187, 188, 190, 199
pathway mapping 35–36
pattern recognition receptors (PRR) **80**, 154, 155, 157
pattern triggered immunity (PTI) 79–81, **80**
PCN Calculator Pallida 110, 111, **111**
PCR (polymerase chain reaction) 4, 147, 150, 189, 216–217, 224, 422, 425, 426–427, **428–429**, 432, 434–435
digital (dPCR) 435
DNA barcoding 433–434, 435
and RAPD/AFLP 432–433
real-time/quantitative (RT-PCR/qPCR) 427, **430**, **431**, 432–433, 434
PCR-RFLP (polymerase chain reaction-restriction fragment length polymorphism) 4, 8, 425–426, **426**
pea, field (*Pisum sativum*) 308
peas 2, 18, 305, **307**, 381
pectate lyases (PELs) 78
Pedaliaceae 10
pepper 219
peptide repellents 220–221, 225–226, **226**, 227–228, 230
allergenic risk of 228
peptides 62, 95–96, 217, 221, 225–226, **226**
signal 30, 32, 34, 35, 80
perivitelline fluid 4, 48, **48**, 57
Peru 3, 10, 13, 389, 411
phagostimulatory response 59–60
Phalaris minor 383
Phalaris paradoxa 383
pharynx 342, 343, **344**, 347, 349
Phaseolus 10
Phaseolus vulgaris 113
phasmids/phasmid openings 57, **340**, 353, 368, 369, 385, 392, 393
PhastSystem 422–423
Phialophora gregata 115, 281, 282–283
phospholipids 91, 164
Phyllanthus maderaspatensis 10
phylogeny/phylogenetic analysis 8, 38, 78, 253, 337, 357, 358, 359, 399–413
and co-evolution 412–413
future prospects for 413
of *Globodera* 410–412
of Heteroderidae 407–410
and mtDNA 403–406, **404**
and nuclear protein-coding genes 400–403
of Punctoderinae 409–410
and rRNA genes 399, 400, **401**, 402

Physalis spp. 386
Physochlaina orientalis 386
Phytophthora capsici 224
Phytophthora infestans 78, 81, 219, 223, 230
Phytophthora sojae 281, 284
phytosanitary status/certification 128, 129–130, **129**, 132, **145**, 419–421
pigeon pea (*Cajanus cajan*) 10, 202, 244, 291, 308, 310
pineapple 253
Pinus 410
Piriformospora indica **239**, 242–243
PIs (proteinase inhibitors) 217, 219–220, 225, 227, 228
Pisum abyssinicum 202
Pisum elatius 202
Pisum sativum var. *arvense* 202
Pisum sativum 10, 308, 381
plant immune system *see* resistance
plant signals 58–60
plantain (*Musa paradisiaca*) 12, 227
Plasmodium spp. 81
Pleurotus spp. **239**
Poaceae 8, **307**
Poales 413
Pochonia chlamydosporia 18, 238, **239**, 241, 251, **255**, 256, 258, **307**, 317–318, 327
Pochonia rubescens 251
Poland 3, 9, 241
Polygonaceae 8, 104, 383
polymerase chain reaction *see* PCR
population dynamics 101–118
  abiotic factors in 105, 106, 115–116, 118
  aggregated distribution 195–196, 285
  and damage thresholds 101, 103, 105–106, 110, 116
  equations for 102–103, 109, 110
  and equilibrium density 102, **102**, 103, 104
  future research areas in 118
  Globodera spp. 107–112, **108**, **111**
  and growing season length/delayed planting 114–115
  Heterodera glycines 103, 112–116
  Heterodera schachtii 103–107, **103**, **104**, **106**, **107**
  Heteroderinae 116–117
  history of 101–103
  and host suitability 104–105
  and management practices 106, **106**, **107**, 110–112, **111**, 114–115, 117–118
  and overwinter survival 112, 114, 116
  and pests/pathogens 115
  and resistant varieties 109, 111, 112–113, 114, 115
  and soil temperature 103, 104, 105, 106, 109, 114
  three phases of 102, **103**
  and virulence 115, 178–179
  and weeds 114, 175, 285, 305, 310–311, 383
population ecology 101, 102, **106**
population genetics/metagenetics 34–35
Portugal 312, 374, 384, 411
pot trials 53, 183–184, 189, 190, **190**, 246, 273
potato root diffusate (PRD) 46–47, 59, 60, 63, 91, 318, 321
potato (*Solanum tuberosum*) 2, 12–13, **15**, 17, 18, 46–47, 62, 81, 94, 107, 129, 130, 131, 161, 165–166, 174, 187–191, 220, 221, 222, 224, 228, 229, 230, 242, 247, **255**, 279, 288–289, 305, **307**, 308, 318, 323, 386, 389, 420, 421
  'Cara' 313
  certification for 139, 388
  'Desiree' 220, 225, 312
  gene introgression into 218–219
  GM 225–226, **226**, 227
  'Golden Wonder' 244
  'Kartel' 313
  late blight disease *see Phytophthora infestans*
  'Maria Huanca' 225, 230
  'Sante' 225, 230
powdery mildew 223, 224
PRD (potato root diffusate) 46–47, 59, 60, 63, 91, 318, 321
protease 94–95, 246, 248
protease inhibitors 94, 156, 163
proteinase inhibitors (PIs) 217, 219–220, 225, 227, 228
proteins 93–96, 116
  identification using 422–425, 434–435
  lipid-binding 94
  neuropeptides 95–96
proteomics 89, 164, 279, 280, 294–295, 399
PRR (pattern recognition receptors) **80**, 154, 155, 157
Pseudomonas spp. 245–246
Pseudomonas aurantiacea 246
Pseudomonas fluorescens 245–246, 319
Pseudomonas oryzihabitans 246
Pseudomonas putida 246
Pseudomonas syringae pv. *glycinea* 284
PTI (pattern triggered immunity) 79–81, **80**
purple deadnettle (*Lamium purpureum*) 310, 381
Purpureocillium lilacinum 238, **239**, 241–242, **255**, 309, 316–317
Pythium aphanidermatum 288
Pythium ultimum 288

qPCR (quantitative PCR) 427, **430**, **431**, 432–433
quarantine 2, 17–18, 112, 128, 129–135, 189, 420
  definitions/categories in 129–130
  and eradication/containment 132–135
  and *Heterodera glycines* 131

and ISPM guidelines 128, 129, **129**, 130, 132, 419
and PCN 130–131, 132–135, **133–134**
and pest free areas (PFA) 132, **133–134**
and phytosanitary status 132
and regulatory control 130, **130**, 135, 144, **145**, 150
and sampling *see* soil sampling
and surveillance 131, 132, **133**, 134
quiescence 53–54

R (resistance) genes 79, 81, 155, 156–159, 161, 162, 163, 165–166, 167, **176**, 178, 191, 199, 218–219, 230
R gene sequence capture (RenSeq) 219
RACE-PCR (rapid amplification of cDNA ends polymerase chain reaction) 27–28
races 10, 11, 13–14, 14, **17**, 179, 194, 377, 384, 422
Ralstonia solanacearum 309
RAPD (random amplified polymorphic DNA) 411, 412, 432
Raphanus sativus 198, 199, **307**, 311, 313, 383
RCRR (Rhizoctonia crown and root rot) 285–286, **287**
Real Trichoderma 319
receptor-like kinases/proteins (RLK/RLP) 157, 159
regional plant protection organizations (RPPOs) 419–420, **420**, 421
regulated non-quarantine pests (RNQPs) 129–130
RenSeq (*R* gene sequence capture) 219
reproductive system 344, 347, 349–350
resistance 7, 13–14, 18, 34, 94, 109, 111, 112–113, 114, 154–168, **158**, 243, 290, 306, **307**
  basal immune responses 155–156, 167
  and biosafety 227–228
  breakdown of 279–280
  and copy number variation/natural mutations 162
  and cystatins 220, 225–226, 227–228, 230
  developing technology for 225–230
  effector recognition 159–160
  effector triggered immunity (ETI) 79–81, **80**, 162
  evaluation of 176, 179–180
  field trials 225–227, **226**
  and genetic engineering/breeding 218–219
  hormone-mediated defence responses 164–166
  host-specific immunity 156–160
  and insecticides 220
  and nematode feeding/digestion 219–220
  partial 175, 188, 199–200, 201, 225, 229, 230, **307**
  and PCR/next generation sequencing 216–217
  and peptide repellents 220–221, 225–226, **226**, 227–228, 230
  quantitative aspects of 161–162
  and RNAi 221–222, **222**, 224, 225, 227
  and susceptibility genes 222–225
  systemic acquired (SAR) 165, 280
  and T-DNA transfer 218
  transcriptional reprogramming/defence gene expression 162–164
resistance breeding 174–203, **203**, 230
  and aggregated population distribution 195–196
  for BCN 197–201
  benefits/costs of 174–176
  for CCN 181–187
  for cereal cyst nematodes 181–187
  Cornell University programme 190, 191
  development of cultivars 185–186, 190–191, 196, 199–200
  essential reading on 186
  future challenges for 186–187, 191, 196–197, 200–201, 202
  and inoculum 177, 180, 182–183
  and interpretation of 'resistance' 185
  and marker assisted selection *see* MAS
  for PCN 187–191
  and pot trials 182, 183–184, 190
  for SCN 191–197
  and screening/phenotyping *see* screening
  and selection of nematode populations 188–190, **190**, 194, 199
  and sources of resistance 178, 179, 181–182, 188, 192–193, 198–199
  and tolerance 176–178, **177**, 183, 187
  and virulence/pathotypes/races 178–179
Resistance (*R*) genes 79, 81, 155, 156–159, 161, 162, 163, 165–166, 167, **176**, 178, 191, 199, 218–219, 230
RFLP (restriction fragment length polymorphism) 399, 425–426, **426**
rhizobacteria 18, 47, 245–247
  Bacillus spp. 246, 252
  plant health/growth promoting (PHPR/PGPR) 245
  Pseudomonas spp. 245–246
  Rhizobium etli 247
rhizobia 292–293
Rhizobium etli 247
Rhizoctonia crown and root rot (RCRR) 285–286, **287**
Rhizoctonia solani 251, 279, 281, 283–284, 285–286, 288–289, 290
rhizosphere 18, 57, 58, 221, **226**, 242, 249, 257, 276, 279, 291, 315, 316, 317
rice *see Oryza sativa*
RLK/RLP (receptor-like kinases/proteins) 157, 159
RNA analysis 215
RNA sequencing (RNAseq) 33–34
RNAi (RNA interference) 35, 78, 197, 217, 221–222, **222**, 224, 225, 227, 228

RNQPs (regulated non-quarantine pests) 129–130
rocket/arugula (*Eruca sativa*) 311, 312
root diffusates/leachates 11, 44, 44–47, **45**, 49, 54, 55, 118, 279
    chemistry of 46–47
root staining 368
Rootrainer system 190, **190**
RPPOs (regional plant protection organizations) 419–420, **420**, 421
Russia 3, **130**, 131, 197, 202, 380
rye 45, 181, **307**, 308, 384

S (susceptibility) genes 222–225
salicylic acid (SA) 164–166, 224, 280, 315, 319
salicylic acid methyl transferase 197, 226
Salpiglossis spp. 386
Saracha jaltomata 386
Saudi Arabia 9, 308, 374
scanning electron microscopy (SEM) 350, 352, 353–355, **355**, 360, 368
SCAR (sequence-characterized amplified region) primers **428**, **431**, 432–433
Schomaker/Been model 137–139
Science and Advice for Scottish Agriculture (SASA) 148
Sclerotionia sclerotiorum 198
screening 176, 179–180, 182–185, 188, 201, 286, 288
    field trials 183
    miniature trials 184–185
    pot trials 53, 183–184, 199
    protocols 189–190, 194–196, 199
    see also bioassays
Scrophulariaceae 8, 381, 383
sDNA (satellite DNA) 430–432
SDS (sudden death syndrome) 280, 281, 282
Secale cereale 382
SEM *see* scanning electron microscopy
semifenestrae **349**, **350**, 351
senescence of host plants 17, 50, 103
sense/antisense gene orientation 221–222, **222**
sensilla 57–58, 221, **341**, 344–345, **345**, 347
sensory perception 57–63
    ascaroside/peptide signals 61–62
    blocking 62–63
    chemoreception 59, 62–63, 344
    plant signals 58–60
    sensilla 57–58, 221, **341**, 344–345, **345**, 347
    sex pheromones 60–61
Sequoia 410
serotonin 63, 221
Sesamum indicum 10, 202
sex pheromones 60–61
sexual dimorphism 346
signal peptides 30, 32, 34, 35, 80
Sinapis alba 59, 198, 199, 374

single-molecule real-time (SMRT) sequencing 219
siRNAs (small interfering RNAs) 221–222, **222**, 227
soil 8, 17, 45, 57, 108–109, 240, 317
    amendments/inundation 11, 116, 134, 309
    enrichment/structural index 227
    food web 227, 319–320
    fumigation 134, 175, 272, 305, **307**, 311–313, **311**, 322, 323, **325**
    microbes 227
    moisture 5, 17, 46, 104, **106**, **107**, 115, 116, 246, 322, 325
    nematode suppressing 238
    organic matter 256
    pH 17, 46, 57, 58, 115, 256, 309, 317, 326
    seals 324, **325**
    temperature *see* temperature factor
soil sampling 129, 134, 135–151, 179
    automated 147, **148**, 151
    challenges with 135
    computer simulations 150
    for detection 136–137, 139–140, **139**, **140**
    EU method 135, 139
    future prospects for 150–151
    and heterogeneity levels 138, **138**
    laboratory diagnosis *see* laboratory diagnosis
    and quantification/distribution 136, 137–139, **138**, 140–144, **141**, **142**, **143**, 148–150
    for quarantine 144
    and Schomaker/Been model 137–139, **139**, 140
Solanaceae 3, 8, 157, **307**, 374, 381, 383, 410, 411, 412
Solanum spp. 12, 386, 389
Solanum lycopersicum see tomato
Solanum melongena 12, 386, 389
Solanum multidissectum 188
Solanum sisymbrifolium **307**, 313, 314
Solanum sparsipilum 188
Solanum spegazzinii **15**, 188, 218
Solanum tuberosum **15**
Solanum tuberosum see potato
Solanum tuberosum spp. *andigena* 188, 218
Solanum vernei **15**, 188, 200
solarization 2, 309–310
sorghum (*Sorghum bicolor*) 306, 308, 374, 384
South Africa 3, 9, 323, 377, 382, 410, 411
South America 2, 8, 10, 12, 13, 14, 130, 131, 159, 187, 197, 380, 386, 388, 389, 405, 410, 411, 412, 429
    see also specific countries
Soviet Union, former 3, 9, 10
soybean *see Glycine max*
soybean aphid (*Aphis glycines*) 115
soybean looper (*Chrysodeixis includens*) 115
soybean stem canker 115, 284
spacer region 221–222, **222**

Spain 3, 9, 374
spatial transcriptomic profiling 33–34
species description, standards for 357–359
spider mite (*Tetranychus urticae*) 166
split root technique 247, 273, 278, 283, 289, 290, 293
SPRY domain/SPRYSEC effectors 30
Staphylococcus aureus 252
Stellaria media 381
Stenotrophomonas maltophilia 247
sterols 92, 242
sticky nightshade (*Solanum sisymbrifolium*) **307**, 313, 314
strawberry 244, 317
Streptomyces 322
Streptomyces scabies 81
Stropharia rugosoannulata **239**, 242
structural index 227
Stylophage spp. 238–239
sucrose 37
sudden death syndrome (SDS) 280, 281, 282
sugar beet 2, 8, 218, 220 *see also Beta vulgaris*
sugarcane 3, 11, 201, 384
sunn hemp (*Crotalaria juncea*) 377
surface coat 243, 250, 252
susceptibility (S) genes 222–225
Sweden 3, 9, 382
sweet potato 220, 308
sweet vetch (*Hedysarum coronarium*) 308
syncytia 1, **5**, 6, 7–8, **7**, 74, 76, 79, 82, 162, 163–164, 166, 167, 178, 198, 223, 228–229, 278, 280, 290, 314
synergistic interactions 247, 257, 276, 277–280, **277**
Syria 3, 9, 383

T-DNA transfer 218
Tagetes patula 59
Tajikistan 3, 9
take-all (*Gaeumannomyces tritici* var. *tritici*) 308–309
TALEN 224
taxonomy 365–393
    Betulodera 392–393
    Cactodera 390–391
    and co-evolution 412–413
    Dolichodera 391–392
    future prospects for 393
    Globodera 385–390
    Heterodera 369–384, **371**
    Heteroderinae 369, **370**
    history of 365–366
    and identification *see* identification of cyst nematodes
    Paradolichodera 393
    Punctodera 390
    Punctoderinae 384–385, 409–410
    reverse 358, 360
    and root-knot nematodes 368–369
    and species description standards 357–359
    Vittatidera 393
Telone/Telone II 321, 323, 324, 326
TEM (transmission electron microscopy) 349, 352, 354–357
temperature factor 5, 6–7, 8, 9–10, 12, 13, 45, 49–50, 51–53, 54–55, 56, 58, 90, 104, 246, 309–310, 313–314, 325
temporal transcriptional profiling 32–33
Tetranychus uricae 166
Thailand 201, 384
thrips (*Frankiniella occidentalis*) 166
tillage 254–256, 257, 308–309
tobacco (*Nicotiana tabacum*) 305, 314, 380, 389
tolerance 176–178, **177**, 183, 187, 294, 314
tomato (*Solanum lycopersicum*) 12, 94, 96, 107, 157–158, **158**, 159, 163, 166, 218–219, 220, 224, **307**, 386, 389
transcriptomics *see* genomics/transcriptomics
transgenesis/transgenic engineering 94, 218–219, 220, 229–230, 258
transmission electron microscopy (TEM) 349, 352, 354–357
trap cropping 2, 106, **133**, 134, 175, **307**, 313–314
trehalose 4, 48, 49, 57, 93
Trichoderma spp. **239**, 242, 251, 318–319
Trichoderma asperellum 318, 319
Trichoderma atroviride 251, 318, 319
Trichoderma harzianum 238, 242, 251, 318, 319
Trichoderma kovingii 318
Trichoderma lignorum 318
Trichoderma longibrachiatum 242, 318–319
Trichoderma virens 242, 318
Trichoderma viride 318
Trichodorus 29
Trichostrongylus colubriformis 248
Trifolium repens 202, 291, 292–293, 383
Trinidad 11, 201
triticale 181, 308
Triticum aestivum 181
Triticum tauschii 117, 181
Triticum triuncialis 181
Triticum variabilis 181
Turkey 3, 9–10, **130**

Umbelliferae 117, 391
United Nations (UN) 230
    see also FAO
United States (USA) 8, 10, 12, 13, 18, 103, 117, 186, 187, 191, 202, 226, 240, 241, **255**, 316, 323, 326, 377, 381, 382, 384, 386, 387, 388, 389, 390, 411, 429
    BCN in 198, 383

United States (USA) (contiuned)
    CCN in 3, 9, 374
    EPA guidelines in 326
    quarantine in **130**, 131
    SCN in 2, 112–113, 114, 115–116, 192–193, **192**, 196, 283, 380
Urtica dioica 382
Urtica urens 382

vagina 342, 344, 347, 349–350, **350**, 352
Vaminoc 244
VAPs (venom allergen-like proteins) 76, 156
Verticillium albo-atrum 289, 382
Verticillium chlamydosporium see Pochonia chlamydosporia
Verticillium dahliae 286–288, 289–290, 309
Verticillium lecanii 319
Verticillium suchlasporium 251
Verticillium wilt 286–288, 289–290
vetiver (*Vetiveria zizanoides*) 384
Vicia faba 308
Vigna spp. 376
Vigna aconitifolia 10
Vigna mungo 10, 202
Vigna radiata 10, 202, 308

Vigna unguiculata 10, 202, 244, 294, **307**
VOTIVO **255**, 318
vulva 342, 348–349, **349**, 365–366, **379**
vulval cone 3–4, 348–349, 353, 366, 368, **371**, **375**, 407
vulval plates **367**, **376**, **387**
Vydate 10G 321, **321**, 322, 323, 326

West Africa 3, 201, 319
wheat 9, **16**, 17, 45, 117, 174, 181, 182, 184, 186, 224, 238, 242, 246, 290–291, **307**, 308, 383, 384, 420
wormwood, sweet (*Artemisia annua*) 314

Xanthomonas oryzae 224

Yursina pestis 250

Zea mays 11, 117, 174, 201, 226, 238, **255**, 306, **307**, 308, 379, 384
zigzag model 79, **80**
Zygomycota 238